WORLD SCIENTIFIC SERIES ON NONLINEAR SCIENCE Series A Vol. 49

Series Editor: Leon O. Chua

NONLINEAR AND PARAMETRIC PHENOMENA

THEORY AND APPLICATIONS IN RADIOPHYSICAL AND MECHANICAL SYSTEMS

Vladimir Damgov

Bulgarian Academy of Sciences

World Scientific

NEW JERSEY • LONDON • SINGAPORE • BEIJING • SHANGHAI • HONG KONG • TAIPEI • BANGALORE

Published by

World Scientific Publishing Co. Pte. Ltd.
5 Toh Tuck Link, Singapore 596224
USA office: 27 Warren Street, Suite 401-402, Hackensack, NJ 07601
UK office: 57 Shelton Street, Covent Garden, London WC2H 9HE

British Library Cataloguing-in-Publication Data
A catalogue record for this book is available from the British Library.

NONLINEAR AND PARAMETRIC PHENOMENA
Theory and Applications in Radiophysical and Mechanical Systems

ISBN 981-02-3051-6

Printed in Singapore by World Scientific Printers (S) Pte Ltd

To TANYA,

NADYA, JORDAN, NIKOLAY,

ELIZA and NYA

PREFACE

This monograph concentrates on radiophysical systems; less attention will be focused on purely mechanical systems, but actually the effects, phenomena and regularities are of a general nature and in most of the cases they are valid for mechanical systems as well. By applying the well-known electromechanical analogues, the resultant mathematical expressions and major conclusions can be successfully used in both areas without substantial adjustments.

Radiophysics is an area within the domain of Physics, which, according to the *Great Soviet Encyclopaedia* [1], studies physical processes related to electromagnetic oscillations and waves with frequencies ranging from a few Hz to 10^{11} and over: the object of exploration is their excitation, propagation, interaction, reception and conversion; issues of signal identification against the background of random (fluctuating) processes are also studied. The subject matter of Radiophysics is closely linked with the *General theory of oscillations, SHF electronics, Quantum electronics,* elements of *Physics of plasma, Remote sensing* [2].

The Theory of nonlinear oscillations is *a science* about the most general regularities in dynamic systems. Its separation and differentiation are mostly related to the common analysis of the phenomena and processes, which resulted in major concepts such as oscillating regularity, stability, dynamic systems, etc., losing their primary physical meaning in the course of their development within this theory and ceased to be physical concepts. Parallel to that, besides its high degree of mathematical formalization in the area of applied mathematics, the Theory of nonlinear oscillations has also „materialized" in natural sciences, mostly in Radiophysics, Mechanics and others.

There is a vast bibliography on the General theory of nonlinear oscillations. The majority of the books, however, deal with the principal analytical methods, while any investigation of the various oscillating processes and phenomena, and of their specificity, for example in particular radiophysical and mechanical systems, is conducted for the sole purpose of illustrating either one or another set of approaches. On the other hand, there are a number of interesting treatises dedicated to the examination of particular classes of systems, for example generating, converting or amplifying systems, which, however, have lost sight of the general oscillations approach and of the general comprehension of the occurring processes from an oscillations perspective. The more general view is indispensable for the modern radiophysicists, specialists in mechanics and other physical-and-engineering areas of today, who should use the most adequate methods of analysis and computation in each particular case considering the oscillation processes in various dynamic systems, and who should be well aware of the category of phenomena to which the object of their research or elaboration belongs.

In recent years modern science has surprised us more than once: it has been proved that chaotic regimes are widespread in strictly determined systems without random external action and with a small number of degrees of freedom; an antipodal structural system has also been developed and a new area of research, dealing with the self-organization and synergetic grouping of systems in steady-state formations, has emerged; the Catastrophe theory has been created, etc.

It has been demonstrated that there is a possibility for the appearence of chaotic oscillations in such a classical dynamical system as is an oscillating circuit under the action of a determined external periodic signal. Hence, it is necessary to focus attention on the nonlinear and parametric phenomena that, in recent years, have resulted in a cardinal revision of our ideas of determinism and radomness of the processes, of general and local stability, of computer simulation and possibility for long-term prediction, etc.

The most general considerations stated above have largely determined the nature of this book.

This book deals with modern research issues related to effects and phenomena in four major classes of radiophysical and physico-technical oscillating systems – linear, parametric, nonlinear and nonlinear-parametric. We present some theoretical generalizations and new theoretical formulations in the following areas: modulation-parametric interactions, interpretation of parametric resonance, generator conversion of signals and chaotic instability of systems with delaying self-action, adaptive (synergetic) grouping of radiophysical and physico-technical oscillating systems in stable formations, the new subclass of modulatory nonlinear-parametric systems with adaptive self-tuning, kick-excitable self-adaptive systems. A suitable analytical technique for investigating the set of nonlinear and modulation-parametric phenomena is provided. We establish classification and generalize the regularities in the manifestation of the different nonlinear and parametric effects and phenomena for the purpose of acquiring a general idea. Some methods of realizing radiophysical and physico-technical systems with new functional capacities are described. Radiophysical and physico-technical devices and systems resting on new effects, phenomena and regularities are presented. A generalized analysis of the possible applications is also offered.

Notwithstanding the monographic nature of the book, it can serve: first, as an introduction to some basic paradigms of the Theory of nonlinear oscillations; second, as a source of initial information for post-graduate students and junior researchers on the methods of and approaches to formulating and solving scientific and research problems; third, as a manual on a definite set of effects and phenomena, and on some methods of their analysis to be used by researchers. The book seeks to assist individual research – the art of creation. Hence one will come across examples drawn from both the first and the second line of the scientific front there. The first (heuristic) line – being the identification and investigation of qualitatively new phenomena, development of new methods and theories. The second („conventional" or „extensive") line – being the utilization of the developed methods for investigating the characteristics and parameters of devices and systems created on the basis of new effects and phenomena found.

A fundamental methodological approach in the structuring of the material is that the presentation of each issue should reflect the overall research cycle – from the general description and classification of effects and phenomena, through the development of an adequate analytical technique, to an analytical, numerical (through computer experiment) and test (natural experiment) investigation of particular devices and systems, and an analysis of the possible applications. The analytical technique is presented in a user-friendly form with the necessary explanations so that the reader can apply it promptly and actively in his creative and research activity.

The book is of an interdisciplinary nature and can serve as a handbook for developing lecture courses such as Fundamentals of Nonlinear Dynamics and Theory of Nonlinear Oscillations, Theory of Nonlinear Circuits and Systems, Fundamentals of Radiophysics and Electronics, Theory of Signals and Theoretical Radiophysics, Theoretical Mechanics and Electrodynamics.

The book is designed for researchers, lecturers, post-graduate students and undergraduates at universities, academies, higher educational institutions, institutes and research centers, where radiophysical and physico-technical subjects are taught. The book can evoke interest in a broad circle of specialists: physicists and mechanicians, specialists in the Theory of nonlinear oscillations, in Radiophysics and Nonlinear Mechanics; engineers, specialists in Radioelectronics, Nonlinear radio engineering, Nonlinear electrical engineering, Microwave engineering, Automation and instrument engineering; specialists involved in Applied Mathematics and dealing with the Theory of nonlinear oscillations with applications in Radiophysics, Mechanics, Biocybernetics, etc.

The author appreciates highly his fertile joint work with Prof. Sc.D. N.D.Biryuk, (Voronezh State University, Russia) on the general analysis of the parametric phenomena in radiophysical systems, Dr.D.B.Duboshinsky and Dr.Y.B.Duboshinsky, (Vladimir Polytechnic Institute, Russia) on the phenomena of adaptive grouping of radiophysical and mechanical systems in stable formations, and Assoc. Prof. Dr. P.G.Georgiev, (Department of Physics, Technical University, Varna) on generator transformation of signals in radiophysical systems.

The author is particularly delighted to underscore the prolific sharing of expertise and ideas with Prof. V.Cimagalli (Rome University, Italy), Dr.A.Holden (Leeds University, UK), Prof. Z.Jasino (Montreal University, Canada), Prof. A. Bossavit (Electricité de France, Paris), Prof. A.Cronin (New Jersey University, USA), Prof. H.Hagedorn (Darmstadt Polytechnic University, Germany), Prof. H. Kawakami (Tokushima University, Japan), Prof. Sc.D. Y.L.Khotuntsev, (Moscow Polytechnic University, Moscow, Russia), Prof.M.Farkas (Budapest Polytechnic University, Hungary), Prof. J.Mavhin (Leuven University, Belgium), Prof. D.R.K. Sastry (Kondapur Research Centre, India), Prof. A.Tondal (National Machine

Research Institute in Prague, Czech Republic), Corresp. member Prof. L.Pust (Institute of Thermomechanics in Prague, the Czech Republic), Acad. Y.A. Mitropolskiy (Institute of Mathematics in Kiev, Ukraine), Prof. Sc.D. M.Marinov, (Institute of Applied Physics at the Technical University, Sofia), Acad. K.Serafimov (Institute of Astronomy, National Astronomic Observatory at the Bulgarian Academy of Sciences, Sofia).

The author expresses his gratitude to the authorities of the Space Research Institute at the Bulgarian Academy of Sciences for providing the environment needed and being able to insure the tests recorded in the book.

The author would like to thank the translator, Mrs. Mariana Stoycheva, the editors, Mrs. Anne-Marie De Grazia and Mrs. Katerina Spasova, and the computer operator Mr. Plamen Chavdarov for their helpful cooperation.

25th May, 2001 Vladimir DAMGOV
Sofia

CONTENTS

INTRODUCTION

(Theory of nonlinear oscillations. Invariance and isomorphism of the oscillating processes in systems of varying physical nature. Nonlinear and parametric phenomena)

The complex natural phenomena that seemed incompatible in the past have been subject to such interpretation in recent years that they have been „brought down" to the level of our day-to-day research practices and have become part and parcel of them. This is mostly valid for Nonlinear Dynamics. Some new scientific disciplines providing invariant description of effects and phenomena, as regards the material carrier, have emerged – such as Nonlinear radio engineering, Nonlinear Radiophysics, Nonlinear Astrophysics, Nonlinear Optics; Astrodynamics is closely intertwined with Radiophysical wave dynamics, etc. Periodical orbits with a large number of harmonics and the dependence of the period on the amplitude have proved to be regularities characteristic both of planetary systems and of „ordinary" radiophysical oscillating systems. Given a relevant choice of coordinates in systems of varying physical nature (electric, mechanical, acoustic, biological, etc.), processes can be described by the same differential equations and their theoretical oscillating characteristics coincide in form. It is in this sense that we view the isomorphism in systems of varying physical nature. In recent years it has become clear that in the case of typical nonlinear situations – both in Radiophysics and Mechanics and in many other disciplines – it is impossible to foresee the dynamic properties over a period of arbitrary duration even in the conditions of slight perturbations, i.e. a possibility has been discovered of creating chaos in the oscillating processes (peculiar generation of chaotic motions in the absence of random external actions) [3–31]. It has been shown that chaos can appear even in oscillating systems with 1.5 degrees of freedom. This has thrown new light on the issues of steadiness, stability and reliability of the oscillating systems.

At the same time, the opposite research scheme – synergetic self-organization of the systems, has also evolved [32–38]. It has been shown that synergetic grouping of oscillating systems in stable formations is possible even in the case of a limited number of degrees of freedom of these systems.

The fundamental research conducted in the past years in the field of the nonlinear theory of oscillations has resulted in the formation of qualitatively new views as regards the oscillating processes and phenomena. They have evolved to give rise to the Catastrophe Theory and Theory of bifurcations [39–41], the doctrine of self-organization of inorganic systems, the new interpretation of the synchronization phenomenon as one of the mechanisms of self-organization [42], the new ideas of determined and chaotic oscillations as a manifestation of different facets of the common nature of dynamic systems. The issue of the self-organization of systems has been recognized as one of the most important scientific problems of the last quarter of the twentieth century [43]. The present-day combination of approaches and the unity of the stochastic and synergetic processes create possibilities for establishing, figuratively speaking, an engineering-and-physical laboratory, where various phenomena can be simulated and studied, and real physical analogues involving virtually arbitrary radiophysical and physico-technical processes can be demonstrated. Nowadays all those phenomena are described with one word: complexity.

The Theory of oscillations, as a separate scientific discipline, deals with the investigation of the most general properties and regularities of the processes and phenomena in dynamic systems.

The theoretical and methodological basis of the Theory of oscillations and, above all, of the modern development of the theory and technology of nonlinear and parametric systems includes, parallel with the classical works of Poincaré, Lord Rayleigh and Melde from the end of the 19th century, the classical works from the 1920s, 1930s, and 1950s written by Van-der-Paul (the Netherlands), Barkhausen and Müller (Germany), Epilton (UK), Koga and Hayashi (Japan), Lenard and Karatanou (France), Mandelshtam, Papaleksi, Andronov, Vitt, Haikin, Migulin, Krilov, Bogolyubov, etc. (Russia) [44–84].

From the viewpoint of the Theory of oscillations, oscillating systems can be classified in several big groups (classes):

1. Linear systems – they are described by linear differential equations, whose parameters (coefficients) are independent of the mode (regime) of work and of time.

2. Parametric systems – they are described by linear differential equations with periodic coefficients, i.e. the parameters are independent of the mode (regime) of work but dependent on time.

3. Nonlinear systems – they are described by nonlinear differential equations, whose parameters (coefficients) are time-independent but dependent on the mode (regime) of work.

4. Nonlinear-parametric systems – they are described by nonlinear differential equations, whose parameters are both mode (regime) and time-dependent.

The book presents an analysis of different oscillating systems belonging to the four major groups of the aforementioned classification. It is noteworthy that if one applies a rigorous approach, all modern oscillating systems should be referred to group four. But this would make the analysis extremely complicated, so in many cases either one or another set of relations and dependencies are ignored and the oscillating systems are referred to the other three groups. Moreover, it should always be remembered that the „linear" (or „linearized") treatment may entail not only inaccuracies in the quantitative evaluations but qualitatively erroneous results as well. Besides, as a rule it is possible to generate and transform different signals only in systems of the latter three groups.

In accordance with the small signal theory and the filter method [78–85], the considered modulation-parametric systems, where small modulating signals and „pumping" high-intensity action are present, are referred to the second group even in the cases when nonlinear elements are used in the circuits.

Chapters 1, 2, 3, and 4 deal with effects and phenomena in modulation-parametric and nonlinear systems belonging to the second and third groups.

Chapters 5, 6, 7, 8, and 9 present a general theory of some phenomena taking place in group one and group two systems.

The systems belonging to group four are the most complicated and least investigated throughout the world. The phenomena occurring in the systems of this group may differ from the phenomena taking place in the linear parametric systems (group two) as drastically as any nonlinear processes, phenomena and systems differ from the linear ones. That is why, in principle, the fourth group of systems is boundless. A peculiar subclass of group four: modulatory nonlinear systems with adaptive self-control (kick-excited self-adaptive systems), is set apart in the book. Besides the typical nonlinear-parametric peculiarities, it also exhibits modulation interactions, whose intensity is small in comparison with definite basic oscillations in the system. The new initial conditions formulated for this subclass allow to qualitatively identify new oscillation phenomena and regularities in macro-systems.

Chapter 10 deals with the properties, regularities and phenomena in the separated subclass of systems, and together with Chapter 9 it tackles the phenomena of synergetic (adaptive) grouping of oscillating systems into stable formations.

Radiophysical systems undergo a new evolutionary stage today. While the first stage saw the development of the theory and practice of receiving and transmitting radio-engineering modules, and of some fundamental issues in the generation, amplification, modulation and manipulation of radio-engineering signals, the modern stage in the development of radio-engineering systems is marked by: the discovery and exploration of new physical effects and phenomena, the evolution of new analytical methods, the optimization of limiting characteristics, the design of qualitatively new precision sensors and radio-engineering means, the development of new methods of radio signal processing on the basis of adaptive and synergetic effects, neuron networks and the artificial intelligence principles.

The development logic of radiophysics can be depicted as a spiral of circles that are attached to each other, while at the same time, as a new turn is made, each subsequent circle rises to a higher level than the previous one.

As an outline this can be presented as the following sequence of stages:

— Discovery and exploration of new physical phenomena and effects, new theoretical generalizations;

— Evolution of analytical methods, synthesis and design of new sensors, devices and systems;

— Analysis and optimization of radio-engineering facilities and of their parameters and characteristics;

— Improvement of the element basis and a new circle of synthesis and analysis of radio-engineering systems and their characteristics;

— Emergence of a new set of physical phenomena and effects bringing about a new circle of synthesis, analysis and optimization of radio-engineering facilities;

— Interaction and mutual enrichment with other areas of science and technology – physics, electronics, mechanics, biology, chemistry, medicine, etc.

Figuratively speaking, the book is positioned on the circle corresponding to the exploration of new effects and phenomena in radiophysical systems, theoretical generalization, the development of analytical methods, and the implementation of general analysis of radiophysical devices and systems with a view to possible practical needs and applications.

The available literature in the field of Theory of oscillations can be conventionally divided into three groups.

Group *one* encompasses mathematical monographs, articles and other publications written by mathematicians, whose expertise is mainly in the area of asymptotic and qualitative methods of analysis. As a rule, these works are notable for the mathematical strictness and for the abstract nature of the methods put forward, while the constructive aspect of the research is only touched upon. For example, when assessing any methodological errors, the authors of this group of works consider only the possibility of elaborating such an assessment in principle, and not the particular procedure of its actual development that could be used in applied problems.

One can refer to group *two* such books, articles and other publications that belong to the other extreme – where authors are involved in the analysis of defined particular problems and come up with clear engineering and technical solutions.

It is obvious that somewhere „in the middle" one could distinguish a group *three* (a „buffer" one), containing works that develop theoretical aspects concerning specific types or classes of systems that can serve as a basis for experimental studies and for conducting the research in group two works. As a rule, the majority of the radiophysical and theoretical engineering elaborations are reflected in group three literary sources.

Due to its nature, the bulk of this book can be referred to group three but it also contains a number of elements belonging to group two.

Linear oscillating systems have been thoroughly studied. The employment of linear differential equations has allowed studying these systems in great detail and conducting an analysis entailing almost automatic application of the suitably developed mathematical technique.

When investigating parametric systems, one could utilize not only the specific technique developed by Floquet, Mathieu, Hill, Gorelik, Yakubovich, Starzhinsky, Taft, etc. [78–89], but almost all approximate methods evolved for analyzing nonlinear systems as well.

It is well-known that we are still in need of general methods for solving nonlinear differential equations, describing nonlinear and nonlinear-parametric systems. Considerable achievements in the theory of nonlinear oscillations were scored in the 1920s and 1930s, particularly in connection with the invention of the electronic (vacuum) valve and the intensive development of radio engineering. It is also worth nothing that the foundations of the mathematical technique adequate to the overall set of issues concerning the Theory of nonlinear oscillations were first laid out in Poincaré's classical works in celestial mechanics and in Lyapounov's elaborations on motion stability at the end of the 19th century [90–95]. A.A.Andronov [49] was the first to draw attention to the relation between these works and the task of exploring continuous oscillations in non-conservative systems. He was the first to introduce the term „self-oscillation".

The rapid development of the modern Theory of nonlinear oscillations has led to the creation of a number of analytical methods that can be divided into two big groups:

A. Methods applicable to systems of both weak and considerable nonlinearity. The following are referred to this group:

1. The methods of the qualitative theory of differential equations developed by Poincaré and presenting geometric images (maps) of the different motions in the system by constructing a family of integral curves on the phase plane [90–92, 94, 96].

2. Methods of piece-wise linearization of nonlinear characteristics with subsequent solutions join in the border-line areas of the individual linearized parts [87].

B. Methods applicable to systems of weak nonlinearity: the small parameter method [54], the method of slowly changing amplitudes [51], asymptotic methods [97], quasi-linear methods [81, 82], the method of the harmonic balance [98], the energy method [99], the method of symbolic equations [63], the stroboscopic method [74], the method of M. Rausher [100], the averaging method [101], the quasi powered method [102], the variable scale method [103], the graph-analytical method [104], numerical methods [105].

The book develops an analytical technique for analyzing nonlinear and modulation-parametric phenomena in radiophysical systems, for investigating the conversion properties of radiophysical systems of a generator type. It also developes a modified complex amplitudes method for analyzing parametric and nonlinear systems. It proposes an adequate analytical technique for studying linear radiophysical systems with almost periodic parameters – linear spatial systems of algebraic equations. The concept of multi-index multiplication of spatial matrices is introduced and methods of conducting such multiplication are presented. A general analytical approach to the implementation of the principle of linear connection to an application of nonlinear radiophysical systems is also offered.

The fundamental research carried out in the past quarter of the century in the field of Radiophysics, Solid state physics and Low temperature physics revealed radically new possibilities for utilizing the electric properties of a solid state for amplifying and transforming electromagnetic oscillations [4, 43].

The creation and utilization of parametric one-ports of different functional designation was a powerful stimulus for the development of modern Radiophysics and Mechanics [2, 78].

Van-der-Ziele [106], while investigating circuits containing nonlinear capacitance, first pointed out that such circuits can be used as low-noise amplifiers. Later on Vulle invented the semi-conductor nonlinear capacitor [107], and Sul offered a microwave solid state amplifier based on ferrite [108].

Nowadays, parametric amplifiers and converters based on the nonlinear capacitance of a semi-conductor diode are widely applied in various areas of science and technology. The developed parametric systems cover an enormous frequency range – from single-digit hertz frequencies to those corresponding to waves in the sub millimeter band [85].

A number of monographs [78, 81–85, 107] and a multitude of articles published in periodicals explore parametric systems.

Most generally, parametric systems can be divided into three major classes: frequency-sustaining systems (of a communicating and reflective type) and systems transforming the frequency „upwards" or „downwards" along the frequency axis. As the frequency is transformed „upwards", it is possible to achieve amplification even without regeneration.

When a weak signal enters parametric systems of a regenerative type, it draws energy from a pumping source and its intensity increases. The mechanism of establishing a positive inverse feed-back in parametric systems is conditioned by the reciprocal properties of the linear-parametric circuits. When the resonance systems serving as a load for the combined frequency components are tuned in a particular way, the pulsation of the reactance with pumping frequency leads to the introduction of energy in the signal circuit. In the course of the analysis, this property of parametric systems is reflected through a negative resistance introduced into the signal circuit. Obviously, when selecting the load of the combined components in a particular way, it can be introduced into the circuit of the signal not only negative or positive resistance but negative reactance (negative capacitance or negative inductance) as well. When a broad-band negative equivalent reactance introduced into the signal circuit is ensured with respect to a signal spectrum, the sensitivity of radiophysical systems can be substantially promoted by compensating the own or parasitic reactive component of their impedance [78].

One of the promising directions in the field of parametrics is that of parametric systems with four and more working frequencies [51, 78, 85, 109] using differential and sum combined frequencies obtained as a result of mixing the signal frequency and the pumping one. We term such systems parametric modulation systems (PMS) since, on the one hand, the manipulations are carried out with a power accumulating (parametric) element and, on the other hand, all regularities of modulation influence are in existence. PMS are based on the principle of modulating forced oscillations in a resonance system containing a nonlinear (or time-dependent) reactive element. It is worth nothing that any nonlinear oscillating circuit, where forced oscillations or self-oscillations (the latter in the case if it is a part of a generator system) have been excited, is a modulation-parametric system with respect to a signal with a spectrum falling far below the transmission band of the circuit.

The fundamentals of the theory and the design of four-frequency parametric systems are presented in the works of Ekhart, Stetser, Bayar, Hauson, Kuh, Mathey, DeJager, M. D. Karasev, M. E. Gertsenstein, V. S. Etkin, Yu. L. Khotuntsev, V. N. Detinko, etc. [78].

An interesting property of four-frequency parametric systems from a physical point of view is that, given definite PMS working modes, the active and reactive parts of the input differential impedance of the system have a negative sign within a broad frequency range of the input signals up to d. c. [78]. This opens up broad vistas for creating regenerative video-amplifying systems without frequency transformation, low-noise parametric frequency correctors, etc.

The terms of „negative capacitance" and „negative inductance" were introduced by Van-der-Paul B. [48], who developed the first circuits for their implementation on the basis of negative resistances. Tube and transistor feed-back circuits also allow realizing negative capacitances and inductances, which, however, come out as narrow-band and frequency dependent ones with relatively high own fluctuation noise.

On the whole, the negative resistances used in science and engineering can be conventionally referred to the following major groups:

1. „Natural" negative resistances typical of the devices and elements with falling volt-ampere characteristic (VAC), in both static and dynamic mode (electric arc, dynatron, transitron, thermistor, dinistor, tunnel diode, etc.).

2. Negative resistances based on the period of electron drift. A falling VAC is realized only in a dynamic mode (diodes with field emission, klystrons, magnetrons, avalanche-drift diodes, Gunn diodes, etc.).

3. Negative resistances based on phase shifting with an external active three-pole element. A falling VAC can exist in both static and dynamic modes.

4. Negative resistances realized with the help of a feed-back. A falling VAC emerges only in a dynamic mode. This method is particularly important from a practical point of view, since it serves as a basis for designing tube and semiconductor generators, regenerative filters, etc.

5. Negative resistances resulting from parametric interactions with the participation of a power-accumulating (reactive) element.

It is obvious that negative resistances are proper not only to individual elements but to devices (semiconductor, ferromagnetic, ferroelectric, etc.) and systems (generator, parametric systems, etc.) as well.

Nonlinear resonance systems with semi-conductor diodes are widely applied in various radiophysical devices – low-noise parametric amplifiers, generators, converters, highly stable parametric frequency multipliers and dividers, frequency modulators, receivers with electronic tuning, trigger circuits, parametrons, etc. [63, 78, 85, 109, 110].

As regards the utilization of modulation-parametric interactions for the effective control of equivalent impedances, a number of issues emerge: first, with respect to the power interchange between pumping circuits and signal in parametric systems in different specific modes and under varying conditions; second, the specificities of the nonlinear and parametric resonance in resonance systems with partially forward-biased (forward-conducting) p – n junction; third, the legality of using small-signal differential parameters and the quasi-static VAC of p – n structures in the conditions of high frequencies and amplitudes, etc.

Generators in autonomous and non-autonomous modes have been widely investigated and used as frequency converters in recent years [78]. This interest has been provoked mostly by the possibility of coming up with high conversion ratios, and it is also related to the development and investigation of autodyne systems for short-range Doppler radars, as well as with the extension of the functional capacities of generator systems. Analysis shows that modulation-parametric phenomena, analogous to those taking place in parametric systems with external pumping, can be manifested in non-autonomous generator systems. A regenerative change in the effective (equivalent) impedance is observed in the synchronization band and, therefore, the synchronized generator can be viewed as a one-port with controllable parameters with respect to external video-signals. A generator nonlinear one-port, set in a conversion mode, may have negative conductance at intermediate frequency, whereas the module of its value reaches its maximum at a power of the basic high-frequency (or SHF) oscillations smaller by far than the possible maximum. The mechanism of this effect is also related to modulation-parametric and detuning effects.

The interest in the investigation of nonlinear parametric interactions of oscillations of different frequencies in complex generator systems that has risen in recent years is also related to the possibility to create, on their basis, selective receiving systems, capable of separating signals of close frequencies and comparable amplitudes [78].

Short-range Doppler radars with self-mixing diodes (Doppler autodynes) are among the most widely developed and applied functional generator systems. They are miniature receiver-and-transmitter modules, widely used in systems for short-range radiolocation [111]. Their major advantage is their compact form and low consumption. Besides, their small transmission power is particularly topical in relation to the problems of electromagnetic compatibility. Certain issues emerge in this connection regarding: a more detailed analysis of the conversion properties, of the spectrum of oscillations, of the stability of Doppler autodynes; the optimization of their characteristics and the development of new functional solutions. For example, one of the ways of increasing their sensitivity is to use one element for generating the oscillations that are to be emitted and for amplifying the received signal, and to use an external converter for the frequency conversion aimed at taking down (recording) the Doppler signal [78]. Moreover, the effects of modulation-parametric amplification can be used for increasing the general sensitivity of the system. The fundamental works on Doppler autodynes (for example [111]) solve the radiolocation problem of continuous signals mainly by taking into account the first harmonic in the Doppler spectrum and the amplitude modulation of the oscillations. Considerable interest in the structure of the signal in Doppler autodynes has been evinced in recent years in connection with the investigation of the utmost parameters of the facilities for short-range radiolocation and extension of their functional capacities [78]. It becomes necessary to analyze in greater detail the spectrum of the oscillations in Doppler autodynes by taking into account both amplitude and phase modulation with the first and second harmonics of the Doppler frequency shift of the reflected signal.

Science and engineering have been using radiophysical systems of growing complexity. One can juxtapose an equivalent circuit to each such system, consisting of a definite number of sources of electromagnetic power, inductances, capacitances, active resistances, connected in a certain way. These circuits with point parameters may be classified in a definite way and the common properties of the separate classes may be worked out.

Above we adopted a classification of the radiophysical systems with point parameters dividing them into four classes – linear, parametric, nonlinear and nonlinear-parametric.

The first class is conditioned by the multitude of systems with time-independent parameters. The first subclass can be identified as containing autonomous linear systems with constant parameters and driving forces (the latter may even be ommitted). Non-autonomous systems with constant parameters and time-dependent driving forces can be referred to a second subclass.

The second class encompasses systems whose parameters are explicitly time-dependent. This class can also be subdivided into two: a subclass of linear systems whose parameters change in time, and a subclass of systems containing explicitly time-dependent nonlinear parameters. We separate the second subclass here, since the phenomena and the typical regularities occurring in numerous practical cases, such as, for example, modulation-parametric systems or systems to which the small signal theory applies, are similar to those in the class of linear-parametric systems.

In the general case, nonlinear-parametric systems constitute the most general and extensive fourth class of systems, including the remaining three classes as particular cases.

The topicality of the developing theory of second-class radiophysical systems in the possibly most general form is confirmed by the present-day condition of science and technology: 1. The success scored by solid state and cryogenic electronics has led to the appearance of new radiophysical elements; a new branch – bioelectronics – has also emerged; 2. A precise analysis of a broad range of universal, generator and converting radiophysical systems is only possible if they are considered as second-class systems with parameters changing in time; 3. The behaviour of the radiophysical systems in the majority of the practical cases is determined by the „resonance" phenomenon which has been insufficiently investigated with respect to second-class systems, including the general problems of stability.

In this connection, the task of developing a general method for analyzing second-class radiophysical systems that would be convenient for practical research, and of assessing the potential capacities of these systems has been formulated. The basic mathematical technique for investigating second-class systems presented in this book is the developed version of the method of complex amplitudes, which ranks closest to engineering practice.

The linear connection principle [112] is used as a basis for offering a method for analyzing nonlinear systems by using linear methods.

Systems described by differential equations of the second or any higher order with periodic coefficients have been tackled in a lot of works: beginning with the classical ones of Lyapounov [93] and ending up with the numerous publications by modern researchers, for example [78–89] and many more. In spite of the considerable number of instructive mathematical publications, the problem of analyzing qualitatively the free processes in a parametric oscillating circuit cannot be regarded as solved. There is an essential difference between the analysis of the abstract mathematical equation and the particular engineering-and-physical system. As a rule the engineering-and-physical problem consists of three parts. The first one allows using the physical properties of the system as a starting point to obtain its schematic and analytical description, as well as the respective mathematical equation. The second part consists in solving and exploring the equation obtained. The third part provides a physical-and-engineering interpretation of the results. The mathematical problem is a component of the engineering-and-physical one and constitutes the latter's second part. The powerful mathematical means used in its solution often allows obtaining in-depth results. Thus, in a certain sense, the engineering-and-physical approach is broader than the mathematical one, but the latter is more profound. When solving the engineering-and-physical problem, it is important to adapt and use adequately a relevant mathematical technique. The book seeks to combine the general formulation of the engineering-and-physical problem concerning the processes in a generalized oscillating circuit with the profundity of the mathematical exploration.

The general analysis of the stationary modes of oscillating systems with periodic parameters is of great theoretical and practical significance. Similar obstacles crop up in connection with many problems related to the Theory of nonlinear oscillations, in particular when investigating parametric amplification and generation of oscillations, frequency modulation, detection and conversion, suppression of undesired oscillations and intermodulation distortions, etc.

The phase plane is of considerable importance for the qualitative analysis of the free processes in a linear oscillating circuit (either autonomous or non-autonomous) with periodic parameters. We shall use the phase plane as a basis for a visual presentation of a dynamic picture of the multitude of free processes in non-autonomous periodic oscillating circuits. The possible transformations of the equations of an oscillating circuit with periodic and almost periodic parameters will be given and the expedience of using different equation forms will be analyzed. A qualitative picture of the free processes in an oscillating circuit will be presented on the basis of the mathematical theory of Hamiltonian systems.

Oscillating circuits with periodic parameters can be divided into two groups. One of them includes oscillating circuits where, given arbitrary initial conditions, free processes are limited. The other group, respectively, encompasses oscillating circuits, whose initial conditions can be selected in such a way as to ensure unlimited free processes. Each group of oscillating circuits, in its turn, is characterized by a set of stability and instability areas. It is particularly important to develop analytical approaches for determining the area of stability or instability to which the specific oscillating circuit belongs.

The „resonance" phenomenon occupies a peculiar position in natural and other sciences, in technology, in civil engineering subjects, in medicine, in the theory of musical instruments, in aeronautics theory, in rocket technics and in astronautics, etc. [4, 44-78, 113–116]. Resonance is often manifested in the world that surrounds us either as a highly useful phenomenon or as an extremely harmful one. Radio communications, radio broadcasting, television and the other radio engineering systems would be absolutely inconceivable without resonance. This book shows that resonance is quite multiaspectual and multiform even in oscillating circuits with constant parameters. Resonance phenomena in nonlinear oscillating systems are virtually boundless.

Linear systems are quite frequently identified as systems with constant parameters not only in textbooks and other teaching aids but in scientific works as well. This approach reduces drastically the class of linear systems, since it excludes linear systems with time-dependent parameters. The principle of linear connection, formulated relatively recently, has boosted the significance of linear systems with variable parameters, since it follows from this principle that if the whole set of linear systems can be studied, this will automatically lead to the establishment of the necessary scientific basis for investigating the processes occurring in nonlinear systems.

The theory of parametric resonance, created mostly by G. S. Gorelik and the school of L. I. Mandelstam and N. D. Papaleksi [46, 47, 50, 86, 115, 118], has been recognized and renowned throughout the world namely because it substantially broadens our notion of the resonance phenomenon. It has been proved that sine functions, to which harmonic resonators respond, are not „the simplest" oscillations. There is quite a broad class of functions which elicit response from resonators whose parameters change periodically in time.

The Theory of nonlinear systems has by now reached a high level of development. The objective difficulties related to the establishment of a general analytical basis uniting nonlinear systems are well-known. While the superposition principle is valid for the Theory of linear systems, as it brings together all possible approaches for their analysis, no similar general concept has been identified with respect to the Theory of nonlinear systems. Due to this, the Theory of nonlinear systems is in a sense a theory of particular cases. The relation between the own, free and forced oscillations in nonlinear systems is quite diluted and cannot always be identified. Even if the process is known under certain initial conditions, one cannot always use it as a basis for stating something definite about the oscillating process under other, even quite close, initial conditions. Hence, qualitative methods of analysis that allow of identifying the essential properties of the oscillating processes without going into great detail acquire particular significance in this situation.

Nonlinear systems have many distinctive features which hamper their classification. There are two large classes that are generally accepted: autonomous and non-autonomous systems. The latter may be non-autonomous due to the action of an external signal, and also due to a time-dependent change in the parameters of the system. The systems of the latter subclass contain the easily distinguished (according to their practical significance) set of nonlinear systems whose parameters are obviously time-dependent and governed by a periodic or almost periodic law.

The method of complex amplitudes can be adapted for analyzing nonlinear systems. Its application allows a considerable simplification of the intermediary analytical transformations. But this method has a specificity: it clearly distinguishes positive from negative frequencies, though they are physically inseparable. Hence, an artificial unification of positive and negative frequencies within the method of complex amplitudes can lead to mistakes. Frequently it is expedient to use the method of complex amplitudes for the intermediate analytical transformations and then to switch to another method for the rest of the analysis.

The most typical property of any oscillating circuit with constant parameters is the fact that a sine action exercised in a stationary mode leads to a sine response, i.e. a sinusoid gives rise to another sinusoid. That is why the amplitude vrs. frequency characteristic of an oscillating circuit together with its phase vrs. frequency characteristic represent a complete description of the circuit adequate to the processes occurring in it. In the case of a nonlinear oscillating circuit, one can imagine that a single sinusoid engenders an infinite set of sinusoids. No characteristic obtained by accounting only for the first harmonic can be regarded as a complete characteristic adequate to the processes in the nonlinear circuit. In principle, there can be nonlinear circuits whose response to sine excitation may not contain a first harmonic at all. Hence, the purpose of developing the theory of resonance in a nonlinear oscillating circuit is to come up with a general approach covering the whole real spectrum.

The processes of self-organization and adaptive grouping in stable formations are an object of analysis not only in the living world but also in technical systems, where they are related with different phase and bifurcation transitions and with a specific response of the system to external action [32–38, 42, 78, 117]. At present these issues are referred to the new scientific domain of Synergetics [32], and more generally to the theory of self-organizing systems. An essential role in the self-organization and adaptive grouping in radiophysical and electromechanical systems is played by the reversibility of the interactions, by the bifurcation nonlinearities and cooperative effects when feedbacks are created [32–38, 78]. A primary form of self-organization in the non-living world is the spatial grouping of separate particles (component parts, subsystems) into more complex systems with definite stability. In addition, the following important situation occurs: the objects grouped together into stable formations are, as a rule, oscillating systems.

Nowadays the interest of an enormous number of researchers from different scientific areas and schools is attracted by the issues of self-organization of matter and the emergence of regular motions, by the processes of adaptive maintenance of definite, virtually unchanging, parameters of the systems grouped in stable formations, by the adaptive adjustment of the systems to external action, etc. [32–38, 42, 78, 117]. The issue of the essence and mechanisms of the grouping of the elementary material objects (microscopic and macroscopic systems) into stable formations is related to one of the most topical and important tasks of modern science. To one extent or another, all of the following can be referred to the „history" of the issue under consideration: the forecast of the atomic composition of matter, the discovery of the periodic system of elements, the rise and evolution of the ideas of electric, oscillating, wave and quantum phenomena, etc. One should mention here the experiments and the scientific treatises of R. Hook (1635-1703), G. Mechtenberg (1744-1799), M. Faraday (1792-1867), V. Berknes (1862-1951), J. Ch. Bousse (1859-1937), P. N. Lebedev (1866-1942), B. B. Golitsin, S. A. Boguslavskiy, M. I. Mandelstam, N. D. Papaleksi [44-53, 118]. The works of the classics listed above lay out the foundations of the theory of periodic motions, of ponderomotive interactions of different oscillating systems. A. Einstein made wide use of oscillating analogues to describe the heat-consuming property and solidity of substances. It is worth noting that in spite of the availability of a number of wonderful models (those of Gins, Tyuring, Eigen, Zhabotinski-Belousov and many more), the modern stage in the development of synergetics as a scientific area is related to the analysis of most simple basic models, which serve as a foundation for exploring some general features of the processes of adaptive grouping and self-organization.

This book has the task of studying definite synergic aspects of oscillating systems with a limited number of degrees of freedom. Some processes of ponderomotive equilibrium and continuous oscillations in grouped systems are examined. A new subclass of systems is formed, i.e. modulation nonlinear parametric systems with adaptive phase self-adjustment. Figuratively speaking, objects that fall into groups and share a property designated as $1 + 1 \neq 2$ are studied. In other words, two grouped objects have new properties that neither of them possesses separately. In the case of radiophysical systems, the properties regarded as most important are the adaptive self-tuning of the system as a whole, as well as the consistency of some internal characteristics and parameters of a separate element under changing external actions.

M. Plank [119] dedicated a cycle of his works to the issue of the absorption and radiation of electromagnetic waves by a system of linear resonators. He believed that in the case of irradiation with an electromagnetic wave the latter changes its resonance properties itself, thereby providing „multiple resonance", i.e. one and the same source of electromagnetic waves excites quite different resonators (as regards their own frequency). A result achieved in this book shows that the „multiple resonance" mechanism lies neither in the properties of the excited resonators, which can be strictly linear, nor in the emitters – it lies in the nonlinearity of the very process of interaction between the electromagnetic wave and a separate resonator. Moreover, a phenomenon of „quantization" of the possible amplitudes of the oscillations excited in each resonator, taken separately, is found.

The basic synergetic system for analytical purposes is the system of two interacting electric oscillating systems with a mechanical degree of freedom. The conditions for adaptive grouping of the system in stable formations are studied.

A class of modulation nonlinear-parametric systems and phenomena with adaptive self-tuning of the phase is created. For the sake of brevity they are termed „argument systems (phenomena, oscillations)" or „an argument method of oscillation excitation". This class corresponds to new initial conditions: nonlinear or linear oscillating systems experience the action of external periodic forces that are nonlinear as regards the coordinate of the excited system. The new initial conditions determine a specific action on the argument of motion in the system, which hunts with respect to the argument of speed and acceleration. For example, a linear differential equation of argument oscillations may be written in the form

$$\ddot{x}(t) + 2\beta\dot{x}(t) + \omega_0^2 x(t + \Delta \sin pt) = 0.$$

Obviously, the argument processes are described by equations that cannot be transformed into equations with periodic coefficients, which are typical of parametric phenomena, i.e. into Mathieu's and Hill's equations. In fact these are equations with constant coefficients and a specific nonlinear right-hand part.

Given a non-zero amplitude Δ of the argument external force, over a certain part of the period $\frac{2\pi}{p}$ the argument of the shift along the coordinate will run ahead of the argument of acceleration, while during the remaining part of the period it will lag behind that argument. Under certain conditions, a predominant delay of the coordinate argument with respect to the argument of acceleration may occur in various oscillating systems (radiophysical, electromechanical, etc.), which may lead to the excitation of continuous oscillations.

The foundation for studying these processes is the analysis, from the general positions of the Theory of nonlinear oscillations, of the processes and phenomena well-known in Radiophysics, Electromechanics, Microwave electronics, Optics, technology for accelerating charged particles, etc. [4, 42, 69, 78, 120–149]. The general regularities identified during the analysis allow for a model consideration of processes with identical properties in relatively simple oscillating systems with a limited number of freedom degrees, for formulating a definite mechanism for exciting stable continuous oscillations, for substantiating – both theoretically and experimentally – a number of new, previously unknown but objectively existing, properties and regularities of the oscillating systems.

The mechanism of exciting natural or close to natural oscillations through the action of an external high frequency force allows obtaining stable periodic regimes with high multiplicity of unitary frequency conversion, high efficiency, a discrete set of possible amplitudes, self-control of the input of energy from the external source, high stabilization coefficient in the case of changing external action amplitude and load of the oscillating system in a very wide range.

The model examination of the processes of wave interaction with oscillators allows to create a method of converting waves from the light band into submilimeter waves. This issue is directly related to the problems of the controlled thermonuclear reaction.

The book has the following structure:

Chapter 1 formulates a principle of reversibility of the modulation-parametric interactions which underlies the effective control of equivalent impedances in radiophysical systems. This serves as a basis for a classification of the radiophysical systems containing forces aimed at changing the effective reactive parameters and the dissipation in these systems. The classical Manley-Row energy relations are generalized for an arbitrary number of signal sources in the conditions of linear and periodic in time reactance, or nonlinear reactance with explicit time dependence. A model of a modulation-parametric modulator and a generalized conversion matrix are presented. An analytical technique for investigating modulation-parametric, conversion, autodyne and other processes and modes in autonomous and non-autonomous generator systems under the action of small external signals is developed.

Chapter 2 presents a theoretical generalization of the modulation-parametric interaction in nonlinear and parametric oscillating systems with external pumping and generator oscillating systems. Nonlinear and modulation-parametric phenomena occurring in several classes of radiophysical systems – four-frequency modulation-parametric systems, generator non-autonomous systems, etc. are described. It is shown that under certain conditions, modulation-parametric interactions lead to an effective change in the value and sign of the frequency-determining (reactive) parameters and of the dissipation, and to the appearance of new conversion properties and regularities, etc. Methods for controlling impedance parameters in modulation-parametric systems with external pumping and non-autonomous generator systems are developed. A number of effects occurring in generator radiophysical systems with synchronizing and asynchronous external action are presented, they are analyzed theoretically and methods for signal amplification, conversion and processing are offered. The peculiarities of the modulation-parametric interactions as reflected in the conversion properties, spectrum and stability of microwave autodyne systems for short-range radars are studied. Autodyne systems described by differential equations of the second and third order are analyzed in theoretical and comparative terms. Different applications of one-ports with negative parameters based on four-frequency parametric systems and non-autonomous generators are presented, i.e. modulation-parametric devices in systems for optical television and radio engineering reconnaissance, vidicons for security purposes, pyroelectric receivers for special fire-warning systems, special sensor systems, including an inductive one for analyzing the vibrospectrum of the body and other elements of machines, aircraft, marine vessels, railway and motor vehicles.

Chapter 3 conveys the investigation of nonlinear and parametric resonance in a generalized radiophysical oscillating circuit. A generalized modified method of complex amplitudes for analyzing processes in nonlinear radiophysical oscillating systems in a general form is offered. Some issues related to the realization of parametric one-ports are considered. A theory of diffusion impedance and detection with a semi-conductor diode in the case of high frequencies and increased signal amplitudes is presented. The influence of the effect of charge accumulation by non-basic carriers on the diffusion conductance and frequency characteristic of the detection with a p-n junction is investigated. So is the latter's work in resonance systems under the real conditions of a periodic signal with large amplitude, at high frequencies and with various loads.

Chapter 4 considers issues of the chaotization of oscillations in radiophysical systems. Some basic concepts, such as bifurcation, chaos, strange attractor, fractal dimension are presented. The conditions and basic mechanisms of transition from determined to chaotic oscillations are described. The basic methods and criteria for identifying chaotic oscillations are provided. The conditions for the emergence of specific instabilities are theoretically considered as stochastization of the oscillations in generator systems with additional delaying feedback without any random actions. It is shown that oscillation chaotization with complete loss of the informative qualities of the system is possible in the case of a system with delayed self-action, such as an autodyne system for short-range radars.

Chapter 5 develops a version of the method of complex amplitudes suitable for analyzing linear systems with periodic and almost periodic parameters. It is shown that this method can be used to analyze not only the stationary mode but the transition process as well – the concept of complex frequency is employed to this end. Properties of the elements of radiophysical systems, such as time-dependent periodic and almost periodic resistances, capacitances, inductances, are considered; their energy characteristics are studied. The mechanism of power transformation with reactive and active elements is investigated.

Chapter 6 presents the resonance properties of a generalized non-conservative oscillating circuit with constant parameters. It is shown that the „resonance" phenomenon refers not to the circuit but to a definite physical value: capacitor charge, magnetic flux of an inductance, etc. When the circuit is considered as a one-port, the concepts of amplitude resonance and phase resonance are used. The energy balance in an oscillating circuit with constant parameters is analyzed.

Chapter 7 contains formulations of general theorems for systems with positive elements concerning the relation between the parameters of the system and the matrix elements of the respective differential vector equation. Infinite systems of algebraic equations for some typical systems with periodic parameters are obtained; the properties of these systems of equations are explored, the general case relation between the complication of the radiophysical systems and the respective alteration of the systems of equations describing them is identified. Solutions of the infinite systems of algebraic equations are constructed in the form of infinite continued fractions. A clear dynamic picture of the multitude of free processes occurring in an oscillating circuit with periodic and almost periodic parameters is given. Theorems concerning the conditions that would be sufficient for the stability or instability of an oscillating circuit with periodic parameters (parametric oscillating circuit) are formulated. The phenomenon „resonance" in a linear oscillating circuit with periodic parameters is analyzed. A generalized interpretation of parametric resonance is provided by using visualizing geometrical categories. The equation of the parametric oscillating circuit is correlated with an expanded hyperplane of the solutions, forming a four-dimensional Euclidean space provided with an oxygonal coordinate system. Linear systems with almost periodic parameters are examined and the concept of „branching" of almost periodic functions is introduced. It is characterized by a positive integer depending on the number of the incommensurable frequencies in the spectrum. An adequate analytical technique for studying linear radiophysical systems with almost periodic parameters is presented. The concept of spatial matrixes is introduced and methods of multi-index multiplication of space matrixes are provided. A classification of linear systems with almost periodic parameters by the respective multi-dimensional infinite systems of linear algebraic equations is offered.

Chapter 8 offers a general analytical approach to the realization of the principle of linear connection as applied to nonlinear radiophysical systems. The phenomena of „strong" and „weak" resonance in a nonlinear system with explicitly time-dependent parameters are considered. It is shown that the output power spectrum of the microwave generators, where nonlinear interelectrode capacitances are connected to resonator systems, largely depends on the number of the degrees of freedom in them. If the number of the degrees of freedom is two or more, the oscillations in the generator are almost periodic and their branching capacity increases.

Chapter 9 presents a generalized theory of adaptive (synergetic) grouping of coupled resonance systems into stable electromechanical formations. An oscillating system with three degrees of freedom is studied. It consists of two linear oscillating systems coupled through a magnetic and electric link and capable of changing the distance between them. The generalized conditions for stability of the coupled oscillating systems are studied on the basis of a system of three differential equations describing the electric and mechanical processes in them. A physical effect consisting in an unambiguous correspondence between the frequency of the exciting generator and the relevant stable state of the system (including configuration and resonance properties) is presented. A system of two linear oscillating circuits linked with a capacitor, one of whose plates can be shifted freely, is separately explored. It is shown that the position of the mobile plate is unambiguously determined by the frequency of the exciting generator connected to one of the coupled oscillating circuits, i.e. that the capacitor „follows" the frequency of the generator by adequately changing the distance between the plates. An expression for the ponderomotive force seeking to change the position of the mobile plate in relation to the fixed one is obtained. The condition for a zero ponderomotive force is used as a starting point for establishing the dependence between the stationary values of the linking capacitance and the distance between its plates as a function of the frequency. A theory of resonance connection frequencies (natural resonance frequencies) of a radiophysical system with 2.5 degrees of freedom in the conditions of adaptive grouping of the system in a stable electromechanical formation is developed. An effect of „frequency attraction" is described. It differs, both in quality and in principle, from the effects of frequency entrainment and synchronization widely known in Radiophysics and Mechanincs.

Chapter 10 presents a theory of the newly-formed class of modulation nonlinear-parametric phenomena and systems with adaptive phase self-tuning (kick-excited self-adapting systems). The theory includes two major elaborations: a general analysis of the dynamic properties and regularities in a linear or nonlinear resonance system under the impact of a nonlinear-on-the coordinate external periodic force and a linear or nonlinear resonance system under the action of a falling electromagnetic wave. A phenomenon of excitation of continuous oscillations with a discrete set of stable amplitudes in linear or nonlinear oscillating systems under the action of an external periodic force, nonlinear on the coordinate of the system subject to excitation, is described. This is a peculiar „quantization" by the parameter intensity in a macro system. The phenomenon has a high degree of generality and besides radiophysical systems, it also occurs in various other material media of an oscillating nature. The novelty of the phenomenon is based on the novelty of the premises presupposing the existence of an external exciting force, nonlinear on the coordinate of the system subject to excitation. The new premises predetermine the qualitatively new characteristics of the phenomenon: „quantization" (discreteness) of the stationary amplitudes, adaptive self-sustenance of the oscillations in the situation of external and internal changes in parameters and impacts, etc. Some analytical dependencies determining the discrete set of stable amplitudes for both nonlinear and linear oscillating systems are derived. It is shown that the key condition for rendering the possible stationary amplitudes discretization is the nonlinear dependence of the external impact force with respect to the coordinate of the system subject to excitation. Furthermore, the system tends towards a stationary mode with a phase belonging to a discrete set of favourable phases. This is the gist of the adaptive nature of system-grouping in a discrete series of stable dynamic states, adaptive tuning and sustenance of oscillations with preset kinematic parameters. The fundamental investigations are based on a model nonlinear differential equation describing a number of real systems, such as simple pendulum, Josephson superconducting junction, oscillating motions of space particles, etc. Special emphasis is laid on the „quantization" of the oscillations in an oscillator under the action of a falling electromagnetic wave. Data from the numerical and natural experiment exploring the characteristic regularities of the phenomenon are quoted. A diagram involving multiple bifurcations for the attractor set of the system under consideration is obtained and analyzed. The complex dynamics, evolution and the energy boundaries of the multiple attractor basins in state space corresponding to energy and initial phase variables are obtained, traced and discussed. An analytic proof is presented showing the existence of „quantized" oscillations for the kick-excited pendulum. An analytic approach is given applicable to the cases of small and large amplitudes (small and large nonlinearity). The spectrum of possible oscillation amplitudes for the pendulum is studied as well as its motion in a rotational regime under the influence of an external non-homogeneous periodic force. Generalized conditions for the excitation of pendulum oscillations under the influence of an external nonlinear force are derived. A generalized model of an oscillator, subjected to the influence of an external wave is considered. It is shown that the systems of diverse physical background, which this model encompasses by their nature, should belong to the broader class mentioned of „kick-excited self-adaptive dynamical systems". Derived also are generalized conditions for the transition of systems of this „oscillator-wave" type to non-regular and chaotic behaviour. For the purpose of demonstrating the heuristic properties of the generalized „oscillator-wave" model from this point of view are considered the relevant systems and phenomena of the kick-rotator, quantized cyclotron resonance and the mega-quantum resonance-wave model of the Solar System. We point to a number of other natural and scientific phenomena, which can be effectively analyzed from the point of view of the developed approach. In particular we stress on the possibility for development and the wide applicability of specific wave influences, for example for the improvement and the speeding up of technological processes. A number of practical applications of the method of exciting „quantized" oscillations in macro systems are presented.

Due to the considerable volume of the material in the book, it is impossible to detail elaborately all the results concerning the issues under consideration. Hence a number of issues as well as some applications are only most broadly outlined and in such cases relevant works of the author are quoted. The purpose is to focus on the theoretical generalizations and on the new theoretical points, presenting them most comprehensively and in a logical sequence. The range of references is sufficiently extensive to give the reader an idea of the scale of the investigations carried out in this and other related areas in recent years, moreover one should bear in mind that modern phenomena and methods are invariant with respect to the material carrier and that they are elaborated in the sources from the perspective of different though close scientific disciplines.

The different paragraphs in the book, as well as all mathematical expressions, tables and figures have two-digit numbers indicating the chapter they belong to.

The denotations used in the book are not listed separately in the beginning since the enormous volume of the material, as well as the relative independence and the peculiarities of the different chapters do not allow treating these denotations in a systematic and uniform manner. Wherever possible, we have used universally accepted denotations.

Not having an usual practice of English, the quality of the English of this book is certainly affected. The reader may excuse this fact.

CHAPTER 1.

PRINCIPLE OF REVERSIBILITY OF MODULATION-PARAMETRIC INTERACTIONS

1.1. Classification of oscillating systems, which give rise to forces seeking to change the effective reactive parameters and dissipation

Under certain conditions one can observe changes in the equivalent frequency determining (reactive) parameters and dissipation in oscillating systems, for example:

— realization of negative or additional positive C (capacitance), L (inductance) and R (resistance) while using non-linear oscillating circuits [78, 151–153];

— alteration of the elasticity parameter (introduction of a positive or negative equivalent elasticity) of a mechanical oscillator in a capacitive or inductive sensor, including a gravitation wave transducer, a piezoelectric resonance sensor, etc. [78, 154, 155];

— excitation of low frequency oscillations in the case of a high frequency non-homogeneous action, including excitation of pendulum oscillations under the action of a high-frequency force [78, 156, 157] or, if the matter is treated differently, division of a frequency with a high coefficient of single conversion [78, 158];

— emergence, in powerful accelerators, of oscillating mechanical instabilities of the diaphragms included in the composition of the electric resonators [125, 127, 143, 147, 159, 160];

— introduction of relatively big differential elasticity from the light flux into the optical indicators of small mechanical shifts (this elasticity can be conventionally termed „light elasticity") both in the case of using diffraction gratings and when employing interferometers [122, 129, 146, 161];

— appearance of rotational instability in artificial space objects as a consequence of the generation of additional differential elasticity [162–165];

— inducement of low-frequency electromagnetic oscillations in a waveguide diode system under the action of a microwave field [142, 144];

— occurrence of a low-frequency wide-band negative resistance in a microwave generator with an avalanche-and-drift diode [166], a Gunn diode, a tunnel diode, and a charge accumulating diode [167, 168], in Josephson SQUIDs [140], etc. [169].

All cases of effective (equivalent) alterations of reactive and active (dissipative) impedance parameters referred to above can be covered by a general classifying scheme [78, 170, 171].

In the general case, there are forces in the oscillating systems seeking to change the equivalent frequency-determining (reactive) parameters and dissipation when:

a) the system is a non-linear oscillating system (an oscillator, a vibrator) where forced oscillations are excited;

b) the system is a linear oscillating system (an oscillator, a vibrator) with an existing possibility for modulatory action on one of the parameters;

c) the system is a parametric oscillating circuit;

d) the system is a generator in a non-autonomous mode.

The forces listed above are manifested both in radiophysical and in mechanical and other oscillating systems. The mechanism of occurrence of these forces is parametric and it consists in the reversibility of the modulation-parametric interactions conditioning the formation of positive and negative feedbacks. In electric modification particularly interesting are C, L and R parametric modulation systems, allowing realization of low-noise wide-band negative capacitances, inductances and resistances in a wide frequency range.

The parametric method of inputting, transforming or transferring energy is mostly based on a change of an energy accumulating (reactive) parameter [78, 85]. The most typical feature of parametric processes is the generation and mutual transformation of combined frequencies. Two opposite processes – a process of generation of combined frequencies in the case of interaction of an input signal with the parametric element, and a process of „reverse" transformation of combined frequencies with the participation of the same parametric element – occur simultaneously and in an inseparable unity. In practice these two processes are inseparable and interdependent. The composition of the combined frequencies generated during the „direct" process as well as their amplitude and phase relations are determined by the nature of the external circuits. The „reverse" process of transformation of the combined frequencies into the initial spectrum of the input signal, and into other combined frequencies, takes place simultaneously. This mutual transformation determines the reaction of the system with respect to the input signal or to the signals from the other combined frequencies, as well as its regenerative or degenerative nature.

It is in this sense that the principle of reversibility of the modulation-parametric interactions is formulated [78, 171] in relation to a mutually reversible transformation and mixing of the signals with the involvement of the parametric element. The process is accompanied by a change in the effective (equivalent) impedances. It is worth noting that the principle of reversibility of the modulation-parametric interactions is formulated by taking into account the property of reciprocity (mutuality) of parametric circuits. This approach gives prominence to the „modulation-parametric" interactions and not to the „parametric" ones, since, on the one hand, the modulation parametric interactions have all properties of manipulations of the type of amplitude or frequency-and-phase modulation, and, on the other hand, the modulation results in the generation of combined frequencies, which, in keeping with the principle of reversibility of the modulation parametric interactions in the presence of an energy accumulating (reactive) parametric element, become energy carriers.

1.2. Generalization of Manley–Row's classical energy relations

The analysis of the interactions in reactive (energy accumulating) elements with varying parameters is an important section of the General theory of oscillations. Such processes underlie parametric amplifiers, generators and converters, which occupy an important position among modern radiophysical devices [78, 85, 107].

Manley–Row's classical energy relations [172, 173] provide a most comprehensive and general description of the generation of an infinite spectrum and of the energy allocation by combined frequencies during the interactions in a nonlinear reactive (energy accumulating) element without losses.

In actual fact, Manley–Row's relations reflect the law of the conservation of energy as applied to a definite class of radiophysical systems.

Within the framework of the principle of reversibility of the modulation parametric interactions, formulated in 1.1, these relations reflect analytically the „direct" process of transformation and production of combined frequencies. By performing the respective mathematical operations with respect to the frequency component we are concerned with, we can reproduce the „reverse" transformation process analytically and thus get a general idea of the reaction of the system and the influence of the overall spectrum on a certain combination component.

Thus Manley–Row's energy relations allow getting a general idea of the operation of the respective device in a stationary mode without analyzing the specific circuit (system), for an arbitrary type of a non-linear reactance and at arbitrary power levels: first, they allow estimating the limiting gain coefficients, and, second, of analyzing stability as a whole.

Manley–Row's relations are derived for a case where two generators with different and incommensurable frequencies f_0 and f_1 act in a circuit with a non-linear capacitance in the situation of guaranteed active load for each frequency in the infinite spectrum of combined frequencies mf_0+nf_1, where $m, n = \pm1, \pm2, \pm3, \ldots \pm \infty$. For this case Manley and Row obtained the following relations:

$$\begin{aligned} \sum_{m=0}^{\infty} \sum_{n=-\infty}^{\infty} \frac{mP_{m,n}}{mf_0 + nf_1} &= 0, \\ \sum_{m=-\infty}^{\infty} \sum_{n=0}^{\infty} \frac{nP_{m,n}}{mf_0 + nf_1} &= 0, \end{aligned} \tag{1.1}$$

where $P_{m,n}$ is the active power of the respective combined frequency $mf_0 + nf_1$.

We set ourselves the task of generalizing Manley–Row's relations for the case, when an arbitrary number of generators of harmonic oscillations act on a linear and time-dependent periodic reactance [174, 175]. Formulated in this way, the task is directly related both to the classical radiophysical systems with mechanically changeable reactances [118] and to the modern parametric amplifiers, modulators and converters, whose input circuits in the approximation of the small signal theory can be regarded as linear circuits with periodically changing reactances [85].

We shall make the analysis by using a special version of the method of complex amplitudes [174] presented in detail in Chapter 5 of this book. We shall demonstrate the major transformations and denotations on the basis of an abstract example:

$$p = P\cos(\omega t + \varphi) \leftarrow \hat{p} = \dot{P}e^{j\omega t}\ , \quad \dot{P} = Pe^{j\varphi}.$$

To the left of the arrow is the function subject to transformation (the original), and to its right is the function obtained as a result of the transformation (the symbol-function or transform). Parallel with the symbol $\hat{p}$, we also use another one, complex conjugate with respect to it: $\check{p} = \overset{*}{P}e^{-j\omega t}$, $\overset{*}{P} = Pe^{-j\varphi}$. The following denotations are introduced, too: „o" – symbolic multiplication, „⊢" – symbolic division. These operations are interpreted in the following way (we use the function $q = Q\cos(\Omega t + \Psi)$, too):

$$p.q \leftarrow \hat{P} \circ \hat{q} = \frac{\hat{P} + \check{P}}{2}\hat{q}, \quad \hat{q} = \dot{Q}e^{j\Omega t}, \quad \dot{Q} = Qe^{j\Psi},$$

$$\frac{p}{q} \leftarrow \vdash\frac{\hat{P}}{\hat{q}} = \frac{2\hat{P}}{\hat{q} + \check{q}}, \quad \check{q} = \overset{*}{Q}e^{-j\Omega t}, \quad \overset{*}{Q} = Qe^{-j\Psi}.$$

Let us consider a linear and time-dependent periodic capacitance $C = C(t)$. The capacitance $C(t) = C(t + T)$, changing in accordance with a periodic law with a period T, can be harmonically expanded with the participation of frequencies multiple of the frequency $\omega = \dfrac{2\pi}{T} = 2\pi f$. Since the superposition principle is valid for the linear capacitance, we shall consider the interaction with just one generator with a frequency f_1, and the result can then be generalized for the case of an arbitrary number of generators.

The symbol-functions of the capacitance charge and current can be presented in the following form:

$$\hat{q} = e^{jx_1} \sum_{k=-\infty}^{\infty} \dot{Q}_k e^{jkx}, \quad x_1 = \omega_1 t = 2\pi f_1 t, \quad x = \omega t, \tag{1.2}$$

$$\hat{i} = \frac{d\hat{q}}{dt} = e^{jx_1} \sum_{k=-\infty}^{\infty} \dot{I}_k e^{jkx}, \quad \dot{I}_k = j(\omega_1 + k\omega)\dot{Q}_k, \tag{1.3}$$

while the symbol-functions of the voltage on the capacitor is expressed as

$$\hat{u} = \frac{\hat{q}}{C(t)}. \tag{1.4}$$

We carry out harmonic expansion of the periodic function $\dfrac{1}{C(t)}$:

$$\frac{1}{C(t)} = \frac{1}{C\left(\dfrac{x}{\omega}\right)} = \sum_{k=-\infty}^{\infty} \dot{S}_k e^{jkx}, \quad \dot{S}_k = \overset{*}{\dot{S}}_{-k}, \tag{1.5}$$

$$\dot{S}_k = \frac{1}{2\pi}\int_0^{2\pi} \frac{1}{C\left(\dfrac{x}{\omega}\right)} e^{-jkx}\,dx.$$

After taking into account (1.3) and (1.5), we present the expression (1.4) in the form

$$\hat{u} = e^{jx_1} \sum_{k=-\infty}^{\infty} U_k e^{jkx}, \quad \dot{U}_k = \sum_{l=-\infty}^{\infty} \dot{Q}_l \dot{S}_{k-l}. \tag{1.6}$$

The following sum is introduced:

$$\begin{aligned}\sum_{k=-\infty}^{\infty} jk\overset{*}{\dot{Q}}_k \dot{U}_k &= \sum_{k=-\infty}^{\infty} jk\overset{*}{\dot{Q}}_k \sum_{l=-\infty}^{\infty} \dot{Q}_l \dot{S}_{k-l} \\ &= \frac{1}{2\pi}\int_0^{2\pi} \frac{1}{C\left(\dfrac{x}{\omega}\right)} \left(\sum_{k=-\infty}^{\infty} jk\dot{Q}_k e^{-jkx}\right)\left(\sum_{l=-\infty}^{\infty} \dot{Q} e^{jlx}\right) dx.\end{aligned} \tag{1.7}$$

After taking into consideration (1.2) and (1.3), the sums in (1.7) can be expressed as

$$\sum_{l=-\infty}^{\infty} \dot{Q} e^{jlx} = \hat{q} e^{-jx_1}, \quad \sum_{k=-\infty}^{\infty} jk\dot{Q}_k e^{-jkx} = -\frac{\partial \check{q}}{\partial x} e^{jx_1}. \tag{1.8}$$

We substitute (1.8) in (1.7) to come up with

$$\begin{aligned}\sum_{k=-\infty}^{\infty} jk\overset{*}{\dot{Q}}_k \dot{U}_k &= -\frac{1}{2\pi}\int_0^{2\pi} \frac{1}{C\left(\dfrac{x}{\omega}\right)} \hat{q}\left(\frac{x}{\omega}\right) \frac{\partial \check{q}\left(\dfrac{x}{\omega}\right)}{\partial x} dx \\ &= -\frac{1}{2\pi}\int_{\check{q}(0)}^{\check{q}(2\pi)} \frac{1}{C\left(\dfrac{x}{\omega}\right)} \hat{q}\left(\frac{x}{\omega}\right) d\check{q}\left(\frac{x}{\omega}\right).\end{aligned} \tag{1.9}$$

Since for $x_1 = \text{const}$ the symbol-functions of the charge $\hat{q}\left(\dfrac{x}{\omega}\right)$ and $\check{q}\left(\dfrac{x}{\omega}\right)$ are periodic functions by x with a period 2π, the integral in (1.9) is equal to zero. Therefore we obtain

$$\sum_{k=-\infty}^{\infty} jk\overset{*}{\dot{Q}}_k \dot{U}_k = 0. \tag{1.10}$$

Bearing in mind that $\overset{*}{Q}_k = \dfrac{\overset{*}{I}_k}{-j(\omega_1 + k\omega)} = \dfrac{\overset{*}{I}_k}{-2\pi j(f_1 + kf)}$, we depart from (1.10) to obtain

$$\sum_{k=-\infty}^{\infty} \frac{k\overset{*}{I}_k U_k}{2\pi j(f_1 + kf)} = 0. \tag{1.11}$$

Denoting the average power, absorbed by the capacitor at frequency $f_1 + kf$, by $W_k = Re\left(\dfrac{\overset{*}{I}_k \overset{*}{U}_k}{4}\right)$, we write

$$\sum_{k=-\infty}^{\infty} \frac{kW_k}{f_1 + kf} = 0. \tag{1.12}$$

The resultant formula (1.12) is the general energy relation for the case of one generator with frequency f_1 and a linear capacitance undergoing a periodic change in time with frequency f.

If n generators with frequencies $f_1, f_2, \ldots, f_n$ are connected to the circuit of the periodic capacitance, each generator interacts independently of the others, in concord with the superposition principle. Hence, in this case, we can write n relations of the type of (1.12):

$$\begin{gathered} \sum_{k=-\infty}^{\infty} \frac{kW_k^{(1)}}{f_1 + kf} = 0, \\ \sum_{k=-\infty}^{\infty} \frac{kW_k^{(2)}}{f_2 + kf} = 0, \\ \cdots\cdots\cdots\cdots\cdots \\ \sum_{k=-\infty}^{\infty} \frac{kW_k^{(n)}}{f_n + kf} = 0, \end{gathered} \tag{1.13}$$

where $W_k^{(i)}$ is the power at frequencies $f_i + kf$, caused by the interaction between the i-th generator and the periodic capacitance.

Relations (1.13) can also be obtained for a linear periodic inductance by applying analogous reasoning, for the relations between the magnetic flux, the voltage, the current and the power of the inductance are analogous to those between the charge, the current, the voltage and the power of the capacitance.

Further on, we set ourselves the task of generalizing Manley–Row's energy relations for the case of an explicit time dependence of a non-linear capacitance governed by a periodic law [176]. Initially we shall analyze the action of two generators, and then we shall make a generalization for n generators.

In a stationary mode the capacitance charge can be presented in the form

$$q = \sum_{n_1=-\infty}^{\infty} \sum_{n_2=-\infty}^{\infty} \sum_{n=-\infty}^{\infty} Q_{n_1,n_2,n} e^{j(n_1 x_1 + n_2 x_2 + nx)}, \tag{1.14}$$

where $x_1 = \omega_1 t$, $x_2 = \omega_2 t$, $x = \omega t$; ω_1 and ω_2 are frequencies of the driving generators, ω is the frequency of explicit change of the non-linear capacitance in time. Since q is real, $Q_{n_1,n_2,n} = \overset{*}{Q}_{-n_1,-n_2,-n}$, where the symbol „*" means a complex conjugate quantity.

The current flowing through the capacitance is determined as

$$i = \frac{dq}{dt} = \sum_{n_1=-\infty}^{\infty} \sum_{n_2=-\infty}^{\infty} \sum_{n=-\infty}^{\infty} I_{n_1,n_2,n} e^{j(n_1 x_1 + n_2 x_2 + nx)}, \tag{1.15}$$

$$I_{n_1,n_2,n} = j(n_1\omega_1 + n_2\omega_2 + n\omega) Q_{n_1,n_2,n} = \overset{*}{I}_{-n_1,-n_2,-n}. \tag{1.16}$$

At this point we shall stress a peculiarity of the expressions (1.14) and (1.15), which is important for our further analysis. If x_2, x = const, then q is a periodic function of t with a period $T_1 = \frac{2\pi}{\omega_1}$, or if x_1, x = const, then q is a periodic function of t with a period $T_2 = \frac{2\pi}{\omega_2}$. The same holds true for current i. The right-hand parts of (1.14) and (1.15) are triple Fourier series. When taking into account the unambiguity of the volt-coulomb characteristic $U = U(q) = U[q(x_1, x_2, x)] = f(x_1, x_2, x)$, it becomes obvious that $f(x_1, x_2, x)$ is unambiguous and periodic by x_1, x_2, x, i.e. a valid presentation will be the following:

$$U = \sum_{n_1=-\infty}^{\infty} \sum_{n_2=-\infty}^{\infty} \sum_{n=-\infty}^{\infty} U_{n_1,n_2,n} e^{j(n_1 x_1 + n_2 x_2 + nx)}, \tag{1.17}$$

$$U_{n_1,n_2,n} = \overset{*}{U}_{-n_1,-n_2,-n},$$

$$U_{n_1,n_2,n} = \frac{1}{(2\pi)^3} \int_0^{2\pi} dx \int_0^{2\pi} dx_2 \int_0^{2\pi} e^{-j(n_1 x_1 + n_2 x_2 + nx)} f(x_1, x_2, x) dx_1.$$

We write out the sum

$$\sum_{n_1=-\infty}^{\infty} \sum_{n_2=-\infty}^{\infty} \sum_{n=-\infty}^{\infty} j n_1 \overset{*}{Q}_{n_1,n_2,n} U_{n_1,n_2,n}$$

$$= \frac{1}{(2\pi)^3} \int_0^{2\pi} dx \int_0^{2\pi} dx_2 \int_0^{2\pi} dx_1 f(x_1, x_2, x)$$

$$\times \sum_{n_1=-\infty}^{\infty} \sum_{n_2=-\infty}^{\infty} \sum_{n=-\infty}^{\infty} j n_1 \overset{*}{Q}_{n_1,n_2,n} e^{-j(n_1 x_1 + n_2 x_2 + nx)}. \tag{1.18}$$

We differentiate the Exp. (1.14) by x_1 and taking into account that $Q_{n_1,n_2,n} = \overset{*}{Q}_{-n_1,-n_2,-n}$, we obtain

$$\frac{\partial q}{\partial x_1} = -\sum_{n_1=-\infty}^{\infty}\sum_{n_2=-\infty}^{\infty}\sum_{n=-\infty}^{\infty} jn_1\overset{*}{Q}_{n_1,n_2,n}e^{-j(n_1x_1+n_2x_2+nx)}. \tag{1.19}$$

Expression (1.19) allows of simplifying Eq. (1.18):

$$\begin{aligned}&\sum_{n_1=-\infty}^{\infty}\sum_{n_2=-\infty}^{\infty}\sum_{n=-\infty}^{\infty} jn_1\overset{*}{Q}_{n_1,n_2,n}U_{n_1,n_2,n}\\ &= \frac{1}{(2\pi)^3}\int_0^{2\pi}dx\int_0^{2\pi}dx_2\int_0^{2\pi}f(x_1,x_2,x)dx_1\frac{\partial q}{\partial x_1}\\ &= -\frac{1}{(2\pi)^3}\int_0^{2\pi}dx\int_0^{2\pi}dx_2\int_{q(0)}^{q(2\pi)}f(x_1,x_2,x)dq.\end{aligned} \tag{1.20}$$

If we assume that $x_1, x_2 =$ const, the charge will be a periodic function with respect to x_1 with a period 2π, i.e. $q(x_1+2\pi) = q(x_1)$, so the last integral in (1.20) will be equal to zero. We obtain

$$\sum_{n_1=0}^{\infty}\sum_{n_2=-\infty}^{\infty}\sum_{n=-\infty}^{\infty} jn_1\overset{*}{Q}_{n_1,n_2,n}U_{n_1,n_2,n} = 0. \tag{1.21}$$

If we introduce the concept average active power

$$W_{n_1,n_2,n} + W_{-n_1,-n_2,n} = \frac{U_{n_1,n_2,n}\overset{*}{I}_{n_1,n_2,n}}{2} + \frac{\overset{*}{U}_{n_1,n_2,n}I_{n_1,n_2,n}}{2},$$

absorbed at frequency $n_1\omega_1+n_2\omega_2+n\omega$, and take into consideration relation (1.16), we can group together the expressions with equal positive and negative frequencies appearing in (1.21) to obtain

$$\sum_{n_1=0}^{\infty}\sum_{n_2=-\infty}^{\infty}\sum_{n=-\infty}^{\infty}\frac{n_1W_{n_1,n_2,n}}{nf_1+n_2f_2+nf} = 0. \tag{1.22}$$

A procedure analogous to that used in the formation and conversion of sum (1.18) can be employed to form and convert the following sums:

$$\sum_{n_1=-\infty}^{\infty}\sum_{n_2=-\infty}^{\infty}\sum_{n=-\infty}^{\infty} jn_2\overset{*}{Q}_{n_1,n_2,n}U_{n_1,n_2,n},$$

$$\sum_{n_1=-\infty}^{\infty}\sum_{n_2=-\infty}^{\infty}\sum_{n=-\infty}^{\infty} jn\overset{*}{Q}_{n_1,n_2,n}U_{n_1,n_2,n}.$$

Then we can obtain two more expressions of type (1.22).

As we generalize that for the case of an explicitly time-dependent and periodic nonlinear capacitance and two exciting generators, the following energy relations are obtained:

$$\begin{gathered}
\sum_{n_1=0}^{\infty}\sum_{n_2=-\infty}^{\infty}\sum_{n=-\infty}^{\infty}\frac{n_1 W_{n_1,n_2,n}}{n_1 f_1+n_2 f_2+nf}=0,\\
\sum_{n_1=-\infty}^{\infty}\sum_{n_2=0}^{\infty}\sum_{n=-\infty}^{\infty}\frac{n_2 W_{n_1,n_2,n}}{nf_1+n_2 f_2+nf}=0,\\
\sum_{n_1=-\infty}^{\infty}\sum_{n_2=-\infty}^{\infty}\sum_{n=0}^{\infty}\frac{n W_{n_1,n_2,n}}{nf_1+n_2 f_2+nf}=0.
\end{gathered}\tag{1.23}$$

Equations (1.23) can be generalized for the case of k generators by using the method of mathematical induction. Then we write $k+1$ equations:

$$\begin{gathered}
\sum_{\underline{n_1},\ldots,n_k,n}\frac{n_1 W_{n_1,n_2,\ldots,n_k,n}}{\sum_{i=1}^{k} n_i f_i+nf}=0,\\
\sum_{n_1,\underline{n_2},\ldots,n_k,n}\frac{n_2 W_{n_1,n_2,\ldots,n_k,n}}{\sum_{i=1}^{k} n_i f_i+nf}=0,\\
\cdots\cdots\cdots\cdots\cdots\cdots\cdots\cdots\cdots\cdots\\
\sum_{n_1,\ldots,\underline{n_i},\ldots,n_k,n}\frac{n_i W_{n_1,n_2,\ldots,n_k,n}}{\sum_{i=1}^{k} n_i f_i+nf}=0,\\
\cdots\cdots\cdots\cdots\cdots\cdots\cdots\cdots\cdots\cdots\\
\sum_{n_1,\ldots,\underline{n_k},n}\frac{n_k W_{n_1,n_2,\ldots,n_k,n}}{\sum_{i=1}^{k} n_i f_i+nf}=0,\\
\sum_{n_1,\ldots,n_k,\underline{n}}\frac{n W_{n_1,n_2,\ldots,n_k,n}}{\sum_{i=1}^{k} n_i f_i+nf}=0.
\end{gathered}\tag{1.24}$$

For the purpose of simplifying the written form, we have introduced an abbreviated denotation of the multiple sums, given that all variables subject to summation fall in the range from $-\infty$ to $+\infty$, except for the underlined variable, whose range is from 0 to ∞. For example:

$$\sum_{n_1,\ldots,\underline{n_i},\ldots,n_k,n}=\sum_{n_1=-\infty}^{\infty}\cdots\sum_{n_{i-1}=-\infty}^{\infty}\sum_{n_i=0}^{\infty}\sum_{n_{i+1}=-\infty}^{\infty}\cdots\sum_{n_k=-\infty}^{\infty}\sum_{n=-\infty}^{\infty}.$$

As expressions (1.23) and (1.24) show, the explicit change of the nonlinear capacitance in accordance with a periodic law is equivalent to an additional generator.

An analogous approach can be used for deriving energy relations for a nonlinear inductance that is explicitly time-dependent and changing in accordance with a periodic law.

1.3. Analytical techniques for investigating modulation-parametric phenomena in resonance systems with external pumping

In accordance with the principle of reversibility of modulation-parametric interactions, formulated in 1.1., the general approach to the analysis of the modulation-parametric phenomena in specific resonance systems consists of two major stages:

1. Formulation of system-describing equations, where a range of combined frequencies are present either explicitly or implicitly.
2. Solution of the equations with respect to an arbitrary spectral frequency from the range of combination spectrum or with respect to the overall spectrum of the input signal.

The first stage includes the „direct" process of generation of the spectrum of combination frequencies, while the second one encompasses the „reverse" conversion of the combined components and the reaction of the system conditioned by them. Thus the two processes of „direct" and „reverse" conversion of signal frequencies, which occur in parallel and are practically inseparable, are divided for analytical purposes.

The specific application of the general analytical approach related above requires the employment of various mathematical procedures and methods. These methods can be divided into quasi-direct methods and methods utilizing certain aspects of the theory of perturbations. The quasi-direct methods or procedures for solving the problem exploring forced oscillations are based on the use of the eigen-functions of the equations, describing the eigen-oscillations in systems with periodically changing parameters. This approach was employed in the investigation of the forced oscillations in circuits with variable parameters, as presented in the classical works of G.S.Gorelik [50, 86], and of the parametric amplifiers [85].

When perturbation theory methods are used, the solution is sought in a temporal or frequency form. In most cases these are asymptotic methods related to the introduction of a small parameter (for example, [97, 177–179]), the quasi-linear method and the harmonic balance method in its different forms [88, 98, 180–183].

From a mathematical point of view, there are substantial differences between the circuits with nonlinear reactance and the circuits with time-dependent linear reactance. The superposition principle is valid for the latter case and the applicable theories include such powerful techniques as Fourier's spectral methods and the theory of Mathieu and Hill's equations [47, 184]. The superposition principle is not valid in the case of nonlinear oscillating circuits and the difficulties associated with any precise analysis increase considerably.

Linear-parametric systems are described by linear differential equations with variable coefficients. The investigation of such equations, which was initiated in the fundamental works of Floquet, A.M. Lyapunov, A. Poincaré and continued by L. I. Mandelstam, N. D. Papaleksi, A. Andronov, is being most intensively conducted at present [89]. It is well-known that the attainment of precise solutions of such equations even of a low order (first and second) is associated with serious calculation difficulties. That is why efforts are focused on a quality investigation of the properties of these equations and on the attainment of analytical solutions by using approximate methods.

From the perspective of the task set in this section, we can group the most popular practical methods of exploring modulation-parametric systems in the following way.

The methods using, in one form or another, the harmonic balance principle in the conditions of a finite number of harmonics can be referred to the *first group*. The advantage of the methods belonging to this type is the general form of the solution, which allows for a general qualitative exploration of the circuit properties. At the same time, the larger number of the harmonics accounted for leads to quite voluminous expressions. Besides, certain difficulties may arise - first, in the estimation of the method error, and second, even in the course of the qualitative analysis, since certain information concerning some subtle properties and effects conditioned by the higher harmonics can be lost as a result of the harmonic linearization.

The asymptotic methods related to the introduction of a small parameter can be referred to the *second group* [97, 177–179]. In spite of the overall universality and effectiveness of these methods, they have a disadvantage: they do not allow investigating modulation-parametric systems, if the modulation depth of the parametric element is considerable. The efforts of a host of authors have endeavoured to eleminate this cardinal shortcoming. Thus, for example, the monograph on the analysis of second order oscillating systems [185] offers a phase plane method with nonlinear conversion of the variables, which, coupled with the asymptotic methods, allows obtaining full information on the investigated system through the new parameters.

The *third group* encompasses any methods employing the overall frequency spectrum of the solution. Such methods are mostly based on the generalization of the method of infinite normal determinants in combination with the transformations of Laplace and Fourier. They allow making qualitative analysis of complex parametric circuits from a single point of view and obtaining sufficiently effective algorithms for computer-aided circuit exploration. We shall dwell on this type of methods in detail in Chapters 7 and 8 of this book.

Numeric methods should be referred to the *fourth group*, e.g. methods of the type of those of Runge-Kutta, applicable in the cases when a qualitative analysis of a specific circuit should be performed. An attempt to investigate the stationary mode directly, by eliminating the transition process, causes difficulties.

The theory and practice of developing and investigating modulation-parametric systems with external pumping indicate that the most appropriate generalized method for such purposes is that of complex amplitudes [78, 85], where relations are directly written in the form of spectral matrix equations linking the complex amplitudes of the combined products with the signal and the response. Spectral methods allow tracing the internal structure of the signal and making effective use of computers for the calculation part.

Since the number of the combined frequencies occurring in a parametric circuit is infinite, the spectral matrixes of such circuits are infinite as well. In practice selective systems are often explored by using what is known as the filter method [85], i.e. the number of the combined frequencies that are read is limited in advance. The filter method is inapplicable, when poorly selective or non-selective circuits are studied. In this case the reduction method can be used for solving the infinite system of equations describing the parametric circuit. The number of the combined frequencies that are accounted for, i.e. the number of the equations, is gradually increased and the calculations are discontinued at that numeric cycle, for which a preset precision is reached. Thus one can also determine the number of the combined oscillations, which describe the parametric circuit with a given accuracy.

Having clarified the major analytical problems associated with the investigation of the modulation-parametric systems, we set ourselves the task of creating an initial model and deriving a generalized matrix equation describing the processes in a four-frequency complex parametric modulator [150], sticking to the small signal theory and the filter hypothesis [85]. We shall account for four components – input signal, sum and difference combined frequencies and pumping frequency.

Fig. 1.1 shows two dual equivalent circuits – of a capacitance complex parametric modulator and of an inductance one, respectively. The following designations have been used: $\dot{I}_1$ and $\dot{V}_1$ are the complex amplitudes of the signal current and voltage with frequency ω_1, entering the parametric modulator; $\dot{V}_1$, $\dot{V}_+$, $\dot{V}_-$ and $\dot{I}_1$, $\dot{I}_+$, $\dot{I}_-$ are the complex amplitudes of the signals and the combined voltages and currents (the frequencies of the combined components are respectively $\omega_\pm = \omega_p \pm \omega_1$, where ω_p is the frequency of periodic change of the parametric elements $C(t)$, $G(t)$ and $L(t)$, $R(t)$). It is assumed that the loading complex conductances and resistances of the parametric modulator ($\dot{Y}_1$, $\dot{Y}_\pm$, $\dot{Z}_1$, $\dot{Z}_\pm$) satisfy the condition $\dot{Y}_i = \infty$ and $\dot{Z}_i = 0$ for all frequencies $\omega \neq \omega_i$, where the index $i \to 1, +, -$.

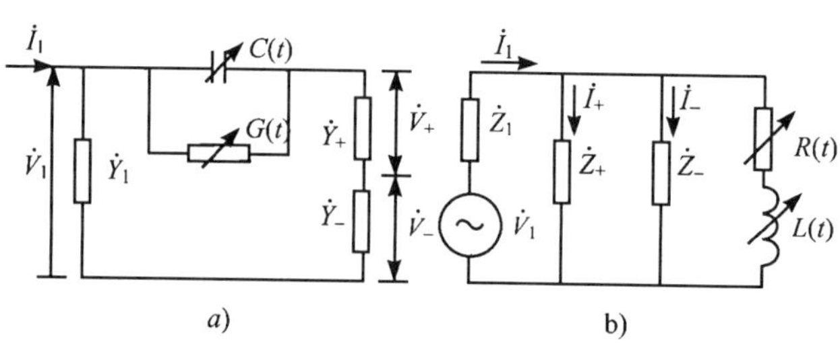

Fig. 1.1. Equivalent circuit of a capacitance (a) and inductance (b) complex parametric modulator

As the parametric elements are presented as Fourier series

$$\begin{Bmatrix} C(t) \\ G(t) \\ L(t) \\ R(t) \end{Bmatrix} = \begin{Bmatrix} C_0 \\ G_0 \\ L_0 \\ R_0 \end{Bmatrix} + 2\sum_{n=1}^{\infty} \begin{Bmatrix} C_n \\ G_n \\ L_n \\ R_n \end{Bmatrix} \cos n\omega_p t, \tag{1.25}$$

$$\begin{Bmatrix} C_n \\ G_n \\ L_n \\ R_n \end{Bmatrix} = \frac{1}{2\pi}\int_0^{\pi} \begin{Bmatrix} C(t) \\ G(t) \\ L(t) \\ R(t) \end{Bmatrix} \cos n\omega_p t \; d(\omega_p t)$$

where and as the spectrum of these elements is limited to the second harmonic, a generalized matrix equation, describing the processes in a parametric complex modulator as in Fig. 1.1 is derived directly:

$$\begin{vmatrix} \dot{\zeta} \\ 0 \\ 0 \end{vmatrix} = \begin{Vmatrix} \dot{H}_1 & D_1 e^{-j\Psi_2} + j\omega_1 S_1 e^{-j\Psi_1} & D_1 e^{j\Psi_2} + j\omega_1 S_1 e^{j\Psi_1} \\ D_1 e^{j\Psi_2} + j\omega_+ S_1 e^{j\Psi_1} & \dot{H}_+ & D_2 e^{j2\Psi_2} + j\omega_+ S_2 e^{j2\Psi_1} \\ D_1 e^{-j\Psi_2} - j\omega_- S_1 e^{-j\Psi_1} & D_2 e^{-j2\Psi_2} - j\omega_- S_2 e^{-j2\Psi_1} & \overset{*}{H}_- \end{Vmatrix} \begin{vmatrix} \dot{\eta}_1 \\ \dot{\eta}_+ \\ \overset{*}{\eta}_- \end{vmatrix}, \tag{1.26}$$

where for Fig. 1.1a: $S_j \equiv C_n$, $D_j \equiv G_n$, $\dot{H}_i \equiv \dot{Y}_i$, $\dot{\zeta}_1 \equiv \dot{I}_1$, $\dot{\eta}_i \equiv \dot{V}_i$; for Fig. 1.1b: $S_j \equiv L_n$, $D_j \equiv R_n$, $\dot{H}_i \equiv \dot{Z}_i$, $\dot{\zeta}_1 \equiv \dot{V}_1$, $\dot{\eta}_i \equiv \dot{I}_i$, $i \to 1,+,-$; the character „*" means a complex conjugate quantity. For the purpose of making the analysis more general, we have also introduced phase shifts of pumping of the complex parametric elements, where Ψ_1 is meant for the parametric capacitance $C(t)$ or parametric inductance $L(t)$, while Ψ_2 is meant respectively for the parametric conductance $G(t)$ or the parametric resistance $R(t)$.

The matrix equation (1.26) is derived in a simple way, by using the following approach. For example, for Fig. 1.1a the relation between the current i and the voltage V on the capacitive complex parametric element can be written in the form: $i = G(t)V + \frac{d}{dt}[C(t)V]$. Obviously each spectral component of the signal voltage V causes the appearance of several combined components in the spectrum of the current i (strictly speaking, when the repeated secondary conversions are accounted for, each spectral component of the signal causes a respective infinite spectrum of combined frequencies in the current flowing through the complex parametric element). The volume of the spectrum and the relation between the amplitudes and the phases of the combined components are determined by the external load, connected to the parametric element (in the equivalent scheme in Fig. 1.1a it is reflected through the complex admittances $\dot{Y}_1$, $\dot{Y}_+$, $\dot{Y}_-$). In reality only the first and the second harmonic of the spectrum of the current and the voltage play an essential part in the resonance modulation-parametric systems. In this case, as we express the voltage in the form

$$V = \dot{V}_1 e^{j\omega_1 t} + \dot{V}_+ e^{j\omega_+ t} + \dot{V}_- e^{j\omega_- t},$$

where $\omega_{\pm} = \omega_p \pm \omega_1$ are, respectively, the sum and difference combined frequencies, and as we make a substitution in the expression for the current i, it will not be difficult for us, after accounting for (1.25), to obtain a complex matrix equation of the type of (1.26).

The matrix equation (1.26) describes in a most general form the conversion processes in a four-frequency parametric system with a complex parametric modulator.

We shall consider some regenerative effects on the basis of Eq. (1.26) in 2.1, and some applications of parametric one-ports with wide-band negative parameters in 2.4.

While concluding this section, we should note that yet another approach to the analysis of four-frequency modulation-parametric systems is developed in publication [186]. The equivalent admittances of a partially open (forward-conducting) $p - n$ junction (one of the possible realizations of a complex parametric element) working in the conditions of a four-frequency parametric system are determined. The analysis is performed for the cases of both small and large signal amplitudes and of high frequencies. In addition, the effect of accumulation and the establishing of a quasi-stationary charge minor carriers in the base of the semiconductor diode is also accounted for (See Chapter 3). Some general expressions for the equivalent conductances in the case of the action of four signals with arbitrary amplitudes on the semiconductor diode with arbitrary volt-ampere and volt-coulomb characteristics are derived. The resultant general expressions also allow computing the equivalent conductances of devices with contacts of heterogeneous material – diodes with Schottky barriers, devices based on structures of the type: metal – dielectric – semiconductor, heterojunctions, etc. with arbitrary volt-ampere and volt-coulomb characteristics.

This approach allows determining the equivalent impedance of the complex parametric element for an arbitrary spectral frequency and, on that basis, calculating the basic parameters and selecting the optimal elements of the four-frequency parametric system, and of computing, with the possibly highest precision, the amplifying and converting characteristics and impedance parameters of the system as a whole.

1.4. Analytical techniques for investigating modulation-parametric phenomena in generator systems

The different methods and approaches, systematized and analyzed in the foregoing section 1.3 are also applicable to the analysis of the modulation-parametric phenomena in generator radiophysical systems.

Here we set ourselves the task of developing analytical techniques, specific for the system class under review. They will combine the use of nonlinear differential equations, describing the generator systems under examination in a most comprehensive way, with the generalized method of complex amplitudes [187, 188].

In its general form, the differential equation of the second order, describing the nonlinear oscillating system, can be written as follows:

$$\frac{d^2x}{dt^2} + \left(a_0 + \sum_{i=1}^{\infty} a_i x^i\right)\frac{dx}{dt} + \left(b_0 + \sum_{i=1}^{\infty} b_i x^i\right)x$$
$$= A\cos(\Omega t + \alpha) + \mu a \cos(\omega t + \gamma), \quad \mu \ll 1. \tag{1.27}$$

The left-hand part of Eq. (1.27) may serve to describe both a passive oscillating system with external forced action and a generator system with synchronizing external action $A\cos(\Omega t + \alpha)$. Under the small parameter μ in equation (1.27) there is an external modulating signal $a\cos(\omega t + \gamma)$.

Equation (1.27) is of an essentially general nature, since the majority of the second order systems can be described by equations, which are particular cases of it. Moreover, the approach recounted below imposes no restrictions on either the order of the differential equation or the functions describing its coefficients. A broad class of functions presupposes a possibility for Taylor series expansion, which leads to the power dependencies of the coefficients used here.

It is assumed, first, that the system is dominated by a certain basic oscillation expressed by function x_0, and, second, that the system has an infinite spectrum of combined frequencies $\sum_{k=-\infty}^{\infty} x_k$ with small intensity as compared with the basic oscillation x_0, generated by the external modulating action $\mu a\cos(\omega t + \gamma)$.

Under these conditions, the solution of Eq. (1.27) can be written in the following form:

$$x = x_0 + \sum_{k=-\infty}^{\infty} x_k. \tag{1.28}$$

Taking into account (1.28), one can break down Eq. (1.27) into two equations:

— regarding the basic oscillations

$$\frac{d^2x_0}{dt^2} + \left(a_0 + \sum_{i=1}^{\infty} a_i x^i\right)\frac{dx}{dt} + \left(b_0 + \sum_{i=1}^{\infty} b_i x_i^i\right)x_0 = A\cos(\Omega t + \alpha); \tag{1.29}$$

— regarding the spectrum of combined frequencies

$$\sum_{k=-\infty}^{\infty} \frac{d^2x_k}{dt^2} + \left(a_0 + \sum_{i=1}^{\infty} a_i x_0^i\right)\sum_{k=-\infty}^{\infty}\frac{dx_k}{dt}$$
$$+ \left(b_0 + 2\sum_{i=1}^{\infty} b_i x_0^i + \sum_{i=1}^{\infty} a_i x_0^{i-1}\frac{dx_0}{dt}\right)\sum_{k=-\infty}^{\infty} x_k = \mu a\cos(\omega t + \gamma). \tag{1.30}$$

If function x_0 is presented in a harmonic form, two shortened equations can be obtained in the first approximation from Eq. (1.29) – concerning the amplitude and concerning the phase of the basic oscillations in the system respectively (the transition to shortened equations concerning the amplitude and the phase is a basic approach, for example, in the method of slowly changing amplitudes [51]. An analogous procedure can be applied to obtain two reduced differential equations concerning the amplitude and the phase of each combined component of Eq. (1.30). In this case a comprehensive exploration of the system necessitates the solution of $2k + 2$ shortened differential equation in a time-dependent form.

The following mixed (combined) analytical approach is useful for practical purposes [170]. Writing the system equations in the form of (1.29) and (1.30), one can avoid the need to solve $2k$ shortened differential equations concerning the amplitudes and phases of the combined components in a time-dependent form. To this end the coefficients and operators in the differential Eq. (1.30) should be written in a complex matrix form. This is done by presenting the coefficients of Eq. (1.30) in a complex form and generating products of the coefficients written in this way, with exponents of the type $\exp[j(m\Omega+\omega)]$, $m = 1, 2, \ldots$ [170]. The same procedure is used with respect to the operators of Eq. (1.30). When this is done Eq. (1.30) is presented in a complex matrix form:

$$\left[\frac{d}{dt}\right]^2 [x] + [A_\infty] \left[\frac{d}{dt}\right] [x] + [B_\infty][x] = [\mu a^0], \tag{1.31}$$

where $\left[\frac{d}{dt}\right] = \mathrm{diag}[-\infty, \ldots, j(\omega - k\Omega), \ldots, j\omega, \ldots, j(\omega + k\Omega), \ldots, \infty]$ is a diagonal matrix of the differentiation operator, $[x] = \mathrm{colon}[-\infty, \ldots, x_{-k}, \ldots, x_k, \ldots, \infty]$ is a column matrix of the spectrum of combined frequencies, $[A_\infty]$ and $[B_\infty]$ are infinite square matrices of the coefficients, $[\mu a^0] = \mathrm{colon}[\ldots 0, \mu a, 0 \ldots]$ is a column matrix of the modulating signal.

Thus in accordance with the approach, the analysis, proposed in the most general case includes the following successive stages:

1. Determining the functions of the basic oscillations x_0 from Eq. (1.29).
2. Choice of the combined frequencies spectrum $[k\Omega+\omega]$ that will be accounted for.
3. Specification of the matrices in Eq. (1.31), taking into consideration the combined frequencies spectrum that has been chosen.
4. Determining the vector of the combined components in Eq. (1.31).
5. Calculation of the basic parameters of the system by using the obtained vector $[x]$ – for example, computation of the coefficients of signal conversion and transmission, calculation of the equivalent impedances, etc.

CHAPTER 2.

CONTROLLING EQUIVALENT IMPEDANCES OF RADIOPHYSICAL SYSTEMS

2.1. Methods of controlling the active and reactive parameters in modulation-parametric systems with external pumping

An analysis of Manley–Row's energy relations (1.1) can be used for formulating the general idea of one of the methods of effective control of the equivalent impedances in a modulation-parametric system.

Let us consider the input impedance of a four-frequency parametric modulator. The response of the parametric modulator to the input signal is presented by the superposition of two factors conditioned respectively by the power absorption of the sum and difference combined frequencies $f_0 \pm f_1$. For the case under consideration, Manley–Row's relations take the form:

$$\begin{aligned} \frac{P_{1,0}}{f_0} + \frac{P_{1,1}}{f_0 + f_1} + \frac{P_{1,-1}}{f_0 - f_1} = 0 \\ \frac{P_{0,1}}{f_1} + \frac{P_{1,1}}{f_0 + f_1} - \frac{P_{1,-1}}{f_0 - f_1} = 0 \end{aligned} , \tag{2.1}$$

where $P_{0,1}$ is the power of the signal, $P_{1,0}$ is the pumping power, $P_{1,-1}$ and $P_{1,1}$ are respectively the powers of the difference and sum combined frequencies.

The ratio between the signal power and the pumping power is determined from (2.1) under the condition $f_1 \ll f_0$:

$$\frac{P_{0,1}}{P_{1,0}} = \frac{f_1}{f_0} \frac{1 - \dfrac{P_{1,-1}}{P_{1,1}}}{1 + \dfrac{P_{1,-1}}{P_{1,1}}} . \tag{2.2}$$

It follows from this ratio that if the powers of the combined frequencies $P_{1,-1}$, $P_{1,1}$ are equal to each other, $P_{0,1} = 0$, i.e. the parametric modulator does not absorb energy at the frequency of the signal. If $P_{1,-1} \neq P_{1,1}$, the direction of the energy flow in the input circuit is determined by the value of the ratio $\dfrac{P_{1,-1}}{P_{1,1}}$. If $\dfrac{P_{1,-1}}{P_{1,1}} > 1$, it follows from Eq. (2.2) that $P_{0,1} < 0$, i.e. the parametric modulator emits energy at the frequency of the signal (it is considered that $P_{1,0} > 0$, $P_{1,-1} > 0$ and $P_{1,1} > 0$, since the power is drawn from the pumping source and dissipated in the passive circuits at the combined frequencies). If $\frac{P_{1,-1}}{P_{1,1}} < 1$, $P_{0,1} > 0$, i.e. in this case the parametric modulator absorbs energy at the frequency of the signal. In the first case the parametric modulator has negative input resistance and in the second one the resistance is positive.

In this way it is concluded, on the basis of Manley–Row's energy relations, that by controlling the detuning of the resonance system of the parametric modulator one can change the ratio $\frac{P_{1,-1}}{P_{1,1}}$, hence – the equivalent impedance at signal frequency. We shall analyze this issue in detail and more precisely on the basis of the generalized parametric modulators presented in Fig.1.1.

2.1.1. Input conductance (impedance) of a cophasal parametric modulator

For the case of a cophasal change of the reactive and active parametric element in the parametric modulator in Fig.1.1, it should be borne in mind the condition $\Psi_1 = \Psi_2 = 0$ in the generalized matrix equation (1.26).

Assuming that the load of the parametric modulator is an oscillating circuit, the values $\dot{H}_+$ and $\overset{*}{H}_-$ can be specified in the following form:

$$\dot{H}_+ = D_p[1 + 2Qj(\eta + \xi)], \quad \overset{*}{H}_- = D_p[1 - 2Qj(\eta - \xi)], \tag{2.3}$$

where $\eta = \frac{\omega_p - \omega_{op}}{\omega_p}$, $\xi = \frac{\omega_1}{\omega_p}$, ω_{op} is the resonance frequency of the oscillating circuit at an infinitesimal pumping amplitude, Q is the quality factor of the oscillating circuit, D_p is the average of the parameter of the active parametric element.

The total input conductance (impedance) of the cophasal modulator at the frequency of the signal is

$$\dot{H}_{in} = \dot{H}_1 + \dot{H}_g, \tag{2.4}$$

where $\dot{H}_g$ is the admittance (impedance) introduced in the input circuit as a result of the reversibility of the modulation-parametric interactions,

$$\dot{H}_g = D_g + jB_g. \tag{2.5}$$

Using equation (1.26) for the active D_g and reactive B_g part of the admittance (impedance) introduced in the input circuit, we obtain the following expressions:

$$D_g = -2D\frac{(M_1 + 2Q^2\xi_0 m_1)\{M_1[1 + Q^2(\varkappa - 4\xi^2)] + 4Q^2\xi^2 m_1 M_0\}}{[1 + Q^2(\varkappa - 4\xi^2)]^2 + (4Q\xi M_0)^2} \cdots \to$$

$$\to \cdots \frac{+M_1 4Q^2\xi^2\{4M_1M_2 - m_1[1 + Q^2(\varkappa - 4\xi^2)]\}}{}, \tag{2.6}$$

$$B_g = -2DQ\xi \frac{(M_1 + 2Q^2\xi_0 m_1)\{m_1[1 + Q^2(\varkappa - 4\xi^2)] - 4M_1M_0\}}{[1 + Q^2(\varkappa - 4\xi^2)]^2 + (4Q\xi M_0)^2} \cdots \to$$

$$\to \cdots \frac{+2M_1\{M_1[1 + Q^2(\varkappa - 4\xi^2)] + 4Q^2\xi^2 m_1 M_0\}}{}, \tag{2.7}$$

where $\xi_0 = \eta - \frac{m_2}{2}$ is the effective relative detuning of the system oscillating circuit; $m_1 - \frac{S_1}{S_0'}$, $m_2 - \frac{S_2}{S_0'}$, $M_1 = \frac{D_1}{D}$ and $M_2 = \frac{D_2}{D}$ are modulation coefficients of the non-linear reactive and active parametric element; $\varkappa = 4(\xi_0^2 + \xi_0 m_2) + \frac{2M_2}{Q}$, $S_0' = S_0 - S_2$, $D = D_0 - D_2$ are effective (average) reactive and active parameters of the system.

It follows from (2.6) and (2.7) that the sign of D_g and B_g depends on the detuning ξ_0:

$$D_g, B_g < 0 \ \text{ at } \ \xi_0 > 0 \,,$$

$$D_g, B_g > 0 \ \text{ at } \ 2Q\xi_0 < -\frac{M_1}{Qm_1} \,.$$

In the first case, there exists negative differential conductance (or negative differential resistance) in a wide frequency range of input signals up to direct current, where

$$D_g(\xi = 0) = -2D\frac{M_1^2 + 2Q^2\xi_0 m_1 M_1}{1 + 2M_2 + 4Q^2(\xi_0^2 + \xi_0 m_2)}. \tag{2.8}$$

The maximum value of the introduced negative conductance (negative resistance)

$$|D_{g-}|_{\max} = D\frac{M_1^2 + Qm_1M_1}{1 + Qm_2} \tag{2.9}$$

and it corresponds to detuning $\xi_0 = \frac{1}{2Q}$.

One can see from (2.9) that under certain conditions the negative conductance (negative resistance) introduced by the four-frequency parametric modulator may substantially exceed the total active loss of the resonance load.

The negative susceptance introduced under certain conditions (2.7) may be perceived as occurrence of effective negative differential capacitance C_g (negative differential inductance L_g) [150]:

$$\left\{ \begin{matrix} C_g \\ L_g \end{matrix} \right\} = -2\left\{ \begin{matrix} C_0' \\ L_0' \end{matrix} \right\} \frac{(M_1 + 2Q^2\xi_0 m_1)\{m_1[1 + Q^2(\varkappa - 4\xi^2)] - 4M_1M_0\}}{[1 + Q^2(\varkappa - 4\xi^2)]^2 + (4Q\xi M_0)^2} \cdots \to$$

$$\to \cdots \frac{+2M_1\{M_1[1 + Q^2(\varkappa - 4\xi^2)] + 4Q^2\xi^2 m_1 M_0\}}{}, \tag{2.10}$$

where $C_0' = C_0 - C_2,\ L_0' = L_0 - L_2$.

The maximum value of the negative capacitance (negative inductance) is

$$\left\{ \begin{matrix} C_{g-} \\ L_{g-} \end{matrix} \right\}_{\max} = \left\{ \begin{matrix} C_0' \\ L_0' \end{matrix} \right\} \frac{M_1 + 2M_1^2 + Qm_1^2}{1 + M_2 + Qm_2}. \tag{2.11}$$

It is obvious that under certain conditions the total input conductance (input resistance) and the total input capacitance (input inductance) of the modulation-parametric system can be negative. For instance, when $Q \sim 100$, $m_1 \sim 0,5$, $M_1 \sim 0,6$, we obtain

$$\frac{\left|D_{g-\max}^{(m)}\right|}{D} = (M_1^2 + Qm_1M_1) \simeq 30,$$

$$\frac{\left\{ \begin{matrix} \left|C_{g-}^{(m)}\right|_{\max} \\ \left|L_{g-}^{(m)}\right|_{\max} \end{matrix} \right\}}{\left\{ \begin{matrix} C_0' \\ L_0' \end{matrix} \right\}} = (M_1 + 2M_1^2 + Qm_1^2) \simeq 25.$$

The stability of the modulation-parametric system under consideration is limited, first, by the requirement $H_{in} \geq 0$ and, second, by the regenerative action of the second harmonic in the spectrum of the parametric reactive element. The stable area is defined by the condition: $\varkappa Q^2 \leq -1$, which, in a particular case, provides evidence that $\varkappa$ plays the part of generalized detuning.

Publication [150] also considers an inertial parametric modulator, which uses a $p-n$ junction in a partially opening regime and high pumping frequencies $\omega_p \gg \frac{1}{\tau}$, where τ is the life-span of the minority carriers in the base of the semiconductor diode (See Chapter 3). In this case $D_1, D_2 = 0$, since the conductance of the $p-n$ junction „fails" to follow the pumping voltage. It is shown that if the system provides conditions for modulation action on an active or reactive parameter, the inertial parametric modulator has all the above-considered peculiarities that are characteristic for the non-inertial parametric modulator.

It follows from the above-recounted analysis that one of the methods of effective control of the active and reactive parameters in a modulation-parametric system with external pumping is the utilization of resonance parametric modulators with controllable detuning of the oscillating system.

2.1.2. Input admittance (impedance) of a complex parametric modulator with „quadratic" pumping of the parametric elements

One of the ways of obtaining higher absolute values of the introduced negative D_g and B_g and more effective control of the active and reactive parameters of the system is to use a phase shift of the pumping of the reactive and active parametric elements. In this case, unlike that of the cophasal parametric modulator, maximum effective changes of the equivalent admittances (impedances) are obtained at zero detunings ξ_0.

The analysis shows that it is most effective to apply „quadratic" phase shift of the pumping of the parametric elements:

i) $\Psi_1 = \pm\frac{\pi}{2}$, $\Psi_2 = 0$ and ii) $\Psi_1 = 0$, $\Psi_2 = \pm\frac{\pi}{2}$.

In the case of i), the following is obtained from Eq.(1.26):

$$D_g^Q = -2D(M_1 \mp m_1 Q)\frac{M_1[1+Q^2(\varkappa-4\xi^2)] + 8Q^2\xi^2 M_0(M_1 \mp Q\xi_0 m_1)}{[1+Q^2(\varkappa-4\xi^2)]^2 + (4Q\xi M_0)^2}, \tag{2.12}$$

$$B_g^Q = -4QD\xi(M_1 \mp m_1 Q)\frac{(M_1 \mp Q\xi_0 m_1)[1+Q^2(\varkappa-4\xi^2)] - 2M_0 M_1}{[1+Q^2(\varkappa-4\xi^2)]^2 + (4Q\xi M_0)^2}, \tag{2.13}$$

(the top sign refers to $\Psi_1 = +\frac{\pi}{2}$, and the bottom one to $\Psi_1 = -\frac{\pi}{2}$).

The effective capacitance (inductance), introduced in the input circuit of the modulator is

$$\begin{Bmatrix} C_g^Q \\ L_g^Q \end{Bmatrix} = -\begin{Bmatrix} C_0' \\ L_0' \end{Bmatrix} 4(M_1 \mp m_1 Q)\frac{(M_1 \mp Q\xi_0 m_1)[1+Q^2(\varkappa-4\xi^2)] - 2M_0 M_1}{[1+Q^2(\varkappa-4\xi^2)]^2 + (4Q\xi M_0)^2}. \tag{2.14}$$

The theoretical frequency dependencies of the relative quantities $\frac{|\dot{D}_{g-}^Q|}{D}$ and $\frac{|C_{g-}^Q|}{C_0'}$ (or $\frac{|L_{g-}^Q|}{L_0'}$) are given in Fig.2.1 for the following values of the parameters: $\Psi_1 = -\frac{\pi}{2}$, $\Psi_2 = 0$, $Q = 15$, $m_1 = 0,42$, $m_2 = 0,2$, $M_0 = 1,69$, $M_1 = 0,63$ and $M_2 = 0,49$. As can be seen from the Figure, the negative D_{g-}^Q and C_{g-}^Q (L_{g-}^Q), inserted into the input circuit, may exceed considerably, in a wide video frequency band of the input signals, respectively the own losses and the own average capacitance (or inductance) of the parametric modulator that generates them.

The parameters D_g^Q, B_g^Q, C_g^Q (L_g^Q) (for case b as well) are determined in an analogous way from Eq. (1.26) [150].

The investigation shows that from the viewpoint of the maximization of the introduced G_g^Q and C_g^Q (L_g^Q) the optimal phases will be provided by the following combinations: $\Psi_1 = -\frac{\pi}{2}$, $\Psi_2 = 0$ and $\Psi_1 = 0$, $\Psi_2 = +\frac{\pi}{2}$.

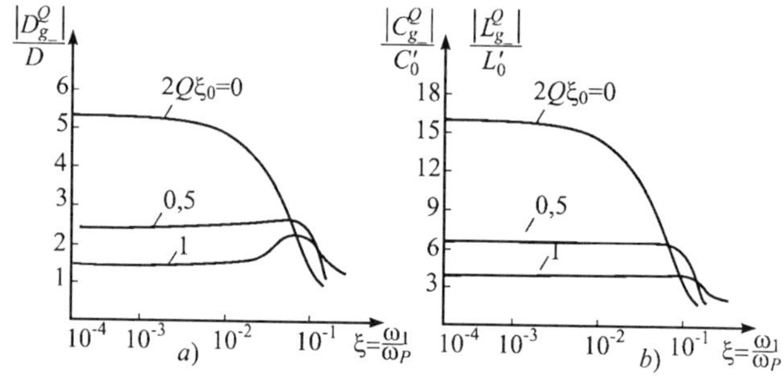

Fig. 2.1. Theoretical frequency dependence of the normalized absolute values of the negative conductance (resistance) – a) and negative capacitance (inductance) – b) introduced by the parametric modulator in the case of „quadratic" pumping of the elements of the complex parametric element

2.1.3. Experimental results concerning the input conductance of a cophasal parametric modulator

As an illustration of the theoretical conclusions drawn above, we shall quote some experimental results concerning the input conductance of a cophasal parametric modulator. Section 2.4 will present the fundamental circuit schemes and describe a number of applications of the complex parametric modulators with „quadratic" pumping.

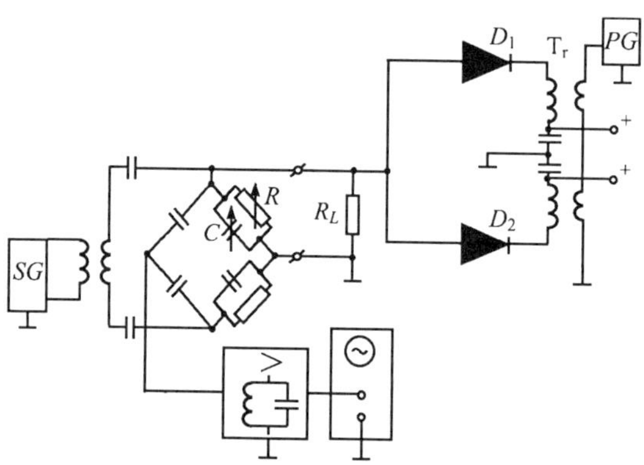

Fig. 2.2. Diagram of the experimental set-up for investigating the input impedance of a cophasal parametric modulator

The schematic diagram of the experimental set-up is shown in Fig.2.2. The cophasal parametric modulator is represented by a balanced circuit consisting of the secondary windings of the transformer T_r and the diodes D_1 and D_2. The diodes D_1 and D_2 have negative bias but they operate in a mode of direct currents obtained as a result of the action of the pumping generator PG (1.5 MHz). The signal comes in from a signal generator SG. The value of the introduced negative (or positive in a different mode) G_{g-} and C_{g-} are determined at the terminals of the load R_L with the help of a bridge circuit.

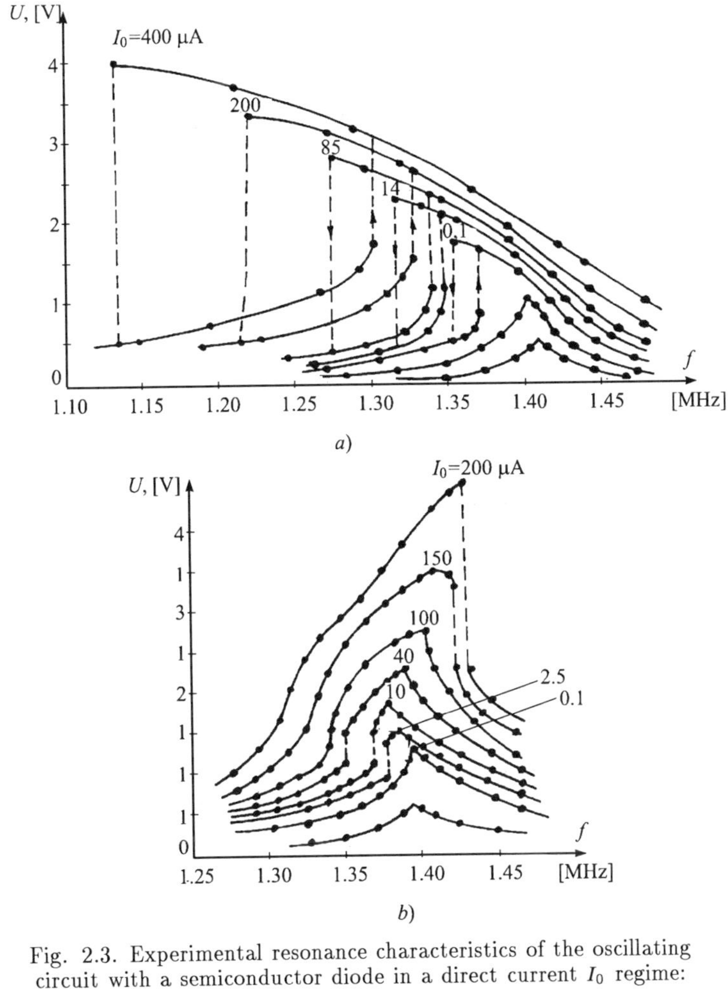

Fig. 2.3. Experimental resonance characteristics of the oscillating circuit with a semiconductor diode in a direct current I_0 regime: a) single oscillating circuit; b) two-stroke balance oscillating circuit

For the sake of comparison, Fig.2.3 presents the experimental resonance curves of a single oscilating circuit with a semiconductor diode in a negative bias mode (a) and in the mode of a two-stroke balance oscillating circuit (b), corresponding

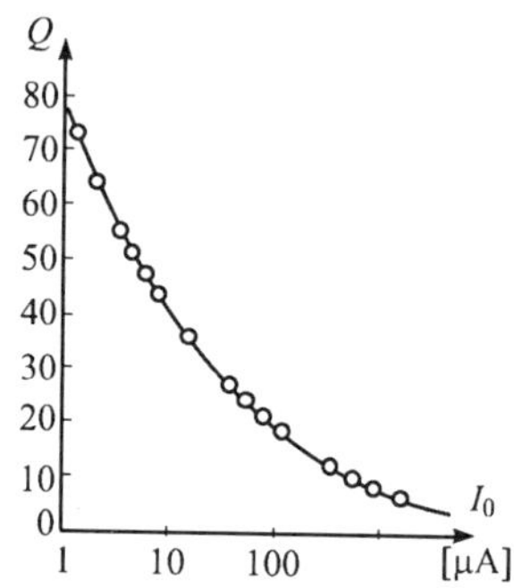

Fig. 2.4. Experimental dependence of the equivalent quality factor of a two-stroke balance oscillating circuit on the value of the direct current flowing through the diodes

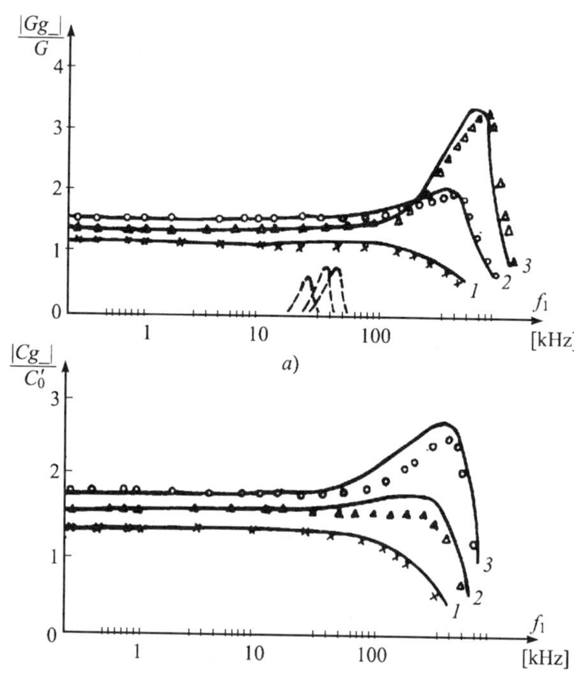

Fig. 2.5. Frequency dependencies of negative conductance (a) and negative capacitance (b) introduced by a cophasal parametric modulator; $I_0 = 5\ \mu A$

to Fig.2.2. Fig.2.4 shows the dependence of the equivalent qualitative factor of the two-stroke balance oscillating circuit on the value of the flowing direct current.

Fig.2.5 a) and b) discloses the experimental (dotted line) and theoretical (continuous lines, formulae (2.6) and (2.10)) dependencies of the introduced negative $G_{g\_}$ and $C_{g\_}$, given the following detunings of the oscillating circuit: curve 1 – 15, 2 – 30 and 3 – 100 kHz. For comparison, Fig.2.5a) shows, with a dash line, the frequency dependence of the introduced negative G_g by a purely reactive (capacitive) modulator. It is clearly seen that the addition of an active parametric element considerably extends the frequency band of the introduced negative conductance up to zero frequencies.

2.1.4. On the existence of a modulation-parametric channel of energy conversion and input in the course of different manipulations with signals

It has been shown that in resonance systems the effect of parametric energy input, resulting from a periodic change in a reactive parameter, increases Q times, where Q is the quality factor of the resonance system. This circumstance is reflected in the formulae (See (2.6) – (2.11)) by the product $m_1 Q$, where m_1 is the modulation depth of an energy accumulative (reactive) parameter.

We shall show that a modulation-parametric energy input channel is existent in non-resonance systems as well, if there are conditions for a modulation-parametric change of an energy accumulative (reactive) parameter. This will be demonstrated on the basis of a system described by a differential equation of the first order with one time-dependent periodic coefficient. Similar systems have been considered in relation to the discovered effect of non-degenerate single-frequency parametric regeneration in Josephson junctions [189], the differential inductance of which assumes negative values over a part of the period of change. Obviously the issue boils down mainly to the modulation depth of the parameter of the reactive element in the system.

In order to illustrate the regenerative capacity of a non-resonance circuit, provided that the modulation coefficient of the parametric element is larger than one, and parallel with that to demonstrate once again the operation of the principle of reversibility of the modulation-parametric interactions (See 1.1), we shall consider the processes occurring in the circuit in Fig.2.6. Since usually the analysis is carried out for a regime of idle running (at no-load) or shorting, we shall assume that the voltage on the parametric element changes in accordance with a harmonic law $V_1(t) = V_1 \cos \omega_1 t$.

As the average capacitance C_0 does not exert direct influence on the conversion processes, it may not be accounted for further on, as we set $C(t) = 2C_1 \cos \omega_p t$.

The low external voltage of the signal $V_1(t)$ creates the following charge on the variable capacitance:

$$q_1 = C(t)V_1(t) = C_1 V_1[\cos(\omega_p - \omega_1)t + \cos(\omega_p + \omega_1)t].$$

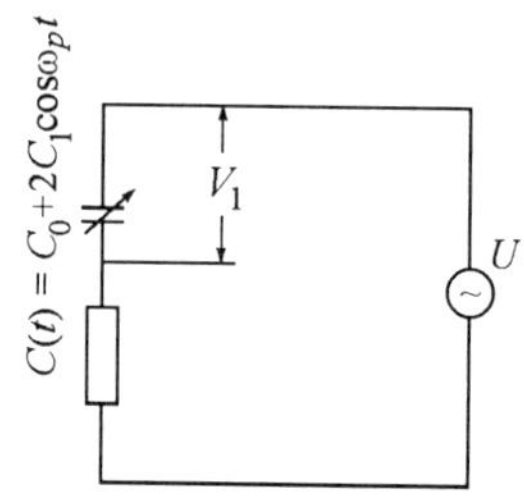

Fig. 2.6. Illustrations of the issue of the regeneration of a non-resonance circuit with a modulation depth of the parametric element larger than unity

Respectively, the current in the circuit will be written as $i_1 = \frac{dq_1}{dt}$, while the induced additional voltage will be expressed in the form

$$V_{1\to 2} = R\frac{dq_1}{dt} = -C_1V_1R[(\omega_p - \omega_1)\sin(\omega_p - \omega_1) + (\omega_p + \omega_1)\sin(\omega_p + \omega_1)t].$$

The voltage $V_{1\to 2}$ is also applied to the non-linear capacitance causing a charge

$$q_2 = C(t)V_{1\to 2}(t) = -C_1^2V_1R\{(\omega_p - \omega_1)[-\sin\omega_1 t + \sin(2\omega_p - \omega_1)t] + (\omega_p + \omega_1)[\sin\omega_1 t + \sin(2\omega_p + \omega_1)t].$$

Respectively, the additional current is $i_2 = \frac{dq_2}{dt}$ and the additional voltage

$$V_{2\omega_1} = -2C_1^2V_1R^2\omega_1^2\cos\omega_1 t.$$

The considered double (direct and reverse) conversion of the frequency creates a feedback loop, whose action has led to the emergence of an additional voltage $V_{2\omega_1}$, antiphasal to the initial voltage $V_1(t)$.

Further on, the matter can be considered repeatedly by using as output voltage $V_1 + V_{2\omega_1}$, then $V_1 + V_{2\omega_1} + V_{4\omega_1}$ and so on, which will yield the respective additional voltages:

$V_{4\omega_1} = 4C_1^4V_1R^4\omega_1^4\cos\omega_1 t$ – cophasal with $V_1(t)$,

$V_{6\omega_1} = -8C_1^6V_1R^6\omega_1^6\cos\omega_1 t$ – antiphasal with $V_1(t)$, etc.

The ultimate expression of the voltage on the parametric element assumes the form

$$V_\Sigma = (1 - \alpha + \alpha^2 - \alpha^3 + \ldots \pm \alpha^n + \ldots)V_1\cos\omega_1 t,$$

where $\alpha = 2C_1^2R^2\omega_1^2$.

As we pass to the limit of the infinite sum in the brackets of the expression above, we obtain:

— for the case of $\alpha < 1$

$$V_\Sigma = \frac{1}{1+\alpha} V_1 \cos\omega_1 t;$$

— for the case of $\alpha > 1$

$$V_\Sigma = \frac{\alpha^{2n} - 1}{\alpha^2 - 1}(1-\alpha) V_1 \cos\omega_1 t, \quad n \to \infty.$$

It is obvious that for $\alpha > 1$ a regenerative mode is possible, even though there is no oscillating circuit. Besides, $\alpha \sim \omega_1^2$, i.e. as the frequency of the signal increases, the conditions for the occurrence of regeneration improve.

For the purpose of exploring the matter in detail, we shall consider the cases of inductive and capacitive parametric elements, presented in Fig.2.7 a, b, where G and R reflect the inevitable losses, L_0 and C_0 are constant, while $\lambda(\tau)$ is a periodic function of the normalized time τ with period p. In order to trace the emergence of regenerative effects during the transition from resonance to non-resonance circuit, we shall complement the parametric elements to obtain a resonance circuit with an antipodal constant reactance (dotted line in Fig.2.7 a, b).

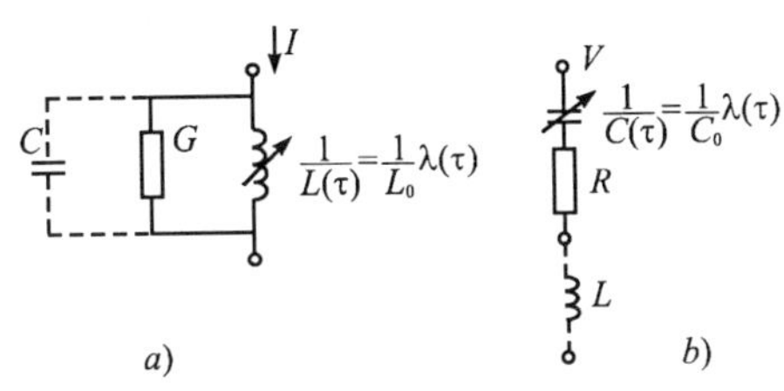

Fig. 2.7. Equivalent circuits for analyzing the effect of single-frequency non-degenerate parametric regeneration

In this case we shall impose no restrictions on the spectrum of combined frequencies or on the modulation depth of the parametric element. The physical prerequisite for the last circumstance is provided by the superconductive Josephson junctions, where the modulation depth of the parametric inductance is larger than one, i.e. the reactance parameter periodically assumes negative values for a part of the period up to half a period from its change [140]. The authors of work [189] associate the effect of non-degenerate single-frequency parametric energy input with dynamic (explosive) instability and with the emergence of specific processes for the time when the reactive parametric element takes negative values. If we use impedance phraseology, the effect consists in the emergence of effective negative resistance with respect to external signals with frequencies exceeding insignificantly the change frequency of the parametric element.

The processes in the systems of both types shown in Fig.2.7 are described by the following differential equation

$$Q^2 \frac{d^2 x}{d\tau^2} + \frac{dx}{d\tau} + \lambda(\tau) x = f(\tau), \quad \overline{\lambda} = 1. \tag{2.15}$$

The following notation has been adopted in (2.15):
— regarding Fig.2.7 a

$$x = G \int V d\tau, \quad \lambda(\tau) = \frac{L_0}{L(\tau)}, \quad f = I, \quad \tau = \frac{t}{GL_0}, \quad Q = \frac{1}{G}\sqrt{\frac{C}{L_0}}; \tag{2.16}$$

- regarding Fig.2.7 b

$$x = R \int I d\tau, \quad \lambda(\tau) = \frac{C_0}{C(\tau)}, \quad f = V, \quad \tau = \frac{t}{RC_0}, \quad Q = \frac{1}{R}\sqrt{\frac{L}{C_0}}. \tag{2.17}$$

In the analysis we assume that the value f is harmonic, i.e.

$$f = a \cos \omega\tau. \tag{2.18}$$

Under this condition we assume that the systems in Fig.2.7 are connected to external electrodynamic circuits in such a way that the oscillations can interact.

The intensity of the interactions (of the energy exchange) can be characterized by the average power flow between the element and the external system:

$$P = \overline{I(t)V(t)}. \tag{2.19}$$

We shall assume that the positive power corresponds to the power flow from the external system to the element, while the negative one indicates power transfer from the element to the external system, i.e. energy input into the acting oscillation. In the latter case, there is regeneration of the external system.

We shall consider the case when the interaction takes place at only one frequency ω. It is assumed that the selective properties of the external system are such that either the current or the voltage on the element change in accordance with a harmonic law. If the power flow – P – from the external system to the element proves to be negative, we shall consider that the element effects single-frequency energy input into the external system.

For the purpose of determining the magnitude and sign of the energy input into the external system as effected by the parametric element, it is sufficient to calculate the real part of the non-dimensional complex value

$$Z = \frac{\overline{\dot{x} e^{-j\omega\tau}}}{a}, \tag{2.20}$$

which characterizes the impedance (or respectively the admittance) of the element at frequency ω. Indeed, in the single-frequency case (2.18) the expression for the energy flux (2.19) takes the form

$$P = \frac{1}{2} a^2 ReZ \times \begin{Bmatrix} R \\ G \end{Bmatrix}. \tag{2.21}$$

Thus, if the real part of the impedance proves to be negative, i.e. $ReZ(\omega) < 0$, this will mean that $P < 0$ and that the external electrodynamic system receives energy from the element at frequency ω, i.e. the effect of the single-frequency parametric energy input can really occur at this frequency.

Adhering to the approach of the authors of work [189], we shall analyze the impedance characteristics of the parametric elements of both types (See Fig.2.7). The processes that occur there are described by one and the same equation

$$\frac{dx}{d\tau} + \lambda(\tau)x = f(\tau), \quad \overline{\lambda} = 1. \tag{2.22}$$

The solution of equation (2.22) is of the form

$$x = e^{-\int \lambda d\tau'} \int f(\tau) e^{-\int \lambda d\tau'} d\tau, \tag{2.23}$$

where $f(\tau)$ is a harmonic function (2.18).

The exponents in (2.23) can be harmonically expanded by the pumping period of the parametric element $\frac{2\pi}{p}$:

$$\exp\{\pm \int \lambda(\tau) d\tau\} = \exp(\pm\tau) \sum_n C_n^{\pm} \exp\{jnp\tau\}. \tag{2.24}$$

As (2.24) is replaced in (2.23) and the components with frequency ω are separated, an expression of the impedance of the element is obtained in accordance with Eq.(2.20):

$$Z = \sum_{n=-\infty}^{\infty} \frac{j\omega}{1 + j(\omega - np)} C_n^- C_{-n}^+. \tag{2.25}$$

The coefficients C_n^- and C_{-n}^+ are determined by the type of the modulation of the reactive parameter λ.

Let us consider the major types of modulations.

In the case of a *rectangular modulation law* the reactive parameter

$$\lambda(\tau) = \begin{cases} 1 + S, & N\frac{2\pi}{p} \le \tau \le \frac{\pi}{p} + N\frac{2\pi}{p} \\ 1 - S, & \frac{\pi}{p} + N\frac{2\pi}{p} \le \tau \le (N+1)\frac{2\pi}{p} \end{cases}. \tag{2.26}$$

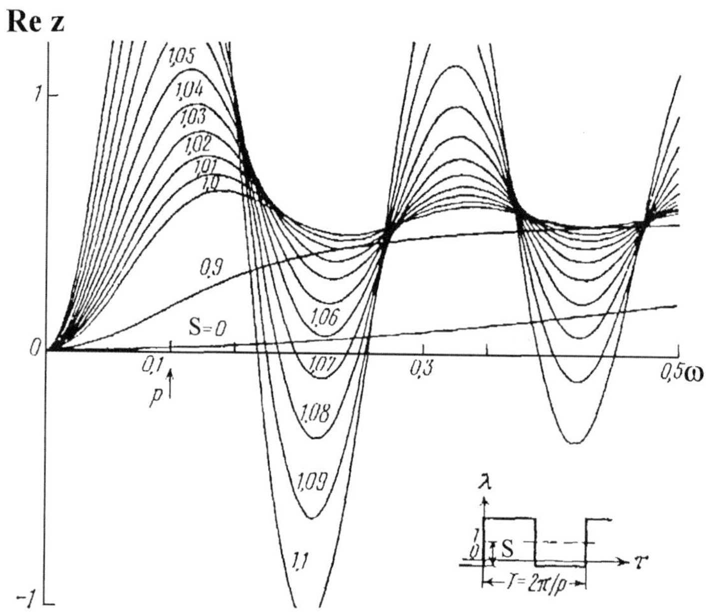

Fig. 2.8. Dependence of the real part of the element impedance Re Z on the signal frequency ω for the case of rectangular modulation of the reactive parameter and different modulation depths. The pumping frequency is small, $p = 0.1$

(the parameter S is clear from the insertion in Fig. 2.8).

As this dependence is substituted in formula (2.24), the following is obtained:

$$\left|\begin{array}{l} C^{+}_{-n} = \dfrac{p}{2\pi}\left(\dfrac{1 - e^{S\frac{\pi}{p} - jn\pi}}{-S + jnp} + \dfrac{e^{S\frac{\pi}{p} + jn\pi} - 1}{S + jnp}\right) \\ C^{-}_{n} = \dfrac{p}{2\pi}\left(\dfrac{1 - e^{-S\frac{\pi}{p} + jn\pi}}{S - jnp} + \dfrac{e^{-S\frac{\pi}{p} - jn\pi} - 1}{-S - jnp}\right) \end{array}\right. . \tag{2.27}$$

As the expressions for coefficients C^{+}_{-n} and C^{-}_{n} are borrowed from (2.27), formula (2.25) assumes, in this case, the following form:

$$Z = \frac{2p^2 S^2}{\pi^2} \sum_{n=-\infty}^{\infty} \frac{j\omega}{[1 + j(\omega - np)]} \frac{1}{(S^2 + n^2 p^2)^2}$$

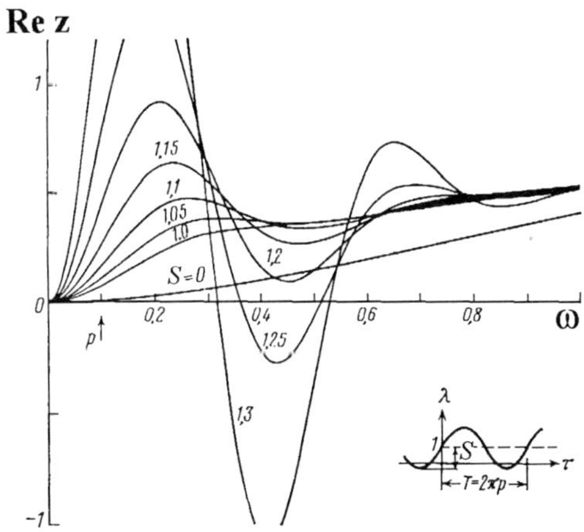

Fig. 2.9. Dependence of the real part of the element impedance Re Z on the signal frequency ω for the case of sine modulation of the reactive parameter. The pumping frequency is small, $p = 0.1$

$$\times \left[-1 + \operatorname{ch}\left(\frac{S\pi}{p}\right)(-1)^n\right]. \tag{2.28}$$

Then we carry out summation to obtain the final expression for the junction impedance:

$$Z = \frac{2p\omega S^2}{\pi} \frac{\operatorname{ch}\left(\frac{S\pi}{p}\right) - \cos\left[\frac{\pi(\omega - j)}{p}\right]}{[S^2 - 1 + \omega^2 - 2j\omega]^2 \sin\left[\frac{\pi(\omega - j)}{p}\right]} + \frac{\omega(j - \omega)}{[1 - S^2 - \omega^2 + 2j\omega]}. \tag{2.29}$$

Typical dependencies of the real part of the impedance, expressed through formula (2.29), are presented in Fig.2.8 for $p = 0,1$. It is obvious that when $S < 1$, the values of ReZ are positive for all ω. But for modulation depths S exceeding one the values of ReZ in certain frequency ranges become negative.

Let us determine the impedance of the element for the case of *sine modulation* of the parameter (See the insertion in Fig.2.9)

$$\lambda(\tau) = 1 + S\cos p\tau. \tag{2.30}$$

As we substitute (2.30) in (2.24), we obtain

$$C_{-n}^{+} = I_n\left(\frac{S}{p}\right), \quad C_n^{-} = I_n\left(\frac{S}{p}\right)(-1)^n, \tag{2.31}$$

where $I_n\left(\frac{S}{p}\right)$ is a modified n-th order Bessel function of the argument $\frac{S}{p}$. Using expressions (2.31), we obtain, from formula (2.25), an expression for the impedance of the element:

$$Z = \sum_{n=-\infty}^{\infty} \frac{j\omega}{1 + j(\omega - np)} I_n^2\left(\frac{S}{p}\right)(-1)^n. \tag{2.32}$$

Fig.2.9 shows typical dependencies of the real part of the impedance ReZ on the signal frequency, derived from formula (2.32) at $p = 0, 1$. It can be seen that, in a way analogous to the rectangular modulation case, the value ReZ for $S > 1$ can take negative values and, respectively, single-frequency non-degenerate parametric energy input can occur.

Fig.2.10 indicates that even for high pumping frequencies, such as for example $p = 10$, the effect still occurs but when the modulation depth is increased respectively (the threshold value for S increases in proportion to $p^{1/2}$).

Now we shall show that besides the hypothetical peculiar dynamic instability discussed by the authors of [189], such a system also contains a modulation-parametric channel for regenerative energy input. Under this condition, the modulation-parametric regenerative effect is manifested under conditions coinciding with those of work [189], when the modulation depth of the parametric element is larger than one and the signal frequency is essentially close to the change frequency of the parametric element. From a physical point of view this is understandable, bearing in mind that the absence of a resonance system may be „compensated" by higher values of the modulation depth of the parametric element, while the necessary difference in the intensity of the sum and difference combined frequencies can be achieved in the situation of a different relation between the signal frequency and change frequency of the parametric element.

It is worth noting that, in accordance with the adopted notation and normalization, $Q \sim \frac{1}{p}$.

The effective impedance and, respectively, the effective conductance of the systems in Fig.2.7 a, b with respect to the external signal I and V with frequency ω is determined as

$$Z_\omega = \frac{V_\omega}{i_\omega} = \frac{\left(\frac{dx}{d\tau}\right)_\omega}{Gf}, \tag{2.33}$$

$$Y_\omega = \frac{i_\omega}{V_\omega} = \frac{\left(\frac{dx}{d\tau}\right)_\omega}{Rf}. \tag{2.34}$$

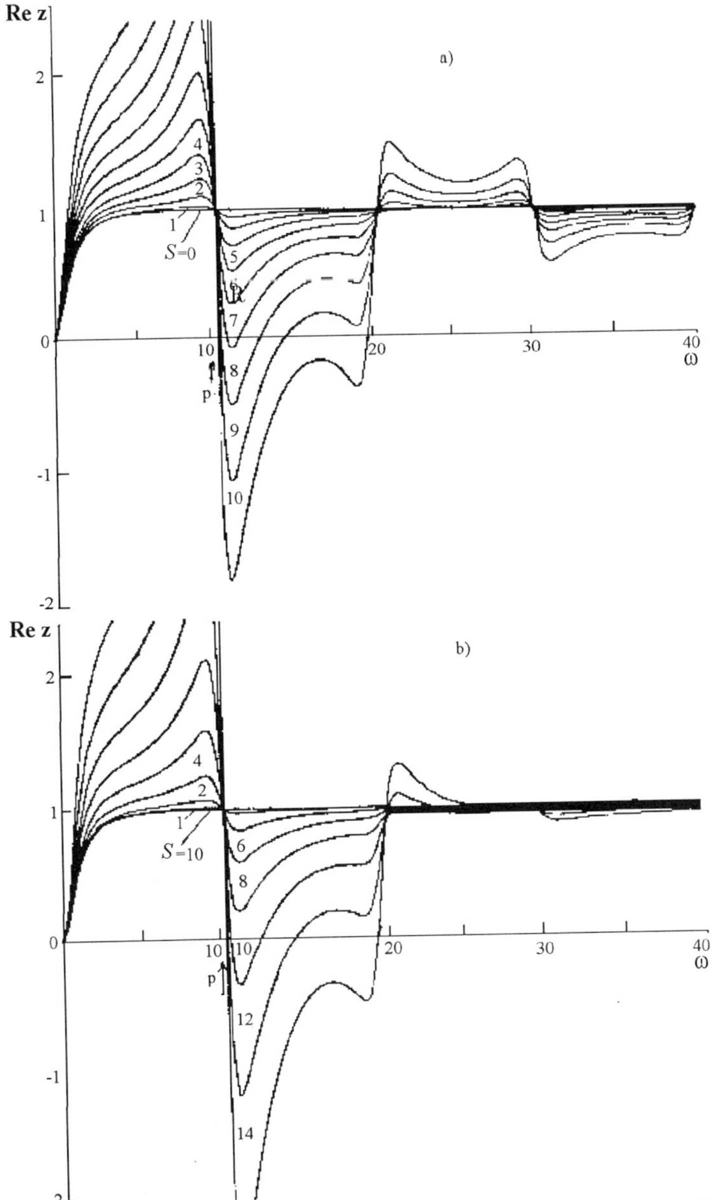

Fig. 2.10. Dependence of the real part of the element impedance Re Z on the signal frequency ω at high pumping frequency ($p = 10$) for the cases of rectangular (a) and sine (b) law of modulation of the reactive parameter

In keeping with the analytical approach recounted in section 1.4, we shall write Eq.(2.15) in a matrix complex form. To this end the coefficient $\lambda(\tau)$ is presented in the form

$$\lambda(\tau) = 1 + \frac{S}{2}\left[e^{jp(\tau-\tau_0)} + e^{-jp(\tau-\tau_0)}\right] \tag{2.35}$$

and the following products are formed: $\left(\frac{d}{d\tau}\right)^2 e^{j(\omega\pm mp)\tau}$, $\left(\frac{d}{d\tau}\right) e^{j(\omega\pm mp)\tau}$, $\lambda(\tau)e^{j(\omega\pm mp)\tau}$, $-\infty < m < \infty$.

The result is equation

$$\left\{Q\left[\frac{d}{d\tau}\right]^2 + \left[\frac{d}{d\tau}\right] + [\lambda]\right\}[x] = [f], \tag{2.36}$$

where $\left[\frac{d}{d\tau}\right]^2 = \mathrm{diag}[\ldots, -(\omega-2p)^2, -(\omega-p)^2, -\omega^2, -(\omega+p)^2, -(\omega+2p)^2, \ldots]$,
$\left[\frac{d}{d\tau}\right] = \mathrm{diag}[\ldots, j(\omega-2p), j(\omega-p), j\omega, j(\omega+p), j(\omega+2p), \ldots]$,

$$[\lambda] = \begin{bmatrix} \ddots & \ddots & \ddots & \\ \ddots & 1 & Se^{-jp\tau_0} & 0 \\ Se^{jp\tau_0} & 1 & Se^{-jp\tau_0} & \\ 0 & Se^{jp\tau_0} & 1 & \ddots \\ \ddots & \ddots & & \ddots \end{bmatrix}$$ – an infinite quadratic triangular matrix,

$[x] = \mathrm{colon}[\ldots, x_{\omega-2p}, x_{\omega-p}, x_\omega, x_{\omega+p}, x_{\omega+2p}, \ldots]$,
$[f] = \mathrm{colon}[\ldots, 0, 0, a, 0, 0, \ldots]$.

In accordance with (2.33) and (2.34) the generalized non-dimensional impedance of the systems can be expressed in a complex form as

$$W_\omega = Z_\omega G = Y_\omega R = \frac{j\omega}{a} x_\omega. \tag{2.37}$$

The solution of Eq.(2.36) concerning the component x_ω, by taking into account (2.37), allows us to obtain the following expression for the generalized impedance:

$$W_\omega = \frac{H_1 H_n \prod\limits_{k=2}^{m-2} [1 - Q^2(\omega+kp)^2 + j(\omega+kp)}{[1-Q^2(\omega+mp)^2+j(\omega+mp)] \prod\limits_{k=0}^{m-1} [1-Q^2(\omega+kp)+j(\omega+kp)} \ldots \rightarrow$$

$$\to \frac{+\overset{1}{\underset{i=k-1}{D}}\frac{-S^2}{1-Q^2(\omega+ip)^2+j(\omega+ip)}\Big]}{+\overset{-m}{\underset{i=k-1}{D}}\frac{-S^2}{1-Q^2(\omega+ip)^2+j(\omega+ip)}\Big]}, \tag{2.38}$$

where $H_1 = j\omega[1 - Q^2(\omega + p)^2 + j(\omega + p)]$, $H_n = [1 - Q^2(\omega + mp)^2 + j(\omega + mp)]$
$\times \left\{1-Q^2[\omega+(m-1)p]^2+j[\omega+(m-1)p]+\overset{1}{\underset{i=m-2}{D}}\frac{-S^2}{1-Q^2(\omega+ip)^2+j(\omega+ip)}\right\}-S^2.$

Expression (2.38) uses short denotations of continued (chain) fractions of the n-th order [190], for example:

$$\overset{1}{\underset{i=m-2}{D}}\frac{-S^2}{1-Q^2(\omega+ip)^2+j(\omega+ip)}=-\frac{S^2}{1-Q^2[\omega+(m-2)p)]^2+[\omega+(m-2)p)]-}\cdots\to$$

$$\to\cdots\frac{}{-\cfrac{S^2}{1-Q^2[\omega+(m-1)p]^2+j[\omega+(m-1)p]-\cfrac{S^2}{\cdots\cdots\cdots\cdots-\cfrac{S^2}{1-Q^2(\omega+p)^2+j(\omega+p)}}}}.$$

If the spectrum is limited only to the sum and difference combined frequencies $(\omega \pm p)$, the generalized impedance is expressed in the form

$$W_\omega^{(\pm)} = \frac{j\omega A}{(1 - Q^2\omega^2 + j\omega)(A - 2S^2) + 2(SQp)^2}, \tag{2.39}$$

where $A = [1 - Q^2(\omega^2 + p^2) + j\omega]^2 - [2\omega Q^2 p - jp]^2$.

It is worth noting that Eq.(2.15) for $p > 1$ is reduced to an equation of the first order:

$$\frac{dx}{d\tau} + \lambda(\tau)x = f(\tau), \quad \text{since} \quad Q \sim \frac{1}{p}. \tag{2.40}$$

As mentioned above, the absence of a resonance circuit in this case can be „compensated" by the higher values of the modulation depth of the parametric element, $S > 1$. Under these conditions however the three-diagonal matrix $[\lambda]$ in (2.39) is transformed into the following completely filled square matrix (the higher powers of the modulation index S are also taken into account):

$$
[\lambda]=\left[\begin{array}{cccccccc}
\cdot & \cdots & \cdots & \cdots & \cdots & \cdots & \cdots & \cdot \\
\vdots & 1+j(\omega-np) & S & S^2 \cdots & \cdots & \cdots & S^{2n} & \vdots \\
\vdots & S & 1+j[\omega-(n-1)p] & S & & & \vdots & \vdots \\
\vdots & S^2 & S & \ddots & \ddots & & \vdots & \vdots \\
\vdots & & \ddots & \ddots & \ddots & & \vdots & \vdots \\
\vdots & S^n & S^{n-1} \cdots S & 1+j\omega & S \cdots & \cdots & S^n & \vdots \\
\vdots & \vdots & & \ddots & \ddots & \ddots & \vdots & \vdots \\
\vdots & \vdots & & & \ddots & \ddots & \ddots & \vdots \\
\vdots & S^{2n-1} & & & S & 1+j[\omega+(n-1)p] & S & \vdots \\
\vdots & S^{2n} & S^{2n-1} \cdots & \cdots & \cdots & \cdots S & 1+j(\omega+np) & \vdots \\
\cdot & \cdots & \cdots & \cdots & \cdots & \cdots & \cdots & \cdot
\end{array}\right].
$$

In the general case the solution of Eq.(2.36), taking into account the reduced matrix, can be presented through branching continued (chain) fractions [190].

The Exps.(2.38) and (2.39) allow assessing the need to account for a certain number of combined components, and also to analyze the two borderline cases:

a) the case when the reactive parametric element has been complemented to yield a resonance circuit by using the opposite reactance, and when it works in the situation of precise resonance;

b) the case of an aperiodic circuit, when $Q \to 0$ in Eq.(2.15). Moreover, in the second case one can trace at what values of the depth of modulation S the lack of an oscillating circuit will be „compensated" for.

The investigation has been conducted on the basis of Exps.(2.38) and (2.39) by employing numeric methods. A reduction method has also been used for the purpose of assessing the role of the combined components and the need to account for a certain number of them.

As an example of the results that have been obtained, Fig.2.11 a, b shows the dependence of the active and imaginary component of the nondimensional impedance W_ω on the frequency ω of the external acting signal at a frequency of change of the parametric element $p = 10$ and the different values of the parameter S, describing the modulation depth of the parametric element. One can clearly see the drastic alteration and the change in the sign of the impedance components in the value range $\omega \sim p$.

The major results from the numeric analysis can be summed up in the following way:

1. For systems described by an equation of the second order of the type of (2.15), 5% precision of the quantitative analysis is ensured by accounting for only

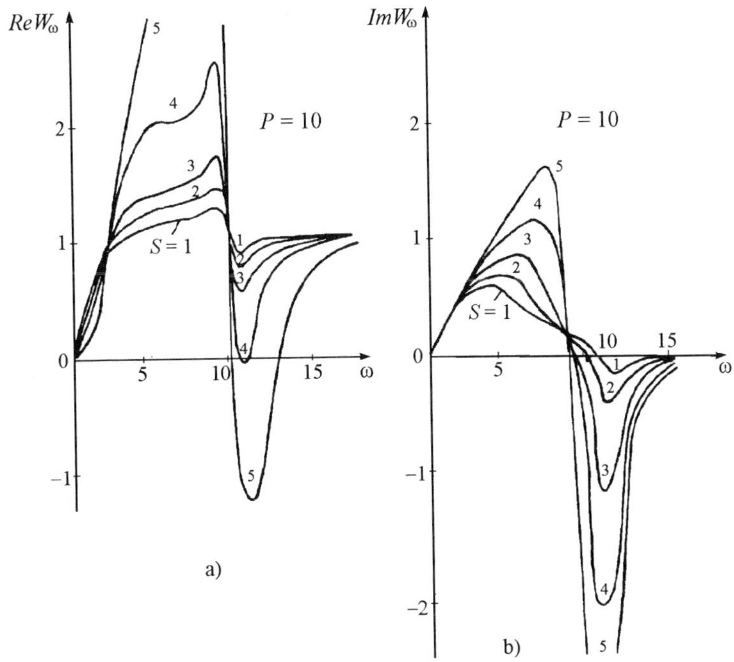

Fig. 2.11. Theoretical dependence of the active (a) and imaginary (b) component of the non-dimensional impedance W_ω on the frequency ω of the external signal at a frequency of change of the parametric element $p = 10$ and different values of the parameter S, describing the modulation depth of the parametric element

two combined frequencies (sum and difference). The precision increases to 1%, when accounting, for example, for 14 combined components.

2. For systems described by an equation of the first order of the type of (2.40), in cases of high values of the modulation depth of the parametric element $S > 1$ and frequencies $\omega, p > 1$, the conclusions drawn above for circuits of the second order hold.

3. For the same systems described by an equation of the first order of the type of (2.40), but under the condition $\omega, p < 1$, a larger number of combined frequencies should be accounted for. In the particular case of $\omega, p \sim 0,1$, satisfactory precision can be obtained only when 22 combined frequencies are accounted for.

2.2. Injection-locked oscillators as one-ports with controllable parameters

In this section we shall consider the following two physical effects manifested in radiophysical oscillator systems [170, 191–195]:

a) Alteration and control of the parameters of the equivalent impedance of an oscillator system viewed as a one-port in relation to an external low-frequency signal;

b) Frequency transformation of a signal by using an injection-locked oscillator, where, in a definite area of the equivalent synchronization band the transformation is accompanied by a regenerative parametric amplification and a simultaneous suppression of the image channel.

The second effect is applied in the solution of one of the key problems related to the design of superheterodyne SHF receivers for special purposes – the task of effectively suppressing the image channel.

First of all, we shall present briefly the method employed in the analysis of non-linear differential equations with strong nonlinearity.

2.2.1. A method for analyzing second order oscillating systems, close to the conservative ones, with strong reactive nonlinearity

The method is applicable to equations of the second order [185] appearing in the form

$$\frac{d^2x}{dt^2} + f(x) = \mu F\left(x, \frac{dx}{dt}, t\right), \tag{2.41}$$

where $\mu \ll 1$ is a small parameter, $f(x)$ is an analytical single-valued function, satisfying the following conditions:

1. The sign of the function $f(x)$ coincides with the sign x: $\operatorname{sign} f(x) = \operatorname{sign} x$.
2. The normalized time t has been selected in such a way that $\left.\dfrac{df(x)}{dx}\right|_{x=0} = 1$.
3. For a certain range $x_1 < x < x_2$ the polynomial approximation $f(x) = x(1 + C_1x + \ldots + C_nx^n)$ is applicable.

A nonlinear conversion of the variable

$$y = \sqrt{2\int_0^x f(x)dx}\ \operatorname{sign} x \tag{2.42}$$

is used and nonlinear time τ is introduced through the ratio

$$\frac{dt}{d\tau} = \frac{dx}{dy} = G(y). \tag{2.43}$$

The new variables y, τ satisfy the condition that the phase trajectory of the system on the plane $(y, \dot{y})$ should be a family of concentric circles, and that the angular motion speed of the point of representation should be equal to one.

The following equation corresponds to the new variables:

$$\frac{d^2y}{d\tau^2} + y = \mu\Phi\left(y, \frac{dy}{d\tau}, \tau\right), \tag{2.44}$$

where $\Phi\left(y, \frac{dy}{d\tau}, \tau\right) = F\left(x, \frac{dx}{dt}, t\right) G(y)$.

The newly introduced function $G(y)$ should possess the following properties:

1) $G(y) > 0$ at $x_1 < x < x_2$; 2) $G(0) = 1$;

3) $G(y)$ is an analytical function that can be presented by an approximating polynomial

$$G(y) = 1 + b_1 y + \ldots + b_m y^m. \tag{2.45}$$

Equation (2.44) is convenient because it can be investigated by using all methods developed to analyze oscillating systems with weak nonlinearity.

The solution in zero-order approximation is usually presented in the form of a harmonic function

$$y = R\cos(\tau + \varphi) = R\cos\Psi. \tag{2.46}$$

Substituting (2.46) in (2.45) we obtain

$$G(y) = B_0 + B_1 \cos\Psi + \ldots + B_n \cos n\Psi, \tag{2.47}$$

where

$$B_0 = \frac{1}{2\pi}\int_0^{2\pi} G(R\cos\Psi)d\Psi,$$

$$B_i = \frac{1}{\pi}\int_0^{2\pi} G(R\cos\Psi)\cos i\Psi d\Psi, \quad i = 1, 2, 3, \ldots,$$

or

$$B_0 = 1 + \frac{b_2}{2}R^2 + \frac{3}{8}b_4 R^4 + \frac{5}{16}b_6 R^6 + \ldots,$$

$$B_1 = b_1 R + \frac{3}{4}b_3 R^3 + \frac{5}{8}b_5 R^5 + \ldots,$$

$$B_2 = \frac{b_2}{2}R^2 + \frac{b_4}{2}R^4 + \frac{15}{32}b_6 R^6 + \ldots,$$

and so on.

Since in the course of the analysis it is convenient to seek the solution at nominal frequency, Eq.(2.44) is presented in the form

$$\frac{d^2y}{d\tau^2} + \beta^2 y = \mu\Phi\left(y, \frac{dy}{d\tau}, \tau\right) + (\beta^2 - 1)y \equiv \mu L\left(y, \frac{dy}{d\tau}, \tau\right), \tag{2.48}$$

where $\beta^2 - 1$ is the relative detuning of the system.

Obviously the „reverse" transition to real time t can take place in the following way:

$$t = \int_0^\Psi G(y)d\Psi = \frac{B_0}{\beta}\Psi + \frac{B_1}{\beta}\sin\Psi + \frac{B_2}{2\beta}\sin 2\Psi + \ldots + \frac{B_n}{n\beta}\sin n\Psi.$$

The oscillation period at time τ is $\tau_0 = 2\pi$, and at the initial normalized time t it is respectively $t_0 = \dfrac{2\pi B_0}{\beta}$.

The coefficient β, which is close to one, is determined in the following way: $\beta = \dfrac{m}{n}B_0\dfrac{\omega_e}{\omega_{ro}}$, where ω_e is the frequency of the external action, ω_{ro} is the respective resonance frequency of the system when the oscillations are infinitesimal, m and n are integers taking into account the possible frequency division and multiplication regimes.

The resonance frequency, conditioned by the non-isochronous character of the system, is determined as $\omega_r = \dfrac{\omega_{ro}}{B_0}$.

The amplitude R and the phase φ in the solution of (2.46) are determined by the system of shortened equations

$$\frac{dR}{d\Psi} = -\frac{\mu}{2\pi\beta^2}\int_0^{2\pi} L[R\cos\Psi, -\beta R\sin\Psi, e(\Psi-\varphi)]\sin\Psi d\Psi, \tag{2.49}$$

$$\frac{d\varphi}{d\Psi} = -\frac{\mu}{2\pi\beta^2 R}\int_0^{2\pi} L[R\cos\Psi, -\beta R\sin\Psi, e(\Psi-\varphi)]\cos\Psi d\Psi, \tag{2.50}$$

where $e(\Psi - \varphi)$ is a function reflecting the external periodic action.

2.2.2. Initial model and analytical techniques for investigating perturbed non-autonomous self-oscillating systems

This section presents an initial model of a non-autonomous oscillator as a functional or converting one-port and develops an adequate analytical technique on the basis of the general analytical approach proposed in 1.4.

The equivalent circuit of a non-autonomous oscillator with a negative resistance can be reduced to one of the circuits displayed in Fig.2.12. The negative differential resistance R_- (or negative conductance G_-) characterizes the falling region of the volt-ampere characteristics (VAC) $U(i)$ (or $i(U)$) of the active non-linear element. The oscillating circuit consists of the nonlinear inductance $L(\Phi)$ (or nonlinear capacitance $C(q)$) and the constant capacitance C (or inductance L). The resistance r (or conductance G) reflects the circuit losses. The oscillator experiences the action of an external harmonic current $i_b = I_b'\cos\omega_b t_r$ (or harmonic voltage $U_b = U_b'\cos\omega_b t_r$), synchronizing the oscillations. The input signal is

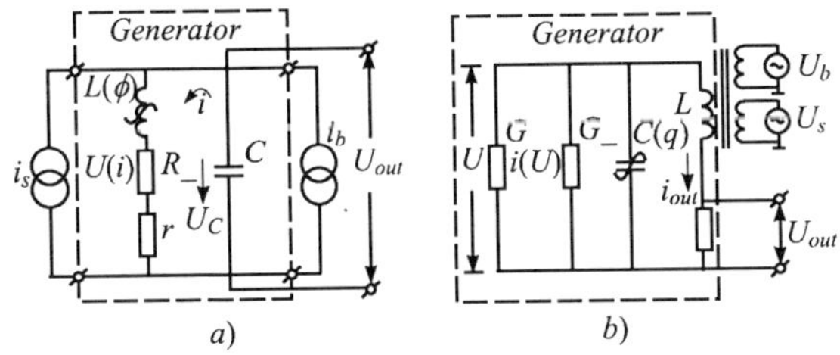

Fig. 2.12. Equivalent circuits of a non-autonomous oscillator with a negative resistance (a) and negative conductance (b)

presented by an external oscillator of current $i_s = I'_s \cos(\omega_s t_r + \varphi_s)$ (or of voltage $U_s = U'_s \cos(\omega_s t_r + \varphi_s)$), Φ and q are the respective total magnetic flux and charge, t_r is real time.

By virtue of the obvious duality of the equivalent circuits shown in Fig.2.12, their functioning can be described by invariant differential equations. The further analysis developed is therefore applicable to both cases. For the purpose of clarity, the mathematical theory presented takes into account only the equivalent circuit in Fig.2.12a.

The differential equation of the circuit is

$$\frac{d^2x}{dt^2} + f(x) = I_s \cos(\Omega_s t + \varphi_s) + I_b \cos \Omega_b t + \frac{d}{dt}\{U[f(x)]\}, \tag{2.51}$$

where $x = \dfrac{\Phi}{\Phi_0}$ (Φ_0 is a constant normalizing magnetic flux), $t = \omega_r t_r$, $\omega_r = \dfrac{1}{\sqrt{L_0 C}}$, ($L_0$ is the inductance at the working point);

$$f(x) = \frac{x}{L(x)C\omega_r^2}, I_s = \frac{I'_s}{I_0}, I_b = \frac{I'_b}{I_0}, I_0 = \Phi_0 \omega_r^2 C, \Omega_s = \frac{\omega_s}{\omega_r}, \Omega_b = \frac{\omega_b}{\omega_r}.$$

The function $U[f(x)]$ corresponds to nondimensional VAC parameters of the active element.

It is assumed that the solution of Eq.(2.51), when the signal is absent ($i_s \equiv 0$) is to be presented by the function x_0, and also that $x = x_0 + \Delta x$, when $i_s \neq 0$. The value Δx is a small perturbation determined by the signal influence. Under these conditions the Eq.(2.51) can be divided into two equations as follows:

$$\frac{d^2x_0}{dt^2} + f(x_0) = I_b \cos \Omega_b t + \frac{d}{dt}\{U[f(x_0)]\}, \tag{2.52}$$

$$\frac{d^2\Delta x}{dt^2} + \frac{df(x_0)}{dx}\Delta x = I_s \cos(\Omega_s t + \varphi_s) + \frac{d}{dt}\left\{\frac{dU[f(x_0)]}{dx}\Delta x\right\}, \tag{2.53}$$

Applying the nonlinear transformation of the variable, i.e. $x_0 \to y_0 = \sqrt{2\int_0^{x_0} f(x)dx}\,\mathrm{sign}\,x_0$ and introducing a new time parameter τ, where $\frac{dt}{d\tau} = \frac{dx_0}{dy_0} = \frac{y_0}{f(x_0)} = G(y_0)$ [185], Eqs.(2.52) and (2.53) can be transformed as follows:

$$\frac{d^2 y_0}{d\tau^2} + \beta^2 y_0 = 2\nu y_0 + I_b \cos \Omega_b t G(y_0) + \frac{d}{d\tau}\{U[f(x_0)]\} = \mu \mathcal{L}_0, \tag{2.54}$$

$$\frac{d}{d\Psi}\left\{\frac{1}{G(y_0)}\frac{d[G(y_0)]}{d\Psi}\Delta y\right\} + \frac{[G(y_0)]^2}{\beta^2}\frac{df(x_0)}{dx}\Delta y$$
$$= \frac{G(y_0)}{\beta^2} I_s \cos(\Omega_s t + \varphi_s) + \frac{1}{\beta}\frac{d}{d\Psi}\left\{\frac{dU[f(x_0)]}{dx} G(y_0)\Delta y\right\}, \tag{2.55}$$

where $\Delta y \simeq \frac{\Delta x}{G(y_0)}$, $d\Psi = \beta d\tau$, $\nu = \beta^2 - 1$ (ν is the relative detuning between the frequency Ω_b of the locking oscillation and the resonance frequency of the oscillator), $\mu \ll 1$ is a small parameter.

Equations (2.54) and (2.55) describe an injection-locked oscillator system, perturbed by a small external signal.

The equivalent low-frequency impedance of the one-port represented by an injection-locked oscillator can be determined as a ratio of complex values:

$$\dot{Z}_{Lf} = \frac{\dot{U}_{Lf}}{\dot{I}_s}, \tag{2.56}$$

where $\dot{U}_{Lf}$ and $\dot{I}_s$ are respectively the complex values of the low-frequency voltage and the input low-frequency current. For the two cases under consideration $\dot{U}_{Lf}$ means:

a) Voltage at the signal frequency Ω_s, when we consider the video-frequency-band impedance of the oscillator one-port, assuming $\Omega_s \ll \Omega_b$.

b) Voltage at the conversion (input or intermediate) frequency $\Delta\Omega = |\Omega_s - \Omega_b|$, when we consider the frequency conversion properties of the oscillator one-port and $\Omega_s \sim \Omega_b$.

In the second case $\dot{Z}_{Lf}$ will be called „transmission impedance of conversion".

In order to determine $\dot{U}_{Lf}$, we shall write an equation for the relative change of the voltage ΔU_c (Fig.2.12a) which takes place under the influence of the signal:

$$\frac{\Delta U_c}{\Phi_0 \omega_p} = \Delta\left\{\frac{dx}{dt} - U[f(x_0)]\right\} = \beta \frac{1}{G(y_0)}\frac{dG(y_0)}{d\Psi}\Delta y - \left[\frac{dU}{dx}G(y_0)\right]\Delta y. \tag{2.57}$$

With respect to a definite spectrum of the oscillations $\{\delta F\}$ in the non-autonomous oscillator, determined by the action of the signal, Eqs.(2.55) and (2.57) can be rewritten in the following matrix form (See 1.4 and [170])

$$[D(\delta F)][G^r][D(\delta F)][G][Y(\delta F)] + \frac{1}{\beta^2}[K][Y(\delta F)]$$

$$-\frac{1}{\beta}[D(\delta F)][H][Y(\delta F)] = \frac{1}{\beta^2}[G][I], \tag{2.58}$$

$$[\dot{U}(\delta F)] = \beta[G^r][D(\delta F)][G][Y(\delta F)] - [H][Y(\delta F)], \tag{2.59}$$

where the following relations are used: $G(y_0)\to[G]$, $\frac{1}{G(y_0)}\to[G^r]$,

$[G(y_0)]^2\frac{df(x_0)}{dx} \to [K]$, $\frac{dU[f(x_0)]}{dx}G(y_0) \to [H]$, $I_s\cos\Psi_s \to [I]$, $\frac{\Delta U_c}{\Phi_0\omega_r} \to [\dot{U}(\delta F)]$,

$\frac{d}{d\Psi} \to [D(\delta F)]$ and $\Delta y \to [Y(\delta F)]$.

The low-frequency voltage $\dot{U}_{Lf}$ can be determined by using (2.59) as follows:

$$\dot{U}_{Lf} = [\dot{E}(\delta F)][\dot{U}(\delta F)], \tag{2.60}$$

where $[\dot{E}(\delta F)]$ is the matrix of the inverse transformation to the initial variables.

The analysis of the impedance and converting properties of a one-port represented by an non-autonomous oscillator can be carried out on the basis of Eqs.(2.56), (2.58), (2.59), (2.60) and the methods recounted in section 1.4 in the following order:.

(i) The considered spectrum $\{\delta F\}$, which is a perturbation of the main oscillation in the system δy_0 under the signal action, should be set.

(ii) The complex matrixes $[G]$, $[G^r]$, $[K]$, $[H]$ and $[I]$ are specified on the basis of the preset type of the functions $f(x)$ and $U[f(x)]$ and by taking into account $\{\delta F\}$.

(iii) The perturbation matrix $[Y(\delta F)]$ is determined from Eq.(2.58) by taking into account the preset spectrum $\{\delta F\}$.

(iv) As $[Y(\delta F)]$ is substituted in (2.59) and the established voltage matrix $[U(\delta F)]$ of the combined frequencies with a set spectrum $\{\delta F\}$ is substituted in (2.60), and as the definition (2.56) is taken into account, one can obtain expressions for the impedance (or for the transmitting impedance of conversion) by means of an injection-locked oscillator.

Further on we shall consider the two specific regimes: a) alteration and control of the parameters of the equivalent impedance of a self-oscillator system as regards the external low-frequency signal; b) frequency conversion of a signal with a frequency close to that of the non-autonomous oscillator used for the purpose.

2.2.3. Equivalent video-impedance of a one-port represented by an injection-locked oscillator

The equivalent impedance of a one-port represented by an injection-locked oscillator is investigated with respect to external low-frequency signals (i_s in Fig.2.12a) in a large video bandwidth $\Omega_s \in [0, \Omega_L]$, when $\Omega_L \ll \Omega_b$.

The following spectral components will be taken into account: the frequency of the input signal Ω_s, and the combined frequencies $(\Omega_b \pm \Omega_s)$.

In the absence of a signal ($i_s \equiv 0$), (2.54) yields the solution for the system's main oscillations

$$y_0 = a\cos(\beta\tau + \varphi) = a\cos\Psi, \tag{2.61}$$

where the shortened equations concerning the amplitude a and the phase φ are written in the form:

$$\left|\begin{aligned} \frac{da}{d\tau} &= -\frac{1}{2\pi\beta}\int_0^{2\pi} \mu\mathcal{L}_0(y_0,\tau)\sin\Psi d\Psi \\ \frac{d\varphi}{d\tau} &= -\frac{1}{2\pi\beta a}\int_0^{2\pi} \mu\mathcal{L}_0(y_0,\tau)\cos\Psi d\Psi \end{aligned}\right. . \tag{2.62}$$

Taking into account that $\Omega_b = \Omega_r = \frac{\beta}{B_0}$, where Ω_r is the frequency of the non-autonomous oscillator, $B_0 = \dfrac{1}{2\pi}\displaystyle\int_0^{2\pi} G(a\cos\Psi)d\Psi$ (See (2.47)) and denoting $\Gamma = \dfrac{\Omega_s B_0}{\beta}$, we shall present the spectrum in the following form:

$$\{\delta F\} = \Gamma, \quad 1+\Gamma, \quad 1-\Gamma$$

(we not only used nondimensional frequencies but also normalized the main spectral components to unity).

We select, for the dependencies $L(\Phi)$ and $U(i)$, the following approximations

$$L = \frac{L_0}{1 + L_1\Phi} \quad \text{and} \quad U(i) = -ki + \gamma' i^2 + \varkappa' i^3,$$

so as to write the functions $f(x)$ and $U[f(x)]$ in the form

$$f(x) = x + Sx^2, \tag{2.63}$$

$$U[f(x)] = \frac{f(x)}{Q_-} - \gamma[f(x)]^2 - \varkappa[f(x)]^3, \tag{2.64}$$

where $S = \dfrac{L_1}{\Phi_0}$, $Q = -\dfrac{(r-k)I_0}{\Phi_0\omega_r}$, $\gamma = \gamma' I_0^2$, $\varkappa = \varkappa' I_0^3$.

Employing (2.63) and (2.64) we also determine the following approximations:

$$x \simeq y - \frac{1}{3}y^2, \quad G(y) \simeq 1 - \frac{2}{3}Sy + \frac{2}{3}S^2y^2,$$
$$[G(y)]^2\frac{df(x)}{dx} \simeq 1 + \frac{2}{3}Sy, \quad \frac{dU[f(x)]}{dx}G(y) = H_0 + H_1 y - H_2 y^2, \tag{2.65}$$

where $H_0 \simeq \frac{1}{Q}$, $H_1 = 4\frac{S}{3}H_0$, $H_2 = 18\left(\frac{S}{3}\right)^2\frac{1}{H_0}$.

In order to determine the matrixes $[I]$ and $[\dot{E}(\delta F)]$, we need to apply the following direct and inverse transformations:

$$\left|\begin{array}{l} \cos(\Omega_s t + \varphi_s) = \cos[(\Omega_s \frac{B_0}{\beta}(\Psi - \varphi) + \varphi_s + \Omega_s \frac{B_1}{\beta}\sin\Psi + \ldots] \\ \qquad = Re\{[-I_1(\xi)e^{j(\Gamma-1)\Psi} + I_0(\xi)e^{-j\Gamma\Psi} + I_1(\xi)e^{j(\Gamma+1)\Psi}\}e^{j\Theta_s} \\ \cos\Gamma\Psi = \cos(\Omega_s t + \Gamma\varphi) + \text{ high-frequency components} \\ \cos(1\pm\Gamma)\Psi = \frac{1}{2}(1\pm\Gamma)\frac{B_1}{B_0}\cos(\Omega_s t + \Gamma\varphi) + \text{ high-frequency components,} \end{array}\right. \tag{2.66}$$

where I_0 and I_1 are modified Bassel functions, $\xi = \Gamma\frac{B_1}{B_0}$, $\Theta_s = \varphi_s - \Gamma\varphi$, $\Psi = \frac{\beta}{B_0}t + \varphi - \frac{B_1}{B_0}\sin\left(\frac{\beta}{B_0}t + \varphi\right)$. Here and henceforth

$$B_n = \frac{1}{\pi}\int_0^{2\pi} G(a\cos\Psi)\cos n\Psi d\Psi, \qquad n = 1, 2 \tag{2.67}$$

Using the Exps.(2.65) and (2.66), Eqs.(2.58), (2.59) and (2.60) can be written in the following complex matrix form:

$$\begin{bmatrix} j(\Gamma-1) & 0 & 0 \\ 0 & j\Gamma & 0 \\ 0 & 0 & j(\Gamma+1) \end{bmatrix} \begin{bmatrix} 1 & \frac{Sa}{3} & 0 \\ \frac{Sa}{3} & 1 & \frac{Sa}{3} \\ 0 & \frac{Sa}{3} & 1 \end{bmatrix} \begin{bmatrix} j(\Gamma-1) & 0 & 0 \\ 0 & j\Gamma & 0 \\ 0 & 0 & j(\Gamma+1) \end{bmatrix}$$

$$\times \begin{bmatrix} 1 & -\frac{Sa}{3} & 0 \\ -\frac{Sa}{3} & 1 & -\frac{Sa}{3} \\ 0 & -\frac{Sa}{3} & 1 \end{bmatrix} [Y(\delta F)] + \frac{1}{\beta^2} \begin{bmatrix} 1 & \frac{Sa}{3} & 0 \\ \frac{Sa}{3} & 1 & \frac{Sa}{3} \\ 0 & \frac{Sa}{3} & 1 \end{bmatrix} [Y(\delta F)]$$

$$-\frac{1}{\beta}\begin{bmatrix} j(\Gamma-1) & 0 & 0 \\ 0 & j\Gamma & 0 \\ 0 & 0 & j(\Gamma+1) \end{bmatrix}\begin{bmatrix} H_0 & \frac{H_1 a}{2} & \frac{H_2 a}{2} \\ \frac{H_1 a}{2} & H_0 & \frac{H_1 a}{2} \\ \frac{H_2 a}{2} & \frac{H_1 a}{2} & H_0 \end{bmatrix}[Y(\delta F)]$$

$$= \frac{1}{\beta^2}\begin{bmatrix} 1 & -\frac{Sa}{3} & 0 \\ -\frac{Sa}{3} & 1 & -\frac{Sa}{3} \\ 0 & -\frac{Sa}{3} & 1 \end{bmatrix}\begin{bmatrix} -I_1(\xi) \\ I_0(\xi) \\ I_1(\xi) \end{bmatrix}, \tag{2.68}$$

$$[\dot{U}(\delta F)] = \beta\begin{bmatrix} 1 & \frac{Sa}{3} & 0 \\ \frac{Sa}{3} & 1 & \frac{Sa}{3} \\ 0 & \frac{Sa}{3} & 1 \end{bmatrix}\begin{bmatrix} j(\Gamma-1) & 0 & 0 \\ 0 & j\Gamma & 0 \\ 0 & 0 & j(\Gamma+1) \end{bmatrix}$$

$$\times\begin{bmatrix} 1 & -\frac{Sa}{3} & 0 \\ -\frac{Sa}{3} & 1 & -\frac{Sa}{3} \\ 0 & -\frac{Sa}{3} & 1 \end{bmatrix}[Y(\delta F)] - \begin{bmatrix} H_0 & \frac{H_1 a}{2} & \frac{H_2 a}{2} \\ \frac{H_1 a}{2} & H_0 & \frac{H_1 a}{2} \\ \frac{H_2 a}{2} & \frac{H_1 a}{2} & H_0 \end{bmatrix}[Y(\delta F)], \tag{2.69}$$

$$\dot{Z}_{Lf} = \begin{bmatrix} \frac{1}{2}(1-\Gamma)\frac{B_1}{B_0} & 1 & \frac{1}{2}(1+\Gamma)\frac{B_1}{B_0} \end{bmatrix}[\dot{U}(\delta F)], \tag{2.70}$$

where

$$[Y(\delta F)] = \begin{bmatrix} \Delta y_{\Gamma-1} \\ \Delta y_{\Gamma} \\ \Delta y_{\Gamma+1} \end{bmatrix}\frac{1}{I_s e^{j\Theta_s}}, \quad [\dot{U}(\delta F)] = \begin{bmatrix} \dot{U}_{\Gamma-1} \\ \dot{U}_{\Gamma} \\ \dot{U}_{\Gamma+1} \end{bmatrix}\frac{1}{I_s e^{j\Theta_s}}.$$

Determining $[Y(\delta F)]$ in Eq.(2.68) and substituting the Eq.(2.69) in Eq.(2.70) and after a series of mathematical transformations, the following expression for the low-frequency „input" impedance of the injection-locked oscillator is obtained:

$$\dot{Z}_{Lf} = (H_0 + j\Gamma)P, \tag{2.71}$$

where $P = 1 + \nu - \frac{(Sa)^2}{3}\frac{\nu}{\nu^2 + \left(\frac{H_0}{2}\right)^2}$.

It is evident that the Eq.(2.71) can be presented as a real and an imaginary part:

$$\dot{Z}_{Lf} = R + jB. \tag{2.72}$$

Fig.2.13 a, b shows the theoretical dependencies of the components of the effective impedance – the resistance R (a) and the inductance $\frac{B}{\Omega_s}$ (b) – on the detuning of the frequency ν at $H_0 = 0,02$, $\Gamma = 0,01$ and for different values of the parameter Sa. It is obvious that the values of R and $\frac{B}{\Omega_s}$ depend essentially on the detuning in the oscillator locking bandwidth. At a certain degree of nonlinearity of the inductance in the oscillator resonance circuit (parameter S), the effective low frequency resistance R and inductance $\frac{B}{\Omega_s}$ can take negative values in a wide video frequency band of the input signals.

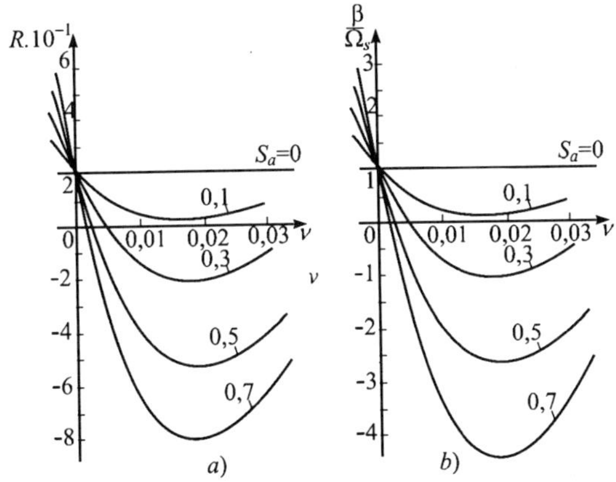

Fig. 2.13. Theoretical dependences of the equivalent resistance R (a) and equivalent inductance $\frac{B}{\Omega_s}$ (b) of an oscillator one-port on the frequency detuning ν

The experimental set-up includes a two-transistor model of a $p-n-p-n$-type device with a falling sector in the V/A characteristics and a nonlinear inductance (See Fig.2.14)

Fig.2.15 exhibits the experimental dependence of the equivalent (effective) low-frequency inductance $\frac{L_{in}}{L_0}$ of an injection-locked self-oscillator with negative resistance and nonlinear inductance on the locking frequency (detuning ν) at different levels of the locking current I_b and a fixed low test-signal frequency ($I_{b1}>I_{b2}>I_{b3}$).

The theory and the experiment described above present an effective method of using the reversibility of the modulation parametric interactions to control the equivalent impedances of radiophysical circuits. It is evident that injection-locked oscillators can be considered, in relation to signals in a wide video frequency bandwidth, as one-ports with controllable parameters.

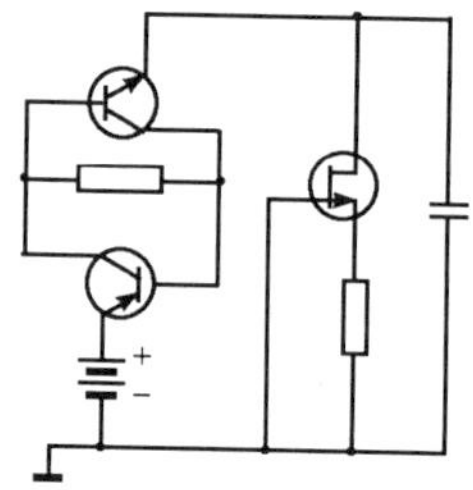

Fig. 2.14. A two-transistor model of a $p-n-p-n$ type device with a falling sector in the Volt-Ampere characteristics and nonlinear inductance

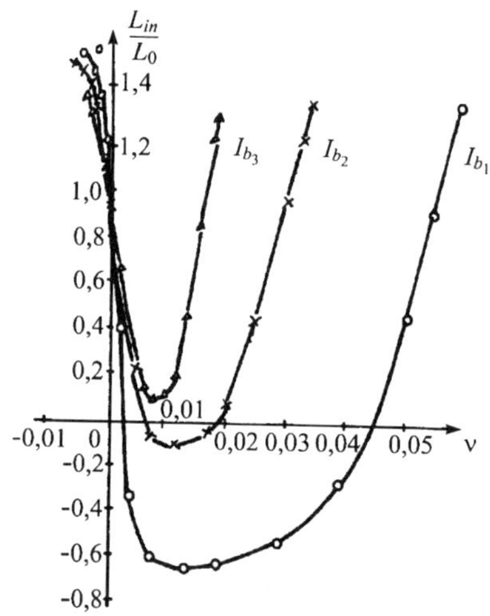

Fig. 2.15. Experimental dependence of the effective (equivalent) low-frequency inductance of a synchronized oscillator on the frequency detuning of the synchronizing action, $I_{b_1} > I_{b_2} > I_{b_3}$

The phenomena associated with a change in the effective low-frequency impedance of an injection-locked oscillator with a nonlinear capacitance, or an oscillator working in a regime of synchronous frequency division and multiplication, in the respective locking bandwidths, have analogous qualities.

The considered effects of alteration of the effective impedances of injection-locked oscillators brought about by the locking conditions highlight on a number of low-frequency instabilities observed in oscillator systems with modern semiconductor devices, particularly in a SHF range (for example the spontaneous low-frequency excitations and the defects occurring in the active element in SHF oscillators with avalanche-drift diodes).

2.2.4. Analysis of the conversion properties of a one-port presented by a non-autonomous oscillator

The conversion properties of a one-port presented by an injection-locked oscillator will also be considered on the basis of Fig.2.12a. The current of frequency Ω_s is the input signal, while the voltage of frequency $\Delta\Omega = |\Omega_b - \Omega_s|$ is the output signal, Ω_b is the oscillating frequency of the injection-locked oscillator, $\Omega_s \sim \Omega_b$.

We shall consider two cases:

a) $\Omega_s < \Omega_b$ (lower-sideband, main channel)

b) $\Omega_s > \Omega_b$ (upper-sideband, image channel)

We shall account for the following spectral components: an output conversion frequency $\Delta\Omega = |\Omega_b - \Omega_s|$ and combined frequencies $(\Omega_b + \Omega_s)$ and $(2\Omega_b - \Omega_s)$. The selected spectrum helps account for the influence of the second harmonic of the time-varying impedance of the injection-locked oscillator.

As the denotation $\alpha = \dfrac{\Omega_s B_0}{\beta}$ is introduced, the spectrum under consideration can be presented in the form:

i) for the main channel: $\{\delta F\} = \Gamma,\ 1+\Gamma,\ 2-\Gamma$

ii) for the image channel: $\{\delta F\} = \Gamma,\ 1-\Gamma,\ 2+\Gamma$,

where $\Gamma = (1-\alpha)$ is the output conversion frequency in a nondimensional (reduced to one) form, $1 \pm \Gamma = 2 - \alpha$ and $2 \mp \Gamma = 1 + \alpha$. Here and henceforth the upper sign corresponds to the spectrum of the main channel, and the lower one to the spectrum of the image channel.

The matrices $[I]$ and $[\dot{E}(\delta F)]$ are determined by using the following direct and inverse transformations:

$$\left|\begin{aligned}
&\cos(\Omega_s t + \varphi_s) = \cos[\alpha(\Psi - \varphi) + \varphi_s + \xi_1 \sin\Psi + \xi_2 \sin 2\Psi + \ldots] \\
&\qquad = Re\left\{\left[-I_1(\xi_1)e^{-j\Gamma\Psi} - I_1(\xi_2)e^{-j(1\pm\Gamma)\Psi} + I_1(\xi_1)e^{j(2\mp\Gamma)\Psi}\right]e^{j\Theta}\right\}, \\
&\cos\Psi\Gamma = \cos[\Delta\Omega t + \Gamma\varphi] + \text{ high-frequency components}, \\
&\cos\Psi(1\pm\Gamma) = -\frac{1}{2}(1\pm\Gamma)\frac{B_1}{B_0}\cos[\Delta\Omega t + \Gamma\varphi] + \text{ high-frequency components}, \\
&\cos\Psi(2\mp\Gamma) = \frac{1}{2}(2\mp\Gamma)\frac{B_2}{2B_0}\cos[\Delta\Omega t + \Gamma\varphi] + \text{ high-frequency components},
\end{aligned}\right. \tag{2.73}$$

where $\xi_1 = \dfrac{\Omega_s B_1}{\beta}$, $\xi_2 = \dfrac{\Omega_s B_2}{2\beta}$, $\Theta = \varphi_s - \alpha\varphi$.

Since the main purpose of the study is to determine the output voltage of conversion frequency $\Delta\Omega$, the HF components in (2.73) should be neglected.

Equations (2.57)–(2.59) can be written as follows:

$$\begin{bmatrix} j\Gamma & 0 & 0 \\ 0 & j(1\pm\Gamma) & 0 \\ 0 & 0 & j(2\mp\Gamma) \end{bmatrix} \begin{bmatrix} 1 & \frac{Sa}{3} & -\frac{(Sa)^2}{6} \\ \frac{Sa}{3} & 1 & 0 \\ -\frac{(Sa)^2}{6} & 0 & 1 \end{bmatrix} \begin{bmatrix} j\Gamma & 0 & 0 \\ 0 & j(1\pm\Gamma) & 0 \\ 0 & 0 & j(2\mp\Gamma) \end{bmatrix}$$

$$\times \begin{bmatrix} 1 & -\frac{Sa}{3} & \frac{(Sa)^2}{6} \\ -\frac{Sa}{3} & 1 & 0 \\ \frac{(Sa)^2}{6} & 0 & 1 \end{bmatrix} [\,Y(\delta F)\,] + \frac{1}{\beta^2} \begin{bmatrix} 1 & \frac{Sa}{3} & -\frac{(Sa)^4}{4} \\ \frac{Sa}{3} & 1 & 0 \\ -\frac{(Sa)^2}{4} & 0 & 1 \end{bmatrix} [\,Y(\delta F)\,]$$

$$-\frac{1}{\beta} \begin{bmatrix} j\Gamma & 0 & 0 \\ 0 & j(1\pm\Gamma) & 0 \\ 0 & 0 & j(2\mp\Gamma) \end{bmatrix} \begin{bmatrix} H_0 & \frac{H_1 a}{2} & \frac{H_2 a}{2} \\ \frac{H_1 a}{2} & H_0 & 0 \\ \frac{H_2 a}{2} & 0 & H_0 \end{bmatrix} [\,Y(\delta F)\,]$$

$$= \frac{1}{\beta^2} \begin{bmatrix} 1 & -\frac{Sa}{3} & \frac{(Sa)^2}{6} \\ -\frac{Sa}{3} & 1 & 0 \\ \frac{(Sa)^2}{6} & 0 & 1 \end{bmatrix} \begin{bmatrix} -I_1(\xi_1) \\ -I_1(\xi_2) \\ +I_1(\xi_1) \end{bmatrix}, \tag{2.74}$$

$$[\,U(\delta F)\,] = \beta \begin{bmatrix} 1 & \frac{Sa}{3} & -\frac{(Sa)^2}{6} \\ \frac{Sa}{3} & 1 & 0 \\ -\frac{(Sa)^2}{6} & 0 & 1 \end{bmatrix} \begin{bmatrix} j\Gamma & 0 & 0 \\ 0 & j(1\pm\Gamma) & 0 \\ 0 & 0 & j(2\mp\Gamma) \end{bmatrix}$$

$$\times \begin{bmatrix} 1 & -\frac{Sa}{3} & \frac{(Sa)^2}{6} \\ -\frac{Sa}{3} & 1 & 0 \\ \frac{(Sa)^2}{6} & 0 & 1 \end{bmatrix} [\,Y(\delta F)\,] - \begin{bmatrix} H_0 & \frac{H_1 a}{2} & \frac{H_2 a}{2} \\ \frac{H_1 a}{2} & H_0 & 0 \\ \frac{H_2 a}{2} & 0 & H_0 \end{bmatrix} [\,Y(\delta F)\,], \tag{2.75}$$

$$\dot{Z}^{(-),(+)} = \begin{bmatrix} 1 & -\frac{1}{2}(1\pm\Gamma)\frac{B_1}{B_0} & \frac{1}{4}(2\mp\Gamma)\frac{B_2}{B_0} \end{bmatrix} [\,U(\delta F)\,], \tag{2.76}$$

where

$$[\,Y(\delta F)\,] = \begin{bmatrix} \Delta y_\Gamma \\ \Delta y_{1\pm\Gamma} \\ \Delta y_{2\mp\Gamma} \end{bmatrix} \frac{1}{I_s e^{j\Theta}}, \quad [\,U(\delta F)\,] = \begin{bmatrix} \dot{U}_\Gamma \\ \dot{U}_{1\pm\Gamma} \\ \dot{U}_{2\mp\Gamma} \end{bmatrix} \frac{1}{I_s e^{j\Theta}}.$$

As $[Y(\delta F)]$ is determined from (2.74) and substituted in (2.75) and (2.76), the following expressions for the transmission impedance of frequency conversion are obtained:

a) for the main (lower-sideband) channel

$$\dot{Z}^{(-)} = R^{(-)} + jB^{(-)}, \tag{2.77}$$

where

$$R^{(-)} = -\frac{\frac{Sa}{3}H_0(1-\Gamma-4\nu)}{D_1} - \frac{12\left(\frac{Sa}{3}\right)^3\Gamma + 2\left(\frac{Sa}{3}\right)^5(2+5\Gamma-14\nu)}{D_2}H_0, \tag{2.77a}$$

$$B^{(-)} = \frac{\frac{Sa}{3}\Gamma}{D_1} - \frac{8\left(\frac{Sa}{3}\right)^5\Gamma + 4\left(\frac{Sa}{3}\right)^7(5\Gamma+9\nu)}{D_2}, \tag{2.77b}$$

b) for the image (upper sideband) channel

$$\dot{Z}^{(+)} = R^{(+)} + jB^{(+)}, \tag{2.78}$$

where

$$R^{(+)} = -\frac{\frac{Sa}{3}H_0(1+\Gamma-4\nu)}{D_1} - \frac{4\left(\frac{Sa}{3}\right)^2\Gamma + 4\left(\frac{Sa}{3}\right)^5(1+5\Gamma-5\nu) + 2\left(\frac{Sa}{3}\right)^6(8-17\Gamma-117\nu)}{D_3}H_0, \tag{2.78a}$$

$$B^{(+)} = \frac{\frac{Sa}{3}\Gamma}{D_1} - \frac{8\left(\frac{Sa}{3}\right)^5(\Gamma-\nu) - 36\left(\frac{Sa}{3}\right)^7(\Gamma-\nu)}{D_3}, \tag{2.78b}$$

$$D_1 = 1 + 2\left(\frac{Sa}{3}\right)^2\Gamma - 4\nu, \quad D_2 = [-2(\Gamma+\nu) - 2\left(\frac{Sa}{3}\right)^2(1-\Gamma-9\nu)]^2$$
$$+ [(1+\Gamma-9\nu)H_0 - 2\left(\frac{Sa}{3}\right)^2 H_0(1+\Gamma-9\nu)]^2, \quad D_3 = [2(\Gamma-\nu)$$
$$-2\left(\frac{Sa}{3}\right)^2(1-3\Gamma-8\nu)]^2 + [(1-\Gamma-8\nu)H_0 - 2\left(\frac{Sa}{3}\right)^2 H_0(1-3\Gamma-8\nu)]^2.$$

The analysis shows that the first terms in Eqs. (2.77a,b) and (2.78a,b) are conditioned by the action of the first harmonic in the time-varying impedance of the injection-locked oscillator, while the second terms are mostly conditioned by the action of its second harmonic.

Fig.2.16 shows the theoretical dependances of the transmission impedance of frequency conversion by main (solid line) and image (dotted line) channel on the detuning ν. They are calculated by using formulae (2.77) and (2.78) at $H_0 = 0,04$, $\Gamma = 0,01$ and different values of the parameter Sa. For comparison, the dash-dot lines show the value of the transmission impedance of frequency conversion of an autonomous oscillator. It can be seen that the oscillator conversion properties change considerably in the locking bandwidth. The low-sideband conversion (main channel) can be carried out with a signal amplification, whereas the upper-sideband conversion (image channel) can be suppressed.

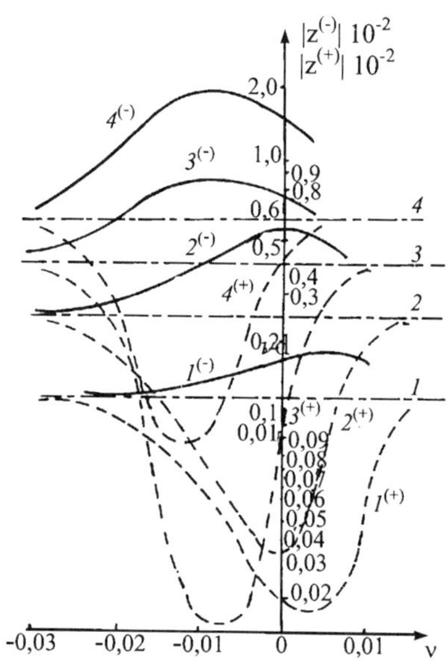

Fig. 2.16. Theoretical dependence of the modulus of the transmission impedance of conversion on the detuning at $H_0 = 0.04$ and $\Gamma = 0.01$ for different values of the parameter Sa: $1^{(-)}, 1^{(+)}$, 1 – 0,1; $2^{(-)}, 2^{(+)}$, 2 – 0,2; $3^{(-)}, 3^{(+)}$, 3 – 0,3; $4^{(-)}, 4^{(+)}$, 4 – 0,4

Fig.2.17 presents the theoretical dependances of the maximum ratio of the transmission impedances of frequency conversion by main and image channel $(|\dot{Z}^{(-)}|/|\dot{Z}^{(+)}|)_{\max}$, reached at certain detuning, on the oscillation amplitude of the injection-locked oscillator at $\Gamma = 0,01$ and for different values of the parameter H_0. The pronounced extremum manifests the specificity of the oscillator-based frequency conversion – the typical dependence of the conversion properties on the oscillation amplitude.

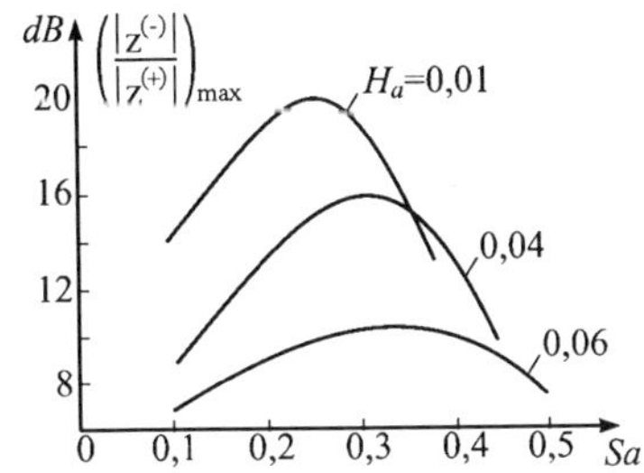

Fig. 2.17. Theoretical dependence of the maximum values of the ratio of the transmission impedances of conversion by main and image channel on the oscillation amplitude in the non-autonomous oscillator at $\Gamma = 0.01$

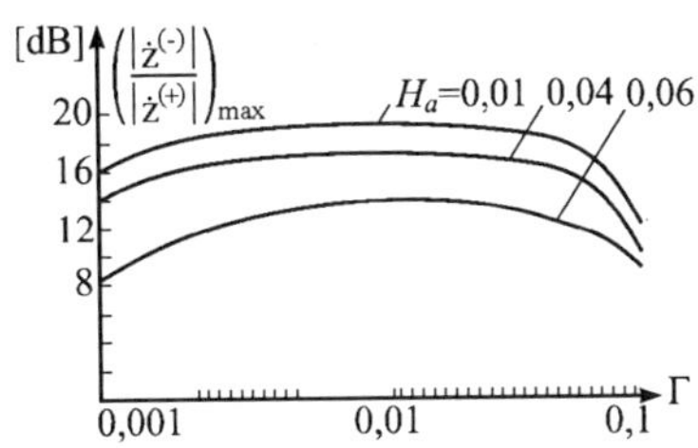

Fig. 2.18. Theoretical dependence of the maximum values of the ratio of the transmission impedances of conversion by main and image channel on the output converted frequency at $Sa = 0.3$

The theoretical dependence of the maximum ratio $(|\dot{Z}^{(-)}|/|\dot{Z}^{(+)}|)_{\max}$ on the output converted frequency for $Sa = 0,3$ and for different values of the parameter H_0 is plotted in Fig.2.18.

Figs.2.17 and 2.18 show that within the locked bandwidth the ratio $(|\dot{Z}^{(-)}|/|\dot{Z}^{(+)}|)_{\max}$ can exceed 20 dB in the frequency bandwidth of the output converted (intermediate frequency) signal from 0 Hz to the borderline frequency, which represents 10–15% of the oscillation frequency of the non-autonomous generator.

The conversion properties of the non-autonomous generator have been studied by using the experimental oscillator shown in Fig.2.14.

Fig.2.19 shows the experimental dependence of the transmission impedances of frequency conversion $|\dot{Z}^{(-)}|$ and $|\dot{Z}^{(+)}|$ on the locking frequency at signal frequency and current of $f_s = 3$ kHz and $I_s = 50$ nA respectively, and bias current $I_0 = 0,11$ mA. The dash-dot line shows the autonomous oscillator conversion level. The experiment corroborates the possibility of frequency conversion on the basis of an injection-locked oscillator with main channel amplification and simultaneous image channel suppression.

The experimental dependence of the maximum ratio of the transmission

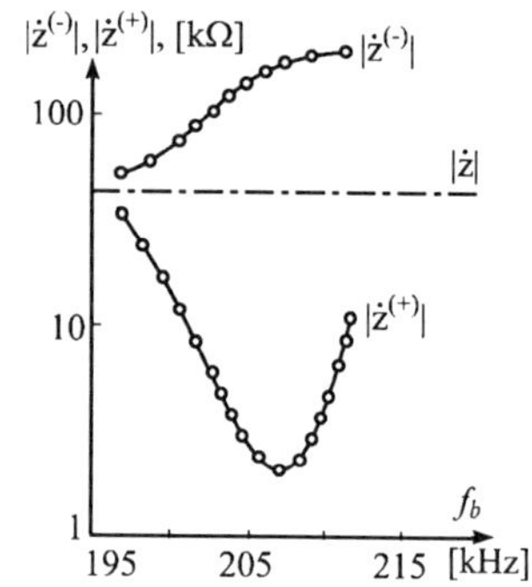

Fig. 2.19. Experimental dependence of the transmission impedances of conversion in the frequency bandwidth of synchronization of the generator

impedances of frequency conversion, achieved at a definite frequency of the locking action, on the output converted (intermediate) frequency is shown in Fig.2.20.

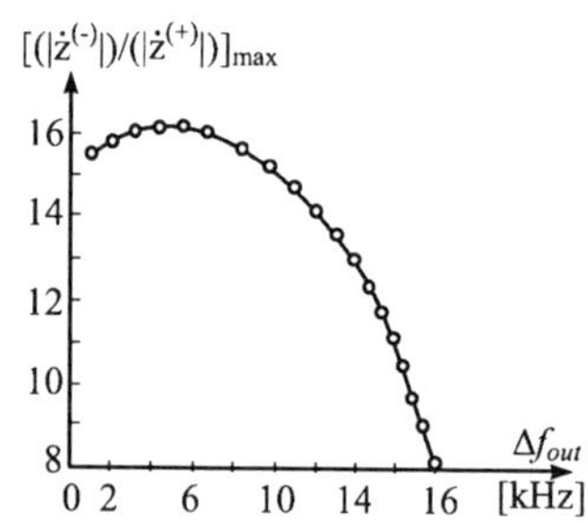

Fig. 2.20. Experimental dependence of the ratio of the transmission impedances of conversion on the output (intermediate) frequency

Figs.2.21 and 2.22 illustrate the dependence of the conversion properties of the non-autonomous generator on the operating mode. By way of comparison, Fig.2.21 displays the experimental V/A characteristic $U_0(I_0)$ and the dependence of the oscillating voltage in the non-autonomous generator U_g on the bias current. Fig.2.22 shows the experimental dependence of the maximum ratio of the transmission impedances of frequency conversion and of the transmission impedances of conversion by the main channel on the bias current. It can be observed that the ratio $(|\dot{Z}^{(-)}|/|\dot{Z}^{(+)}|)_{\max}$ reaches its maximum at a voltage amplitude in the self-oscillator considerably lower than the maximum value. The fact that the position of the maximum values of the dependencies $(|\dot{Z}^{(-)}|/|\dot{Z}^{(+)}|)_{\max}$ and $|\dot{Z}^{(-)}|$ on the bias current coincide is characteristic, i.e. the conditions for attaining the maximum of the transmission impedances of frequency conversion through the main

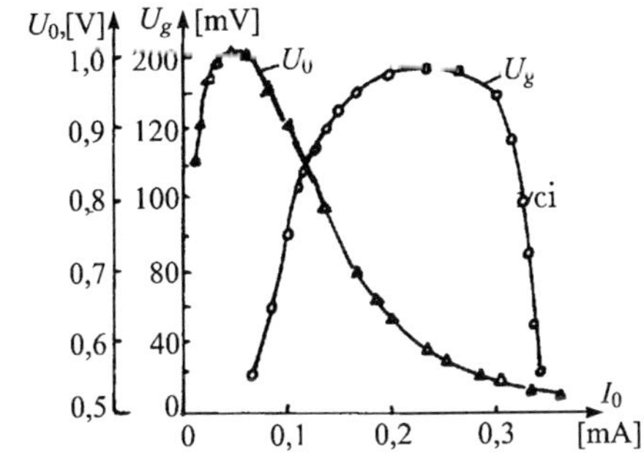

Fig. 2.21. Experimental Volt-Ampere characteristic $U_0(I_0)$ and dependence of the amplitude of the oscillating voltage in the generator U_g on the bias current I_0

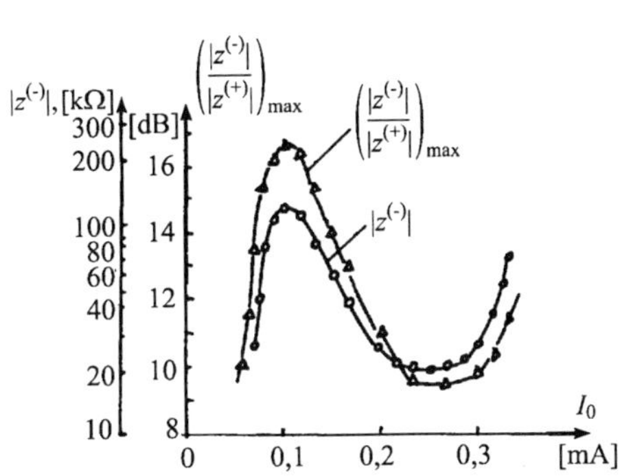

Fig. 2.22. Experimental dependence of the maximum ratio of the transmission impedances of conversion and of the transmission impedances of conversion by main channel on the bias current

channel overlap with the conditions for providing maximum suppression of the image channel.

Qualitatively analogous changes occur in the transmission impedances of conversion by the main and image channel in the respective locking bandwidths in the cases when the oscillator operates in a mode of synchronous frequency multiplication or division.

2.2.5. Selective properties of an oscillator with asynchronous action

The theoretical and experimental results of a study on the effect of a selective change (with respect to an external signal) in the effective impedance of a one-port, represented by an oscillator with weak asynchronous external action are outlined. Unlike the case considered above, now the frequency of the external action does not fall within the effective frequency bandwidth of synchronization, i.e. the external impact is asynchronous. The effect related to a drastic change in the effective impedance of the oscillator one-port is manifested, when the frequency of the external signal is equal to the absolute difference between the own frequency of the oscillations in the generator and the frequency of the external asynchronous action. The analysis shows that the mechanism of the effect is of a regenerative-parametric character and that it is related to the appearance of a very characteristic peculiarity – synchronization of the oscillations in the generator with a combined tone, whose frequency is equal to the sum or difference between the frequencies of the external asynchronous action and the signal. As a result of this synchronization, the oscillations in the system acquire the necessary phase for the occurrence of regenerative parametric effects. The effect is manifested in a very narrow frequency bandwidth, where the above mentioned synchronization and the favourable phase are realized. This conditions the high selectivity of the system, which is actually a highly selective electrically tunable filter with a very high qualitative factor. A filter obtained in such a way is controlled by changing the frequency of the external asynchronous action.

A detailed theoretical analysis of the effect has been carried out in [192] on the basis of a circuit equivalent to that in Fig.2.12a and Eq.(2.51), studied by the procedure presented above. The following expression for the equivalent impedance of the one-port, represented by an oscillator with asynchronous action with respect to the signal [192] has been obtained:

$$\dot{Z} = -\left[(\Gamma+1)\frac{Sa}{3}H_0 + j(-\Gamma^2+2\Gamma-1)\left(1+\frac{\nu}{2}\right)\right]\dot{A}_{\Gamma-1} + \left\{\left[2\left(\frac{Sa}{3}\right)^2 - 1\right]H_0 \right.$$

$$\left. + j\Gamma\left(1+\frac{\nu}{2}\right)\right\}\dot{A}_{\Gamma} + \left[(\Gamma-1)\frac{Sa}{3}H_0 - j(1-\Gamma)^2\left(1+\frac{\nu}{2}\right)\frac{Sa}{3}\right]\dot{A}_{\Gamma+1}, \qquad (2.79)$$

where

$$\dot{A}_{\Gamma-1} = \frac{(1-\nu)(\Gamma-1)\frac{Sa}{3}+C_1}{\Gamma(2-\Gamma)-j(\Gamma-1)\left(1-\frac{\nu}{2}\right)H_0} + \frac{j(\Gamma-1)\left(1-\frac{\nu}{2}\right)2\frac{Sa}{3}H_0\left[1-\nu-C_1\Gamma\frac{Sa}{3}\right]}{D_1},$$

$$\dot{A}_{\Gamma} = \frac{C_2\left[(1-\nu)(\Gamma-1)\frac{Sa}{3}+C_1\right]}{D_1} + \frac{(1-\nu)-C_1\Gamma\frac{Sa}{3}}{1-\Gamma^2-j\Gamma\left(1-\frac{\nu}{2}\right)H_0} - \frac{C_2(1-\nu)(\Gamma+1)\frac{Sa}{3}}{D_2},$$

$$\dot{A}_{\Gamma+1} = \frac{\left(1+\frac{1}{\Gamma}\right)C_2\left[(1-\nu)-C_1\Gamma\frac{Sa}{3}\right]}{D_2} + \frac{(1-\nu)(\Gamma+1)\frac{Sa}{3}}{\Gamma(2+\Gamma)+j(\Gamma+1)\left(1-\frac{\nu}{2}\right)H_0},$$

$$C_1 = \frac{I_b}{I_s}(1-\nu)e^{j(\varphi-\varphi_s)}, \quad C_2 = j\Gamma\left(1-\frac{\nu}{2}\right)2\frac{Sa}{3}H_0,$$

$$D_1 = \Gamma(1-\Gamma^2)(2-\Gamma) - j\left(1-\frac{\nu}{2}\right)H_0(-2\Gamma^3+3\Gamma^2+\Gamma-1),$$

$$D_2 = \Gamma(\Gamma^2-1)(2+\Gamma) + j\left(1-\frac{\nu}{2}\right)H_0(2\Gamma^3+3\Gamma^2-\Gamma-1).$$

Fig.2.23 shows the theoretical dependencies of the impedance ratio $\frac{|\dot{Z}|}{|\dot{Z}|_{\max}}$, calculated according to the formula (2.79), on the detuning $\nu = \frac{(\omega_s+\omega_b)^2-\omega_r^2}{\omega_r^2}$ at $H_0 = 0,02$ and different values of the parameter Sa. The substantial change in the equivalent impedance of the oscillator with asynchronous action within a narrow frequency bandwidth can be clearly seen.

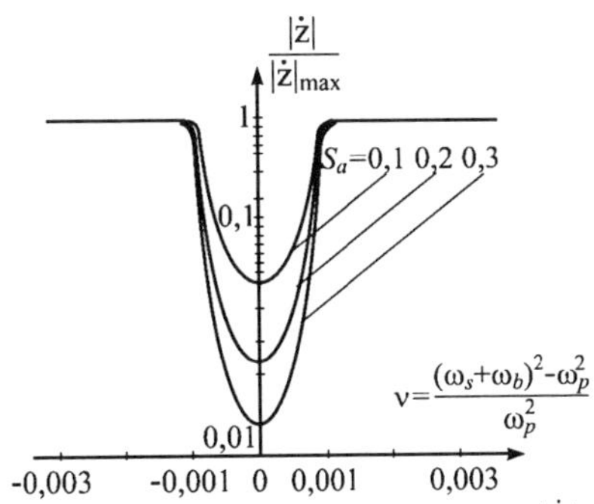

Fig. 2.23. Theoretical dependence of the impedance ratio $\frac{|\dot{Z}|}{|\dot{Z}|_{\max}}$ of a one-port, represented by an oscillator with asynchronous action, on the frequency detuning; $H_0 = 0,02$

Fig.2.24 reveals the experimental dependence of the same impedance ratio $\frac{|\dot{Z}|}{|\dot{Z}|_{\max}}$ (i.e. dependence of the relative change in the impedance) on the signal frequency f_s. The oscillator used in the experiment has a two-transistor $p-n-p-n$ structure (Fig.2.14) and its own generation frequency $f_g = 200$ kHz, while the frequency of the external asynchronous action is $f_b = 151$ kHz. A pronounced change in the equivalent impedance of the oscillator one-port in relation to external signals with frequency $f_s = 49$ kHz is observed. The impedance of the oscillator with asynchronous action alters within a very narrow frequency bandwidth of the input signals f_s, which can shift, within a wide range, along the frequency axis by simply changing the frequency of the external asynchronous action f_b.

The effect of the selective change (with respect to external signals) in the effective impedance of a one-port representing an oscillator with asynchronous action is manifested in a broad class of generator systems with modern semiconductor devices, particularly in the SHF range and is widely applicable. Some applications, important for special purposes, are presented in 2.4.

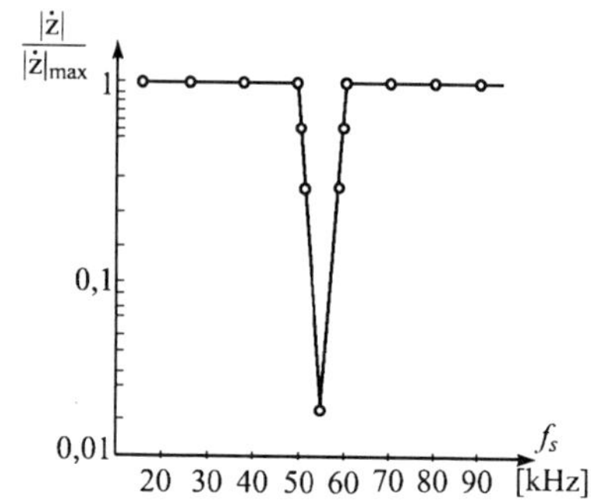

Fig. 2.24. Experimental dependence of the same impedance ratio $\frac{|\dot{Z}|}{|\dot{Z}|_{max}}$ of a one-port, represented by an oscillator with asynchronous action, on the signal frequency f_s; generation frequency $f_s = 200$ kHz, frequency of the asynchronous action $f_b = 151$ kHz

2.3. Converting properties of receiving and transmitting SHF oscillator modules

2.3.1. Short-range self-detecting Doppler radars (autodyne systems) described by differential equations of the second order

The generation of the oscillations to be transmitted, the high-frequency amplification of the received signal reflected by a moving object, the frequency conversion and low-frequency amplification of the Doppler component are all carried out by using one active element in the autodyne systems. The signal with Doppler frequency ω_D is received directly in the LF circuit of the autodyne (AD) – Fig.2.25. In the case of a homodyne scheme of construction of the facilities for short-range Doppler radars (Fig.2.26) the oscillations to be transmitted are generated by an oscillator (G), while the reflected signal is transmitted through a circulator to an external converter (CV) without amplification. Improved characteristics of the autodyne facilities for short-range Doppler radars can be obtained, if the generation and HF amplification of the reflected signal are carried out by using one active element, while the Doppler component is separated by means of an external converter – Fig.2.27.

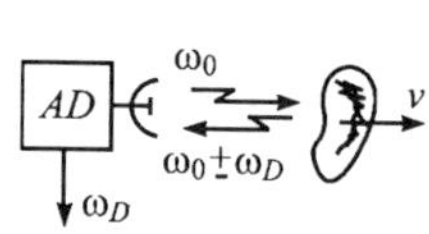

Fig. 2.25. Classical short-distance Doppler radar (autodyne); AD – autodyne

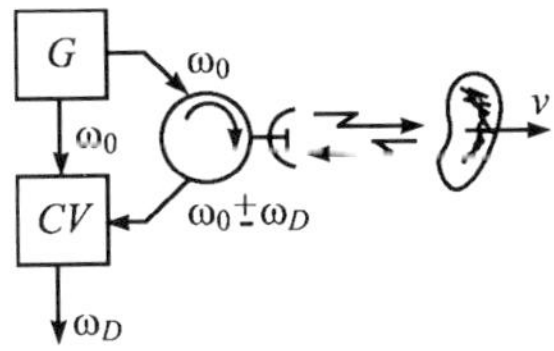

Fig. 2.26. A homodyne Doppler scheme for the construction of a system of short-distance Doppler radars; G – oscillator, CV – frequency converter

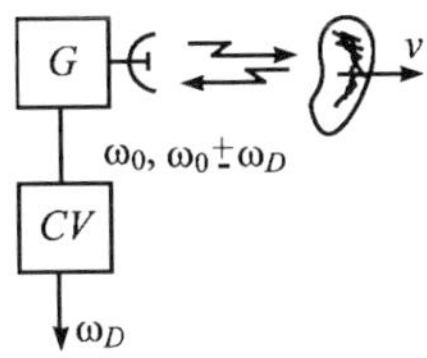

Fig. 2.27. A Doppler autodyne system with external converter; G – oscillator, CV – frequency converter

The systems in Fig.2.25, 2.26 and 2.27 are studied in the works [111, 196–205]. The autodyne and heterodyne processing of the reflected Doppler signal (Fig.2.25 and Fig.2.26) are compared in [201] and it is concluded that autodyne processing ensures a higher conversion coefficient than heterodyne handling. [202] indicates that from the viewpoint of its energy potential (the ratio between the dynamic range and the sensitivity of the system) the autodyne method is in no way inferior to the heterodyne one, while [203] specifies that the potential of the system, where the oscillator and the converter are coupled with a circulator, exceeds – by 20 dB – the potential of the autodyne, where the generation of the oscillations to be transmitted, the conversion and amplification of the small reflected signal are conducted by using a single active element.

This section provides a detailed analysis and comparison of the conversion characteristics of the systems in Figs. 2.25 and 2.27.

The analysis is carried out on the basis of the equivalent circuit of the autodyne system, including an oscillator with negative resistance – Fig.2.28, where G_{D-} is negative differential conductance characterizing the falling section of the volt-ampere characteristic $i_D(U)$ and of the active nonlinear element. L_pC_p represents the resonator of the oscillator, C_D and $L_D(\Phi)$ are reactive parameters of the active device, Φ is a total magnetic flux, R_L is the equivalent load of the generator, $i_b = I_b \cos(\omega_b t_r + \varphi_b)$ is the signal reflected by a moving object, $\omega_b = \omega_0 \pm \omega_D$, ω_0 is the frequency of the oscillations in the generator, ω_D is the Doppler shift of the frequency of the reflected signal, $i_s = I_s \cos(\omega_0 t_r + \varphi_s)$ is a synchronizing signal.

The processes in the system shown in Fig.2.28 are described by the following

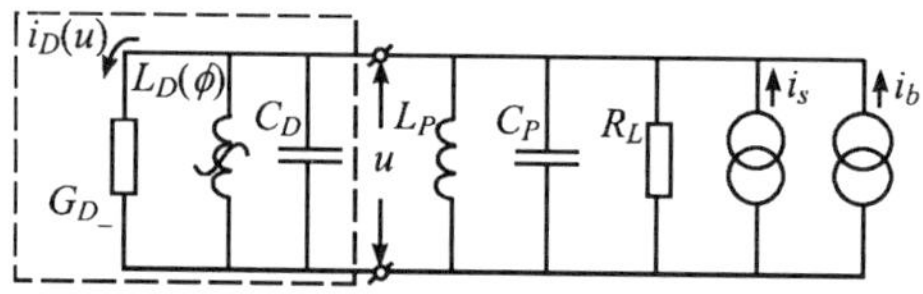

Fig. 2.28. An equivalent circuit of an autodyne system containing an oscillator with negative resistance

differential equation:

$$(C_D + C_p)\frac{d^2\Phi}{dt_r^2} + \frac{1}{R_L}\frac{d\Phi}{dt_r} + \frac{\Phi}{\dfrac{L_D(\Phi)L_p}{L_D(\Phi)+L_p}} + i_D - i_s - i_b = 0, \tag{2.80}$$

$U = \dfrac{d\Phi}{dt_r}$ is the voltage of the oscillator circuit.

We use the following approximating expressions:

$$i_D(U) = -G_1\frac{d\Phi}{dt_r} + \sum_{n=2}^{m} G_n\left(\frac{d\Phi}{dt_r}\right)^n, \quad L_D(\Phi) = \frac{L_0}{1+\sum_{n=1}^{k} L'_n\Phi^n},$$

where G_n, L'_n are coefficients, L_0 is the inductance at the operating point.

As non-dimensional parameters and symbols are introduced – $x = \dfrac{\omega_{ro}\Phi}{U_{oo}}$, ($U_{oo}$ is normalizing voltage), $t = \omega_{ro}t_r$, $\omega_{ro} = \dfrac{1}{\sqrt{\dfrac{L_0L_p}{L_0+L_p}C}}$, $C = C_D + C_p$,

$Q{=}\omega_{ro}CR_L$, $f(x){=}x\left(1+\sum_{n=1}^{k} L_nx^n\right)$, $L_n = \dfrac{L_pL'_n}{L_0+L_p}\left(\dfrac{U_{oo}}{\omega_{ro}}\right)^n$, $\dfrac{\omega_b}{\omega_{ro}}{=}\Omega_b$, $\dfrac{\omega_0}{\omega_{ro}}{=}\Omega_0$, $U_s = \dfrac{I_sR_L}{U_{oo}}$, $U_b = \dfrac{I_bR_L}{U_{oo}}$, Eq.(2.80) takes the form

$$\frac{d^2x}{dt^2} + f(x) = \frac{1}{Q}(G_1R_L - 1)\frac{dx}{dt} - \frac{R_L}{Q}\sum_{n=2}^{m} G_nU_{oo}^{n-1}\left(\frac{dx}{dt}\right)^n + \frac{U_s}{Q}\cos(\Omega_0t + \varphi_s) + \frac{U_b}{Q}\cos(\Omega_bt + \varphi_b). \tag{2.81}$$

In accordance with the analytical approach developed in 1.4, we shall present the solution of Eq.(2.81) in the form $x = x_0 + \Delta x$, where Δx is a small perturbation conditioned by the action of the reflected signal. Then Eq.(2.81) is expressed in the form of two equations concerning the major generator oscillations x_0 and the small variation in these oscillations Δx:

$$\frac{d^2 x_0}{dt^2} + f(x_0) = \frac{1}{Q}(G_1 R_L - 1)\frac{dx_0}{dt} - \frac{R_L}{Q}\sum_{n=2}^{m} G_n U_{oo}^{n-1}\left(\frac{dx_0}{dt}\right)^n + \frac{U_s}{Q}\cos(\Omega_0 t + \varphi_s), \tag{2.82}$$

$$\frac{d^2 \Delta x}{dt^2} + \frac{df(x_0)}{dx}\Delta x = \frac{1}{Q}(G_1 R_L - 1)\frac{d\Delta x}{dt} - \frac{R_L}{Q}\sum_{n=2}^{m} G_n U_{oo}^{n-1} n\left(\frac{dx_0}{dt}\right)^{n-1}\frac{d\Delta x}{dt} + \frac{U_b}{Q}\cos(\Omega_b t + \varphi_b). \tag{2.83}$$

The nonlinear transformation of the variables [185] and the respective denotations, described in 1.4, are used to rewrite Eqs.(2.82) and (2.83) as follows:

$$\frac{d^2 y_0}{d\Psi^2} + y_0 = 2\nu y_0 + \frac{1}{\beta Q}(G_1 R_L - 1)\frac{dy_0}{d\Psi} - \frac{R_L}{Q}\sum_{n=2}^{m} G_n U_{oo}^{n-1}\beta^{n-2}\left(\frac{dy_0}{d\Psi}\right)^n + \frac{U_s}{\beta^2 Q}G(y_0)\cos(\Omega_0 t + \varphi_s), \tag{2.84}$$

$$\frac{d}{d\Psi}\left\{\frac{1}{G(y_0)}\frac{d}{d\Psi}[G(y_0)\Delta y]\right\} + \frac{[G(y_0)]^2}{\beta^2}\frac{df(y_0)}{dx}\Delta y = \frac{1}{\beta Q}(G_1 R_L - 1)\frac{d}{d\Psi}[G(y_0)\Delta y] - \frac{R_L}{Q}\sum_{n=2}^{m} G_n U_{oo}^{n-1} n\beta^{n-2}\left(\frac{dy_0}{d\Psi}\right)^{n-1}\frac{d}{d\Psi}[G(y_0)\Delta y] + \frac{U_b}{\beta^2 Q}G(y_0)\cos(\Omega_b t + \varphi_b), \tag{2.85}$$

where $\Delta y \simeq \frac{\Delta x}{G(y_0)}$, $2\nu = \beta^2 - 1 = \frac{2(\omega_0 - \omega_{ro})}{\omega_{ro}}$ – relative detuning in the case of a non-autonomous regime in the oscillator system of the autodyne, $\mu \ll 1$.

We shall characterize the converting properties of the autodyne systems by using the conversion coefficients

$$K_{\pm\Omega_0\pm\Omega_D} = \frac{V_{\pm\Omega_0\pm\Omega_D}}{U_b} \quad \text{and} \quad K_{\pm\Omega_D} = \frac{V_{\pm\Omega_D}}{U_b}, \tag{2.86}$$

where $V_{\pm\Omega_0\pm\Omega_D}$ is the voltage of the combined frequencies $(\pm\Omega_0 \pm \Omega_D)$ fed at the external converter (Fig.2.27); $V_{\pm\Omega_D}$ is the output converted voltage with Doppler frequency $\Omega = \frac{\omega_D}{\omega_{ro}}$ (Fig.2.25).

In (2.80) and further on in the text the sign „±" before Ω_D corresponds to a relative moving away or approaching of an object in relation to the autodyne system, while the sign „±" coming before the own frequency Ω_0 of the oscillator determines the respective sum or difference combined frequencies $(\pm\Omega_0 \pm \Omega_D)$.

For the purpose of determining $V_{\pm\Omega_0 \pm\Omega_D}$, we derive an expression for the relative change of the voltage of the oscillating circuit of the generator under the action of the reflected signal (Fig.2.28):

$$\frac{\Delta U}{U_{oo}} = \frac{d\Delta x}{dt} = \frac{\beta}{G(y_0)} \frac{d}{d\Psi}[G(y_0)\Delta y]. \tag{2.87}$$

As a certain spectral volume $\{\delta F\}$ of the generator oscillations, conditioned by the action of the reflected signal, is set, (2.85) and (2.87) will be rewritten in a complex matrix form:

$$[D(\delta F)][G^r][D(\delta F)][G][Y(\delta F)] + \frac{1}{\beta^2}[K][Y(\delta F)] - \frac{\delta}{\beta}[D(\delta F)][G][Y(\delta F)]$$
$$+ \sum_{n=2}^{m} \delta_n \beta^{n-2}[D(\delta F)]^{n-1}[Y]^{n-1}[D(\delta F)][G][Y(\delta F)] = \frac{U_b}{\beta^2 Q}[G][I], \tag{2.88}$$

$$[U(\delta F)] = \beta[G^r][D(\delta F)][G][Y(\delta F)], \tag{2.89}$$

where $\delta = \dfrac{1}{Q}(G_1 R_1 - 1)$, $\delta_n = \dfrac{R_L}{Q} n G_n U_{oo}^{n-1}$, $n = 2, 3, \ldots$; $G(y) \to [G]$, $\dfrac{1}{G(y_0)} \to [G^r]$, $[G(y_0)]^2 \dfrac{df(x_0)}{dx} \to [K]$; $\Delta y \to [Y(\delta F)]$; $\dfrac{d}{d\Psi} \to [D(\delta F)]$; $\cos(\Omega_b t + \varphi_b) \to [I]$; $\dfrac{\Delta U}{U_{oo}} \to [U(\delta F)]$; $y_0 \to [Y]$.

The output voltage of frequencies $(\pm\Omega_0 \pm \Omega_D)$ is

$$V_{\pm\Omega_0 \pm \Omega_D} = [E(\delta F)][U(\delta F)], \tag{2.90}$$

where $[E(\delta F)]$ is the matrix of the reciprocal conversion to the output variables, accounting for the contribution of each spectral component from the selected spectrum $\{\delta F\}$ in the output signal at frequency $(\pm\Omega_0 \pm \Omega_D)$.

In the absence of a reflected signal $(i_b \equiv 0)$ (2.84) yields a solution for the basic oscillations in the system

$$y_0 = a\cos(\beta\tau + \varphi) = a\cos\Psi, \tag{2.91}$$

where the equations concerning the amplitude a and the phase φ are written in a way analogous to (2.62).

The analysis shows that when the output signal of the system presented in Fig.2.27 is taken off at the level of combined frequencies $(\pm\Omega_0 \pm \Omega_D)$, the largest contribution comes from the spectral components $\pm\Omega_D$, $(\pm\Omega_0 \pm \Omega_D)$, $(\pm 2\Omega_0 \pm \Omega_D)$, which also constitute the selected range of the oscillation spectrum. The latter can be presented in the form $\{\Delta F\} = \pm\Gamma, \pm 1 \pm \Gamma, \pm 2 \pm \Gamma$, where $\Gamma = \dfrac{\Omega_D B_0}{\beta}$. If we restrict the approximating polynom of the function $f(x) = x + Lx^2$ to the second power, we can derive the following approximating expressions:

$$y = x + \frac{Lx^2}{2}, \quad G(x) = 1 - \frac{2}{3}Lx, \quad G(y) = 1 - \frac{2}{3}Ly,$$
$$[G(y)]^2 \frac{df(y)}{dx} = 1 + \frac{2}{3}Ly. \tag{2.92}$$

When determining matrixes $[I]$ and $[E(\delta F)]$, the following direct and inverse transformations are made:

$$\left|\begin{aligned}
&\cos(\Omega_b t + \varphi_b) = Re\left[\left\{\mp I_1\left[(\pm 1 \pm \Gamma)\frac{B_1}{B_0}\right] e^{j(\pm\Gamma)\Psi} + e^{j(\pm 1 \pm \Gamma)\Psi}\right.\right.\\
&\qquad\left.\left. \pm\; I_1\left[(\pm 1 \pm \Gamma)\frac{B_1}{B_0}\right] e^{j(\pm 2 \pm \Gamma)}\right\} e^{j[\varphi_b - (\pm 1 \pm \Gamma)\varphi]}\right],\\
&\cos(\pm\Gamma\Psi) = \mp I_1\left[\pm\Gamma\frac{B_1}{\beta}\right]\cos[(\pm\Omega_0 \pm \Omega_D)t + (\mp 1 \pm \Gamma)\varphi] + \text{HF components},\\
&\cos[(\pm 1 \pm \Gamma)\Psi] = \cos[(\pm\Omega_0 \pm \Omega_D)t + (\pm 1 \pm \Gamma)\varphi] + \text{HF components},\\
&\cos[(\pm 2 \pm \Gamma)\Psi] = \mp\, I_1\left[(\pm 2 \pm \Gamma)\frac{B_1}{\beta}\right]\cos[(\pm\Omega_0 \pm \Omega_D)t\\
&\qquad + (\pm 1 \pm \Gamma)\varphi] \;+\; \text{HF components},
\end{aligned}\right. \tag{2.93}$$

where I_1 is a modified Bessel function, $t = \displaystyle\int_0^T G(a\cos\Psi)d\tau = \frac{B_0}{\beta}(\Psi - \varphi) + \frac{B_1}{\beta}\sin\Psi + \ldots$, B_0 and B_1 are determined by (2.47) and (2.67).

As (2.91) – (2.93) are taken into account, the square matrixes in (2.88) and

(2.89) can be specified for the case under consideration as follows:

$$[D(\delta F)] = \begin{bmatrix} \pm\Gamma & 0 & 0 \\ 0 & j(\pm 1 \pm \Gamma) & 0 \\ 0 & 0 & j(\pm 2 \pm \Gamma) \end{bmatrix}, \quad [G] = \begin{bmatrix} 1 & -\frac{La}{3} & \frac{(La)^2}{3} \\ -\frac{La}{3} & 1 & -\frac{La}{3} \\ \frac{(La)^2}{3} & -\frac{La}{3} & 1 \end{bmatrix},$$

$$[G^r] = \begin{bmatrix} 1 & \frac{La}{3} & -\frac{(La)^2}{3} \\ \frac{La}{3} & 1 & \frac{La}{3} \\ -\frac{(La)^2}{3} & \frac{La}{3} & 1 \end{bmatrix}, \quad [K] = \begin{bmatrix} 1 & \frac{La}{3} & -\frac{(La)^2}{4} \\ \frac{La}{3} & 1 & \frac{La}{3} \\ -\frac{(La)^2}{4} & \frac{La}{3} & 1 \end{bmatrix}, \quad (2.94)$$

$$[I] = \begin{bmatrix} \mp I_1[(\pm 1 \pm \Gamma)\frac{B_1}{B_0}] \\ \pm I_1[(\pm 1 \pm \Gamma)\frac{B_1}{B_0}] \end{bmatrix}, \quad \delta_3\beta[D(\delta F)]^2[Y]^2 = \begin{bmatrix} \frac{\delta_3\beta a^2}{2} & 0 & -\frac{\delta_3\beta a^2}{2} \\ 0 & \frac{\delta_3\beta a^2}{2} & 0 \\ -\frac{\delta_3\beta a^2}{2} & 0 & \frac{\delta_3\beta a^2}{2} \end{bmatrix},$$

$$[E(\delta F)] = \left[\mp I_1[(\pm\Gamma)\frac{B_1}{\beta}] \quad 1 \quad \mp I_1[2 \pm 2 \pm \Gamma)\frac{B_1}{\beta}] \right].$$

As (2.94) is taken into account and Eq.(2.88) is used as a starting point, the following is determined:

$$[Y(\delta F)] = [A]^{-1} \frac{U_b}{\beta^2 Q} [G][I], \tag{2.95}$$

where

$$[A] = [D(\delta F)][G^r][D(\delta f)][G] + \frac{1}{\beta^2}[K] - \frac{\delta}{\beta}[D(\delta F)][G] + \delta_3\beta[D(\delta F)]^2[Y]^2[D(\delta F)][G].$$

Then (2.89) yields

$$[U(\delta F)] = [V(\delta F)][Y(\delta F)] = U_b \begin{bmatrix} V_{\pm\Gamma} \\ V_{\pm 1 \pm \Gamma} \\ V_{\pm 2 \pm \Gamma} \end{bmatrix}. \tag{2.96}$$

As (2.96) is substituted in (2.90) and (2.94) is taken into account, an expression for the conversion coefficient is obtained:

$$K_{\pm\Omega_0 \pm \Omega_D} = \frac{V_{\pm\Omega_0 \pm \Omega_D}}{U_b} = V_{\pm 1 \pm \Gamma} \pm \frac{La}{3} \{\pm\Gamma V_{\pm\Gamma} + [\pm 2 - (\pm 2\nu) \pm \Gamma] V_{\pm 2 \pm \Gamma}\}. \tag{2.97}$$

Fig.2.29 reveals the theoretical dependence of $|K_{\Omega_0 + \Omega_D}|$ on the parameter δ, reflecting the regeneration degree in the oscillating circuit of the generator at $\nu = 0$

and $\Gamma = 10^{-7}$. A common parameter of the dependence is La, characterizing both the nonlinearity of the reactive characteristic of the active element and the level of the oscillations in the generator. As the value of δ goes up, the conversion coefficient $|K_{\Omega_0+\Omega_D}|$ falls, furthermore, the degree of the decrease depends on the amplitude a of the own oscillations of the generator. Given a constant δ, the coefficient $|K_{\Omega_0+\Omega_D}|$ tends to its maximum value, when the amplitude of the oscillations in the generator tends to zero – this is one of the characteristic peculiarities of the system in Fig.2.27.

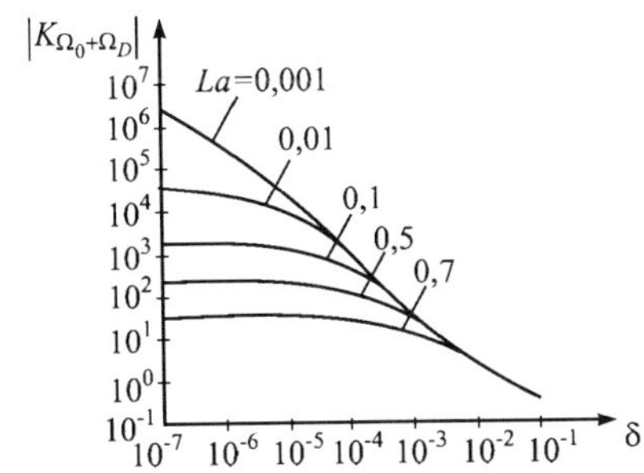

Fig. 2.29. Theoretical dependence of the conversion coefficient up to the sum combined frequency on the regeneration degree in the oscillating circuit of the autodyne oscillator; $\nu = 0$, $\Gamma = 10^{-7}$

The analysis shows that the expressions obtained for $K_{\pm\Omega_0\pm\Omega_D}$ (2.97) display certain symmetry, where the following conversion coefficients are identical:

$$K_{+\Omega_0+\Omega_D} = K^*_{-\Omega_0-\Omega_D}, \quad K_{+\Omega_0-\Omega_D} = K^*_{-\Omega_0+\Omega_D},$$

where „*" is a symbol of a complex conjugate quantity.

Owing to this property, the total conversion coefficient of the system in the case of an approaching object is identical to the total coefficient in the case of an object moving away, all other conditions being equal. This is another essential peculiarity of the system in Fig.2.27.

Fig.2.30 presents the theoretical dependence of the module of the conversion coefficient of the system in Fig.2.27 to the total combined frequency $(\Omega_0 + \Omega_D)$: $|K_{\Omega_0+\Omega_D}|$ (or $|K_{-\Omega_0-\Omega_D}|$), on the non-dimensional Doppler frequency Γ and various detunings in the locking frequency bandwidth (non-autonomous mode of the autodyne generator), $\delta = 10^{-7}$ and $La = 10^{-3}$. In the case of zero detuning ($\nu = 0$) the coefficient $|K_{\Omega_0+\Omega_D}|$ is at its maximum, when $\Gamma \to 0$ and it drops as the Doppler frequency increases. When the detunings in the locking frequency bandwidth are negative ($\nu = -10^{-3}$, $\nu = -10^{-5}$), high selectivity with respect to the Doppler frequencies, approximately equal to the respective detunings, $\Gamma = \nu$, is observed. There is no such selectivity in the case of positive detunings.

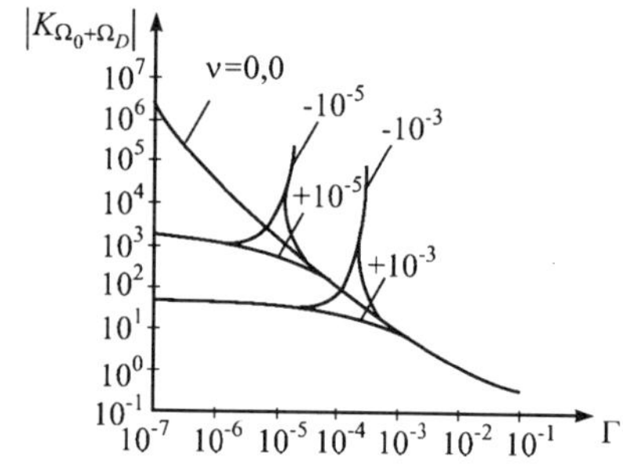

Fig. 2.30. Theoretical dependence of the conversion coefficient up to the sum combined frequency on the reduced non-dimensional Doppler frequency at various detunings (non-autonomous mode of the autodyne generator); $\delta = 10^{-7}$, $La = 10^{-3}$

Fig.2.31 presents the theoretical dependence of the coefficient $|K_{\Omega_0-\Omega_D}|$ (or $|K_{-\Omega_0+\Omega_D}|$) on Γ for $\delta = 10^{-7}$, $La = 10^{-3}$. At $\nu = 0$ the coefficient $|K_{\Omega_0-\Omega_D}|$ may have considerable values for small magnitudes of Γ. Monotonic decrease of $|K_{\Omega_0-\Omega_D}|$ is observed in the cases of both negative and positive detunings: there is no selectivity to the Doppler frequency.

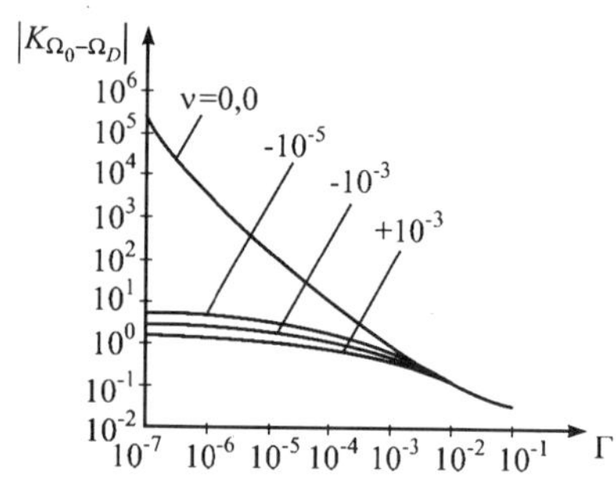

Fig. 2.31. Theoretical dependence of the conversion coefficient up to the difference combined frequency on the reduced non-dimensional Doppler frequency at various detunings (non-autonomous mode of the autodyne generator); $\delta = 10^{-7}$, $La = 10^{-3}$

The dependence in Fig.2.29 reveals the contradiction in the requirements concerning the choice of the amplitude and oscillations in the autodyne generator presented in Fig.2.27. As the amplitude decreases, the conversion coefficient is increased but parallel with that the level of the reflected signal $U_b = \gamma a$ drops. It should be expected that there is an optimal amplitude a_{opt}, at which the sensitivity of the system reaches its peak.

Fig.2.32 presents the theoretical dependence of $\dfrac{La}{\gamma}|K_{-\Omega_0-\Omega_D}|$ on the parameter La. Curve 1 illustrates the case of $\nu = \pm 10^{-3}$, $\Gamma = 10^{-7}$. A pronounced extremum at relatively high amplitudes is observed ($La = 0,1 \ldots 0,3$). For $\nu = 0$, or in a selection mode $\nu = \Gamma$, when the conversion coefficient $|K_{-\Omega_0 - \Omega_D}|$ soars, the extremum is shifted to the area of small values of the parameter La (curve 2 in Fig.2.32, $\Gamma = 10^{-7}$). The effective realization of these modes is highly dependent on the level of the system's own fluctuation noises.

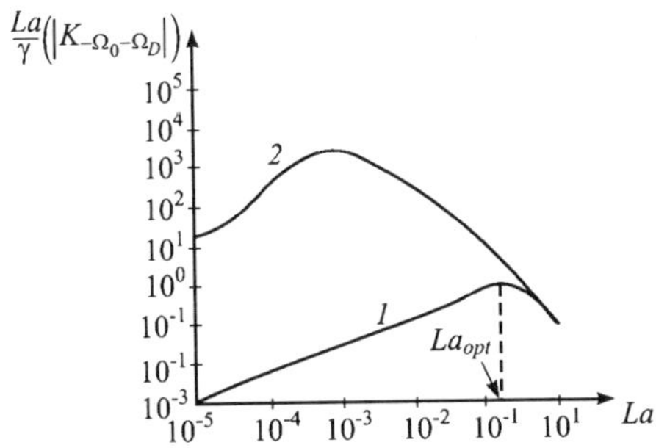

Fig. 2.32. Theoretical dependence of the reduced conversion coefficient up to the difference combined frequency on the parameter La at $\delta = 10^{-7}$, curve *1* – $\nu = \pm 10^{-3}$, $\Gamma = 10^{-7}$, curve *2* – $\nu = 0$, $\Gamma = 10^{-7}$

It can be concluded that for the purpose of a most comprehensive utilization of the mode set by a high conversion coefficient at zero detunings, or of the mode of selective amplification, the functions of generating the probing signal and receiving the reflected signal should be separated and that two different oscillators should be used – a powerful and a low-powered one.

In the system shown in Fig.2.27 the oscillation spectrum of the autodyne $(\Omega_0 \pm \Omega_D)$, Ω_0, $(-\Omega_0 \pm \Omega_D)$ is communicated to an external converter CV. The conversion coefficients of a passive frequency converter with respect to the sum and difference frequencies are determined as

$$K_+ = \frac{V^D_{\pm\Omega_D}}{V_{\Omega_0 \pm \Omega_D}} \quad \text{and} \quad K_- = \frac{V^D_{\pm\Omega_D}}{V_{-\Omega_0 \pm \Omega_D}},$$

where $V^D_{\pm\Omega_D}$ is the Doppler frequency voltage at the exit of the converter CV. Then, as (2.86) is accounted for, the cumulative transmission coefficient of the autodyne system presented in Fig.2.27 is

$$K_\Sigma = \frac{V^D_{\pm\Omega_D}}{U_b} = K_+ K_{\Omega_0 \pm \Omega_D} + K_- K_{-\Omega_0 \pm \Omega_D}.$$

The coefficients K_+ and K_- reflect the losses in the specific passive converter.

The conversion coefficient of the system in Fig.2.25 can be determined in a way analogous to the one recounted above. Since a similar problem was solved in section 2.2 we shall abstain from deriving expressions for the coefficients $K_{\pm\Omega_D}$ here. The dependence of $K_{\pm\Omega_D}$ on the parameter δ characterizing the degree of regeneration is qualitatively analogous to the respective dependence in the case of a system with an external converter – Fig.2.29. However, within a certain range of the locking frequency bandwidth, the conversion coefficients of an object that is moving away ($K_{+\Omega_D}$) may differ substantially from those of an approaching one ($K_{-\Omega_D}$). This is a disadvantage of the short-range self-detecting Doppler radar system with an injection-locked oscillator shown in Fig.2.25 as compared to the system in Fig.2.27 under an analogous mode of operation. It is also essential that the dependence of the conversion coefficient $K_{\pm\Omega_D}$ on the amplitude of the self-oscillator exhibits a pronounced extremum at $La = 0,1 \ldots 0,3$. An important distinction in this case is that when the amplitude of the oscillations in the generator tends to zero, $K_{\pm\Omega_D}$ also tends to zero.

The above-described analytical approach based on the method of a phase plane with nonlinear conversion of the variables and the perturbation method allows to conduct an effective analysis of autodyne, conversion, modulation-parametric and other processes and modes in autonomous and non-autonomous oscillator systems, experiencing the action of small external signals.

2.3.2. Short-range self-detecting Doppler radar systems described by differential equations of the third order

The analysis will be carried out on the basis of the equivalent circuit of the short-range self-detecting Doppler radar system (autodyne system) presented in Fig.2.33, containing an oscillator with negative resistance. The bias circuit will be accounted for as it sets the mode of the active device, and parallel with that, it serves as a load in a mode of immediate Doppler frequency output (Fig.2.25). The active device in the circuit is represented by the active G and reactive (capacitive) element linked with a parallel connection and described by the characteristics $i(U)$ and $C(q)$, where $i(U)$ is an active component of the current, U is the voltage of the generator oscillating circuit, $C(q)$ is the nonlinear capacitance of the active device, q is the flowing charge.

The influence of the bias circuit is accounted for by introducing the pair R_TC into the equivalent circuit; it is assumed that $R_TC \ll \dfrac{1}{\omega_D}$ and $R_TC \gg \dfrac{1}{\omega}$, ω is the frequency of the oscillations in the autodyne generator. The equivalent HF load is represented by the resistance R, the signal reflected by a moving object is given through the voltage source E_r.

The processes occurring in the circuit in Fig.2.33 can be described by a system

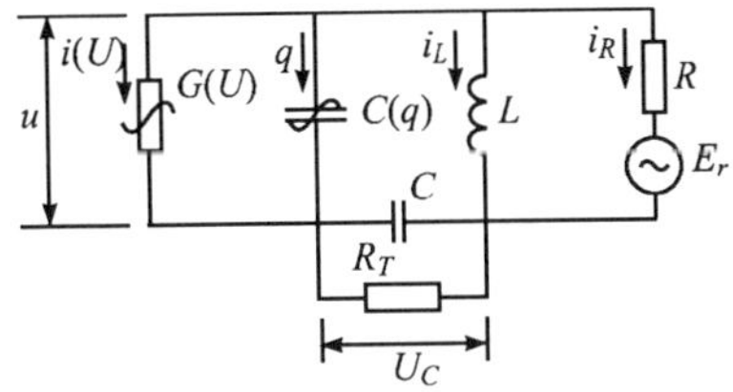

Fig. 2.33. Equivalent circuit of an autodyne system containing an oscillator with negative resistance and accounting for the direct current load circuit

of three differential equations of the first order

$$\left|\begin{array}{l} \dfrac{dq}{dt_r} = -i(U) - i_L - i_R \\ E_r = -i_R R + \dfrac{q}{C(q)} U_C \\ C\dfrac{dU_C}{dt_r} + \dfrac{U_C}{R_T} = \dfrac{dq}{dt_r} + i(U) \end{array}\right. , \tag{2.98}$$

where U_C is the voltage of the circuit $R_T C$, t_r is real time.

The characteristics of the nonlinear active device are presented in the form

$$i(U) = -G_1 U + G_2 U^2, \quad C(q) = \frac{C_0}{1 - C_1' q}, \tag{2.99}$$

where G_1, G_2, C_1' are coefficients, C_0 is the capacitance at the operating point.

The following non-dimensional parameters and notations are introduced: $x = \dfrac{q}{C_0 U_{oo}}$ (U_{oo} is normalizing voltage), $t = t_r \omega_0$, $\xi = \dfrac{\omega - \omega_0}{\omega}$ is a relative detuning, $\omega_0 = \dfrac{1}{\sqrt{LC_0}}$ is a resonance frequency of oscillation with an infinitesimal amplitude, $Q = \omega_0 R C_0$, $\zeta = \dfrac{C_0}{C}$, $\delta_1 = \dfrac{1}{\omega_0 C R_T}$, $\delta = \dfrac{1}{Q}(G_1 R - 1)$ is a parameter characterizing the degree of regeneration in the generator oscillation circuit.

As (2.99) and the notations introduced above are accounted for, (2.98) is reduced to the following equation of the third order:

$$\begin{aligned} &\frac{d^3 x}{dt^3} + (1-\xi)\frac{d^2}{dt^2}\{\alpha x - \delta f(x) + \beta[f(x)]^2\} \\ &+ (1-\xi)^2 \frac{d}{dt}\{\zeta x + \varkappa f(x) + \alpha\beta[f(x)]^2\} \\ &+ (1-\xi)^3\{\gamma f(x) + \zeta\beta[f(x)]^2\} = \frac{1-\xi}{Q}\frac{d^2 \mathcal{E}_r}{dt^2} + (1-\xi)^2 \frac{\delta_1}{Q}\frac{d\mathcal{E}_r}{dt}, \end{aligned} \tag{2.100}$$

where $f(x) = \dfrac{q}{U_{oo}C(q)} = x(1 - C_1 x)$, $C_1 = C_1' U_{oo} C_0$, $\alpha = \delta_1 + \dfrac{\zeta}{Q}$, $\beta = \dfrac{RG_2}{Q} U_{oo}$, $\varkappa = 1 + \dfrac{\delta_1}{Q} - \alpha\left(\delta + \dfrac{1}{Q}\right)$, $\gamma = \delta_1 - \delta\xi - \dfrac{\zeta}{Q}$, $\mathcal{E}_r = \dfrac{E_r}{U_{oo}}$.

The solution of equation (2.100) is presented in the form

$$x = x_0 + \Delta x, \tag{2.101}$$

where x_0 is a function, corresponding to the basic oscillations in the autodyne generator, given the absence of an external signal ($\mathcal{E}_r \equiv 0$), Δx is the variational spectrum of the oscillations under the action of a small reflected signal $\mathcal{E}_r$.

As (2.101) is substituted in (2.100) and the fact that Δx is a small variation of x_0 is taken into account, the following are obtained:

— an equation for the basic oscillations

$$\begin{aligned} &\frac{d^3 x_0}{dt^3} + (1-\xi)\frac{d^2}{dt}\{\alpha x_0 - \delta f(x_0) + \beta[f(x_0)]^2\} \\ &\quad + (1-\xi)^2 \frac{d}{dt}\{\zeta x_0 + \varkappa f(x_0) + \alpha\beta[f(x_0)]^2\} \\ &\quad + (1-\xi)^3\{\gamma f(x_0) + \zeta\beta[f(x_0)]^2\} = 0; \end{aligned} \tag{2.102}$$

— an equation concerning the variational spectrum

$$\begin{aligned} &\frac{d^3 \Delta x}{dt^3} + (1-\xi)\frac{d^2}{dt^2}\left\{\left[\alpha - \delta\frac{df(x_0)}{dx} + 2\beta f(x_0)\frac{df(x_0)}{dx}\right]\Delta x\right\} \\ &\quad + (1-\xi)^2\frac{d}{dt}\left\{\left[\zeta + \varkappa\frac{df(x_0)}{dx} + 2\alpha\beta f(x_0)\frac{df(x_0)}{dx}\right]\Delta x\right\} \\ &\quad + (1-\xi)^3\left\{[\gamma + 2\zeta\beta f(x_0)]\frac{df(x_0)}{dx}\Delta x\right\} \\ &= \frac{1-\xi}{Q}\frac{d^2\mathcal{E}_r}{dt^2} + (1-\xi)^2\frac{\delta_1}{Q}\frac{d\mathcal{E}_r}{dt}. \end{aligned} \tag{2.103}$$

As the resonance nature of the system is accounted for, the solution for the basic oscillations is presented in the form

$$x_0 = D_0 \cos(t + \Theta) - E. \tag{2.104}$$

If Eq.(2.102) is written out in the form $\dfrac{d^3 x}{dt^3} + \dfrac{dx}{dt} = \mu F\left(\dfrac{d^2 x}{dt^2}, \dfrac{dx}{dt}, x, t\right)$, where $\mu \ll 1$ is a small parameter, it would not be difficult to show that the shortened equations concerning the amplitude D_0, the phase Θ, and the additional small charge E, appearing in (2.104) can be presented in the form

$$\frac{dD_0}{dt} = -\frac{1}{2\pi}\int_0^{2\pi} \mu F\left(\frac{d^2x}{dt^2}, \frac{dx}{dt}, x, t\right)\cos(t+\Theta)dt,$$

$$\frac{d\Theta}{dt} = \frac{1}{2\pi D_0}\int_0^{2\pi} \mu F\left(\frac{d^2x}{dt^2}, \frac{dx}{dt}, x, t\right)\sin(t+\Theta)dt,$$

$$\frac{dE}{dt} = \frac{1}{2\pi}\int_0^{2\pi} \mu F\left(\frac{d^2x}{dt^2}, \frac{dx}{dt}, x, t\right)dt.$$

For the purpose of solving Eq.(2.103) we need to set a certain volume of a spectrum accounting for $\{\delta F\}$. The already familiar two cases are considered again – first, when the initial signal has a Doppler component $\varepsilon = \dfrac{\omega_D}{\omega}$ (Fig.2.25), and second, when the initial signal has a sum or difference combined frequency $(\pm 1+\varepsilon)$. As we take into account the possibility for the occurrence of modulation-parametric regenerative phenomena in the case of mutually reversible conversion of the frequencies ε and $(\pm 1+\varepsilon)$, as well as the conversion of the latter in the presence of the second harmonics in the spectrum of the pulsatory nonlinear elements $G(U)$ and $C(q)$ of the active device (2.99), we determine the following spectrum:

$$\{\delta F\} = \varepsilon,\quad 1+\varepsilon,\quad -1+\varepsilon,\quad 2+\varepsilon,\quad -2+\varepsilon. \tag{2.105}$$

As (2.105) is taken into account, the variational spectrum of the basic oscillations in the autodyne generator can be presented in the form of a matrix column

$$[\Delta x] = \text{colon}[x_{2+\varepsilon}, x_{1+\varepsilon}, x_{\varepsilon}, x_{-1+\varepsilon}, x_{-2+\varepsilon}. \tag{2.106}$$

Given conditions (2.105) and (2.106), Eq.(2.103) is written in a complex matrix form:

$$\begin{aligned}
&[D]^3[\Delta x] + (1-\xi)[D]^2\{[\alpha - \delta[DF] + 2\beta[FDF]][\Delta x]\} \\
&\quad + (1-\xi)^2[D]\{[\zeta + \varkappa[DF] + 2\alpha\beta[FDF]][\Delta x]\} \\
&\quad + (1-\xi)^3\{[\gamma[DF] + 2\zeta\beta[FDF]][\Delta x]\} \\
&\quad = \frac{1-\xi}{Q}[D]^2[E^{(\pm)}] + (1-\xi)^2\frac{\delta_1}{Q}[D][E^{(\pm)}],
\end{aligned} \tag{2.107}$$

where

$$\frac{d}{dt} \to [D] = \begin{bmatrix} j(2+\varepsilon) & 0 & 0 & 0 & 0 \\ 0 & j(1+\varepsilon) & 0 & 0 & 0 \\ 0 & 0 & j\varepsilon & 0 & 0 \\ 0 & 0 & 0 & j(-1+\varepsilon) & 0 \\ 0 & 0 & 0 & 0 & j(-2+\varepsilon) \end{bmatrix}, \tag{2.108}$$

$$\frac{df(x_0)}{dx} \to [DF] = \begin{bmatrix} 1 & -C_1D_0 & 0 & 0 & 0 \\ -C_1D_0 & 1 & -C_1D_0 & 0 & 0 \\ 0 & -C_1D_0 & 1 & -C_1D_0 & 0 \\ 0 & 0 & -C_1D_0 & 1 & -C_1D_0 \\ 0 & 0 & 0 & -C_1D_0 & 1 \end{bmatrix}, \qquad (2.109)$$

$$f(x_0)\frac{df(x_0)}{dx} \to [FDF] = \begin{bmatrix} k_{11} & k_{12} & k_{13} & k_{14} & 0 \\ k_{12} & k_{11} & k_{12} & k_{13} & k_{14} \\ k_{13} & k_{12} & k_{11} & k_{12} & k_{13} \\ k_{14} & k_{13} & k_{12} & k_{11} & k_{12} \\ 0 & k_{14} & k_{13} & k_{12} & k_{11} \end{bmatrix}, \qquad (2.110)$$

$$k_{11} = -\frac{3}{2}C_1D_0^2, \quad k_{12} = \frac{D_0}{2}\left(1 + \frac{3}{2}C_1^2D_0^2\right), \quad k_{13} = -\frac{3}{4}C_1D_0^2, \quad k_{14} = -\frac{1}{4}C_1^2D_0^3;$$

$$\mathcal{E}_r \to [E^{(+)}] = \mathcal{E}_0[E_0^{(+)}] = \mathcal{E}_0 \begin{bmatrix} 0 \\ 1 \\ 0 \\ 0 \\ 0 \end{bmatrix} \quad \text{– for an approaching object,} \qquad (2.111)$$

$$\mathcal{E}_r \to [E^{(-)}] = \mathcal{E}_0[E_0^{(-)}] = \mathcal{E}_0 \begin{bmatrix} 0 \\ 0 \\ 0 \\ 1 \\ 0 \end{bmatrix} \quad \text{– for an object that is moving away,} \qquad (2.112)$$

$\mathcal{E}_0 = \dfrac{E_0}{U_{oo}}$ is a relative amplitude of the signal reflected by a moving object.

As (2.108) – (2.112) are taken into account, Eq.(2.107) is presented in the form

$$[A][\Delta x] = \mathcal{E}_0\frac{1-\xi}{Q}\{[D]^2 + (1-\xi)\delta_1[D]\}[E_0^{(\pm)}], \qquad (2.113)$$

where $[A]$ is a common matrix of the coefficients on the left-hand side of (2.107), $[A] = [a_{i,m}], \quad i, m = 1, \ldots, 5$.

After denoting

$$W_{11} = (1-\xi)(-\alpha + \delta + 3\beta C_1D_0^2), \quad W_{12} = (1-\xi)(\gamma - 3\zeta\beta C_1D_0^2),$$

$$W_{13} = (1-\xi)(\zeta + \varkappa - 3\alpha\beta C_1D_0^2), \quad S = 1 + \frac{3}{2}C_1^2D_0^2,$$

$$W_{21} = (1-\xi)D_0(-\delta C_1 - \beta S), \quad W_{22} = (1-\xi)^3D_0(-\gamma C_1 + \zeta\beta S),$$

$$W_{23} = (1-\xi)^2D_0(-\varkappa C_1 + \alpha\beta S), \quad W_{31} = \frac{3}{2}\beta(1-\xi)C_1D_0^2,$$

$$W_{32} = -W_{31}\zeta(1-\xi)^2, \quad W_{33} = -\frac{3}{2}\alpha\beta C_1 D_0^2(1-\xi)^2,$$

$$W_{41} = -\frac{1}{2}\beta(1-\xi)C_1^2 D_0^3, \quad W_{42} = -W_{41}\zeta(1-\xi)^2,$$

$$W_{43} = -W_{41}\alpha(1-\xi)^2, \quad \{Z^n\} = \left\{\begin{matrix} (2+\varepsilon)^n \\ (1+\varepsilon)^n \\ \varepsilon^n \\ (-1+\varepsilon)^n \\ (-2+\varepsilon)^n \end{matrix}\right\}, \qquad n = 1,2,3,$$

we obtain the following expressions for the elements $a_{i,m}$ of matrix $[A]$:

$$\left\{\begin{matrix} a_{11} \\ a_{22} \\ a_{33} \\ a_{44} \\ a_{55} \end{matrix}\right\} = W_{11}\{Z^2\} + W_{12} + j(W_{13}\{Z^1\} - \{Z^3\}),$$

$$\left\{\begin{matrix} a_{12} \\ a_{21},\ a_{23} \\ a_{32},\ a_{33} \\ a_{43},\ a_{45} \\ a_{54} \end{matrix}\right\} = W_{21}\{Z^2\} + W_{22} + jW_{23}\{Z^1\},$$

$$\left\{\begin{matrix} a_{13} \\ a_{24} \\ a_{31},\ a_{35} \\ a_{42} \\ a_{53} \end{matrix}\right\} = W_{31}\{Z^2\} + W_{32} + jW_{33}\{Z^1\},$$

$$\left\{\begin{matrix} a_{14} \\ a_{25} \\ a_{41} \\ a_{52} \end{matrix}\right\} = W_{41}\{Z^2\} + W_{42} + jW_{43}\{Z^1\},$$

$$a_{15} = a_{51} = 0.$$

Using (2.111) and (2.112) we arrive at equations concerning the variational spectrum of the oscillations in the autodyne:

— for an approaching object

$$[\Delta x] = \frac{1-\xi}{Q}(1+\varepsilon)[A]^{-1}\mathcal{E}_0 \begin{bmatrix} 0 \\ -(1+\varepsilon) + j\delta_1(1-\xi) \\ 0 \\ 0 \\ 0 \end{bmatrix};$$

— for an object that is moving away

$$[\Delta x] = \frac{1-\xi}{Q}(-1+\varepsilon)[A]^{-1}\mathcal{E}_0 \begin{bmatrix} 0 \\ 0 \\ 0 \\ -(-1+\varepsilon)+j\delta_1(1-\xi) \\ 0 \end{bmatrix}.$$

The conversion properties of the systems in Fig.2.25 and Fig.2.27 will be characterized through the conversion coefficients determined, respectively, by the variation of the voltage ΔU_C of the low-frequency load $R_T C$ and the variation of the voltage ΔU of the autodyne generator oscillating circuit.

The coefficient of conversion into Doppler frequency ε in the system shown in Fig.2.25 is

$$K_\varepsilon = \frac{\Delta U_C}{E_0} = \frac{1}{\mathcal{E}_0}\frac{\Delta U_C}{U_{oo}} = \frac{1}{\mathcal{E}_0}\frac{\zeta}{\delta_1}\left\{(1-\xi)\frac{d\Delta x}{dt} + \left[-\frac{1}{Q}(1-\delta Q)+2\beta f(x_0)\right]\frac{df(x_0)}{dx}\Delta x\right\}_\varepsilon . \tag{2.114}$$

The coefficient of conversion into combined frequencies from the spectrum $(\pm 1+\varepsilon)$ in the system shown in Fig.2.27 is

$$K_{\pm 1+\varepsilon} = \frac{\Delta U}{E_0} = \frac{1}{\mathcal{E}_0}\left\{\frac{df(x_0)}{dx}\Delta x\right\}_{(\pm 1+\varepsilon)} . \tag{2.115}$$

As the inverse matrix $[A]^{-1}$ is presented in the form $[A]^{-1} = [q_{i,m}]$, $i,m = 1,\ldots,5$ and as (2.106) and (2.107) are taken into consideration, (2.114) and (2.115) are used as points of departure to obtain expressions for the complex transmission coefficients:

— for an approaching object

$$\dot{K}_\varepsilon^{(+)} = \frac{1-\xi}{Q}(1+\varepsilon)b_2\frac{\zeta}{\delta_1}\{j(1-\xi)\varepsilon q_{32} + \frac{1}{Q}(1-\delta Q)[C_1 D_0(q_{22}+q_{42}) - q_{32}] + 2\beta[k_{13}(q_{12}+2q_{32}+q_{52}) + k_{12}(q_{22}+q_{42})]\}, \tag{2.116}$$

$$\dot{K}_{\pm 1+\varepsilon}^{(+)} = \frac{1-\xi}{Q}(1+\varepsilon)[-C_1 D_0(q_{12}+2q_{32}+q_{52}) + q_{22}+q_{42}]b_2; \tag{2.117}$$

— for an object that is moving away

$$\dot{K}_\varepsilon^{(-)} = \frac{1-\xi}{Q}(-1+\varepsilon)b_4\frac{\zeta}{\delta_1}\{j(1-\xi)\varepsilon q_{34} + \frac{1}{Q}(1-\delta Q)[C_1 D_0(q_{24}+q_{44}) - q_{34}] + 2\beta[k_{13}(q_{14}+2q_{34}+q_{54}) + k_{12}(q_{24}+q_{44})]\}, \tag{2.118}$$

$$\dot{K}^{(-)}_{\pm 1+\varepsilon} = \frac{1-\xi}{Q}(-1+\varepsilon)[-C_1 D_0(q_{14}+2q_{34}+q_{54})+q_{24}+q_{44}]b_4, \tag{2.119}$$

where $b_2 = -(1+\varepsilon)+j\delta_1(1-\xi)$, $b_4 = -(-1+\varepsilon)+j\delta_1(1-\zeta)$.

A numerical experiment has been made on the basis of (2.116) – (2.119) and the transmission characteristics of the autodyne systems in Fig.2.25 and Fig.2.27, described by differential equations of the third order, have been studied. The numerical example has been conducted for the typical properties of the parameters:

$$\varepsilon = 10^{-7}, \delta = 10^{-7}, \delta_1 = 10^{-7}, D_0 = 10^{-2}, \xi = 0, C_1 = 0,1, G_2 = 0,01, Q = 50,$$

in addition, the influence of each parameter within a set range of alteration of its values has been examined. The illustrations presented here cover mainly the case of an approaching object. The dependencies for the case of an object that is moving away are described on a comparative basis only when presenting the selectivity effect of the autodyne, related to the detuning of the oscillations with respect to the resonance frequency ω_0. In practice the detuning of the oscillation frequency is achieved by communicating the respective synchronizing signal.

The dependence of the conversion coefficients $K^{(\pm)}_{\varepsilon}$ and $K^{(\pm)}_{\pm 1+\varepsilon}$ on the parameter δ characterizing the degree of regeneration in the autodyne generator is qualitatively analogous to that in Fig.2.30.

An interesting point in the case under consideration is the influence of the load in the bias circuit on the conversion properties. Fig.2.34 a, b presents the dependencies of the conversion coefficients $K^{(+)}_{\varepsilon}$ and $K^{(+)}_{\pm 1+\varepsilon}$ on the parameter δ_1, which is inversely proportional on the time constant $R_T C$. The figure shows clearly that the conversion properties of the autodyne system in Fig.2.25 can be controlled by changing the load R_T in the bias circuit. The dependence of the conversion coefficient $K^{(+)}_{\pm 1+\varepsilon}$ of the system in Fig.2.27 on δ_1 is by far smaller.

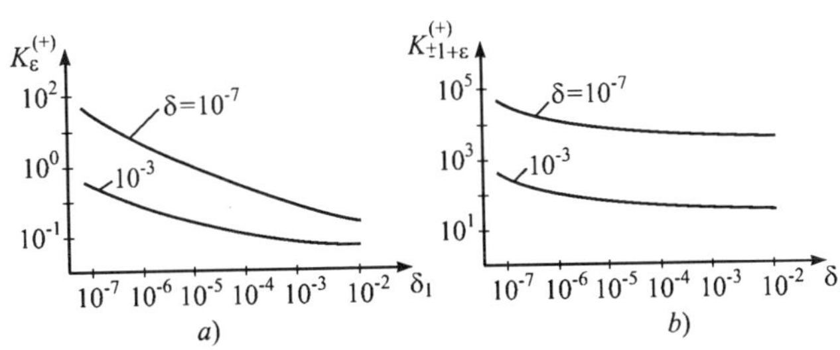

Fig. 2.34. Theoretical dependence of the conversion coefficients in an autodyne system on the time constant of the direct current load circuit

The dependence of the conversion coefficients on the amplitude of the oscillations D_0 in the autodyne oscillator is illustrated in Fig.2.35. As with the autodyne system of the second order, the conversion properties in the case of an immediate output at a Doppler frequency (Fig.(2.25)) have an optimum by the value of the amplitude D_0, while the conversion coefficient $K^{(+)}_{\pm 1+\varepsilon}$ of the system in Fig.2.27 is the larger, the smaller the value of D_0 is. As the regeneration degree of the circuit in the system presented in Fig.2.25 grows, the extremum of the dependence shown in Fig.2.35 is shifted in the direction of the larger values of D_0. This is clear, bearing in mind that the system is quite sensitive for small values of δ and, respectively, small values of D_0, when the oscillations do not essentially cover nonlinear sections of the characteristics of the active element. As the system gets „tougher" when δ increases, there should be a corresponding increase in the pulsations of the impedance of the active element, which is achieved by increasing D.

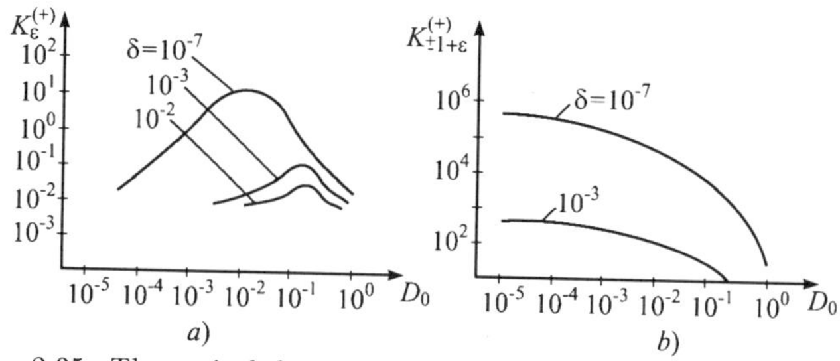

Fig. 2.35. Theoretical dependence of the conversion coefficients in an autodyne system on the oscillation amplitude in the autodyne oscillator

Fig.2.36 a,b demonstrates the „thin structure" of the dependence of the conversion coefficients K_ε in the area of zero detuning. For the sake of comparison we have provided the coefficients of the dependencies both for an approaching object ($K^{(+)}_\varepsilon$) and for an object that is moving away ($K^{(-)}_\varepsilon$). The plots in Fig.2.36a have been built for relative values of the Doppler frequency $\varepsilon = 10^{-7}$, while the plots in Fig.2.36b – for $\varepsilon = 10^{-4}$. As in the case of an autodyne of the second order, the dependencies are characterized by pronounced selectivity at $\varepsilon = |\xi|$. The extreme values of the coefficients $K^{(+)}_\varepsilon$ and $K^{(-)}_\varepsilon$ are determined by the sign of the detuning – in the case of negative detunings the coefficient $K^{(+)}_\varepsilon$ reaches the corresponding extreme value, and in the case of positive detuning the coefficient $K^{(-)}_\varepsilon$ does so. When each of the coefficients is at its extreme value, the value of the other coefficient is substantially lower. In principle we can imagine that the existence of such a peculiarity as the „thin structure" of the dependence of the coefficients $K^{(+)}_\varepsilon$ and $K^{(-)}_\varepsilon$ on the sign of the detuning can be used for determining the direction in which the object moves by way of periodical detuning of the oscillations in the autodyne within a certain narrow bandwidth with the help of an external swept-frequency synchronizing generator. While accounting for the relatively small values of the Doppler frequencies in comparison with the frequency of the signal emitted by the autodyne, in practice one should primarily consider the possibility to change the coefficient of conversion due to the unstable frequency of the self-oscillator within the range of deviations comparable to the range of the Doppler frequencies. This should also be taken into consideration in the cases when the performance of the autodyne generator is stabilized through an external synchronizing action, which shifts the frequency of the emitted signal insignificantly. The striking selectivity already demonstrated in Fig.2.36 is only manifested for relatively small values of the regeneration parameter δ.

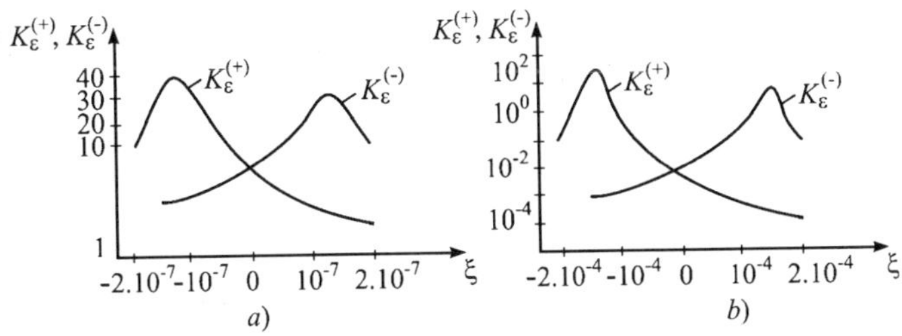

Fig. 2.36. Illustration of the „thin structure" in the dependence of the conversion coefficients in an autodyne system in the area of zero detuning: a – $\varepsilon = 10^{-7}$; b – $\varepsilon = 10^{-4}$

The dependence of the coefficients K_ε and $K_{\pm1+\varepsilon}$ on the value of the Doppler shift of the frequency ε is qualitatively analogous to that in Fig.2.31, and it is substantially reduced as the value of δ increases.

The dependence of the conversion coefficient $K_{\pm1+\varepsilon}^{(+)}$ on the parameter C_1, characterizing the degree of nonlinearity of the capacitive characteristic of the active device is illustrated in Fig.2.37. When the values of δ are small (for example $\delta = 10^{-7}$), the role of C_1 in the process of the initial spectrum formation $(\pm1+\varepsilon)$ is relatively limited. The extreme value of the dependence corresponds to the relatively small values of C_1, and then as C_1 goes up, the coefficient $K_{\pm1+\varepsilon}$ goes down. As the parameter δ increases (the case that is interesting from a practical point of view), the coefficient $K_{\pm1+\varepsilon}^{(+)}$ drops significantly, and its increase can be secured only by a more powerful „switching on" of the modulation-parametric mechanism that forms the initial spectrum $(\pm1+\varepsilon)$, conditioned by the nonlinear capacitance of the reactive element. In addition, as Fig.2.37 shows, the higher the value of δ, the more the extreme value of the dependence shifts in the direction of higher values for C_1, i.e. the more significant is the role of the modulation-parametric mechanism using a power accumulating parameter.

The nonlinear active conductance, which is characterized by the parameter G_2, also exerts considerable influence on the conversion properties of the autodyne systems (Fig.2.38). The coefficient $K_\varepsilon^{(+)}$ (a system presented in Fig.2.25) is highly dependent on G_2. The dependence is determined by the important general regularity that in the case of modulation-parametric interactions, the „upward" conversion of the frequency is more effective when nonlinear reactance is used, while the „downward" conversion is more effective when carried out with a nonlinear

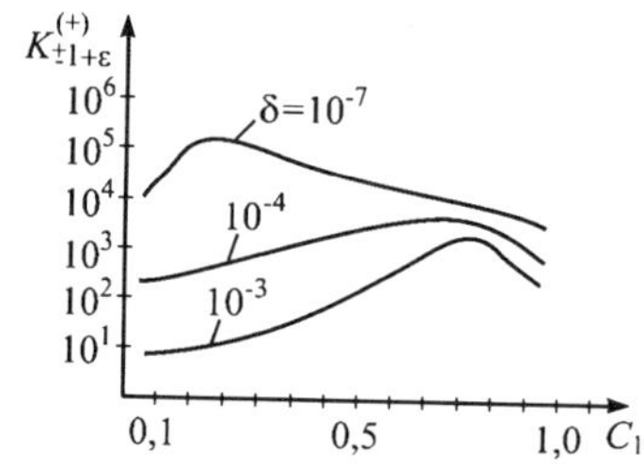

Fig. 2.37. Theoretical dependence of the conversion coefficients in an autodyne system on the degree of nonlinearity of the capacitive characteristic of the active device

active element. That is why the coefficient $K_\varepsilon^{(+)}$ is highly dependent on G_2, and for a broad range of values $K_\varepsilon^{(+)}$ goes up as G_2 increases. The dependence shown in Fig.2.38a has its extreme values but in practice the values of the coefficients G_2 for the modern SHF semiconductor devices generally correspond to the left-hand slope of the graph.

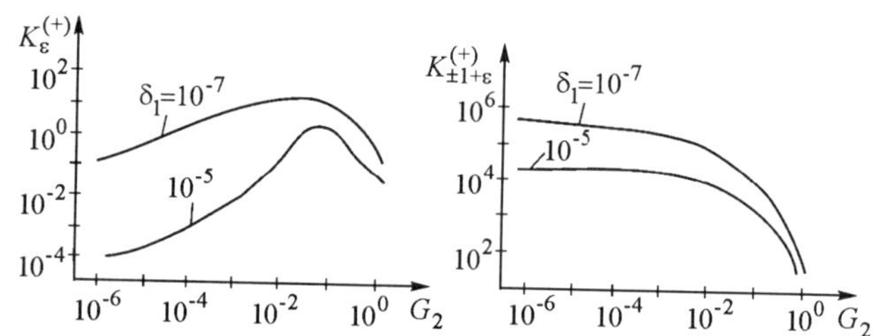

Fig. 2.38. Theoretical dependencies of the conversion coefficients in an autodyne system on the degree of nonlinearity of the conductance of the device

In accordance with the general regularity, the frequency conversion conditions related to the system in Fig.2.27 deteriorate as the value of G_2 goes up (the dependence of $K^{(+)}_{\pm1+\varepsilon}$ in Fig.2.38b). This is due to the fact that as G_2 increases, the role of the modulation-parametric conversion, effected with the aid of a power accumulating element, weakens.

The analysis of two basic autodyne systems, illustrated in Fig.2.25 and Fig.2.27, allows to detect the general potential for controlling and optimizing their conversion properties. In practice, the possible area of parameter variation is normally narrower by far than the broad range of alteration considered above. That is why the optimization task should be viewed in mini-max terms. At the same time, the already conducted general analysis of the conversion properties can also prompt qualitatively different solutions in designing short-range radar systems. For instance, one of the major conclusions of the theoretical analysis presented above is the fact that the conversion properties are strongly influenced by the parameter δ

characterizing the degree of regeneration of the autodyne oscillating circuit. Since for the purpose of achieving high amplitudes of the emitted signal the value of the parameter δ should grow, which, in turn, would cause a drop in the conversion coefficient and in the sensitivity of the system with respect to the reflected signal, one readily arrives at the conclusion that the function of generating and emitting a probing signal should be separated from the function of receiving, amplifying and transforming the signal reflected by a moving object. Under these conditions, two generators should be used – a high powered and a low-powered one (Fig.2.39).

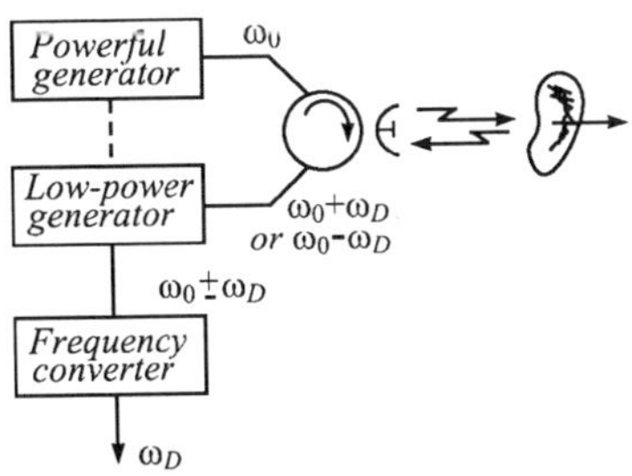

Fig. 2.39. A system of a short-distance Doppler radar with two oscillators

The analysis indicates that the autodyne systems can be highly sensitive receiving and transmitting modules among the short-range radar facilities. It is noteworthy that the attainment of the potential sensitivity, as well as the realization of the frequency selectivity effects, can be limited by the natural fluctuation processes.

Publication [207] analyzes a short-range self-detecting Doppler radar system with focusing the emitted field in the short-range zone. Such a system has a large resolution. Moreover, it is selective with respect to the parameter of distance to a certain object. The latter is of considerable interest with a view to some special applications. As a moving object passes through the focal spot zone, the phase of the incoming reflected signal is further shifted by π radians. While the object passes through this zone, the frequency of the reflected signal grows in a leap-like manner. It is interesting that Adler's equation [208], describing the behaviour of the phase difference between the external action and the natural oscillations of the autodyne generator has no solution.

The analysis shows that self-synchronization in the autodyne systems in the classical sense is impossible. Under definite conditions, however, certain relaxation instability may occur as a result of a peculiar „instantaneous" self-synchronization. The issue regarding the possibility for self-synchronization in such a peculiar radiophysical system as the short-range self-detecting Doppler radar with delaying self-action has been debated in the published sources for years on end. We shall refer to this issue once again in Chapter 4, where we shall address the problem from a different angle by applying the general criteria of oscillation chaotization in generator systems.

Works [205, 206, 247] analyze the oscillation spectrum in short-range self-detecting Doppler radar systems described by differential equations of the second and third order. The investigation has been carried out by taking into account both the amplitude and the phase modulation of the oscillations in the autodyne generator under the action of a signal reflected by a moving object. It has been established that in such a system the amplitude of the basic oscillations in the autodyne generator and their detuning, as well as the spectral components of the amplitude and the phase modulation conditioned by the action of a reflected signal, change slowly in time at a frequency $\sim \dfrac{\omega_D}{\omega_0}$, where ω_D is the Doppler shift of frequency, and ω_0 is the frequency of the unperturbed oscillations in the autodyne generator. A number of other unexpected effects concern the dependence of the amplitudes and phases of the first and second harmonic in the spectrums of the amplitude and phase modulation on the parameters of the oscillating system, the time constant of the bias circuit, the load, the coefficient of the reflection by a moving object and the length of the signal lag, the motion parameters of the object (speed and direction). It is indicated in what cases the phase of the second harmonics in the spectra can be reliably informative with respect to the direction in which the object moves.

2.4. Application of the principle of reversibility of the modulation-parametric interactions in radiophysical systems with external pumping

In this section we shall outline some of the possibilities for applying the modulation-parametric systems for the purpose of increasing the sensitivity of radiophysical systems.

The expediency of using four-frequency parametric systems with negative parameters to increase the sensitivity of radiophysical systems can be assessed on the basis of the general noise parameters of the system.

2.4.1. Noise parameters and properties of one-ports with negative parameters built up on the basis of four-frequency parametric systems

As already shown in 2.1, the reversibility of the modulation-parametric interactions allows using the four-frequency parametric systems successfully to create one-ports with wide-band negative G_-, C_- and R_-, L_- (conductance, capacitance and resistance, inductance), applicable in the cases of regenerative amplification and frequency correction. A statement valid for all these applications is that the load noises cannot be referred to the next stage of the system, since, just as the noise in the signal circuit, they are regenerated by the parametric system. The circuit of the parametric one-port with negative parameters, coupled with the next amplifying stage, make up an overall system. It would be expedient to consider this system as a single (joint) stage.

Having taken these considerations into account, we have presented in Fig.2.40 the equivalent circuit of a video frequency regenerative modulation-parametric amplifier and frequency corrector, which accounts for the respective noise sources [210, 211]. The signal circuit is represented by the conductance $\dot{Y}_s = G_s + jB_s$. Respectively, the circuit of the load (the next amplifying stage) is represented by the equivalent input admittance $\dot{Y}_T = G_T + jB_T$. The noise properties of the signal circuit and the amplifying two-port of the load are characterized by the effective values of the noise currents $\sqrt{i^2_{NS}}$, $\sqrt{i^2_{NT}}$, the noise voltage $\sqrt{u^2}$ and the noise correlation $\dot{Y}_{cor}$, which are determined by the expressions [106, 209]:

$$\overline{i^2_{NS}} = G_s N, \quad \overline{i^2_{NT}} = G_{NT} N, \quad \overline{u^2_{NT}} = R_{NT} N, \quad Y_{cor} = \frac{\overline{\sqrt{i^2_{NT}}\sqrt{u^2_{NT}}}^*}{|\overline{u^2_{NT}}|},$$

where $N = 4kT\Delta f$, k is Boltzmann's constant, T is the absolute temperature, Δf is the frequency bandwidth, G_{NT} and R_{NT} are, respectively, the equivalent noise conductance and the equivalent noise resistance of the load two-port. Both here and further on, the sign „*" will mean a complex conjugate value. The four-frequency modulation-parametric system (MPS) is represented by an equivalent circuit analogous to that in Fig.1.1a. The natural noises of MPS have been taken into account by including the equivalent generators of noise current $\overline{i^2_{Nn}} = NReY_n$, reflecting the thermal noises, as well as equivalent generators $\overline{i^2_{Dn}} = 2qI_0\Delta f_n$, representing the current fluctuation noises of the complex parametric element, $n \to 1, +, -$, q is the electron charge, I_0 is the direct current flowing through the parametric element. Regardless of the fact that there is just one source of shot noise (the complex parametric element), there is no correlation between the equivalent noise generators $\overline{i^2_{Dn}}$, since they are spread along the frequency axis. And the only parameter accounted for with respect to them is the frequency bandwidth f_n, determined by the filtering circuits. Parallel to that, it should be pointed out that both the signals and the noises in the MPS are mutually converted in the three spectral bandwidths ω_s, ω_+, ω_- as a result of the action of the pulsatory parametric element and as a consequence of the reversivility of the modulation-parametric interaction. The flicker noises (of the type of „$\frac{1}{f}$") can be ignored and it can be considered that their relative weight is negligibly small, since in the parametric complex systems they are manifested only in the frequency area from one to several scores of Hz [108]. The given system is considered with regard to a broad video frequency bandwidth of the input signals, which is substantially influenced by the other sources of noise.

As shown in 2.1, as a consequence of the reversibility of the modulation-parametric interaction, the input admittance of MPS in Fig.2.40 can be presented

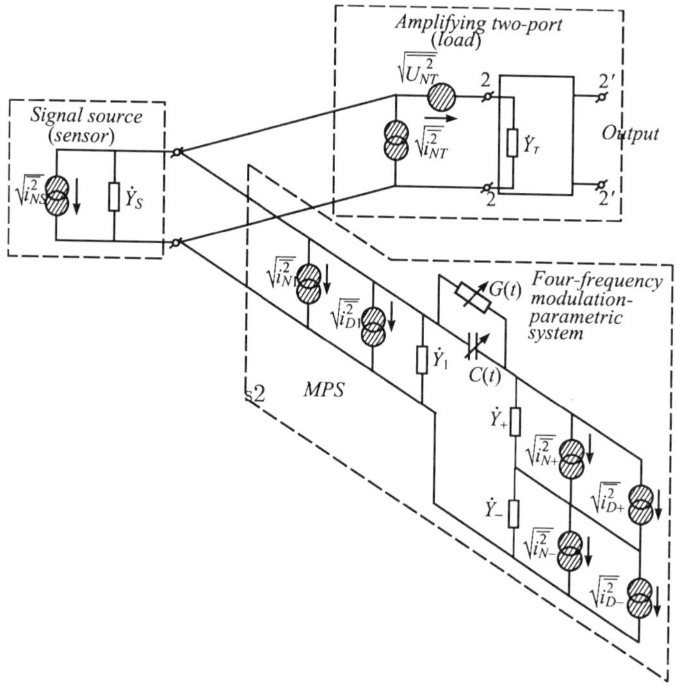

Fig. 2.40. Equivalent circuit of a video-frequency regenerative modulation-parametric amplifier and frequency corrector, which accounts for the signal source, the load and all noise sources

in the form $\dot{Y}_{in} = \dot{Y}_1 + \dot{Y}_g$, where $\dot{Y}_g = -G_g - jB_g$ is the negative admittance introduced into the input circuit of MPS. The case when $|\dot{Y}_g| \gg |\dot{Y}_1|$ is considered. Along with that certain additional current $\overline{i^2_{Ng}}$ is introduced into the input circuit.

As the ideas recounted above are taken into account, the equivalent circuit in Fig.2.40 can be presented in the form given in Fig.2.41.

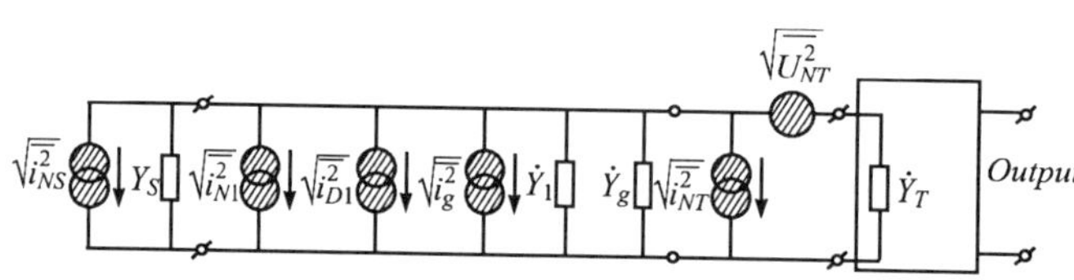

Fig. 2.41. Generalized equivalent circuit for the noise sources of the system „sensor–modulation parametric one-port–amplifying two-port (load)"

Since there is no impedance match in the system, the noise coefficient of the circuit in Fig.2.40 can be defined as

$$F = \frac{P_{N\Sigma}}{P_{NS}}, \tag{2.120}$$

where $P_{N\Sigma}$ is the total power of the noises emitted in the load, P_{NS} is the power of the noises in the load conditioned only by the noises of the signal circuit under normal temperature $T_0 = 300$ K.

As the denotations in Fig.2.40 are accounted for, the expression for the noise coefficient can be presented in the form

$$F = 1 + \frac{\overline{i^2_{N\Sigma}}}{\overline{i^2_{NS}}}, \tag{2.121}$$

where $\overline{i^2_{N\Sigma}} = \overline{i^2_{N1}} + \overline{i^2_{D1}} + \overline{i^2_{Ng}} + \overline{i^2_{NT}} + \overline{u^2_{NT}}\left|\frac{\dot{Y}_T(\dot{Y}_s + \dot{Y}_{in})}{\dot{Y}_T + \dot{Y}_s + \dot{Y}_{in}}\right|^2$

$-2\overline{u^2_{NT}}|Y_{cor}(\overset{*}{Y}_s + \overset{*}{Y}_{in})|\left|\frac{\dot{Y}_T}{\dot{Y}_s + \dot{Y}_{in} + \dot{Y}_T}\right|^2.$

The Exp.(2.121) can also be rewritten as

$$F=1+\frac{G_{NT}}{G_s}+\frac{\overline{u^2_{N1}}|\dot{Y}_s+\dot{Y}_T+\dot{Y}_{in}|^2}{G_sN}+\frac{R_{NT}}{G_s}\left[|\dot{Y}_s+\dot{Y}_{in}|^2-2|Y_{cor}(\dot{Y}_s+\overset{*}{Y}_{in})|\right], \tag{2.122}$$

where $\overline{u^2_{N1}} = \dfrac{\overline{i^2_{N1}} + \overline{i^2_{D1}} + \overline{i^2_{Ng}}}{|\dot{Y}_s + \dot{Y}_T + \dot{Y}_{in}|^2}$ is the noise voltage introduced at the input clamps 1–1 of MPS (Fig.2.40).

The employment of MPS as a one-port with negative differential conductance will be expedient under the condition that

$$M < F_{Nf} - 1, \tag{2.123}$$

where M is the noise number of the system including the signal circuit of MPS and the amplifying two-port (the load); F_{Nf} is the noise figure of the same circuit but in the absence of MPS.

The noise number, in its general form, is defined by the expression [106]

$$M = \frac{F - 1}{1 - \dfrac{1}{K}}, \tag{2.124}$$

where K is the coefficient of amplification by power.

As a result of the presence of MPS, there is a realization of a coefficient of amplification by power

$$K=\gamma\frac{G_s+G_T}{G_s+G_T-G'_g},\ \text{where}\ G'_g=G_g-G_1^0,\ \gamma=\sqrt{1+\left(\frac{B_s+B_T}{G_s+G_T}\right)^2},\dot{Y}_1=G_1^0+jB_1. \tag{2.125}$$

It should be pointed out that in reality the MPS under consideration is used in the conditions of overall offsetting of the system reactances through the introduced broad-band negative reactive conductance B_g, i.e.

$$B_s+B_1+B_T-B_g\simeq 0. \tag{2.126}$$

As (2.121) and (2.124) are substituted in (2.123), and (2.126) is taken into consideration, we obtain

$$M=\left\{G_{NT}+\frac{\overline{u_{N1}^2}(G_s+G_T-G'_g)^2}{4kT\Delta f}+R_{NT}[(G_s-G'_g)^2-2|Y_{cor}(\overset{*}{Y}_s+\dot{Y}_{in})|]\right\}$$
$$\times\frac{1}{G_s}\frac{\gamma(G_s+G_T)}{(\gamma-1)(G_s+G_T)+G'_g}. \tag{2.127}$$

It is not difficult to show that the noise coefficient F_{Nf} is defined by the expression

$$F_{Nf}=1+\frac{G_{NT}}{G_s}+\frac{R_{NT}}{G_s}[G_s^2-2|Y_{cor}\overset{*}{Y}_s|]. \tag{2.128}$$

As (2.127) and (2.128) are combined, the condition (2.123) concerning the expedience of using MPS as a one-port with negative differential conductance, when estimated on the basis of a complex indicator including the amplifying and noise properties, acquires the form

$$\{G_{NT}+\frac{\overline{u_{N1}^2}(G_s+G_T-G'_g)^2}{4kT\Delta f}+R_{NT}[(G_s-G'_g)^2-2|Y_{cor}(\overset{*}{Y}_s$$
$$+\overset{*}{Y}_{in})|]\}\frac{\gamma(G_s+G_T)}{(\gamma-1)(G_s+G_T)+G'_g}<G_{NT}+R_{NT}[G_s^2-2|Y_{cor}\overset{*}{Y}_s|]. \tag{2.129}$$

If the noise coefficient of MPS is defined as

$$F_{MPS}=\frac{G_{MPS}}{G'_g}=\frac{\overline{i_{N1\ MPS}^2}}{4kT\Delta fG'_g}, \tag{2.130}$$

where $\overline{i^2_{N1\ MPS}} = \overline{i^2_{N1}} + \overline{i^2_{Ng}}$, G_{MPS} is the equivalent noise conductance of MPS, the term containing $\overline{u^2_{N1}}$ in (2.129) can be written as $\dfrac{\overline{u^2_{N1}}(G_s + G_T - G'_g)^2}{4kT\Delta f} = F_{MPS}G'_g$. Then the final condition (2.129) will be presented in the form

$$\{G_{NT} + F_{MPS}G'_g + R_{NT}[(G_s - G'_g)^2 - 2|Y_{cor}(\overset{*}{Y}_s + \overset{*}{Y}_{in})|]\}$$
$$\times \frac{\gamma(G_s + G_T)}{(\gamma - 1)(G + G_T) + G'_g} < G_{NT} + R_{NT}[G^2_s - 2|Y_{cor}\overset{*}{Y}_s|]. \tag{2.131}$$

The noise coefficient F_{MPS} will be determined from the matrix equation (1.26) reflecting the processes in MPS. For the case under consideration it is in the form

$$\begin{Vmatrix} \dot{I}_1 \\ \dot{I}_+ \\ \overset{*}{I}_- \end{Vmatrix} \begin{Vmatrix} \dot{Y}_1 & G_1e^{-j\psi_2} + j\omega_sC_1e^{-j\psi_1} & G_1e^{j\psi_2} + j\omega_sC_1e^{j\psi_1} \\ G_1e^{j\psi_2} + j\omega_+C_1e^{j\psi_1} & \dot{Y}_+ & G_2e^{j\psi_2} + j\omega_+C_2e^{j\psi_1} \\ G_1e^{-j\psi_2} - j\omega_-C_1e^{-j\psi_1} & G_2e^{-j2\psi_2} - j\omega_-C_2e^{-j2\psi_1} & \overset{*}{Y}_- \end{Vmatrix} \begin{Vmatrix} \dot{U}_1 \\ \dot{U}_+ \\ \overset{*}{U}_- \end{Vmatrix}. \tag{2.132}$$

As the currents $\dot{I}_1$, $\dot{I}_+$, $\dot{I}_-$ and the voltage $\dot{U}_1$ in the matrix equation (2.132) are presented as noise currents $\dot{I}_1 \to \sqrt{\overline{i^2_{N1\Sigma}}} = \sqrt{\overline{i^2_{N1}} + \overline{i^2_{D1}}}$, $\dot{I}_+ \to \sqrt{\overline{i^2_{N+\Sigma}}} = \sqrt{\overline{i^2_{N+}} + \overline{i^2_{D+}}}$, $\dot{I}_- \to \sqrt{\overline{i^2_{N-\Sigma}}} = \sqrt{\overline{i^2_{N-}} + \overline{i^2_{D-}}}$ and noise voltage $\dot{U}_1 \to \sqrt{\overline{U^2_{N1}}}$, we obtain

$$\overline{U^2_{N1}} = \overline{i^2_{N1\Sigma}}\left|\frac{D_1}{D}\right|^2 + \overline{i^2_{N+\Sigma}}\left|\frac{D_+}{D}\right|^2 + \overline{i^2_{N-\Sigma}}\left|\frac{D_-}{D}\right|^2, \tag{2.133}$$

where D is the determinant of the matrix in (2.132) and D_1, D_+, and D_- are the respective algebraic adjuncts.

Taking into account the fact that the input conductance of MPS is expressed by the matrix equation (2.132) as $\dot{Y}_{in} = \dfrac{\dot{I}_1}{\dot{U}_1} = \dfrac{D}{D_1}$ (See 1.3), the following expression for the total input noise current $\overline{i^2_{N1\ MPS}}$ of MPS is arrived at:

$$\overline{i^2_{N1\ MPS}} = \overline{U^2_{N1}}|\dot{Y}_{in}|^2 = \overline{i^2_{N1\Sigma}} + \overline{i^2_{N+\Sigma}}\left|\frac{D_+}{D_1}\right|^2 + \overline{i^2_{N-\Sigma}}\left|\frac{D_-}{D_1}\right|^2, \tag{2.134}$$

i.e. the inserting by MPS additional noise current is

$$\overline{i^2_{Ng}} = \overline{i^2_{N+\Sigma}}\left|\frac{D_+}{D_1}\right|^2 + \overline{i^2_{N-\Sigma}}\left|\frac{D_-}{D_1}\right|, \tag{2.135}$$

On the basis of (2.134) we obtain the following expression for the product $F_{MPS}G'_g$:

$$F_{MPS}G'_g = \frac{\overline{i^2_{N1\ MPS}}}{4kT\Delta f} = G^0_1+G\left(\left|\frac{D_+}{D_1}\right|^2+\left|\frac{D_-}{D_1}\right|^2\right)+\frac{qI_0}{2kT}\left(1+\left|\frac{D_+}{D_1}\right|^2+\left|\frac{D_-}{D_1}\right|^2\right), \tag{2.136}$$

where G is the conductance of the MPS oscillating circuit (See 2.1).

From (2.132) we obtain the following expressions for the reciprocal conversion coefficients in (2.136) at $\Psi_1 = -\frac{\pi}{2}$, $\Psi_2 = 0$:

$$K_+=\frac{D_+}{D_1}=\frac{M_1-m_1\xi Q(1+2M_2)+j2Q[m_1\xi(\xi_0-\xi)-M_1(\xi_0+m_2-\xi)]}{1+2M_2+4Q^2(\xi_0^2+\xi_0 m_2-\xi^2)}, \tag{2.137}$$

$$K_-=\frac{D_-}{D_1}=-\frac{M_1+m_1\xi Q(1+2M_2)+j2Q[m_1\xi(\xi_0+\xi)+M_1(\xi_0+m_2+\xi)]}{1+2M_2+4Q^2(\xi_0^2+\xi_0 m_2-\xi^2)}. \tag{2.138}$$

Expressions for the negative parameters G_g and B_g introduced by MPS were obtained in 2.1.

In reality the coefficient of power amplification is considerable as a result of the regenerative action of MPS. This is achieved at $G'_g \to G_s + G_T$ and $\gamma \to 1$. Under these conditions (2.131) is reduced to

$$F_{MPS} < \frac{R_{NT}}{G_s+G_T}[G_s^2 - 2|Y_{cor}\overset{*}{Y}_s|]. \tag{2.139}$$

Given that the conductance of the signal circuit G_s is by far larger than the conductance of the noise correlation $|Y_{cor}|$ and the noise conductance of the load two-port G_T, the condition (2.139) assumes the form

$$F_{MPS} < R_{NT}G_s. \tag{2.140}$$

The inequality (2.140) is convenient for an initial practical engineering estimate. It shows that from the viewpoint of noise, it is most convenient to work with low-resistant signal circuits or input oscillating circuits with a low qualitative factor (and this is most frequently encountered in the practical applications).

The analysis shows that in the case of quadrature pumping of the reactive ($C(t)$) and active ($G(t)$) part of the parametric element, i.e. when $\Psi_1 = -\frac{\pi}{2}$ and $\Psi_2 = 0$, MPS has the property of unidirectional transmission of the signals, and the reciprocal transmission coefficients K_+ and K_- (formulae (2.137) and (2.138)) can be negligibly small.

The numerical estimates of real MPS parameters, where (2.132) – (2.137) are used, indicate that

$$F_{MPS}G'_g \sim 1,5G^0_1. \tag{2.141}$$

Taking into account that MPS has been designed to create negative active conductance, whose absolute value exceeds by far the natural input conductance, i.e. $\frac{G_1^0}{G'_g} \ll 1$, it becomes clear that the noise coefficient of MPS $F_{MPS} \ll 1$ (actually the already formulated noise coefficient $F_{MPS} \sim 0,1 \div 0,3$).

Going back to inequality (2.141) and bearing in mind that the modern HF transistor amplifiers have an equivalent noise resistance $R_{NT} \sim 100 \div 300\ \Omega$, we can draw the general conclusion that the application of MPS as one-ports with negative conductance is expedient (from the viewpoint of noise) in the case of signal sources with active conductance $G_s > 10^{-3}\ \Omega^{-1}$.

An analogous estimate has been carried out and it has shown that in the cases of high-resistance signal sources, where $G_s < 10^{-3}\ \Omega^{-1}$, it is expedient to use purely reactive MPS in the form of one-ports, mainly with effective negative reactance. The latter helps carry out the required frequency correction and the additional amplification.

The source [212] provides a comprehensive theoretical analysis of the steadiness and stability of one-ports with negative parameters designed on the basis of four-frequency parametric systems.

The theoretical analysis of the noise properties, of the steadiness and stability of the parametric one-ports with broad-band low-noise negative parameters proves the expedience of their wide application.

2.4.2. Increasing the sensitivity of receiving systems of the type of a capacitive or inductive video sensor

The system of a capacitive or inductive video sensor, combined with a video amplifier, may have very high sensitivity, if a four-frequency parametric system with a negative input capacitance or inductance – Fig.2.42 – is used. The total input capacitance of the video amplifier in Fig.2.42a can be presented as $C_{in} = C_- + C_{in}^a = \mu C_C$, where C_- is negative capacitance, created by a four-frequency parametric system, C_{in}^a is the positive input capacitance in the absence of compensation, μ is the coefficient reflecting the degree of compensation. The situation displayed in Fig.2.42b is analogous – $L_{in} = L_- + L_{in}^a = \mu L_L$. The input conductance G_{in} and resistance R_{in} of the four-frequency reactive parametric system can be presented as: $G_{in} = G - A\omega^2$ and $R_{in} = R - B\omega^2$, where G and R reflect the natural losses, while the coefficients A and B depend on the circuit of the system and on its mode of operation.

For the systems in Fig.2.42 the generalized complex transmission coefficient is equal to:

$$K = \frac{1}{1 + \mu + j\left[\omega\left(\tau_{CL}\mu + \dfrac{F}{S_{CL}}\right) - \dfrac{D_{CL}}{\omega S_{CL}}\right]},$$

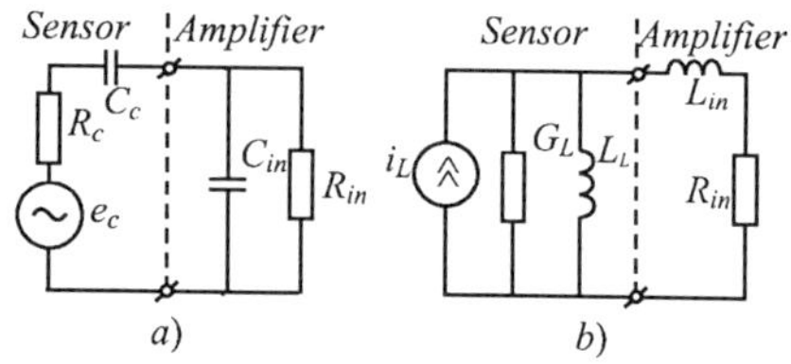

Fig. 2.42. Equivalent circuits of a capacitive (a) and inductive (b) sensor and video amplifier

where, respectively, for Fig.2.42a: $\tau_{CL} = R_C C_C$, $F \equiv A$, $S_{CL} = C_C$, $D_{CL} = D$ and for Fig.2.42b: $\tau_{CL} = G_L L_L$, $F \equiv B$, $S_{CL} = L_C$, $D_{CL} = R$. It is assumed that $R_C G_{in} \ll 1$, $G_L R_{in} \ll 1$.

The module of $\dot{K}$, as a function of the frequency at different degrees of compensation μ is presented in the form of a graph in Fig.2.43. The value of $|\dot{K}|$ in the middle of the frequency bandwidth at $\mu = 0$ has been accepted as a zero level. It can be seen that within a broad frequency range the signal is amplified at $-1 < \mu < 0$, at that the general transmission coefficient of the „reactive video sensor – parametric video amplifier" system has increased $\dfrac{1}{1+\mu}$ times.

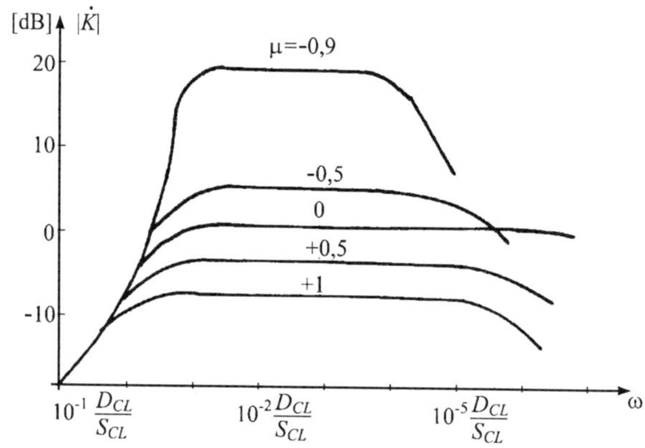

Fig. 2.43. Dependence of the generalized transmission coefficient of a capacitive or inductive sensor operating jointly with a parametric video amplifier

2.4.3. Inductive sensor

Inertia systems, which lack in reactive element changes in time with HF oscillations excited in the system, but have a possibility for modulation action on a reactive element on the part of an external, determined or fluctuation signal, display modulation-parametric interactions analogous to those in the inertia-free case (this issue is elaborately analyzed in the publication [150]). The electromechanical analogues of the four-frequency parametric systems – the highly sensitive capacitive [154, 213] and inductive (Fig.2.44) [214] sensor of small mechanical oscillations – are also referred to these systems. The latter sensor (Fig.2.44) consists of a mechanical oscillator with mass m, elasticity coefficient K_M and friction coefficient H_M. The mass m is the mobile core of a high frequency coil, which forms an electric oscillating circuit with a linear capacitor C. The resistance r reflects the active losses of the circuit, where forced electric oscillations are excited by an external high-frequency source of current $I'_m \cos\omega t$. The core with mass m can oscillate freely around its point of equilibrium under the action of external mechanical vibrations (external periodic forces). Using the principle of reversibility of the modulation-parametric interactions, we can analyze the processes in the system. The response of the electric oscillating circuit with respect to the mechanical oscillations in the system is manifested in the introduction of effective additional friction and additional elasticity in the mechanical oscillator.

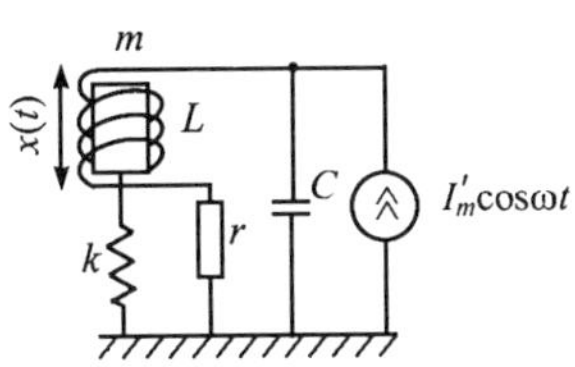

Fig. 2.44. Inductive sensor and analyzer of small mechanical vibrations

The processes in the electromechanical system (Fig.2.44) are described by the following system of differential equations:

$$\left|\begin{array}{l} L(x)\dfrac{d^2 i_L}{dt^2} + r\dfrac{di_L}{dt} + \dfrac{1}{C} i_L = I'_m \cos\omega t \\ \\ m\dfrac{d^2 x}{dt^2} + H_M \dfrac{dx}{dt} + K_M x = F_{el} \quad , \end{array}\right. \tag{2.142}$$

where $L(x)$ is the coil inductance depending on the position of the ferromagnetic core, $F_{el} = A_L i_L^2$ is electric power, A_L – a coefficient accounting for the geometric dimensions of the coil, the type of winding, the diameter of the wire and the number of the windings.

The motion in the mechanical oscillation around the equilibrium x_0 state can be presented as

$$x = X e^{\delta_M t} \cos(\Omega_M t + \varphi_0), \tag{2.143}$$

where $\delta_M = \dfrac{H_M + H_e}{2m} = \delta_{M_0} + \delta'_e$, $\Omega_M = \sqrt{\frac{K_M + K_e}{m}} = \sqrt{\Omega^2_{M_0} + \Omega^2_e}$.

The additional friction coefficient H_e (corresponding to the additional damping δ'_e) and the additional differential elasticity coefficient K_e (corresponding to the change in the natural resonance frequency Ω_e) determined by the response of the electric oscillating circuit to the oscillations of the mechanical oscillator should be determined.

For small deviations from the equilibrium state, the inductance of the coil can be presented in the form $L = \dfrac{L_0}{1+\varepsilon|x|}$, where $\varepsilon \ll 1$ and x is normalized by the x_0 coordinate. Then the system of equations (2.142) can be written in the following form:

$$\left|\begin{aligned} &\frac{d^2 i_L}{dt^2} + 2\alpha_e \frac{di_L}{dt} + \omega_0^2(1+\varepsilon|x|)i_L = I_m \cos\omega t \\ &\frac{d^2 x}{dt^2} + 2\delta_M \frac{dx}{dt} + \Omega_M^2 x = A_e i_L^2 \quad , \end{aligned}\right. \tag{2.144}$$

where $2\alpha_e = \dfrac{r}{L_0}$, $\omega_0^2 = \dfrac{1}{L_0 C}$, $L_0 = L(x_0)$, $I_m = \dfrac{I'_m}{L_0}$.

The solution of the first equation of the system (2.144) is presented in a series by the small parameter

$$i_L = i_{L_0} + \varepsilon i_{L_1} + \varepsilon^2 i_{L_2} + \ldots \tag{2.145}$$

As (2.145) is substituted in the second equation of the system (2.144) and certain transformations are made, the following expressions are obtained for the parameters sought:

$$\delta_M = \frac{H_M + H_e}{2m} = \delta_{M_0} + \delta_e = \delta_{M_0} + \frac{\mp\Delta\omega\alpha_e A_e I_m^2}{8\omega_0[(\Delta\omega)^2 + \alpha_e^2]^3} \tag{2.146}$$

$$\Omega_M^2 = \frac{K_M + K_e}{m} = \Omega_{M_0}^2 + \Omega_e^2 = \Omega_{M_0}^2 + \frac{\pm\Delta\omega A_e I_m^2}{8\omega_0[(\Delta\omega)^2 + \alpha_e^2]^2}, \tag{2.147}$$

where $\Delta\omega$ is the absolute frequency detuning of the electric circuit (the upper sign corresponds to positive detunings and the lower one to negative detunings), ω_0 and α_e are the resonance frequency and damping of the oscillating electric circuit, I_m is the amplitude of the source current.

It can be seen from (2.146) and (2.147) that the damping decrement and the resonance frequency of the mechanical oscillator depend on the current amplitude and on the oscillation detuning in the oscillating electric circuit. When the circuit is tuned to the right-hand slope of the resonance curve (positive values for $\Delta\omega$), an equivalent negative damping decrement δ'_e and negative elasticity are introduced in the mechanical oscillator (the resonance frequency of the mechanical oscillator increases by Ω_e). The parameters δ_e and Ω_e reach their maximum values at detunings corresponding to the maximum slope of the resonance characteristic of the circuit

$$[\delta_e]_{\max} = \frac{A_e Q_e^4 I_m^2}{4\omega_0^5} = \frac{A_e}{4\omega_0}\left(\frac{Q_e}{\omega_0}\right)^4 I_m^2,$$

$$[\Omega_e]_{\max} = \frac{A_e Q_e^3 I_m^2}{4\omega_0^4} = \frac{A_e}{4\omega_0}\left(\frac{Q_e}{\omega_0}\right)^3 I_m^2,$$

where Q_e is the quality factor of the electric circuit.

A potential for controlling the parameters of the mechanical oscillating system through electric means is created. The amplitude of the source current I'_m is changed to regulate the input of effective positive or negative elasticity (altering the resonance frequency of the mechanical oscillator) and of effective positive or negative friction in the mechanical oscillating system. It is also possible to excite continuous mechanical oscillations if the following condition is satisfied: $I_m > F - \sqrt{\frac{4\delta_{M_0}\omega_0^5}{A_e Q_e^3}}$. At $I_m < F$ the mechanical oscillator is in a state of rest, yet it is regenerated, i.e. its quality factor is much higher and this conditions the higher sensitivity of the inductive sensor to external mechanical influences. Given a periodic sweeping of I'_m within definite boundaries, the system in Fig.2.44 can be used as a sensitive sensor – a spectrum analyzer of mechanical oscillations in a relatively broad frequency bandwidth [215]. Devices of this kind can be applied when studying the oscillation spectrum of aircraft and other objects tested in aerodynamic tunnels, the vibrations of the body and other elements of diferent machines, etc.

2.4.4. Capacitive sensor

An analogous compensation of the elasticity of a mechanical oscillator creates additional potential for enhancing the sensitivity of receiving systems of the „capacitive sensor – video amplifier" type. Thus, for example, when tackling the impressive scientific task of practically discovering and proving the existence of the gravitation waves predicted by A. Einstein's General theory of relativity, a high-quality capacitive sensor in combination with a four-frequency parametric system in an optimal noise mode can be used. In this case the compensation of the sensor elasticity through a parametric system can ensure a frequency-independent transmission coefficient of the sensor in a broad frequency bandwidth without any deterioration in its sensitivity (this issue is elaborately examined in [215, 216]). The gravitation receivers (the gravitational resonance bar detectors) that are currently in a process of development in the USA, France, Italy, Germany, Russia and other countries, have input sapphire oscillators with a Q-factor in the order of $10^8 - 10^{10}$. Therefore they have an extremely narrow reception frequency bandwidth – to the order of $10^{-7} - 10^{-8}$ Hz, i.e. it is believed that the frequency of the unknown space gravitation waves will coincide „absolutely precisely" with the tuning frequency of the gravitation sensor. To put it a different way, the response of an arbitrary gravitation wave signal will be a sinusoid with the resonance tuning frequency of the mechanical oscillator. So the question is whether the energy of this narrow spectral range will be sufficient to detect the signal in a reliable manner.

Let us show the way in which the high sensitivity of the capacitive sensor (Fig.2.45) to small mechanical oscillations [109] is brought about. The mechanical oscillator consists of mass M, suspended on a spring with elasticity K. The sensor is an oscillating circuit made up of an inductive coil L and capacitance C, one of whose plates is connected to the mass M. The resistance R reflects the energy losses in the oscillating circuit. The damper h represents the energy losses in the mechanical oscillating system. There is an inductive connection between the coil L and the generator e, which excites forced oscillations in the oscillating circuit. Their frequency is close to the resonance frequency of the circuit. As the mass M oscillates, the capacitance C changes. Such oscillation retunes the oscillating circuit, altering the amplitude and the phase of its forced oscillations. In this way the mechanical oscillation is converted into amplitude and phase modulation of a variable electric signal. Combined frequencies also occur, i.e. $\omega_p \pm n\Omega_s$, where ω_p is the frequency of the generator, Ω_s is the oscillation frequency of the mass M, $n = 0, 1, 2, \ldots$ Given a small amplitude of the mass M, the combined components with frequency $\omega_p \pm \Omega_s$ show the highest intensity. Besides, at $\Omega_s \ll \omega_p$ both combined oscillations $\omega_p \pm \Omega_s$ are separated by the common oscillating circuit of the capacitive sensor. The mechanical oscillating system can exchange energy with the electric oscillations in the circuit, furthermore it can be shown that this energy exchange is in accordance with Manley-Row's energy relations (1.1)

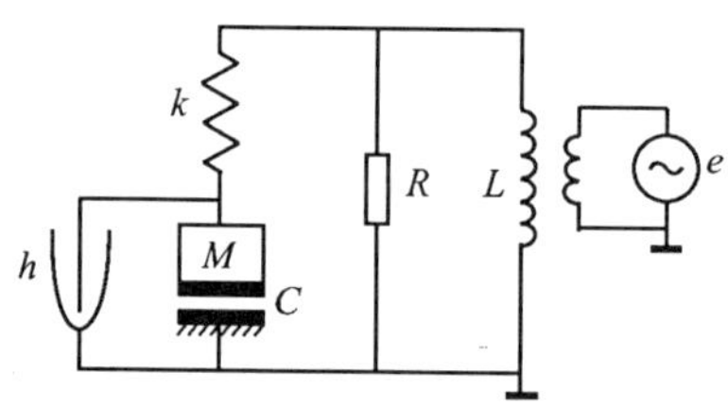

Fig. 2.45. Equivalent scheme of a capacitive sensor of small mechanical vibrations

If the oscillation of the mass M is governed by a harmonic law $x = X_0 \cos \Omega_s t$, the capacitance of the circuit changes as $C = C_0(1 + 2m \cos \Omega_s t)$, where $m = \dfrac{\varkappa X_0}{2C_0}$, $\varkappa = -\dfrac{C_0}{d_0}$, C_0 is the value of the capacitance, if the distance between its plates is d_0.

The voltage and the charge of the capacitor can be presented as

$$U_C = U_p \cos \omega_p t + U_- \cos(\omega_- t + \varphi_-) + U_+ \cos(\omega_+ t + \varphi_+),$$
$$q_C = U_C C_0 (1 + 2m \cos \Omega_s t),$$

where $\omega_\pm = \omega_p \pm \Omega_s$. Then the combined components of the load and the complex amplitudes of the respective currents flowing through the capacitor at frequencies

ω_- and ω_+ are determined in the form:

$$q_{\omega_-} = C_0 U_- \cos(\omega_- t + \varphi_-) + C_0 m U_p \cos \omega_- t,$$

$$q_{\omega_+} = C_0 U_+ \cos(\omega_+ t + \varphi_+) + C_0 m U_p \cos \omega_+ t,$$

$$\dot{I}_{\omega_-} = j\omega_- C_0(\dot{U}_- + mU_p), \quad \dot{I}_{\omega_+} = j\omega_+ C_0(\dot{U}_+ + mU_p).$$

Using as our starting point the condition that for each of the frequencies $\omega_\pm$ the sum of the currents flowing through R and L should be zero, we can write down the following two equations:

$$\left| \begin{array}{l} \dfrac{\dot{U}_-}{R} + \dfrac{\dot{U}_-}{j\omega_- L} + j\omega_- C_0(\dot{U}_- + mU_p) = 0 \\ \dfrac{\dot{U}_+}{R} + \dfrac{\dot{U}_+}{j\omega_+ L} + j\omega_+ C_0(\dot{U}_+ + mU_p) = 0 \end{array} \right. . \tag{2.148}$$

The complex combined amplitudes are determined from (2.148)

$$\dot{U}_- = \frac{-j\omega_- m C_0 R U_p}{1 + 2jQ(\xi_0 - \xi)}, \quad \dot{U}_+ = \frac{-j\omega_+ m C_0 R U_p}{1 + 2jQ(\xi_0 + \xi)}, \tag{2.149}$$

where $\xi_0 = \dfrac{\omega_p - \omega_0}{\omega_p}$, $\omega_0 = \dfrac{1}{\sqrt{LC_0}}$, $\xi = \dfrac{\Omega_s}{\omega_p}$, $Q = \omega_0 C_0 R$.

For the purpose of calculating the amplitude and phase change, it is convenient to present the combined components in (2.149) in the following form:

$$\dot{U}_- = \dot{U}_{-a} + \dot{U}_{-ph}, \quad \dot{U}_+ = \dot{U}_{+a} + \dot{U}_{+ph}, \tag{2.150}$$

where

$$\begin{aligned} \dot{U}_{-a} &= \frac{-2Q^2 m \xi_0 U_p}{(1 - 2jQ\xi)^2 + 4Q^2\xi_0^2}, \quad \dot{U}_{-ph} = \frac{-jQmU_p(1 - 2jQ\xi)}{(1 - 2jQ\xi)^2 + 4Q^2\xi_0^2}, \\ \dot{U}_{+a} &= \frac{-2Q^2 m \xi_0 U_p}{(1 + 2jQ\xi)^2 + 4Q^2\xi_0^2}, \quad \dot{U}_{+ph} = \frac{-jQmU_p(1 + 2jQ\xi)}{(1 + 2jQ\xi)^2 + 4Q^2\xi_0^2}. \end{aligned} \tag{2.151}$$

Further on we put down the following sum of voltages

$$\begin{aligned} U_p e^{j\omega_p t} + |\dot{U}_{-a}| e^{j(\omega_- t + \varphi_{-a})} + |\dot{U}_{+a}| e^{j(\omega_+ t + \varphi_{+a})} &= \\ = [U_p + 2|\dot{U}_{+a}| \cos(\Omega_s t + \varphi_{+a})] e^{j\omega_p t}&, \end{aligned} \tag{2.152}$$

where

$$\varphi_{+a} = -\varphi_{-a} = -\operatorname{arctg} \frac{4Q\xi}{1 - 4Q^2\xi^2 + 4Q^2\xi_0^2}. \tag{2.153}$$

Therefore, the voltage components of combined frequencies with complex amplitudes $\dot{U}_{-a}$ and $\dot{U}_{+a}$ cause only amplitude modulation of the voltage in the oscillating circuit; furthermore it is obvious from (2.152) that the change in the voltage amplitude $\delta A(t)$ is

$$\delta A(t) = 2|\dot{U}_{+a}|\cos(\Omega_s t + \varphi_{+a}) \tag{2.154}$$

Drawing an analogy with (2.151), the following can be written (taking into account that $|\dot{U}_-| \ll U_p$ and $|U_+| \ll U_p$)

$$\begin{aligned} &U_p e^{j\omega_p t} + |\dot{U}_{-ph}|e^{j(\omega_- t + \varphi_{-ph})} + |\dot{U}_{+ph}|e^{j(\omega_+ t + \varphi_{+ph})} \\ &\simeq U_p e^{j[\omega_p t - \frac{2|\dot{U}_{+ph}|}{U_p}\cos(\Omega_s t + \varphi_{+ph})]}, \end{aligned} \tag{2.155}$$

where $\varphi_{+ph} = -\varphi_{-ph} = \operatorname{arctg} 2Q\xi - \operatorname{arctg}\dfrac{4Q\xi}{1 - 4Q^2\xi^2 + 4Q^2\xi_0^2}$.

It follows from (2.155) that the voltage components of the combined frequencies with complex amplitudes $\dot{U}_{-ph}$ and $\dot{U}_{+ph}$, in a first approximation, cause only phase modulation of the voltage in the oscillating circuit, where, given an oscillating mass M, the change in the phase $\delta\Phi(t)$ is determined from the relation

$$\delta\Phi(t) = -\frac{2|\dot{U}_{+ph}|}{U_p}\cos(\Omega_s t + \varphi_{+ph}).$$

In practice the motion of the mass M can be registered by considering the amplitude or phase deviations of the voltage in the oscillating circuit. Let us estimate the minimum shift of the mass M that can be registered, for instance, on the basis of the voltage amplitude change in the oscillating circuit. As (2.151) and (2.153) are substituted in (2.154), the following is obtained for the deviation $\delta A_{\max}$ of the amplitude from the average value ($2Q\xi \ll 1$):

$$\delta A_{\max} = \frac{2Q^2\xi_0 U_p \dfrac{\varkappa X_0}{C_0}}{|(1 + 2jQ\xi)^2 + 4Q^2\xi_0^2|} \simeq \frac{-2Q^2\xi_0 U_p \dfrac{\varkappa X_0}{C_0}}{1 + 4Q^2\xi_0^2}. \tag{2.156}$$

In (2.156) X_0 is the oscillation amplitude of the mass M, $\delta A_{\max}$ has the meaning of an envelope amplitude of the voltage in the oscillating circuit, which reaches its maximum value under the condition $2Q\xi_0 = 1$. The latter condition corresponds to the steepest slope point on the resonance curve of the oscillating circuit. Under this condition, (2.156) can be written in the form

$$|X_0| = \frac{2C_0\delta A_{\max}}{\varkappa Q U_p}. \tag{2.157}$$

For the parallel-plate capacitor $\varkappa = -\dfrac{C_0}{d_0}$. For example, given the trivial conditions $d_0=10^{-1}$ cm, $Q=10^2$, $U_p=10$ V and assuming that the sensitivity of the voltage deviation meter is $\delta A_{\min} = 10^{-6}$ V (which is also an elementary condition), (2.157) yields the average value of the threshold of sensitivity – $X_{0_{\min}} = 2.10^{-10}$ cm. This is a rather small value, yet, in reality, it is absolutely attainable. It should be stressed here that if such high sensitivity is to be realized in practice, special actions should be taken to protect the capacitive sensor against mechanical vibrations and acoustic noise. The mechanical oscillators used to solve the problem of identifying space gravitation waves should have sapphire monocrystals, they should be cooled down to helium temperatures (4.2 K) and kept in special shielded premises. These conditions have yielded record-setting threshold sensitivities $X_{0_{\min}} = 10^{-17}$ cm.

Let us consider the response of the capacitive sensor as a consequence of the reversibility of the modulation-parametric interactions.

The force $\mathcal{F}$ of attraction between the plates of the capacitive sensor is

$$\mathcal{F} = \frac{1}{2}Eq,$$

where E is the intensity of the electric field in the gap, q is the charge of the capacitor. For the parallel-plate capacitor $E = \frac{U}{d}$ and $q = CU$, where U is the voltage of the capacitor, d is the distance between its plates and C is its capacitance. Then the force $\mathcal{F}$ is expressed as

$$\mathcal{F} = \frac{CU^2}{2d}.$$

Using the expressions $C = C_0(1 + 2m\cos\Omega_s t)$, $d = d_0 + X_0\cos\Omega_s t$, $U = [U_p + \delta A(t)]\cos[\omega_p t + \delta\varphi(t)]$ in this formula and calculating the average for a period of high-frequency voltage in the circuit $\left(\dfrac{2\pi}{\omega_p}\right)$, we obtain the average value of the force $\overline{\mathcal{F}}$ in the form:

$$\overline{\mathcal{F}} = \frac{C_0(1+2m\cos\Omega_s t)}{4d_0(1 - \dfrac{X_0}{d_0}\cos\Omega_s t)}[U_p + \delta A(t)]^2.$$

As we retain only the small terms of the first order in this expression and take into account (2.154), we obtain

$$\overline{\mathcal{F}} = \mathcal{F}_0 + \overline{\delta\mathcal{F}},$$

where $\mathcal{F}_0 = \dfrac{C_0U_p^2}{2d_0}$ is the constant component of the force of attraction of the capacitor plates,

$$\overline{\delta\mathcal{F}} = \frac{C_0}{d_0}U_p|U_{+a}|\cos(\Omega_s t + \varphi_{+a}) \tag{2.158}$$

is the variable component of the force of attraction of the capacitor plates.

Let us assume that the motion of the mass in the mechanical oscillator is caused by the action of a periodic external force $f(t) = F_0 e^{j\Omega_s t}$. Then the equation regarding the motion of the mass M is of the form

$$M\frac{d^2x}{dt^2} + h\frac{dx}{dt} + kx = F_0 e^{j\Omega_s t} + \overline{\delta\mathcal{F}}.$$

We use the method of complex amplitudes to present the solution of this equation in the form $x = \dot{X}_0 e^{j\Omega_s t}$. Then, as we take into account (2.158), we obtain the following expression for the complex amplitude $\dot{X}_0$:

$$\dot{X}_0 = \frac{F_0}{-M\Omega_s^2 + j\Omega_s(h + \delta h) + k + \delta k},$$

where the effective additions δh and δk can be expressed, after taking into account the inequality $\omega_s \ll \frac{\omega_p}{2Q}$, as follows:

$$\delta h = -\frac{4Q^3\xi_0 \varkappa U_p^2}{\omega_s d_0(1 + 4Q^2\xi_0^2)^2}, \qquad \delta k = \frac{Q\xi_0 \varkappa U_p^2}{d_0(1 + 4Q^2\xi_0^2)}.$$

As a result of the reversibility of the modulation-parametric interactions effective (equivalent) additions δh and δk occur, which are dually analogous to those of the inductive capacitor considered above.

In the case of $\xi_0 = \dfrac{\omega_p - \omega_0}{\omega_p} < 0$, $\delta k < 0$, $\delta h > 0$ and the resonance frequency of the mechanical oscillator also decreases effectively, while the damping goes up. When $\xi_0 > 0$, respectively, $\delta k > 0$, $\delta h < 0$ and the mechanical frequency of the mechanical oscillator increases effectively, while the damping goes down.

In practice, the modulation-parametric interaction between mechanical and electric oscillations, considered above, should be taken into consideration quite often. An example that can be cited here is the gravimeter – an apparatus for relative measurement of the acceleration of the weight force used in gravimetric surveys for ores and minerals. A gravity meter with a capacitive shift sensor is a particular case of the device in Fig.2.45 and the results of the analysis presented above are fully applicable to it. The practical conclusion is that the satisfaction of the condition $\xi_0 = 0$ should be strictly controlled for each measurement.

Let us once again discuss the mechanism of excitation of mechanical oscillations in a capacitive sensor related to the emergence and interaction of combined frequencies by using the more general and visual approach close to that used under 2.4.3. above during the analysis of the inductive sensor.

The equations of the system under consideration (Fig.2.45) are of the form:

$$\left| \begin{array}{l} \ddot{q} + 2\delta_1 \dot{q} + \dfrac{C_0}{C(x)} \Omega_0^2 q = \dfrac{U_0}{L} \cos vt \\ \ddot{x} + 2\delta_2 \dot{x} + \omega_0^2 x = -\dfrac{1}{m} F(x,q), \end{array} \right.$$

where q is the charge on the capacitor, $C(x) = C_0 \left(1 + \dfrac{x}{d_0}\right)^{-1}$ is its capacitance, $\Omega_0 - \dfrac{1}{\sqrt{LC_0}}$ is the own resonance frequency of the electric oscillating circuit for a value of the mechanical coordinate $x = 0$, d_0 is the initial distance between the plates of the capacitor in the case of undeformed spring, ω_0 is the own resonance frequency of the mechanical oscillating unit, δ_1 and δ_2 are the dissipation coefficients, C_0 and L are constants, U_0 is the amplitude of the acting voltage, $F(x,q) = \dfrac{Sq^2\varepsilon_0\varepsilon}{2C^2(x)(d_0+x)^2} = \dfrac{q^2}{2C_0 d_0}$ is the force of attraction between the plates of the capacitor, m is the mass (Fig.2.45). (All conclusions and derivations are laid out under the condition that a flat capacitor is considered).

In the case of zero-order approximation the solution of the system of equations concerning the mechanical coordinate x can be written down in the form $x = B\cos(\omega t + \varphi_1)$. After substitution, the short equations for the amplitude B and phase φ_1 are obtained in the following form:

$$\left| \begin{array}{l} \dot{B} = (\delta_0 - \delta_2)B_2 - \beta_1 B^3 \\ \dot{\varphi}_1 = -\Delta + KQ_m^2 - \beta_2 B^2, \end{array} \right.$$

where $\delta_0 = \dfrac{\Omega_0^4 \delta_1 [(v^2 - \Omega_0^2)(3v^2 + \Omega_0^2) + 4\delta_1^2 v^2]}{2mLd_0^2, [(v^2 - \Omega_0^2)^2 + 4\delta_1^2 v^2]^3} U_0$, $\Delta = \omega - \omega_0$, K, β_1 and β_2 are coefficients depending on the parameters of the system, Q_m is the amplitude of the alternating charge on the capacitor.

It can be seen that the excitation of oscillations is possible mostly when $v > \Omega_0$, i.e., on the right-hand slope of the resonance curve. Therefore, the condition for exciting self-oscillations is $v \simeq \Omega_0 + \dfrac{\delta_1}{\sqrt{5}}$ and its corresponding threshold value for the voltage amplitude is: $U_{0,\min} = \dfrac{24 d_0 \delta_1}{5} \left(\dfrac{6mL\delta_2}{\Omega_0 \sqrt{5}} \right)^{1/2}$. This voltage of the electric source has the following corresponding voltage on the capacitor: $U_c = 12 d_0 \delta_1 \left(\dfrac{m\delta_2\sqrt{5}}{\Omega_0 C_0} \right)^{1/2}$.

The analysis of the non-linearity coefficient β_1 indicates that the excitation of the oscillations is a „soft excitation" (without any special requirements to the initial conditions).

An analogous effect of self-excitation of mechanical oscillations is observed in the optic sensors of small shifts, where self-oscillations occur at the expense of light pressure. The same effect was also observed experimentally during the action of a SHF field on a rotating pendulum. An analogous mechanism is that leading to oscillations in the resonators used in the powerful accelerators of counter flows. The emergence of a generation of elastic waves in dielectric resonators pumped with a high frequency electromagnetic field is also described.

2.4.5. Implementation and utilization of modulation-parametric one-ports with negative conductance (negative resistance)

The devices with negative conductance (negative resistance) are regarded as functional, since due to the negative conductance they can perform the functions of amplifiers, generators, converters, etc. The Introduction to this work provided a short classification of the negative resistances used in science and technology.

The negative resistances, obtained on the basis of parametric systems, where the pulsation of an energy accumulating (reactive) parametric element leads, under certain conditions, to the input of energy in the signal circuit, occupy a particular position from a theoretical point of view. This property of the parametric system is described through a negative resistance introduced into the signal circuit and occurring only in a dynamic mode.

Section 2.1 indicated in theoretical terms that the usage of RC four-frequency parametric systems, where the pumping of the reactive parametric element goes with a phase shift of $-\frac{\pi}{2}$ with respect to the pumping of the active parametric element, allows of implementing one-ports with broad-band negative resistance. This section describes practical methods and devices for obtaining one-ports with broadband negative resistances [217, 218].

Fig.2.46 shows a circuit of a parametric device for obtaining broadband negative conductance. The oscillating circuit is formed by the inductance L and a reactive parametric element represented by „capacitive" diodes D_1 and D_2, connected in a balanced two-cycle circuit. The diodes operate in a turned-off condition, they are pumped by the pumping generator PG. The diodes (D_3) and (D_4), representing the active parametric element, are connected in parallel with the diodes (D_1) and (D_2) also in a balanced two-cycle circuit. The diodes D_3 and D_4 have a negligibly small natural capacitance, and they work in a mode with direct currents of the order of scores of hundreds of μA. Their pumping is effected by a PG with a phase shift of $\frac{\pi}{2}$. The frequency of the change in the parametric elements is equal to the resonance frequency of the oscillating circuit. The negative conductance, together with a parallel negative capacitance, is implemented on the clamps of the load Z_L. Depending on the mode, it is possible to get negative conductance with an absolute value of $10^{-8} - 10^{-2}\ \Omega^{-1}$. As the frequency of the

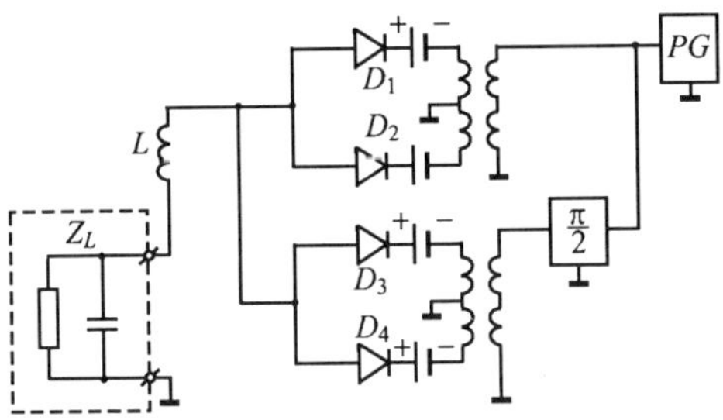

Fig. 2.46. Parametric device for obtaining broadband negative conductance

pumping voltage is changed, the upper boundary frequency of negative parameter existence can also be altered from single-digit Hz to hundreds of MHz.

Fig.2.47 displays modified variants of the circuit in Fig.2.46 for obtaining broadband negative conductance, where FET transistors are used as active parametric elements. These circuits allow ensuring a lower level of the natural (own) fluctuation noises.

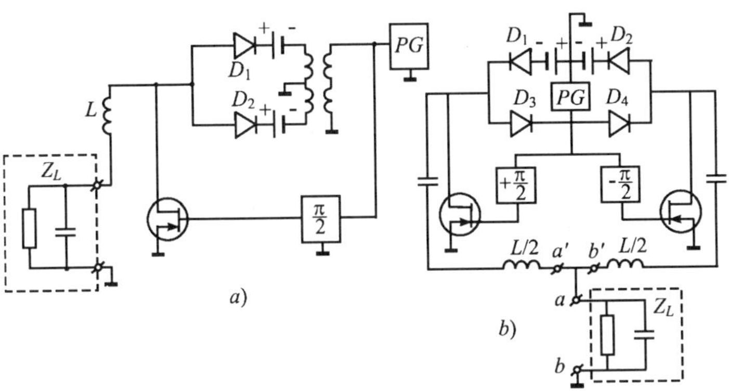

Fig. 2.47. Modified variants of a parametric device for obtaining broadband negative conductance, where field-effect transistors are used

The implementation of a broad-band negative conductance can be effectively applied in the development of new types of parametric devices – regenerative parametric video amplifiers without conversion of the signal frequency. The development of such devices can contribute substantially to the solution of current modern problems of amplifying signals from video sensors with infinitesimal power, such as: high-ohm T.V. camera tubes of the type of „vidicon", „plumbicon", etc. used in reconnaissance and security systems; vidicons for infra-red radiation used for reconnaissance, cancer diagnostics, electronic circuit control, fire extinguishing, in industry, etc.; in pyroelectric receivers used in spectral flame fire-warning systems; impulse photosensors, special encephalographs recording the impulse activeness of neurons, semiconductor particle detectors registering narrow-band lines in cases of radiation reconnaissance, etc. The implementation of a broad-band negative capacitance in parallel with the negative conductance allows of a substantial increase in the sensitivity of video systems by offsetting the parasitic capacitance of the signal source. The level of the useful signal in the T.V. camera tubes can be raised by 10 – 20 dB compared to the case of no compensation for the parasitic capacitance of the tube. For example, if the conductance of the signal source is $\dot{Y}_s = G_s + j\omega C_s = G_s\dot{\Psi}$, where $\dot{\Psi} = 1 + j\omega\frac{C_s}{G_s}$, the coefficient of the signal attenuation caused by the existence of the capacitance C_s is equal to $|\dot{\Psi}|^2 = 1 + \left(\omega_s\frac{C_s}{G_s}\right)^2$, even when the conductance is matched by the maximum of the transmitted power. For T.V. tubes with $C_s \sim 20$ pF, $G_s \sim 10^{-7}\ \Omega^{-1}$ and frequency $\omega_s = 10^{-5}\ \text{s}^{-1}$ the coefficient of attenuation is $|\dot{\Psi}| \simeq 400$.

2.4.6. Implementation and utilization of modulation-parametric one-ports with negative capacitance

A consequence of the theoretical analysis carried out in 2.1 is that the major problem related to the implementation of one-ports with negative capacitance on the basis of four-frequency parametric systems is compensation for the action of the second harmonic in the spectrum of the capacitive parametric element. A practical solution of this problem can be achieved by using „circuit" compensation or non-harmonic pumping. This section recounts methods and devices for obtaining low-noise broad-band negative capacitances [150, 217, 218].

A broad-band low-noise negative capacitance can be realized by employing the circuit in Fig.2.46, with the only difference that both the diodes D_1 and D_2, and the diodes D_3 and D_4 are „capacitive" and operate in a turned-off state. In this case the pumping of the diodes D_3 and D_4 with a phase shift of $\frac{\pi}{2}$ allows of compensating the action of the second harmonic from the spectrum of the parametric element. The maximum absolute value of the negative capacitance is obtained in the case of a positive relative detuning of the oscillating circuit equal to $\frac{1}{2Q}$, where Q is the Q-factor of the circuit.

The action of the second harmonic can be compensated by using two multiple harmonic voltages of pumping – Fig.2.48. The first harmonic of the pumping voltage is obtained through double division of the frequency of the pumping generator PG. The advantage of the circuit in Fig.2.48 in comparison with the circuit in Fig.2.46 is the fact that the operation of the device is independent of the load.

Fig.2.49 shows yet another circuit of a device for an implementation of a low-noise broad-band negative capacitance with circuit compensation of the action of the second harmonic in the spectrum of the parametric element. The parametric

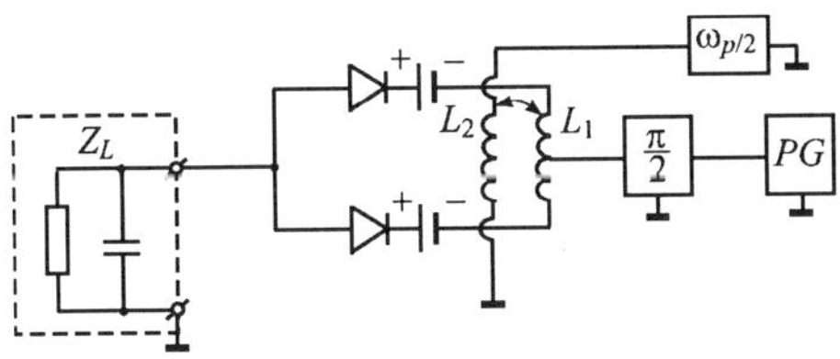

Fig. 2.48. Parametric device for obtaining broadband low-noise negative capacitance by using non-harmonic pumping to compensate for the second harmonic in the spectrum of the parametric element

diodes are connected in a bridge-balanced circuit. For the purpose of reducing the noise coming from the pumping generator, low-noise parametric generators are used for pumping.

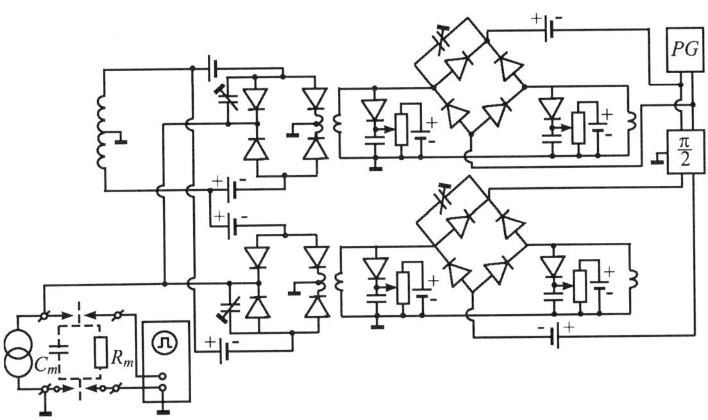

Fig. 2.49. Device for obtaining broadband low-noise negative capacitance with circuit compensation of the action of the second harmonic in the spectrum of the parametric element. It is used to measure the capacitance of biological membranes, for frequency corrections, etc.

A major peculiarity of the implemented negative capacitances, besides the frequency bandwidth liable to regulation and control, is the rather low level of the natural fluctuation noise (the equivalent input noise voltage changes from 5.10^{-3} μV at $f = 20$ Hz to 5.10^{-4} μV at $f = 10^4$ Hz).

Realization of absolute values of the negative capacitance from single-digit pF to 10^6 pF is achieved by selecting the parametric diodes with natural capacitance from single-digit pF to 10^4 pF, for which a parallel connection of single-digit and even scores of diodes is also used. Moreover, the employment of bridge and balanced circuits allows of increasing, the respective number of times, the absolute value of the negative capacitance introduced into the signal circuit.

The low-noise broad-band negative capacitances implemented on the basis of four-frequency parametric systems are effectively applied in the following cases: for broad-band frequency correction seeking to increase the sensitivity of receiving systems of the types of capacitive video sensors, photosensors with video amplifiers, piezoelectric converters, pyroelectric receivers for thermal radiation, etc.; in seismic station receivers, also for the purposes of special reconnaissance; in cases of indication of slight mechanical shifts; in precise medical equipment; when simulating physical, biological and other objects. For example, if we seek to achieve the sensitivity of photodetectors to be limited only by the thermal noise of the current I_T flowing in an unlit state, the capacitance of the photodiode C_{ph} should meet the requirement: $C_{ph} < \dfrac{I_T}{2kT\Delta f}$ [219]. In practice the reduction of the capacitance C_{ph} to the admissible values is possible only through its compensation in the video frequency bandwidth Δf. The sensitivity of the photodetector can be increased by using a photoparametric amplifier permitting the realization of threshold sensitivity in the order of $10^{-14} - 10^{-15}$ A (See Fig.2.50).

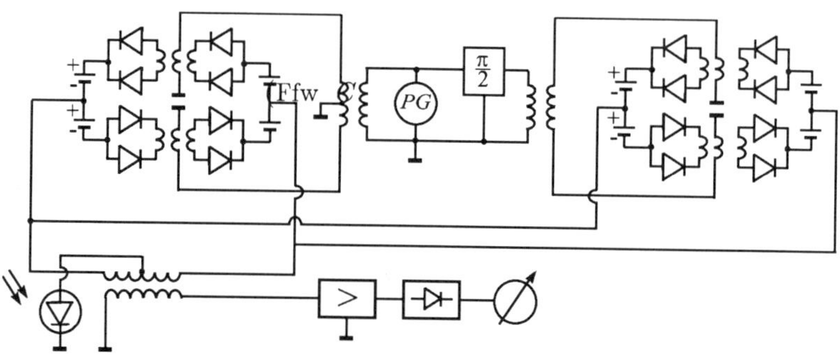

Fig. 2.50. Photoparametric amplifier

Negative capacitances can also be applied in biological membrane research [220]. Biological membranes have natural capacitance, which plays the part of an indicator of their filtering functions, hence its non-destructive measurement evokes considerable scientific interest on the part of bio-cybernetics and microbiologists. Any measurement of such capacitance in a harmonic mode is undesirable due to the inevitable side effect on the biological object and the detecting properties of some membranes, while an obstacle to impulse measurements is the sizable time constant of the circuit. The compensation of membrane capacitance with negative capacitance allows determining, quickly and precisely, its value, as well as the conductance of the membrane in an arbitrary direction. Fig.2.49 illustrates the method of measuring the capacitance of biological membranes by using microelectrodes [220].

The general principles recounted above are used as a basis for increasing the sensitivity of a sensor for interplanetary plasma (solar wind) – a multielectrode modulation measurer of the differential energy spectrum of solar wind [221, 222]. Measurements of the parameters of interplanetary plasma allow forecasting a large number of phenomena in the ionosphere and magnetosphere of the Earth. As the fine structure of the solar wind is studied, the precision of the weather forecasts is improved substantially.

It is a widely known fact that the sensitivity of the interplanetary plasma sensor is highly restricted by the inevitable capacitance of its collector. The reduction of the modulating frequency is undesirable due to the rapid increase in the flicker noise. Neither is it possible to enhance the sensitivity by increasing the area of the collector, due to the proportional increase in the shunting capacitance. The sensitivity of the multielectrode modulation solar wind measurer rises considerably as its natural capacitance and the conductance of the electrometric preamplifier are compensated by using parametric one-ports with low-noise broadband C_- and G_-. The electric circuit of the solar wind sensor is shown in Fig.2.51a [221]. The main pumping channel ensures a wide range time-dependent change in the diode capacitances, while the additional pumping channel is necessary for eliminating the action of the second harmonic in the spectrum of the parametric element.

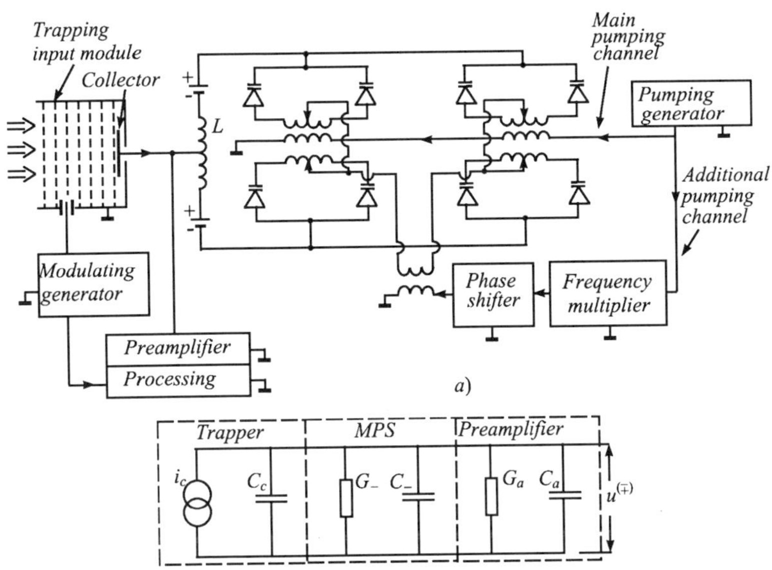

Fig. 2.51. Multielectrode modulation trap of interplanetary plasmas (solar wind) using a modulation-parametric system: a) fundamental scheme; b) equivalent electric circuit

The equivalent electric circuit of the sensor is presented in Fig.2.51 b, where i_C and C_C are, respectively, the current and the capacitance of the collector of the trapping input module, C_- and G_- are the broad-band low-noise negative capacitance and conductance implemented by a four-frequency modulation-parametric system (MPS), C_a and G_a are the input conductance and capacitance of the electrometric preamplifier, $U^{(\mp)}$ is the input voltage of the preamplifier without and with MPS regenerative action (the index „–" is used for the regenerative case, when MPS introduces equivalent negative C_- and G_-, while the index „+" is used for the ordinary case without MPS).

As the regeneration coefficients $\mu_C = \dfrac{C_C + C_a}{C_C + C_a - C_-}$ and $\mu_G = \dfrac{G_a}{G_a - G_-}$ are introduced, one can obtain an expression for the coefficient of the gain in the resultant input voltage at the clamps of the preamplifier in the form

$$K = \frac{U^{(-)}}{U^{(+)}} = \frac{\mu_C}{\left(\dfrac{\mu_C}{\mu_G}\right)^2 + \omega_s^2 \tau_{ca}^2} \sqrt{\left(\omega_s^2 \tau_{ca}^2 + \frac{\mu_C}{\mu_G}\right)^2 + \omega_s^2 \tau_{ca}^2 \left(\frac{\mu_c}{\mu_G} - 1\right)^2},$$

where $\tau_{ca} = \dfrac{C_C + C_a}{G_a}$ is the time constant of the system without regeneration, ω_s is the respective spectral frequency of the collector current.

In the case of complete compensation of the collector capacitance C_C, i.e. when $\mu_C \to \infty$, the coefficient K takes the form

$$K_\infty = \mu_G \sqrt{1 + \omega_s^2 \tau_{ca}^2}.$$

It is easy to see that the regeneration effectiveness increases as the initial value of τ_{ca} increases.

The integral effect of using MPS can be expressed and analyzed on the basis of the mean values of the input voltage of the electrometric preamplifier $U_{a,a}^{(\mp)}$, obtained by averaging the modulation period T_0:

$$U_{a,a}^{(-)} = \frac{i_C \mu_G}{G_a} \left\{ \frac{1}{2} - \frac{C_C + C_a}{\mu_C G_a T_0} \mu_G \left[\frac{1}{2} - \frac{\mu_C G_a T_0}{2\mu_G (C_C + C_a)} \right] \right\} \simeq \frac{i_C}{2G_a} \mu_G,$$

because $\dfrac{C_C + C_a}{\mu_C G_a T_0} \mu_G \ll 1$,

$$U_{a,a}^{(+)} = \frac{i_C}{G_a} \left\{ \frac{1}{2} - \frac{C_C + C_a}{G_a T_0} \left[\frac{1}{2} - e^{-\frac{G_a T_0}{2(C_C + C_a)}} \right] \right\} \simeq \frac{i_c}{4(C_C + C_a)} T_0,$$

where $\dfrac{C_C + C_a}{G_a} \gg T_0$.

The ratio of these expressions yields the coefficient indicating the effectiveness of using MPS:

$$K_a = \frac{U_{a,a}^{(-)}}{U_{a,a}^{(+)}} \simeq \frac{2(C_C + C_a)}{G_a T_0} \mu_G.$$

For all practical cases $K_a \gg 1$.

The effect of sensitivity promotion due to the employment of MPS can be proved even more thoroughly as the noise properties are analyzed. As shown above, the signal/noise ratio N does not deteriorate as an MPS is added. In this case the value N can be expressed as:

— for the regeneration case:

$$N^{(-)} = \frac{U_{a,a}^{(-)}}{\sqrt{\overline{U_{N\Sigma}^2}}} \simeq \frac{i_C \mu_G}{2G_a\sqrt{\overline{U_{N\Sigma}^2}}} = \frac{nS_C\phi(W_\beta)\mu_G}{2G_a\sqrt{\overline{U_{N\Sigma}^2}}};$$

— for the non-regeneration case:

$$N^{(+)} = \frac{U_{a,a}^{(+)}}{\sqrt{\overline{U_{Na}^2}}} \simeq \frac{i_C T_0}{4(C_C + C_a)\sqrt{\overline{U_{Na}^2}}} = \frac{nS_C\phi(W_\beta)T_0}{4(C_C + C_a)\sqrt{\overline{U_{Na}^2}}},$$

where $i = nS_C\phi(W_\beta)$, n is the plasma concentration, S_C is the area of the collector of the plasma trapper, $\phi(W_\beta)$ is a function characterizing the distribution of the charged particles by energy, $\overline{U^2}_{N\Sigma}$ is the common quadratic mean of the noise voltage in the regenerative case (when an MPS is in place), $\overline{U^2}_{Na}$ is the quadratic mean of the noise voltage of the preamplifier.

This can be used as a starting point to obtain expressions for the minimum plasma concentration and minimum collector current of the plasma trapper, which can be registered in each of the cases:

$$n_{\min}^{(-)} = \frac{2N^{(-)}G_a\sqrt{\overline{U_{N\Sigma}^2}}}{\mu_G S_C\phi(W_\beta)}, \quad n_{\min}^{(+)} = \frac{4N^{(+)}\sqrt{\overline{U_{Na}^2}}}{S_C\phi(W_\beta)},$$

$$i_{C\min}^{(+)} = \frac{2N^{(+)}(C_C + C_a)}{T_0}\sqrt{\overline{U_{Na}^2}}, \quad i_{C\min}^{(-)} = \frac{2N^{(-)}G_a}{\mu_G}\sqrt{\overline{U_{N\Sigma}^2}}.$$

The noise voltage values $\overline{U_{N\Sigma}^2}$ and $\overline{U_{Na}^2}$ are determined in accordance with the method detailed above (See Exps.(2.121) – (2.138)). The estimates made there can be used to show that there is reliable satisfaction of the following conditions for effectiveness of the utilization of MPS in a solar wind sensor:

$$N^{(-)} > N^{(+)}, \quad n_{\min}^{(-)} < n_{\min}^{(+)}, \quad i_{C\min}^{(-)} < i_{C\min}^{(+)}.$$

Works [222] contains an elaborate theoretical study of the noise properties of the „amplifier-MPS-electrometric amplifier" system.

2.4.7. Implementation and utilization of modulation-parametric one-ports with negative inductance

A one-port with broad-band negative inductance can be implemented in accordance with the theory presented in 2.1 and on the basis of a four-frequency inductive parametric system [153, 217, 218]. There, the maximum absolute value of the negative inductance is equal to $|L_{g-}|_{\max} = L_{eff}\dfrac{Qm_1^2}{1+Qm_2^2}$, where Q is the Q-factor of the oscillating circuit of the inductive parametric modulator, $m_1 = \dfrac{L_1}{L_{eff}}$ and $m_2 = \dfrac{L_2}{L_{eff}}$ are modulation coefficients of the parametric inductance (L_1 and L_2 are the first and second coefficient of the harmonic expansion of the parametric inductance, see (1.25)), L_{eff} is the effective inductance of the parametric element, which is determined as

$$L_{eff} = \frac{\Psi_1}{I_p} = \frac{1}{I_p}\frac{2}{\pi}\int_0^{\Psi} \Psi(t)\cos\omega_p t \; d(\omega_p t), \tag{2.159}$$

where Ψ_1 is the amplitude of the first harmonic of the magnetic flux, which runs through the parametric inductance – $\Psi(t) = \Psi_0 + \sum_{n=1}^{\infty} \Psi_\pi \cos n\omega_p t$, I_p is the amplitude of the pumping current flowing through the parametric inductance – $i_p = I_0 + I_p\cos\omega_p t$, ω_p is the pumping frequency.

Taking into account that $\dfrac{\partial\Psi}{\partial I_0} = \dfrac{\partial\Psi}{\partial i}$ and $\dfrac{\partial\Psi}{\partial I_p} = \dfrac{\partial\Psi}{\partial i_p}\cos\omega_p t$ and using (2.159), we can write down $\dfrac{\partial(I_p L_{eff})}{\partial I_p} = \dfrac{2}{\pi}\displaystyle\int_0^{\pi} \dfrac{d\Psi}{di_p}(t)\cos^2\omega_p t \; d(\omega_p t)$, which yields

$$L_{eff} + I_p\frac{\partial L_{eff}}{\partial I_p} = L_0 + L_2, \tag{2.160}$$

where

$$L_n = \frac{1}{\pi}\int_0^{\pi} \frac{d\Psi}{di_p}(t)\cos n\omega_p t \; d(\omega_p t), \qquad n = 0, 1, 2, \ldots \tag{2.161}$$

Using (2.159) as a point of departure and taking into account (2.160) and (2.161), we determine $I_p\dfrac{\partial L_{eff}}{\partial I_p} = 2L_2$. Ultimately the following formula is obtained for the effective inductance of the parametric element: $L_{eff} = L_0 - L_2$. The result is rather interesting in that the acting resonance inductance of the parametric element (i.e. the equivalent inductance determining the resonance properties of the pumping oscillating circuit) is equal to the average inductance at the operating point L_0 minus the amplitude of the second harmonic in the spectrum of the parametric element L_2. The effective inductance determining the resonance properties of the parametric modulator is highly dependent on the amplitude of the pumping current and on the degree of non-linearity of the parametric inductance. Moreover, this dependence is realized mostly through the properties of the second harmonic in the spectrum of the parametric element.

The characteristic of the ferromagnetic element in the absence of constant external magnetization can be approximated through the polynomial $L = \sum_{n=0}^{\infty} a_n I^{2n}$ [108]; I – current, a_n – coefficients. In the presence of an external magnetizing current I_0, the approximating polynomial assumes the form

$$L_{I_0} = \sum_{r=0}^{\infty} b_r I^r, \quad \text{where} \quad b_r = \sum_{n=r}^{\infty} a_n C_{2n}^r I_0^{2n-r}. \tag{2.162}$$

(2.162) can be used to express the average inductance at the operating point and the amplitude of the second harmonic as:

$$L_0 = \sum_{r=0}^{\infty} \sum_{n=2r}^{\infty} a_n C_{2n}^{2r} I_0^{2(n-r)} I_p^{2r} \frac{1}{2^{2r}} C_{2r}^r, L_2 = \sum_{r=1}^{\infty} \sum_{n=2r}^{\infty} a_n C_{2n}^{2r} I_0^{2(n-r)} I_p^{2r} \frac{C_{2r}^{r-1}}{2^{2r-1}}.$$

The formulae obtained above are used to estimate the possible values of the negative inductance, which is implemented on the basis of a specific inductive parametric element.

Fig.2.52 shows a circuit of a device for obtaining broad-band negative inductance by using ferromagnetic parametric elements [218]. The pumping circuit consists of the capacitance C and the inductive parametric element, represented by two inductances L' and L'', connected in parallel, with ferromagnetic cores in a quasi-saturated condition and coils in parallel planes. The controlling winding L_p of the pumping generator PG is situated in a perpendicular plane with respect to the inductances L' and L''. A significant reduction of the hysteresis and linearization of the characteristic of the ferromagnetic core as a result of the orthogonal polarization are realized in the direction of the signal field [108]. Negative inductance is obtained at clamps 1–1 of the load, and its equivalent noise voltage is 2–3 orders down from the equivalent noise voltage of a negative inductance obtained with the aid of an impedance converter with vacuum tubes or transistors, and its absolute value is frequency-independent in a broad video-frequency band. A miniature ferromagnetic toroid can be used for implementation of an inductive parametric element. Owing to the orthogonal polarization of the ferromagnetic toroid with respect to the signal field, a substantial suppression of the Barkhausen noise occurs and so does an almost harmonic change in the inductance of the parametric element.

Fig.2.53 shows a circuit of a device for obtaining broad-band negative inductance, where the parametric element is a $p-n-p-n$ structure simulated by a complementary pair of bipolar transistors [218]. The circuit differs from that in Fig.2.14 in that the capacitance used in the pumping oscillating circuit

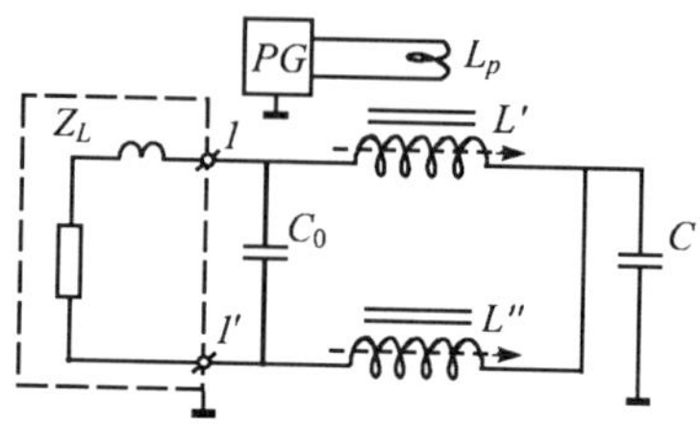

Fig. 2.52. Parametric device for obtaining broadband negative inductance by using ferromagnetic parametric elements

is a semiconductor diode with automatic bias meant „to rectify" the amplitude frequency characteristic of the circuit and thereby to improve considerably the conditions for introducing energy into the signal circuit. (This issue is tackled theoretically in 3.2.) The volt-ampere characteristic of the $p-n-p-n$ structure and the dependence of its natural differential inductance on the mode current are shown in Fig.2.54. The system as a whole is a one-port with a broad-band negative inductance, whose absolute inductance exceeds many times the natural effective inductance of the generating system in an even frequency band from $\sim$ 0 Hz to the maximum frequency equal to 5 – 10% of the pumping frequency. The circuit in Fig.2.53 is also marked, first, by the fact that due to the existence of a negative differential resistance of the $p-n-p-n$ structure a high Q-factor of the oscillating circuit, hence a sizable absolute value of the obtained negative inductance can be ensured, and, second, by the possibility to build the device by employing microelectronic means and uniform technology.

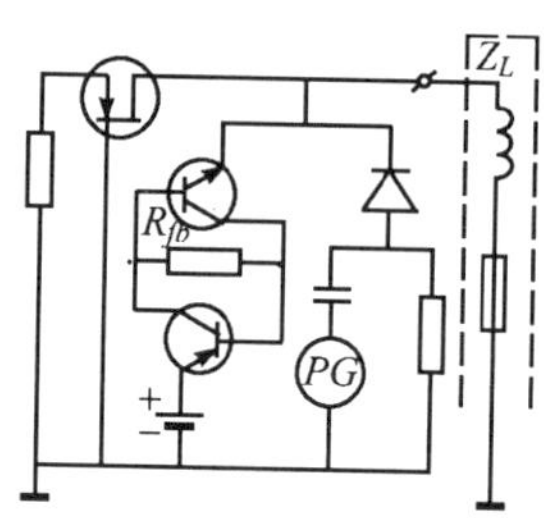

Fig. 2.53. Parametric device for obtaining broadband negative inductance by using a two-transistor $p-n-p-n$ structure as a parametric inductive element

Fig.2.55 shows the experimental frequency characteristic of the negative inductance introduced into the input circuit (in series with the load Z_L) of the four-frequency parametric system from Fig.2.53 at different relative detunings ν. The frequency independence of the negative inductance in a broad frequency bandwidth of the input signals can be seen. The maximum absolute value can be achieved at positive detuning $\nu = \dfrac{1}{2Q}$.

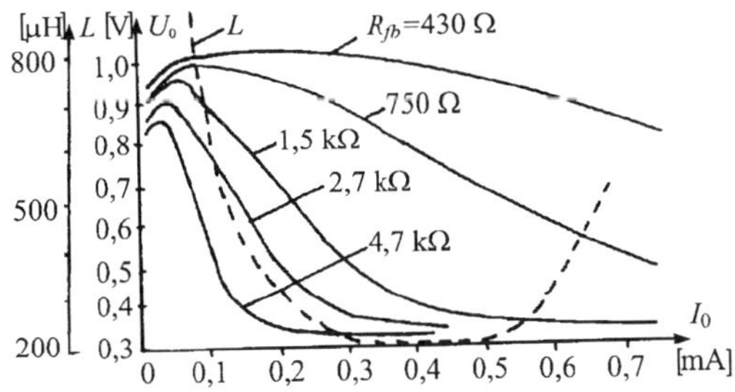

Fig. 2.54. Experimental volt-ampere characteristic of the $p-n-p-n$ structure shown in Fig.2.53 and dependence of its natural differential inductance on the mode current

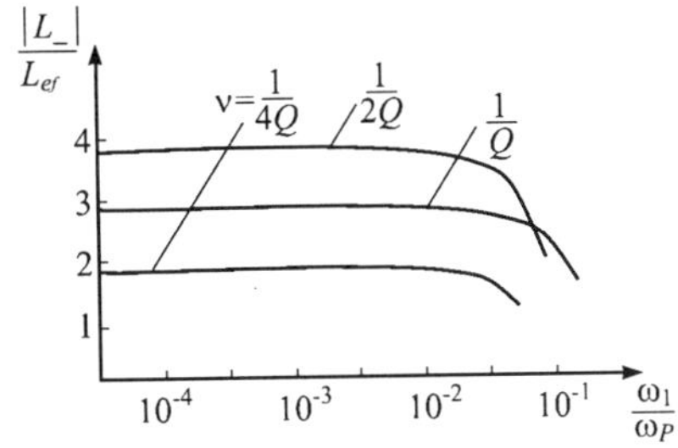

Fig. 2.55. Experimental frequency characteristic of the implemented negative inductance obtained by using the device in Fig. 2.53

The obtained broad-band negative inductances can be applied in various electronic and magnetic devices for broad-band frequency correction. As the input inductances of receiving systems of the type of the inductive video sensor are compensated, the sensitivity of these systems can be increased only as a result of the compensation by 10 to 20 dB and the transmission frequency bandwidth of the system can be expanded by 1 to 2 orders. The sensitivity of these systems can be additionally increased, if the four-frequency parametric system is simultaneously used for low-noise parametric amplification.

2.5. Application of self-oscillating one-ports with controllable parameters

The possibility to control the effective impedances in a non-autonomous generator, presented in 2.2 and 2.3, can be applied in the development of a number of methods and original devices for signal conversion and processing.

The established effect of conversion of a frequency with a different conversion coefficient through main and image channel can be used to develop a method of frequency conversion with image channel suppression. The input signal is communicated to a generator with a non-linear oscillating circuit, whose generation frequency has been shifted with the help of an external synchronizing source of electric oscillations in a definite section of the equivalent synchronization band. The initial converted signal is removed from the generator.

Fig.2.56 shows an electric circuit of a frequency converter with image channel suppression.

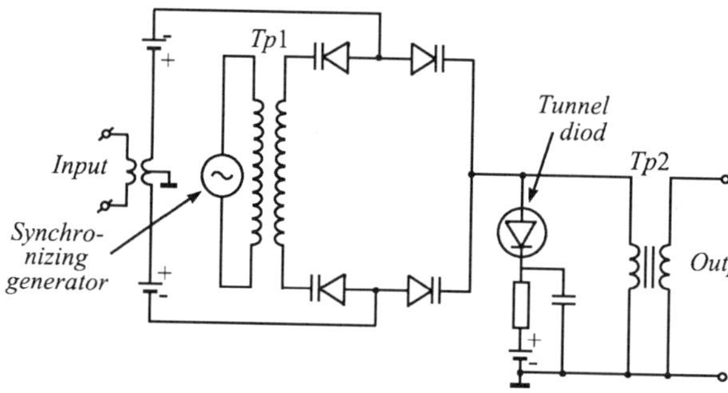

Fig. 2.56. Generator frequency converter with image channel suppression

A frequency converter with image channel suppression, whose circuit is provided in Fig.2.57, can be applied in the SHF centimeter and millimeter range.

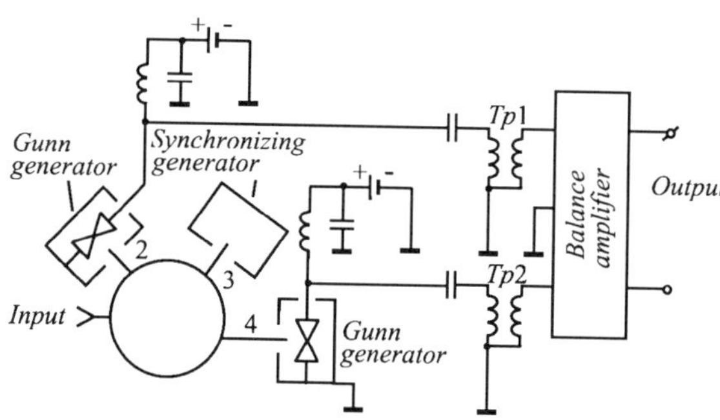

Fig. 2.57. SHF generator frequency converter with image channel suppression

The properties of a non-autonomous generator are used to develop a method and offer original devices for the implementation of parametric elements, for modulation and frequency correction, which can be applied in parametric video amplifiers, in modulators of electric signals in broad frequency bandwidth, for broad-band frequency correction seeking to increase the sensitivity of receiver systems of the type of a capacitive or inductive video sensor, etc. A generator with a nonlinear oscillating circuit is used. A signal, coming from a local source of electric oscillations, is transmitted to it . The frequency of the signal is either close to the frequencies of the self-oscillations or a variable one (frequency-modulated signal) and the signal intensity is sufficient to synchronize the oscillations in the generator.

Fig.2.58 exhibits a circuit of a device for parametric capacitance implementation consisting of a transistor generator with a tapped-coil coupling. The natural generation frequency is determined by the inductance L and the effective capacitance of a balanced bridge circuit including four semiconductor diodes $D_1 - D_2$, operating in a turned-off condition. When the local source sets a voltage that is frequency-modulated in accordance with a sinusoid law (with inconsiderable modulation depth), with a carrier frequency close to the natural frequency of the generator en9uring its synchronization, a capacitance changing in accordance with a sinusoid law occurs at the input clamps of device 1–2 with respect to an external low-frequency signal. As the equivalent capacitance can assume negative values as well, the major indicator of the parametric element – the modulation depth of the parameter – can assume quite high values, and it can even be higher than unity. When the external source of alternating voltage is not frequency modulated, but synchronizes the oscillations in the generator, its frequency can be controlled so as to retune the equivalent capacitance of the device in a broad range and thereby to effect frequency correction of the characteristics of receiving systems of the type of a capacitive video sensor. In practice the equivalent capacitance can undergo a relative change in the range $\frac{\Delta C}{C_0} = 2 \div 5$, where C_0 is the capacitance at the operating point.

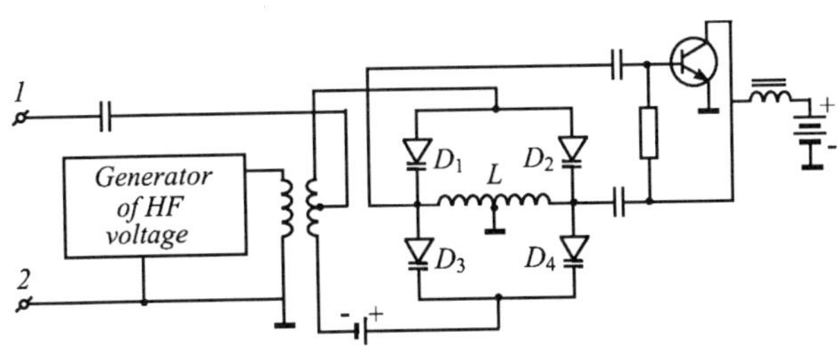

Fig. 2.58. Device for implementation of parametric capacitance with considerable modulation depth

As the parametric element obtained in this way is used with familiar circuits, low-frequency parametric amplifiers with considerable sensitivity and high amplifying coefficient in the video frequency bandwidth can be realized. If the proffered device is used as a modulator of electric signals, it may result in high values of the modulation depth, practically restricted only by the requirement that there should be no self-excitation in the general circuit.

It is noteworthy that a parametric element of one type can serve as a basis for obtaining a parametric element of the opposite type. This possibility is illustrated in Fig.2.59, which shows how a non-autonomous generator with a double transistor simulation of a $p-n-p-n$ structure can be used to obtain a capacitive parametric

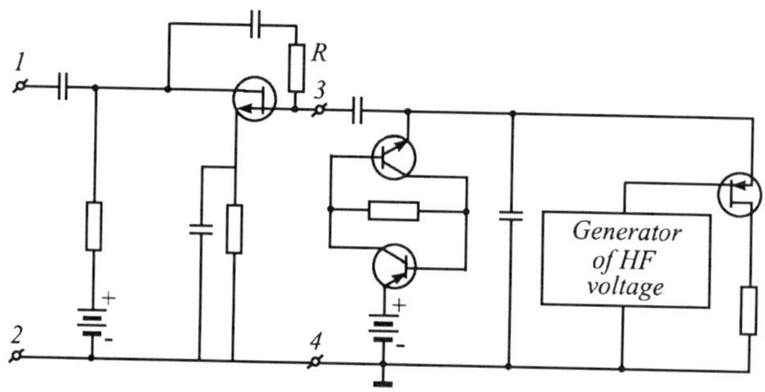

Fig. 2.59. Non-autonomous generator with a double transistor simulation of a $p-n-p-n$ structure used to obtain a capacitive parametric element

element. The equivalent capacitance at clamps 1 – 2 changes in step with the alteration of the equivalent inductance at clamps 3 – 4.

The presentation under 2.2 can serve as a basis for developing a method of converting and amplifying frequency modulated signals, consisting in the following. The frequency modulated signal is fed to a generator with a nonlinear oscillating circuit. The generator also receives a signal from a local source of alternating electric oscillations. The initial amplitude-modulated signal, already converted and amplified, is removed from the same nonlinear oscillating circuit of the generator; besides, its carrier frequency corresponds to the frequency of the local source of alternating electric oscillations. The advantages of the offered method of conversion and amplification of frequency modulated signals are: first, provision of high sensitivity to the level of the frequency modulated signal subject to conversion; second, a possibility for a substantial downward shift of the carrier frequency along the frequency axis; and third, amplification of the signal with a simultaneous transformation of its spectrum.

Fig.2.60 shows an electric circuit of a generator converting frequency-modulated signals into amplitude-modulated ones. The circuit is designed by using the two-transistor model of a $p-n-p-n$ structure (Fig.2.14). Here the input frequency modulated signal plays the part of a synchronizing action, while the low-frequency signal, with respect to which the generator changes its impedance, is the carrier oscillation of the output amplitude-modulated signal. The practically equivalent inductance of the $p-n-p-n$ structure can change within a 5% frequency bandwidth of synchronization from -2 mH to $+7$ mH. Since the signal fed to the input clamps 1 – 2 is frequency-modulated, the equivalent impedance of the $p-n-p-n$ structure changes in step with the changes in the frequency of the input signal. The amplitude-modulated signal occurring at clamps 3 – 4 has a carrier frequency equal to the frequency of the external source of electric low-frequency oscillations. The high sensitivity and amplification of the signal are realized and ensured as a consequence of the high sensitivity and drastic change

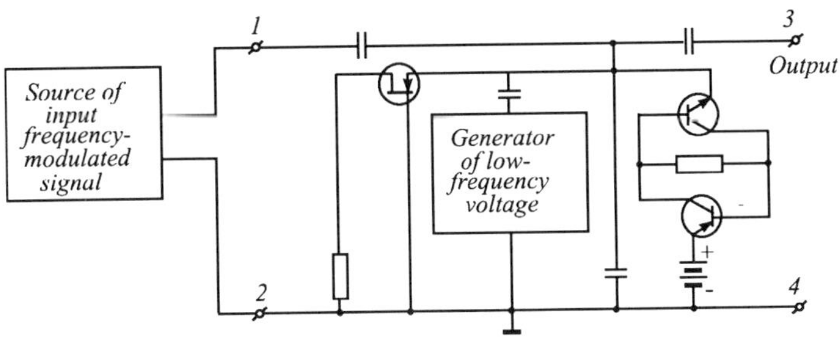

Fig. 2.60. Generator converter of frequency-modulated signals with a double transistor $p-n-p-n$ structure

in the equivalent impedance of the $p-n-p-n$ structure in the synchronization bandwidth.

Fig.2.61 displays another circuit of the converter of frequency-modulated signals. The generator is built by using a tunnel diode. The nonlinear oscillating circuit of the generator is made up of the inductance L and the effective capacitance of the balanced bridge circuit consisting of four „capacitive" semiconductor diodes $D_1 - D_4$, functioning in a turned-off state. For example, when semiconductor diodes with $C(0) = 400$ pF are used, the equivalent capacitance of the generator with respect to the voltage of the high-frequency signal can vary within the 5% frequency bandwidth of synchronization, for example from -250 pF to $+1300$ pF.

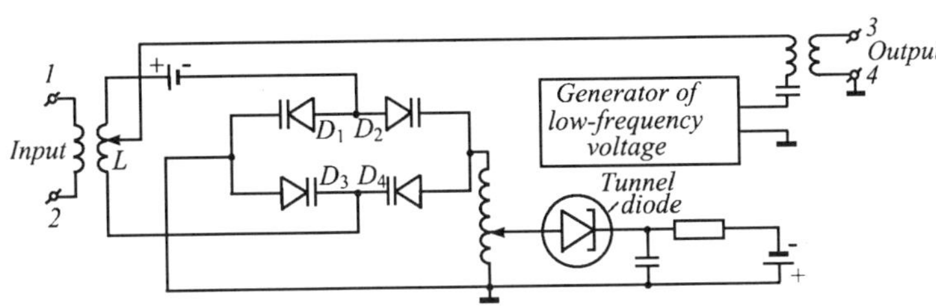

Fig. 2.61. Generator converter of frequency-modulated signals with a tunnel diode and a four-diode bridge balanced circuit

The established effect of frequency selectivity in the case of equivalent impedance changes in a generator one-port with asynchronous action, is used as a basis for proposing a method of spectral selectivity, a method of processing Doppler signals and spectrums, and for developing analyzers of the speed of moving objects.

The spectral selectivity method consists in that the electromagnetic signal with frequency (spectral) shift from a given carrier frequency, which is under investigation, is mixed with a base monochromatic electromagnetic signal with the same carrier frequency, and simultaneously the two signals are mixed with a third harmonic electromagnetic signal, whose frequency changes in accordance with a sawtooth law within the frequency range of the expected frequency (spectral) shift of the electromagnetic signal under examination. Additionally, the instant intensity of the third electromagnetic signal is indicated and compared with its instant frequency.

The method of Doppler spectrum processing renders the spectral selectivity method concrete in its applications to Doppler systems. The Doppler spectrum can be studied by acting on a SHF generator with a swept low-frequency signal in the frequency area of the expected Doppler frequencies.

This method is visually presented in Fig.2.62, showing an autodyne analyzer of the speed of moving objects [223]. Let us assume that in this case ω_g is the frequency of the signal emitted by the autodyne, and ω_b is the frequency of the signal reflected by a moving object. The sweep-frequency generator sets a signal with alternating frequency ω_s in the range of the expected Doppler shifts of the SHF probing signal. At the moment, when the condition $\omega_s = |\omega_g - \omega_b|$ is realized, i.e. when the instant frequency of the swept-generator coincides with the Doppler frequency, a considerable change occurs in the equivalent impedance of the active element of the autodyne (in Fig.2.62 – SHF avalanche-drift diode), which causes a negative output voltage impulse. The latter is recorded by an cathode-ray indicator, to whose horizontal plates a sweep from the swept-generator is fed. The cathode-ray indicator can be graduated directly in Doppler frequencies, so as to reflect the input Doppler spectrum. Owing to the high frequency selectivity of the effect, there is reliable reflection of each spectral Doppler component on the screen. This is a way to realize a tunable Doppler frequency filter with an equivalent quality factor $Q > 1000$.

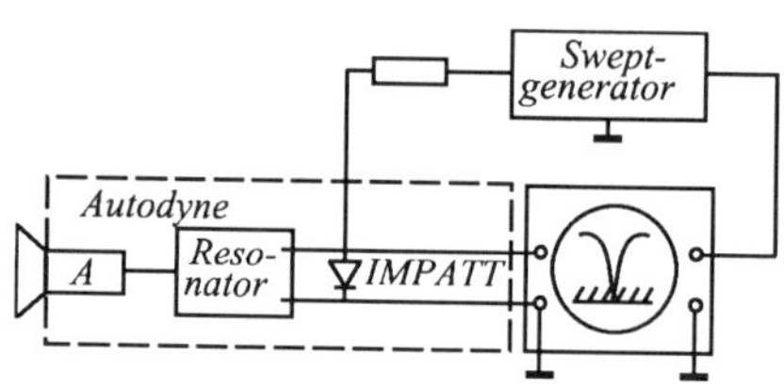

Fig. 2.62. Autodyne analyzer of the speed of moving objects

Fig.2.63 presents an analogous autodyne analyzer of the speed of moving objects but it indicates the Doppler frequency in a digital way [224].

For the purpose of actual modeling, tuning and testing of Doppler autodyne analyzers of the speed of moving objects and of other Doppler signals and systems, a one-band SHF modulator was invented. Its schematic diagram is shown in Fig.2.64 [225]. At this point we shall illustrate the fact of frequency bands separation from modulation products and the second harmonic suppression.

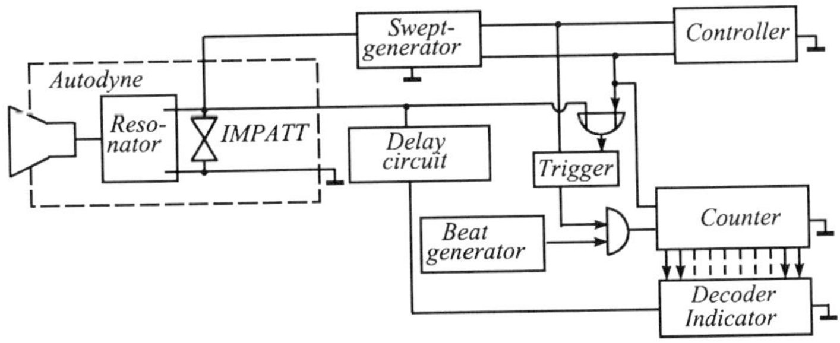

Fig. 2.63. Autodyne analyzer of the speed of moving objects with digital indication

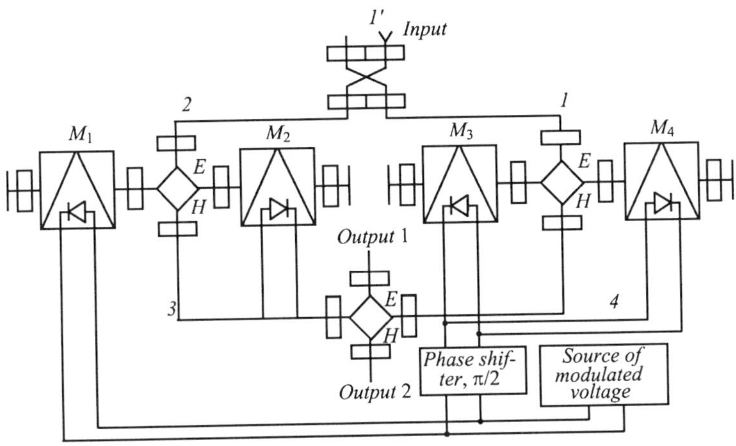

Fig. 2.64. Single-sideband SHF modulator

If we assume that the carrier SHF oscillation reaching signal arm *1'* (Fig.2.64) is $\sin\omega_0 t$ (the amplitude is of no importance for this analysis, so it is omitted), the oscillations at the different points marked on the diagram in Fig.2.64 will be as follows:

Point *1* – $\sin\omega_0 t$; point *2* – $\sin\left(\omega_0 t - \frac{\pi}{2}\right)$ – a consequence of the properties of the slot bridge hybrid.

Points M_1 and M_2 of one of the T-bridges are subjected to the following oscillations:

$$\text{Point } M_1 \begin{cases} \sin\left(\omega_0 t - \pi - \frac{\pi}{2}\right) & \text{– As a consequence of the } T\text{-bridge properties;} \\ \sin\omega_D t & \text{– The low-frequency modulating voltage;} \end{cases}$$

Point M_2 $\begin{cases} \sin\left(\omega_0 t - \dfrac{\pi}{2}\right) & \text{– As a consequence of the } T\text{-bridge properties;} \\ \sin(\omega_D t + \pi) & \text{– The low-frequency modulating voltage with a phase shift of } \pi \text{ compared to that of point } M_1. \end{cases}$

As a result of the modulation, the following oscillations occur at points M_1 and M_2:

$$\text{Point } M_1 \begin{cases} \cos\left(\omega_0 t - \omega_D t - \pi - \dfrac{\pi}{2}\right) \\ \cos\left(\omega_0 t + \omega_D t - \pi - \dfrac{\pi}{2}\right); \end{cases}$$

$$\text{Point } M_2 \begin{cases} \cos\left(\omega_0 t - \omega_D t - \pi - \dfrac{\pi}{2}\right) \\ \cos\left(\omega_0 t + \omega_D t - \dfrac{\pi}{2} + \pi\right); \end{cases}$$

Analogously for points M_3 and M_4:

They are subjected to the following oscillations:

$$\text{Point } M_3 \begin{cases} \sin(\omega_0 t - \pi) \\ \sin\left(\omega_D t - \dfrac{\pi}{2}\right) \end{cases}; \quad \text{Point } M_4 \begin{cases} \sin \omega_0 t \\ \sin\left(\omega_D t + \dfrac{\pi}{2}\right) \end{cases}.$$

As a result of the modulation, oscillations of the following type occur at points M_3 and M_4.

$$\text{Point } M_3 \begin{cases} \cos\left(\omega_0 t - \omega_D t - \pi + \dfrac{\pi}{2}\right) \\ \cos\left(\omega_0 t + \omega_D t - \pi - \dfrac{\pi}{2}\right); \end{cases}$$

$$\text{Point } M_4 \begin{cases} \cos\left(\omega_0 t - \omega_D t - \dfrac{\pi}{2}\right) \\ \cos\left(\omega_0 t + \omega_D t + \dfrac{\pi}{2}\right). \end{cases}$$

The frequency consistent oscillations in the H-arms of the T-bridges are summed up and at the output there is no phase shift:

$$\text{At point } 3 \begin{cases} \cos\left(\omega_0 t - \omega_D t + \dfrac{\pi}{2}\right) \\ \cos\left(\omega_0 t + \omega_D t + \dfrac{\pi}{2}\right); \end{cases}$$

$$\text{At point } 4 \begin{cases} \cos\left(\omega_0 t - \omega_D t - \dfrac{\pi}{2}\right) \\ \cos\left(\omega_0 t + \omega_D t + \frac{\pi}{2}\right). \end{cases}$$

A separation of the frequency bands takes place at the output in the E- and H-planes of the output T-bridge:

Output *1* – $\cos\left[(\omega_0 + \omega_D)t + \dfrac{\pi}{2}\right]$;

Output *2* – $\cos(\omega_0 - \omega_D)t$.

We shall show in an analogous way that the second (parasitic, undesired) modulation band is suppressed by the single-band modulator:

Point *1* – $\sin\omega_0 t$, point M_1 $\begin{cases} \sin\left(\omega_0 t - \pi - \dfrac{\pi}{2}\right) \\ \sin(2\omega_D t); \end{cases}$

Point *2* – $\sin\left(\omega_0 t - \pi - \dfrac{\pi}{2}\right)$, point M_2 $\begin{cases} \sin\left(\omega_0 t - \dfrac{\pi}{2}\right) \\ \sin(2\omega_D t + 2\pi). \end{cases}$

As an outcome of the modulation:

Point M_1 $\begin{cases} \cos\left(\omega_0 t - 2\omega_D t - \pi - \dfrac{\pi}{2}\right) \\ \cos\left(\omega_0 t + 2\omega_D t - \pi - \dfrac{\pi}{2}\right); \end{cases}$

Point M_2 $\begin{cases} \cos\left(\omega_0 t - 2\omega_D t - \dfrac{\pi}{2} - 2\pi\right) \\ \cos\left(\omega_0 t + 2\omega_D t - \dfrac{\pi}{2} + 2\pi\right). \end{cases}$

Analogously -

Point M_3 $\begin{cases} \sin(\omega_0 t - \pi) \\ \sin(2\omega_D t - \pi) \end{cases}$, Point M_4 $\begin{cases} \sin\omega_0 t \\ \sin(2\omega_D t + 2\pi). \end{cases}$

As a result of the modulation:

Point M_3 $\begin{cases} \cos(\omega_0 t - 2\omega_D t - \pi + \pi) \\ \cos(\omega_0 t + 2\omega_D t - \pi - \pi); \end{cases}$

Point M_4 $\begin{cases} \cos(\omega_0 t - 2\omega_D t - \pi) \\ \cos(\omega_0 t + 2\omega_D t + \pi). \end{cases}$

In the H-arms of the T-bridges:

Point *3* $\begin{cases} \cos\left(\omega_0 t - 2\omega_D t + \dfrac{\pi}{2}\right) + \cos\left(\omega_0 t - 2\omega_D t - \dfrac{\pi}{2}\right) = 0 \\ \cos\left(\omega_0 t + 2\omega_D t + \dfrac{\pi}{2}\right) + \cos\left(\omega_0 t + 2\omega_D t - \dfrac{\pi}{2}\right) = 0; \end{cases}$

Point *4* $\begin{cases} \cos(\omega_0 t - 2\omega_D t) + \cos(\omega_0 t - 2\omega_D t - \pi) = 0 \\ \cos(\omega_0 t + 2\omega_D t - 2\pi) + \cos(\omega_0 t + 2\omega_D t + \pi) = 0. \end{cases}$

It is evident from the presentation above that there are no parasitic oscillations from the second modulation band that penetrate the H-arms of the T-bridges (Point *3* and Point *4*), and, respectively, there are none that pass on to the output T-bridge.

The experimental study indicates that in practice the interpenetration between the separate frequency bands has suppression in the order of 40 – 60 dB. The suppression of the parasitic second frequency modulation band in comparison with its intensity when it originates in the modulator is of the same order.

The single-band modulator can be used in special Doppler trainers, in special single-band radars, in the systems for radio counteraction, etc.

CHAPTER 3.

NONLINEAR RESONANCE IN RADIOPHYSICAL SYSTEMS. IMPLEMENTATION OF PARAMETRIC ONE-PORTS. PECULIARITIES OF THE UTILIZATION OF SEMICONDUCTOR STRUCTURES IN RADIOPHYSICAL SYSTEMS

3.1. Nonlinear resonance in an oscillating circuit with a p-n junction of a semiconductor diode

Nonlinear circuits where the role of a nonlinear reactance is played by a semiconductor diode are widely applied in various radiophysical devices: low-noise parametric amplifiers and converters, generators, highly stable frequency multipliers and dividers, frequency modulators, receivers with electronic tuning, trigger circuits, parametrons, for creating non-autonomous phase-impulse multistable elements, etc. [25, 68, 78, 85, 107, 168, 223].

Nonlinear resonance in an oscillating circuit with a $p-n$ junction in a mode of forced external negative bias has been widely studied and properly described in the literature [51, 78, 227]. This section will dwell on the peculiarities of the nonlinear resonance in an oscillating circuit with a $p-n$ junction in an automatic bias regime, i.e. when the negative constant bias is obtained automatically with the assistance of a RC-filter at the expense of inconsiderable direct currents flowing through the diode (Fig.3.1) [78, 228, 230]. This type of an oscillating circuit is frequently used in practice for a number of reasons – stable characteristics of the radiophysical devices, compact design, energy and noise optimization, expansion of the functional capacities, etc.

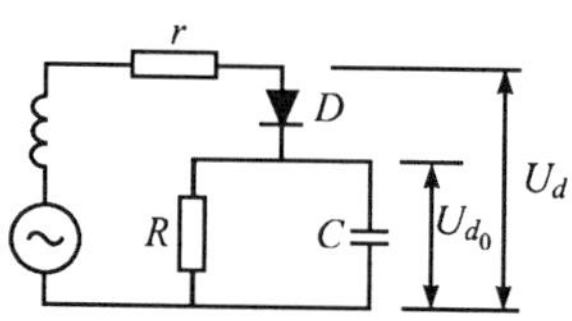

Fig. 3.1. Series oscillating circuit containing a semiconductor diode in an automatic bias regime

Usually, when nonlinear resonance is studied, only the barrier (layer) capacitance is taken into account, and in the case of automatic bias only the static differential capacitance at the operating point is accounted for, but this holds only for infinitesimal voltage amplitudes. Moreover, the volt-ampere and volt-coulomb characteristic of the $p-n$ junction is approximated with a polynomial including a small number of terms, which, in this case, provides rather a crude approximation.

This section will present a qualitative investigation of the peculiarities of a nonlinear resonance in a series oscillating circuit as shown in Fig.3.1. The equivalent linearization method will be used [97]. This method is convenient for the case under review, since we shall account for both the barrier capacitance and the diffusive one by employing the major expressions stemming from the fundamental diffusion theory and reflecting, in a most comprehensive way, the real properties of the $p-n$ junction. As we generalize the equivalent linearization method, we shall study the influence of the higher harmonics on the nonlinear resonance. We shall note that in qualitative and quantitative terms the peculiarities of the nonlinear resonance in a series and parallel oscillating circuit are fully identical with an accuracy to the extent of small quantities of the second gradation order.

3.1.1. Effective (equivalent) parameters of the nonlinear oscillating circuit

It is assumed that the voltage at the semiconductor diode changes in accordance with a harmonic law

$$U = U_{dm} \cos \omega t. \tag{3.1}$$

The analysis is based on the idea that for a given amplitude of the alternating voltage U_{dm} and invariable other conditions, the resonance frequency of the circuit is determined by the effective (equivalent) capacitance of the diode, defined as

$$C_{ef} = \frac{Q_1}{U_{dm}}, \tag{3.2}$$

where Q_1 is the amplitude of the first harmonic of the charge in the circuit.

The equivalent circuit of the semiconductor diode is shown in Fig.3.2, where R_s is the resistance of the semiconductor volume, while the other components are expressed in accordance with the fundamental diffusion theory [168] in the following manner:

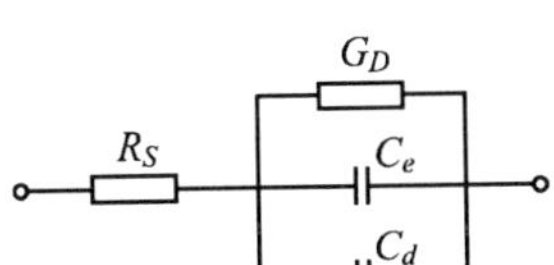

Fig. 3.2. Equivalent circuit of a semiconductor diode

active differential conductance of the junction

$$G_D = \frac{e(I+I_s)}{mkT} \frac{\sqrt{\sqrt{1+\omega^2\tau^2}+1}}{\sqrt{2}}; \tag{3.3}$$

diffusive capacitance

$$C_d = \frac{e(I+I_s)}{mkT}\frac{\tau}{\sqrt{2}\sqrt{\sqrt{1+\omega^2\tau^2}+1}}; \tag{3.4}$$

barrier capacitance of the junction

$$C_e = \frac{C(0)}{\left(1-\dfrac{U}{\varphi_k}\right)^{\gamma}} = C(0)\left[1-\frac{mkT}{e\varphi_k}\ln\left(\frac{I}{I_s}+1\right)\right]^{-\gamma}, \tag{3.5}$$

where e is the electron charge, m is a coefficient reflecting the influence of the generation-recombination processes of the carriers in the $p-n$ junction [231], $C(0)$ is the value of the barrier capacitance at zero bias, I is the direct current, I_s, φ_k, and τ are respectively saturation current, contact difference of potentials and lifetime of the minor carriers, $\gamma = 1, \frac{1}{2}$ or $\frac{1}{3}$ [168].

(3.4) and (3.5) are used for obtaining an expression for the charge at the $p-n$ junction

$$q = \frac{C(0)\varphi_k^{\gamma}(\varphi_k+U_{d_0})^{1-\gamma}}{1-\gamma}\left[1-\left(1-\frac{U}{\varphi_k+U_{d_0}}\right)^{1-\gamma}\right] + \frac{C_D}{\alpha}\left(e^{\alpha U_d}-1\right), \tag{3.6}$$

where

$$\alpha = \frac{e}{mkT},\quad C_D = \frac{eI_s}{mkT}\frac{\tau}{\sqrt{2}\sqrt{\sqrt{1+\omega^2\tau^2}+1}},\quad U_d = -U_{d_0} + U_{dm}\cos(\omega t).$$

As (3.6) is expanded harmonically and (3.2), (3.4) and (3.5) are taken into account, the effective capacitance is determined in the following form

$$C_{ef} = C_0 - C_2 + \frac{2C_D}{\alpha U_{dm}}e^{-\alpha U_{d_0}}J_1(\alpha U_{dm}), \tag{3.7}$$

where C_0 and C_2 are the constant component and the amplitude of the second harmonic in the Fourier spectrum of the barrier capacitance (3.5), expressed in the general case as

$$C_n = \frac{C(0)}{2\pi}\int_{-\pi}^{\pi}\left(1-\frac{U_{dm}\cos\omega t}{\varphi_k+U_{d_0}}\right)^{-\gamma}\cos(n\omega t)d(\omega t), \tag{3.8}$$

$J_1(\alpha U_{dm})$ is a modified Bessel function of the first order (the designation normally accepted in the literature is $I_n(x)$, but here we use the symbol I_n for the amplitudes of the spectral current components).

The constant bias of the $p-n$ junction is determined by the transcendental equation

$$U_{d_0} = RI_s e^{-\alpha U_{d_0}} J_0(\alpha U_{dm}), \tag{3.9}$$

where $J_0(\alpha U_{dm})$ is a modified Bessel function of the zero order.

As the amplitude of the alternating voltage increases, the quantity C_{ef} as a whole decreases, since the operating point U_{d_0} shifts along the volt-farad characteristic in the direction of smaller values of the differential capacitance (Fig.3.3). Under these conditions the transmission coefficient of the diode increases

$$K_D = \frac{U_{d_0}}{U_{dm}}, \tag{3.10}$$

and tends to unity, which entails a reduction in the angle of the segment Θ (Fig.3.3). At the same time, however, there is an increase in the amplitude of the direct current impulse, whose constant component maintains the level of the automatic bias. Non-basic carriers, conditioning an effective modulation of the diffusion capacitance of the $p-n$ junction, are injected in the base of the diode. In this way, as the amplitude of the alternating voltage at the diode increases, the "participation" of the diffusion capacitance of the p-n junction also goes up, which, as we shall see later, results in a number of peculiarities.

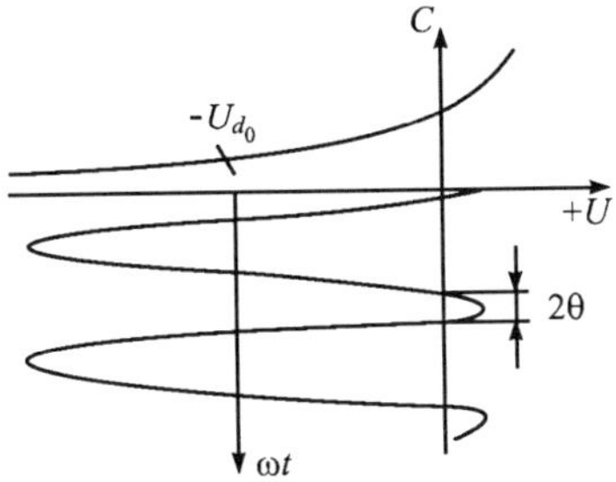

Fig. 3.3. Volt-farad characteristic of a $p-n$ junction

In an automatic bias mode the semiconductor diode represents not only a pulsative capacitance, but a pulsative conductance as well. The effective (equivalent) resistance of the $p-n$ junction in this mode is defined as

$$R_{def} = \frac{\sqrt{2}}{\sqrt{\sqrt{1+\omega^2\tau^2}+1}} \frac{RU_{dm}J_0(\alpha U_{dm})}{2U_{d_0}J_1(\alpha U_{dm})}. \tag{3.11}$$

3.1.2. The influence of higher harmonics on nonlinear resonance

The analysis will be conducted in a more general form.

Let us assume that there is a resonance system containing a nonlinear capacitive element with a volt-coulomb characteristic

$$q = f(U). \tag{3.12}$$

It can be assumed that in the first approximation, the voltage at the nonlinear capacitive element is in the form

$$U = U_m \cos \omega t. \tag{3.13}$$

As (3.13) is substituted in (3.12) and a harmonic expansion is carried out, the following is obtained:

$$q = Q_0 + \sum_{n=1}^{\infty} q_n = Q_0 + \sum_{n=1}^{\infty} Q_n \cos(n\omega t), \tag{3.14}$$

where

$$Q_0 = \frac{1}{2\pi} \int_{-\pi}^{\pi} f(U) d(\omega t), \quad Q_n = \frac{1}{\pi} \int_{-\pi}^{\pi} f(U) \cos(n\omega t) d(\omega t).$$

The effective capacitance in the first approximation can be determined by using a formula analogous to (3.2).

The following is done for the purpose of determining the second approximation: As (3.14) is taken into account, the expression for the voltage at the capacitive element is specified

$$U_{II} = U_m \cos(\omega t) - \sum_{n=2}^{\infty} \frac{dq_n}{dt} Z(n\omega), \tag{3.15}$$

where $Z(n\omega)$ is the impedance of the entire remaining (linear) part of the resonance system. This leads to a criterion for the applicability of the recounted method. The term $U_m \cos(\omega t)$ is basic among the variable components in a system with a sufficiently high quality factor, while the remaining terms are only of a correcting character. Hence the applicability of the method is subject to the following sufficient condition:

$$n\omega Q_n |Z(n\omega)| \ll U_m, \quad (n = 2, 3, \ldots). \tag{3.16}$$

Further on, expression (3.15) will be used for specifying the functional dependence (3.12), given the condition provided by (3.13).

We have

$$q_{II} = f[U_m \cos(\omega t) - \xi] \simeq f(U) - \xi f'(U), \tag{3.17}$$

where $\xi = \sum_{n=2}^{\infty} \frac{dq_n}{dt} Z(n\omega)$, $f'(U)$ is a derivative of the charge function (3.12) with respect to the voltage U at $U = U_m \cos \omega t$.

As a harmonic expansion is conducted, the function $f'(U)$ is expressed in the following way:

$$f'(U) = A_0 + \sum_{n=1}^{\infty} A_n \cos(n\omega t), \tag{3.18}$$

which results in

$$\begin{aligned} \xi f'(U) &= \sum_{n=2}^{\infty} \frac{dq_n}{dt} Z(n\omega) \left[A_0 + \sum_{n=1}^{\infty} A_n \cos(n\omega t) \right] \\ &= \sum_{n=2}^{\infty} I_n |Z(n\omega)| \sin[(n\omega t) + \varphi(n\omega)][A_0 + \sum_{n=1}^{\infty} A_n \cos(n\omega t)], \end{aligned} \tag{3.19}$$

where

$$I_n = n\omega Q_n, \tag{3.20}$$

$|Z(n\omega)|$ and $\varphi(n\omega)$ are, respectively, the module and the phase of the impedance of the linear part of the resonance system.

After certain transformations, the expression (3.19) is presented in the following way:

$$\begin{aligned} \xi f'(U) = &\frac{1}{2} \sum_{k=1}^{\infty} \left\{ \sum_{n=1}^{\infty} I_{n+k} A_n |Z[(n+k)\omega]| \sin[\varphi(n+k)\omega] \right. \\ &\left. + \sum_{n=2}^{\infty} I_n A_{n+k} |Z(n\omega)| \sin[\varphi(n\omega)] \right\} \cos(k\omega t) \\ &+ \frac{1}{2} \sum_{k=1}^{\infty} \left\{ \sum_{n=1}^{\infty} I_{n+k} A_n |Z[(n+k)\omega]| \cos[\varphi(n+k)\omega] \right. \\ &\left. - \sum_{n=2}^{\infty} I_n A_{n+k} |Z(n\omega)| \cos[\varphi(n\omega)] \right\} \sin(k\omega t) + \text{harmonics}. \end{aligned} \tag{3.21}$$

Now we can determine the effective capacitance of the resonance system in the second approximation. As (3.14) and (3.21) are taken into account, the following is obtained

$$C_{ef}^{II} = \frac{1}{U_m} \sqrt{F^2 + G^2}, \tag{3.22}$$

where

$$F = Q_1 - \frac{1}{2} \sum_{m=2}^{\infty} I_n |Z(n\omega)| \sin[\varphi(n\omega)](A_{n-1} + A_{n+1}), \tag{3.22a}$$

$$G = \frac{1}{2}\sum_{n=2}^{\infty} I_n|Z(n\omega)|\cos[\varphi(n\omega)](A_{n-1} + A_{n+1}). \tag{3.22b}$$

Hence, it follows that the influence of the higher harmonics of the capacitive current is manifested in that the effective capacitance of the system becomes a complex quantity,

$$\varphi_{C_{ef}} = \operatorname{arctg}\frac{G}{F}, \tag{3.23}$$

even though it is assumed that the nonlinear element in the system is of a purely capacitive nature and there are no losses in it. The nonlinear element „transfers" energy from the basic tone to the higher harmonics and thus it acts as a generator of harmonics, which is taken into account in the second approximation. The response of the system, as the active losses of the higher harmonic frequencies are accounted for, consists in a reverse action on the effective parameter of the system, which is demonstrated by the appearance of a phase in the effective capacitance. This effect can be qualified as one of the realizations of the general principle of reversibility of the modulation parametric interactions formulated in 1.1. One can see from Exp. (3.23), after taking into account (3.22a) and (3.22b), that the condition $\varphi_{C_{ef}} > 0$ is always satisfied in real systems.

Let us use the methods stated above to identify the respective relations for the effective capacitance of a $p-n$ junction of a semiconductor diode.

The following expression is obtained for the amplitudes of the harmonics of the charge Q_n from (3.6):

$$Q = \frac{U_{dm}}{n}(C_{n-1} - C_{n+1}) + \frac{2C_D}{\alpha}e^{-\alpha U_{d0}} J_n(\alpha U_{dm}), \quad n = 1, 2, \ldots, \tag{3.24}$$

where C_n are coefficients in the harmonic expansion of the barrier capacitance (3.8).

The amplitudes of the harmonics of current I_n can be determined from (3.20) by using (3.24).

The only unidentified elements needed for the calculation of C_{ef} are the coefficients A_n in (3.18). The first derivative from the charge function (3.12), when taking into account (3.6), is the sum of the barrier and diffusion capacitance of the p-n junction. As a harmonic expansion is carried out, the result is

$$f'(U_d) = C_0 + C_D e^{-\alpha U_{d0}} J(\alpha U_{dm}) + 2\sum_{n=1}^{\infty}[C_n + C_D e^{-\alpha U_{d0}} J_n(\alpha U_{dm})], \tag{3.25}$$

that is

$$\begin{aligned} A_0 &= C_0 + C_D e^{-\alpha U_{d0}} J_0(\alpha U_{dm}), \\ A_n &= 2[C_n - C_D e^{-\alpha U_{d0}} J_n(\alpha U_{dm})]. \end{aligned} \tag{3.26}$$

3.1.3. Numerical analysis of the resonance properties of the oscillating circuit and experimental illustrations

The numerical example will be simplified if the effective capacitance is defined in the form:

$$C_{ef} = \frac{I_1}{\omega U_{dm}}, \tag{3.27}$$

where I_1 is the amplitude of the first harmonic in the current spectrum.

The calculation will be made in accordance with the equivalent circuit of a serial oscillating circuit – Fig.3.4. We take as a given the amplitude of the sine voltage (the basic harmonic) U_{dm} at the nonlinear capacitance of the diode C_d, shunted by the linear part of the circuit (L and r), the effective active resistance of the $p-n$ junction R_{def} and the spurious capacitance of the diode body and the wiring C_p (moreover it should be pointed out that as the calculations indicate, the pulsative active conductance of the $p-n$ junction in the conditions of a resonance circuit takes a negligibly small part in the energy conversion between the harmonics, hence R_{def} can be referred to the linear part of the circuit). For the purpose of determining R_{def}, the voltage is averaged with respect to the first harmonic $U_{dm}\cos(\omega t)$ (3.11) but the specific characteristic of the family R_{def} (for example Fig.3.5), which is averaged with respect to a preset higher harmonic, is determined by the frequency of this harmonic. The latter is accounted for by substituting ω in (3.11) with $n\omega$.

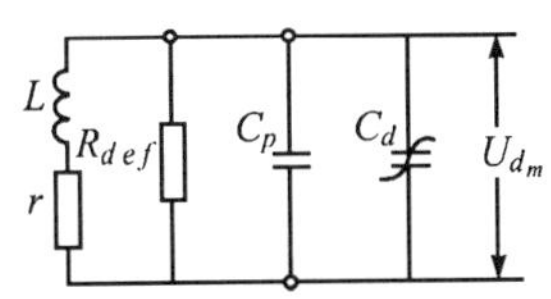

Fig. 3.4. Equivalent circuit, by the alternating signal, of a series oscillating circuit

The capacitive current is written down in the following way:

$$i_c = f'(U_d)\frac{dU_d}{dt} = \left[C(0)\left(\frac{\varphi_k}{\varphi_k - U_{d_0}}\right)^{\gamma} + C_D e^{\alpha U_d}\right]\frac{dU_d}{dt}. \tag{3.28}$$

As the voltage is presented as $U_d = -U_{d_0} + U_{dm}\cos\omega t$, i_c in (3.28) is expressed in the form of an infinite sum of harmonics

$$i_c = \sum_{n=1}^{\infty} I_n \sin(n\omega t). \tag{3.29}$$

The specified expression of the voltage at the diode is

$$U_d = -U_{d_0} + U_{dm}\cos(\omega t) - \sum_{n=2}^{\infty} I_n |Z(n\omega)| \sin[n\omega t + \varphi(n\omega)], \tag{3.30}$$

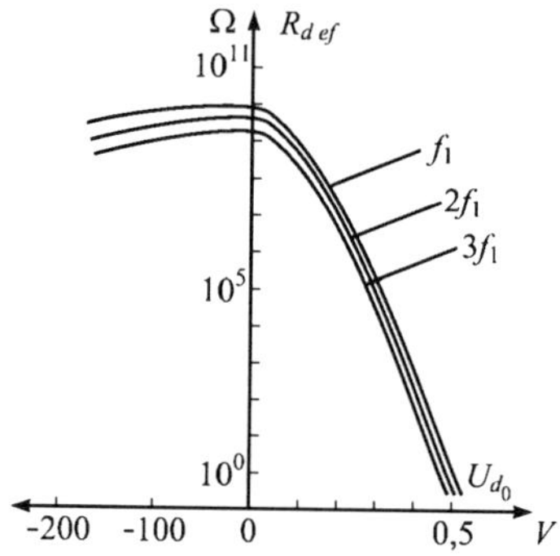

Fig. 3.5. Experimental dependence of the active differential resistance of a $p-n$ junction of a semiconductor diode of the D208 type on the constant voltage applied (the measurement signal has a frequency of 10 kHz)

where the module and the impedance phase of the linearized (by taking into account R_{def}) part of the circuit is put down as:

$$|Z(n\omega)| = \sqrt{H^2 + N^2}, \quad \varphi(n\omega) = \operatorname{arctg}\frac{N}{H}, \tag{3.31}$$

$$H = \frac{rR_{def}(r + R_{def}) + \omega^2 LR_{def}}{(r + R_{def})^2 + (\omega L)^2},$$

$$N = \frac{\omega LR_{def}^2}{(r + R_{def})^2 + (\omega L)^2}.$$

The expression for the capacitive current assumes the form

$$I_c \simeq f'(U_d)\frac{dU_d}{dt} - \xi f''(U_d)\frac{dU_d}{dt}, \tag{3.32}$$

where

$$\xi = \sum_{n=2}^{\infty} I_n |Z(n\omega)| \sin[n\omega t + \varphi(n\omega)].$$

The second derivative of the charge function can be subjected to series expansion as follows

$$f''(U_d)\frac{dU_d}{dt} = \sum_{n=1}^{\infty} V_n \sin(n\omega t), \tag{3.33}$$

which leads to

$$\xi f''(U_d)\frac{dU_d}{dt} = \sum_{n=2}^{\infty} I_n |Z(n\omega)| \sin[n\omega t + \varphi(n\omega)] \sum_{n=1}^{\infty} V_n \sin(n\omega t). \tag{3.34}$$

After the respective transformations, the expression (3.34) takes the form

$$\begin{aligned}
\xi f''(U_d)\frac{dU_d}{dt} &= \frac{1}{2}\sum_{k=1}^{\infty}\left\{-\sum_{n=1}^{\infty} V_n I_{n+k}|Z[(n+k)\omega]|\sin\varphi[(n+k)\omega]\right.\\
&\quad\left.+\sum_{n=2}^{\infty} V_{n+k} I_n|Z(n\omega)|\sin\varphi(n\omega)\right\}\sin(k\omega t)\\
&\quad+\frac{1}{2}\sum_{k=1}^{\infty}\left\{\sum_{n=1}^{\infty} V_n I_{n+k}|Z[(n+k)\omega]|\cos[(n+k)\omega]\right.\\
&\quad\left.+\sum_{n=2}^{\infty} V_{n+k} I_n|Z(n\omega)|\cos\varphi(n\omega)\right\}\cos(k\omega t)+\text{harmonics}. \quad (3.35)
\end{aligned}$$

The module and the phase of the effective capacitance are determined as

$$C_{ef} = \frac{1}{\omega U_{dm}}\sqrt{S^2+W^2}, \tag{3.36}$$

$$\varphi_{C_{ef}} = \operatorname{arctg}\frac{W}{S}, \tag{3.37}$$

where

$$S = I_1 + \frac{1}{2}\sum_{n=2}^{\infty} I_n|Z(n\omega)|\sin\varphi(n\omega)(V_{n-1}-V_{n+1}), \tag{3.37a}$$

$$W = \frac{1}{2}\sum_{n=2}^{\infty} I_n|Z(n\omega)|\cos\varphi(n\omega)(V_{n-1}+V_{n+1}). \tag{3.37b}$$

Computation requires working formulae for the coefficients and parameters. Equation (3.28) yields

$$\begin{aligned}
i_c &= \omega U_{dm}\left[C(0)\left(\frac{\varphi_k}{\varphi_k+U_{d_0}}\right)^{\gamma}\left(1-\frac{U_d}{\varphi_k+U_{d_0}}\right)^{-\gamma}+C_D e^{\alpha U_d}\right]\sin\omega t\\
&=\omega U_{dm}\sum_{n=1}^{\infty}\left[C(0)\left(\frac{\varphi_k}{\varphi_k+U_{d_0}}\right)B_n+\frac{2C_D}{\alpha U_{dm}}e^{-\alpha U_{d_0}}nJ_n(\alpha U_{dm})\right]\sin(n\omega t), \quad (3.38)
\end{aligned}$$

where the expressions for the coefficients B_n are of the form

$$\begin{aligned}
B_1 &= 1+\frac{\gamma(\gamma-1)}{8}\beta^2+\frac{\gamma(\gamma-1)(\gamma-2)(\gamma-3)}{192}\beta^4+\dots,\\
B_2 &= -\frac{\gamma}{2}\beta-\frac{\gamma(\gamma-1)(\gamma-2)}{16}\beta^3-\frac{\gamma(\gamma-1)(\gamma-2)(\gamma-3)(\gamma-4)}{426}\beta^5-\dots,
\end{aligned} \tag{3.39}$$

etc.,

$$\beta = -\frac{U_{dm}}{\varphi_k + U_{d_0}}. \tag{3.40}$$

In order to determine C_{ef}, we should also disclose the coefficients V_n in (3.33) and (3.34).

The second derivative of the load function is expressed in the following way:

$$\begin{aligned} f''(U_d)\frac{dU_d}{dt} &= \sum_{n=1}^{\infty} V_n \sin(n\omega t) \\ &= \omega U_{dm}\left[\gamma C(0)\frac{\varphi_k^{\gamma}}{(\varphi_k + U_{d_0})^{\gamma+1}}\left(1 - \frac{U_d}{\varphi_k + U_{d_0}}\right) + \alpha C_D e^{-\alpha U_d}\right]\sin(\omega t) \\ &= \omega U_{dm}\sum_{n=1}^{\infty}\left[\gamma C(0)\frac{\varphi_k^{\gamma}}{(\varphi_k + U_{d_0})^{\gamma+1}}D_n + \frac{2C_D}{U_{dm}}e^{-\alpha U_{d_0}} n J_n(\alpha U_{dm})\right]\sin(n\omega t). \end{aligned} \tag{3.41}$$

The coefficients D_n in (3.41) are determined in a way analogous to that used for the coefficients B (3.39), if γ in the latter is substituted by $\gamma + 1$.

Fig.3.6 presents the results of the computation of the effective capacitance C_{ef} of the oscillating circuit with a semiconductor diode in an automatic bias regime without accounting for the influence of the higher harmonics (dot-and-dash lines) and by taking into consideration the influence of the second and third harmonic (the solid line) for four different values of the resistance R in the RC-filter of the automatic bias (See Fig.3.1).

The chart in Fig.3.6 shows that given relatively small amplitudes of the voltage, C_{ef} falls as U_{dm} increases. In the case of high amplitudes the rate of C_{ef} decline goes down not only because the operating point projects on a flatter section of the volt-coulomb characteristic (See Fig.3.3), but also at the expense of the considerable growth in the effective diffusive capacitance (the hatch-dotted section of Fig.3.6 singles out only the dependence of the effective diffusive capacitance on the amplitude of the alternating voltage at the diode).

The strong influence of the higher harmonics on the effective capacitance of the diode in the case of a low Q-factor of the oscillating circuit (curves 3 and 4 in Fig.3.6) can be explained by the response of the overtones to the high modulation of the diffusive capacitance at the moments of passage into the area of direct currents through the $p-n$ junction.

The dots in Fig.3.6 indicate the experimental data obtained by measuring the effective capacitance of a semiconductor diode.

Fig.3.7 shows the dependence of the phase of the effective capacitance $\varphi_{C_{ef}}$ on the amplitude of the voltage for the same four values of the resistance R in the RC-filter of the automatic bias.

The equivalent circuit of the oscillating circuit in Fig.3.4 can be used to obtain, without difficulty, the following expressions for the amplitude and the phase of the

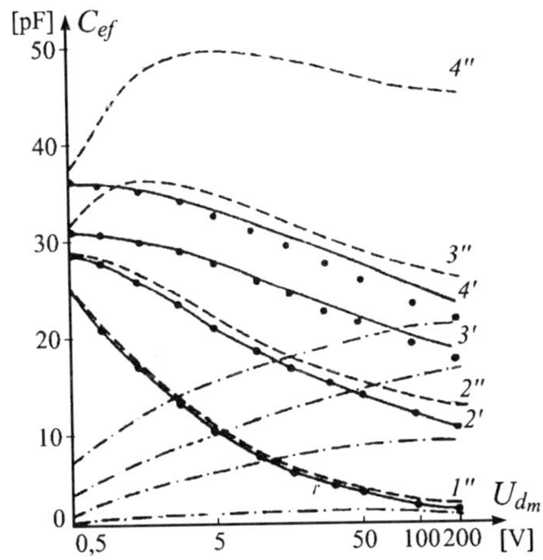

Fig. 3.6. Dependence of the effective (equivalent) capacitance on the amplitude of the circuit voltage without accounting for the influence of the higher harmonics (dotted lines), while accounting for the influence of the second and third harmonic (solid line) for four values of the resistance R in the RC-filter of the automatic bias. *1'*, *1''* – 30 MΩ ($Q \sim 50$); *2'*, *2''* – 18 MΩ ($Q \sim 25$); *3'*, *3''* – 3 MΩ ($Q \sim 15$); *4'*, *4''* – 0,33 MΩ ($Q \sim 7$). Dots are used to denote the experimental data

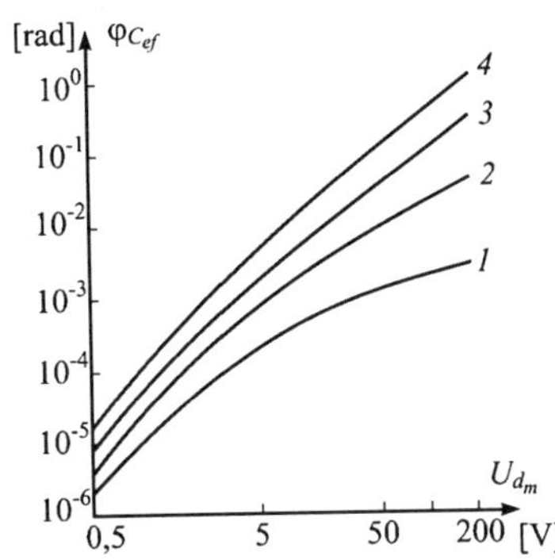

Fig. 3.7. Dependence of the phase of the effective (equivalent) capacitance on the amplitude of the oscillating circuit voltage for four values of the resistance R in the RC-filter of the automatic bias: *1* – 30 MΩ ($Q \sim 50$); *2* – 18 MΩ ($Q \sim 25$); *3* – 3 MΩ ($Q \sim 15$); *4* – 0,33 MΩ ($Q \sim 7$)

oscillations:

$$U_{dm} = \frac{U_m}{\sqrt{(1 - \omega^2 L C_{ef})^2 + (\omega C_{ef} R_p)^2}}, \tag{3.42}$$

$$\varphi_d = -\operatorname{arctg}\frac{\omega C_{ef} R_p}{1-\omega^2 L C_{ef}}, \tag{3.43}$$

where

$$R_p = r + R_s + \frac{R_{def}}{1+(\omega R_{def} C_{ef})^2}. \tag{3.44}$$

The computation of the resonance characteristic of the oscillating circuit has been carried out by using equation (3.42) and by accounting for the expressions (3.36) – (3.41). The computation error does not exceed 5% due to the substitution of the coefficients C_n (3.8) with series with a finite number of terms. In the calculations, the modified Bessel functions are replaced with asymptotic series. The roots of equation (3.42) are sought in the amplitude range $0,1U_m \div (Q+10)U_m$, where Q is the Q-factor of the circuit defined as

$$\frac{1}{Q} = \frac{1}{Q_L} + \frac{1}{Q_d}, \tag{3.45}$$

Q_L is the Q-factor of the linear part of the circuit, Q_d is the Q-factor of the diode for a given amplitude U_{dm},

$$Q_d = \frac{\omega C_{ef} R_{def}}{1 + \dfrac{R_s}{R_{def}} + R_s R_{def} \omega^2 C_{ef}^2}. \tag{3.46}$$

A simple iteration procedure is used, and the change in the sign of the difference between U_{dm} and the right-hand side of equation (3.42) means that a root has been passed over. The value of this root, with preset accuracy (1%), is determined through successive approximations at „half a step". At each stage of the root search, the respective value of the automatic bias U_{d_0} is calculated through an iteration cycle of equation (3.8). Two „external" circles set the variation of the amplitude of the exciting generator U_m and the frequency ω.

Fig.3.8 presents the results of the calculation of the resonance characteristic of an oscillating circuit with automatic bias. It can be seen that as the voltage amplitude increases, the resonance frequency of the circuit keeps growing and an unstable hysteretic section appears on the right-hand slope. As the voltage amplitude goes further up, the growth rate of the resonance frequency decreases and at a certain amplitude the increase stops, the resonance curve „straightens out". The decrease in the value of the resistance R in the RC-filter of the automatic bias leads to a shift of the family of amplitude-frequency characteristics in the direction of smaller frequencies, furthermore the hysteretic instability on the right-hand slope disappears at lower values of the amplitude, while the effect of „straightening out" of the resonance curve is more pronounced.

The analysis stated above shows that, as the diffusive capacitance is accounted for, the nonlinear resonance acquires a number of peculiarities. The typical

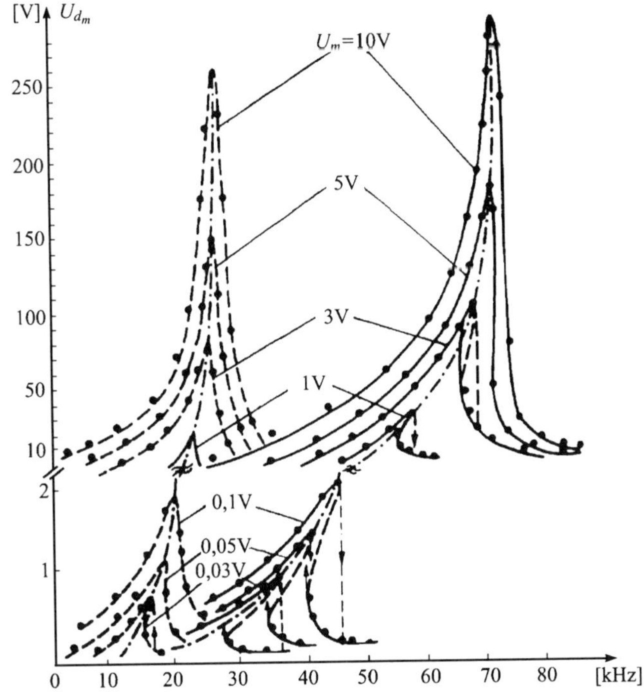

Fig. 3.8. Resonance characteristics of an oscillating circuit with a semiconductor diode in an automatic bias regime for two different values of the resistance R in the RC-filter of the automatic bias: dotted lines – $R = 0,33$ MΩ, solid lines – $R = 30$ MΩ. Dots are used to denote the experimental data

hysteretic instability disappears from the right-hand slope of the resonance characteristic. Such a behaviour of the amplitude-frequency characteristic can be applied in the detectors for reducing intermodulation distortions, in the creation of highly effective modulation-parametric amplifiers, etc.

The presented approach for obtaining equations concerning the higher approximations on the basis of the method of equivalent linearization, allows conducting a sufficiently precise quantitative study of resonance phenomena in nonlinear systems, particularly of those with a relatively low quality factor. Even the second approximation reveals the complex nature of the effective parameter of the resonance system with a nonlinear reactive element, which is determined by the influence of the active losses at the frequencies of the higher harmonics. Given a sufficiently high Q-factor ($Q > 100$), the influence of the higher harmonics on the effective parameter is insignificant and the quantitative analysis of the oscillating processes in the case of a nonlinear resonance can be effected only by accounting for the basic tone. As the Q-factor goes down, the influence of the overtones leads to a considerable reduction in the values of the effective parameters, compared with the value estimates with respect to the first harmonic, and it is imperative that they should be taken into account.

3.1.4. Excitation of periodic oscillations on the basis of a nonlinear oscillating circuit with a pronounced hysteretic area on the resonance characteristic

In this section we shall demonstrate how a typically radiophysical effect can be used to create a method for exciting periodic mechanical motion and to develop the respective facilities. The idea is to use the hysteretic area of the resonance curve of a nonlinear oscillating circuit for an effective energy input in a mechanical motion. This can be achieved by creating the useful acceleration at high resonance amplitudes, while the inevitable reaction against the motion occurs at low amplitudes as a result of a relevant leap within the hysteretic area of the resonance curve.

3.1.4.1. An inductive case (asynchronous inductive motor)

The basic idea formulated above will be illustrated by using the system shown in Fig.3.9 [155]. The latter is a radiophysical oscillating circuit periodically interacting with a physical pendulum. We shall demonstrate that it is possible to excite continuous oscillations or rotary motion of the pendulum in such a system. The power supply for the oscillating circuit is a source of alternating voltage E. The pendulum is a freely suspended ferromagnetic disk, which experiences the action of the magnetic field of the solenoid L. Simultaneously, there is also „reverse" action manifested in periodic modulation of the equivalent inductance of the solenoid.

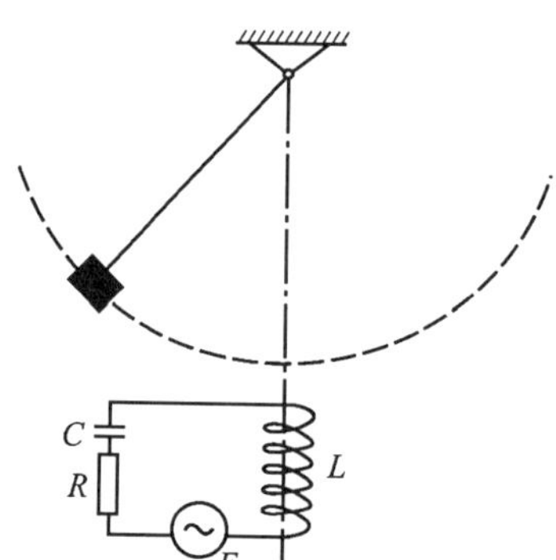

Fig. 3.9. Electric oscillating circuit interacting with a physical pendulum

Fig.3.10 shows the qualitative dependence of the inductance of the solenoid on the angle of pendulum deviation y from the equilibrium position $y = 0$. The periodic change in the inductance of the solenoid during the motion of the pendulum results in a substantial nonlinearity of the resonance characteristic of the electric oscillating circuit. Fig.3.11 is a qualitative demonstration of the dependence of the voltage U at the electric oscillating circuit on the same angle of deviation y. As the pendulum moves, the characteristics with a pronounced hysteretic zone (known as „beak-like" characteristics), typical of nonlinear resonance and symmetric with respect to the point $y = 0$, occur. Since an alternating current is flowing through the solenoid, a ponderomotive attractive force F is acting on the ferromagnetic disk of the pendulum, which always points at the equilibrium position of the pendulum. The outcome is a very specific interaction.

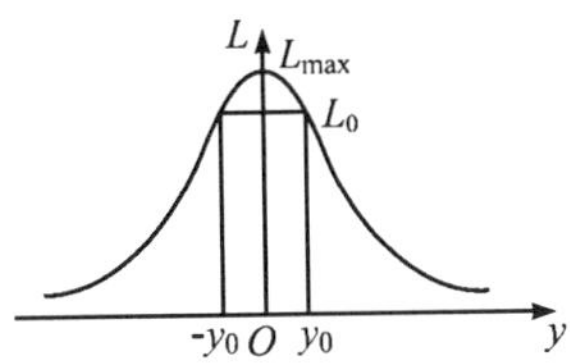

Fig. 3.10. Dependence of the inductance of the solenoid on the position of the ferromagnetic disk

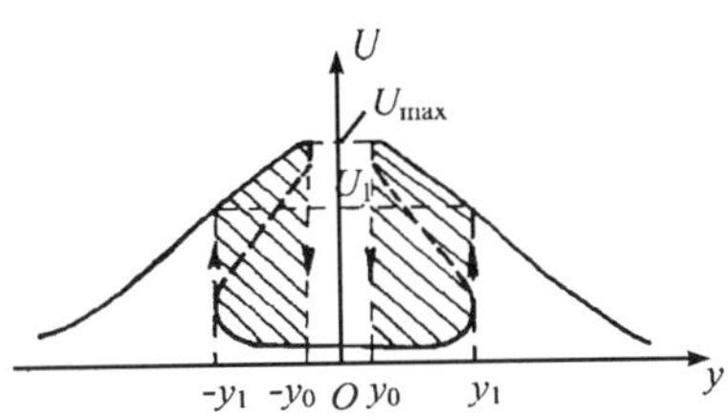

Fig. 3.11. Resonance characteristics of the oscillating circuit in the case of periodic motion of the ferromagnetic disk over the solenoid

The solenoid attracts the ferromagnetic disk, which, as it draws closer to it, changes its inductance in such a way that the electric oscillating circuit „tends" to resonance. As a result of this motion towards resonance, the voltage in the oscillating circuit increases, which causes an increase of the current in the solenoid and a stronger attraction of the disk. When the body draws close to the resonance position $y = 0$, the system „jumps over" the resonance and as a result of the nonlinearity the amplitude at the oscillating circuit changes in a leap-like manner and its value experiences a multiple decline. When the disk leaves the area $y = 0$, the solenoid already exerts a retarding influence on it, but the amplitudes of the voltage at the oscillating circuit are by far lower. Thus a positive energy input occurs for a period. The process exposed above is also illustrated by Fig.3.12. For example, when the pendulum moves from the left to the right (Fig.3.12a) up to $y = 0$, the force F will exert accelerating action, and further on, for $y > 0$, its action will be retarding. It can be seen from Fig.3.11 that the pendulum is accelerated at high amplitudes of the voltage (up to $U_{\max}$) on the solenoid, and that its motion is retarded at relatively small amplitudes, which do not exceed U_1. The correspondence between the amplitudes of the voltage U and the acting force F can be established by comparing Fig.3.11 and Fig.3.12a. As the pendulum moves from right to left (Fig.3.12b), the process is completely analogous – the acceleration exceeds by far the retarding action, since due to the symmetry, on the one hand, and the dynamic ambiguity, on the other, the dependence of the force on the coordinate is turned into its mirror image. In the case of oscillating motion of the pendulum, the useful acceleration is twice per period, while in the situation of rotary motion of the pendulum the useful acceleration is realized only once per period. Thus the effect is an effective input of energy in the pendulum motion, which compensates for its own losses due to friction. The energy input is proportional to the area of the hysteretic section of the resonance curve (the hatched part in Fig.3.11). Asynchronous excitation of stationary oscillations or a rotary motion of the pendulum are realized. It is obvious that the physical mechanism of the described effect is related to the reversibility of the modulation-parametric interactions in the system (See Chapter 1).

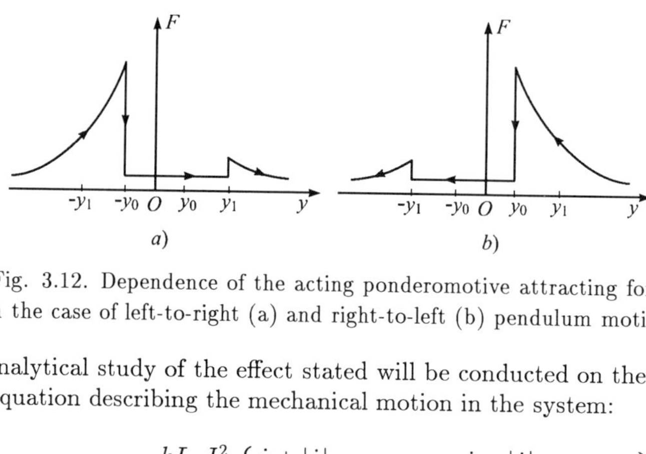

Fig. 3.12. Dependence of the acting ponderomotive attracting force in the case of left-to-right (a) and right-to-left (b) pendulum motion

The analytical study of the effect stated will be conducted on the basis of the following equation describing the mechanical motion in the system:

$$\ddot{y} + 2\beta\dot{y} + \omega_0^2 \sin y = \frac{kL_0 J_0^2}{2ml^2}\left\{\frac{\dot{y}+|\dot{y}|}{2}\delta(y+y_0) + \frac{\dot{y}-|\dot{y}|}{2}\delta(y-y_0)\right\}, \tag{3.47}$$

where β is a coefficient characterizing the energy losses of the pendulum due to friction, ω_0 is the frequency of small natural oscillations of the pendulum, ml^2 is the inertia moment of the pendulum, l is the length of the pendulum, m is the mass of the ferromagnetic disk, L_0 and y_0 are elucidated through Figs 3.10, 3.11

and 3.12, J_0 is the average effective value of the power of the electric current, k is a coefficient accounting for the width of the ponderomotive impulse of the acting force, $\delta(y \pm y_0)$ is Dirac's delta-function.

The solution of equation (3.47) can be presented in the first approximation as follows:

$$y = a\sin\Theta, \quad \Theta = \omega t + \alpha. \tag{3.48}$$

As (3.48) is substituted in (3.47) and given the assumptions $\dot{x} = a\omega\cos\Theta$, $\frac{da}{dt}\sin\Theta + a\frac{d\alpha}{dt}\cos\Theta = 0$, and $a > y_0$, the following shortened equations for the amplitude and phase of the pendulum oscillations are obtained:

$$\frac{da}{dt} = -\beta a + 2B\sqrt{1 - \frac{y_0^2}{a^2}}, \tag{3.49}$$

$$\frac{d\alpha}{dt} = \omega(a) - 2B\frac{y_0}{a^2}, \tag{3.50}$$

where $\omega(a)$ is the frequency of the pendulum oscillations, which depends on its amplitude, $B = \frac{kL_0J_0^2}{2ml^2}$.

(3.49) yields the following solutions for the stationary oscillations ($\frac{da}{dt} = 0$ and $\frac{d\alpha}{dt} = 0$):

$$a_1^2 = \frac{2B^2}{\beta^2}\left[1 - \sqrt{1 - \frac{\beta^2}{B^2}y_0^2}\right], \tag{3.51}$$

$$a_2^2 = \frac{2B^2}{\beta^2}\left[1 + \sqrt{1 - \frac{\beta^2}{B^2}y_0^2}\right]. \tag{3.52}$$

The stability analysis shows that solution (3.51) is unstable, while solution (3.52) is stable. Hence, when continuous oscillations are excited, the pendulum should receive such an initial impulse that its amplitude should be bigger than a_1.

A possible application of the effect of asynchronous excitation of stationary oscillations presented above is the development of a linear reciprocating electric motor [232]. An experimental version of such a motor is shown in Fig.3.13. It is made up of two oscillating systems – electrical and mechanical. The electrical one is a series oscillating circuit consisting of a series connection of an electromagnetic coil (1) of the stator (2), a capacitor (3) and a power source of electric oscillations (4). The mechanical system consists of a steel ball (5) with a diameter of 5 mm, freely positioned in a guiding groove (6) bounded by supports, (7) and (8) with limiting springs (9) and (10) attached to them. The resonance frequency of small

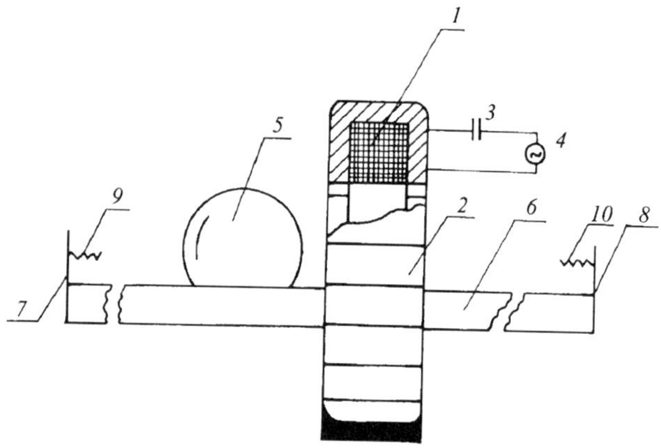

Fig. 3.13. Experimental structure of a linear reciprocating electric motor

oscillations in the electrical oscillating system (given a symmetric positioning of the steel ball (5) in the stator (2)) is approximately 5 kHz.

The interaction of the stator with the secondary element of the motor is due to the dynamic ambiguity of the force acting on the secondary element, which is manifested in the dependence on the coordinate and the direction in which the steel ball moves. The magnetic field of the coil acts on the secondary element – the steel ball. Simultaneously, as the ball passes through the opening of the stator, periodic modulation of the coil inductance occurs. The process is analogous to that described for the system in Fig.3.9. In accordance with the position of the steel ball as related to the stator, certain dependencies analogous to those in Figs.3.10, 3.11, 3.12 are realized. In this case y is the linear coordinate of the steel ball motion in relation to the stator. The frequency of the mechanical oscillations of the steel ball is determined by its mass and the other parameters of the oscillating system (elasticity of the limiting springs, friction coefficient, etc.) as well as by the voltage of the power supplying generator 4 in Fig.3.13. At a definite tuning, the oscillations occur at generator voltage $7 \div 22$ V. When the voltage is less than 7 V no mechanical oscillations are observed due to the lack of a pronounced hysteretic zone at the characteristic (Fig.3.11) required for an energy input exceeding the respective threshold value. When the voltage exceeds 22 V, the left-hand and right-hand hysteretic zones of the characteristic (Fig.3.11) overlap, which violates the conditions for effective energy input in the secondary element. It is worth noting that certain tuning can be applied to shift the range of power supplying voltages in the direction of higher values, which increases the power of the linear motor. The only condition is that the dependence of the voltage of the electric oscillating system on the coordinate along which the steel ball moves should be of the type shown in Fig.3.11.

The dynamic ambiguity and the symmetric nature of the characteristics with respect to the motion coordinate allow creating a linear reciprocating electric motor, which requires no special control systems – commutators, contacts, feedback circuits, hence this is a way to boost its reliability, useful life and efficiency and to reduce its mass per unit of power. The asynchronous nature of the energy input, when the excitation mechanism acts independently of the frequency of periodic motion, enables the attainment of a reciprocating motion of the secondary element with a frequency by far lower than the frequency of the power supplying grid, without any use of additional equipment to convert the frequency of the power supplying current and to exercise control.

In practice the described effect and the basic configuration in Fig.3.9 can serve as a basis for the development of various linear reciprocating motors.

For the sake of completeness, we shall consider here yet another system, which converts high frequency electric energy into low frequency mechanical energy – Fig.3.14 [139].

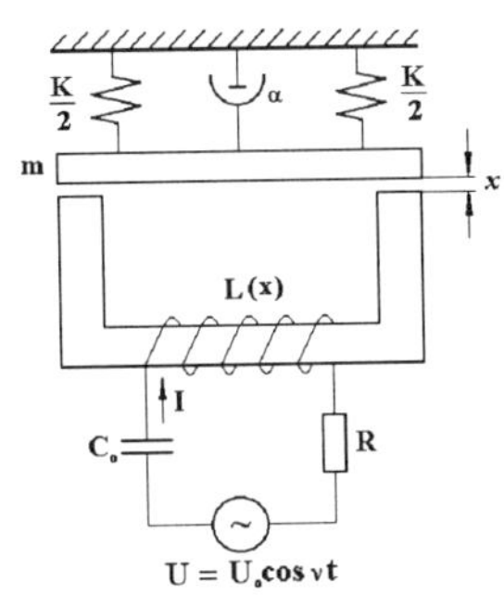

Fig. 3.14. A dynamic system transforming high-frequency electric power into low-frequency mechanical energy

The system is an electromechanical vibrator and the power circuit of its electromagnet includes an oscillating circuit. The oscillation frequency of the mechanical part of the vibrator is by far lower than the frequency of the power source voltage. Under certain conditions the vibrator performs undamping oscillations with a frequency close to its own, whereat the amplitude and frequency of these oscillations are weakly dependent on the frequency of the feeding voltage.

This fact, unlike the previous case discussed above, will be explained on the basis of the combined interaction of the frequency of the power supply current ν and the frequency of the mechanical vibrations ω. A similar approach was used in Section 2.4.3. in the analysis of an inductive sensor.

The equations of the system under consideration are of the following form:

$$\left|\begin{array}{l} \dfrac{d^2}{dt^2}[L(x)I] + R\dfrac{dI}{dt} + \dfrac{I}{C_0} = U_0 v \sin vt \\[2ex] m\dfrac{d^2x}{dt^2} + \alpha\dfrac{dx}{dt} + Kx = F(x,t), \end{array}\right.$$

where $L(x)$ is the inductance of the coil with a magnetic core depending on the slot between the plate with mass m and the core, $F(x,I) = \frac{1}{2}\frac{d\Phi}{dx}$, $\Phi = IL(x)$ is a magnetic flux, I is the current strength, R, C_0, m, α and K are coefficients made clear by Fig.3.14. With a sufficient degree of accuracy we can set $\frac{d\Phi}{dx} = I\frac{dL}{dx}$, [139].

For small values of x the inductance $L(x)$ can be approximated by means of a polynomial: $L(x) = L_0(1 + a_1 x - s_2 x^2)$. Then $F(x,I) = L_0 = \left(\frac{a_1}{2} - a_2 x\right) I^2$.

Let us set $I{=}y{+}A\sin(vt{+}\varphi)$, $x = z{+}x_0$, where $A{=}\frac{U_0 v C_0}{\sqrt{R^2v^2C_0^2{+}(L_0C_0v^2-1)^2}}$, $\operatorname{tg}\varphi = \frac{RvC_0}{L_0C_0v^2-1}$, $x_0 = \frac{1}{2}\frac{L_0 a_1 \overline{I^2}}{K + L_0 a_2 \overline{I^2}}$ is the constant shift of the plate caused by the average square of the current strength $\overline{I^2}$.

The equations regarding the variables y and z, given sufficiently small y, z, A, R and α, can be presented in the following form:

$$
\left|\begin{aligned}
&\ddot{y} + \Omega^2 y = -\mu[2\delta_1\dot{y} - 2\Omega\Delta_1 y + (a_1 - a_2x_0 - a_2z)(x_0 + z)(\ddot{y} - Av^2\sin\Psi) \\
&\qquad + 2(a_1 - 2a_2x_0 - 2a_2z)\dot{z}(\dot{y} - Av\cos\Psi) + (a_1\ddot{z} - 2a_2\dot{z}^2 - 2a_2x_0\ddot{z} \\
&\qquad - 2a_2z\ddot{z})(y + A\sin\Psi)] \\
&\ddot{z}{+}\omega^2 z{=}{-}\mu[2\delta_2\dot{z}{-}2\omega\Delta_2 z{+}\frac{L_0a_2}{m}\overline{I^2}z{-}\frac{L_0}{m}\left(\frac{a_1}{2}{-}a_2x_0{-}a_2z\right)(y{+}A\sin\Psi)^2{-}\overline{I^2}],
\end{aligned}\right.
$$

where, in this case, μ is a small parameter, which should be set as equal to unity in the final expressions; $\delta_1 = \frac{R}{2L_0}$, $\delta_2 = \frac{\alpha}{2m}$, $\Omega_0 = \frac{1}{\sqrt{L_0C_0}}$ is the own frequency of the oscillating circuit, when $x = 0$; $\omega_0 = \sqrt{\frac{K}{m}}$ is the own frequency of the mechanical system; $\Psi = vt + \varphi$, Ω is the frequency of the variable component of the current y; ω is the frequency of the mechanical oscillations; $\Delta_1 = \frac{\Omega^2 - \Omega_0^2}{2\Omega}$ is the detuning between the frequency of the oscillations Ω and the own resonance frequency Ω_0; $\Delta_2 = \omega - \omega_0$ is the detuning between the frequency ω and the own resonance frequency ω_0 (Δ_2 is a small quantity).

With regard to the unknown frequencies Ω and ω, we shall impose the condition: $\Omega + \omega = v$.

Setting $y = B\cos(\omega t + \varphi_1)$, $z = C\cos(\omega t + \varphi_2)$, to a first approximation while using the averaging method [97] yields the following equations for B, C, φ_1 and φ_2:

$$\left|\begin{array}{l} \dot{B} = -\delta_1 B - \dfrac{(a_1 - 2a_2 x_0(v+\omega)^2}{4\Omega} AC\cos v \\ \dot{\varphi}_1 = -\Delta_1 - \dfrac{(a_1 - 2a_2 x_0)x_0\Omega}{2} + \dfrac{a_2\Omega C^2}{4} - \dfrac{(a_1 - 2a_2 x_0)(v+\omega)^2}{4\Omega B} AC\sin v \\ \dot{C} = -\delta_2 C - \dfrac{a_1 - 2a_2 x_0}{2a_2}\gamma\omega AB\cos v \\ \dot{\varphi}_2 = -\Delta_2 + \dfrac{\omega}{2}\gamma(A^2+B^2) - \dfrac{a_1 - 2a_2 x_0}{2Ca_2}\omega\gamma AB\sin v, \end{array}\right.$$

where $\gamma = \dfrac{L_0 a_2}{2m\omega^2}$, $v = \varphi - \varphi_1 - \varphi_2$, $x_0 = \dfrac{a_1}{2a_2}\dfrac{\varphi(A^2+B^2)}{1+\gamma(A^2+B^2)}$.

For the sake of simplicity in any further calculations it is assumed that $\gamma(A^2+B^2) \ll 1$, i.e. $a_2 x_0 \ll a_1$. Apart from that, the following are set: $\delta_1\Omega^2 \gg \delta_2(v+\omega)^2$ and $a_1^2\delta_2(v+\omega)^2 \ll a_2\omega\delta_1\Omega$. Given these assumptions, the stationary solution of the equations is of the following form:

$$\left|\begin{array}{l} \cos v = -\dfrac{4}{\omega+v}\sqrt{\dfrac{\Omega\delta_1\delta_2 a_2}{2\omega a_1^2\gamma A^2}} \\ \operatorname{tg} v = \pm\sqrt{\dfrac{a_1^2\omega(v+\omega)^2\gamma A^2}{8\Omega\delta_1\delta_2 a_2} - 1} \\ \omega - \omega_0 = \delta_2 \operatorname{tg} v \\ C^2 = \dfrac{2\delta_1\Omega\omega}{a_2\delta_2(v+\omega)^2}\gamma B^2 \\ \gamma B^2 = \dfrac{\delta_2(v+\omega)^2}{\delta_1\Omega^2}\left(2\dfrac{\Delta}{\omega} - 2\dfrac{\delta_1}{\omega}\operatorname{tg} v + \dfrac{a_1^2}{2a_2}\dfrac{\Omega}{\omega}\gamma A^2\right), \end{array}\right.$$

where $\Delta = \Delta_1 + \Delta_2$.

It follows from the conditions for the existence of the stationary solutions, $|\cos\theta| \leq 1$, $B^2 + 0$, that:

$$\left|\begin{array}{l} \dfrac{a_1^2\omega(v+\omega)^2\gamma A^2}{8\Omega\delta_1\delta_2 a_2} \geq 1 \\ \Delta \geq \delta_1 \operatorname{tg} v - \dfrac{a_1^2}{4a_2}\Omega\gamma A^2. \end{array}\right.$$

Since γA^2 depends on the frequency v, the first condition out of those listed above, establishes a top limit to the detunings, while the second condition sets a bottom limit.

Let us now examine the stability of the obtained solution.

The equations regarding the deviations from the stationary condition $B - B^{(st)} = \xi_1$, $C - C^{(st)} = \xi_2$, $v - v^{(st)} = \eta$ are of the following form:

$$\dot{\xi}_1 = \delta_1 \left(-\xi_1 + \frac{B}{C}\xi_2 - B \operatorname{tg} v\eta \right)$$

$$\dot{\xi}_2 = \delta_2 \left(\frac{C}{B}\xi_1 - \xi_2 - C \operatorname{tg} v\eta \right)$$

$$\dot{\eta} = \delta_1 \left[\left(\operatorname{tg} v + \frac{\Omega}{2\delta_1} \frac{a_1}{a_2} \gamma B^2 \right) \frac{\xi_1}{B} + \left(\frac{\omega}{\delta_2} \frac{\Omega^2 \gamma B^2}{(\omega + v)^2} - \operatorname{tg} v \right) \frac{\xi_2}{C} - \eta \right].$$

It is established from the characteristic equation that the solution is stable in the case of $\operatorname{tg} v > 0$.

The analysis indicates that the obtained solution can exist and can be stable within a broad range of the detunings Δ. This range is determined by the conditions derived above. Moreover, the frequency of the mechanical oscillations hardly cross the boundary of the band δ_2, corresponding to the resonance curve, while the frequency of the oscillations of the component y of the current in the circuit can differ substantially from the own one. The amplitude of the mechanical fluctuations within a broad range detunings may be weakly dependent on the frequency of the power source v. By way of illustration, Fig.3.15 shows the dependencies of the amplitude C on $\dfrac{v}{\Omega_0}$ for the following values of the parameters: $\dfrac{\omega_0}{\delta_2} = 100$, $\dfrac{\Omega_0}{\delta_1} = 10$, $\dfrac{a_1^2}{a_2} = 1$, $\Omega_0 \gg \omega_0$, $\gamma A_0 = \dfrac{\gamma U^2}{R^2} = 1$ (curve 1) and $\gamma A_0^2 = 0,3$ (curve 2). The value C in the figure is related to $|C_0|$, which is the absolute value of C in the case of $\Delta = 0$ (it follows from the solution obtained above that at $\Delta = 0$ the value C is imaginary, i.e. such a regime cannot exist). As it can be seen from Fig.3.15, the reduction in the amplitude of the power source voltage entails a substantial shrinkage of the existence area of the oscillations under consideration. It is not difficult to determine the condition of self-excitation of oscillations in the system, which is of the following form:

$$\frac{a_1^2 v^4 [(v^2 - \Omega_0^2)(v^2 + 3\Omega_0^2) - 4\delta_1^2 \gamma^2]}{2mL_0[(v^2 - \Omega_0^2)^2 + 4\delta_1^2 \gamma^2]^3} \geq \delta_2.$$

This condition determines the critical value of the amplitude of the power source $U_{0,cr}$ as dependent on the frequency v and the other parameters of the system. The dependence $U_{0,cr}(x)$ is of a non-monotonous nature and it is represented in Fig.3.16. for two private cases: a) $\Omega_0 \gg \delta_1$ and b) $\Omega_0 \ll \delta_1$. In both cases, for a certain value of the frequency $v = v_m$, the quantity $U_{0,cr}$ reaches its

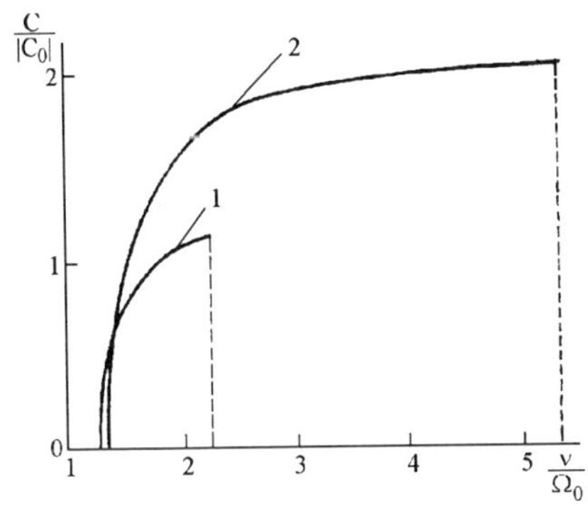

Fig. 3.15. Amplitude frequency characteristic of excited mechanical oscillations in the case of $\gamma A_0^2 = 1$ (curve 1) and $\gamma A_0^2 = 0,3$ (curve 2)

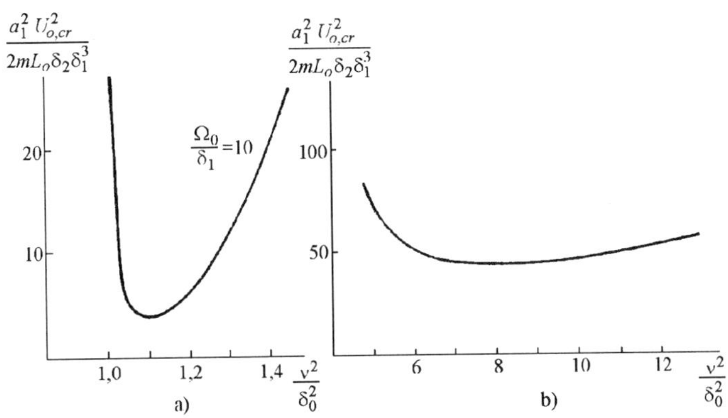

Fig. 3.16. Dependence of the critical amplitude of the pumping voltage on its frequency, when the other parameters of the system are given

minimum level. If $\Omega_0 \ll \delta_1$, then $\omega_m = \sqrt{8}\delta_1$; in the opposite case, when $\Omega_0 \gg \delta_1$, the result is $\omega_m = \Omega_0 + \dfrac{\delta_1}{\sqrt{5}}$.

The frequency of oscillations ω is close to the own frequency ω_0. The correction Δ_2 of the own frequency in a stationary regime is determined by the last equation in the system of equations, given $\varphi_2 = 0$.

It should be pointed out that the mechanism of excitation of mechanical oscillations described above is manifested in various systems: for example, excitation of elastic oscillations of resonators filled in with high-frequency electromagnetic radiation, occurrence of negative friction in mechanical constructions situated in electromagnetic fields (phenomena kindred to the inductive and capacitive sensor considered above – See Sections 2.4.3. and 2.4.4.).

3.1.4.2. A capacitive case (an asynchronous capacitive motor)

Another possibility is to excite rotary motions (in accordance with the mechanism described above) in a dual circuit – a nonlinear capacitive oscillating circuit with a rotating secondary „capacitive" element – Fig.3.17. The oscillating circuit with a semiconductor diode referred to above is used in an automatic bias regime, which creates the required nonlinearity and dynamic ambiguity. The lower plate of the „whirligig" is immobile, while the upper one is mobile. The dependence of the voltage U in the oscillating circuit on the angle α between the two plates of the „whirligig" is analogous to the dependence in Fig.3.11, i.e. there is a „beak-like" characteristic with a broad hysteresis section. There are forces of attraction between the plates of the „capacitive motor". They try to bring the mobile plate to a position, when $\alpha = 0$, i.e. to a position of complete overlapping with the immobile plate. During this motion there is a drive towards resonance and the amplitude in the oscillating circuit increases, and, respectively, the attraction force between the plates also goes up drastically. When the mobile plate passes through the $\alpha = 0$ position as a consequence of the inertia in its motion, the attraction force already acts as a brake. At that moment, however, a leap has occurred in the hysteresis section of the characteristic, and the amplitude in the oscillating circuit has experienced a multiple decline. Respectively, the braking action is by far weaker than the preceding accelerating action and there is effective energy input in the system, which takes an average of one rotation period of the mobile plate. Thus asynchronous excitation of rotary motion takes place. Its frequency is not directly related to the frequency of the electric oscillations in the circuit (which may fall in the kilohertz or megahertz range). Instead it is determined by the most favourable conditions for optimal energy input in the mechanical motion.

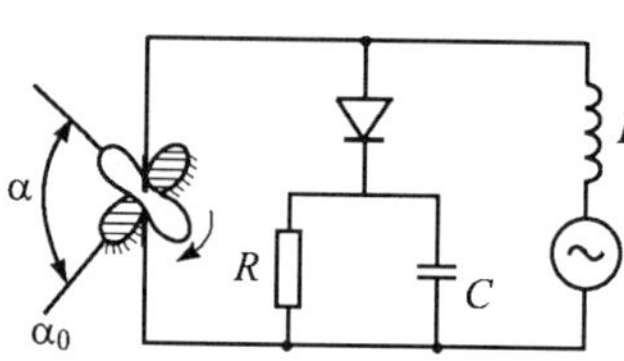

Fig. 3.17. Basic diagram of an asynchronous capacitive motor

Let us present analytically the possibility of effective energy input into the asynchronous capacitive motor described above.

Let q be the charge on the capacitor plates. The torque, conditioned by the electric forces and applied to the mobile plate (in this case playing the part of a motor rotor), can be expressed as

$$M = -\frac{\partial W}{\partial \alpha},$$

where $W = \dfrac{q^2}{2C}$ is the energy accumulated in the capacitor, α is the angle of rotation.

The voltage between the capacitor plates changes in accordance with a periodic law $U = A\cos(\omega t + \varphi)$, where ω is the frequency of the oscillations in the electric circuit.

Since the rotation frequency of the rotor, as well as the frequency of leap-like alteration in the amplitude of the oscillations, are much lower than the frequency of these oscillations in the electric circuit, it would be appropriate to use the concept of average energy accumulated in the capacitance of the mechanical part of the system:

$$\overline{W} = \frac{\overline{q}^2}{2C}.$$

Taking into account that the torque is $M = -\dfrac{q^2}{2C^2}\dfrac{dC}{dt}$, we can determine the capacitance energy as $W = -\displaystyle\int M d\alpha = \int \frac{A^2}{4} dC$.

We shall approximate the amplitude of the voltage at the capacitor by using the expression

$$A = A_{\min} + \frac{A_{\max} - A_{\min}}{2\pi}\alpha = A_{\min} + \frac{\Delta A}{2\pi}\alpha.$$

We shall proceed to approximate, in an analogous way, the law of capacitance alteration taking into account the fact that it falls in the range $0 < \alpha < \pi$, and, respectively, goes up when $\pi < \alpha < 2\pi$ ($\alpha = 0$ corresponds to the situation, when the two capacitor plates overlap completely – Fig.3.17).

The following expression is obtained for the total capacitance energy:

$$W = W^{(+)} + W^{(-)} = \left[\frac{1}{8}(\Delta A)^2 + \frac{1}{4}(\Delta A)A_{\min}\right]\Delta C,$$

where $W^{(+)} = \displaystyle\int_{\pi}^{2\pi} \frac{\left(A_{\min} + \frac{\Delta A}{2\pi}\alpha\right)^2}{4\pi} \Delta C d\alpha$ is the energy obtained by the rotor as the capacitance increases, $W^{(-)} = -\displaystyle\int_{0}^{\pi} \frac{\left(A_{\min} + \frac{\Delta A}{2\pi}\alpha\right)^2}{4\pi} \Delta C d\alpha$ is the energy dissipated as the value of the capacitance decreases as a result of the braking action of the coulomb forces of attraction, ΔC is the capacitance change in the oscillating circuit.

It is obvious that $W > 0$, i.e. the energy input takes place for one period of high-frequency oscillations. Simultaneously, the energy is dissipated due to friction,

$$W_f = \int_0^t M_f d\alpha,$$

where W_f is the energy dissipated because of the action of the moment M_f of the friction force.

Let us assume that $M_f = \mathcal{L}\Omega$, where $\mathcal{L}$ is the proportionality factor, while Ω is the angular rotational frequency of the rotor ($\Omega \ll \omega$). Then over the period $T = \dfrac{2\pi}{\Omega}$

$$W_f = \int_0^T \mathcal{L}\Omega d\alpha = \int_0^T \mathcal{L}\Omega^2 dt = \mathcal{L}\Omega_{st}^2 T = 2\pi\mathcal{L}\Omega_{st},$$

where Ω_{st} is the stationary angular rotational frequency of the rotor.

The factor $\mathcal{L}$ can be determined from the equation:

$$J\frac{d\Omega}{dt} + \mathcal{L}\Omega = 0,$$

where J is the moment of the inertia of the rotor.

If at $t = 0$ the rotor turns with frequency Ω_0, the solution of this equation can be written down as $\Omega = \Omega_0 e^{-\Delta t}$, where $\Delta = \dfrac{\mathcal{L}}{J}$ is the damping decrement. From here we can determine the time T_e over which Ω falls e times – $\Delta T_e = 1$, therefore $\mathcal{L} = \dfrac{J}{T_e}$. The ultimate result that we get for the energy dissipation due to friction is $W_f = 2\pi\dfrac{J}{T_e}\Omega_{st}$.

In order to determine the stationary rotational speed, we equalize the accumulated and dissipated energy over one period $W = W_f$. We obtain

$$f_{st} = \frac{T_e W}{4\pi^2 J} = \frac{T_e \Delta C}{4\pi^2 J}\left[\frac{1}{8}(\Delta A)^2 + \frac{1}{4}(\Delta A)A_{\min}\right].$$

Since the parameters ΔC and ΔA are related to the width of the hysteretic area on the characteristic (Fig.3.11), we can draw the obvious conclusion that the effect of asynchronous excitation of stable stationary motions is determined by the hysteresis area of the highly nonlinear amplitude-frequency and phase-frequency characteristics of the electric oscillating circuit in the system.

It is worth noting that the area of the hysteretic section is not dependent on the angular rotational frequency of the rotor, when satisfying the inequality

$$\Omega \ll \frac{2\pi}{\tau_R},$$

where $\tau_R = \dfrac{2Q}{\omega}$ is the relaxation time of the oscillation amplitude in the resonance circuit, Q is the Q-factor of the oscillating circuit. The average moment of rotation under this condition is

$$\overline{M} = \frac{\Delta C A^2}{8\pi},$$

while the average power obtained by the rotor from the electric field is determined as

$$\overline{W} = \frac{\Delta C \Omega Q W_f}{4\pi C\omega},$$

where $W_f = \dfrac{A^2 \omega C}{2Q}$.

This yields the efficiency of the motor in the form

$$\eta = \frac{\zeta}{\zeta + 1}, \quad \text{where} \quad \zeta = \frac{\Delta C \Omega Q}{4\pi\omega C}.$$

It is important to note that when the conditions referred to above are satisfied, the power of such a capacitive motor can exceed by far the limit set by Manley-Row's energy relations (See 1.2).

This was an illustration of the possibilities for extensive utilization of the peculiarities of nonlinear radio engineering oscillating circuits in the case of excitation of periodic mechanical motions. It can be shown that the same peculiarities of radio engineering oscillating circuits can also be used on the „molecular" level for exciting specific precession in ensembles of molecules, which will contribute to the acceleration of other processes, for example specific chemical interactions.

3.2. Nonlinear and parametric resonance in a generalized oscillating circuit

Many modern resonance systems can be represented as oscillating circuits with two opposite nonlinear reactive elements: thin-film contacts of superconductors with semiconductors [140], SHF transistor resonance amplifiers, resonance systems tunable by means of JIG spheres, LC systems with complex nonlinear circuits and filters [78], etc. These systems can be used to construct different radiophysical amplifiers, generators and transformers. The investigation of such systems is therefore of theoretical and practical significance. However, great difficulties arise in the theoretical analysis of such systems. The presence of two opposite nonlinear reactive elements considerably complicates the respective calculations.

The examination concerns the most general case of a resonance system with nonlinear inductance, nonlinear resistance, nonlinear capacitance and nonlinear conductance. An example of such a system is a two-transistor model of a $p-n-p-n$ inductive structure, complemented to a resonance oscillating circuit by a self-bias $p-n$ junction – Fig.3.18 [233, 234]. The system is compared with oscillating circuits containing but one nonlinear reactive element. For the sake of brevity, the circuit with nonlinear capacitance is conventionally termed circuit of type I, while the circuit with nonlinear inductance is called circuit of type II.

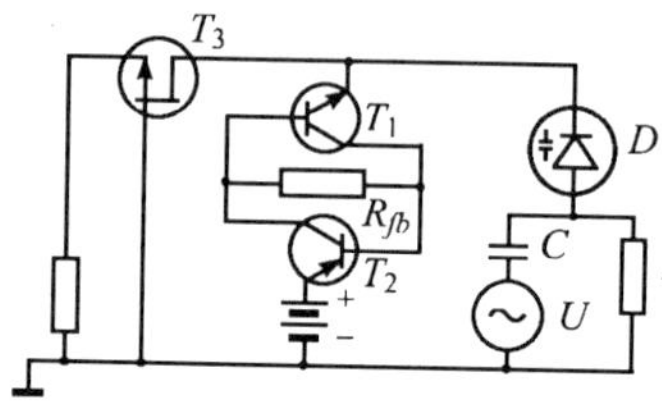

Fig. 3.18. Complex oscillating circuit represented by a double transistor simulation of a $p-n-p-n$ structure with nonlinear inductance and negative resistance, and $p-n$ junction with automatic bias representing nonlinear capacitance and nonlinear conductance

3.2.1. Equivalent circuit and approximation of the nonlinear characteristics

The equivalent circuit of the resonance system under consideration is shown in Fig.3.19, where $R(i)$ and $L(\Psi)$ are the nonlinear negative resistance and the nonlinear negative inductance of the $p-n-p-n$ structure, $C_d(q)$ and $G_d(q)$ are the nonlinear capacitance and nonlinear conductance of the self-biased $p-n$ junction, r is the resistance of the losses, $U = U_m \cos(\omega t_r - \varphi)$ is the external harmonic voltage exciting oscillations in the system.

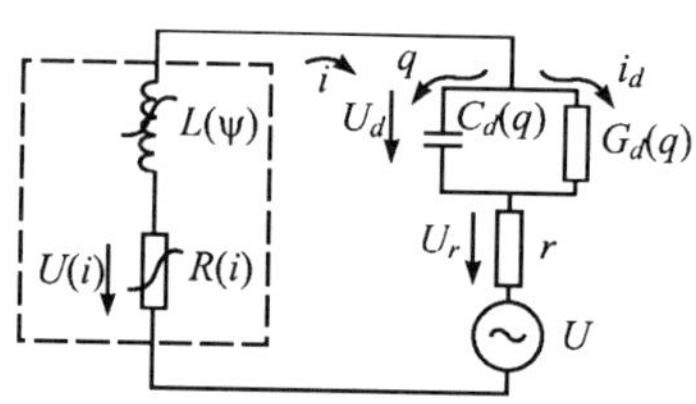

Fig. 3.19. Equivalent circuit of a complex oscillating system with negative resistance and four nonlinear elements

The characteristics of the elements used in the system in Fig.3.19 are represented as polynomial functions:

$$U_d = d_1'q - d_2'q^2 - d_3'q^3, \quad i_d = g_1'q - g_2'q^2 - g_3'q^3, \quad L(\Psi) = \frac{L_0}{1 + L_1'\Psi + L_2'\Psi^2 + L_3'\Psi^3},$$

$U(i) = -k_1'i + k_2'i^2 + k_3'i^3$, where U_d and q are the voltage and the charge on the diode D (Fig.3.18), Ψ is the magnetic flux, L_n' and k_n' $(n = 0, 1, 2, 3)$ are constant coefficients. The coefficients d_n' and g_n':

$$d_1' = \frac{\left(1 + \frac{E_0}{\phi_k}\right)^{\gamma}}{C_0}, \quad d_2' = \frac{\gamma\left(1 + \frac{E_0}{\phi_k}\right)^{2\gamma-1}}{2C_0^2\phi_k}, \quad d_3' = \frac{\gamma(2\gamma-1)\left(1 + \frac{E_0}{\phi_k}\right)^{3\gamma-2}}{6C_0^3\phi_k^2},$$

$$g_1' = \alpha I_s e^{-\alpha E_0} d_1', \quad g_2' = \alpha I_s e^{-\alpha E_0}\left(d_2' - \frac{\alpha d_1'^2}{2}\right),$$

$$g_3' = \alpha I_s e^{-\alpha E_0}\left(d_3' + \alpha d_1' d_2' - \frac{\alpha^2 d_1'^3}{6}\right),$$

are determined by using the dependence of the $p-n$ junction capacitance on the applied voltage and V/A characteristics given by

$$C_d(U) = \frac{C_0}{\left(1 - \dfrac{U}{\phi_k}\right)^{\gamma}}, \quad i_d = I_s(e^{\alpha U} - 1),$$

where $U = U_d - E_0$ and $E_0 = E_0(Q_m)$ is the constant self-bias, which depends on the charge amplitude Q_m, ϕ_k is the contact potential and $\gamma = 1, \dfrac{1}{2}$ or $\dfrac{1}{3}$.

Fig.3.20 presents the analytical and experimental dependence $E_0(Q_m)$ of a semiconductor stabilizer diode at $R = 0,33\ M\Omega$ and $C = 0,1\ \mu$F (See also Fig.3.18). It is clear that in the case of small Q_m the dependence is approximated as $E_0 = -(\varkappa_1 Q_m + \varkappa_2 Q_m^2)$, and when Q_m is large as $E_0 = -\varkappa(Q_m - Q_c)$, where $\varkappa_1$, $\varkappa_2$, $\varkappa_3$, and Q_c are constant coefficients.

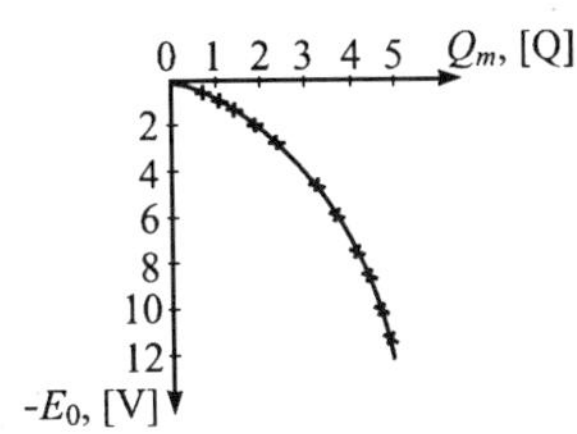

Fig. 3.20. Analytical and experimental dependences of the constant automatic bias of a $p-n$ junction on the amplitude of the flowing alternating charge

Further on, the polynomials approximating the characteristics are used in the nondimensional form:

$$f_1(x_1) \equiv U_d(x) = x_1(1 - d_2 x_1 - d_3 x_1^2), \; H(x_1) \equiv i_d(x_1) = x_1(g_1 - g_2 x_1 - g_3 x_1^2),$$

$$f_2(x_2) \equiv \frac{1}{q_0\omega_0}\frac{x_2}{L(x_2)} = x_2(1 + L_1 x_2 + L_2 x_2^2),$$

$$U(x_2) = -\frac{1}{Q_-} f_2(x_2) + k_2[f_2(x_2)]^2 + k_3[f_2(x_2)]^3,$$

where $x_1 = \dfrac{q}{q_0} = \dfrac{q}{\dfrac{U_0 C_0}{\left(1+\dfrac{E_0}{\phi_k}\right)^\gamma}}$, $x_2 = \dfrac{\Psi}{\omega_0 q_0 L_0}$, $\omega_0 = \dfrac{1}{\sqrt{\dfrac{L_0 C_0}{\left(1+\dfrac{E_0}{\phi_k}\right)^\gamma}}}$,

U_0 and q are, respectively, normalizing voltage and normalizing charge,

$d_2 = \dfrac{d_2' C_0 q_0}{\left(1+\dfrac{E_0}{\phi_k}\right)^\gamma}$, $d_3 = \dfrac{d_3' C_0 q_0^2}{\left(1+\dfrac{E_0}{\phi_k}\right)^\gamma}$, $L_1 = L_1'\omega_0 q_0 L_0$, $L_2 = L_2'(\omega_0 q_0 L_0)^2$,

$g_1 = \dfrac{g_1'}{\omega_0}$, $\quad q_2 = g_2' \dfrac{q_0}{\omega_0}$, $\quad g_3 = g_3' \dfrac{q_0^2}{\omega_0}$, $\quad \dfrac{1}{Q_-} = \dfrac{k_1'}{\omega_0 L_0}$, $k_2 = k_2' \dfrac{q_0}{\omega_0}$, $k_3 = k_3' \dfrac{\omega_0 q_0}{L_0}$.

3.2.2. Equations

The differential equations of the system in Fig.3.19 are

$$\left|\begin{array}{l} \dfrac{d\Psi}{dt_r} + U_d + r\dfrac{\Psi}{L(\Psi)} + U(i) = U_m' \cos(\omega t_r - \varphi) \\ \dfrac{\Psi}{L(\Psi)} = i_d(q) + \dfrac{dq}{dt_r}, cr \end{array}\right.$$

where $\dfrac{\Psi}{L(\Psi)} = i$, t_r is a real time.

Using dimensionless parameters, the equations can be rewritten:

$$\left|\begin{array}{l} \dfrac{dx_1}{dt} = f_2(x_2) - H(x_1) \\ \dfrac{dx_2}{dt} = -f_1(x_1) - U(x_2) - \dfrac{1}{Q} f_2(x_2) + U_m \cos\left(\dfrac{\omega}{\omega_0} t - \varphi\right), \end{array}\right. \tag{3.53}$$

where $t = \omega_0 t_r$, $\dfrac{1}{Q} = \dfrac{r}{\omega_0 L_0}$, $U_m = \dfrac{U_m'}{\omega_0^2 q_0 L_0}$.

The essence of the method applied ([185], See 2.2) is the introduction of new variables

$$y_1 = \left[2\int_0^{x_1} f_1(x_1)dx_1'\right]^{1/2} \operatorname{sign} x_1, \quad y_2 = \left[2\int_0^{x_2} f_2(x_2)dx_2'\right]^{1/2} \operatorname{sign} x_2$$

and a new nonlinear time variable τ,

$$\frac{d\tau}{dt} = \frac{f_1(x_1)}{y_1}\frac{f_2(x_2)}{y_2} = \frac{1}{G_1(y_1)}\frac{1}{G_2(y_2)}.$$

In terms of the newly introduced quantities, Eqs.(3.53) can be rewritten:

$$\left| \begin{array}{l} \dfrac{d^2 y_1}{d\tau^2} + \beta^2 y_1 = \mu F_1[y_1, y_2, \dot{y}_1, \dot{y}_2, \tau] \\ \dfrac{d^2 y_2}{d\tau^2} + \beta^2 y_2 = \mu F_2[y_1, y_2, \dot{y}_1, \dot{y}_2, \tau] \end{array} \right. , \tag{3.54}$$

where

$$F_1[y_1, y_2, \dot{y}_1, \dot{y}_2, \tau] = \frac{\nu}{\mu} y_1 + \frac{1}{\mu} \left\{ \left[-U(x_2) - \frac{1}{Q} f_2(x_2) \right.\right.$$
$$\left.\left. + U_m \cos\left(\frac{\omega}{\omega_0} t - \varphi \right) \right] G_1(y_1) - \frac{d}{d\tau}[H(x_1) G_2(y_2)] \right\},$$

$$F_2[y_1, y_2, \dot{y}_1, \dot{y}_2, \tau] = \frac{\nu}{\mu} y_2 + \frac{1}{\mu} H(x_1) G_2(y_2) + \frac{1}{\mu}\frac{d}{d\tau}\{[-U(x_2)$$
$$- \frac{1}{Q} f_2(x_2) + U_m \cos\left(\frac{\omega}{\omega_0} t - \varphi \right)] G_1(y_1)\},$$

$\nu = 2\dfrac{\omega_n - \omega_r}{\omega_r} = \beta^2 - 1$, ω_r is the resonance frequency, $\omega_n = \dfrac{m}{n}\omega$, $m, n = 1, 2, 3 \ldots$, $\mu \ll 1$ is a small parameter.

The functions $G_1(y_1)$ and $G_2(y_2)$ can be represented as polynomials:

$$G_1(y_1) = b_0^I + b_1^I y + \ldots + b_n y^n, \quad G_2(y_2) = b_0^{II} + b_1^{II} y + \ldots + b_n^{II} y^n.$$

An analogous procedure is applied to obtain

— for the circuit of type I

$$\frac{d^2 y_1}{d\tau^2} + \beta^2 y_1 = \mu F_{11}(y_1, \dot{y}_1, \tau), \tag{3.55}$$

where $\mu F_{11}(y_1, \dot{y}_1, \tau) = \nu y_1 - \dfrac{1}{Q}\dfrac{dy_1}{d\tau} + \dfrac{1}{Q} H(y_1) G_1(y_1) + \dfrac{d}{d\tau}[H(y_1)]$
$-U_m \cos\left(\frac{\omega}{\omega_0} t - \varphi\right) G_1(y_1)$, $Q = \dfrac{\left(1 + \dfrac{E_0}{\phi_k}\right)^{\gamma}}{\omega_0 c_0 r}$, $U_m = U'_m \dfrac{C_0}{q_0 \left(1 + \dfrac{E_0}{\phi_k}\right)^{\gamma}}$,

— for the circuit of type II

$$\frac{d^2 y_2}{d\tau^2} + \beta^2 y_2 = \mu F_{22}(y_2, \dot{y}_2, \tau), \tag{3.56}$$

where $\mu F_{22}(y_2, \dot{y}_2, \tau) = \nu y_2 - \frac{d}{dt}\left[\frac{1}{Q}f(x_2) + U(x_2) - U_m \cos\left(\frac{\omega}{\omega_0}t - \varphi\right)\right]$,
$Q = \frac{\omega L_0}{r}$, $U_m = \frac{U'_m}{\omega_0^2 q_0 L_0}$.

3.2.3. Forced oscillations

The solution of Eqs.(3.54) can be represented as

$$\left|\begin{array}{ll} y_1 = R\cos\beta\tau = R\cos\Psi, & \frac{dy_1}{d\tau} = -\beta R\sin\Psi \\ y_2 = R\sin\beta\tau = R\sin\Psi, & \frac{dy_2}{d\tau} = \beta R\cos\Psi \end{array}\right. , \tag{3.57}$$

where the differential equations for the establishment of the amplitude R and the phase φ of the oscillations are:

$$\left|\begin{array}{rl} 2\beta\frac{dR}{d\tau} & = -\frac{\mu}{\pi}\int_0^{2\pi} F_1[y_1, y_2, \dot{y}_1, \dot{y}_2, \tau]\sin\Psi d\Psi \\ & = \left(-\frac{1}{Q_\Sigma}I_1 + 2k_2 I_0 I_1 + 3k_3 I_0^2 I_1 - U_m\sin\varphi\right)B_0^I - \frac{S_1 B_0^{II}}{2}, \\ 2\beta\frac{d\Psi}{d\tau} & = -\frac{\mu}{\pi}\int_0^{2\pi} F_1[y_1, y_2, \dot{y}_1, \dot{y}_2, \tau]\cos\Psi d\Psi \\ & = -\nu R - \frac{1}{Q_\Sigma}I_0 B_1^I + k_2[I_0^2 B_1^I + I_1 I_2 B_0^I] \\ & + k_3[I_0^3 B_1^I + 3I_0 I_2 B_0^I] - U_m\cos\varphi B_0^I + S_0 B_1^{II}, \end{array}\right. \tag{3.58}$$

$\frac{1}{Q_\Sigma} = \frac{1}{Q} - \frac{1}{Q_-}$.
(3.58) takes into account the following equalities, obtained under condition (3.57):

$$\begin{aligned} H(y_1) &= S_0 + S_1\cos\Psi + S_2\cos 2\Psi + \ldots, \\ f_2(y_2) &= I_0 - I_1\sin\Psi + I_2\sin 2\Psi + \ldots, \\ G_1(y_1) &= B_0^I + B_1^I\cos\Psi + B_2^I\cos 2\Psi + \ldots, \\ G_2(y_2) &= B_0^{II} - B_1^{II}\sin\Psi + B_2^{II}\sin 2\Psi + \ldots \end{aligned}$$

An analogous procedure is applied to obtain
— for the circuit of type I

$$\left|\begin{array}{l} 2\frac{dR}{d\tau} = -\frac{R}{Q} - S_1 - U_m\sin\varphi \\ 2\beta R\frac{d\varphi}{d\tau} = \frac{S_0 B_1^I + S_1 B_0^I}{Q} - U_m\cos\varphi - \nu R \end{array}\right. ; \tag{3.59}$$

— for the circuit of type II

$$\left|\begin{array}{l} 2\dfrac{dR}{d\tau}=-\dfrac{I_1}{Q_\Sigma}+2k_2I_0I_1+3k_3I_0^2I_1+U_m\cos\varphi \\ 2\dfrac{d\varphi}{d\tau}=-\nu-\dfrac{U_m}{R}\sin\varphi. \end{array}\right. \tag{3.60}$$

The amplitude-frequency and phase-frequency characteristics of the complex oscillating circuit are derived from (3.58)

$$\left|\begin{array}{l} \nu=\nu_0^g\pm\dfrac{1}{R}\sqrt{(U_mB_0^I)^2-\left[\left(-\dfrac{1}{Q_\Sigma}I_1+2k_2I_0I_1+3k_3I_0^2I_1\right)B_0^I-\dfrac{S_1B_0^{II}}{2}\right]^2} \\ \varphi=\operatorname{arctg}\dfrac{1}{U_mB_0^I}\dfrac{\left(-\dfrac{1}{Q_\Sigma}I_1+2k_2I_0I_1+3k_3I_0^2I_1\right)B_0^I-S_1B_0^{II}}{\nu_0R+\nu_0^gR}, \end{array}\right. \tag{3.61}$$

where $\nu_0=\dfrac{2(\omega-\omega_{00})}{\omega_{00}}, \omega_{00}=\dfrac{1}{\sqrt{LC_0}}, \nu_0^g=-\dfrac{2\omega}{\omega_{00}}(B_0^IB_0^{II}-1)+\dfrac{1}{R}\Big\{-\dfrac{1}{Q_\Sigma}I_0B_1^I$ $+k_2[I_0^2B_1^I+I_1I_2B_1^I]+k_3[I_0^3B_1^I+3I_0I_2B_0^I]+S_0B_1^{II}\Big\}$ is the equation of the so called „skeleton curve" [51].

The condition for the „linearization" of the amplitude-frequency characteristics is

$$\frac{d\nu_0^g}{dR}=2\left[-\frac{\omega}{\omega_{00}}(b_2^I+b_2^{II})+b_1^{II}\left(\frac{1}{Q_\Sigma}\frac{b_1^I}{2}+\frac{k_2}{2}+\frac{g_1b_1^I}{2}\right)\right]R=0.$$

In this case the condition for the absolute stability of the complex resonance system is

$$\left|\begin{array}{l} \left(\dfrac{1}{Q_\Sigma}+g_1\right)\left(1+\dfrac{R}{1+\dfrac{b_2^I}{2}R^2}\right)+\left(\dfrac{b_2^I}{2Q_\Sigma}+k_2b_1^{II}+g_1\dfrac{b_1^{II}}{2}\right)\left(3+\dfrac{R}{1+\dfrac{b_2^I}{2}R^2}R^2\right)>0 \\ \left[\dfrac{1}{Q_\Sigma}+g_1+\left(\dfrac{b_2^I}{2Q_\Sigma}+k_2b_1^{II}+g_1\dfrac{b_1^{II}}{2}\right)R^2\right]^2+\nu^2>0. \end{array}\right.$$

The analysis of these inequalities shows that the lack of a slope on the skeleton curve $\left(\dfrac{d\nu_0^g}{dR}=0\right)$ is a sufficient condition for the absolute stability of the stationary oscillations.

An analogous procedure is applied to obtain

— for the circuit of type I

$$\nu_0 = \nu_{0s} \pm \sqrt{\left(\frac{U_m}{R}\right)^2 - \left(\frac{1}{Q} + \frac{S_1}{R}\right)^2}, \quad \varphi = -\operatorname{arctg}\frac{\dfrac{1}{Q} + \dfrac{S_1}{R}}{\nu_{0s} - \nu_0}, \tag{3.62}$$

where $\nu_{0s} = -\dfrac{2\omega}{\omega_0}(B_0^I - 1) + \dfrac{S_0 B_1^I + S_1 B_0^I}{QR}$;

— for the circuit of type II

$$\left|\begin{aligned} &\nu_0 = -\frac{2\omega}{\omega_{00}}(B_0^{II} - 1) \pm \sqrt{\left(\frac{U_m}{R}\right)^2 - \frac{1}{R}\left(\frac{I_1}{Q_\Sigma} - 2k_1 I_0 I_1 - 3k_3 I_0^2 I_1\right)^2} \\ &\varphi = \operatorname{arctg}\frac{\nu_0\left(\dfrac{\omega}{\omega_{00}}\right)(B_0^{II} - 1)}{\dfrac{1}{R}\left(\dfrac{I_1}{Q_\Sigma} - 2k_2 I_0 I_1 - 3k_3 I_0^2 I_1\right)}. \end{aligned}\right. \tag{3.63}$$

To determine the area of bistable work – for example of a circuit of type I, equation (3.62) is written in an implicit form

$$F(R, \nu_0) = \left(\frac{R}{Q} + S_1\right)^2 \left[\frac{S_0 B_1^I + S_1 B_0^I}{Q} - \nu_0 R - \frac{2\omega}{\omega_{00}} R(B_0^I - 1)\right]^2 - U_m^2 = 0.$$

The boundaries of the area with bifurcational instability is determined by the condition

$$\frac{d\nu_0}{dR} = \frac{dF(R,\nu_0)/dR}{dF(R,\nu_0)/d\nu} = \frac{dF(R,\nu_0)}{dR} = 0.$$

After certain transformations the following parabola, delineating the area of bifurcational instability, is obtained

$$\begin{aligned} &\left[\frac{3}{4}\left(\frac{g_1}{Q} - \nu_0\right)\left(\frac{3}{2}b_1 b_3 + b_2^2\right)\frac{g_1}{Q} + 3\left(\frac{g_1}{Q}\frac{b_1^2}{2} + \frac{g_1 b_2}{Q} - b_2\nu_0\right.\right. \\ &\left.\left. - 2b_2\right)\left(\frac{1}{4}\frac{g_1}{Q}b_1^2 + \frac{3}{4}\frac{g_1}{Q}b_2 - \frac{1}{2}b_2\nu_0 - b_2\right)\right] R^4 \\ &+ \left[\frac{3}{2}g_1 b_2\left(\frac{1}{Q} + g_1\right) + 2\left(\frac{g_1}{Q} - \nu_0\right)\left(\frac{5}{4}\frac{g_1}{Q}b_1^2 + \frac{13}{4}\frac{g_1 b_2}{Q}\right.\right. \\ &\left.\left. - \frac{5}{2}b_2\nu_0 - 5b_2\right)\right] R^2 + 2\left[\left(\frac{1}{Q} + g_1\right)^2 + \left(\frac{g_1}{Q} + \nu_0\right)^2\right] = 0, \end{aligned} \tag{3.64}$$

where the coefficients b_n $(n = 1, 2, 3, \ldots)$ are obtained from the characteristic of the nonlinear capacitance $C_d(q)$ by using the method presented in [185].

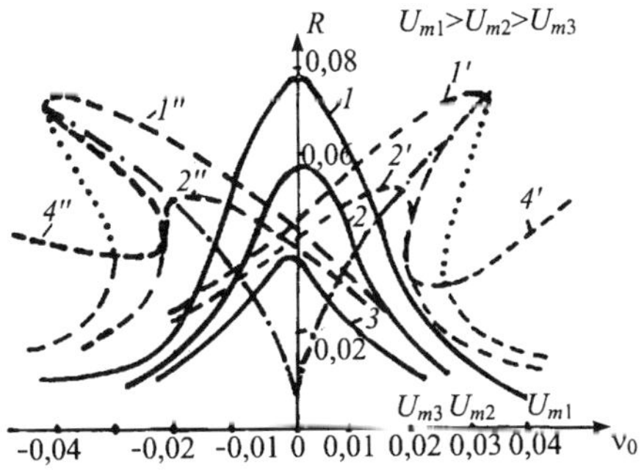

Fig. 3.21. Theoretical amplitude-frequency characteristic of a complex oscillating circuit (continuous curves 1, 2, 3) and types I and II oscillating circuits (dotted curves 1^I, 2^I, 1^{II}, 2^{II}). Curves 4^I and 4^{II} delineate the bifurcational area of instability

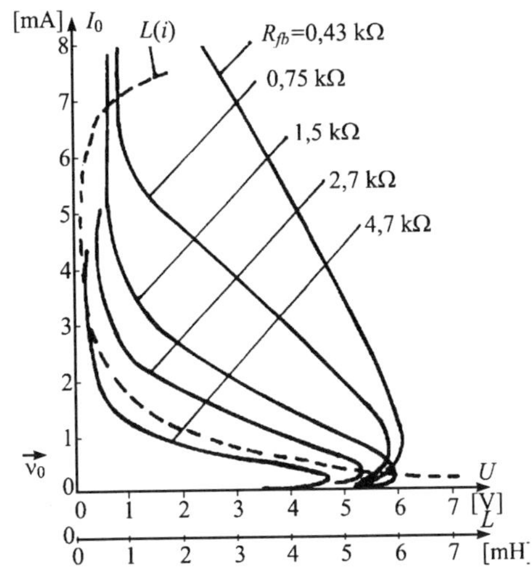

Fig. 3.22. Experimental volt-ampere characteristic of a $p-n-p-n$ structure and dependence of its natural differential inductance on the regime current

Fig.3.21 shows the theoretical amplitude-frequency characteristics of the complex oscillating circuit (curves 1, 2, 3) and of the circuits of types I and II (respectively curves $1^I, 2^I, 3^I$ and $1^{II}, 2^{II}, 3^{II}$), calculated according to formulae (3.61), (3.62) and (3.63) for three values of the exciting voltage. The following elements have been taken into account during the computation: the real characteristics of the nonlinear elements used in the experiment, a quadratic law governing the change in the differential capacitance of the diode with C_0=430 pF and the characteristics of the p–n–p–n structure $R(i)$ and $L(\Psi)$ for R_{fb}=1,5 kΩ and 2,7 kΩ (See Fig.3.22); $\frac{1}{Q_\Sigma} = \frac{1}{Q} = 0,02$, $R = 3,0$ MΩ and $0,33$ MΩ. It is clear that it is possible „to straighten" the resonance curves in such a system, i.e. to eliminate the area of bifurcational instability. The parabola (3.64) describing the area of bifurcational instability of the circuit of type I is given by the dash-line curve 4^I in Fig.3.21.

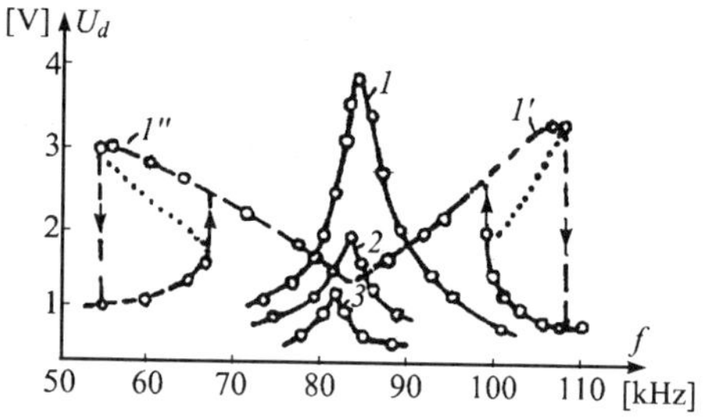

Fig. 3.23. Experimental amplitude-frequency characteristic of a complex oscillating circuit (continuous curves 1 and 2) and of types I and II oscillating circuits (dotted curves 1^I and 1^{II})

Fig.3.23 presents the respective experimental amplitude-frequency characteristics for the parameters given above. There is no area of bifurcational instability on the resonance curve of the complicated oscillating circuit, whose type is close to that of a linear oscillating circuit.

3.2.4. Parametric resonance

The differential equations for parametric resonance in a complex oscillating circuit ($\omega \simeq 2\omega_0$) are:

$$\left|\begin{aligned} 2\beta\frac{dR}{d\tau} &= \left(-\frac{1}{Q_\Sigma}I_1 + 2k_2 I_0 I_1 + 3k_3 I_0^2 I_1\right)B_0^I + U_m B_1^I \Bigg[1 \\ &\quad + \left(\frac{B_1^{II}}{B_0^{II}}\right)^2\Bigg]\sin\varphi - U_m B_0^I\left[1 + \left(\frac{B_1^I}{B_0^I}\right)^2\right]\frac{B_1^{II}}{B_0^{II}}\cos\varphi - \frac{S_1 B_0^{II}}{2}, \\ 2\beta\frac{d\varphi}{d\tau} &= -\nu R - \frac{1}{Q_\Sigma}I_0 B_1^I + k_2[I_0^2 B_1^I + I_1 I_2 B_0^I] \\ &\quad + k_3[I_0^3 B_1^I + 3I_0 I_2 I_3 B_0^I] + U_m B_1^I\left[1 + \left(\frac{B_1^{II}}{B_0^{II}}\right)^2\right]\cos\varphi \\ &\quad + U_m B_0^I\left[1 + \left(\frac{B_1^I}{B_0^I}\right)^2\right]\frac{B_1^{II}}{B_0^{II}}\sin\varphi. \end{aligned}\right. \qquad (3.65)$$

The amplitude-frequency characteristics for this mode are presented as

$$\nu_0 = \nu_{oh} \pm \frac{1}{R}\sqrt{U_m^2\left\{B_1^{I^2}\left[1+\left(\frac{B_1^{II}}{B_0^{II}}\right)^2\right]^2 + \left(\frac{B_0^I B_1^{II}}{B_0^{II}}\right)^2\left[1+\left(\frac{B_1^I}{B_0^I}\right)^2\right]^2\right\} \cdots \rightarrow}$$

$$\rightarrow \cdots \overline{-\left[\left(-\frac{1}{Q_\Sigma}I_1 + 2k_2 I_0 I_1 + 3k_3 I_0^2 I_1\right)B_0^I + \frac{S_1 B_0^{II}}{2}\right]^2}, \tag{3.66}$$

where

$$\nu_{oh} = -\frac{2\omega}{\omega_{00}}(B_0^I B_0^{II} - 1) - \frac{1}{RQ_\Sigma}I_0 B_1^I + \frac{k_2}{R}[I_0^2 B_1^I + I_1 I_2 B_0^I] + \frac{k_3}{R}[I_0^3 B_1^I + 3I_0 I_2 I_3 B_0^I].$$

For the resonance of the third subharmonic ($\omega \simeq 3\omega_0$) one obtains

$$\nu_0 = \nu_{oh} + \frac{S_0 B_1^{II}}{R} \pm \frac{1}{R}\sqrt{U_m^2 B_1^{I^2}\left\{\left(\frac{27}{16}\right)^2\left[\frac{3}{8}\frac{B_1^I}{B_0^I}\left(\frac{B_2^I}{B_0^I}\right)^2 - \frac{B_1^{II}B_2^{II}}{B_0^{II^2}}\right]^2 \cdots \rightarrow}$$

$$\rightarrow \cdots \overline{+\left[\frac{9}{4}\frac{B_1^{II}}{B_0^{II}}\left(1+\frac{3}{4}\frac{B_2^I}{B_0^I}\right)\right]^2\right\} - \left[\left(-\frac{1}{Q_\Sigma}I_1 + 2k_2 I_0 I_1 + 3k_3 I_0^2 I_1\right)B_0^I - \frac{S_1 B_0^{II}}{2}\right]^2}. \tag{3.67}$$

An analogous procedure can be applied to examine the parametric phenomena in oscillating circuits of types I and II. For example, the amplitude-frequency characteristic of the parametric resonance and the third subharmonic resonance for the circuit of type I is obtained in the following form:

— for $\omega \simeq 2\omega_0$

$$\nu_0 = \nu_{oc} \pm \sqrt{\left(\frac{3}{4}\frac{U_m}{R}\frac{B_2^I}{B_0^I}\right)^2 - \left(\frac{1}{Q} + \frac{S_1}{R}\right)^2}, \tag{3.68}$$

— for $\omega \simeq 3\omega_0$

$$\nu_0 = \nu_{oc} \pm \sqrt{\left[\frac{9}{4}\frac{U_m}{R}\left(\frac{B_1^I}{B_0^I}\right)^2\right]^2 - \left(\frac{1}{Q} + \frac{S_1}{R}\right)^2}, \tag{3.69}$$

where $\nu_{oc} = -\frac{2\omega}{\omega_{00}}(B_0^I - 1) + \frac{S_0 B_1^I + S_1 B_0^I}{QR}$.

Figs.3.24 and 3.25 show the theoretical (Eqs.(3.66) and (3.68)) and the experimental amplitude-frequency characteristics of the parametric resonance in

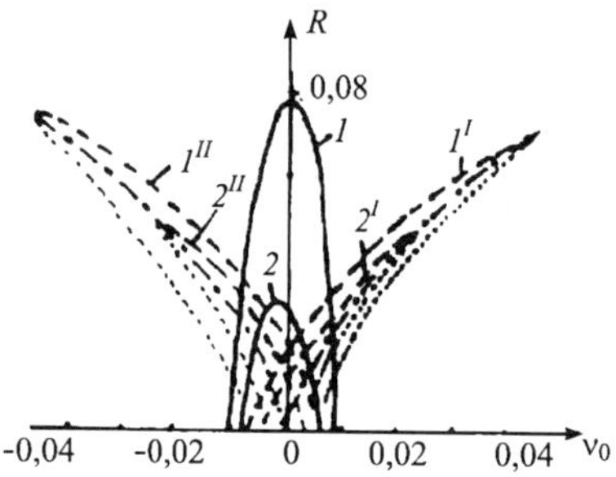

Fig. 3.24. Theoretical amplitude-frequency characteristics of parametric resonance in a complex oscillating circuit (continuous curves 1, 2) and in types I and II oscillating circuits (dotted curves 1^I, 2^I and 1^{II}, 2^{II})

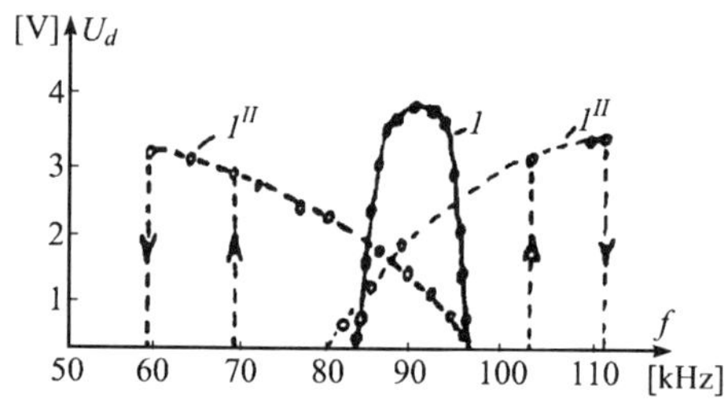

Fig. 3.25. Experimental amplitude-frequency characteristics of parametric resonance in a complex oscillating circuit (continuous curve 1) and in types I and II oscillating circuits (dotted curves 1^I and 1^{II})

a complicated oscillating circuit (curves 1 and 2) and in circuits of types I and II (respectively curves 1^I, 2^I and 1^{II}, 2^{II}).

Equations (3.66) and (3.68) yield formulas for the critical amplitudes necessary for the excitement of parametric resonance:

— for the complicated oscillating circuit

$$U_{mcr}^2 = \frac{R^2\left(\nu_0^2 + \frac{1}{Q_\Sigma^2}\right)}{B_1^{I^2}\left[1 + \left(\frac{B_1^{II}}{B_0^{II}}\right)^2\right]^2 + \left(\frac{B_0^I B_1^{II}}{B_0^{II}}\right)^2\left[1 + \left(\frac{B_1^I}{B_0^I}\right)^2\right]^2}, \tag{3.70}$$

— for the circuit of type I

$$U_{mcr}^{I^2} = \left(\frac{RB_0^I}{B_1^I}\right)^2\left(\nu_0^2 + \frac{1}{Q^2}\right). \tag{3.71}$$

The dependencies (3.70) and (3.71) on the detuning ν_0 are shown in Fig.3.26 (for the complicated oscillating circuit – solid-line curve, for the circuit of type I – dashed line). The conclusion is that the excitation threshold of parametric oscillations in a complicated oscillating circuit can be considerably lower than in a circuit possessing one nonlinear reactive element. This fact can be explained by the improved conditions for parametric energy accumulation in the circuit of two parametric elements. The best conditions for parametric energy input are realized for equal depth of the capacitive and inductive parametric element modulation. If, for example, $B_0^I = B_0^{II}$ and $\dfrac{B_1^I}{B_0^I} = \dfrac{B_1^{II}}{B_0^{II}} = 0,8$, then $\left(\dfrac{U_{mcr}^I}{U_{mcr}}\right)^2 \simeq 5B_0^{I^2}$.

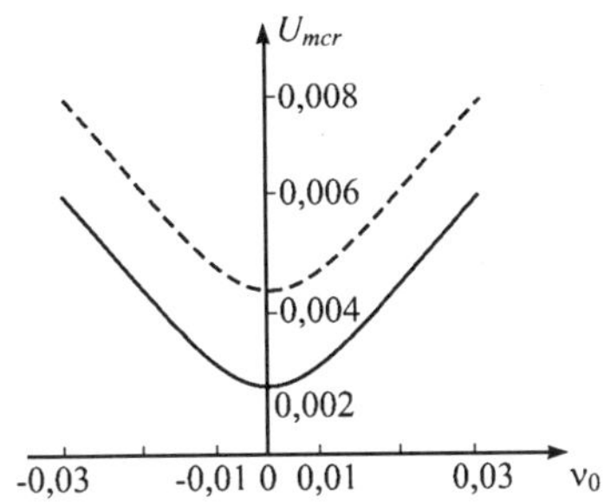

Fig. 3.26. Theoretical dependencies of the critical amplitudes of parametric excitation on the detuning for a complex oscillating circuit (continuous line) and type I circuit (dotted line)

The analytical expressions obtained provide opportunities to examine the influence of each element on the resonance and parametric phenomena in a complicated oscillating circuit.

The complicated oscillating circuit considered can be applied in various amplifiers, generators and convertors, active quality filters. Their construction using microelectronic technology is of great importance.

The analytical approach outlined above can be used to examine the characteristics of any device (particularly those possessing the resonance systems listed in the beginning of the section), which can be presented in the form of a complicated oscillating circuit with two nonlinear reactive elements of antipodal nature.

3.3. Generalized modified method of complex amplitudes for analyzing processes in nonlinear oscillating systems

Consider a nonlinear oscillating system of second order described by two nonlinear differential equations, which can be presented in the following vector form:

$$\frac{d}{d\tau}\mathbf{x} = \mathbf{A}(\mathbf{x})\mathbf{x} + \mathbf{f}(\tau), \tag{3.72}$$

$$\mathbf{A}(\mathbf{x}) = \{a_{ij}(\mathbf{x})\}, \quad i,j = 1,2, \quad \mathbf{x} = \begin{bmatrix} x_1 \\ x_2 \end{bmatrix}, \tag{3.73}$$

$$\mathbf{f}(\tau) = \begin{bmatrix} 0 \\ 1 \end{bmatrix} \sum_{k=0}^{\infty} U_k \cos(k\nu\tau + \varphi_k).$$

System (3.72) can be solved by using the method of successive approximations presented below, where, for the sake of expediency, it will be rewritten as

$$\frac{d}{d\tau}\mathbf{x}^{(l)} = \mathbf{A}(\mathbf{x}^{(l-1)})\mathbf{x}^{(l)} + \mathbf{f}(\tau), \tag{3.74}$$

$\mathbf{x}^{(l-1)}$, $\mathbf{x}^{(l)}$ are the solutions of system (3.72), respectively in the $(l-1)$-th and l-th approximation. Initially we shall determine $\mathbf{x}^{(l)}$ assuming that $\mathbf{x}^{(l-1)}$ is given.

During the analysis we shall use a special version of the method of complex amplitudes. Some major transformations related to this method were explained in section 1.2, while its overall presentation is given in section 5.1 of the book.

It follows from some general considerations that the solutions concerning approximations of arbitrary order can be presented in the form

$$\left| \begin{array}{l} \mathbf{x}^{(l-1)} = \sum\limits_{k=-\infty}^{\infty} \mathbf{x}_k^{(l-1)} \cos(k\nu\tau + \varphi_k^{(l-1)}) \leftarrow \hat{\mathbf{x}}^{(l-1)} = \sum\limits_{k=-\infty}^{\infty} \dot{\mathbf{x}}_k^{(l-1)} e^{jk\nu\tau} \\ \mathbf{x}^{(l)} = \sum\limits_{k=-\infty}^{\infty} \mathbf{x}_k^{(l)} \cos(k\nu\tau + \varphi_k^{(l)}) \leftarrow \hat{\mathbf{x}}^{(l)} = \sum\limits_{k=-\infty}^{\infty} \dot{\mathbf{x}}_k^{(l)} e^{jk\nu\tau}. \end{array} \right. \tag{3.75}$$

Since it is regarded that $\mathbf{x}^{(l-1)}$ is a given quantity, as we substitute it, in the capacity of an argument, in matrix (3.73), we linearize the latter. Now matrix (3.73) can be regarded as directly dependent on time.

$$\mathbf{A}(\mathbf{x}^{(l-1)}) = \mathbf{A}^{(l-1)}(\tau) = \sum_{k=-\infty}^{\infty} 2\mathbf{A}_k^{(l-1)} \cos[k\nu\tau + \varphi_k^{(l-1)}]. \tag{3.76}$$

In this case the periodic matrix function $\mathbf{A}$ is represented in a harmonic Fourier series.

Using (3.75) and (3.76) we rewrite equation (3.74)

$$(jk\nu\mathbf{I} - \dot{\mathbf{A}}_0^{(l-1)})\dot{\mathbf{x}}_k^{(l)} - \sum_{\substack{n=-\infty \\ n\neq k}}^{\infty} \dot{\mathbf{A}}_{k-n}^{(l-1)} \dot{\mathbf{x}}_n^{(l)} = \dot{\mathbf{f}}_k, \tag{3.77}$$

$k = 0, \pm 1, \pm 2, \ldots, \pm\infty$, for negative k $\dot{\mathbf{f}}_k = 0$.

In this case $\mathbf{I}$ is a unit matrix of the second order.

A compact presentation of the latter system can take a matrix form:

$$\bar{\dot{\mathbf{A}}}^{(l-1)}\bar{\dot{\mathbf{x}}} = \bar{\dot{\mathbf{f}}}. \tag{3.78}$$

The bar above the symbols means that the infinite matrices and column vectors are of a block type, i.e. that their elements consist, respectively, of matrixes and column vectors of the second order. The matrix of system (3.78) is of the following type:

$$\begin{array}{c|c|c|c|c}
\cdots & \cdots & \cdots & \cdots & \cdots \\
\hline
 & -j\nu\mathbf{I} - \dot{\mathbf{A}}_0^{(l-1)} & -\dot{\mathbf{A}}_{-1}^{(l-1)} & -\dot{\mathbf{A}}_{-2}^{(l-1)} & \\
\hline
 & -\dot{\mathbf{A}}_1^{(l-1)} & -\dot{\mathbf{A}}_0^{(l-1)} & -\dot{\mathbf{A}}_{-1}^{(l-1)} & \\
\hline
 & -\dot{\mathbf{A}}_2^{(l-1)} & -\dot{\mathbf{A}}_1^{(l-1)} & j\nu\mathbf{I} - \dot{\mathbf{A}}_0^{(l-1)} & \\
\hline
\cdots & \cdots & \cdots & \cdots & \cdots
\end{array} \tag{3.79}$$

The frequency multipliers of the first terms in the main diagonal of the matrix are of the type $jk\nu$, $k = 0, \pm 1, \pm 2, \ldots, \pm\infty$.

The infinite algebraic system (3.79) can be solved approximately by using the method of reduction. A system of three equations can be solved in the first approximation, five equations in the second one, etc. Continuing this process we obtain the q-th approximation as a result of solving a system of $2q + 1$ equations,

$$\hat{x}_0^{(l)} = \sum_{k=-q}^{q} \dot{x}_k^{(l)} e^{jk\Omega t}. \tag{3.80}$$

The approximate solution of the initial equation is obtained from (3.80):

$$x_0^{(l)} = Re\hat{x}_0^{(l)} = \sum_{k=-q}^{q} x_k^{(l)} \cos(k\Omega t + \varphi_k^{(l)}). \tag{3.81}$$

The l-th approximate solution can be estimated in terms of the scalar square of the solutions in the l-th and $(l-1)$-th approximations:

$$\begin{aligned}
\Delta_l^2 &= (x_0^{(l)} - x_0^{(l-1)},\ x_0^{(l)} - x^{(l-1)}) \\
&= \frac{1}{2} \sum_{k=-q}^{q} [x_k^{(l)^2} + x_k^{(l-1)^2} - 2x_k^{(l)} x_k^{(l-1)} \cos(\varphi_k^{(l)} - \varphi_k^{(l-1)})].
\end{aligned}$$

The inequality $\Delta_l \ll \varepsilon$, where ε is the tolerated error, should be satisfied.

If one knows nothing about the solution of Eq.(3.72), he should start with the zero approximation. It should be assumed that in (3.72) $k = 0$. For this quantity system (3.72) is integrated without difficulty. Then, following the procedure described above, the first approximation is obtained, then the second one, and so on until the l-th approximation is obtained with the required precision. It is worth noting that the problems for obtaining all approximations of the solution of system (3.74), except the zero one, are of one and the same type. Coupled with numerical (computation) methods, the proposed approach can ensure a virtually arbitrary preset precision of the solution. The analytical approach proposed allows examining systems of nonlinear differential equations of an arbitrary type.

We shall apply the analytical approach exposed above to the investigation of a serial resonance circuit with nonlinear capacitance $C = \dfrac{C_0}{1+\lambda q^2}$ excited by sine voltage $U = U\cos\omega t$. This is a typical problem, which appears whenever one uses semiconductor parametric diodes (varactors, varicaps), ferroelectric variconds, etc. as well as in the air capacitor cases, when a powerful source of electric energy is used for excitation purposes. Enormous forces undergoing periodic time-dependent changes with a double charge-altering frequency are applied to the capacitor plates. Such a phenomenon occurs, for example, at the output stages of the powerful radio transmitters for special purposes.

The oscillating circuit equation concerning the capacitor charge is

$$\frac{d^2q}{dt^2} + 2\alpha\frac{dq}{dt} + \omega_0^2(1+\lambda q^2)q = \frac{U}{L}\cos\omega t, \tag{3.82}$$

where α is a damping decrement, $\omega_0 = \dfrac{1}{\sqrt{LC_0}}$, λ is a small number, $\lambda q^2 \ll 1$.

Equation (3.82) is a particular case of Eq.(1.27) under section 1.4, if $a = 0$, $A = UL^{-1}$, $\Omega = \omega$.

In our case the approximation we have used presupposes that the coefficient before q changes at frequency 2ω, i.e. $\Omega = 2\omega$. This yields the spectral composition of the charge in the form

$$\omega + k\Omega = \omega(2k+1), \quad k = 0, \pm1, \pm2, \ldots, \pm\infty,$$

i.e. in a stationary regime the charge can contain only odd harmonics of the frequency ω. The solution of Eq.(3.82) and its transform are given in the form

$$q = \sum_{k=-\infty}^{\infty} Q_{2k+1}\cos[(2k+1)\omega t + \varphi_{2k+1}] \leftarrow \hat{q} = \sum_{k=-\infty}^{\infty} \dot{Q}_{2k+1}e^{j(2k+1)},$$

$$\dot{Q}_{2k+1} = Q_{2k+1}e^{j\varphi_{2k+1}}, \quad k = 0, \pm1, \ldots, \pm\infty.$$

The oscillating circuit equation concerning the charge transform is

$$\frac{d^2\hat{q}}{dt^2}+2\alpha\frac{d\hat{q}}{dt}+\omega_0^2\left(1+\lambda\frac{\hat{q}+\check{q}}{2}\right)\hat{q}=\frac{U}{L}c^{j\omega t}, \tag{3.83}$$

where $\check{q}$ is the time function that is complex-conjugate with respect to $\hat{q}$. For $\lambda=0$ the result is a zero approximation of Eq.(3.83):

$$\frac{d^2\hat{q}^{(0)}}{dt^2}+2\alpha\frac{d\hat{q}^{(0)}}{dt}+\omega_0^2\hat{q}^{(0)}=\frac{U}{L}e^{j\omega t}, \tag{3.84}$$

The precise solution of (3.84) is

$$\hat{q}^{(0)}=\dot{Q}^{(0)}e^{j\omega t},$$

$$Q^{(0)}=\frac{U}{L\sqrt{(\omega_0^2-\omega^2)^2+4\alpha^2\omega^2}},\qquad \varphi^{(0)}=-\operatorname{arctg}\frac{2\alpha\omega}{\omega_0^2-\omega^2}.$$

As we know the zero approximation of the solution, we obtain the first approximation of Eq.(3.84):

$$\frac{d^2\hat{q}^{(1)}}{dt^2}+2\alpha\frac{d\hat{q}^{(1)}}{dt}+\omega_0^2\left[\left(1+\frac{\lambda}{2}Q^{(0)^2}\right)+\frac{\lambda}{4}\overset{*}{Q}{}^{(0)}e^{-j2\omega t}\right.$$

$$\left.+\frac{\lambda}{4}\dot{Q}^{(0)}e^{j2\omega t}\right]\hat{q}^{(1)}=\frac{U}{L}e^{j\omega t} \tag{3.85}$$

(the sign „*" denotes the complex-conjugate quantity).

The precise solution of this equation can be presented in the form of an infinite sum. We shall limit our presentation to only four terms, which allow us to estimate the first and the third harmonic:

$$\hat{q}^{(1)}=\sum_{k=-2}^{1}\dot{Q}_{2k+1}e^{j(2k+1)\omega t},\quad \dot{Q}^{(1)}_{2k+1}=Q^{(1)}_{2k+1}e^{j\varphi^{(1)}_{2k+1}}.$$

As $\hat{q}^{(1)}$ is substituted in (3.85) and the terms with equal frequencies are selected, a system of four equations is obtained, which is conveniently represented in the following matrix form.

$$\begin{bmatrix} a_{-3-3} & a_{-3-1} & 0 & 0\\ a_{-1-3} & a_{-1-1} & a_{-11} & 0\\ 0 & a_{1-1} & a_{11} & a_{13}\\ 0 & 0 & a_{31} & a_{33}\end{bmatrix}\begin{bmatrix}\dot{Q}^{(1)}_{-3}\\ \dot{Q}^{(1)}_{-1}\\ \dot{Q}^{(1)}_{1}\\ \dot{Q}^{(1)}_{3}\end{bmatrix}=\begin{bmatrix}0\\0\\ \frac{U}{L}\\0\end{bmatrix}, \tag{3.86}$$

where

$$a_{-3-3} = -9\omega^2 - \omega_0^2\left(1 + \frac{\lambda}{2}\dot{Q}^{(0)^2}\right) - j6\alpha\omega,$$

$$a_{-3-1} = \frac{\lambda\omega_0^2}{4}\overset{*}{Q}^{(0)}, \quad a_{-1-3} = \frac{\lambda\omega_0^2}{4}\dot{Q}^{(0)},$$

$$a_{-1-1} = -\omega^2 + \omega_0^2\left(1 + \frac{\lambda}{2}\dot{Q}^{(0)^2}\right) - j2\alpha\omega,$$

$$a_{-11} = \frac{\lambda\omega_0^2}{4}\overset{*}{Q}^{(0)}, \quad a_{1-1} = \frac{\lambda\omega_0^2}{4}\dot{Q}^{(0)},$$

$$a_{11} = -\omega^2 + \omega_0^2\left(1 + \frac{\lambda}{2}\dot{Q}^{(0)^2}\right) + j2\alpha\omega, \quad a_{13} = \frac{\lambda\omega_0^2}{4}\overset{*}{Q}^{(0)},$$

$$a_{31} = \frac{\lambda\omega_0^2}{4}\dot{Q}^{(0)}, \quad a_{33} = -9\omega^2 + \omega_0^2\left(1 + \frac{\lambda}{2}\dot{Q}^{(0)^2}\right) + j6\alpha\omega.$$

In order to solve system of Eqs. (3.86), we obtain the general D and particular determinants:

$$D = a_{-3-3}a_{33}(a_{-1-1}a_{11} - a_{-11}a_{1-1}) - a_{-1-3}a_{-3-1}(a_{11}a_{33} - a_{13}a_{31})$$
$$-a_{-3-3}a_{-11}a_{13}a_{31}, \; D_{-3} = UL^{-1}a_{-3-1}a_{-11}a_{33}, \; D_{-1} = -UL^{-1}a_{-3-1}a_{-11}a_{33},$$
$$D_1 = UL^{-1}a_{33}(a_{-3-3}a_{-1-1} - a_{-3-1}a_{-1-3}),$$
$$D_3 = -UL^{-1}a_{31}(a_{-3-3}a_{-1-1} - a_{-3-1}a_{-1-3}).$$

The complex amplitudes of the solution are expressed as

$$\dot{Q}_{-3}^{(1)} = \frac{D_{-3}}{D}, \quad \dot{Q}_{-1}^{(1)} = \frac{D_{-1}}{D}, \quad \dot{Q}_1^{(1)} = \frac{D_1}{D}, \quad \dot{Q}_3^{(1)} = \frac{D_3}{D}.$$

The transform of the first and third harmonics of the solution are obtained by means of the complex amplitudes:

$$\hat{\overline{q}}_1^{(1)} = \dot{Q}_1^{(1)}e^{j\omega t} + \dot{Q}_{-1}^{(1)}e^{-j\omega t}, \quad \hat{\overline{q}}_3^{(1)} = \dot{Q}_3^{(1)}e^{j3\omega t} + \dot{Q}_{-3}e^{-j3\omega t}.$$

As the transforms are known, the harmonics themselves can be put down

$$\overline{q}_1^{(1)} = Re\hat{\overline{q}}_1^{(1)} = Re[(\dot{Q}_1^{(1)} + \overset{*}{Q}_{-1}^{(1)})e^{j\omega t}] = \overline{Q}_1^{(1)}\cos(\omega t + \overline{\varphi}_1^{(1)}),$$

$$\overline{q}_3(1) = Re\hat{\overline{q}}_3^{(1)} = Re[(\dot{Q}_3^{(1)} + \overset{*}{Q}_{-3}^{(1)})e^{j3\omega t}] = \overline{Q}_3^{(1)}\cos(3\omega t + \overline{\varphi}_3^{(1)}),$$

where

$$\overline{Q}_1^{(1)} = \sqrt{(Q_1^{(1)}\cos\varphi_1^{(1)} + Q_{-1}^{(1)}\cos\varphi_{-1}^{(1)})^2 + (Q_1^{(1)}\sin\varphi_1^{(1)} - Q_{-1}^{(1)}\sin\varphi_{-1}^{(1)})^2},$$

$$\overline{Q}_3^{(1)} = \sqrt{(Q_3^{(1)}\cos\varphi_3^{(1)} + Q_{-3}^{(1)}\cos\varphi_{-3}^{(1)})^2 + (Q_3^{(1)}\sin\varphi_3^{(1)} - Q_{-3}^{(1)}\sin\varphi_{-3}^{(1)})^2},$$

$$\operatorname{tg}\overline{\varphi}_1^{(1)} = \frac{Q_1^{(1)}\sin\varphi_1^{(1)} - Q_{-1}^{(1)}\sin\varphi_{-1}^{(1)}}{Q_1^{(1)}\cos\varphi_1^{(1)} + Q_{-1}^{(1)}\cos\varphi_{-1}^{(1)}},$$

$$\operatorname{tg}\overline{\varphi}_3^{(1)} = \frac{Q_3^{(1)}\sin\varphi_3^{(1)} - Q_{-3}^{(1)}\sin\varphi_{-3}^{(1)}}{Q_3^{(1)}\cos\varphi_3^{(1)} + Q_{-3}^{(1)}\cos\varphi_{-3}^{(1)}}.$$

In the case of small values of λ in (3.82) this can be the end of the analysis. Otherwise a second approximation equation is drafted with respect to the transforms

$$\frac{d^2\hat{q}^{(2)}}{dt^2} + 2\alpha\frac{d\hat{q}^{(2)}}{dt} + \omega_0^2\left\{1 + \lambda\frac{\overset{\cdot}{\overline{Q}}{}_1^{(1)^2} + \overset{\cdot}{\overline{Q}}{}_3^{(1)^2}}{2}\right.$$

$$+\frac{\lambda}{2}\left[\left(\frac{\overset{\cdot}{\overline{Q}}{}_1^{(1)^2}}{2} + \overset{*}{\overline{Q}}{}_1^{(1)}\overset{\cdot}{\overline{Q}}{}_3^{(1)}\right)e^{j2\omega t} + \left(\frac{\overset{*}{\overline{Q}}{}_1^{(1)^2}}{2} + \overset{\cdot}{\overline{Q}}{}_1^{(1)}\overset{*}{\overline{Q}}{}_3^{(1)}\right)e^{-j2\omega t}\right.$$

$$\left.\left.+\frac{\overset{\cdot}{\overline{Q}}{}_1^{(1)}\overset{\cdot}{\overline{Q}}{}_3^{(1)}}{2}e^{j4\omega t} + \frac{\overset{*}{\overline{Q}}{}_1^{(1)}\overset{*}{\overline{Q}}{}_3^{(1)}}{2}e^{-j4\omega t}\right]\right\}\hat{q}^{(2)} = \frac{U}{L}e^{j\omega t}. \qquad (3.87)$$

The approximate solution of this equation is obtained by using the methods presented above. Approximations of a higher order can be obtained in an analogous manner.

It is notable that the coefficients of Eq.(3.87) and the following approximations do not change the sign and that they are perfectly smooth. It follows from the theory of Fourier series that in this case the amplitude value of the harmonics desreases as the order increases, i.e. the sum of the Fourier series displays absolute and uniform convergence. In such cases it is not necessary to provide special evidence on the convergence in the solution of the infinite system of equations.

3.4. Implementation of parametric one-ports

Parametric one-ports with a high-frequency change in the parameter are implemented mostly through the action of voltage or high-intensity current on nonlinear reactive and resistive elements and devices [78, 109]. Parametric one-ports are obtained mainly by using nonlinear capacitive and active conductance of a p-n junction of a semiconductor (parametric) diode. Varicond and reactance valve capacitances, nonlinear inductances of ferrite cores and thin magnetic films, tunnel diode, Gunn diode, avalanche transit time diode conductances and transistor collector junction [78, 85, 235] are used less frequently. Surface semiconductor capacitances, ferroelectric structures, impedance converters with modulation in the reactive feed-back circuit, Josephson superconductor contacts, special capacitors with a light-beam controller capacitance, etc. [109] can also be used as parametric reactances.

The major difference between reactive and active elements consists in that oscillation energy in an electric and magnetic form is accumulated and exchanged in reactive elements; at the same time resistances (positive or negative) are elements, which serve as a means of dispersing or increasing the energy in the system. The reactive and active one-ports with time-dependent parameters (parametric one-ports) have both common and distinctive features. For example, the ability to generate harmonics is proper both to nonlinear and parametric capacitances and inductances, and to nonlinear and parametric resistances. A nonlinear element is also required, when alternating current is converted into direct current, but in this case the nonlinear element can only be a nonlinear resistance. A generalization of the basic properties of nonlinear reactive elements is provided by the theorem formulated by M. D. Karasev [236]: „It is not possible to obtain direct current from alternating current with the aid of a purely reactive element, even if this element is nonlinear. It is not possible to convert direct current into periodic alternating current by using a system consisting of an arbitrary combination of nonlinear reactive elements and constant resistances." A direct consequence of these principles is the postulate that it is impossible to design a negative reactance, which, having only its internal feedbacks and devoid of external sources of alternating electromotive force could impart energy to an external circuit. Negative resistance can be obtained artificially at the expense of a direct current source and a feedback controlling the conductance only under the condition that the total energy, which can be communicated by the negative resistance sector, exceeds the energy needed to control the feedback.

The nonlinear capacitance of the $p-n$ junction has become particularly common as a nonlinear reactive element in semiconductor parametric systems owing to its simplicity of design, compactness, low natural losses and the actual inertia-free barrier capacitance up to 10^{12} Hz [85].

It is a well-known fact already used in section 3.2, that the barrier capacitance complies with the following law of external voltage alteration [168]:

$$C_b = \frac{dq}{dU} = \frac{C(0)}{\left(1 - \frac{U}{\varphi_k}\right)^{\gamma}}, \tag{3.88}$$

where $C(0)$ is the differential capacitance at external voltage equal to zero, $\gamma = 1, \frac{1}{2}$ or $\frac{1}{3}$ is a coefficient depending on the type of the $p-n$ junction, and φ_k is the contact difference of the potentials.

Fig.3.27 shows an example of a volt-farad and volt-ampere characteristic of the semiconductor diode. The value U_f limits the range of the direct voltage values for which the current flowing through the diode is virtually zero. In the case of silicon diodes the values of U_f can fall in the range from 0.2 to 0.5 V, while in the case of gallium arsenide diodes U_f has higher values. From the viewpoint of the realization of parametric one-ports, the voltage U_f can be regarded as one of the parameters of the semiconductor diode.

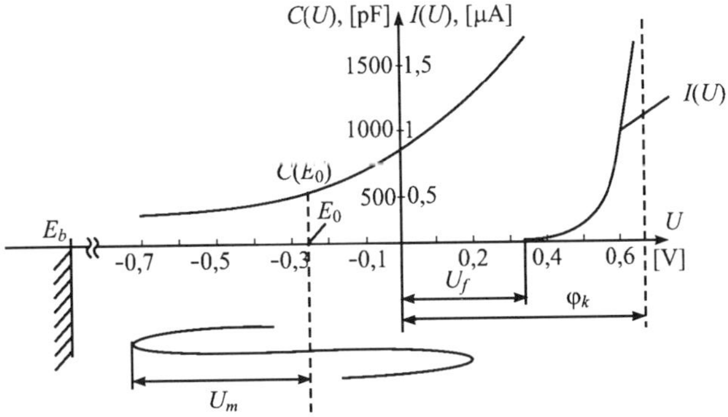

Fig. 3.27. Volt-farad and volt-ampere characteristics of a semiconductor diode of the D817A type

The parameter U_f, together with the parameter E_b (the inverse voltage of $p-n$ junction break-down) determine the possible maximum amplitude of the voltage U_m at the diode in the absence of direct currents.

We shall term the relation $\alpha_V = \dfrac{U_m}{\varphi_k - E_0}$ coefficient of utilization of the characteristic of the semiconductor diode. The maximum value of the coefficient α_V for a given negative bias E_0 in the absence of direct currents is:

$$\alpha_{Vm} = \frac{U_f - E_0}{\varphi_k - E_0}.$$

The coefficient α_{Vm} grows with the increase of the negative bias to the maximum possible value

$$\alpha_{Vm}^{(m)} = \frac{U_f - E_b}{2\varphi_k - U_f - E_b}.$$

Fig.3.27 shows a regime of pumping with harmonic voltage (HVP). The other extreme case is harmonic current pumping (HCP). This regime requires a dependence of the diode barrier capacitance on the charge applied.

As (3.88) is integrated at initial conditions $U = 0$ and $q = 0$, the following expression is obtained for the volt-coulomb characteristics of the parametric diode

$$q = Q_k \left[1 - \left(1 - \frac{U}{\varphi_k}\right)^{1-\gamma}\right], \tag{3.89}$$

where $Q_k = \dfrac{C(0)\varphi_k}{1-\gamma}$.

As U in (3.89) is expressed through q and substituted in (3.88), the required relationship is obtained:

$$C(q) = C(0)\left(1 - \frac{q}{Q_k}\right)^{\frac{-\gamma}{1-\gamma}}. \tag{3.90}$$

Fig.3.28 provides examples of a coulomb-farad and volt-coulomb characteristic of a semiconductor diode corresponding to (3.89) and (3.90).

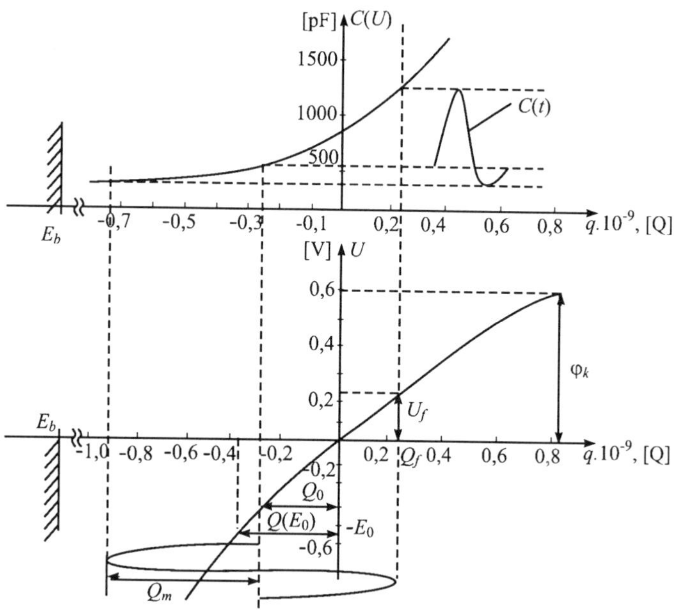

Fig. 3.28. Coulomb-farad and volt-coulomb characteristic of a semiconductor diode of the D817A type

There is a peculiar aspect that is worth mentioning. It follows from the volt-coulomb characteristic shown in Fig.3.28 that additional constant voltage should appear as a result of the nonlinearity of this characteristic. But in accordance with M.D. Karasev's general theorem [236], it is impossible to achieve detection (rectification) by using a nonlinear reactive (in this case capacitive) element. This apparent paradox is resolved by assuming that the high-frequency pumping results in the occurrence of an additional constant charge on the nonlinear capacitor that meets the requirement of zero additional constant voltage. The diode retains only the external constant bias E_0, while the constant charge at its capacitance changes to $Q_0 + \Delta Q_0$, where ΔQ_0 is the additional charge caused by the action of the high frequency pumping.

An analogy can be drawn with the HVP regime, when identifying the maximum value of the factor characterizing the utilization of the parametric diode characteristic in the HCP regime, i.e.

$$\alpha_{qm} = \frac{Q_f - Q_0}{Q_k - Q_0}. \tag{3.91}$$

It follows from the expressions for α_{Vm} and α_{qm}, that at $U_f = \varphi_k$ the utilization factor of the diode characteristic is equal to unity and does not depend on the choice of the operating point and the amplitude of the pumping voltage. In the case of real diodes, it is always true that $U_f < \varphi_k$ and that the maximum coefficient α_{Vm} or α_{qm} tends to unity only in a direct current mode and at high voltage (or current) pumping amplitudes.

Most frequently, the investigation of the parametric systems takes place within the framework of the small signal theory. The limitations imposed by this theory come to the requirement that the intensity of the signal and the combination components at the nonlinear element should be small in comparison with the intensity of the pumping generator. When this condition is met, the nonlinear element with respect to the signal and the combined oscillations can be presented as a linear one with time-dependent periodic pulsation with the pumping frequency and its harmonics. As we saw in sections 1.2 and 1.3, such a presentation simplifies by far the mathematical description of the conversion processes; moreover, it makes it possible to take into account the phase difference in the pumping of the active and reactive elements.

As we set $U = -E_0 + U_m \cos\omega_p t$ in (3.88), the following is obtained:

$$C(t) = C(E_0)(1 - \alpha_V \cos\omega_p t)^{-\gamma}. \tag{3.92}$$

Analogously for the opposite (inverse) capacitance (that is elastance)

$$S(t) = \frac{1}{C(E_0)}(1 - \alpha_V \cos\omega_p t)^{\gamma}.$$

Similarly, for the HCP mode we set $q = -Q_{E_0} + Q_m \cos\omega_p t$ in (3.90) to obtain

$$\left| \begin{aligned} C(t) &= C(Q_{E_0})(1 - \alpha_q \cos\omega_p t)^{\frac{-\gamma}{1-\gamma}} \\ S(t) &= \frac{1}{C(E_0)}(1 - \alpha_q \cos\omega_p t)^{\frac{\gamma}{1-\gamma}} \end{aligned} \right. . \tag{3.93}$$

Regardless of the power supply mode of the diode, its capacitance and elastance can be presented through the following Fourier series (See also (1.25)):

$$\left| \begin{aligned} C(t) &= C_0 + 2\sum_{n=1}^{\infty} C_n \cos n\omega_p t \\ S(t) &= S_0 + 2\sum_{n=1}^{\infty} S_n \cos n\omega_p t. \end{aligned} \right. \tag{3.94}$$

The coefficients of the series (for example in (3.94)) will differ at $\alpha_V = \alpha_q$, as follows from (3.92) and (3.93), hence the energy transformation in the devices employing nonlinear reactances will depend on the power-supply mode of the diode. Some studies show that at high pumping levels the HVP mode ensures more effective energy conversion in the broad-band modulation-parametric converters.

In a mode of a partially forward-biased $p-n$ junction, the sum total of the diffusion current and the current of the barrier capacitance flows through the diode. The diffusion theory of the flat $p-n$ junction developed by Shokly is mostly applicable to lab germanium diodes. The distinctions from this model can be explained by the generation-recombination processes in the space charge layer, which can be actually accounted for by introducing a coefficient m in the analytical expression of the $p-n$ junction volt-ampere characteristic [78, 231]:

$$I = I_S\left(e^{\frac{qU}{mkT}} - 1\right), \tag{3.95}$$

where I_S is the saturation current, U is the external voltage applied to the $p-n$ junction, q is the electron charge, k is Boltzmann's constant, T is the absolute temperature. In addition, the components of the parallel equivalent circuit of the $p-n$ junction are expressed in the following way:

— differential conductance

$$G = \frac{q(I+I_S)}{mkT}\frac{\sqrt{\sqrt{1+\omega^2\tau^2}+1}}{\sqrt{2}}; \tag{3.96}$$

— diffusion capacitance

$$C_d = \frac{q(I+I_S)}{mkT}\frac{\tau}{\sqrt{2}\sqrt{\sqrt{1+\omega^2\tau^2}+1}}; \tag{3.97}$$

— barrier capacitance

$$C_e = \frac{C_0}{\left(1+\dfrac{U}{\varphi_k}\right)^{\gamma}} = C_0\left[1 - \frac{mkT}{q\varphi_k}\ln\left(\frac{I}{I_S}-1\right)\right]^{-\gamma}, \tag{3.98}$$

where τ is the life-time of the minority carriers in the diode base, ω is the circular frequency of the small alternating signal.

Figs.3.29 and 3.30 present, in a double logarithmic scale [78, 231], some typical experimental dependencies of a $p-n$ junction conductance and capacitance for a silicon and geranium diode in the current range of 1 – 100 μA and a frequency range of 5 – 500 kHz. The solid lines show the theoretical dependence of the conductance G and the sum of capacitances $C = C_d + C_e$. For the sake of convenience during the comparison, there are additional entries along the x-axis: the direct voltage for the silicon diodes and the direct diffusion current $I_{dif} = I_S e^{\frac{qU}{mkT}}$ for the germanium diodes.

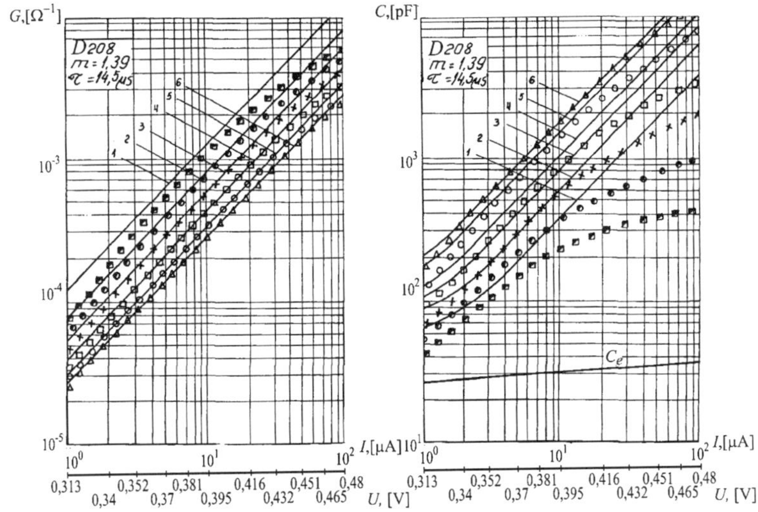

Fig. 3.29. Experimental (with different dots) and theoretical (continuous lines) dependencies of the conductance (a) and capacitance (b) of a silicic semiconductor diode of the D208 type: *1* – 500, *2* – 200, *3* – 100, *4* – 50, *5* – 15, and *6* – 5 kHz; $m = 1,39$, $\tau = 14$ μs

Fig.3.31 provides an answer to the logical question of how the dependencies of the capacitance C and conductance G of the $p-n$ junction are going to change, given a further increase in the direct current. Fig.3.31 provides a clear picture of the values of the inverse currents I_i, for which the capacitive nature of the differential reactance is changed to inductive. The dotted line shows the value of the „negative" capacitance obtained experimentally, while the dash-dot line indicates the equivalent inductance of the diode at frequency 500 kHz.

When working with high pumping voltage amplitudes ($\gg \frac{kT}{q}$) and periodic

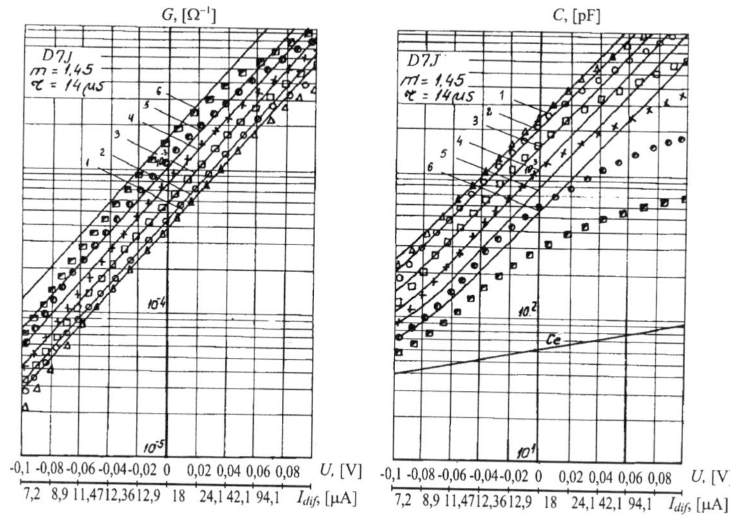

Fig. 3.30. Experimental (with different dots) and theoretical (continuous lines) dependencies of the conductance (a) and capacitance (b) of a germanium semiconductor diode of the D7J type ($m = 1,45$ $\tau = 14\ \mu s$); *1* – 500, *2* – 200, *3* – 100, *4* – 50, *5* – 15, and *6* – 5 kHz

forward-conducting of the $p - n$ junction, one should bear in mind the following typical aspects:

1. In the case of low frequencies $\left(\omega \leq \dfrac{1}{\tau}\right)$, the small signal differential characteristics $G(I, \omega)$ and $C_d(I, \omega)$ can be used, e.g. those shown in Fig.3.29 and 3.30, and as the frequency dependencies are taken into account a spectral analysis of the pulsative complex one-port can be made.

2. In the case of higher frequencies $\left(\omega \gg \dfrac{1}{\tau}\right)$, the effect of accumulating quasi-stationary charge of minority carriers in the diode base should be taken into account.

Indeed, within the action time of the positive half-wave of the current, a charge of minority carriers is accumulated in the diode base. It fails to attain complete recombination over that time – $t = \dfrac{T}{2}$ (T is the period of the high-frequency pumping voltage). Therefore, at time $t = \dfrac{T}{2}$ the diode is in imbalance, i.e. it is a source of voltage $\Delta U(t)$, which causes the emergence of an additional current component, whose value is not set by an external source – it depends on the diode parameters, on the circuit, operation mode and frequency of the signal. The presence of a residual charge in the base is equivalent to a reduction of the

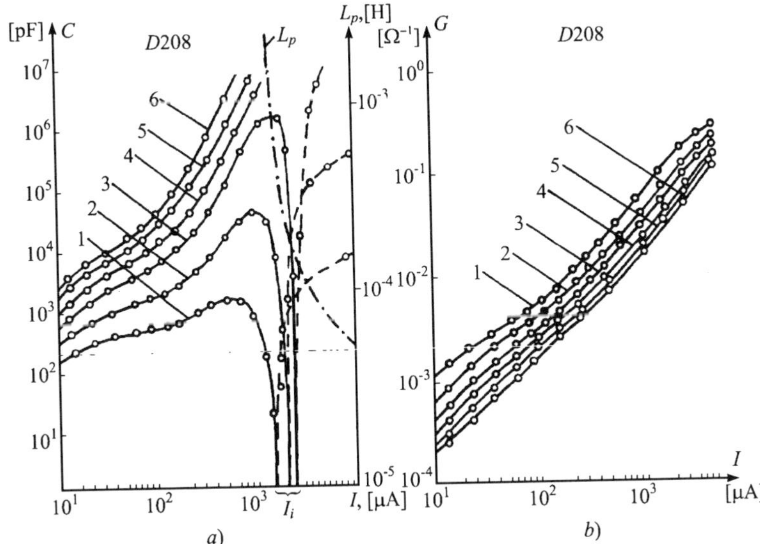

Fig. 3.31. Experimental dependencies of the differential capacitance (a) and conductance (b) of a silicic diode of the D208 type in the case of high injection levels: *1* – 500, *2* – 200, *3* – 100, *4* – 50, *5* – 15, and *6* – 5 kHz

equilibrium potential difference of the $p-n$ junction by $\Delta U(t)$. This results in a change in the constant current component both in value and in time, as well as in a bias shift. The influence of the effect of accumulation and establishment of a quasi-stationary charge of minority carriers on the performance of the semiconductor diodes in radio physical systems will be considered in detail in the following section.

When the semiconductor diode operates in a direct current mode, its nonlinear differential conductance, rendered pulsative under the action of the pumping voltage, can be presented in terms of the small signal theory through a Fourier series (See also 1.25):

$$G_d = G_0 + 2\sum_{n=1}^{\infty} G_n \cos n\omega t, \tag{3.99}$$

where

$$G_n = \frac{1}{2\pi}\int_{-\pi}^{\pi} G_d \cos n\omega t d(\omega t).$$

If the high-frequency voltage at the diode is relatively high, for example several volts, then in practice it is often sufficient to present the volt-ampere characteristic in the following simplified way:

$$I = \begin{cases} G_i U, & \text{U>0} \\ 0, & \text{U<0} \end{cases}. \tag{3.100}$$

The operation mode of the nonlinear conductance can also be convincingly described by using the concept „line segment angle":

$$\Theta_{om} = \arccos \frac{U_0}{U_m},$$

where U_0 is the direct voltage at the load connected in the circuit of the nonlinear conductance, U_m is the amplitude of the alternating pumping voltage. In this case, the convenience stems from the fact that the line segment angle is unambiguously set by the ratio of the load resistance at the signal frequency (R_T) and the internal (natural) resistance of the diode ($R_i = \frac{1}{G_i}$) through the formula:

$$\frac{\operatorname{tg}\Theta_{om} - \Theta_{om}}{\pi} = \frac{R_i}{R_T} = \frac{Y_T}{G_i}.$$

The real values of the ratio $\frac{R_i}{R_T} - 0,2 \geq \frac{R_i}{R_T} \geq 0,001$ encountered in practice have the following corresponding range of values for the line segment angle: $45° \geq \Theta_{om} \geq 15°$.

As regards the parametric diode with external bias, the line segment angle is related to the parameter α_V in the following way:

$$\cos\Theta_{om} = \frac{1}{\alpha_V\left(1 + \frac{\varphi_k}{U}\right)}.$$

The concept of „line segment angle" can also be used when the parametric diode operates in an automatic bias mode. In this case, in order to determine the line segment angle, one has to measure the direct current I_0, which can be related to the amplitude of the alternating pumping voltage U_m and Θ_{om} in the following way:

$$I_0 = \frac{U_m}{\pi R_i}(\sin\Theta_{om} - \Theta_{om}\cos\Theta),$$

hence

$$\Theta_{om} \simeq \sqrt[3]{\frac{3\pi R_i I_0}{U_m}}.$$

The Fourier coefficients in (3.99) can be expressed through the angle of the line segment Θ_{om} in the following manner:

$$G_0 = \frac{G_i}{2\pi}2\Theta_{om}, \quad G_1 = \frac{G_i}{2\pi}2\sin\Theta_{om}, \quad G_2 = \frac{G_i}{2\pi}\sin\Theta_{om},$$

etc.

For the purpose of ensuring a high Q-factor in the resonance radiophysical systems, one usually uses a mode of inconsiderable transition into the direct current range of the diode characteristic. In this case the approximation of the volt-ampere characteristic (3.100) is quite crude, so the analytical expression (3.95) will be used.

In this case the Fourier coefficients in (3.99) can be expressed as follows:

$$G_0 = \alpha I_s e^{-\alpha E_0} J_0(\alpha U_m) = \alpha I_0,$$

$$G_1 = \frac{1}{2}\alpha I_s e^{-\alpha E_0} J_1(\alpha U_m) = \frac{1}{2}\alpha I_0 \frac{J_1(\alpha U_m)}{J_0(\alpha U_m)},$$

$$G_2 = \frac{1}{2}\alpha I_s e^{-\alpha E_0} J_2(\alpha U_m) = \frac{1}{2}\alpha I_0 \frac{J_2(\alpha U_m)}{J_0(\alpha U_m)},$$

etc., where $J_n(x)$ are the modified Bassel functions of the n-th order.

Studies of Josephson's quantum-mechanical effect conducted in recent years have shown that the Josephson element has nonlinear inductance, which can be considered as a parametric element [140].

As first shown by Josephson, the phase difference φ of the wave functions describing superconductive electrodes is related to the junction voltage V through the relation

$\frac{d\varphi}{dt} = \frac{2\pi}{\phi_0}V$, $\phi_0 = \frac{h}{2q_e} = 2.0710^{-15}$ Vb, q_e is the electron charge, h is Planck's constant.

The junction current may have a „superconductive" component (supercurrent) I_s, which depends on the phase φ: $I_s = I_c \sin\varphi$, where I_c is the so-called critical junction current.

Provided that small alternating current $\tilde{I}_s$ flows through the junction, it conditions the appearance of an additional change in the phase $\tilde{\varphi}$. Then we can write: $I_s + \tilde{I}_s = I_c \sin(\varphi + \tilde{\varphi})$.

As we expand $\sin(\varphi + \tilde{\varphi})$ and retain only the first order terms with respect to φ, we obtain

$$\tilde{I}_s = \frac{1}{L}\int \tilde{V}\,dt, \qquad \tilde{\varphi} = \frac{2q_e}{h}\int \tilde{V}\,dt,$$

where L has the dimensions of an inductance: $\frac{1}{L} = \frac{1}{L_c}\cos\varphi$, $L_c = \frac{\phi_0}{2\pi I_c}$.

With respect to small external signals the occurring supercurrent with its properties is equivalent to a reactive element with differential inductance L.

The inductances that we are aware of are divided into geometric and kinetic according to the type of the energy stored. The geometric (magnetic) inductances are characterized by the capacity of current carriers to accumulate magnetic energy. Kinetic inductance has to do with the inertia of the current carriers. In reality, kinetic inductance does not occur in ordinary metals as a consequence of a relatively large dispersion of the current carriers.

In the case of the Josephson effect, the inductance of the contacts is of a kinetic nature. Its nonlinearity is associated with a quantum-mechanical interferential effect of the wave functions. The typical dependence of the Josephson contact inductance on the phase φ determines the two basic modes of its utilization in parametric systems: with external pumping and with self-pumping.

A major peculiarity of the Josephson supercurrent as a parametric element is the possibility to achieve considerable depth of parametric modulation. The Josephson inductance can take negative values for certain phases. On this basis a generic presentation of the effect of single-frequency non-degenerate parametric regeneration has been attempted [189]. The occurrence of this effect is associated with the possibility for the reactive parametric element to take negative values periodically in the process of its alteration. Section 2.1 has shown that similar phenomena are observed in modulation-parametric systems, where strong regeneration effects are manifested as a result of the reversibility of the interactions. Given a greater modulation depth, as in the case of a Josephson inductance with periodic transition to the area of negative values, there is no further need for a resonance circuit and the regenerative effects are also manifested in an aperiodic system.

3.5. Influence of the effect of accumulation of a minority carrier charge on the performance of semiconductor diodes in radiophysical systems

The classical $p-n$ junction diffusion theory developed by Shokli [168] assumes that the $p-n$ junction has no resistive losses. Moreover, as diffusive conductance is calculated, it is regarded that the $p-n$ junction is connected to a voltage source with zero internal resistance. Given this formulation of the problem, the charge of minority carriers in the diode base is „obliged" to follow the voltage applied, since the recharge of the diffusive capacitance conditioned by this charge should take place „instantly" through the zero resistance of the source. Consequently, this charge is always equal to zero at the beginning of each subsequent period of high frequency voltage and it is not accumulated at the diode base.

Since the real diode always has finite resistance of losses, the mechanism of signal detecting and frequency dependence of the diffusive conductance of the p-n junction should be determined by the accumulation process. In addition, the well-known definitions of „quadratic" and „linear" signal detection become meaningless.

Usually, when the inertia properties of the semiconductor diodes are analyzed in the written sources, the transient characteristics, for which a voltage impulse is taken as an initial signal, are considered. But in the case of such radiophysical devices as detectors, rectifiers, frequency multipliers, parametric systems with semiconductor diodes in a direct current mode, etc., a quasi-harmonic signal is fed at the input and the issue of the choice of a diode cannot be unequivocally resolved on the basis of its transient characteristic, for in this case there is no unambiguous correspondence between the transient and the frequency characteristic.

Both theory and practice indicate that the semiconductor diodes in the resonance circuits of radiophysical devices can operate within quite a broad frequency range, and that they are also used at frequencies for which $\omega\tau \gg 1$, where τ is the life-time of the minority carriers at the diode base. This possibility is mostly conditioned by the actual lack of inertia of the barrier capacitance of the $p-n$ junction, up to frequencies of 10^{12} Hz, and the „top" limit is necessitated by the resistance of the semiconductor volume R_s.

As the semiconductor diode operates in a transfer mode, for a part of the period the inertia of the diffusive capacitance and active conductance influence tangibly the frequency dependence of the detected voltage in the direct current area. In this connection, there arises a number of questions concerning, first, the legitimacy of using small signal differential parameters of the impedance and quasi-static volt-ampere characteristic of the p-n junction, when carrying out the computations in the conditions of high frequencies and considerable amplitudes of the voltage ($\gg \frac{kT}{q}$, q is the charge of the electron), and, second, the peculiarities of using partially forward-biased $p-n$ junctions in resonance circuits of radiophysical devices: the influence of the accumulation of a minority carrier charge on the nonlinear resonance in the oscillating circuit with a $p-n$ junction in a direct current mode and on the exchange, in the parametric modulator, between the pumping circuit and the signal one in the situation of a parametric complex one-port, which shows inertia with respect to the pumping but is liable to modulation by the signal, etc.

At this point we shall examine the process of accumulation and establishment of a quasi-stationary minority carrier charge at the diode base given a different nature of the load, the peculiarities of the diffusive detection mechanism at high amplitudes of the alternating voltage ($\gg \frac{kT}{q}$) and higher frequencies ($\omega \gg \frac{1}{\tau}$), and the influence of the effect on the diffusive conductance of the semiconductor diode.

3.5.1. *Phenomenological model of the process of charge accumulation*

As we build a phenomenological model by using the charge method [78], we shall use as a basis the equations describing the behaviour of the carriers in the base area of the diode and providing a mathematically adequate description of the accumulation process. We shall carry out the calculations for the well-known model of a semiconductor diode with a plane $p-n$ junction, subject to certain limiting conditions [168, 237].

Fig.3.32 shows the form of the current and the voltage at a semiconductor diode operating in a mode with an active load at high frequencies ($\omega \gg \frac{1}{\tau}$). We shall divide the task into two parts: the first one – from the beginning of the sine up to point ξ_π (Fig.3.32), as long as the current has a sine form, and the second one – from point ξ_π up to the beginning of the next positive period.

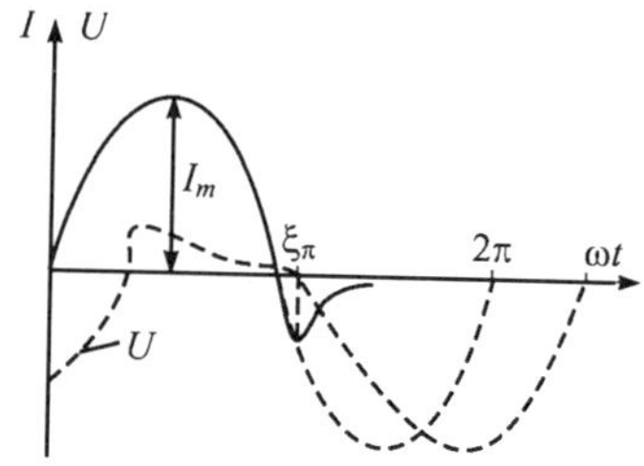

Fig. 3.32. Form of the current (I) and of the voltage (U) on a semiconductor diode operating in a regime with an active load at high frequency ($\omega\tau = 50$)

Taking into account the familiar limiting conditions [168, 237], we can write an equation concerning the diffusion of the holes, the boundary and initial conditions in the following form:

$$\frac{\partial P}{\partial T_c} = \frac{\partial^2 P}{\partial x^2} - P, \tag{3.101}$$

$$P(x,0) = 0, \quad P(\infty, T_c) = 0,$$

$$\frac{\partial P}{\partial x}(0, T_c) = -\sin \Omega T_c \ \text{ at } \ \frac{2\pi n}{\Omega} < T_c < T_\pi + \frac{2\pi n}{\Omega},$$

$$P(0, T_c) = 0 \text{ at } \frac{2\pi n}{\Omega} + T_\pi \le T_c \le \frac{2\pi(n+1)}{\Omega}, n = 0, 1, 2, \ldots$$

In Eq.(3.101) the concentration of the holes P is normalized to $P_0 = \frac{j_m}{q}$, where j_m is the amplitude of the current density, the time T_c and the coordinate x are taken in dimensionless units, $\Omega = 2\pi F$, F is the frequency in units $\frac{1}{\tau}$. The boundary conditions are changed at moment T_π, corresponding to ξ_π (Fig.3.32).

Equation (3.101) is solved by using a grid method and an implicit scheme, with a step of $\Delta x = 0,01$ along the coordinate and a time step $\Delta T_c = \frac{(\Delta x)^2}{2}$, which differentiates the optimal speed of convergence. The calculation error has been established and it does not exceed 5%.

Fig.3.33 reveals the dependencies of the charge concentration at the diode base at different points in the course of the first period. Curves 1, 2 and 3 correspond to different moments during the action of the positive half-wave. In the case of a negative half-wave the holes „collapse" back through the $p-n$ junction, with a part of them recombinating. At the boundary of the $p-n$ junction ($x = 0$), the concentration falls to zero but a certain charge remains to be accumulated at the base in depth. Curve 6 illustrates the distribution of the concentration at the end of the first period.

Fig.3.34 presents the dependencies of the residual charge at the diode base during the non-stationary accumulation process at the end of 1, 2, 3, 5, 8, 16

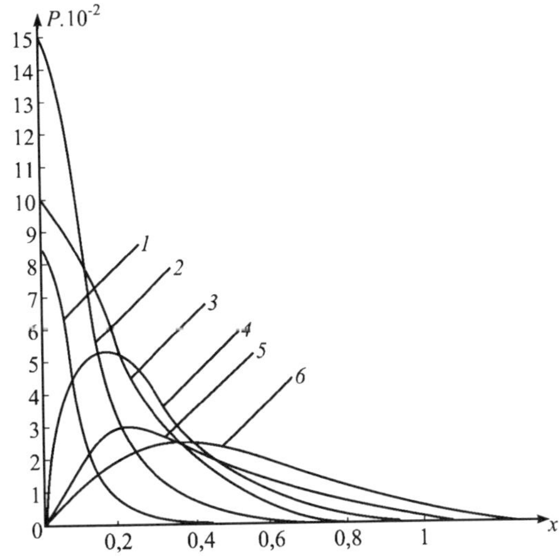

Fig. 3.33. Distribution of the concentration of the non-basic charge in the base of the diode at different moments in the course of the first period ($\omega\tau = 50$)

and 40 periods for frequency $F = 8$ ($\omega\tau = 50$). The process of accumulation and establishment of a quasi-stationary charge of minority carriers at the diode base is clearly seen.

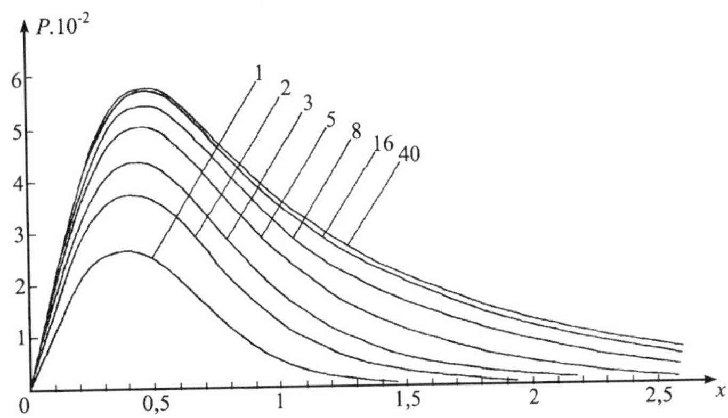

Fig. 3.34. Distribution of the residual charge from non-basic carriers at the base of the diode in the case of non-stationary accumulation process ($\omega\tau = 50$)

The dependence of the non-dimensional quasi-stationary charge Q_0 on the reduced frequency $\theta = \omega\tau$ is represented in Fig.3.35. The charge Q_0 reaches

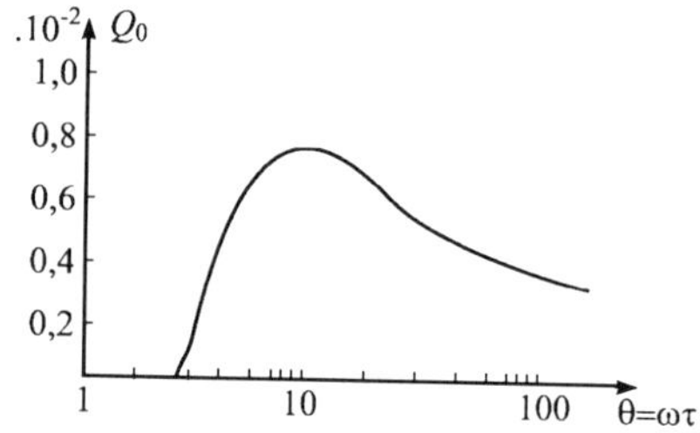

Fig. 3.35. Dependence of the non-dimensional quasi-stationary accumulated charge in the base of the diode on the reduced frequency

maximum value at frequency $\theta = 12$. At a further increase of the frequency, Q_0 falls gradually and at $\omega \to \infty$ it tends to zero asymptotically.

Later on it will be shown that the charge method can serve as a basis for conducting a sufficiently thorough study of the frequency detection characteristic and the diffusive admittance of a semiconductor diode in an extreme mode – high amplitudes ($\gg \dfrac{kT}{q}$) and higher frequencies ($\omega \gg \dfrac{1}{\tau}$).

The process of accumulation and establishment of a quasi-stationary charge at the base of a semiconductor diode occurs at $\omega \geq \dfrac{2,7}{\tau}$ in the following way [237].

The charge left over at the base after the first period at zero initial condition $Q(0) = 0$ is

$$Q_1 = Q_\pi e^{-\frac{\pi}{\theta}} - Q_{1r},$$

where Q_π is the unbalanced charge at time $\omega t = \pi$, Q_{1r} is the charge that has run out during the negative half-wave of the first period, $\theta = \omega\tau$ is the frequency reduced in accordance with the lifetime τ of the minority carriers.

After the second period, the charge left over at the base

$$Q_2 = Q_1\left(1 + e^{-\frac{2\pi}{\theta}}\right) - (Q_{2r} - Q_{1r}).$$

Analogous reasoning is applied to obtain the following recurrent formula for the n-th period:

$$Q_n = \frac{Q_\pi e^{-\frac{\pi}{\theta}}}{1 - e^{-\frac{2\pi}{\theta}}}\left(1 + e^{-n\frac{2\pi}{\theta}}\right) - Q_{nr}\left[1 + \frac{Q_{(n-1)r}}{Q_{nr}}e^{-\frac{2\pi}{\theta}} + \frac{Q_{(n-2)r}}{Q_{nr}}e^{-\frac{4\pi}{\theta}} + \ldots + \frac{Q_{1r}}{Q_{nr}}e^{-(n-1)\frac{2\pi}{\theta}}\right],$$

or

$$Q_n = \frac{Q_\pi e^{-\frac{\pi}{\theta}} - Q_{nr}}{1 - e^{-\frac{2\pi}{\theta}}}\left(1 - e^{-n\frac{2\pi}{\theta}}\right), \qquad (3.102)$$

where $Q_\pi = \dfrac{I_m \tau \theta}{1+\theta^2}\left(1+e^{-\frac{\pi}{\theta}}\right)$, I_m is the current amplitude.

It follows from (3.102) that the process of establishing quasi-stationary charge Q_0 at the diode base is determined by the lifetime of the minority carriers τ according to the law $\left(1-e^{-\frac{t}{\tau}}\right)$. At $n \to \infty$ (3.102) yields

$$Q_0 = \frac{Q_\pi e^{-\frac{\pi}{\theta}} - Q_r}{1-e^{-\frac{2\pi}{\theta}}}. \tag{3.103}$$

3.5.2. Frequency characteristics of detection

The rate of change in the detected current I_0, corresponding to the change in the frequency $\left|\dfrac{dI_0}{d\omega}\right|$, is proportional to the charge Q_0, accumulated by the diffusive capacitance of the semiconductor diode [238, 239, 240]. In the general case

$$\left|\frac{dI_0}{d\omega}\right| = \frac{dI_m}{d\omega}\frac{K(\theta)}{\pi} + \frac{I_m\tau}{\pi}\left|\frac{dK(\theta)}{d\theta}\right| = \frac{1}{2\pi}\theta_0, \tag{3.104}$$

where $K(\theta) = 1 - \dfrac{Q_r}{Q_W}$ is the frequency characteristic of detection, $Q_W = \displaystyle\int_0^{\frac{T}{2}} I(t)dt$ is the charge of minority carriers injected into the base of the positive half-period, $Q_r(\theta) = \dfrac{1}{\omega}\displaystyle\int_0^{2\pi} I(\eta)e^{-\frac{\pi-\eta}{\theta}}d\eta$ is the part of this charge, which runs back through the $p-n$ junction during the negative half-period, T is the period.

If the diode operates with an active charge R, $I_m = \dfrac{U_m}{R}$ and $\dfrac{dI_m}{d\omega} = 0$. Then

$$Q_0 = 2I_m\tau\left|\frac{dK(\theta)}{d\theta}\right|. \tag{3.105}$$

In the case of a complex charge y_L, the following can be written:

$$U_m\frac{d|y_L|}{d\omega}\frac{K(\theta)}{\pi} + \frac{U_m|y_L|\tau}{\pi}\left|\frac{dK(\theta)}{d\theta}\right| = \frac{1}{2\pi}Q_0. \tag{3.106}$$

In the particular case of an automatic bias, the load represents a RC filter, and Eq.(3.106) takes the form

$$K(\theta) + \theta\left|\frac{dK(\theta)}{d\theta}\right| = \frac{Q_0}{2CU_m} = \frac{Q_0}{Q_W}, \tag{3.107}$$

where $Q_W = \dfrac{2I_m\tau}{\theta} = 2CU_m$, $U_m = \dfrac{I_m}{\omega C}$.

As (3.103) and (3.105) are equated and the denotation $\dfrac{Q_r}{Q_W} = Z_R$ is introduced, the following equation is obtained

$$\frac{dZ_R}{d\theta} + \frac{1}{\theta\left(1 - e^{-\frac{2\pi}{\theta}}\right)} Z_R = \frac{\theta\left(e^{-\frac{\pi}{\theta}} + e^{-\frac{2\pi}{\theta}}\right)}{2(1+\theta^2)\left(1 - e^{-\frac{2\pi}{\theta}}\right)}.$$

Its solution is of the form

$$Z_R = \frac{e^{-f(\theta)}}{2}\int_0^\theta \frac{\theta\left(e^{-\frac{\pi}{\theta}} + e^{-\frac{2\pi}{\theta}}\right)e^{f(\theta)}}{(1+\theta^2)\left(1 - e^{-\frac{2\pi}{\theta}}\right)}d\theta, \tag{3.108}$$

where $f(\theta) = \displaystyle\int_0^\theta \frac{d\theta}{\theta\left(1 - e^{-\frac{2\pi}{\theta}}\right)}$.

For the case of external forced bias E, the charge equation $\left(\dfrac{dQ}{dt} + \dfrac{Q}{\tau} = I(t)\right)$ yields:

$$Q_0 = Q_{\varphi W}\frac{\dfrac{Q_{\pi-2\varphi}}{Q_{\varphi W}}e^{\frac{-\pi+2\varphi}{\theta}} - \dfrac{Q_{\varphi_r}}{Q_{\varphi W}}}{1 - e^{-\frac{2\pi}{\theta}}}, \tag{3.109}$$

where $Q_{\pi-2\varphi} = \dfrac{I_m\tau\theta}{1+\theta^2}\left(1 + e^{-\frac{\pi-2\varphi}{\theta}}\right)\left[\cos\varphi - \theta\beta\dfrac{1 - e^{-\frac{\pi-2\varphi}{\theta}}}{1 + e^{-\frac{\pi-2\varphi}{\theta}}}\right]$ is the charge at moment $\omega t = \pi - 2\varphi$ for a zero initial condition, the angle φ is determined by the parameter $\beta = \dfrac{E}{U_m}$ according to the formula $\varphi = \arcsin\beta$, $Q_{\varphi W} = \dfrac{2I_m\tau}{\theta}[\cos\varphi$ $-0,5\beta(\pi - 2\varphi)]$, Q_{φ_r} is the charge flowing out during the negative half-wave.

As (3.103) and (3.109) are equated and the equation is solved with respect to $\dfrac{Q_{\varphi_r}}{Q_{\varphi W}} = Z_E$, the following is obtained

$$Z_E = \frac{e^{-F(\theta)}}{2}\int_0^\theta H(\theta)e^{F(\theta)}d\theta, \tag{3.110}$$

where

$$F(\theta) = [\cos\varphi - \frac{1}{2}\beta(\pi - 2\varphi)]\int_0^\theta \frac{d\theta}{\theta\left(1 - e^{-\frac{2\pi}{\theta}}\right)},$$

$$H(\theta) = \frac{\theta\left(e^{-\frac{\pi+2\varphi}{\theta}} + e^{-\frac{2\pi}{\theta}}\right)\left[\cos -\theta\beta\dfrac{1-e^{-\frac{\pi-2\varphi}{\theta}}}{1 \mid c^{-\frac{\pi-2\varphi}{\theta}}}\right]}{(1+\theta^2)\left(1-e^{-\frac{2\pi}{\theta}}\right)}.$$

For the case of an automatic bias mode the following is obtained for the relation between the direct and inverse charge Z_{RC}:

$$Z_{RC} = \frac{e^{-F(\theta)}}{2}\int_0^\theta \left[H(\theta) - \frac{1}{|y_L|}\frac{d|y_L|}{d\theta}\right] e^{F(\theta)} d\theta, \tag{3.111}$$

where

$$H(\theta) = \frac{Q_0}{Q_{\varphi W}}\frac{2}{\theta}\left[\cos\varphi - \theta\beta\frac{1-e^{-\frac{\pi-2\varphi}{\theta}}}{1+e^{-\frac{\pi-2\varphi}{\theta}}}\right].$$

The expressions (3.108), (3.110) and (3.111) are used to obtain analytical expressions for the frequency characteristic of detection and the quasi-stationary accumulated charge Q_0 for the different modes in the form

$$K(\theta) = 1 - Z(\theta), \quad Q_0(\theta) = \frac{Q_W}{1-e^{-\frac{2\pi}{\theta}}}\left[\frac{Q_\pi}{Q_W}e^{-\frac{\pi}{\theta}} - Z(\theta)\right].$$

Fig.3.36 shows the theoretical characteristics of detection in a circuit with a semiconductor diode at different external negative biases. The dots indicate the experimental data.

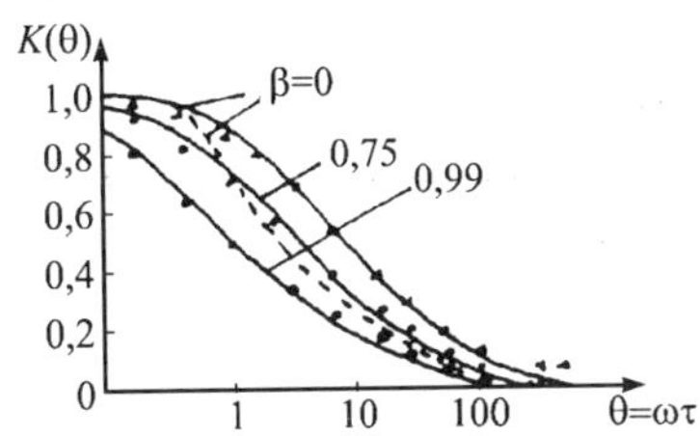

Fig. 3.36. Theoretical frequency characteristics of detection with a semiconductor diode in an external negative bias regime. The experimental data are plotted by means of dots

It is clear from the analysis shown that at high frequencies the detection mechanism is associated with the accumulation of additional quasi-stationary charge at the diode base. The emerging direct current $I_0^\theta = \dfrac{Q_0}{\tau}$ flowing along the active resistance of the load R, creates an automatic negative bias $-RI_0^\theta$, which leads to the reduction of the potential difference of the $p-n$ junction.

Let us estimate the relation between the constant voltage u_0, corresponding to I_0^θ, and the amplitude of the variable component of the voltage U_j at the $p-n$ junction itself. The direct voltage at the $p-n$ junction is obtained from the volt-ampere characteristic:

$$u_0 = \frac{1}{\lambda} \ln \left(\frac{K_R(\theta) I_m}{\pi I_s} + 1 \right), \quad \text{where } \lambda = \frac{1}{kT}. \tag{3.112}$$

The amplitude of the variable component of the voltage at the $p-n$ junction is $U_j = I_m |Z_d|$, where $|Z_d|$ is the module of the diode resistance at high frequency,

$$|Z_d| = \left[I_s \lambda \left(\frac{K_R(\theta) I_m}{\pi I_s} + 1 \right) \sqrt{\theta} \right]^{-1}.$$

Then

$$\frac{u_0}{U_j} \simeq \frac{K_R(\theta)}{\pi} \sqrt{\theta} \ \ln \frac{I_0}{I_s}. \tag{3.113}$$

The latter ratio shows that U_j does not depend on the amplitude of the alternating current I_m and that it falls off as the frequency increases.

At $\dfrac{I_0^\theta}{I_s} = 10^2$, $Q = 3$ and $K_R(\theta) = 0,8$, (3.113) yields $\dfrac{u_0}{U_j} \simeq 1,17$, i.e. starting at frequencies $Q = 2,7$, the ratio $\dfrac{u_0}{U_j} > 1$ increases as the frequency goes up.

It is obvious that at high frequencies $U_j < u_0$, and the diode is virtually always „forward-biased" for both half-waves (positive and negative) of the alternating signal.

The analytical expressions for $K_{RC}(\theta)$ and $K_E(\theta)$, corresponding to the mode of automatic and forced external bias, can be checked through a numerical solution of the diffusive equation (3.101) for the adopted model of a $p-n$ structure and for the conditions of a stationary regime. To this end, however, the overall transition process should be calculated many times, which, apart from the error accumulation, will require too much computer time.

In order to avoid these inconveniences, we can work out an equation on the function of distribution of the minority carriers $\varphi(x)$ at the diode base in a stationary regime.

We present the solution at moment T_π (corresponding to moment ξ_π in Fig.3.32) as a sum: 1) of the solution with initial zero condition and boundary conditions corresponding to the sine current $h(x)$, where x is once again a coordinate determining the distance in depth of the diode base, and 2) a solution with homogeneous boundary conditions and an initial condition coming from the previous moment of switching $H(x)$.

The expression for $h(x)$ is obtained by solving the diffusive equation (3.101) at zero initial conditions by using an operator method on the basis of the Carson-Laplace transformations:

$$h(x, T_\pi) = \frac{2}{\sqrt{\pi}} \int_0^{\sqrt{T_\pi}} \sin[\Omega(T_\pi - Z^2)] e^{-z^2 + \frac{x^2}{4z^2}} dZ, \tag{3.114}$$

where $\Omega = 2\pi F$, z is the impedance, which is external with respect to the diode.

The following can be written for $H(x)$ at the switching point T_π:

$$H(x, T_\pi) = \int_0^\infty G_2(x, \eta) \left\{ \int_0^\infty \varphi(\xi) G_1(u, \xi) d\xi \right\} d\eta, \tag{3.115}$$

where $\varphi(\xi)$ is the distribution of the minority carriers at the base and at the moment of previous switching, $G_1(\eta, \xi)$ and $G_2(x, \eta)$ are Green's functions for the first and second boundary-valu problems [96, 100, 104]. The multiplier in the figure brackets represents the distribution of the holes at the end of the previous period determined on the basis of the certain distribution at moment T_π of the same period.

(3.115) is rewritten in the following way:

$$H(x, T_\pi) = \int_0^\infty \varphi(\xi) G(x, \xi) d\xi, \tag{3.116}$$

where $G(x, \xi)$ can be defined as Green's function for the problem with conditions switching at moment T_π. After certain transformations, the following is obtained for $G(x, \xi)$:

$$G(x, \xi) = \frac{1}{2\sqrt{\pi T_0}} \left\{ \operatorname{erf}\left(\frac{\xi T_\pi + x T_2}{2\sqrt{T_\pi T_2 T_0}} \right) e^{-\frac{(x-\xi)^2}{4T_0}} + \operatorname{erf}\left(\frac{\xi T_\pi - x T_2}{2\sqrt{T_\pi T_2 T_0}} \right) e^{-\frac{(x+\xi)^2}{4T_0}} \right\} e^{-T_0},$$

where T_0 is the voltage period in units τ, $T_2 = T_0 - T_\pi$.

For the stationary process the solution obtained for the switching moment T_π is identical with the distribution $\varphi(x)$ of the holes at the switching moment during the previous period, so

$$\varphi(x) = h(x) + \int_0^\infty \varphi(\xi) G(x, \xi) d\xi. \tag{3.117}$$

The expression (3.117) is Fredholm's equation of the second order, which can be used to establish $\varphi(x)$. Then $\varphi(x)$ is used to obtain the value of the unknown reverse current at an arbitrary moment after T_π, the direct current I_0^Θ and the quantity of Q_0, the current and charge harmonics, etc.

In order to obtain the solution in the case of an automatic bias mode, it is sufficient to simply transform the function $h(x)$:

$$h(x,T_\pi)=\frac{2}{\sqrt{\pi}}\int_0^{\sqrt{T_\pi}}\{\sin[\Omega(T_\pi-z^2)+\varphi]-\beta\}e^{-z^2+\frac{x^2}{4z^2}}dz. \tag{3.118}$$

The current at an arbitrary moment after T_π is determined according to the formula

$$I(T_c)=-\frac{\partial}{\partial x}P(x,T_c)_{x=0}=\frac{e^{-T_c}}{2T_c'\sqrt{\pi T_c}}\int_0^\infty\varphi(\xi)\xi e^{-\frac{\xi^2}{4T_c'}}d\xi, \tag{3.119}$$

$T_c'=T_c-\left(T_\pi+2\pi\frac{S}{\Theta}\right)$, $S=1,2,3,\ldots$, T_c is the current time in units τ.

The harmonics of the current can be calculated by simply using the Fourier formulae.

Knowing the charge at the base of the diode Q_{T_π} at moment T_π, we can resort to the charge equation

$$\frac{dQ}{dT_c}+\frac{Q}{\tau}=I(T_c), \tag{3.120}$$

to determine the charge at an arbitrary moment as

$$Q(T_c)=C_ie^{-T_c}+e^{-T_c}\int_0^{T_c}e^\eta I(\eta)d\eta,$$

where C_i is determined by the equation

$$C_i=Q_{T_\pi}e^{T_\pi}-\int_0^{T_\pi}e^\eta\sin\Omega\eta d\eta.$$

For the purpose of calculating the charge harmonics, a relation can be established between them and the current harmonics.

Multiplying the left-hand and right-hand parts of Eq.(3.120) by $\sin n\Omega$ and $\cos n\Omega$ successively, and averaging for a period, we obtain

$$Q_{ns}=\frac{I_{ns}+n\Omega I_{nc}}{1+n^2\Omega^2},\quad Q_{nc}=\frac{I_{nc}-n\Omega I_{ns}}{1+n^2\Omega^2}, \tag{3.121}$$

where I_{nc}, Q_{nc} are the cosine and I_{ns}, Q_{ns} the sine components of the current and the charge, $n=1,2,3,\ldots$ is the number of the harmonic.

Equation (3.117) has been investigated via numerical methods and its error does not exceed 5%.

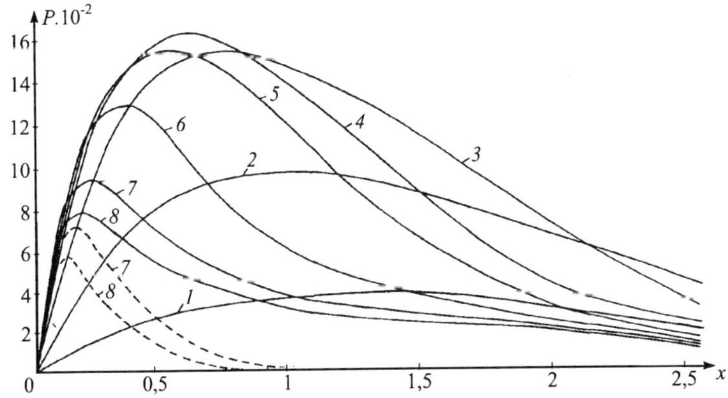

Fig. 3.37. Distribution of the non-basic carriers in the base of the diode at the moment of switching T_π in the case of a stationary process, for frequencies: *1* – $F = 0,03$; *2* – 0,1; *3* – 0,3; *4* – 0,7; *5* – 1; *6* – 3; *7* – 8 and *8* – 12

Fig.3.37 shows the distribution of the minority carriers at the diode base at the moment of switching T_π for a stationary process and for different frequencies. The charge reaches its maximum at frequency $F = 0,3$ ($\Theta = \omega\tau \simeq 2$). At $F = 0,7$ ($\Theta \simeq 4$) the maximum concentration of the minority carriers is higher but the charge is smaller than in the case of $F = 0,3$. When the frequency increases, the charge decreases, and the switching current remains virtually unchanged.

The dotted lines in Fig.3.37 indicate the distribution of the minority carriers at frequency $F = 3$ and 8 at the moment of switching during the first period.

Fig.3.38 shows the calculated reverse current of the diode in a stationary regime for different frequencies. There is a reverse current impulse even at $F = 0,01$ ($\Theta \simeq 0,06$) but, as the Figure reveals, the accumulation of additional charge from period to period begins only at frequency $F = 0,45$ ($\Theta \simeq 2,7$). The experimental study is fully in concord with these conclusions [239].

It is worth noting that the theoretical frequency characteristics of detection, obtained on the basis of a numerical solution of Eq.(3.117) and by using the analytical expressions (3.103), (3.108), (3.109), and (3.110) (Fig.3.36) coincide with at least 5% precision.

3.5.3. Diffusive impedance of the p–n junction at high signal amplitudes and higher frequencies

In order to determine the diffusive components of the impedance of the $p-n$ junction in the situation of accumulated quasi-stationary charge, we shall express the amplitude U_j of the alternating voltage at the $p-n$ junction in the following

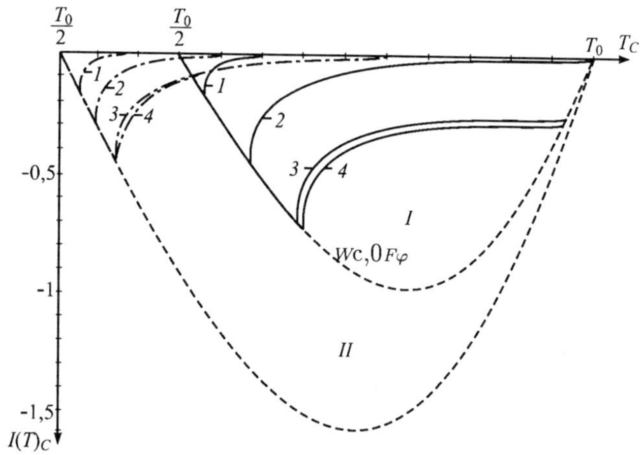

Fig. 3.38. Reverse current of the diode in stationary mode; *I* – without bias, *II* – in the case of external forced bias $\beta_E = 0,75$, for frequencies: *1* – $F = 0,03$ ($\Theta \approx 0,2$); *2* – 0,3 (2); *3* – 3 (20); *4* – 8 (50)

way [241]:

$$U_j = \frac{I_m}{\left[K(\Theta)\frac{I_m}{\pi} + I_s\right]\Lambda\sqrt{\Theta}} \simeq \frac{\pi\Theta_0}{\Lambda\sqrt{\Theta}K(\Theta)(Q_0 + I_s\tau)}, \tag{3.122}$$

where $\Lambda = \frac{1}{\Theta_k}$, $K(\Theta)$ and Q_0 are the frequency characteristic of detection and the quasi-stationary charge for the different modes considered above, I_s is the reverse conductance current of the diode.

It can be seen from (3.122) that the charge at the base of diode Q_0 is related to the voltage at the $p-n$ junction by a relationship different from that known in the classical diffusion theory [168].

The diffusive capacitance of the $p-n$ junction is determined as

$$C^{\Theta}_{dp-n} = \frac{\partial Q_0}{\partial U_j} = I_s\tau\Lambda K(\Theta)\sqrt{\Theta}\frac{1}{\pi\left[1 - U_j\sqrt{\Theta}K(\Theta)\frac{\Lambda}{\pi}\right]^2}. \tag{3.123}$$

Given $U_j\sqrt{\Theta}K(\Theta)\frac{\Lambda}{\pi} \ll 1$, i.e., when $I_0^{\Theta} \ll 1$, where $I_0^{\Theta} = \frac{Q_0}{\tau}$, the diffusive capacitance of the $p-n$ junction is $C^{\Theta}_{dp-n} = I_s\Lambda\tau\frac{K(\Theta)}{\pi}\sqrt{\Theta}$ and it does not depend on the amplitude of the alternating voltage $U_{j\Theta}$ (for example at $\Theta = 50$, $I_s = 10^{-5}$ A and $\tau = 10^{-5}$ s we determine $C^{\Theta}_{dp-n} \simeq 280$ pF).

When $I_m \gg I_s \dfrac{\pi}{K(\Theta)}$, i.e. $I_0^\Theta \gg I_s$,

$$C^\Theta_{dp-n} = I_m^2 \tau \Lambda \frac{\sqrt{\Theta} K^3(\Theta)}{\pi^3 I_s}. \tag{3.124}$$

In this way, in the case of high signals, C^Θ_{dp-n} depends on I_m as $\sim I_m^2$ and it goes up quickly as the amplitude of the current increases.

The frequency dependence of C^Θ_{dp-n} is not identical with the classical $(C_d \sim \dfrac{1}{\sqrt{\Theta}})$ and, as follows from (3.123) and (3.124), it is more complex in character: in the case of small signals $C^\Theta_{dp-n} \sim K(\Theta)\sqrt{\Theta} = a\Theta^{\frac{1}{2}} + b\Theta^{-\frac{1}{2}}$, and at higher signals $C^\Theta_{dp-n} \sim K^3(\Theta)\sqrt{\Theta} = \left(a\Theta^{\frac{1}{6}} + b\Theta^{-\frac{5}{6}}\right)^3$.

In an analogous way, the diffusive resistance of the $p-n$ junction can be determined as

$$R^\Theta_{dp-n} = \frac{\partial U_j}{\partial I_m} = \frac{I_s}{\left[I_m \frac{K(\Theta)}{\pi} + I_s\right]^2 \Lambda \sqrt{\Theta}}. \tag{3.125}$$

If $I_m < I_s \dfrac{\pi}{K(\Theta)}$, i.e. $I_0^\Theta \ll I_s$, then $R^\Theta_{dp-n} = \dfrac{1}{I_s \Lambda \sqrt{\Theta}}$.

Obviously, at small signals, R^Θ_{dp-n} coincides exactly with the classical formula for differential resistance for an ideal $p-n$ junction.

Given $I_m > \dfrac{R_s \pi}{K(\Theta)}$, i.e. $I_0^\Theta \gg I_s$, $R^\Theta_{dp-n} = \dfrac{I_s}{I_0^\Theta} \dfrac{1}{I_0^\Theta \Lambda \sqrt{\Theta}}$, and the frequency resistance of R^Θ_{dp-n} is retained within the classical framework $(R_d \sim \dfrac{1}{\sqrt{\Theta}})$, while the absolute value goes down $\dfrac{I_0^\Theta}{I_s}$ times, which is indicative of the existence of strong modulation of the conductance at the expense of the accumulated charge.

The dependencies obtained yield an expression for the diffusive time constant at high frequencies

$$(R^\Theta_{dp-n} C^\Theta_{dp-n}) = \frac{K(\Theta)}{\pi} \tau, \tag{3.126}$$

while the classical diffusion theory gives $(R_d C_d) = \dfrac{1}{\omega}$, which does not depend on τ.

When the $p-n$ junction is used in an automatic bias mode as a capacitor in oscillating resonance systems, as the analysis shows, the diffusive capacitance C^Θ_{dp-n} largely determines the characteristic bifurcational peculiarities of the nonlinear resonance at high amplitudes and higher frequencies $\omega \gg \dfrac{1}{\tau}$.

Fig.3.39 shows the dependence of the effective capacitance of a semiconductor diode in an automatic bias mode (the experimental data are marked with different

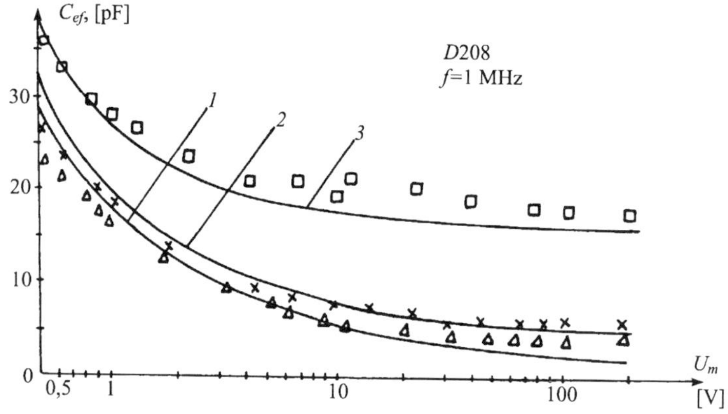

Fig. 3.39. Dependence of the effective capacitance of a semiconductor diode (D208) on the voltage amplitude at a high frequency ($\Theta = \omega\tau = 120$) for the values of the resistance R of the RC-filter automatic bias, as follows: *1* – 30 MΩ; *2* – 3 MΩ; *3* – 0,33 MΩ

dots). A comparison of the diagram in Fig.3.39 with the dependencies of C_{ef} at low frequencies $\left(\omega < \dfrac{1}{\tau}\right)$ – for example Fig.3.6 – shows their qualitative similarity.

As shown in Chapter 1 and Chapter 2, when the purpose is to promote the effectiveness and expand the functional capacities, the parametric diodes operate in a partially forward-conducting regime with small direct currents. At first glance it may seem that when the pumping frequency $\omega_p \gg \dfrac{1}{\tau}$, the $p-n$ junction cannot play the part of a complex parametric one-port, since the conductance (as well as the diffusive capacitance) fail to change in accordance with the pumping frequency. It is not difficult to show, however, that the modulation of the quasi-stationary charge with signal frequency (in the modulation-parametric systems the signal frequency is much smaller than the pumping frequency) provides a possibility for obtaining all functional regularities typical of the four-frequency parametric systems with inertia-free complex parametric elements (See Chapter 1 and Chapter 2).

Thus, for example, the charge of the signal frequency can be defined as a variation of the constant component of the charge at the p-n junction. We have

$$q_1 = \delta(Q_E + Q_0) = \frac{\partial(Q_E + Q_0)}{\partial U_0} U_1 + \frac{\partial(Q_E + Q_0)}{\partial U_m} \delta U_p, \tag{3.127}$$

where Q_E is the constant term in the Fourier expansion of the barrier capacitance charge, U_0 is the constant voltage at the $p-n$ junction, U_m is the amplitude of the pumping voltage, δU_p is the variation of the pumping voltage, which is essentially the sum of the voltages of the combined frequencies.

The expression (3.127) can be rewritten as

$$q_1 = (C_E + C_0^\Theta)U_1 + (C_1 + C_d^\Theta)\delta U_p, \tag{3.128}$$

where C_E and C_1 are the constant terms and the first coefficient in the Fourier expansion of the barrier capacitance, which is subjected to the action of the pumping voltage, $C_0^\Theta = \frac{Q_0}{U_0}$ is an addition to the average capacitance conditioned by the accumulated charge Q_0, $C_d^\Theta = \frac{\partial Q_0}{\partial U_m}$ is the diffusive capacitance of the semiconductor diode at high amplitudes of the pumping voltage and higher frequencies.

In an analogous way, the diffusive current of the signal frequency can be presented through the variation of the constant components of the current

$$i_1 = \delta I_0^\Theta = \frac{\partial I_0^\Theta}{\partial U_0} U_1 + \frac{\partial I_0^\Theta}{\partial U_m} \delta U_p,$$

or

$$i = G_0^\Theta U_1 + G_d^\Theta \delta U_p, \tag{3.129}$$

where $G_0^\Theta = \frac{Q_0}{\tau U_0}$ is the average conductance of the $p-n$ junction, $G_d^\Theta = \frac{\partial I_0^\Theta}{\partial U_m}$ is the diffusive conductance of the semiconductor diode at high amplitudes of the pumping voltage and at higher frequencies.

Taking into account the active losses in the oscillating circuit of the pumping section expressed as series active resistance R_p, as well as the formula for the impedance module of the $p-n$ junction $|Z_d|$, the amplitude of the pumping voltage can be determined as

$$U_m = \frac{Q_0 \pi}{\tau K(\Theta)} \frac{\left(\frac{Q_0}{\tau} + I_s\right) \Lambda \sqrt{\Theta} R_p + 1}{\left(\frac{Q_0}{\tau} + I_s\right) \Lambda \sqrt{\Theta}}$$

In the case of a high-frequency pumping corresponding to the condition $\omega_p \gg \frac{1}{\tau}$, the conductance of the forward-biased $p-n$ junction virtually fails to change with the pumping frequency and its harmonics, so the coefficients G_1 and G_2 in the matrix equation of the parametric modulator (1.26) can be regarded as equal to zero. Taking into account (3.127) and (3.129), the parametric modulator in this case can be described by the following matrix equation [231]:

$$\begin{Vmatrix} \dot{I}_1 \\ 0 \\ 0 \end{Vmatrix} = \begin{Vmatrix} \dot{y}_1 & G_d^\Theta + j\omega_1(C_1 + C_d^\Theta) & G_d^\Theta + j\omega_1(C_1 + C_d^\Theta) \\ j\omega_+ C_1 & \dot{y}_+ & j\omega_+ C_2 \\ -j\omega_- C_1 & -j\omega_- C_2 & \overset{*}{y}_- \end{Vmatrix} \begin{Vmatrix} \dot{U}_1 \\ \dot{U}_+ \\ \overset{*}{U}_- \end{Vmatrix}. \tag{3.130}$$

If we conduct a study of the input impedance of the parametric modulator – in a way analogous to that developed in 2.1 – and if we use (3.130), we obtain for the conductance ($G_d^\ominus$) and susceptance ($B_g^\ominus$) introduced in the input circuit:

$$G_g^\ominus = -4Q^2 G^\ominus m_1 \xi_0 \frac{4Q^2(m_1+m_d)\xi^2 + M_d[1+Q^2(\varkappa' - 4\xi^2)]}{[1+Q^2(\varkappa'-4\xi^2)]^2 + (4Q\xi)^2}, \tag{3.131}$$

$$B_g^\ominus = -4Q^3 G^\ominus m_1 \xi_0 \xi \frac{(m_1+m_d)[1+Q^2(\varkappa'-4\xi^2)] - 4M_d}{[1+Q^2(\varkappa'-4\xi^2)]^2 + (4Q\xi)^2}, \tag{3.132}$$

where $m_d = \frac{C_d^\ominus}{C_0^\ominus}$, $M_d = \frac{G_d^\ominus}{G^\ominus}$, $C_0^\ominus = C_E + C_0 - C_2$, $G^\ominus = G_0^\ominus + G_0 - G_2$, $\varkappa' = 4(\xi_0^2 + \xi_0 m_2)$.

The effective (equivalent) negative capacitance introduced into the signal circuit, will obviously be written as

$$C_g^\ominus = -4Q^2 m_1 \xi_0 C_0^\ominus \frac{(m_1+m_d)[1+Q^2(\varkappa'-4\xi^2)] - 4M_d}{[1+Q^2(\varkappa'-4\xi^2)]^2 + (4Q\xi)^2}, \tag{3.133}$$

In the bandwidth ($\varkappa' \gg 4\xi_m$, ξ_m is the upper boundary spectral component of the signal)

$$G_g^\ominus = -\frac{4Q^2 G^\ominus m_1 M_d \xi_0}{1+Q\varkappa'}, \tag{3.134}$$

$$C_g^\ominus = -4Q^2 m_1 \xi_0 C_0^\ominus \frac{(m_1+m_d)(1+Q^2\varkappa) - 4M_d}{(1+Q^2\varkappa)^2}. \tag{3.135}$$

The maximum value $G_g^\ominus$ and $C_g^\ominus$ realize at detuning $\xi_0 = \frac{1}{2Q}$:

$$\left|G_{g-}^\ominus\right|_{\max} = \frac{QG^\ominus m_1 M_d}{1+Qm_1}, \tag{3.136}$$

$$\left|C_{g-}^\ominus\right|_{\max} = Qm_1 C_0^\ominus \left[\frac{m_1+m_d}{1+Qm_2} - \frac{2M_d}{(1+Qm_2)^2}\right], \tag{3.137}$$

hence, at $m_2 \ll \xi_0$

$$\left|G_{g-}^\ominus\right|_{\max}^{(m)} = QG^\ominus m_1 M_d, \tag{3.138}$$

$$\left|C_{g-}^\ominus\right|_{\max}^{(m)} = QC_0^\ominus m_1(m_1 + m_d - 2M_d). \tag{3.139}$$

Dependencies $G_g^\ominus$ and $C_g^\ominus$ from (3.131) and (3.133) repeat qualitatively the respective dependencies (2.6) and (2.10) – Fig.2.5. This result shows that owing to the effect of accumulation and establishment of a quasi-stationary charge at the base of the parametric diode it is possible to use a parametric one-port, which is of an inertia type as regards pumping but liable to modulation as regards the signal, for modulation-parametric amplification.

CHAPTER 4.

CHAOTIC OSCILLATIONS IN RADIOPHYSICAL SYSTEMS

4.1. Radiophysical systems with natural complex dynamics. Research methods

4.1.1. Major concepts: bifurcations, chaos, strange attractor, fractal dimension. Conditions for the manifestation of chaotic behaviour in radiophysical systems

The introduction of the concept „chaos" and its discovery and investigation in a broad class of dynamic systems, including those in Mechanics, Radiophysics and Electronics, has been one of the major achievements of the physics of nonlinear oscillations in recent years [3–31].

A dynamic system usually implies a physical object, a mathematical model or a process, whose condition is determined by the totality of some quantities at the initial moment; furthermore, the state of the system at an arbitrary point of time, set in advance, can be established with the help of a given functional operator.

The chaotization of oscillations can be defined as the occurrence of peculiar non-homogeneous (irregular) motions in absolutely determined oscillating systems. The describing equations have limited non-periodic solutions of a random nature even though there are no random parameters in these equations.

Apart from the heuristic nature of the problems related to the investigation of chaotic oscillations, the latter is important for the engineering and research practice due to a number of other reasons as well. First, the chaos or noise in the radiophysical systems impedes the prediction of the length of their operation and their reliability, since the exact dependencies of the voltages and currents on time prove to be unknown. Second, realizing that the „ordinary" nonlinearities can lead to the development of chaotic modes, we are faced with issues related to the value and authenticity of the numerical modeling of nonlinear radiophysical systems in general. It is usually believed that the more powerful supercomputers are used, the more precisely the behaviour of the system can be predicted. In nonlinear problems involving chaotization, however, the behaviour of the system is highly sensitive to the initial conditions and the precise calculation of the future behaviour may prove impossible even in the case of periodic modes.

Mechanics, Radiophysics and Electronics are those areas of human knowledge where the phenomena of complex dynamics are most widespread. At the same time, these scientific areas provide the best possibilities for complex analytical, numerical and experimental research. Besides their general scientific and world-outlook value, the investigations of the chaotic oscillations in the mechanical, radiophysical and electronic systems specifically are of great practical significance in the analysis of regime stability and reliability in generators, amplifiers, SHF and Josephson systems, bistable dynamic systems, in the development of sources of stochastic and noise signals, etc.

While the possibility for chaotic behaviour in the absence of external fluctuational (random) action is regarded as more or less obvious for the systems with a large number of degrees of freedom, in the case of systems with a small number of degrees of freedom (for example with 1.5 degrees of freedom) this possibility has proved to be unexpected for the researchers. Hence, the most extensively studied issue in recent years has been the chaotic behaviour of systems with an ultimately small dimension of the phase space $N = 3$ (both autonomous systems described by differential equations of the third order and non-autonomous systems with one degree of freedom, subjected to external determined periodic action).

Until recently, it was believed that the behaviour of the simple systems of the type of a radiophysical nonlinear oscillating system or a generator of radiophysical signals is completely predictable for a given arbitrary time interval by virtue of the determinateness of the equations that describe them. It turns out that chaotic oscillations may occur in such systems without external random action as a result of complex natural behaviour of the system brought about by the instability of the phase trajectories.

The chaotic oscillations can be qualified as non-homogeneous motions characterized by irregularity, unpredictability and randomness.

A well-known fact from the Theory of nonlinear oscillations [44–48] is that the dynamics of oscillating systems is reflected through the totality of phase trajectories in the phase space determining the phase portrait of the system. Each attracting or repelling set in the phase space is called an attractor. Such attractors can be the stable and unstable peculiar points, limit cycles, separatrices and other sets.

In the case of chaotic oscillations, certain phase trajectories prove to be unrealizable as a consequence of a peculiar type of instability but the totality of phase trajectories as a whole is realized in a definite limited area of the phase space. Since it is not simple for one to imagine all this, the geometrical image, consisting of such a totality of phase trajectories, has been termed a „strange attractor" – a strange attracting set. Phase trajectories are internally unstable, the proximity between them, typical of determined systems, is not retained, most frequently they „diverge" exponentially. Yet, for an arbitrarily long interval of time the totality of phase trajectories „shrinks" in the direction of the set that attracts it – the strange attractor.

A strange attractor can be defined as a geometrical object in the phase space, where all phase trajectories of the system are tending to, and which, at the same time, repels them as a result of their instability in the area of this geometrical object. Obviously it is not possible to realize simultaneously exponential instability and „return" of the phase trajectories typical of chaotic oscillations, on the phase plane, since the phase trajectories do not cross each other on this plane. That is why the minimum dimension N of the phase space at which chaotic oscillations can exist is $N = 3$, which corresponds to an oscillating system with 1.5 degrees of freedom. Chaotic oscillations can also be excited in an ordinary passive nonlinear oscillating circuit with 1.0 degree of freedom, when the circuit is subjected to the action of an external determined periodic signal. In this case it is also regarded that the system has three-dimensional phase space, with time playing the part of a third coordinate.

The behaviour of the systems with chaotic dynamics is sensitive to the precision of setting the initial conditions and this is indeed one of the reasons for its unpredictability, i.e. regardless of the fact that the description of the system is fully determined, it is not possible to predict its evolution in time. It is necessary to develop a special statistic theory of the dynamic systems with a strange attractor.

A typical concept of the Theory of nonlinear oscillations is bifurcation defined as a change (most often leap-like) in the equilibrium state and stability of the system, or a change in the number and stability of the solutions of the describing equation. Bifurcation is a typical nonlinear phenomenon.

In the classical Theory of probability, the existence of randomness and respective probabilities is postulated and the issue of their origin is not raised. The canonized axiomatics of the classical Theory of probability operates with the concept of stochasticism and it is accepted that the latter is engendered in open dynamic systems in a fluctuating medium. Until recently it was accepted that all phenomena in our surrounding world were divided into determined and random ones. The former were studied by the Theory of differential equations, classical Electrodynamics, Mechanics and Mathematical Physics, while the latter were considered by the Theory of probability, the Theory of random processes, Statistical Physics and Quantum Mechanics. Such a strict division seemed indisputable and raised no doubts. At present, views are radically changing. Moreover, the development of the Theory of nonlinear oscillations and the Theory of dynamic systems naturally leads to the issue of the origin of randomness and the essence of probability, and to the design of respective new methods for resolving the emerging problems. At present, stochastics in continuous dynamic systems is mostly associated with the impossibility to achieve an exact repetition of the same initial conditions and the considerable influence that can be exerted even by the smallest perturbation. In this sense we can maintain that the stochastic and chaotic motions of determined dynamic systems are the „real" randomness in our surrounding world. These new ideas are taking shape only due to the discovered possibility for stochastization and chaotization of the motions in determined dynamic systems. Obviously, this is a colossal set of problems, whose elaboration has just been initiated. Every day, the modern scientific world is flooded by a huge amount of new facts and events in this area. The mechanisms of the occurrence of stochastic behaviour in the simplest determined dynamic systems are subjected to particularly precise and comprehensive consideration, and new initial paradigms affecting our deepest intuitive notions are worked out. The term „chaotic", as used in modern publications, applies to those motions in determined mechanical, physical and mathematical systems, in which lack random or unpredictable forces or parameters, while at the same time their phase trajectories demonstrate great dependence on the initial conditions.

In order to grasp more profoundly the meaning of the stochastic behaviour of oscillating systems, one should first answer the question in what way chaotic motion, i.e. a random process, may arise in systems without fluctuations and with a minimum number of freedom degrees, described by determined differential equations. It follows from the theorem of the existence and uniqueness of the solution, that in a set of strictly given initial conditions the differential equation has a definite determined solution. It is not possible to obtain random function after solving the equation. It turns out that the answer lies in the instability of all or almost all solutions. But even if all solutions of the equation are unstable, for each specific initial condition one invariably obtains a solution in the form of a determined function. Then how can the instability of the solutions cause stochasticism? The point is how one understands stochasticism. Currently the concept „determinism" is rationalized in the following aspect. It is believed that the state of the system cannot be specified with infinite accuracy, hence the initial conditions will contain certain indeterminacy ε. It is clear that the solution of the differential equation will also contain indeterminacy (finite dispersion) depending on ε and on the time t. If initially ε is set as tending to zero and then the equation is solved, this will imply that it will be solved under initial conditions set rigorously, identified with infinite accuracy. In this case, the result is always utmost predictability. From all appearences, this track is not practicable and should be replaced with the following rule: first, given a set ε, identify the behaviour of the trajectories and areas of finite dispersion for an arbitrary t, and then set ε as tending to zero. In addition, the change in the final dispersion at $t \to \infty$ should be traced. If at $\varepsilon \to 0$ the extreme dispersion tends to zero (and such will be the case whenever the solution is stable), the behaviour of the system will be determined. If, on the other hand, at $\varepsilon \to 0$ the dispersion of the trajectories does not tend to zero for $t \to 0$ (and this is observed, if there are unstable solutions), the behaviour of the system will be „unpredictable" and it will be described by a random temporal function. This understanding of determinedness differs from the traditional one in that the order of the limit transitions $\varepsilon \to 0$ and $t \to 0$ is changed.

Since the evolution in chaotic systems is unpredictable, the question whether it is relevant even to talk about any dynamic structures stands out. The answer to the question on dissipative systems is affirmative: chaotic motions have a new geometric property called fractal structure. A definite internal order has been discovered in the chaotic dynamics. It conditions the possibility to predict, study, and, when necessary, suppress the chaotic noises in radiophysical and other systems.

It is believed that just as dynamics, the new science of the 17th century, called to life differential calculus, nowadays nonlinear dynamics calls forth such new ideas from the areas of geometry and topology as fractals which should a 21th century researcher fail to grasp them, he would remain unable to understand the subject of his own investigations comprehensively.

Currently there are two most popular qualitative indicators of the existence of chaos: a positive exponential indicator of Lyapounov and a fractal structure of the orbit in the phase space. The peculiarities characteristic of the chaotic oscillations can be most readily observed by using the cross section of Poincaré. It provides a picture of the cross section of the attractor in whose area the motion in the phase space is focused. When the motion is chaotic, a complex labyrinth-like multi-leaf structure occurs. The set of points delineating definite lines have a more subtle structure as well, which can be analyzed given a greater magnification of a certain local section. Such Poincaré cross sections are said to have fractal structure. What is known as fractal dimension is used to characterize the strange attractors and measure the fractal boundaries in the areas of initial data and parameters. Currently, there are a number of fractal dimension definitions: capacitive (or geometric), mean score, correlational, informational, Lyapounov dimension. In all cases, the calculation of the fractal dimension is based on digitalization of the signals in equal time intervals. The fractal dimension is meaningful only with respect to dissipative systems. It is typical of them that orbits tend to fill only a part of the phase space. Fractal dimension is a measure of the extent to which, a certain space is filled by the orbit. In a more general perspective, the dimension determines the amount of information necessary for setting the coordinates of a point belonging to the attractor. In the case of classical (regular) attractors, the dimension is an integer: the dimension of a fixed point is zero, that of the limit cycle is unity, while that of the two-dimensional torus is two, etc. The existence of a fractional (non-integer) dimension is indicative of the presence of a strange attractor.

The general cause for the occurrence of chaotic motion in dynamic systems is the loss of stability and the exponential „scattering" of the close phase trajectories combined with their general limitation and general shrinkage. The exponential increase in the deviation followed by shutting down can serve as the simplest model of instability coupled with general shrinkage. One can imagine that there is an area in the phase space G, where the phase trajectories diverge and as a result leave the area. When the phase trajectories leave the area G, the nature of the motion changes in such a way that the phase trajectories go back to the initial area G. This return to the area G is the very phase of „shutting down", when the deviations that have occurred are restored. This cycle is repeated many times. It is noteworthy that the controversy between these two tendencies - instability with a rapid process of increase in the oscillations and a limitation of that increase – is also typical of the strictly periodic oscillations (in resonance or generator systems). The causes for the instability and the increase in the oscillations may be the same in both cases of determined and chaotic oscillations as well. The difference lies in the nature of the limitation: in the first case, this is gradual slowing down and restriction of the oscillations, most frequently leading to the formation of a stable limit cycle in the phase space, while in the second case it is the „abrupt" shutting down.

4.1.2. Basic mechanisms of transition from determined to chaotic oscillations

As far back as the end of the 19th century, the great French researcher A. Poincaré suggested [91] that randomness (indeterminacy) may occur as a consequence of the instability of the very motion in the system. In such an unstable system, „an absolutely trifling cause, escaping our notice due to its insignificance, can bring about actions that we cannot predict in advance".

Chaotic oscillations were first discovered in 1963 in the Lorentz system, already regarded as classical. The latter describes the thermal convection in a horizontal liquid layer heated from below. Lorentz's work, whose significance is generally acknowledged today, was hardly known for many years after its publication (till 1975). Today the system of equations examined by Lorentz is called „the Lorentz model". Nowadays it has been shown that the system of Lorentz equations describes a number of radiophysical generator and converting systems, where analogous phenomena occur.

Chaotic behaviour of systems, described by means of a strange attractor, has been encountered in numerous radiophysical and electronic systems, as well as in mechanics and biology, in economy and in the organization of society, in ecology, hydrodynamics and many other areas of science and technology.

Several models of bifurcations [17, 19] are known for their ability to cause, for example, the emergence of chaotic oscillations in a radiophysical nonlinear resonance system.

The first mechanism has been proposed by Reuel and Takens and it consists in a three-dimensional dynamic system (described in a three-dimensional phase space) that passes from a state of stable equilibrium through three successive bifurcations: a limit cycle – a two-dimensional torus – a three-dimensional torus. There is a considerable probability that a strange attractor (chaos) will occur afterwards.

The second mechanism has been developed by Feigenbaum. When a certain bifurcation parameter, K, is changed, the limit cycle becomes unstable. A new, two-revolution stable cycle appears. Given a further change in K, the following bifurcation occurs: the two-revolution limit cycle loses its stability and a new four-revolution cycle comes about. If parameter K is changed further, such a critical value, $K = K_{cr}$, is reached, beyond which an infinite number of unstable cycles with different periods emerge in a limited area of the phase space. Besides, it has been established that in spite of the specific peculiarities of the system and, in many cases, despite the dimension of the phase space, the value of the bifurcation parameter is governed by the Feigenbaum relation

$$\lim \frac{K_{n+1} - K_n}{K_n - K_{n-1}} = \frac{1}{\delta}, \qquad \delta = 4,6692... ,$$

where n is the number of the bifurcation.

The third mechanism can be called an intermittency mechanism. The stability of the limit cycle is lost as a consequence of its merger with the unstable one.

Periodic oscillations gradually disappear as a result of their interception by chaotic ones. A lasting laminar process, alternating with chaotic oscillations, is observed under this mechanism.

The fourth mechanism is associated with firm occurrence of chaos around a stable singular point. Such a bifurcation is termed „reverse Hopf bifurcation". Its essence consists in that the unstable limit cycles, which occur around the stable singular point, penetrate into these points during the next bifurcation. It is in such a way, for instance, that the chaotic oscillations emerge in the classical Lorentz system.

We can also imagine that a complex, non-homogeneous motion, can be caused to happen by resonance overlapping. If an external force, $\rho(t)$, with a period $\frac{2\pi}{\Omega}$ acts on a linear oscillating system, and if it is described by

$$\ddot{x} + \omega_0^2 x = \rho(t) \ , \tag{4.1}$$

where ω_0 is the natural resonance frequency, then, as it is well known, the solution of the homogeneous equation is

$$x_0 = A\cos(\omega_0 t) + B\sin(\omega_0 t). \tag{4.2}$$

If the external periodic force is presented in a Fourier series,

$$\rho(t) = \frac{a_0}{2} + \sum_{n=1}^{\infty}[a_n \cos(n\Omega t) + b_n \sin(n\Omega t),$$

then, as we carry out a harmonic expansion of the particular solution of the describing equation (4.1) as well, we obtain for the nth term

$$\lambda_n = \frac{a_n \cos(n\Omega t) + b_n \sin(n\Omega t)}{\omega^2 - n^2\Omega^2}.$$

Given $\omega_0^2 \simeq n^2\Omega^2$, there are resonances causing the increase in the amplitudes of the respective spectral components. If the oscillating system is nonlinear, all harmonics, multiple to the base frequency, ω_0, will be present in the homogeneous solution (4.2). In this case resonance effects appear whenever the ratio $\frac{\omega_0}{\Omega}$ is a rational number, thus a dense system of resonances emerges. Since the natural frequency of the oscillations in nonlinear systems is a function of their amplitude, while the resonance phenomena change this amplitude and, therefore, the frequency of the spectral components, a peculiar counteracting process of violating the conditions of resonance manifestation occurs. This intricate interdependence obviously changes the nature of the phase trajectories and causes a complex non-homogeneous (irregular) motion in the system.

4.1.3. Methods and criteria for identifying chaotic oscillations

A number of methods are used for investigating the nature of the motion in specific oscillation systems for the purpose of establishing the presence or absence of a strange attractor.

1. Numerical integration of the differential equation describing the system except for the transient process. The result is the picture of the time-dependent change of one or several variables – $x(t)$, $\dot{x}(t)$, etc.

2. Investigation of the Fourier spectrum of the oscillations, most frequently by using a quick Fourier transformation. It is possible to trace, for example, the evolution of the oscillations in a radiophysical system from periodic to chaotic ones, when changing one or several controlling parameters, such as amplitude and frequency of the external action, coefficient of nonlinearity, damping decrement (increasing increment), etc. Initially an increase in the level of the harmonics is observed in the spectrum, then subharmonic components appear (bifurcation of the period of oscillations). A continuous spectrum takes shape in the chaos, and only several characteristic frequencies of the system can be discerned.

3. Investigation of the correlation functions of the chaotic oscillations. Given a continuous spectrum, the correlation function for an arbitrary phase variable $x_i(t)$ is of a declining nature by virtue of the Wiener-Hinchin theorem. The correlation function can be calculated directly from the formula

$$k(\tau) = < x_i x_{i+\tau} - m^2 > ,$$

where m is the average of the variable x_i ($m = \overline{x}_i$).

4. One of the criteria for the existence of chaos is the positive sign of the entropy K of the dynamic system, which is characterized by the average speed of divergence of the phase trajectories in the strange attractor,

$$K = \lim_{\varepsilon \to 0} \lim_{t \to \infty} \frac{M(\varepsilon, t)}{t} ,$$

where ε is the inaccuracy of the measurements, t is the time of observation, $M(\varepsilon, t)$ is the maximum number of trajectories the distance between which is larger than ε. If the trajectories are stable according to Lyapounov's criteria, i.e. the close initial representing points do not diverge considerably, then $K = 0$.

5. Another widely applied criterion for the existence of a state of chaos is Lyapounov's characteristic index λ, which also characterizes the average speed of exponential divergence of close trajectories:

$$\lambda = \lim_{t \to \infty} \frac{1}{t} \ln |x(t)|.$$

The fact that the parameter λ is positive implies that the phase trajectories are diverging, while the value λ characterizes the chaos depth.

6. Determination of the dispersion $D^2 = (x - \overline{x})^2$ and plotting of histograms on the basis of process realization. As the strange attractor cuts across a plane known as Poincaré plane, the dimension of the phase space can be reduced by unity and a picture of the evolution of the oscillations from periodic to chaotic when changing the controlling parameter (two-dimensional map of the shift) can be obtained on the plane.

As an example, we shall present the evolution of the map of the Van-der-Paul equation with nonlinear reactance (also known as Van-der-Paul – Duffing equation):

$$\ddot{x} - \varepsilon(1 - x^2)\dot{x} + x + \gamma x^3 = \rho \cos(\omega_1 t). \tag{4.3}$$

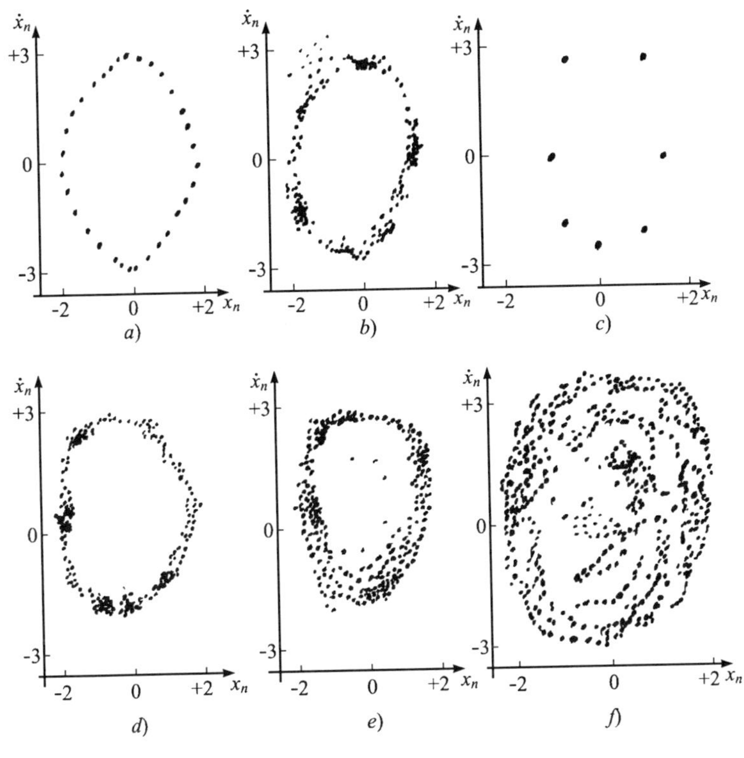

Fig. 4.1. Two-dimensional Poincare map given an amplitude increase of an external harmonic signal acting upon a Van-der-Paul generator with nonlinear reactance (Van-der-Paul – Duffing equation)

In the case of small values of ρ ($0 < \rho < 1,7$ at $\{\varepsilon, \gamma, \omega\} = \{0.33; 0.7; 2.46\}$), equation (4.3) may have a periodic or quasi periodic solution. The periodic solution corresponds to a closed limit cycle at the phase plane $(x, \dot{x})$. The Poincaré section, represented by a stroboscopic sample $x(t_n)$ with an interval $t_n = \dfrac{n2\pi}{\omega_1}$ is represented by a single point on the same plane.

When there are substantial distinctions between the frequency of the external acting signal ω_1 and the frequency of the natural oscillations in the system, ω_0, almost periodic oscillations are realized

$$x = b_1 \cos \omega_1 t + b_2 \cos \omega_2 t \ ,$$

where the frequency ω_2 may differ from ω_0 as a consequence of the non-isochronous nature of the oscillations. Respectively, the following can be written out for the stroboscopic sample

$$x_n = b_1 + b_2 \cos \frac{2\pi n \omega_2}{\omega_1} \ , \qquad \dot{x}_n = -\omega_2 b_2 \sin \frac{2\pi n}{\omega_1} .$$

Figure 4.1 shows the Poincaré section (two-dimensional map) for the case of almost periodic oscillations and finite n. When $n \to \infty$, the set of points $\{x_n, \dot{x}_n\}$ fill in the closed curve completely.

As the amplitude ρ increases, there is a transition to a seven-stroke cycle (Fig.4.1 b, c), which is stable at $1,9 < \rho < 2,65$. Given a further increase in ρ, the seven-stroke cycle is disrupted (Fig.4.1 d, e), and in the case of $\rho > 2,75$, chaotic oscillations occur. When $\rho = 5$, developed chaos can be observed in the system (Fig.4.1 f)

4.1.4. Experimental and numerical methods for investigating chaotic oscillations

The experimental investigation of determined and stochastic oscillations in radiophysical systems can be conducted by using standard measuring equipment for analyzing time-dependent, spectral and statistical characteristics of signals [17].

The time-dependent characteristics include the very realization of the oscillation process, the phase portraits of the system obtained by feeding in signals from various points of the system under investigation – x_i, as well as their derivatives, at the inputs of an oscillograph. By choosing different variables $\{x_i, \dot{x}_i\}$ or $\{x_i, \dot{x}_j\}$ different projections of the phase portrait of the system can be obtained. Different orientation of the phase portraits is achieved by turning the respective axes, while adding phase shifts along the coordinates. Data storing oscillographs are particularly useful, when studying complex signals, including chaotic ones. The time-dependent realizations and particularly the phase portraits of the system recorded by means of them not only make it possible to establish the existence of a non-periodic oscillation process, but also reflect the general structure of the attractors emerging in the phase space of the system (e.g. Fig.4.1). For the purpose of obtaining statistical information concerning the motion of a certain dynamic system, such as the distribution function, the correlation function, the power spectrum, etc. standard measures of correlation characteristics, spectrum analyzers, including panoramic spectrum analyzers and the like, can be used.

One of the fundamental methods of studying chaotic oscillations in a system with a small number of degrees of freedom is to obtain the Poincaré map. Experimentally this is achieved by raising the level of brightness and presenting, on the screen of the data storing oscillograph, only certain points of the phase trajectories $\{x_i, x_j\}$ at preset points of time (Fig.4.2). The brightness of the oscillograph beam is changed by communicating the respective strobing impulses to input z.

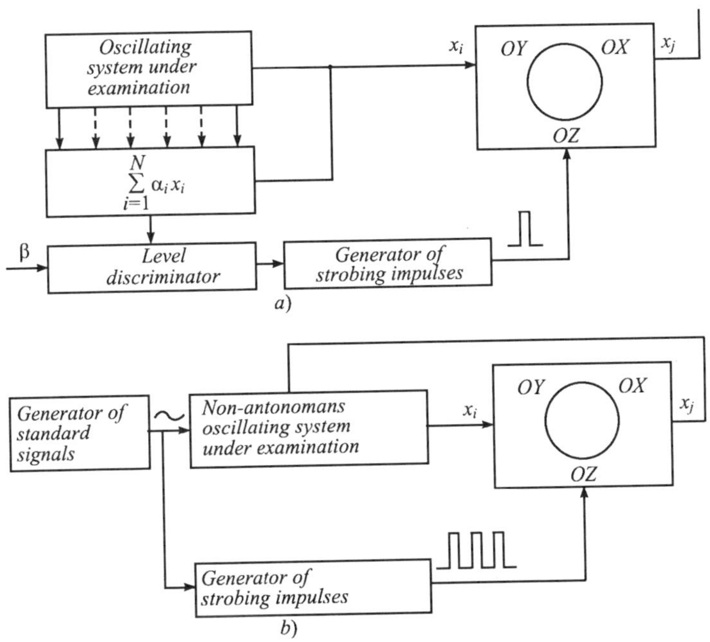

Fig. 4.2. Block diagram of set-ups for visualization of Poincaré map: a. Map of a radio engineering system with N-dimensional phase space; b. Map of a non-autonomous radio engineering system

In the general case, when investigating a radiophysical system with phase space dimension N, a two-dimensional projection of the Poincaré map is obtained as a two-dimensional section of the phase portrait with a hyperplane with dimension

$N-1$, presented through a linear equation

$$\sum_{i=1}^{N} \alpha_i x_i - \beta = 0.$$

For the purpose of registering the Poincaré map, the signals x_i, with weight coefficients α_i are conveyed to an adder (Fig.4.2a). The moment of the cross-section with the hyperplane is determined by the condition $\sum \alpha_i x_i = \beta$. At that point of time, a certain voltage comes in from the generator of strobing impulses, which increases the brightness of the respective point of the phase trajectory. A secant hyperplane is selected by choosing the weight coefficients α_i and the level β of operation of the discriminator. This methodology makes it possible to plot projections of maps corresponding to the maximum values of an arbitrary coordinate – x_i. In this case, the discriminator determines the maximum value of the coordinate under the condition $\dot{x}_i = 0$, while two other arbitrary coordinates of the process $\{x_j, x_k\}$ are brought up on the oscillograph screen.

When studying non-autonomous radiophysical systems, it is convenient for the Poincaré map to be presented by those points of the phase trajectory which are separated in time by the period T of the external action (Fig.4.2b). As in the previous case, the brightness of the oscillograph beam is modulated by the generator of a series of strobing impulses with period T.

One of the major issues arising in the course of investigating the oscillation systems is connected with the nature of the evolution of the dynamic regimes, when a set (controlling) parameter is changed. Visualization of single-parametric bifurcational diagrams on an oscillograph screen can be achieved, for example, in the experimental set presented in Fig.4.3. When the evolution of the dynamic regime is examined in its dependence on the amplitude of the external harmonic signal, the following steps should be undertaken (Fig.4.3a). The signal x_i is fed at input OY of the oscillograph and saw-tooth voltage is supplied at input OX, which also helps modulate the amplitude of the external harmonic signal. The synchronization of the map on the screen is carried out with the assistance of the generator of strobing impulses in a way analogous to that in Fig.4.2b.

Yet another type of bifurcation diagrams comes about, when investigating the frequency properties of the oscillating systems. When visualizing such a type of diagrams, one can use a standard device for analyzing amplitude-frequency characteristics (Fig.4.3b). The bifurcational diagram obtained in this way reflects the response of the system to an external action with a constant amplitude, depending on its frequency. The change in the amplitude of the output signal makes it possible to trace the dynamism in the development of the chaotic regimes.

When analyzing the dynamics of nonlinear radiophysical systems, it is hard to give preference to either solely experimental or exclusively numerical methods of investigation. In the course of an experiment, it is simpler and faster to conduct

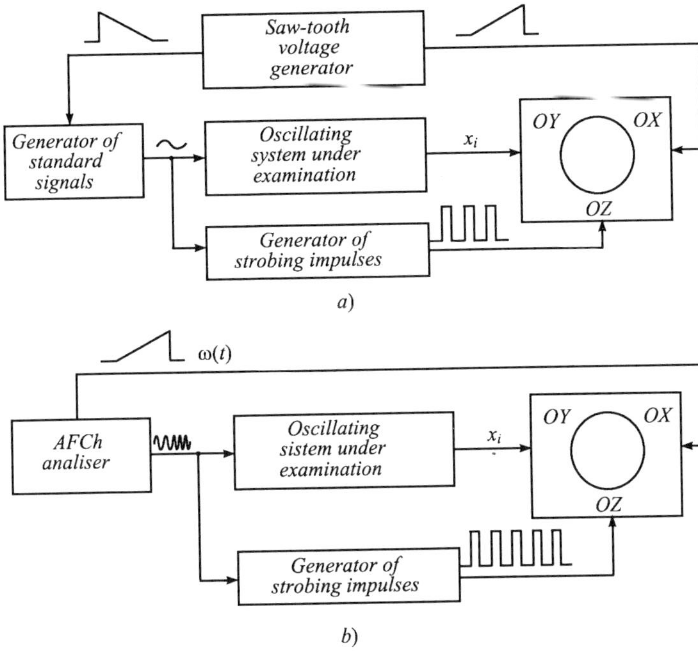

Fig. 4.3. Block diagram of set-ups for visualization of the evolution of dynamic systems (development of bifurcational diagrams): a. depending on the amplitude of the external harmonic signal, b. depending on the frequency of the external harmonic signal

the analysis in the event of variance of the parameters of the system; there is a higher degree of visualization involved in experimental studies, and frequently they are less costly. Yet, when altering the parameters of the system, there is a risk of disrupting the adequacy between the natural and the mathematical model, hence, in all cases it is necessary to control the results numerically, at least with respect to certain typical modes. The latter is also linked with the important circumstance of the dependence, as a matter of principle, of the realizations on the initial conditions. During the experiment, it is by far harder to control the initial conditions of the process under investigation, when a realization may lose stability and be replaced by another one. The experimenter may fail to notice all this and it may lead to wrong conclusions. The numerical investigations of chaos require special algorithms and software, but as a rule they build upon well-known methods of numeric integration of systems of ordinary differential equations. The important aspect with the numeric investigations is the possibility to analyze regular unstable solutions that are not realized in the natural experiment, but whose evolution under changing parameters may result in the occurrence of structurally stable modes, which can be observed experimentally.

Research practice has shown that the investigation of the dynamics of the radiophysical systems under laboratory conditions is normally conducted by applying the experimental and numeric approaches in parallel. Research software packages have been developed, mainly orientated towards personal computers, which make it possible to analyze the basic characteristics of the radiophysical oscillating systems in both stochastic and determined modes. There are algorithms for calculating the oscillation process, the phase portraits and their projections, the spectral power of the process, the spectrum of the Lyapounov characteristic exponents, the Lyapounov dimension of the attractors, the Poincaré map, the single-parameter bifurcational diagrams. The realization of the oscillation process is the simplest, but sufficiently informative characteristic. The Rhunge-Kutta methods of numerical integration of the systems of ordinary differential equations are the most frequently used methods.

The phase portraits and their projections provide a visual idea of the stream of trajectories of the system in the phase space and make it possible to identify such important phenomena as the emergence of the limit cycle, the doubling of the period of the limit cycle, the appearance and character of two-dimensional tori, etc.

The spectrum of the process power is one of the basic and universal characteristics of the multi-frequency and stochastic oscillations in radiophysical systems. It determines the distribution of the dispersion (the value of the average variance) of the time-dependent realization by frequencies. A quick Fourier transformation followed by an averaging of the set of spectrograms is most frequently used.

The spectrum of the Lyapounov characteristic indices and the Lyapounov dimension are among the most important characteristics of the attractor. The Lyapounov indices determine the average stability of the orbit of the attractor in quantitative terms.

The concept of Lyapounov dimension is introduced in the following way. Let us arrange the Lyapounov indices as follows: $\lambda_1 \geq \lambda_2 \geq \ldots \geq \lambda_p$. A typical feature of strange attractors is the divergence of the trajectories in accordance with an exponential law. Hence, at least one of the Lyapounov indices is positive. This introduces quantitative understanding in the concept sensitivity to the initial conditions, which exists in the different attempts to define the strange attractor.

Currently a Lyapounov dimension is taken to mean a value defined as:

$$D_L = j + \frac{1}{|\lambda_{j+1}|} \sum_{i=1}^{j} \lambda_i ,$$

where j is the maximum integer, for which $\lambda + \ldots + \lambda_j \geq 0$.

The correlational dimension along with the Lyapounov dimension is an accessible characteristic from the viewpoint of the volume of computations required to estimate the fractal dimension of the attractor. The correlational dimension is estimated on the basis of the correlations between the random points of the attractor.

Consider the set of attractor points $\{x_i\}$, $i = 1, \ldots, n$, obtained from the time series, i.e. $x_i = x(t + i\tau)$, τ is a constant time interval between the measurements. Due to the exponential divergence of the trajectories, most of the point couples (x_i, x_j) at $i \neq j$ will be dynamic uncorrelated couples of essentially random points. These points, however, lie on the attractor. Hence, they will be spatially correlated. This spatial correlation is measured with the help of the integral of the correlation $C(l)$, defined as [17]

$$C(l) = \lim_{n\to\infty} \frac{1}{n^2} \left\{ \begin{array}{c} \text{number of point couples } \{i, j\} \text{ the distance between} \\ \text{which } (||x_i - x_j||) \text{ is less than } l \end{array} \right\}.$$

The link between the correlational integral $C(l)$ and the correlational dimension D_ν is subject to the relation $C(l) \sim e^{D_\nu}$, which makes it possible to define D_ν on the basis of $C(l)$.

The Poincaré map is a universal informative characteristics of the oscillations in the different dynamic systems, which is convenient thanks to its visual nature and the possibility to identify a broad range of dynamic processes. In order to obtain it, however, one needs a relatively large volume of computer time.

The one-parameter bifurcational diagram yields the dependence of the extreme points in the oscillating process for one of the components in the case of an adiabatically slow change in a parameter. These dependencies create possibilities to identify oscillating modes with multiple periods, as well as quasi-periodic, synchronized and chaotic oscillations. A major advantage of this method of identifying oscillating modes is its efficiency from the viewpoint of the necessary computer resources.

4.2. Chaotic oscillations in non-autonomous radiophysical systems with nonlinear reactance and parametric systems

A classical paradigm of the Theory of oscillations is the oscillating circuit (linear or nonlinear). The theory of nonlinear resonance (including that exposed in Chapter 3) is usually built on the assumption that a periodic action evokes a periodic response. The new theory of chaotic oscillations precludes this postulate. One of the principles of chaotic oscillations is connected with the appearance of a broad continuous frequency spectrum, when a harmonic signal is used to act on a system.

An important class of systems, playing an essential role in radiophysics, is that of the systems of second order, and in particular the non-autonomous oscillating systems with nonlinear reactance. These are mostly amplifiers, converter and generator resonance systems, Josephson superconductive SHF systems and quantum interferometers, semiconductor and ferroelectric resonator systems, parametric resonance filters and the like.

A major problem when examining the stability of radiophysical systems with respect to the development of chaotic oscillations is the identification of the areas of their emergence in the parameter space of the system. Chaos occurs in non-autonomous radiophysical systems with nonlinear reactance given certain parameter combinations determining the type of the reactive nonlinearity and the nature of the external periodic signal.

At this point, we shall focus on the non-autonomous systems most widely used in radiophysics and described by nonlinear differential equations of the second order. When investigating the nonlinear oscillating and generator systems in the previous chapters, we assumed that only strictly periodical oscillations existed in them. It stands to reason that since these systems are described by three independent variables (the third coordinate is time) and since as a rule these systems are nonlinear, chaotic oscillations can be expected to occur in them given certain correlations of the amplitude and frequency of the external force, the damping decrement (growth increment), the coefficient of nonlinearity, etc.

Let us consider nonlinear radiophysical oscillating systems influenced by an external harmonic signal, whose motion is described by the following equations [51, 242]:

$$\ddot{x} + 2\delta\dot{x} + x + x^3 = \rho\cos\omega t, \tag{4.4}$$

$$\ddot{x} + 2\delta\dot{x} + x - x^3 + \gamma x^5 = \rho\cos\omega t, \tag{4.5}$$

$$\ddot{x} + 2\delta\dot{x} - x + x^3 = \rho\cos\omega t. \tag{4.6}$$

Equations (4.4) – (4.6) represent a very broad class of radiophysical systems, as well as optical and mechanical systems, etc. Equations (4.4) – (4.5) are known in the literature as Duffing equations [51]. Equation (4.6) describes systems with „a deep potential well" that have a broad declining sector in the characteristic of the nonlinear reactive element corresponding, for example, to a system with a negative differential capacitance. Fig.4.4 shows characteristics of the nonlinear reactances in equations (4.4 – 4.6).

A general picture of the areas of chaotic and periodic motions obtained through a numeric analysis of Eqs.(4.4) – (4.6) is shown in Fig.4.5, Fig.4.6 and Fig.4.7, at $2\delta = 0,1$ and $\gamma = 0,1$. The same figures also show the bifurcational areas characterized by multiplication of the period of oscillations. A complex picture of the location of the different areas can be seen. A more powerful magnification reveals the fine structure of the various areas presented in these figures.

Both the chaotic and the periodic areas are sloping to the right, i.e. in the direction of the skeleton curve of the radiophysical system (See sections 3.1 and 3.2). The stationary mode of both periodic and chaotic oscillations is established within an interval in the order of 20 periods of the acting external signal (for Eq.(4.4)), and within approximately 400–600 periods for equations (4.5) and (4.6). The relative share of the plane of parameters $\{\rho, \omega\}$, which corresponds to chaotic oscillations, is the smallest in the case of Eq.(4.4): 13%, and the largest in the situation described by Eq.(4.5): 49%. According to this indicator, Eq.(4.6) occupies an intermediate position: 28%.

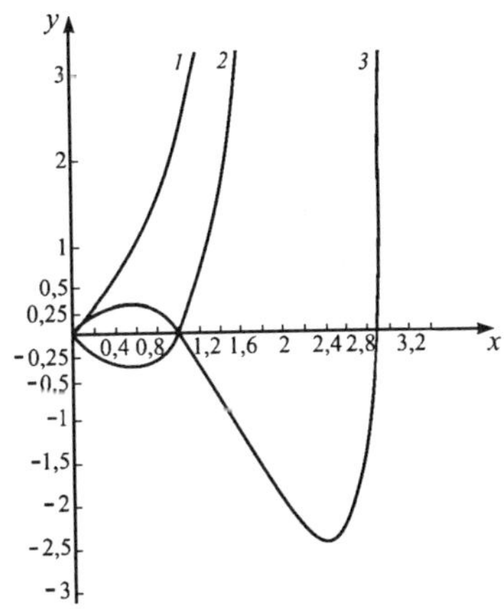

Fig. 4.4. Plots of the nonlinear reactances in equations (4.4), (4.5), (4.6): 1– $y = x + x^3$, 2 – $y = x - x^3 + 0,1x^5$, 3 – $y = -x + x^3$

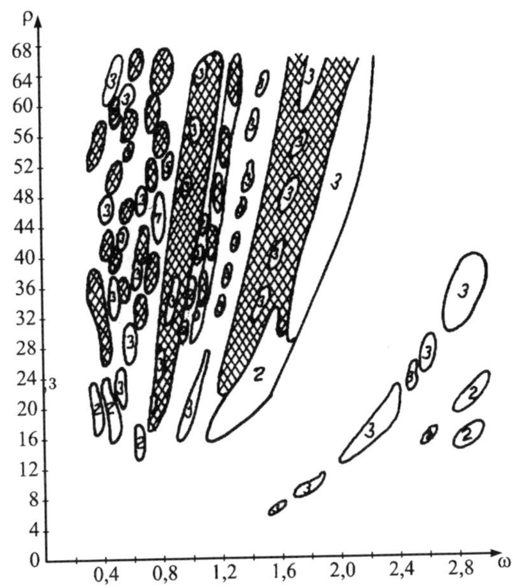

Fig. 4.5. A picture of the position of the areas of chaotic oscillations (double hatching), of periodic oscillations (blank areas), and of periodic oscillations with the respective bifurcation of period multiplication (2, 3, 4, 5 ....) in the plane of parameters $\{\rho, \omega\}$. Equation (4.4).

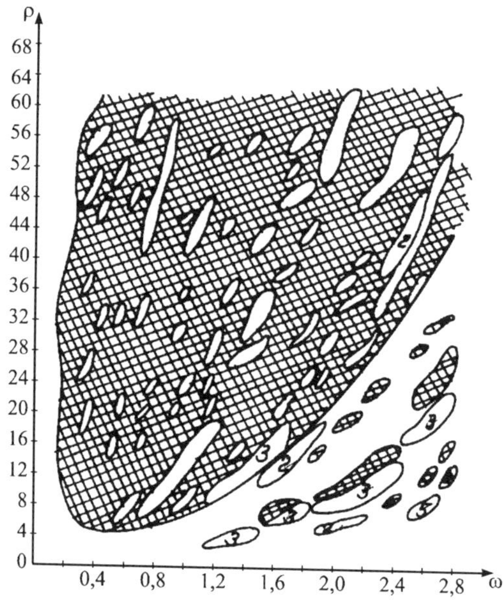

Fig. 4.6. A picture of the position of the areas of chaotic oscillations (double hatching), of periodic oscillations (blank areas), and of periodic oscillations with the respective bifurcation of period multiplication (2, 3,....) in the plane of parameters $\{\rho, \omega\}$. Equation (4.5).

By way of an example of a non-autonomous resonance system, we shall quote the data of an experimental study of a radiophysical oscillating circuit containing a varactor diode: Fig.4.8a. In line with the presentation in Chapter 3 the processes in the circuit are described by the following system of equations:

$$\left|\begin{array}{l} L\dfrac{di}{dt} = -Ri - U + U_0 \sin\omega t \\ \dfrac{d}{dt}[C(U)U] = i \\ C(U) = C_0(1-\alpha U)^{-\gamma}. \end{array}\right.$$

After an experimental study of the behaviour of the system, the areas of subharmonic and chaotic response are identified on the plane of parameters $\{U_0, \omega\}$: Fig.4.8b. The magnified section of the diagram discloses that islands of chaotic behaviour can exist in the plane of parameters $\{U_0, \omega\}$.

We shall consider yet another example – a parametric amplifier with a Josephson junction described by the following equation:

$$\ddot{\varphi} + \frac{1}{\sqrt{\beta_c}}\dot{\varphi} + \sin\varphi = \rho\cos(\omega t) , \qquad (4.7)$$

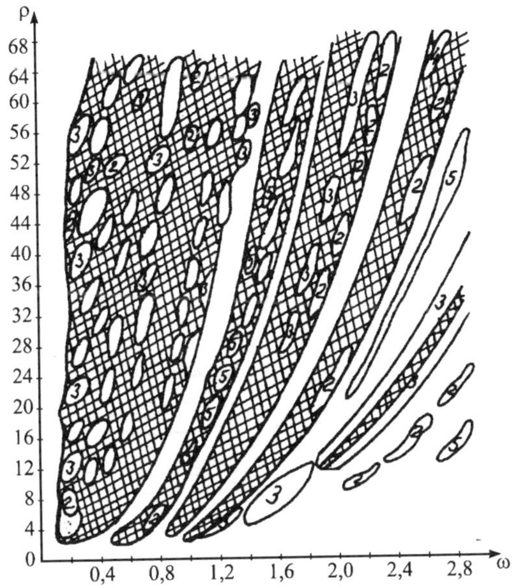

Fig. 4.7. A picture of the position of the areas of chaotic oscillations (double hatching), of periodic oscillations (blank areas), and of periodic oscillations with the respective bifurcation of period multiplication (2, 3,....) in the plane of parameters $\{\rho, \omega\}$. Equation (4.6).

where φ is the Josephson phase difference, ρ and ω are the amplitude and frequency of the acting SHF signal, β_c is a parameter equal to the external frequency normalized with respect to the plasma frequency.

The numeric analysis of equation (4.7) shows that chaos occurs and the amplification in the parametric amplifier with a Josephson contact is enhanced at an approximately equal correlation of the parameters. This prompts the conclusion that the growing noise observed in almost all experiments with Josephson parametric amplifiers with external pumping is due to the chaotic oscillations occurring in the system.

Fig.4.9 reveals the spectrum evolution in the system at $\beta_c = 25$. It can be seen (curve 1) that in a mode of periodic oscillations the noise background is 15 orders lower than the desired signal. As the amplitude of external action increases (curve 2) the background noise grows inconsiderably. As chaotic oscillations occur, the background noise soars by some 14 orders (curve 3), so the desired signal exceeds this noise by only 1–2 orders. The areas of periodic and chaotic oscillations in the system depending on the parameters ρ and ω are shown in Fig.4.10.

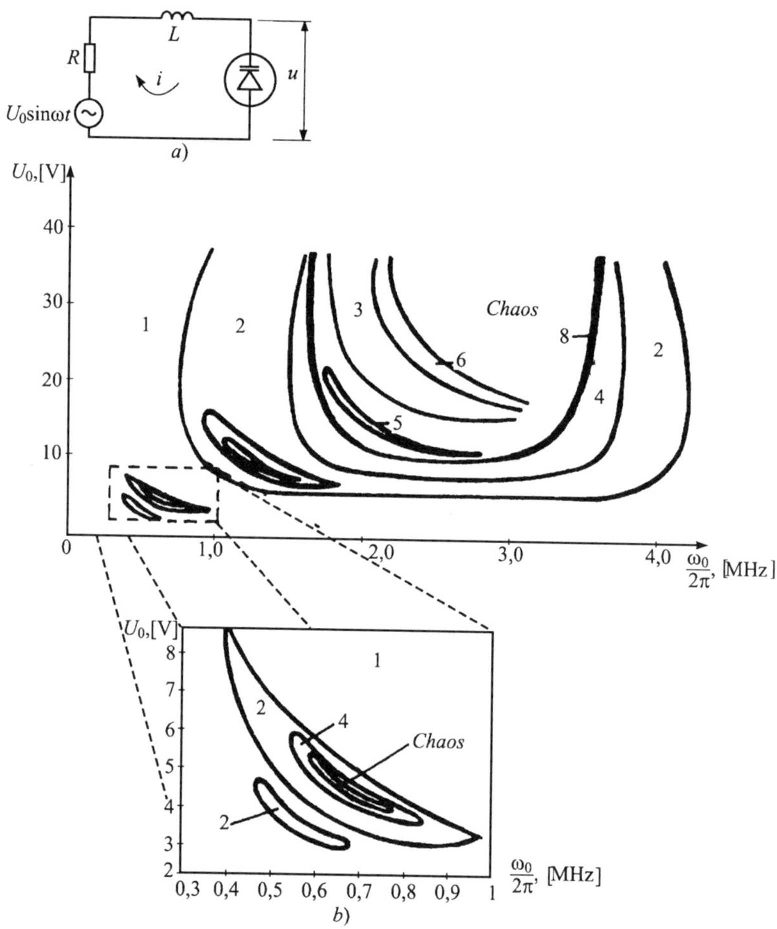

Fig. 4.8. Oscillating circuit with a varactor diode (a) and a diagram of the subharmonic and chaotic oscillations in the plane of the parameters $\{U_0, \omega_0\}$ (b)

It is noteworthy that the existence of reactive nonlinearity is a necessary condition for the development of chaotic oscillations in the parametrically excited oscillating systems as well. For example, developed chaos can be observed in the „ordinary" nonlinear parametric oscillating circuit described by the equation

$$\ddot{x} + 2\delta\dot{x} + (1 + b\cos\Omega t)x + x^3 = 0.$$

It is amazing that this relatively simple radiophysical system displays regularities of transition to chaos that are absolutely identical to those in the

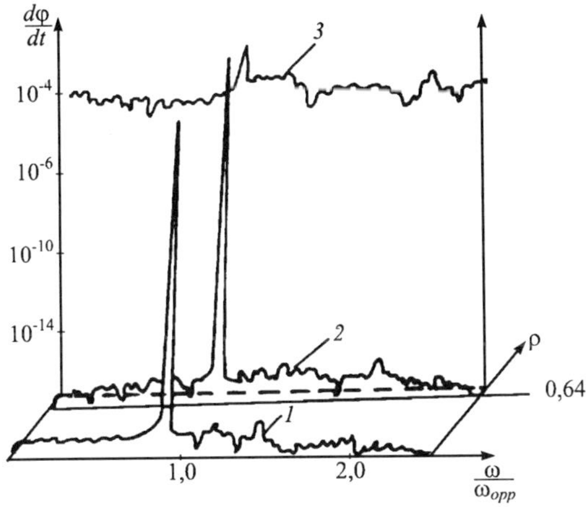

Fig. 4.9. Evolution of the spectrum in a parametric amplifier with a Josephson junction; *1, 2* – periodic oscillation mode, *3* – chaotic oscillation mode

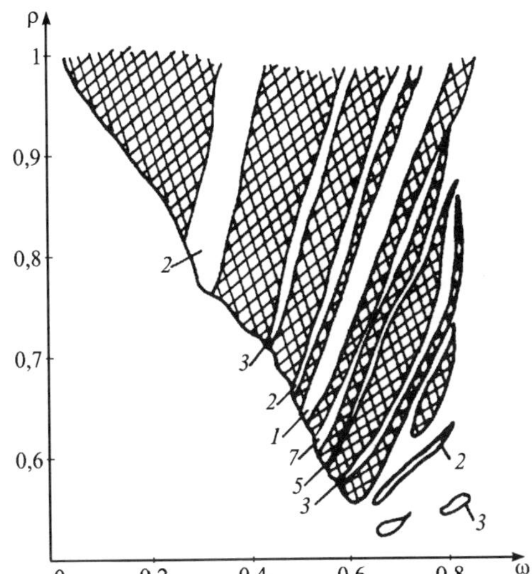

Fig. 4.10. A picture of the position of the areas of chaotic oscillations (double hatching), of periodic oscillations (1), and of periodic oscillations with the respective bifurcation of period multiplication (2, 3,....) in the plane of parameters $\{\rho, \omega\}$. Equation (4.7).

classical Lorentz system describing the processes occurring in a heated volume of liquid by means of equations in partial derivatives. This phenomenon is conditioned by the fact that as the bifurcational point of transition to chaos is approached, the multidimensional (and even infinitely dimensional) dissipative system behaves as a small-sized one. This mechanism is another illustration of the unity of the dynamic laws in nature.

4.3. Chaotic oscillations in generator radiophysical systems and SHF short-range self-detecting Doppler radars (autodyne systems) for close radiolocation

4.3.1. Conditions for chaotization of the oscillations in generator systems

From the perspective of the new ideas of chaos and stochastization of the oscillations, the following classification of the generator systems widely used in Radiophysics emerges:

1. Radiophysical systems, where oscillations are excited by compensating own losses with negative resistance, or through positive feedback.

2. Resonance radiophysical systems with forced excitation achieved through the action of an amplified signal that has passed through an inertia unit.

3. Resonance radiophysical systems where parametric excitation takes place at the expense of an inertia action of a signal from the resonance system upon elements of this resonance system.

There are models of stochastic dynamics in each of the three subclasses. For example, the first subclass can be represented by a modified LC-generator with inertia type of nonlinearity described by the following equations:

$$\left| \begin{array}{l} L\dfrac{di}{dt} + Ri + U = E\cos\omega t, \\ i = i_1 + i_2, \quad i_1 = C\dfrac{dU}{dt}, \quad i_2 = f(U), \end{array} \right.$$

where i, i_1, i_2 are currents, U is voltage, $f(U)$ is a nonlinear function.

The second subclass is represented by a model described by the equations

$$\ddot{x} + 2\beta\dot{x} + x = \eta, \quad \dot{\eta} = f_\mu(x)$$

and the third one by a mathematical model in the following form:

$$\ddot{y} + 2\varepsilon\dot{y} + y^3 + zy = 0, \quad \dot{z} = -\varepsilon\alpha z + \varepsilon\beta y.$$

The first dynamic system, where chaotic oscillations have been numerically established and investigated, is the following system of Lorentz equations:

$$\dot{x} = -\sigma(x - y), \quad \dot{y} = rx - y - xz, \quad z = -bz + xy.$$

The Lorentz equations describe a system of the second subclass.

The system of Lorentz equations can be presented in a form more relevant to radiophysics as follows:

$$\left| \begin{array}{l} \ddot{x} + 2\delta\dot{x} + \omega_0^2 x = -ky + f(x, \dot{x}, y), \\ \dot{y} + \gamma y = ax + \varphi(x, \dot{x}, y), \end{array} \right.$$

where δ is the friction coefficient, γ is an inertia parameter, k and a are coefficients of the linear relation between the variables, f and φ are nonlinear functions containing no linear terms. In this case the variable y has been excluded from the system of Lorentz equations and the following substitutions have been made:

$$x = X - \sqrt{b(r-1)}, \quad y = \sigma\left[z - \frac{1}{2\sigma}X^2 - (r-1)\left(1 - \frac{b}{2\sigma}\right)\right],$$

$$2\delta = \sigma + 1, \quad \omega_0^2 = b(r-1), \quad k = \sqrt{b(r-1)}, \quad \gamma = b,$$

$$a = (2\sigma - b)\omega_0, \quad f(x, y) = -\left(\frac{3}{2}\omega_0 x + \frac{x^2}{2} + y\right)x,$$

$$\varphi(x) = \left(\sigma - \frac{b}{2}\right)x^2.$$

The general conditions for the occurrence of chaotic behaviour in radiophysical systems, including generator systems, were presented in section 4.1.

The investigations of a number of authors show that stochastic modes do not occur in the classical model of a Van-der-Paul generator placed under external action and described by the equation:

$$\ddot{x} + \varepsilon(1 - x^2)\dot{x} + x = B\cos\omega t, \quad \varepsilon > 0,$$

(this equation was first used to describe the classical valve generator) owing to the lack of reactive nonlinearity. A series of studies demonstrates that the emergence of developed chaos is possible in a system with weak nonlinearity [243].

An interesting point that deserves to be mentioned is the fact that it is possible to suppress the chaotic behaviour of a radiophysical generator by means of external synchronizing action. In this case the strange attractor that describes it is converted to a limit cycle. Furthermore, as different from the phenomenon synchronization of determined generators, when there is no threshold value for its occurrence as regards the intensity of the synchronizing action, in the case of synchronization of chaotic oscillations, the amplitude of the external action should exceed a certain threshold value. In stochastic generator systems, the amplitude threshold of synchronization is directly related to the average positive Lyapounov index characterizing the speed of divergence of closely lying trajectories in the phase space of the system. By virtue of this fact, the amplitude threshold of synchronization can serve as a quantitative measure of chaos in the stochastic generator system, which is of particular importance in the experimental investigation of such radiophysical systems.

As two generators, respectively of periodic and chaotic oscillations, interact, the possible outcome is either periodization or chaotization of the oscillations. The different modes, all other conditions being equal, depend on the coefficient of the relation between the generators. An interesting point worth mentioning is that the oscillations in a Van-der-Paul generator, which are generally determined in an autonomous mode, can be rendered chaotic given a definite coefficient of a relation with a generator of chaotic oscillations.

Interesting phenomena are also observed when two generators of chaotic oscillations interact. This is first and foremost the effect of stochastic synchronization. The latter consists in the following: when a dissipative coupling is established between stochastic generators demonstrating qualitatively different stochastic behaviour, the generators can be synchronized. Given a sufficiently powerful coupling, they start generating chaotic realizations that are identical in their average characteristics (entropy, dimension, spectrum). Moreover, it turns out that even the topology of the strange attractors in the respective partial phase spaces is the same.

4.3.2. Bifurcations and chaos in autonomous and non-autonomous generator systems

So far bifurcational phenomena and the conditions for rendering oscillations chaotic have been studied in a large number of diverse radiophysical generator systems operating practically in all the frequency bands available. The main R&D thrust is made into generator systems with a feedback lag, models of a stochastic generator with a tunnel diode, an avelanch-drift diode, bipolar transistors etc., models of a generator with an inertia type of nonlinearity and circular structure, models of a generator with nonlinear resistance [17, 19, 25]. Some of the major methods of analysis were described in the foregoing sections. At this point, we shall use some more typical examples to demonstrate various research approaches.

Let us consider the generator of chaotic oscillations shown in Fig.4.11 [20, 25, 67]. The generator is an LC oscillating circuit with a tunnel diode (TD) in a series connection and a negative conductance in a parallel connection, constituting the active element (for example an amplifying triode in the respective regime). The capacitance C_1, in a parallel connection with the TD, upgrades the freedom degree of the system up to 1.5. The oscillating process in the system is described by the following system of equations:

$$\dot{x} = y - \delta z, \quad \dot{y} = -x + 2\gamma y + \alpha z + \beta, \quad \mu\dot{z} = x - f(z), \tag{4.8}$$

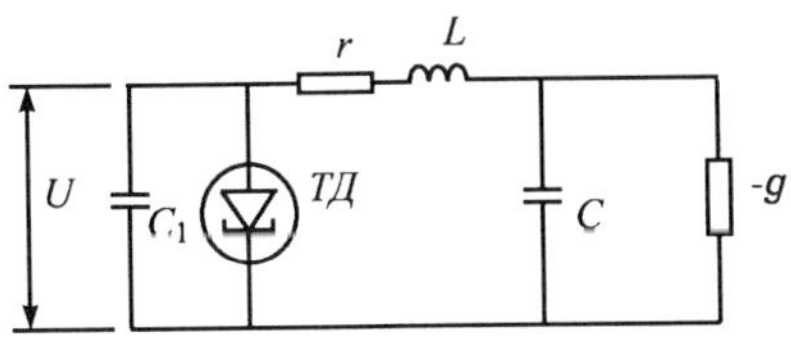

Fig. 4.11. Basic diagram of a generator of chaotic oscillations

where $x = \frac{I}{I_0} - 1$, $y = \frac{1}{I_0}\left[(U_0 - U)\sqrt{\frac{C}{L} + \frac{1}{r^2}}\right]$, $z = \frac{U}{U_0} - 1$, $\tau = \frac{t}{\sqrt{LC}}$,

$\delta = \frac{U_0}{I_0}\sqrt{\frac{C}{L}}$, $\gamma = \frac{1}{2\sqrt{LC}}(gL - Cr)$, $\beta = g\frac{U_0}{I_0} - 1$, $\mu = \delta\frac{C_1}{C}$, $\alpha = 1 + \beta - 2\gamma\delta$,

$f(z)$ reflects the N-shaped volt-ampere characteristic (VAC) of the TD in non-dimensional parameters, U_0 and I_0 determine the coordinates of the central point in the falling section of VAC.

At $\mu \ll 1$ all motions in the phase space of the system are either slow, with related phase trajectories on the surface $x = f(z)$ (in the planes $z = -1$, $x < 1$ and $z = 1$, $x > -1$), and quick (along the straight lines $x =$ const and $y =$ const). Given $\mu \to 0$, the phase space of the system (4.8) degenerates into two semi-planes overlapping in the zone $-1 < x < 1$ (Fig.4.12). The transition of the representing point from one of them to the other takes place along the straight lines $x = 1$ and $x = -1$. In line with the strange attractor condition, there are three equilibrium states in the system (one at the origin of the coordinates and two symmetrical to the slow motions plane); all states are unstable, moreover.

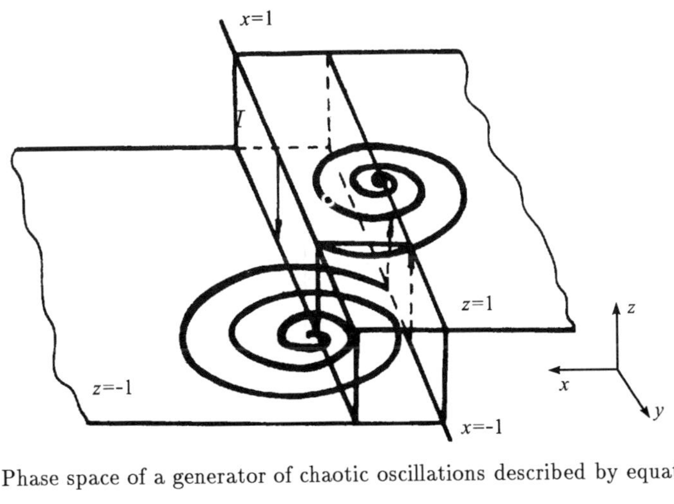

Fig. 4.12. Phase space of a generator of chaotic oscillations described by equation (4.8)

Due to the exponential instability of the system, the deviation of the initial conditions entails both a shift of the unstable foci and an indefinite phase of the transition from one plane to the other, hence, an indefinite number of the oscillations in the voltage trains generated by such a system. Indeed, two arbitrary close phase trajectories around the leap boundary may behave quite differently. The trajectories that remain within the area limited by the tangent to the straight line $|x| = \pm 1$, remain in the slow motions plane to perform almost another entire loop unlike those that happen to be beyond that area and, therefore, move to the symmetrical unstable equilibrium of the other plane. It can be shown that the voltage values generated by the system at discrete points of time represent a series of independent random quantities. Since these oscillations occur under the action of non-random energy sources and, within certain finite ranges, are independent of the initial conditions, it is natural that they should be termed stochastic self-oscillations (on the analogy of the regular self-oscillations, whose map is the limit cycle).

One of the typical configurations of stochastic generators with a delayed feedback can be presented as a closed circle consisting of a nonlinear element, a linear amplifier, a low-frequency RC-filter, an RLC-filter. Various nonlinear elements with a falling VAC section are used in the different frequency ranges – SHF devices of the O-type, dinistors, thyristors, unijunction avalanche-and-injection transistors, tunnel diodes, avalanche thermo- and photo-transistors, etc.

$P-n-p-n$ structures are particularly promising with respect to the creation of circular generator systems with stochastic behaviour. Fig.4.13a shows the structure of two coplanar field-effect transistors, which, due to its VAC (Fig.4.13b), has been called a λ-diode. A $p-n-p-n$ structure (Fig.4.14), where one of the transistors is an MOS-transistor with an induced channel, is used to create a possibility for electronic control of the VAC parameters and for extending the range of the dynamic negative resistance. The structure is used in frequency ranges, where it is an inertia-free element. As the structure is formed in a common semiconductor crystal, it can be applied not only in the low-frequency and radio range, but in the SHF range as well.

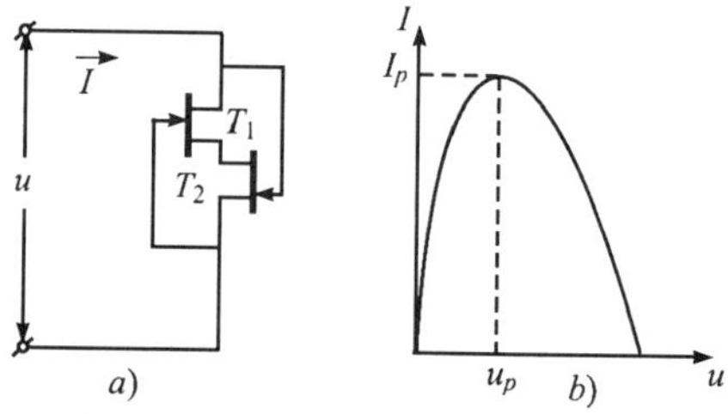

Fig. 4.13. Basic diagram of a λ-diode (a) and its typical Volt-Ampere characteristic (b)

Given two varying parameters in the stochastic generator system with a delayed feedback (e.g. the time constant τ_{RC} of the low-frequency RC-filter and

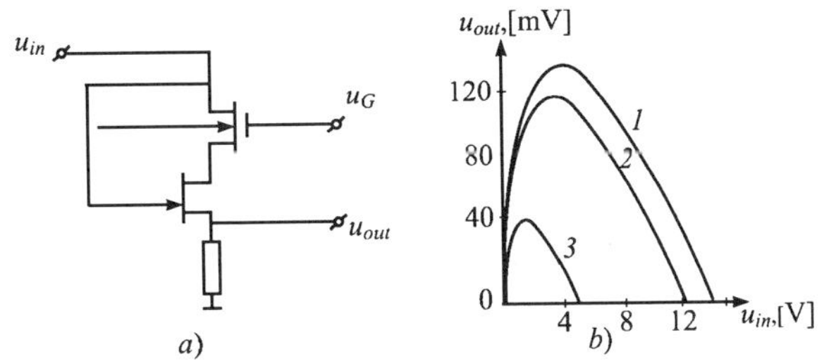

Fig. 4.14. Diagram of a nonlinear element (a) and its Volt-Ampere characteristic for different values of the gate voltage U_G: –15 V (1), –12 V (2) and –5 V (3)

the constant voltage U_r, applied at the input of the nonlinear element – Fig.4.14a – and determining the position of the operating point of the falling VAC section – Fig.4.14b) the result is a two-parameter diagram of the oscillating modes in the generator system – Fig.4.15. The diagram has been obtained by a top-down alteration of the parameter U_r for each fixed value of τ_{RC}. There are one-revolution limit cycles in area 1 of the diagram. As the boundary between areas 1 and 3 is transcended, the oscillations lose stability and a two-revolution limit cycle occurs. A transition to this oscillating mode can also be made via area 2, characterized by a definite blurring of the phase portrait of the output one-revolution limit cycle. This blurring is conditioned by the appearance of a very slow modulation of the limit cycle. A further development of the oscillation process in the system, provided that U_r goes down, is related to the loss of stability of the two-revolution limit cycle (passage across the boundary between areas 3 and 4 in the diagram). Area 5 corresponds to four-stroke cycles based on a two-revolution cycle. And finally, an oscillating mode with a strange attractor is realized in the area CA_1. A transition from stochastic motion to a three-revolution stable limit cycle takes place at the boundary between areas CA_1 and 6. The transition from the three-revolution cycle to chaos CA_2, given a further decrease in the parameter U_r, also begins with bifurcation of doubling of the period (area 7). Area p is characterized by stable equilibrium positions.

A number of radiophysical and electronic tasks entail the establishment of generator systems with an external harmonic signal acting on them. The non-autonomous generator system also serves as a simple model of generator interaction. This model is used as a basis to study the response, to an input signal, of amplifiers prone to self-oscillation, as well as the processes in superregenerative receiver systems. We have already discussed the classical model of a generator with external harmonic action (4.3) described by a Van-der-Paul's equation with nonlinear reactance. The interest in Van-der-Paul's non-autonomous equation as in a system possible with stochastic behaviour is due to several reasons. Firstly, this equation serves as a model for a large number of phenomena and processes occurring not only in radiophysical systems but in other areas of science and

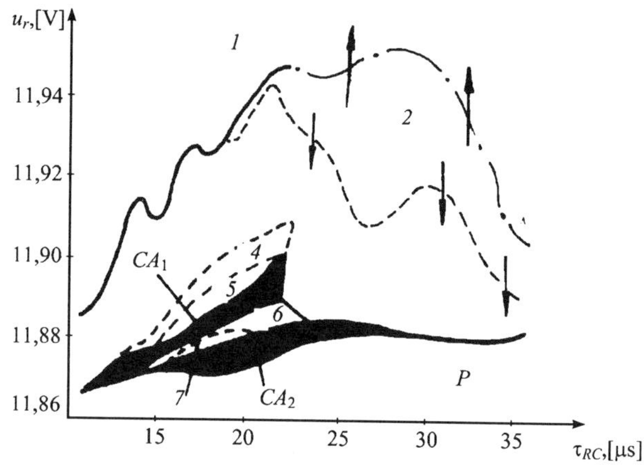

Fig. 4.15. Two-parameter diagram of the oscillating modes in a stochastic circular generator system

technology as well. Secondly, this very equation has provided the foundation for the development and improvement of the major analytical methods of the Theory of nonlinear oscillations: the asymptotic methods, the small parameter methods, the perturbation methods, etc. However, chaotic oscillations require nonlinear reactance. Hence, the reactive term of the equation is modified to yield an equation of Van-der-Paul – Duffing (4.3).

4.3.3. Creating chaotic oscillations in generator systems with a delayed feedback

In this section we shall discuss, from a phenomenological perspective, the creation of stochastic oscillations in a generator with an additional delayed feedback [244]. We shall analyze the mechanisms of chaos appearance both in the case of a constant delay and, in the typical autodyne mode of slow change, when a delay occurs in the additional feedback.

The analysis will be conducted on the basis of the following differential equation describing the generator with a delayed feedback from a phenomenological perspective:

$$\ddot{x} - \alpha(1 - x^2)\dot{x} + \omega_0^2 x + f(x) = k\dot{x}[t - \tau(t)], \tag{4.9}$$

where $f(x)$ is a nonlinear function, $\alpha > 0$, k is a coefficient of the additional delayed feedback, $\tau(t)$ is the delay in the feedback, which can also be time dependent, ω_0 is the „linear" part of the own frequency of the generator.

Equation (4.9) may also describe a Doppler autodyne system, and in this case $\tau = \tau_0 + at$, where $\tau_0 = \dfrac{2l_0}{C_L}$, $a = \dfrac{2V}{C_L}$. Moreover, Eq.(4.9) may characterize one-mode generation in a gas laser with an additional resonator, one of whose mirrors moves at a definite speed.

Equation (4.9) is studied by means of numerical methods at $f(x) = Gx^3$, $\omega_0 = 1$ and $\alpha = 0,1$ in two cases: $\tau = \tau_0 = \text{const}$ and $\tau = \tau_0 + at$. The numerical experiment shows that Eq.(4.9) predetermines stochastic modes in a broad space area of parameters.

In this case of stochastic oscillation investigation, the researcher's interest is aroused by the mechanism of chaos generation, given a change in the determining parameters of the system – the feedback coefficient, the nonlinearity coefficient, and the time of the delay.

Let us consider the case of an invariable delay ($a \equiv 0$), which corresponds to a stationary object in the autodyne model (Fig.2.25) and to a stationary mirror in the model of a laser with an additional resonator. We assume $G = 0,1$ and $\tau_0 = 30$. At low feedback factors ($0 < k < k_0 = 0,09$) periodic oscillations are excited in the system (4.9). Their frequency f_0 is close to its natural frequency $f_0 = 0,17 = \frac{1}{2\pi}$. As the feedback factor goes up, the limit cycle of the phase plane, corresponding to the periodic oscillations, loses stability. A two-dimensional torus emerges in the phase space of the system. It is consistent with quasi-periodic oscillations with two basic frequencies: f_0 and $f_1 = 0,029 \simeq \frac{1}{\tau_0}$ (Fig.4.16a). This torus is stable for values of the feedback factors in the range $k_0 < k < k_1$, $k_1 = 0,42$. As the feedback factor is increased further, two pronounced bifurcations of doubling of the torus quasi-period can be observed (Fig.4.16b): at $k = k_1 = 0,42$ and $k = k_2 = 1,2$ respectively. At $k = k_3 = 1,3$ a third, less pronounced bifurcation of doubling of the torus quasi-period can be observed, since the subharmonics are already comparable with the level of the background noise. As the factor k goes further up ($k > k_4 = 1,7$), the background noise augments and the individual spectral frequencies (f_1, $\frac{f_1}{2}$, $\frac{f_1}{4}$, $\frac{f_1}{8}$,...) become indiscernible (Fig.4.16c). This moment can be regarded as the birth moment of chaos. It is worth noting that as the feedback factor k changes, the frequencies f_0 and f_1 experience inconsiderable alterations.

The dependence of the frequency f_1 on the time of the delay τ_0, all other parameters being fixed, is shown in Fig.4.17. The specific values of the feedback factor at which bifurcations of torus birth and doubling of its period occur, depend largely on the value of the constant delay τ_0 and the nonlinearity G. As the time of the delay increases, the bifurcational values of the feedback delays decrease. Fig.4.18 shows the dependencies of the coefficients $k_0(1)$ and $k_1(2)$ corresponding to a torus birth and the first bifurcation of doubling on the time delay τ_0 ($G = 0,1$). As the nonlinearity increases, the bifurcational values of the feedback factor also decrease, i.e. chaos occurs at smaller quantities of k. For example, when $\tau_0 = 40$, $G = 0.1$ the following bifurcational values are obtained: $k_0 = 0.03$, $k_1 = 0.3$, and

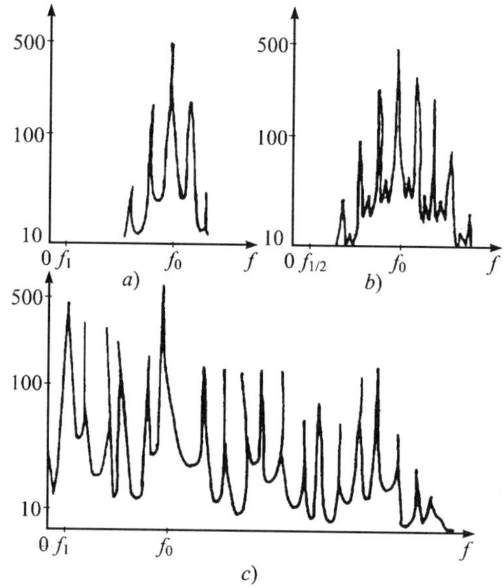

Fig. 4.16. Spectrum of the process $x(t)$ described by the model equation of a generator with an additional delay feedback (4.9) at k: 0.4 (a), 0.75 (b) and 1.75 (c)

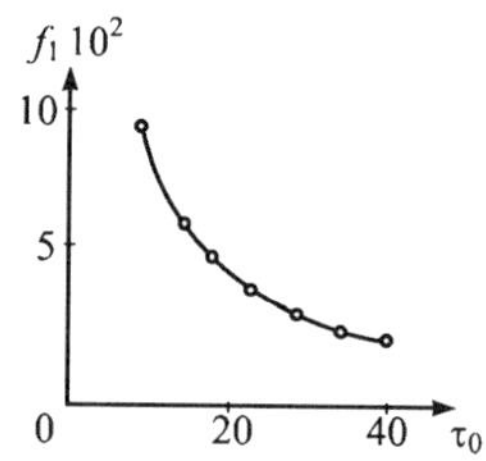

Fig. 4.17. Dependence of the bifurcational frequency on the time of the delay in a generator system with an additional delay feedback at $a = 0$, $G = 0$, $k = 1$

when $\tau_0 = 40$, $G = 1$, respectively $k_0 = 0.003$, $k_1 = 0.025$ and there is chaos at $k > 0.09$.

The nature of the transition to chaos remains invariable, if the time delay τ_0 or the nonlinearity coefficient G is taken as a controlling parameter, rather than k. The difference consists in that fact that as τ_0 changes, the frequency $f_1 = \dfrac{1}{\tau_0}$ is modified substantially, while a change in G entails mainly an alteration of the frequency f_0.

For the purpose of establishing the influence of the speed on the reflecting object Eq.(4.9) is examined for $a = 0.01$. The rest of the parameters are fixed: $G = 0.1$ and $\tau_0 = 20$. In this case the transition to chaos is quite different from that

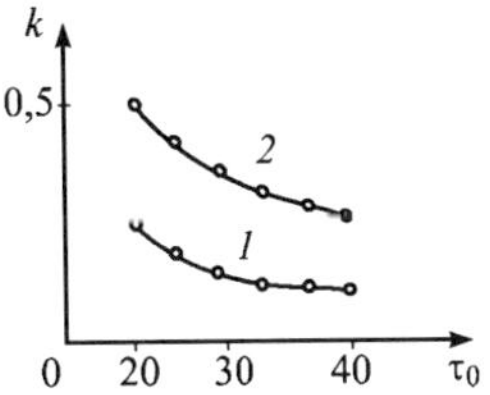

Fig. 4.18. Dependence of the bifurcational values of the delay feedback coefficients k_0 (1) and k_1 (2), corresponding to a torus birth and its first bifurcation of doubling, on the time of delay

in the $a = 0$ mode. Even at infinitesimal values of k, quasi-periodic oscillations with two basic frequencies are generated in the system: $f_0 = 0.17$ and $f_D = 0.002 \simeq af_0$ (f_D – Doppler frequency), i.e. there is a two-dimensional torus in the phase space. As k goes up ($k > k_1 = 0.3$), the two-dimensional torus loses stability and a three dimensional torus occurs, i.e. quasi-periodic oscillations with three basic frequencies: f_0, f_D, and $f_1 = 0.05$. The frequency f_1 is close to the frequency f_1 in the case of $a = 0$. A further increase in the feedback factor boosts the background noise and the result is chaos in the system.

The observed transitions to chaos can be characterized by means of the attractor dimension. Here, seeking to assess the dimension of the attractor, we use the correlational index ν, which is calculated in the following way. The temporary realization of the process $x(t)$ is used as a basis to establish a sample $x_i = x(t+iT)$, $i = 0, \dots, N$ and to build d-dimensional phase spaces, whose points are set by vectors $\zeta = (x_i, x_{i+1}, \dots, x_{i+d})$. The correlational index for each d-dimensional phase space ($d = 1, 2, 3, \ldots$) is calculated according to the formula

$$\nu_d = \lim_{\varepsilon \to 0} \frac{\ln C(\varepsilon)}{\ln \varepsilon},$$

where $C(\varepsilon) = \lim \dfrac{M(\varepsilon)}{N^2}$, $M(\varepsilon)$ is the number of the point pairs (ζ_i, ζ_j), $i \neq j$, the distance between which is less than ε. As d increases, initially ν_d increases, but later on it tends to a definite value of ν. This value of ν is close to the Hausdorf's dimension of the strange attractor. Fig.4.19 discloses the dependence of the correlational index ν on the quantity of the feedback factor k for the case of $G = 1$, $\tau_0 = 40$, $a = 0$. In the area of periodic oscillations $\nu = 1$, in the area of quasi-periodic oscillations $\nu = 2$, and in the area of stochastic oscillations, ν rises gradually as k increases, starting at $\nu = 2$.

The major conclusion of the analysis is that the chaos in the system under review occurs as a result of the bifurcational appearance of a torus in the phase space and its doubling in the case of a standing delay, or as a result of a consecutive occurrence of two-dimensional and three-dimensional torus in the case of a time-dependent delay. The possibility for a limit cycle, a torus or a strange attractor to exist in the phase space of the system under consideration at different values of the parameters is confirmed by means of a statistical measurement of their dimensions.

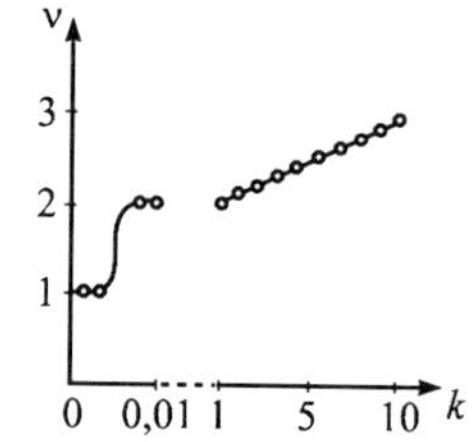

Fig. 4.19. Dependence of the dimension of the attractor on the additional feedback coefficient

4.3.4. Chaos in SHF short-range self-detecting Doppler radars

Autodyne systems can be defined as a specific class of systems with a delayed self-action. Since, as a rule, these systems operate at short distance, given the different varying conditions in the range, the intensity of the reflected signal can change by several orders. The reflected signals obtained in practice are often very strong, their intensity being comparable to that of the emitted probing SHF electromagnetic oscillations. In this case, typical relaxation instabilities may emerge in the master oscillator of the autodyne system. Under certain conditions they convert it into chaos with a total loss of its informative qualities.

At this point we shall investigate, from the perspective of the stochastic conversion of determined oscillations, the possibilities for the occurrence of instabilities of a fundamental nature that can be observed in modern SHF semiconductor autodyne systems with negative resistance [245–248]. The conditions for the emergence of chaos in autodyne systems described by a differential equation of the third order (2.100) are compared with such described by a differential equation of the second order (2.81). For the sake of convenience, we shall hereinafter call them Type I systems and Type II systems. The controlling parameter selected for the investigation is the coefficient of reflection of the signal by a moving object k. The closest reflection of the processes in the real autodyne systems is provided by the differential equations (2.100) and (2.81). To make the numerical analysis more convenient, we present equation (2.100) in the form of the following system of first-order equations:

$$\left|\begin{aligned} &\frac{dx}{dt} = -(1-\xi)\left\{F(x) + y - \frac{k}{Q}[f(x_\tau) + z_\tau]\right\} - \frac{1}{Q}\frac{dy}{dt}, \\ &\frac{dy}{dt} = (1-\xi)[f(x) + z], \\ &\frac{dz}{dt} = (1-\xi)[F(x) - \delta_1 z] + \zeta\frac{dx}{dt} \quad , \end{aligned}\right. \qquad (4.10)$$

where $x = \dfrac{q}{C_0U_{00}}$, $y = \dfrac{i_L}{\omega_0 C_0 U_{00}}$, $z = \dfrac{U_C}{U_{00}}$, $F(x) = f(x)[-G_1 + G_2 f(x) + G_3 f^2(x)]$,

$x_\tau = x(t-\tau)$, $z_\tau = z(t-\tau)$, $f(x) = x(1 - C_1 x)$, (See Figs.2.33 and 2.3).

As for the coefficients in system (4.10) and equation (2.81), we take typical values, characterizing respectively the SHF Gunn oscillators and the SHF avalanche-and-drift (IMPATT) oscillators. The time of the delay τ is assumed to be $\tau_0 = 30$. For the purpose of achieving more general analysis, values for $k > 1$ are also considered. The latter is also of practical significance, bearing in mind the potential of special applications to exert radio counteraction by receiving the probing signal of the autodyne system, amplifying it coherently and emitting it back to it for the purpose of disturbing its operation.

The major numeric results are grouped in Table 4.1, which presents the obtained typical bifurcational values of the reflection factor k.

Table 4.1

Bifurcational values of the coefficients of reflection from a moving object

Koefficients k_i, $i = 0, 1, 2, ..., 4$	k_0	k_1	k_2	k_3	k_4
System I	0,07	0,19	0,43	0,47	0,53
System II	0,17	0,43	0,97	1,11	1,19
Characteristic spectral frequencies	f_0	f_0, f_1	f_0, f_1 $\frac{f_1}{2}$	f_0, f_1 $\frac{f_1}{2}$ $\frac{f_1}{4}$	CHAOS

In this case, the technique of chaos generation is absolutely analogous to that of a generator with an additional delayed feedback, considered above from a phenomenological point of view.

Given a relatively small increase in the time of delay to the value $\tau_0 = 40$ and all other conditions unchanged, the bifurcational values of the coefficient k change respectively to $k_0^I = 0.015$, $k_1^I = 0.11$ and $k_0^{II} = 0.025$, $k_1^{II} = 0.27$. In the case of an increase in the nonlinearity of the function $f(x)$ the bifurcational values of the reflection factor also go down, i.e. chaos occurs at lower values for k. For example, at $\tau_0 = 40$ and one degree higher values of the coefficients C_2 and L_2 preceding the cubic term in $f(x)$, the following are obtained: $k_0^I = 0.0015$, $k_1^I = 0.01$ and $k_0^{II} = 0.0023$, $k_1^{II} = 0.025$.

The numeric investigation that has been conducted (Table 4.1) shows that the bifurcational values for k are by far lower for the system (4.10) than for the system (2.81). This is obviously due to the fact that the system (4.10) has an extra half a degree of freedom. From a physical point of view, this phenomenon can be explained by the regenerative action of the direct-current circuit R_LC in the system shown in Fig.2.33 which leads to a potential relaxation (bifurcational) instability. For the sake of verifying this conclusion, the system in Fig.2.33 (the system of equations

(4.10)) was examined for the same values of the parameters but in the absence of the R_LC circuit. In this case the system of equations assumes the form:

$$\left|\begin{aligned} \frac{dx}{dt} &= -(1-\xi)\left[F(x) + y - \frac{k}{Q}f(x_\tau)\right] - \frac{1}{Q}\frac{dy}{dt}, \\ \frac{dy}{dt} &= (1-\xi)f(x). \end{aligned}\right. \tag{4.11}$$

The results of the numeric experiment indicate that the bifurcational values for k (k_1, k_2, k_3) of the system (4.11) are close to those for a system of type II with no influence of the direct current circuit $(k_1^{II}, k_2^{II}, k_3^{II})$. This confirms the hypothesis of the regenerative influence of the R_LC circuit in the system of Fig.2.33 leading to the establishment of more favourable conditions for the occurrence of chaos.

The process of chaos occurrence and the gradual disappearance of the Doppler component ω_D from the spectrum of the autodyne system (loss of the informative qualities of the system) was studied through an experiment involving the set-up shown in Fig.4.20, where: AD – an autodyne, PM – a measure of power, RA – a SHF reflective generative amplifier, DL – delay line, DFI – a simulator of a Doppler frequency shift utilizing a mechanically rotating capacitive pin in a waveguide, SA – a spectrum analyzer. The time of signal delay is set either by using a high-frequency cable of the respective length or through a couple of receiving and emitting aerials positioned at a certain distance from each other. The damping is offset by the reflective regenerative amplifiers. The experiment with an autodyne system based on a SHF Gunn oscillator and a SHF avalanche-and-drift (IMPATT) oscillator shows that for parameter $\tau = 20\varepsilon = 10^{-6}$ the bifurcational values of the reflection factor at which developed chaos occurs and the Doppler frequency ω_D disappears from the spectrum, are respectively $k_4^I = 0,35$ and $k_4^{II} = 0,95$.

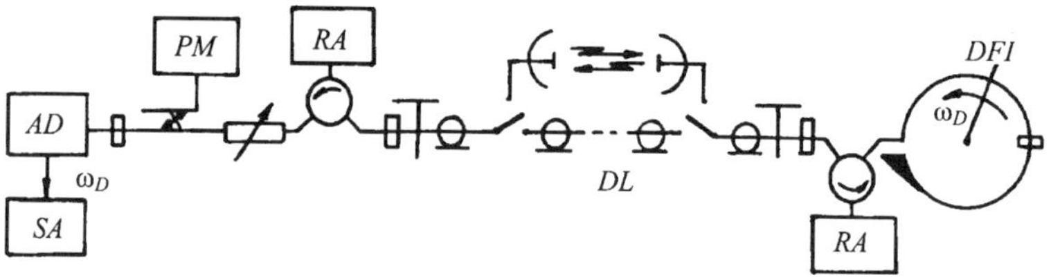

Fig. 4.20. Diagram of an experimental set-up for investigating the chaotic behaviour of a SHF autodyne system

The results that have been obtained are of heuristic significance for the more comprehensive characteristics of the behaviour of short-range self-detecting Doppler radars, which constitute a specific class of radio systems of self-action. Taking into account the peculiarities of the operation of these systems at close distances, it follows from the analysis of the numerical experiment that certain conditions (signal delay plus relatively high values of the reflection factor) can be realized, that are sufficient for the occurrence of instabilities as a result of the oscillation chaotization. From the perspective of engineering practice, it should be borne in mind that these instabilities are of fundamental nature and that they are not the outcome of stray feedbacks, defects or other instabilities in the equipment. In certain cases the acquired stochastic behaviour of the oscillations may entail partial or total loss of the informative properties of the radio system.

CHAPTER 5.

ELEMENTS OF RADIOPHYSICAL SYSTEMS

5.1. Generalization of the method of complex amplitudes for linear oscillating systems with periodic parameters and nonlinear systems

A variant of the complex amplitudes method, specially adapted for analyzing linear periodic and nonlinear oscillating systems, is described below (See also 1.2 & 4.3 and [187, 188]).

Let us set four different sine functions and construct their transforms in the complex plane through the following correspondence:

$$\begin{aligned} a_1 &= A_1 \cos(\omega_1 t + \varphi_1) \leftarrow \hat{a}_1 \equiv \dot{A}_1 e^{j\omega_1 t} = A_1 e^{j\varphi_1} e^{j\omega_1 t}, \\ a_2 &= A_2 \cos(\omega_2 t + \varphi_2) \leftarrow \hat{a}_2 = \dot{A}_2 e^{j\omega_2 t} = A_2 e^{j\varphi_2} e^{j\omega_2 t}, \\ a_3 &= A_3 \cos(\omega_3 t + \varphi_3) \leftarrow \hat{a}_3 = \dot{A}_3 e^{j\omega_3 t} = A_3 e^{j\varphi_3} e^{j\omega_3 t}, \\ a_4 &= A_4 \cos(\omega_4 t + \varphi_4) \leftarrow \hat{a}_4 = \dot{A}_4 e^{j\omega_4 t} = A_4 e^{j\varphi_4} e^{j\omega_4 t}. \end{aligned} \tag{5.1}$$

The arrow, pointing at the original above, stands for the relative correspondence.

Several mathematical operations typical of the method will be presented. They need no comment:

$$\left| \begin{aligned} & a_1 + a_2 \leftarrow \hat{a}_1 + \hat{a}_2, \quad ka_1 \leftarrow k\hat{a}_1, \quad k = \text{const}, \\ & \frac{da_1}{dt} \leftarrow \frac{d\hat{a}_1}{dt} = j\omega_1 \hat{a}_1, \quad \int a_1 dt \leftarrow \int \hat{a}_1 dt = \frac{\hat{a}_1}{j\omega_1}. \end{aligned} \right. \tag{5.2}$$

When the classical method of complex amplitudes is used, it is rare to come across products and quotients of harmonic functions, and, in such cases, they are specially regulated (for example, when computing instantaneous powers or modules of the Umov-Pointing vector).

Products and quotients of harmonic functions are widely used when analyzing linear periodic and nonlinear systems. That is why their symbols have to be formalized. For this purpose we introduce mathematical operations suitable for such cases: symbolic multiplication "$\circ$" and symbolic division "$\vdash$", whose meaning is revealed by the following examples:

$$\left|\begin{array}{l} a_1.a_2 \leftarrow \hat{a}_1 \circ \hat{a}_2 = \dfrac{\hat{a}_1 + \breve{a}_1}{2}\hat{a}_2 \text{ or } \dfrac{\hat{a}_2 + \breve{a}_2}{2}\hat{a}_1, \\ a_1.a_2.a_3.a_4 \leftarrow \hat{a}_1 \circ \hat{a}_2 \circ \hat{a}_3 \circ \hat{a}_4 = \dfrac{\hat{a}_1 + \breve{a}_1}{2}\dfrac{\hat{a}_2 + \breve{a}_2}{2}\dfrac{\hat{a}_3 + \breve{a}_3}{2}\hat{a}_4 \\ \qquad \text{or } \dfrac{\hat{a}_4 + \breve{a}_4}{2}\dfrac{\hat{a}_1 + \breve{a}_1}{2}\dfrac{\hat{a}_2 + \breve{a}_2}{2}\hat{a}_3, \\ \dfrac{a_1}{a_2} \leftarrow \dfrac{\hat{a}_1}{\hat{a}_2} = \dfrac{2}{\hat{a}_2 + \breve{a}_2}\hat{a}_1, \\ \dfrac{a_1 a_2}{a_3 a_4} \leftarrow \dfrac{\hat{a}_1 \circ \hat{a}_2}{\hat{a}_3 \circ \hat{a}_4} = \dfrac{2}{\hat{a}_3 + \breve{a}_3}\dfrac{2}{\hat{a}_4 + \breve{a}_4}\dfrac{\hat{a}_1 + \breve{a}_1}{2}\hat{a}_2. \end{array}\right. \tag{5.3}$$

Mutually conjugated functions are used here, such as $\hat{a}_1$ and $\breve{a}_1$, where

$$\breve{a}_1 \overset{*}{A}_1 e^{-j\omega_1 t} = a_1 e^{-j(\omega_1 t + \varphi_1)}, \quad \overset{*}{A}_1 = A_1 e^{-j\varphi_1}$$

is a complex conjugate amplitude.

It follows from (5.3) that several transforms correspond to the original of the product of harmonic time functions. These transforms differ from each other only with respect to their imaginary parts. It is typical of the symbolic method recounted here that the imaginary functions existing in the intermediate transformations can influence only the imaginary part of the finally achieved transform, which is rejected when the final solution of the problem is obtained.

If the original is the sum of harmonic functions, symbolic multiplication and division follow the same rules. For example, if $a_{12} = a_1 + a_2$ and $a_{34} = a_3 + a_4$, then

$$a_{12}.a_{34} \leftarrow \hat{a}_{12} \circ \hat{a}_{34} = \frac{\hat{a}_{12} + \breve{a}_{12}}{2}\hat{a}_{34}, \quad \frac{a_{12}}{a_{34}} \leftarrow \frac{\hat{a}_{12}}{\hat{a}_{34}} = \frac{2}{\hat{a}_{34} + \breve{a}_{34}}\hat{a}_{12}.$$

The proposed variant of the complex amplitudes method is convenient for studying the processes in linear periodic systems and non-linear systems with polynomial nonlinearity.

5.2. Linear periodic and almost periodic elements of radiophysical systems

5.2.1. Periodic and almost periodic resistance and conductance

The objects to be considered are linear temporally periodic and almost periodic resistances and conductance.

Let us assume that there is resistance

$$R = R_0[1 + m_R \cos(\Omega t + \varphi_R)], \tag{5.4}$$

linked up to an ideal source of harmonic voltage $U_g = U_g \cos(\omega t + \varphi)$. When Ohm's law is applied, the following equation is obtained:

$$R_0[1 + m_R \cos(\Omega t + \varphi_R)]i = U_g \cos(\omega t + \varphi). \tag{5.5}$$

In order to make use of the complex amplitudes method, we should set the type of the current as a function of time. It is not difficult to verify that Eq.(5.5) will be correct (with the required correspondence between the frequencies in the left-hand and right-hand parts of the equation) only when the current is set as an infinite series

$$i = \sum_{k=-\infty}^{\infty} I_k \cos[(\omega + k\Omega)t + \varphi_k] \leftarrow \hat{i} = \sum_{k=-\infty}^{\infty} \dot{I}_k e^{j(\omega+k\Omega)t}, \ \dot{I}_k = I_k e^{j\varphi_k}. \tag{5.6}$$

By using (5.6) Eq.(5.5) can be „moved" in the complex plane following the method of complex amplitudes:

$$R_0 \frac{\dot{m}_R}{2} \sum_{k=-\infty}^{\infty} \dot{I}_k e^{j[\omega+(k+1)\Omega]t} + R_0 \sum_{k=-\infty}^{\infty} \dot{I}_k e^{j(\omega+k\Omega)t}$$
$$+ R_0 \frac{\overset{*}{m}_R}{2} \sum_{k=-\infty}^{\infty} \dot{I}_k e^{j[\omega+(k-1)\Omega]t} = \dot{U}_g e^{j\omega t}. \tag{5.7}$$

The sign $*$ is used here and further on to denote a complex-conjugate amplitude, for instance $\overset{*}{m}$.

Any equation of the type of (5.7) is of great importance when analyzing radiophysical systems by applying the complex amplitudes method, so we shall term such equations *determining equations.*

Determining Eq.(5.7) can be presented in the following extended matrix form:

$$R_0 \cdot \begin{bmatrix} 1 & \frac{\overset{*}{m}_R}{2} & & & \\ \frac{\dot{m}_R}{2} & 1 & \frac{\overset{*}{m}_R}{2} & & \\ & \frac{\dot{m}_R}{2} & \boxed{1} & \frac{\overset{*}{m}_R}{2} & \\ & & \frac{\dot{m}_R}{2} & 1 & \frac{\overset{*}{m}_R}{2} \\ & & & \frac{\dot{m}_R}{2} & 1 \end{bmatrix} \begin{bmatrix} \dot{I}_{-2} \\ \dot{I}_{-1} \\ \boxed{\dot{I}_0} \\ \dot{I}_1 \\ \dot{I}_2 \end{bmatrix} = \begin{bmatrix} \\ \\ \boxed{\dot{U}_g} \\ \\ \\ \end{bmatrix}. \tag{5.8}$$

or in brief

$$\dot{\mathbf{R}}\dot{\mathbf{I}} = \dot{\mathbf{U}}. \tag{5.9}$$

Matrix boxes with zero elements are left empty both here and further on in our presentation.

Bold symbols will be used to denote matrixes or vectors, and the dot above will be a reminder that the respective matrix or vector contains complex elements.

The matrix in Eq.(5.8) has only three non-zero diagonals, the elements of each of the diagonals are moreover identical. This allows obtaining a result of a unique nature – a precise solution of the infinite system (5.8).

Indeed, let us present the system in the following way:

$$\left|\begin{array}{l} \cdots\cdots\cdots\cdots \\ \dfrac{\dot I_{-2}}{\dot I_{-1}} = -\dfrac{\overset{*}{m}_R}{2}\dfrac{1}{1+\dfrac{\dot m_R}{2}\dfrac{\dot I_{-3}}{\dot I_{-2}}}, \quad \dfrac{\dot I_{-1}}{\dot I_0} = -\dfrac{\overset{*}{m}_R}{2}\dfrac{1}{1+\dfrac{\dot m_R}{2}\dfrac{\dot I_{-2}}{\dot I_{-1}}}, \\ \dfrac{R_0 \dot I_0}{U_g} = \dfrac{1}{1+\dfrac{\dot m_R}{2}\dfrac{\dot I_{-1}}{\dot I_0}+\dfrac{\overset{*}{m}_R}{2}\dfrac{\dot I_1}{\dot I_0}}, \\ \dfrac{\dot I_1}{\dot I_0} = -\dfrac{\dot m_R}{2}\dfrac{1}{1+\dfrac{\overset{*}{m}_R}{2}\dfrac{\dot I_2}{\dot I_1}}, \quad \dfrac{\dot I_2}{\dot I_1} = -\dfrac{\dot m_R}{2}\dfrac{1}{1+\dfrac{\overset{*}{m}_R}{2}\dfrac{\dot I_3}{\dot I_2}}, \\ \cdots\cdots\cdots\cdots \end{array}\right. \tag{5.10}$$

Equations (5.10) yield the following dependence:

$$\frac{\dot m_R}{2}\frac{\dot I_{-1}}{\dot I_0} = \frac{\overset{*}{m}_R}{2}\frac{\dot I_1}{\dot I_0} = \cfrac{\frac{m_R^2}{4}}{1-\cfrac{\frac{m_R^2}{4}}{1-\frac{4}{1-\ldots}}}. \tag{5.11}$$

The right-hand side of Eq.(5.11) has turned out to be an infinite continuous fraction with equal elements which is denoted as K. It is known that such a fraction is convergent and has an estimate of $|K| \leq \dfrac{m_R^2}{2}$, where $m_R \leq 1$. Observing that $K = \dfrac{-\frac{m_R^2}{4}}{1+K}$, we can determine the values of the continuous fraction as:

$K_{1,2} = \dfrac{-1 \pm \sqrt{1-m_R^2}}{2}.$

Some physical considerations obviously require the satisfaction of the condition that given $m_R = 0$, there should be $K = 0$, which allows of dropping the „–" sign in front of the radical and finally determining: $K = \dfrac{-1+\sqrt{1-m_R^2}}{2}$.

This result allows obtaining the following expressions as regards the complex amplitudes of all spectral harmonics of the current.

$$\left|\begin{array}{l}\dot{I}_0 = \dfrac{\dot{U}_g}{R_0}\dfrac{1}{\sqrt{1-m_R^2}},\\ \dot{I}_{-k} = (-1)^k \dfrac{\dot{U}_g}{\dot{R}_0}\dfrac{1}{\sqrt{1-m_R^2}}\left(\dfrac{\overset{*}{m}_R}{1+\sqrt{1-m_R^2}}\right)^k,\\ \dot{I}_k = (-1)^k \dfrac{\dot{U}_g}{\dot{R}_0}\dfrac{1}{\sqrt{1-m_R^2}}\left(\dfrac{\dot{m}_R}{1+\sqrt{1-m_R^2}}\right)^k, \quad k = 1, 2, \ldots, \infty.\end{array}\right. \tag{5.12}$$

The vector diagrams of the voltages and currents are plotted in Fig.5.1. The vector diagram in Fig.5.1a is trivial for the case when $R = \text{const}$. Fig.5.1b and 5.1c correspond to expressions (5.12). Fig.5.1b shows vectors I_0; I_1, I_{-1}; I_2, I_{-2}, etc. which turn, in the respective different directions, with frequency Ω. The symmetric character of the diagram in Fig.5.1b allows presenting it in a more compact form – Fig.5.1c – by using the complex function $\dot{I}(t)$ that changes in time in compliance with a complex periodic law with a period of $2\pi/\Omega$.

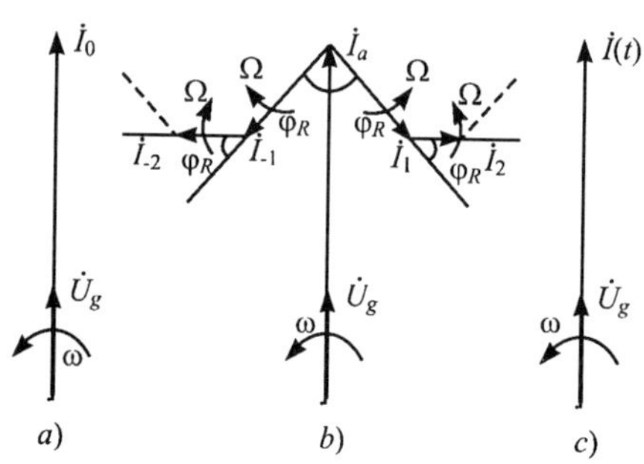

Fig. 5.1. A vector diagram of the voltage and current flowing through a resistance connected to a source of voltage: a) a case of constant resistance; b) and c) time-dependent periodic linear resistance with a harmonic variable part

Given that the periodic resistance contains two harmonics

$$R(t) = R_0[1 + m_1 \cos(\Omega t + \varphi_1) + m_2 \cos(2\Omega t + \varphi_2)], \tag{5.13}$$

a matrix equation of the following type can be obtained in an analogical manner

$$\begin{bmatrix} 1 & \frac{\overset{*}{m}_1}{2} & \frac{\overset{*}{m}_2}{2} & & \\ \frac{\dot{m}_1}{2} & 1 & \frac{\overset{*}{m}_1}{2} & \frac{\overset{*}{m}_2}{2} & \\ \frac{\dot{m}_2}{2} & \frac{\dot{m}_1}{2} & 1 & \frac{\overset{*}{m}_1}{2} & \frac{\overset{*}{m}_2}{2} \\ & \frac{\dot{m}_2}{2} & \frac{\dot{m}_1}{2} & 1 & \frac{\overset{*}{m}_1}{2} \\ & & \frac{\dot{m}_2}{2} & \frac{\dot{m}_1}{2} & 1 \end{bmatrix} \cdot \begin{bmatrix} \dot{I}_{-2} \\ \dot{I}_{-1} \\ \dot{I}_0 \\ \dot{I}_1 \\ \dot{I}_2 \end{bmatrix} = \begin{bmatrix} \\ \\ \dot{U}_g \\ \\ \\ \end{bmatrix} \tag{5.14}$$

If the periodic resistance is expressed by k harmonics, then the respective matrix $\dot{\mathbf{R}}$ will contain $2k+1$ non-zero diagonals situated symmetrically with respect to the main diagonal. In the general case of a periodic function $R(t)$, represented by an infinite harmonic series, matrix $\dot{\mathbf{R}}$ will not contain zero elements.

The matter is much more complex when the function is almost periodic. In this case instead of (5.14), the resulting systems are multidimensional and infinite, and they are not specifically studied by the Theory of oscillations. The multidimensional systems of algebraic equations below are used for classifying the respective radiophysical systems.

Let us assume that resistance changes in time according to the law

$$R(t) = R_0[1 + m_1 \cos(\Omega_1 t + \varphi_1) + m_2(\Omega_2 t + \varphi_2)], \tag{5.15}$$

where Ω_1 and Ω_2 are incommensurable frequencies, i.e. $R(t)$ is an almost periodic function, whose branching is equal to two (we introduce the concept of „branching", which is related to the number of incommensurable frequencies; obviously, a periodic function has a branching equal to one).

The equation concerning the current flowing through the resistance, given an ideal source of voltage, can be written down in a form analogical to that of (5.5):

$$R_0[1 + m_1(\Omega_1 t + \varphi_1) + m_2 \cos(\Omega_2 t + \varphi_2)]i = U_g \cos \omega t. \tag{5.16}$$

The analysis indicates that Eq.(5.16) will be correct only if the current is given in the form of a double sum

$$i = \sum_{k=-\infty}^{\infty} \sum_{l=-\infty}^{\infty} I_{kl} \cos[(\omega + k\Omega_1 + l\Omega_2)t + \varphi_{kl}],$$

$$\leftarrow \hat{i} = \sum_{k=-\infty}^{\infty} \sum_{l=-\infty}^{\infty} I_{kl} e^{j(\omega + k\Omega_1 + l\Omega_2)t}, \quad \dot{I}_{kl} = I_{kl} e^{j\varphi_{kl}}.$$

In this case the determining equation is obtained in the form

$$R_0 \sum_{k=-\infty}^{\infty} \sum_{l=-\infty}^{\infty} \left(\frac{\dot{m}_2}{2} \dot{I}_{k,l-1} + \frac{\dot{m}_1}{2} \dot{I}_{k-1,l} + \dot{I}_{k,l} + \frac{\overset{*}{\dot{m}}_1}{2} \dot{I}_{k+1,l} + \frac{\overset{*}{\dot{m}}_2}{2} I_{k,l+1} \right)$$

$$\times e^{j(\omega + k\Omega_1 + l\Omega_2)t} = \dot{U}_g e^{j\omega t}. \tag{5.17}$$

When the components of equal frequencies are selected, an infinite linear algebraic system of an extraordinary form is obtained. Since we are not in a position to write it out in full, we shall confine ourselves to the following short presentation:

$$\dot{\mathbf{R}} \bullet \bullet \dot{\mathbf{I}} = \dot{\mathbf{U}}. \tag{5.18}$$

The two horizontal dots in (5.18) denote a peculiar double-index multiplication of matrixes; $\dot{\mathbf{I}}$ and $\dot{\mathbf{U}}$ are infinite square matrixes with respective indices (k, l) and (i, j) and with the following structures:

$$\dot{\mathbf{I}} = \overset{l \rightarrow}{k \downarrow \begin{array}{|c|c|c|c|c|} \hline \dot{I}_{-2,-2} & \dot{I}_{-2,-1} & \dot{I}_{-2,0} & \dot{I}_{-2,1} & \dot{I}_{-2,2} \\ \hline \dot{I}_{-1,-2} & \dot{I}_{-1,-1} & \dot{I}_{-1,0} & \dot{I}_{-1,1} & \dot{I}_{-1,2} \\ \hline \dot{I}_{0,-2} & \dot{I}_{0,-1} & \boxed{\dot{I}_{0,0}} & \dot{I}_{0,1} & \dot{I}_{0,2} \\ \hline \dot{I}_{1,-2} & \dot{I}_{1,-1} & \dot{I}_{1,0} & \dot{I}_{1,1} & \dot{I}_{1,2} \\ \hline \dot{I}_{2,-2} & \dot{I}_{2,-1} & \dot{I}_{2,0} & \dot{I}_{2,1} & \dot{I}_{2,2} \\ \hline \end{array}} , \quad \dot{\mathbf{U}} = \overset{j \rightarrow}{i \downarrow \begin{array}{|c|c|c|c|c|} \hline \phantom{\dot{U}_g} & \phantom{\dot{U}_g} & \phantom{\dot{U}_g} & \phantom{\dot{U}_g} & \phantom{\dot{U}_g} \\ \hline & & & & \\ \hline & & & \boxed{\dot{U}_g} & \\ \hline & & & & \\ \hline & & & & \\ \hline \end{array}} ,$$

$\dot{\mathbf{R}}=\{R_{i,j,k,l}\}$ is a four-dimensional infinite matrix, $i, j, k, l=-\infty, \ldots, -1, 0, 1, \ldots, \infty$. In the case under consideration, matrix $\dot{\mathbf{R}}$ has proved to be quite sparse, since its non-zero elements are only the following ones:

$$\left| \begin{array}{l} \dot{R}_{k,l,k,l} = R_0, \\ \dot{R}_{k,l,k-1,l} = \frac{1}{2} \dot{m}_1 R_0, \quad \dot{R}_{k,l,k,l-1} = \frac{1}{2} \dot{m}_2 R_0, \\ \dot{R}_{k,l,k+1,l} = \frac{1}{2} \overset{*}{\dot{m}}_1 R_0, \quad \dot{R}_{k,l,k,l+1} = \frac{1}{2} \overset{*}{\dot{m}}_2 R_0. \end{array} \right. \tag{5.19}$$

Let us explain the product of matrixes $\dot{\mathbf{R}}$, $\dot{\mathbf{I}}$, $\dot{\mathbf{U}}$ in (5.18) as compared to the ordinary product of matrixes. Let us assume that we have three ordinary (two-dimensional) rectangular matrixes:

$$\begin{array}{llll} \mathbf{A} = \{a_{i,j}\} , & i = 1, 2, \ldots, p , & j = 1, 2, \ldots, q, \\ \mathbf{B} = \{b_{j,k}\} , & k = 1, 2, \ldots, r , & \mathbf{C} = \{c_{i,k}\}. \end{array}$$

If these matrixes are related through the ordinary product $\mathbf{A}.\mathbf{B} = \mathbf{C}$, this implies that $c_{ik} = \sum_{j=1}^{q} a_{ij} b_{jk}$.

The ordinary product of matrices can be termed single-index product since it leads to a summing by a single index (in this case by j).

Let us assume now that the first matrix is four-dimensional, i.e.

$$\mathbf{A} = \{a_{\varepsilon,\mu,i,k}\}\ , \quad \varepsilon = 1,2,\ldots,s\ , \quad \mu = 1,2,\ldots,t,$$
$$\mathbf{B} = \{b_{j,k}\}\ , \quad \mathbf{C} = \{c_{\varepsilon,\mu}\}.$$

In this case the relation between the matrices is realized through a peculiar product $\mathbf{A} \bullet \bullet \mathbf{B} = \mathbf{C}$, where

$$c_{\varepsilon,\mu} = \sum_{i=1}^{p} \sum_{j=1}^{q} a_{\varepsilon,\mu,j,k} b_{j,k}.$$

Obviously this product differs from the ordinary one by the existence of a double sum. It is natural to qualify such multiplication as a two-index one.

If another harmonic component with incommensurable frequency is added to (5.15), the equation complying to the generalized Ohm's law will acquire the following form

$$\dot{\mathbf{R}} \bullet \bullet \bullet \dot{\mathbf{I}} = \dot{\mathbf{U}}, \tag{5.20}$$

where three–index multiplication appears. Moreover, the matrix of the system is six-dimensional, while the other matrixes are three-dimensional.

Following this logic, a whole family of multidimensional matrixes can be introduced: two–index, three–index, four–index ones, etc. They are all of practical importance. As a rule, the matrix of the system has an even number of dimensions, while the other matrixes have half the dimensions of the system one.

Let us now consider the other typical problem – the variable resistance is attached to an ideal source of harmonic current $i_g = I_g \cos(\omega t + \varphi)$, where

$$U = R_0[1 + m_R \cos(\Omega t + \varphi_R)] I_g \cos(\omega t + \varphi). \tag{5.21}$$

The same two cases tackled above will be considered here (See (5.13) and (5.15)):

1/ $R = R_0[1 + m_1 \cos(\Omega t + \varphi_1) + m_2 \cos(2\Omega t + \varphi_2)]$;

2/ $R = R_0[1 + m_1 \cos(\Omega_1 t + \varphi_1) + m_2 \cos(\Omega_2 t + \varphi_2)]$.

In the first case the voltage symbol is obtained in the form of a finite sum

$$\left| \begin{array}{ll} \hat{U} = \sum\limits_{k=-2}^{2} \dot{U}_k e^{j(\omega + k\Omega)t}\ , & \dot{U}_0 = R_0 \dot{I}_g, \\ \dot{U}_1 = \frac{1}{2} \dot{m}_1 R_0 \dot{I}_g\ , & \dot{U}_2 = \frac{1}{2} \dot{m}_2 R_0 \dot{I}_g, \\ \dot{U}_{-1} = \frac{1}{2} \overset{*}{m}_1 R_0 \dot{I}_g\ , & \dot{U}_{-2} = \frac{1}{2} \overset{*}{m}_2 R_0 \dot{I}_g. \end{array} \right. \tag{5.22}$$

This sum can be regarded as an almost periodic time function with a branching equal to one.

In the second case we obtain:

$$\left|\begin{array}{ll} \hat{U} = \sum_{k=-1}^{1} \sum_{l=-1}^{1} \dot{U}_{kl} e^{j(\omega+k\Omega_1+l\Omega_2)t}, & \dot{U}_{00} = R_0 \dot{I}_g, \\ \dot{U}_{1,0} = \frac{1}{2}\dot{m}_1 R_0 \dot{I}_g, & \dot{U}_{0,1} = \frac{1}{2}\dot{m}_2 R_0 \dot{I}_g, \\ \dot{U}_{-1,0} = \frac{1}{2}\overset{*}{m}_1 R_0 \dot{I}_g, & \dot{U}_{0,-1} = \frac{1}{2}\overset{*}{m}_2 R_0 \dot{I}_g. \end{array}\right. \tag{5.23}$$

The sum written down in (5.21) is an almost periodic function with a branching equal to two.

Obviously the considered problem with supply from a source of current is incomparably simpler than that with supply from a source of voltage.

It is not difficult to be convinced that the problems related to the investigation of the properties of periodic or almost periodic conductance are analogical to the problems related to periodic and almost periodic resistance considered above.

Indeed, if a linear periodic conductance, presented as

$$G = G_0[1 + m_G \cos(\Omega t + \varphi_G)], \tag{5.24}$$

is linked up to an ideal voltage source $U_g \cos(\omega t + \varphi)$, then the determining equation will be

$$i = G_0[1 + m_G \cos(\Omega t + \varphi_G)] U_g \cos(\omega t + \varphi). \tag{5.25}$$

When the conductance is linked up to a source of harmonic current $I_g \cos(\omega t + \varphi)$, the basic equation will be

$$G_0[1 + m_G \cos(\Omega t + \varphi_G)] U = I_g \cos(\omega t + \varphi). \tag{5.26}$$

Comparing (5.25) and (5.26) with (5.21) and (5.5), we become convinced that these equations are equivalent to an approximation of the denotations.

In practice it may be relevant to make use of the choice of a source (current or voltage) and of the value (resistance or conductance) for simplifying the analysis.

5.2.2. Periodic inductance and magnetic susceptibility

When a periodic inductance of the type

$$L = L_0[1 + m_L \cos(\Omega t + \varphi_L)] \tag{5.27}$$

is attached to an ideal source of harmonic voltage U_g, the following determining equation is obtained

$$j\omega L_0 \frac{\dot{m}_1}{2} \sum_{k=-\infty}^{\infty} \dot{I}_{k-1} e^{j(\omega+k\Omega)t} + j\omega L_0 \sum_{k=-\infty}^{\infty} \dot{I}_k e^{j(\omega+k\Omega)t}$$

$$+j\omega L_0 \frac{\overset{*}{m}_1}{2} \sum_{k=-\infty}^{\infty} \dot{I}_{k+1} e^{j(\omega+k\Omega)t} = \dot{U}_g e^{j\omega t}. \tag{5.28}$$

This equation is analogous to Eq.(5.7) – the difference is that here multiplier R_0 is replaced with $j\omega L_0$. Hence, all conclusions concerning (5.8) will be applicable here as well. In this case, the vector diagram differs from that in Fig.5.1 only in that vectors $\dot{I}_0$ and $\dot{U}_g$ are moved apart from each other at an angle of $\frac{\pi}{2}$.

When the periodic inductance is connected to an ideal source of harmonic current $i_g = I_g \cos(\omega t + \varphi)$, the following is obtained:

$$\left| \begin{array}{ll} \hat{U} = \sum_{k=-1}^{1} \dot{U}_k e^{j(\omega+k\Omega)t}\ , & \dot{U}_0 = j\omega L_0 I_g, \\ \dot{U}_1 = j(\omega+\Omega) L_0 \frac{\dot{m}_L}{2} \dot{I}_g\ , & \dot{U}_{-1} = j(\omega-\Omega) L_0 \frac{\overset{*}{m}_L}{2} \dot{I}_g. \end{array} \right. \tag{5.29}$$

This case is analogous to the one considered above, where a periodic resistance is connected to a source of current, with the important difference that here $U_{-1} \neq U_1$. This difference can be visualized in the vector diagram – Fig.5.2, plotted in correspondence with (5.29). It is obvious that the instantaneous complex voltage amplitude $\dot{U}(t)$ leads the complex current amplitude with an angle, which is periodically either larger or smaller than 90°, but its average is equal to 90°. If vector $\dot{I}_g$ is regarded as motionless, then vector $\dot{U}(t)$ will periodically change its module and direction with angular frequency Ω, constantly remaining in a definite sector wide 2α (See Fig.5.2b). By way of visualization, one can imagine that $\dot{U}(t)$ is a harmonic oscillation modulated both by amplitude and by phase with a modulation frequency Ω. As a rule, the „modulation frequency" Ω in modulation-parametric systems is much higher than the „carrier" frequency ω, i.e. $\Omega \gg \omega$.

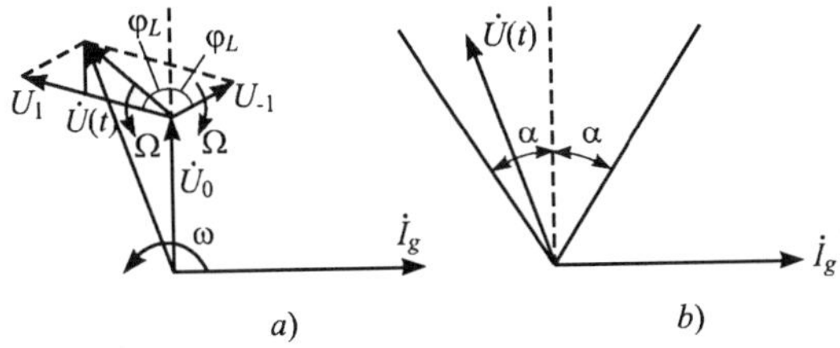

Fig. 5.2. A vector diagram of the voltage of a linear periodic inductance with a harmonic variable part connected to a source of harmonic current: a) in full form. b) in compact form

The considered case of connection of periodic inductance to a source of current is characterized by this specific peculiarity – the appearance of phase modulation. Such phase modulation fails to appear when a periodic inductance is connected to a source of ideally harmonic voltage.

If the inductance changes in time in accordance with an almost periodic law, then the modulation that comes about is characterized by an almost periodic law and a multidimensional quality emerges. Then sector 2α is defined in multidimensional space.

Consider a periodic magnetic susceptibility (a value opposite to inductance) $\Gamma = \Gamma_0[1 + m_\Gamma \cos(\Omega t + \varphi_\Gamma)]$. In the case of a connection to an ideal source of harmonic voltage U_g, the following complex form of an expression is obtained for the flowing current

$$\left| \begin{array}{l} \hat{i} = \dot{I}_{-1} e^{j(\omega-\Omega)t} + \dot{I}_0 e^{j\omega t} + \dot{I}_1 e^{j(\omega+\Omega)t}, \\ \dot{I}_0 = \dfrac{\Gamma_0 \dot{U}_g}{j\omega}, \ \dot{I}_1 = \dfrac{\dot{m}_\Gamma}{2}\dfrac{\Gamma_0 \dot{U}_g}{j\omega}, \ \dot{I}_{-1} = \dfrac{\overset{*}{m}_\Gamma}{2}\dfrac{\Gamma_0 \dot{U}_g}{j\omega}. \end{array} \right. \tag{5.30}$$

This case is close to the one considered above (See (5.29)), with the only difference that no phase modulation emerges here.

When the magnetic susceptibility is linked up to an ideal source of harmonic current i_g, the following determining equation is obtained

$$-j\Gamma_0 \sum_{k=-\infty}^{\infty} \left[\frac{U_k}{\omega + k\Omega} + \frac{\dot{m}_\Gamma}{2} \frac{\dot{U}_{k-1}}{\omega + (k-1)\Omega} + \frac{\overset{*}{m}_\Gamma}{2} \frac{\dot{U}_{k+1}}{\omega + (k+1)\Omega} \right]$$

$$\times e^{j(\omega + k\Omega)t} = \dot{I}_g e^{j\omega t}, \tag{5.31}$$

which describes an infinite system of the type

$$-j\Gamma_0 \begin{bmatrix} 1 & \dfrac{\overset{*}{m}_\Gamma}{2} & \\ \dfrac{\dot{m}_\Gamma}{2} & \boxed{1} & \dfrac{\overset{*}{m}_\Gamma}{2} \\ & \dfrac{\dot{m}_\Gamma}{2} & 1 \end{bmatrix} \begin{bmatrix} \omega - \Omega & & \\ & \boxed{\omega} & \\ & & \omega + \Omega \end{bmatrix} \begin{bmatrix} \dot{U}_{-1} \\ \boxed{\dot{U}_0} \\ \dot{U}_1 \end{bmatrix} = \begin{bmatrix} \\ \boxed{\dot{I}_g} \\ \\ \end{bmatrix}. \tag{5.32}$$

or in brief: $\dot{\Gamma}(j\omega)^{-1}\dot{\mathbf{U}} = \dot{\mathbf{I}}$, where $\omega = \mathrm{diag}(\ldots, \omega - k\Omega, \ldots, \omega - \Omega, \omega, \omega + \Omega, \ldots, \omega + k\Omega, \ldots)$ is an infinite diagonal matrix of frequencies. The emergence of a matrix of frequencies is indicative of the existence of phase modulation.

Phase modulation appears when the inductance or magnetic susceptibility are tied to a source of harmonic current and fails to appear when they are connected to a source of harmonic voltage. This is obvious, since in the first case an infinite sum of voltage harmonics appears and each term is divided into its frequency with the respective phase shift. As a result, a matrix frequency multiplier appears which conditions an angular modulation.

5.2.3. Periodic capacitance and electric elastance

When a variable capacitance $C(t) = C_0[1 + m_C \cos(\Omega t + \varphi_C)]$ is hooked to an ideal source of harmonic voltage U_g, the following can be determined:

$$\left|\begin{array}{l} \hat{i} = \dot{I}_{-1} e^{j(\omega-\Omega)t} + \dot{I}_0 e^{j\omega t} + \dot{I}_1 e^{j(\omega+\Omega)t}, \ \dot{I}_0 = j\omega C_0 \dot{U}_g, \\ \dot{I}_1 = j(\omega+\Omega) C_0 \frac{\dot{m}_C}{2} U_g, \ \dot{I}_{-1} = j(\omega-\Omega) C_0 \frac{\overset{*}{\dot{m}}_C}{2} U_g. \end{array}\right. \tag{5.33}$$

The vector diagram for this case is presented in Fig.5.3.

This case is dual to the case of inductance with a source of current (Fig.5.2).

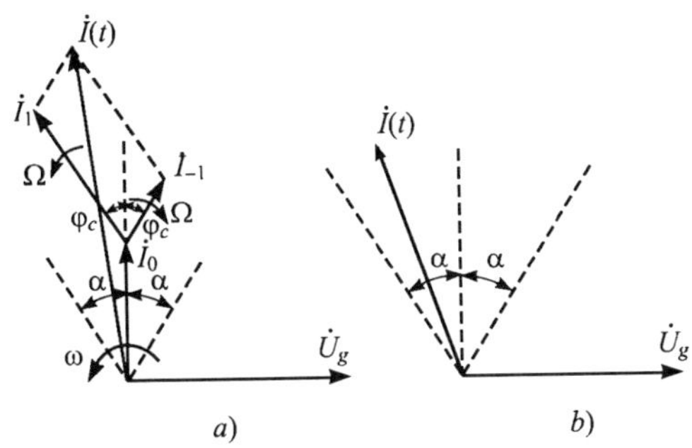

Fig. 5.3. A vector diagram of the voltage and currents flowing through periodic capacitance with a harmonic variable part connected to a source of harmonic voltage: a) in full form. b) in compact form

When the capacitance is connected to an ideal source of harmonic current, the following determining equation is obtained

$$j\omega C_0 \sum_{k=-\infty}^{\infty} \left(\frac{\dot{m}_C}{2} \dot{U}_{k-1} + \dot{U}_k + \frac{\overset{*}{\dot{m}}_C}{2} \dot{U}_{k+1} \right) e^{j(\omega+k\Omega)t} = \dot{I}_g e^{j\omega t}, \tag{5.34}$$

which structurally coincides with (5.7), given the substitution $j\omega C_0 \to R, m_C \to m_R$.

When a periodic electric elastance $S(t) = S_0[1 + m_S \cos(\Omega t + \varphi_S)]$ (a quantity opposite to capacitance) is connected up to an ideal source of harmonic voltage, the result is a determining equation

$$-jS_0 \sum_{k=-\infty}^{\infty} \left[\frac{\dot{I}_k}{\omega + k\Omega} + \frac{\dot{m}_S}{2} \frac{\dot{I}_{k-1}}{\omega + (k-1)\Omega} + \frac{\overset{*}{\dot{m}}_S}{2} \frac{\dot{I}_{k+1}}{\omega + (k+1)\Omega} \right]$$

$$\times e^{j(\omega+k\Omega)t} = U_g e^{j\omega t}, \tag{5.35}$$

which differs from (5.31) only by the denotations. The respective infinite system is of the type

$$-jS_0 \begin{bmatrix} 1 & \frac{\overset{*}{m}_s}{2} & \\ \frac{\dot{m}_s}{2} & \boxed{} & \frac{\overset{*}{m}_s}{2} \\ & \frac{\dot{m}_s}{2} & \end{bmatrix} \cdot \begin{bmatrix} \omega - \Omega & & \\ & \boxed{\omega} & \\ & & \omega + \Omega \end{bmatrix} \begin{bmatrix} \dot{I}_{-1} \\ \boxed{\dot{I}_0} \\ \dot{I}_1 \end{bmatrix} = \begin{bmatrix} \\ \boxed{\dot{U}_g} \\ \end{bmatrix}, \tag{5.36}$$

or in brief: $\dot{\mathbf{S}}(j\omega)^{-1}\dot{\mathbf{I}} = \dot{\mathbf{U}}$.

Given a linkage of the electric elastance to an ideal source of harmonic current, the result is

$$\left| \begin{array}{ll} \hat{U} = \sum\limits_{k=-1}^{1} \dot{U}_k e^{j(\omega+k\Omega)t} , & \dot{U}_0 = \dfrac{S_0 \dot{I}_g}{j\omega}, \\ \dot{U}_1 = \dfrac{\dot{m}_S}{2}\dfrac{S_0}{j\omega}\dot{I}_g , & \dot{U}_{-1} = \dfrac{\overset{*}{m}_S}{2}\dfrac{S_0}{j\omega}\dot{I}_g . \end{array} \right. \tag{5.37}$$

The general conclusion is that the capacitance and the electric elastance linked up to an ideal source of harmonic voltage yield a current modulated by amplitude and phase, while their connection to an ideal source of harmonic current results only in amplitude modulated voltage.

5.2.4. Power consumed by periodic elements of the radio circuits

An instantaneous power consumed by a periodic resistance connected to an ideal source of harmonic voltage is represented as

$$p = U_g i \leftarrow \hat{p} = \frac{\hat{U}_g + \check{U}_g}{2} i = \frac{1}{2} \sum_{k=-\infty}^{\infty} \dot{U}_g \dot{I}_k e^{j(2\omega+k\Omega)t} + \frac{1}{2} \sum_{k=-\infty}^{\infty} \overset{*}{U}_g \dot{I}_k e^{jk\Omega t}. \tag{5.38}$$

In the special case when one can find such $k = k_a < 0$, where the equation $2\omega + k_a\Omega = 0$ would be fair, the average power is equal to

$$P_{av} = Re\left(\frac{\overset{*}{U}_g \dot{I}_0}{2} + \frac{\dot{U}_g \dot{I}_{ka}}{2} \right) = \frac{1}{2} U_g I_0 + \frac{1}{2} U_g I_{ka} \cos(\varphi + \varphi_{ka}). \tag{5.39}$$

Fig.5.4 shows a typical vector diagram of the symbol of instantaneous power $\hat{p}$ plotted on the basis of expression (5.38), as well as the qualitative time dependence of the instantaneous power $p(t)$ consumed by a periodic resistance.

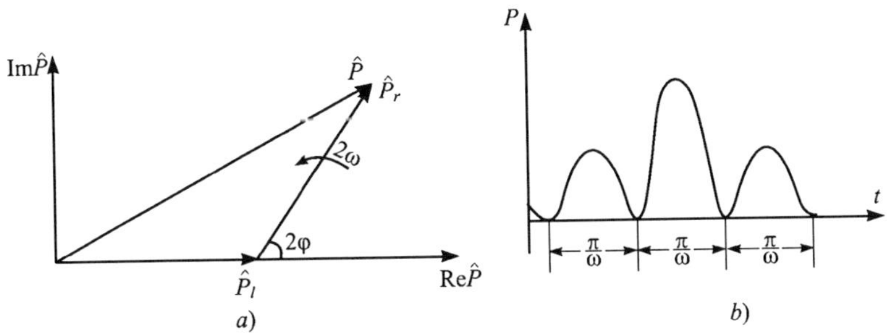

Fig. 5.4. Illustration of the instantaneous power consumed by periodic resistance connected to an ideal source of harmonic voltage:
a) A typical vector diagram of the symbol of the instantaneous power;
b) Quality dependence on the time of the instantaneous power

The vector diagram of the symbol of instantaneous power in dynamics (Fig.5.4a) can be explained by using the vector diagram in Fig.5.1c. It follows from (5.38) that the „lying" vector

$$\hat{P}_l = \frac{1}{2} \sum_{k=-\infty}^{\infty} \overset{*}{\dot{U}}_g \dot{I}_k e^{jk\Omega t}$$

is always situated in the direction of the real axis and that it undergoes modular changes with frequency Ω according to a periodic law, and its value cannot be negative. The „rotating" vector

$$\hat{P}_r = \frac{1}{2} \sum_{k=-\infty}^{\infty} U_g I_k e^{j(2\omega + k\Omega)t}$$

turns regularly, with an angular speed 2ω, round the end of vector $\hat{P}_l$ undergoing modular changes analogous to vector $\hat{P}_l$.

Fig.5.4b is a consequence of Fig.5.4a. The almost periodic function $p(t)$ is built on the basis of two frequencies: frequency Ω, at which modules P_l and P_r change in synchronism, and frequency ω, when vector $\hat{P}_r$ turns round the end of vector $\hat{P}_l$ with angular speed 2ω. Given fixed frequencies ω and Ω, the form of function $P(t)$ may be changed in a wide range by changing the initial phases φ and φ_R of the source voltage and the variable part of the resistance correspondingly, as well as the amplitudes of both functions.

In the case when a periodic resistance is connected to an ideal source of harmonic current i_g, the symbol of the instantaneous power consumed is determined as

$$\hat{p} = \frac{\hat{i}_g + \check{i}_g}{2} \hat{U} = \frac{1}{2} \sum_{k=-1}^{1} \dot{I}_g U_k e^{j(2\omega + k\Omega)t} + \frac{1}{2} \sum_{k=-1}^{1} \overset{*}{\dot{I}}_g \dot{U}_k e^{jk\Omega t}. \tag{5.40}$$

All qualitative peculiarities of the previous case are preserved (Fig.5.4).

The matter with the two regimes of linkage to periodic conductance stands analogically by virtue of the obvious duality of the systems.

From a formal point of view, the expression of the instantaneous power consumed by a periodic inductance connected to an ideal source of harmonic voltage does not differ from that of a periodic resistance (5.38). But as the phase relations between voltage and current are taken into account, the result is a vector diagram of the power symbol and a graph of the instantaneous power as a function of time, as shown in Fig.5.5. In this case the „standing" power vector is

$$\hat{P}_S = \frac{1}{2}\sum_{k=-\infty}^{\infty} \overset{*}{\dot{U}}_g \dot{I}_k e^{jk\Omega t},$$

while the „rotating" power vector is

$$\hat{P}_r = \frac{1}{2}\sum_{k=-\infty}^{\infty} \dot{U}_g \dot{I}_k e^{j(2\omega+k\Omega)t}.$$

The instantaneous power, as in the previous case, is a periodic function of the time with fixed interval between the zeroes. Yet, unlike the previous case, this function has a zero average value only for a large time interval. An exception from the general rule is the case when the condition that $2\omega + k\Omega = 0$ is met. Then the average power consumed by the periodic inductance depends on the sign of $\cos(\varphi + \varphi_k)$ and can be both positive and negative. From a physical point of view, this fact is easy to explain, bearing in mind the generalization of the Manley-Row energy relations (See section 1.2) for the case of linear periodic reactance, as well as the effect of the single-frequency non-degenerate parametric regeneration considered in section 2.1 from the position of the four-frequency modulation-parametric systems.

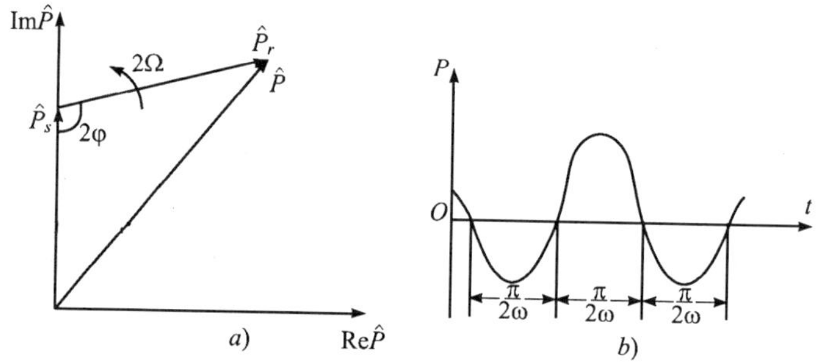

Fig. 5.5. Illustration of the instantaneous power consumed by periodic inductance connected to an ideal source of harmonic voltage:
a) A typical vector diagram of the symbol of the instantaneous power;
b) Quality dependence on the time of the instantaneous power

In the case of a periodic inductance linked up to an ideal source of harmonic current, the instantaneous consumed power is expressed by an equation analogous to (5.35). Then the first term yields no constant component, while the second one can yield such a component provided that $2\omega + k\Omega = 0$ and $P_{av} = \frac{1}{2} U_g I_k \cos(\varphi + \varphi_k)$. Obviously, the constant component can be both larger than zero and smaller than zero. Fig.5.6 displays a vector diagram of the power symbol and the qualitative dependence of instantaneous power on time in such a case. The difference from Fig.5.5a consists in that instead of the „standing" vector $\hat{P}_S$ we have an oscillating „suspended" vector – $\hat{P}_h$, that changes its direction but remains, in the average, turned opposite to the direction of the imaginary axis. As a consequence of the modulation of components $\dot{U}_1$ and $\dot{U}_{-1}$ not only by amplitude but by phase as well, as a rule, the interval between the zeros in the diagram shown in Fig.5.6b differs from $\frac{\pi}{2\omega}$, but its average over a long time interval is equal to that quantity.

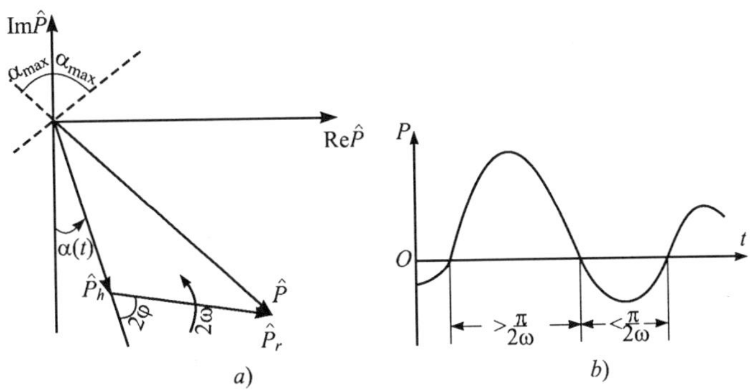

Fig. 5.6. Illustration of the instantaneous power consumed by periodic inductance connected to an ideal source of harmonic current:
a) A typical vector diagram of the symbol of the instantaneous power;
b) Quality dependence on the time of the instantaneous power

The cases of the two possible linkages of periodic magnetic susceptibility are dual to the periodic inductance cases considered above.

Essentially new versions fail to appear when a periodic capacitance and periodic elastance are connected to ideal sources of harmonic voltage and currency, too.

Periodic reactance, unlike constant reactance, extends the signal spectrum to infinity, thus imparting a completely definite structure to it – furthermore, at some spectral components the active power can be absorbed, analogically to the case of positive resistances, while at other spectral components this power can be generated and transmitted to the circuit, as is the case with negative resistances. The consumption and generation of active power is a peculiarity of the periodic reactances.

By way of comparison, we shall consider the power consumed by a constant capacitance C. The instantaneous power and its transform are presented as

$$P = iU = \frac{\omega}{2} CU^2 \sin(2\omega t + 2\varphi) \leftarrow \hat{P} = \frac{\hat{U} + \check{U}}{2} i = j\frac{\omega C}{2} U^2 + j\frac{\omega C}{2} U^2 e^{j2\omega t}.$$

The original P is an oscillating power changing in time according to a harmonic law with double frequency. The transform $\hat{P}$ is presented by the sum of a constant component and a complex-valued periodic time function.

The constant component $\dot{P}_0 = j\frac{\dot{\omega} C}{2} U^2$ is conditioned by the complex amplitudes method and it is a convenient object of study. It is an analogue of the average power absorbed by an active resistance. From the perspective of the complex amplitudes method, a capacitor, like an active resistance, „absorbs" power, but this „absorbed" power is imaginary, reactive. The „absorbed" power $\dot{P}_0$ provides important information on the oscillating power since its module is equal to the latter.

In the case of periodic capacitance, the constant „absorbed" power is expressed through the following infinite matrix

$$\mathbf{P}_0 = \frac{1}{2} \overset{*}{\mathbf{U}} j\omega \dot{\mathbf{C}} \dot{\mathbf{U}} = \left\{ \dot{P}_{kl} \right\}, \tag{5.41}$$

where $\dot{\mathbf{C}}$ is an infinite square three-diagonal matrix of the type of matrix $\dot{\mathbf{R}}$ in (5.8), given the following substitutions: $m_R \to m_C$, $\omega = \mathrm{diag}(\ldots, \omega - k\Omega, \ldots, \omega - \Omega, \omega, \omega + \Omega, \ldots, \omega + k\Omega, \ldots)$, $\dot{\mathbf{U}} = \mathrm{diag}(\ldots, \dot{U}_{-k}, \ldots, \dot{U}_{-1}, \dot{U}_0, \dot{U}_1, \ldots, \dot{U}_k, \ldots)$, $\dot{P}_{k,k} = \frac{1}{2} j(\omega + k\Omega) C_0 U_k^2$, $\dot{P}_{k,k-1} = j(\omega + k\Omega)\frac{\dot{m}_C}{4} C_0 \overset{*}{U}_k \dot{U}_{k-1}$, $\dot{P}_{k,k+1} = j(\omega + k\Omega)\frac{\overset{*}{m}_C}{4} C_0 \overset{*}{U}_k \dot{U}_{k+1}$.

It is convenient to present Exp.(5.36) in the form

$$\dot{\mathbf{P}}_0 = \omega \dot{\mathbf{K}}, \tag{5.42}$$

where $\mathbf{K} = \{k_{pq}\}$ is a complex energetic coefficient of the capacitance represented by an infinite slanting Hermitian matrix, which, in the case under consideration, consists, similarly to $\dot{\mathbf{C}}$, of three diagonals

$$\dot{k}_{p,p} = j\frac{1}{2} C_0 U_p^2, \quad \dot{k}_{p,p-1} = j\frac{\dot{m}_C}{4} C_0 \overset{*}{U}_p \dot{U}_{p-1},$$

$$\dot{k}_{p,p+1} = j\frac{\overset{*}{m}_C}{4} C_0 \overset{*}{U}_p \dot{U}_{p+1}.$$

Unlike $\dot{P}_0$, matrix $\dot{\mathbf{P}}_0$ is not an imaginary one and consists of complex elements. This implies that, as it changes in time, the capacitance absorbs a certain part of the source energy. In the case under consideration, the exchange of energy is by far more complicated and involves three objects: the signal source, the source of capacitance change in time, and the capacitance. If a definite harmonic component is considered, it may happen that the power is absorbed at this component and a part of the absorbed power is transformed into generated power at other harmonics, while another part is directed into the source of capacitance change. The opposite transformation may also take place, i.e. power is generated at a certain harmonic component due to the fact that a part of this power is obtained from other harmonics, while another part has come from the source of capacitance change. These two opposite tendencies are combined in a complex way and in a stationary state the energy of the two sources is balanced against the infinite set of spectral components of the capacitance current and voltage. In all cases, the linear periodic capacitance consumes active power which is lacking in the constant capacitance. Thus matrix $\dot{\mathbf{P}}_0$ is a characteristic of the complex power of the capacitor with periodically changing capacitance. The real part of this matrix – $\mathbf{P}_A = Re\mathbf{P}_0$ – can be defined as a characteristic of the active power consumed or transmitted by all harmonic components of the capacitance. Obviously, it is correct to write $\mathbf{P}_A = \omega\mathbf{K}_D$, where $\mathbf{K}_D = Re\dot{\mathbf{K}}$ is the real energetic coefficient of the power. This is an infinite real slanting symmetric matrix. In the case under consideration, it has the following structure

$$K_D = \frac{1}{2}C_0 \begin{array}{|c|c|c|c|c|} \hline & k_{-2,-1} & & & \\ \hline -k_{-2,-1} & & k_{-1,0} & & \\ \hline & -k_{-1,0} & & k_{0,1} & \\ \hline & & -k_{0,1} & & k_{1,2} \\ \hline & & & -k_{1,2} & \\ \hline \end{array},$$

where $k_{p,p+1} = \dfrac{1}{4}C_0 Re(j\overset{*}{m}_C\overset{*}{\dot{U}}_p\dot{U}_{p+1})$, $k_{p,p-1} = \dfrac{1}{4}C_0 Re(jm_C\overset{*}{\dot{U}}_p\dot{U}_{p-1})$.

The real energy coefficient $\mathbf{K}_D$ allows creating a clear mental picture of the exchange of energy between the individual harmonics, the signal source and the source of capacitance change. Indeed, it is correct to write the following expressions for matrix $\mathbf{P}_A = \left\{\dot{P}_{l,m}\right\}$: $P_{l,l-1} = (\omega + l\Omega)k_{l,l-1}$, $P_{l-1,l} = [\omega + (l-1)\Omega]k_{l-1,l}$ $= -[\omega + (l-1)\Omega]k_{l,l-1}$.

Let us assume that $k_{l,l-1} > 0$, $l > 1$. Then $P_{l,l-1} > 0$, $P_{l-1,l} < 0$. The power $P_{l,l-1}$ is directed from the signal source to the capacitance by the harmonic component with frequency $(\omega + l\Omega)$; the capacitance, on its part, transmits power $P_{l-1,l}$ to the harmonic component with frequency $[\omega + (l-1)\Omega]$, which brings this power back to the signal source. The algebraic sum of these powers $P_{l,l-1} + P_{l-1,l} = \Omega k_{l,l-1}$ is absorbed by the capacitance, i.e. it is transmitted to the source of capacitance change.

In the more general case of periodic linear capacitance (or nonlinear capacitance with harmonically varying argument), its parameter is presented by an infinite Fourier series: $C(t) = C_0\left[1 + \sum_{k=1}^{\infty} m_k \cos(k\Omega t + \varphi_k)\right]$. In this case matrix $\dot{\mathbf{C}}$ is completely full, and its major diagonal consists of unities, while all other diagonals parallel to the major one are made up of identical elements. The diagonals above the major one consist, respectively, of the elements $\frac{\overset{*}{m}_1}{2}, \frac{\overset{*}{m}_2}{2}, \ldots, \frac{\overset{*}{m}_k}{2}, \ldots$, while those under it encompass $\frac{\dot{m}_1}{2}, \frac{\dot{m}_2}{2}, \ldots, \frac{\dot{m}_k}{2}, \ldots$.

The theory recounted here is generalized for any linear reactive circuits whose elements change periodically in time with equal periods. In all cases the infinite complex matrix $\dot{\mathbf{P}}_0$ and the complex matrix energy coefficient $\dot{\mathbf{K}}$ are characteristics of the complex powers of the reactances with periodically changing parameters and they determine the energy balance. The real matrix energy coefficient $\mathbf{K}_D$ reveals the mechanism of converting active power through periodic reactance. The results give an idea of the transformation dynamism of the active power by periodic reactance and provide an algorithm for calculating the average absorbed or generated power of the individual harmonics.

5.3. Nonlinear elements of radiophysical systems

The nonlinear elements of radiophysical systems are usually given in the following forms: $R(U)$ and $R(i)$, $C(U)$ and $C(q)$, $L(i)$ and $L(\Phi)$. In most of the practical cases, the radiophysical systems with nonlinear elements can be considered as systems with periodic time parameters, since voltage U, current i, charge q and magnetic flow Φ are periodic time functions.

When analyzed, nonlinear characteristics are most often presented in a Taylor's series, such as for example

$$R = \sum_{k=0}^{\infty} R_k U^k \tag{5.43}$$

while in the process of problem solving they appear in a Fourier series, for example

$$R = R_0 + 2\sum_{n=1}^{\infty} R_n \cos(n\omega t + \varphi_n). \tag{5.44}$$

Let us consider a resistance with the following dependence on voltage

$$R = R_0 + R_1 U, \tag{5.45}$$

linked up to an ideal source of harmonic voltage $U_g = U_0 + U_1 \cos(\omega t + \varphi)$. The determining equation of the system is written in the form

$$R_0 \sum_{k=-\infty}^{\infty} \left[\left(1 + \frac{U_0}{U_{00}}\right) \dot{I}_k + \frac{1}{2}\frac{\dot{U}_1}{U_{00}} \dot{I}_{k-1} + \frac{1}{2}\frac{\overset{*}{\dot{U}}_1}{U_{00}} \dot{I}_{k+1} \right] e^{jk\omega t} = U_0 + \dot{U}_1 e^{j\omega t},$$

where $U_{00} = \dfrac{R_0}{R_1}$ and corresponds to the following infinite system of algebraic equations

$$R_0 \begin{bmatrix} 1 + \dfrac{U_0}{U_{00}} & \dfrac{1}{2}\dfrac{\overset{*}{\dot{U}}_1}{U_{00}} & \\ \dfrac{1}{2}\dfrac{\dot{U}_1}{U_{00}} & 1 + \dfrac{U_0}{U_{00}} & \dfrac{1}{2}\dfrac{\overset{*}{\dot{U}}_1}{U_{00}} \\ & \dfrac{1}{2}\dfrac{\dot{U}_1}{U_{00}} & 1 + \dfrac{U_0}{U_{00}} \end{bmatrix} \begin{bmatrix} \dot{I}_{-1} \\ \dot{I}_0 \\ \dot{I}_1 \end{bmatrix} = \begin{bmatrix} \dot{U}_{-1} \\ \dot{U}_0 \\ \dot{U}_1 \end{bmatrix}, \tag{5.46}$$

or $\dot{\mathbf{R}}(\mathbf{U}).\dot{\mathbf{I}} = \dot{\mathbf{U}}$.

It is obvious that the infinite system (5.46) has much in common with the infinite system (5.8) in the case of a periodic resistance. Here the infinite matrix $\dot{\mathbf{R}}$ is obtained on the basis of the functional dependence $U_g(t)$ which is typical for nonlinear systems. The infinite system (5.46) is solved with the help of the infinite chain fraction K, which, in this case, yields a precise solution:

$$K = \frac{1}{2}\left[1 + \sqrt{1 - \left(\frac{U_1}{U_{00} + U_0}\right)^2}\,\right].$$

The solution of system (5.46) is presented in the form

$$i(t) = I_0 + i_1 + i_2 + \ldots + i_k + \ldots, \tag{5.47}$$

where

$$i_1 = Re\left(\dot{I}_1 e^{j\omega t} + \dot{I}_{-1} e^{-j\omega t}\right) = Re(\dot{I}_1 + \overset{*}{\dot{I}}_{-1}) e^{j\omega t} = Re\dot{\bar{I}} e^{j\omega t},$$

$$i_2 = Re\dot{\bar{I}}_2 e^{j2\omega t}, \ldots, \; i_k = Re\dot{\bar{I}}_k e^{jk\omega t}, \ldots,$$

$$\dot{\bar{I}}_1 = \dot{I}_1 + \overset{*}{\dot{I}}_{-1}, \; \dot{\bar{I}}_2 = \dot{I}_2 + \dot{I}_{-2}, \ldots, \; \dot{\bar{I}}_k = \dot{I}_k + \overset{*}{\dot{I}}_{-k}, \ldots,$$

$$\dot{I}_0 = \frac{\left(a - \dfrac{b^2}{aK}\right) U_0 - \overset{*}{\dot{b}}\dot{U}_1}{\left(a - \dfrac{b^2}{aK}\right)^2 \dot{U}_1 - b^2}, \quad \dot{I}_1 = \frac{\left(a - \dfrac{b^2}{aK}\right) \dot{U}_1 - bU_0}{\left(a - \dfrac{b^2}{aK}\right)^2 - b^2},$$

$$\dot{I}_{-1} = \frac{-\dfrac{\overset{*}{b}}{aK}\left(a - \dfrac{b^2}{aK}\right) U_0 + \dfrac{b^{*2}}{aK}\dot{U}_1}{\left(a - \dfrac{b^2}{aK}\right)^2 - b^2}, \quad a = R_0\left(1 + \frac{U_0}{U_{00}}\right), \ \dot{b} = \frac{R_0}{2}\frac{\dot{U}_1}{U_{00}}.$$

The rest of the complex amplitudes are calculated by applying the following recurrent ratios:

$$\frac{\dot{I}_{k+1}}{\dot{I}_k} = -\frac{\dot{b}}{aK}, \quad k = 1, 2, 3, \ldots, \infty,$$

$$\frac{\dot{I}_{-(k+1)}}{\dot{I}_{-k}} = -\frac{\overset{*}{\dot{b}}}{aK}, \quad k = 0, 1, 2, 3, \ldots, \infty.$$

If instead of (5.45) the square term of Taylor's series is taken into account, matrix $\dot{\mathbf{R}}$ will contain not three but five non-zero diagonals. In the general case, if the non-linear resistance (5.43) contains k terms from Taylor's series, the matrix of the system will contain $2k+1$ non-zero diagonals.

An analogical result is obtained if we use resistance of the type of (5.45) but the input voltage U is a periodic function represented by a Fourier series. If we use the first k harmonics of the harmonic expansion of U, then the system matrix $\dot{\mathbf{R}}$ will contain $2k+1$ non-zero diagonals.

In the most general case, when the nonlinear resistance (5.43) contains k terms of Taylor's series and the forcing voltage has n harmonic terms with commensurable frequencies, the system matrix $\dot{\mathbf{R}}$ will have $2(k+n)+1$ non-zero diagonals and the vector column will have respectively $n+1$ non-zero elements.

If the forcing voltage is an almost periodic time function, the resultant system (5.46) will be multi-dimensional. If the almost periodic function has branching k, the current and voltage matrixes in (5.46) will be k-dimensional and the resistance matrix will be $2k$-dimensional; the multiplication of the resistance matrix and current matrix will be k-index.

Let us consider yet another generalization for the case when the nonlinear resistance explicitly depends on time according to the law

$$R = R_0 + R_1 U + R^{(1)} \cos(\Omega t + \Psi_1) \tag{5.48}$$

and is connected up to an ideal source of harmonic voltage with frequency ω.

The current symbol is presented in the form $\hat{i} = \sum_{k=-\infty}^{\infty} \sum_{l=-\infty}^{\infty} \dot{I}_{kl} e^{j(k\omega + l\Omega)t}$, while a system of the type of (5.46) is written in the following succinct way:

$$\dot{\mathbf{R}}(\dot{\mathbf{U}}) \bullet \bullet \dot{\mathbf{I}} = \dot{\mathbf{U}},$$

where $\dot{\mathbf{R}}(\mathbf{U})$ is a four-dimensional matrix, $\dot{\mathbf{I}}$ and $\dot{\mathbf{U}}$ are two-dimensional matrixes, while the two horizontally located dots denote double-index multiplication. If the explicit dependence on time is almost periodical with a k-branched almost periodic function, the current and voltage matrices will be $k+1$-dimensional, while the resistance matrix will be $2(k+1)$-dimensional.

There is a more general case when nonlinear resistance with almost periodic and explicit dependence on time, containing m basic (incommensurable) frequencies is connected to a source of almost periodic voltage characterized by n basic frequencies. Out of the $m+n$ set of frequencies, p basic frequencies are identified so that $\max(m,n) \leq p \leq m+n$. Here the current and voltage matrices are $(p+1)$-dimensional, while the resistance matrix is, respectively, $2(p+1)$-dimensional.

It is worth noting that analyses such as that presented above are also typical of the cases when $C(U)$ and $L(\Phi)$ are connected to a source of voltage, or $C(q)$ and $L(i)$ are connected to a source of current.

So far we have considered problems that can be defined as *first type* problems with nonlinear elements. This type of problems are characterized by the fact that the argument of the nonlinear elements is a known time function. For example, when $R(U)$ is connected to a voltage source, the law of resistance dynamism in time can be determined at once.

Problems where the function sought is the argument of the nonlinear element are referred to as *second type* problems with nonlinear elements. For example, if the same resistance $R(U)$ is linked up to a source of current, this resistance, as a function of time, will not be known in advance, for its argument U is the value sought.

Let us consider the following case that can be referred to the second type of problems: Nonlinear resistance $R = R_0 + R_1 U + R_2 U^2$ is connected to an ideal source of harmonic current $i_g = I_0 + I_1 \cos(\omega t + \varphi)$. The determining equation is obtained in the following form:

$$\sum_{k=-\infty}^{\infty} \left[\frac{\dot{I}_1}{2U_{02}^2} \sum_{l=-\infty}^{\infty} \dot{U}_{k+l-1}\dot{U}_{-l} + \frac{\dot{I}_1}{2U_{01}}\dot{U}_{k-1} + \frac{\dot{I}_0}{U_{02}^2} \sum_{l=-\infty}^{\infty} \dot{U}_{k+l}\dot{U}_{-l} \right.$$
$$\left. + \left(\frac{I_0}{U_{01}} - \frac{1}{R_0}\right)\dot{U}_k + \frac{\overset{*}{I}_1}{2U_{02}^2} \sum_{l=-\infty}^{\infty} U_{k+l+1}U_{-l} + \frac{\overset{*}{I}_1}{2U_{01}}\dot{U}_{k+1} \right] e^{jk\omega t}$$
$$= I_0 + \frac{1}{2}\dot{I}_1 e^{j\omega t} + \frac{1}{2}\overset{*}{I}_1 e^{-j\omega t}, \tag{5.49}$$

where $U_{01} = \dfrac{R_1}{R_0}$, $U_{02} = \dfrac{R_2}{R_0}$. Equation (5.49) determines an infinite system of nonlinear algebraic equations which can be written in the following concise vector form

$$\mathbf{G}(i,U).\dot{\mathbf{U}} = \dot{\mathbf{I}}, \tag{5.50}$$

where

$$\dot{\mathbf{U}} = \mathrm{colon}(\ldots,\dot{U}_{-k},\ldots,\dot{U}_{-1},\dot{U}_0,\dot{U}_1,\ldots,\dot{U}_k,\ldots),$$

$$\dot{\mathbf{I}} = \mathrm{colon}(\ldots,0,\ldots,0,\frac{1}{2}\overset{*}{I}_1,\dot{I}_0,\frac{1}{2}\dot{I}_1,0,\ldots,0,\ldots).$$

Matrix $\mathbf{G}(i,U) = \{\dot{g}_{i,j}\}, i,j = -\infty,\ldots,-1,0,1,\ldots,\infty$ contains no non-zero elements, all of its elements depend on the complex amplitudes of both current and voltage:

$$\dot{g}_{k,k} = \frac{I_0}{U_{01}} - \frac{1}{R_0} + \frac{1}{U_{02}^2}\left(I_0U_0 + \frac{\dot{I}_1\dot{U}_{-1} + \overset{*}{I}_1\dot{U}_1}{2}\right),$$

$$\dot{g}_{k,k+1} = \frac{\overset{*}{I}_1}{2U_{01}} + \frac{1}{U_{02}^2}\left(I_0\dot{U}_{-1} + \frac{\dot{I}_1\dot{U}_{-2} + \overset{*}{I}_1U_0}{2}\right),$$

$$\dot{g}_{k,k-1} = \frac{\dot{I}_1}{2U_{01}} + \frac{1}{U_{02}^2}\left(I_0\dot{U}_1 + \frac{\overset{*}{I}_1\dot{U}_{-2} + \dot{I}_1U_0}{2}\right),$$

$$\dot{g}_{k,k+l} = \frac{1}{U_{02}^2}\left(I_0\dot{U}_{-l} + \frac{\dot{I}_1\dot{U}_{-l-1} + \overset{*}{I}_1U_{-l+1}}{2}\right),$$

$$\dot{g}_{k,k-l} = \frac{1}{U_{02}^2}\left(I_0\dot{U}_l + \frac{\overset{*}{I}_1\dot{U}_{-l-1} + \dot{I}_1U_{-l+1}}{2}\right).$$

The specificities and exacerbated analytical difficulties that go with the second type of problems are evident.

5.4. The law of energy conservation in oscillating systems

The general law of energy conservation was first used by Manley and Row, in their classical articles [172, 173], for analyzing radiophysical systems while accounting for their specificity.

In section 1.2 Manley–Row's classical energy relations were developed and generalized for an infinite number of forcing generators in the case of linear periodic reactance or nonlinear reactance with explicit dependence on time. The resultant relations are used for investigating parametric systems with several pumping generators. In systems of this type it is possible to suppress the detrimental influence of the second harmonic from the spectrum of the parametric element

(See section 2.3), to achieve a high degree of unidirectional (not reciprocal) signal transfer, to promote reliability by transmitting one and the same message via different frequency channels with following transformation into one and the same intermediate frequency at the receptor point, to increase the signal/noise ratio of radiophysical receiving systems, etc.

The Manley–Row relations can be viewed as one of the bridges thrown between the radiophysical macroscopic and quantum systems [78, 249]. For example, if all terms in Manley–Row's relations are divided by Planck's constant $\hbar$, then the expression in the denominator of each term of the sum will be the energy of the oscillations quantum with the respective frequency. In this case, the first Manley–Row relation assumes the following form:

$$\sum_{m=1}^{\infty}\sum_{n=-\infty}^{\infty}\frac{mW_{mn}}{\hbar(m\omega_0+n\omega_1)}=0. \tag{5.51}$$

The quantity $\dfrac{W_{10}}{\hbar\omega_0}$ obviously determines the number of the quanta with pumping frequency ω_0 sent by the pumping generator to the nonlinear capacitance per unit of time.

Let us consider one of the terms of the sum (5.51), for example: $\dfrac{mW_{mn}}{\hbar(|n|\omega_1-m\omega_0)}$. It is evident that n quanta with frequency ω_1 are required for producing one quantum of combined frequency $(|n|\omega_1 - m\omega_0)$. Besides, this process yields m quanta with frequency ω_0.

It is obvious that relation (5.51) can be interpreted as a law of conservation of the number of quanta with pumping frequency. Similarly, the second of Manley–Row's relations,

$$\sum_{m=-\infty}^{\infty}\sum_{n=1}^{\infty}\frac{nW_{mn}}{\hbar(m\omega_0+n\omega_1)}=0,$$

can be viewed as a law of conservation of the number of quanta with the frequency of the signal.

It is worth noting that Manley–Row's general energy relations can be applied in an infinite number of ways and this will bring about changes in the relative weight of the different harmonic components in the respective sums. The distribution of power by combined frequencies is determined by the magnitude and sign of the combined frequencies, as well as by the nonlinearity of the reactance and the type of the energy consumer.

If we introduce the denotation $N_{mn} = \dfrac{W_{mn}}{\hbar(m\omega_0+n\omega_1)}$, where N_{mn} is the number of the power quanta absorbed by the reactance at frequency $m\omega_0+n\omega_1$, then Manley–Row's relations can be written as follows:

$$\sum_{m=0}^{\infty}\sum_{n=-\infty}^{\infty} mN_{mn}=0,$$

$$\sum_{m=-\infty}^{\infty}\sum_{n=0}^{\infty} nN_{mn} = 0.$$

Using these relations for describing the energy transformation in a quantum-mechanical system with four working levels, we obtain

$$\left| \begin{array}{l} N_{13} + N_{24} + N_{14} + N_{23} = 0 \\ N_{12} + N_{34} + N_{14} - N_{23} = 0. \end{array} \right. \tag{5.52}$$

If we take into account that $W_{01} = (N_{13} + N_{24})\hbar\omega_0$, $W_{10} = (N_{12} + N_{34})\hbar\omega_1$, $W_{11} = N_{14}\hbar\omega_+$, $W_{-11} = N_{23}\hbar\omega_-$, we can use (5.52) to derive

$$\left| \begin{array}{l} \dfrac{W_{10}}{f_1} + \dfrac{W_{11}}{f_+} - \dfrac{W_{-11}}{f_-} = 0 \\ \dfrac{W_{01}}{f_0} + \dfrac{W_{11}}{f_+} + \dfrac{W_{-11}}{f_-} = 0. \end{array} \right.$$

The resultant system presents Manley–Row's energy relations for the four-frequency parametric system considered in section 1.3. Therefore, from an energy point of view, the quantum mechanical system with four working levels is equivalent to a four-frequency parametric modulator. The classical and quantum-mechanical presentation of Manley–Row's energy relations have proved to be equivalent.

Viewed as a consequence of the general law of energy conservation, Manley–Row's energy relations have clear physical meaning. Yet, they treat the transformations of energy by an ideal nonlinear or parametric reactive element. An interesting point worth mentioning is that the issue of the conversion of active power in the parametric reactances is analogical to that of the transformation of reactive power in the parametric active elements. If complex power is set as $S = W + jX$, the active power flow W will obviously be directed to the points where it will be converted into other types of energy, while the reactive power flow X will either be channelled towards elements accumulating reactive energy or derive from such elements. On the whole, a just condition for the active element in stationary mode is $W \neq 0$, $X = 0$, while for the reactances it is vice versa: $W = 0$, $X \neq 0$. These conditions are indeed just as a whole, and they may not be satisfied for certain separate combined frequencies. In the case of nonlinear reactance, the active power at certain combined frequencies $m\omega_0 + n\omega_1$ may be absorbed, i.e. $W_{mn} > 0$, while that at others may be transferred to the circuit, i.e. $W_{mn} < 0$. Yet, as a whole, the average active power of all combined frequencies is zero. It can be shown that the situation with the energy aspect of the parametric active element is the same – at some combined frequencies the average reactive power is directed towards the active element, i.e. $X_{mn} > 0$, while at others it is driven off it, i.e. $X_{mn} < 0$. One could say that the parametric active element stores electric energy at certain frequencies, but only instantaneously, and then transfers it to the circuit at other frequencies. This perspective can be employed when explaining specific phenomena. For example, let us assume that a nonlinear active element is linked up to a source of harmonic voltage. There will be currents and voltages of the upper harmonics in the circuit. From the viewpoint of the stored energy, their appearance may be explained in the following way: the nonlinear active element „stores" electric power with basic frequency, which is immediately transferred to the circuit at the frequencies of the upper harmonics. It is evident that while there is definite awareness of the distinctions between the constant (linear) active and reactive elements, the boundary between the active and reactive parametric (or nonlinear) elements is quite blurred.

Manley–Row's relations for the case of a nonlinear resistor can be written in the form:

$$\left| \begin{aligned} &\sum_{m=0}^{\infty} \sum_{n=-\infty}^{\infty} m X_{m,n} = 0 \\ &\sum_{m=-\infty}^{\infty} \sum_{n=0}^{\infty} n X_{m,n} = 0 \ , \end{aligned} \right. \tag{5.53}$$

where X_{mn} is the reactive power consumed by the nonlinear resistor at frequency $m\omega_0 + n\omega_1$. The nonlinear resistor absorbs energy continuously, while at the same time a part of the energy of the main harmonic is transformed into upper harmonics and vice versa. While active power is the amount of energy absorbed per unit of time, reactive power characterizes the amount of energy involved in these transformations per unit of time. It is obvious that in the stationary regime the reactive power is stored in the resistor, since in this regime the relative weight of the basic and upper harmonics remains unchanged. This condition is met only when the power of the basic harmonic, which is converted into upper harmonics, is equal to the power of the upper harmonics, which is converted into the basic frequency component, i.e. the total reactive power of the resistor is equal to zero. Hence, there is an evident analogy between the reactive power of the parametric (nonlinear) resistor and the active power of the parametric (nonlinear) reactance.

The energy relations for a nonlinear resistor (5.53) can be generalized in an analogical way, similar to that used in section 1.2 for nonlinear reactance, for the case of explicit dependence on time under the action of an arbitrary number of generators.

CHAPTER 6.

OSCILLATING CIRCUIT WITH CONSTANT PARAMETERS

Before analyzing the properties of resonant systems with periodic parameters, it would be logical from a methodological point of view to consider the processes occurring in an oscillating circuit with constant parameters [78, 250]. Despite the widely popular view that oscillating circuits with constant parameters have been studied in detail, our analysis has shown that a number of their properties have escaped the notice of researchers. This chapter presents a theory of a generalized oscillating circuit with constant parameters. The relevant equations are derived, the phenomenon resonance is considered elaborately, the energy balance in a resonant circuit is investigated, a number of tuning characteristics are presented.

6.1. Free and forced oscillations in a generalized oscillating circuit

The oscillating circuit shown in Fig.6.1. will be termed a generalized oscillating circuit. Given $i_g \equiv 0$, this is a generalized series circuit, while in the case of $i_g \equiv 0$ and $G \equiv 0$ it is an ordinary series circuit. Analogously, when $U_g \equiv 0$, the outcome is a parallel oscillating circuit, and when $U_g \equiv 0$ and $R \equiv 0$ it is an ordinary parallel circuit.

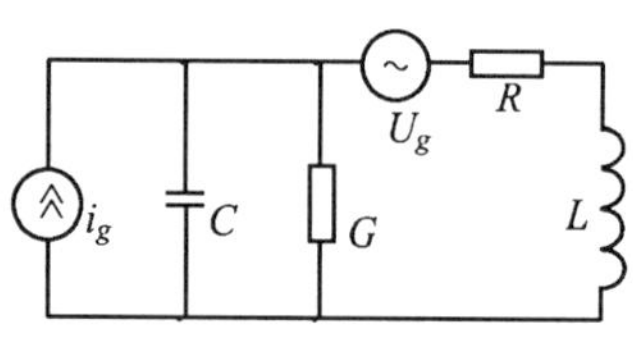

Fig. 6.1. Equivalent circuit of a generalized oscillating circuit

The processes taking place in the generalized oscillating circuit (Fig.6.1) are described by the following basic system of differential equations:

$$\left| \begin{aligned} \frac{dq}{dt} &= -\frac{G}{C}q - \frac{1}{L}\phi + i_g \\ \frac{d\phi}{dt} &= \frac{1}{C}q - \frac{R}{L}\phi + U_g, \end{aligned} \right. \tag{6.1}$$

where i_g and U_g are current and voltage sources, q stands for charge and ϕ for magnetic flow.

The unknown quantity in system (6.1) is the two-dimensional vector-function colon$[q(t), \phi(t)]$.

System (6.1) can be transformed into two independent differential equations of second order:

$$\frac{d^2q}{dt^2}+\left(\frac{R}{L}+\frac{G}{C}\right)\frac{dq}{dt}+\frac{1+RG}{LC}q=-\frac{U_y}{L}-\frac{Ri_y}{L}+\frac{di_y}{dt}, \tag{6.2}$$

$$\frac{d^2\phi}{dt^2}+\left(\frac{R}{L}+\frac{G}{C}\right)\frac{d\phi}{dt}+\frac{1+RG}{LC}\phi=\frac{i_g}{C}+\frac{GU_g}{C}+\frac{dU_g}{dt}. \tag{6.3}$$

It is evident that the left-hand parts of equations (6.2) and (6.3) are identical, while the right-hand ones are dually symmetrical with an accuracy up to the signs.

As we pass on to non-dimensional parameters $x_1=\frac{q}{q_{00}}$, $x_2=\frac{\phi}{\phi_{00}}$, $\tau=\frac{t}{t_0}$, where q_{00}, ϕ_{00} and t_{00} are scaling (normalizing) constants, system (6.1) takes the form

$$\frac{d}{d\tau}\begin{bmatrix}X_1\\X_2\end{bmatrix}=\begin{bmatrix}-t_{00}\dfrac{G}{C} & -\dfrac{t_{00}}{q_{00}}\dfrac{\phi_{00}}{L}\\ \dfrac{t_{00}}{\phi_{00}}\dfrac{q_{00}}{C} & -t_{00}\dfrac{R}{L}\end{bmatrix}\begin{bmatrix}X_1\\X_2\end{bmatrix}+\begin{bmatrix}\dfrac{t_{00}}{q_{00}}i_g\\ \dfrac{t_{00}}{\phi_{00}}U_g\end{bmatrix}. \tag{6.4}$$

or written in a succinct vector form

$$\frac{d}{d\tau}\mathbf{x}=\mathbf{A}\mathbf{x}+\mathbf{f}(\tau). \tag{6.5}$$

For the purpose of analyzing free processes we shall make use of the scalar Eq.(6.2) in the case of $i_g\equiv 0$ and $U_g\equiv 0$:

$$\frac{d^2q}{dt^2}+2\alpha\frac{dq}{dt}+\Omega_0^2q=0, \tag{6.6}$$

where $\alpha=\frac{1}{2}\left(\frac{R}{L}+\frac{G}{C}\right)$, $\Omega_0^2=\frac{1+RG}{LC}$.

Under initial conditions $t=0$, $q=q_0$ and $\frac{dq}{dt}=i_0$ the free process is described by the expression

$$q=e^{-\alpha t}q_m\cos(\omega_f t-\varphi_0), \tag{6.7}$$

where $q_m=\sqrt{q_0^2+\frac{1}{\omega_f^2}(i_0+\alpha q_0)^2}$, $\operatorname{tg}\varphi_0=\frac{1}{\omega_f}\left(\frac{i_0}{q_0}+\alpha\right)$, $\omega_f=\sqrt{\omega_0^2-\beta^2}$,

$\omega_0=\frac{1}{\sqrt{LC}}$, $\beta=\frac{1}{2}\left|\frac{R}{L}-\frac{G}{C}\right|$.

The expressions given above highlight a very typical property of the generalized oscillating circuit: given arbitrary losses in the circuit, the resonance frequency of

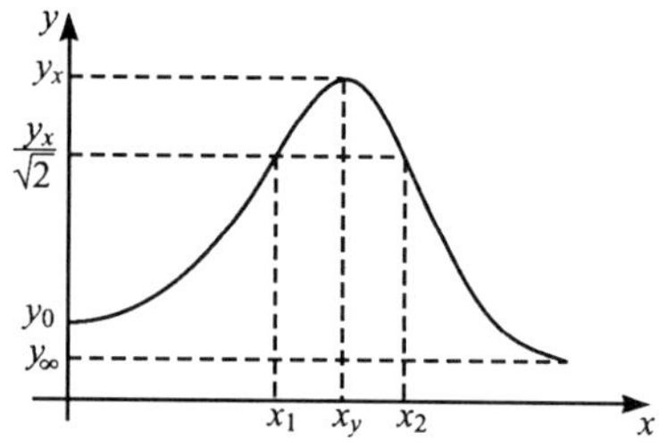

Fig. 6.2. Generalized tuning characteristic of an oscillating circuit

the free oscillations will be exactly equal to the natural resonance frequency of the circuit, provided that the following condition is met:

$$\frac{R}{L} = \frac{G}{C}. \tag{6.8}$$

The established property can be successfully used in measuring practices – for instance, when measuring the natural frequency of an oscillating circuit with a relatively low quality factor by applying the kick-excitation method.

Let us consider the forced oscillations in the generalized resonance circuit.

We „shift" Eq.(6.2) to the complex plane and, taking into account the denotations introduced above, in the case of $U_g = U\cos(\omega t + \varkappa)$ and $i_g \equiv 0$ we obtain

$$\frac{d^2\hat{q}}{dt^2} + 2\alpha\frac{d\hat{q}}{dt} + \Omega^2\hat{q} = \frac{1}{L}\dot{U}e^{j\omega t}. \tag{6.9}$$

We substitute $\hat{q} = \dot{q}_m e^{j\omega t}$ in (6.9) to obtain expressions for the charge amplitude and phase:

$$q_m = \frac{U}{L}[(\Omega_0^2 - \omega^2)^2 + 4\alpha^2\omega^2]^{-\frac{1}{2}}, \quad \varphi = \varkappa - \operatorname{arctg}\frac{2\alpha\omega}{\Omega^2 - \omega^2}. \tag{6.10}$$

Discussions usually focus on a „resonance adjusted oscillating circuit". Strictly speaking this is quite inaccurate. The resonance concept does not refer to the circuit but instead to the physical quantity determining the process in the oscillating circuit. One could talk about resonance of a capacitor charge, or of the magnetic flux permeating the coil, about resonance of the voltage at a certain resistance, etc. Hence, each generalized oscillating circuit has a definite family of tuning characteristics which can be divided into four groups: a) resonance characteristics determined by the frequency of the excitation source; b) capacitance tuning characteristics; c) inductance tuning characteristics; d) active tuning characteristics.

Fig.6.2 provides a generalized tuning characteristic and its typical parameters. Variable x can be the frequency of the external source, capacitance, inductance or active resistance, while function y can be the physical quantity determining the process in the circuit.

As regards the charge of the generalized series oscillating circuit, we have the following (we use (6.10)): $x = \omega$, $y = q_m$, $y_0 = q_m(0) = \dfrac{U}{\Omega^2 L}$, $y_\infty = q_m(\infty) = 0$,

$$x_y = \omega_r = \omega_0\sqrt{1 - \frac{1}{2}\left(\frac{R^2}{\rho^2} + \rho^2 G^2\right)},\ \rho = \sqrt{\frac{L}{C}},$$

$$y_x = q_{mr} = \frac{U}{\omega_0(R+\rho^2 G)\sqrt{1-\dfrac{1}{4}\left(\dfrac{R}{\rho}-\rho G\right)}},\ x_{1,2}=\omega_{1,2}=\sqrt{\Omega_0^2 - 2\alpha(\alpha \pm \sqrt{\Omega^2-\alpha^2})}.$$

The frequency bandwidth of the oscillating circuit is $2\Delta\omega = \omega_2 - \omega_1$.

The resonance characteristic of the generalized series oscillating circuit is asymmetrical, i.e. $\omega_2 - \omega_r \neq \omega_r - \omega_1$.

The typical parameters of the capacitance tuning characteristic of the charge are as follows: $x = c$, $y = q_m$, $y_0 = q_m(0) = 0$,

$$y_\infty = q_m(\infty) = \frac{U}{\omega\sqrt{R^2+\omega^2 L^2}},\quad x_y = C_r = \frac{G}{\omega}\left[\frac{(1+RG)^2}{\omega LG} + \omega LG\right],$$

$$y_x = q_{mr} = \frac{U}{\omega^2 L}\left[1 - \frac{1}{(1+RG)^2+\omega^2 L^2 G^2} + \frac{R^2}{\omega^2 L^2}\right]^{-\frac{1}{2}},$$

$$\frac{1}{x_{1,2}} = \frac{1}{C_{1,2}} = \frac{\omega^2 L}{(1+RG)^2+\omega^2 L^2 G^2}\left\{1 \pm \sqrt{\left(1+\frac{R^2}{\omega^2 L^2}\right)[(1+RG)^2+\omega^2 L^2 G^2]-1}\right\}.$$

The width of the tuning curve is: $2\Delta C = C_2 - C_1$.

The inductive tuning characteristic of the charge is characterized by the following parameters: $x = L$, $y = q_m$,

$$y_0 = q_m(0) = CU[(1+RG)^2 + \omega^2 L^2 R^2]^{-\frac{1}{2}},\quad y_\infty = q_m(\infty) = 0,$$

$$x_y = L_r = C(G^2+\omega^2 C^2)^{-1},\quad y_x = q_{mr} = CU\sqrt{1+\frac{\omega^2 C^2}{G^2}}\left(1 + RG + \frac{\omega^2 C^2 R}{G}\right)^{-1},$$

$$x_{1,2} = L_{1,2} = \frac{C}{G^2+\omega^2 C^2}\left[1 \mp \sqrt{\frac{G^2}{\omega^2 C^2}(1+RG)^2 + 2RG(1+RG) + \omega^2 C^2 R^2}\right],$$

$2\Delta L = L_2 - L_1$ is the width of the tuning curve.

The active tuning characteristic of the charge is characterized by the following parameters: $x = R$, $y = q_m$, $y_\infty = q_m(\infty) = 0$,

$$y_0 = q_m(0) = \frac{U}{L\sqrt{(\omega_0^2-\omega^2)^2 + \dfrac{\omega^2 G^2}{C^2}}},\quad x_y = R_r = -\frac{G}{G^2+\omega^2 C^2},$$

$$y_x = q_{mr} = \frac{U}{\omega\sqrt{\omega^2 + \dfrac{G^2}{C^2}\left(L - \dfrac{C}{G^2+\omega^2C^2}\right)}},$$

$$x_{1,2} = R_{1,2} = -\frac{G}{G^2+\omega^2C^2} \mp \omega\left(\frac{C}{G^2+\omega^2C^2} - L\right).$$

The expression for x_y shows that in the case of the active tuning characteristic, the charge amplitude reaches its maximum when the active resistance and the conductance have different signs. This gives rise to a problem related to the stability of the oscillating circuit in the respective frequency bandwidth $2\Delta x = R_2 - R_1$. The following differential equations for the free oscillations in the generalized oscillating circuit are obtained for the three typical points x_y, x_1 and x_2 respectively:

$$\frac{d^2q}{dt^2} + \frac{G}{C}\left(\frac{C}{L}\frac{1}{G^2+\omega^2C^2} - 1\right)\frac{dq}{dt} + \frac{\omega^2C^2}{G^2+\omega^2C^2}\frac{1}{LC}q = 0,$$

$$\frac{d^2q}{dt^2} + \left(\frac{G}{C} \mp \omega\right)\left(\frac{C}{L}\frac{1}{G^2+\omega^2C^2} - 1\right)\frac{dq}{dt} + \left[1 \pm G\left(\frac{\omega C \mp G}{G^2+\omega^2C^2} - \omega L\right)\right]\frac{1}{LC}q = 0.$$

The upper sign in the second equation refers to x_1, while the lower one refers to x_2.

Provided that $G < 0$, the sufficient conditions for asymptotic stability of the oscillating circuit can be derived from the differential equations referred to in the following form:

$$\left|\begin{array}{l} |G| > \omega C \\ (G^2+\omega^2C^2)\dfrac{L}{C} - \dfrac{\omega C}{|G|} < 1 < (G^2+\omega^2C^2)\dfrac{L}{C} + \dfrac{\omega C}{|G|}. \end{array}\right. \tag{6.11}$$

The active tuning characteristic of the charge can also be determined by the conductance. Then: $x = G$, $y = q_m$, $y_\infty = q_m(\infty) = 0$,

$y_0 = q_m(0) = CU[(1-\omega^2LC)^2 + \omega^2C^2R^2]^{-\frac{1}{2}}$, $x_y = G_r = -R(R^2+\omega^2L^2)^{-1}$,

$$y_x = q_{mr} = CU\sqrt{1+\frac{R^2}{\omega^2L^2}}[1 - \frac{C}{L}(R^2+\omega^2L^2)]^{-1},$$

$x_{1,2} = G_{1,2} = -\dfrac{R}{R^2+\omega^2L^2} \mp \omega\left(\dfrac{L}{R^2+\omega^2L^2} - C\right)$, $2\Delta x = x_2 - x_1$.

Provided that we opt in this case for $R < 0$, the sufficient conditions for asymptotic stability of the oscillating circuit will take the following form:

$$\left|\begin{array}{l} |R| > \omega L \\ (R^2+\omega^2L^2)\dfrac{C}{L} - \dfrac{\omega L}{|R|} < 1 < (R^2+\omega^2L^2)\dfrac{C}{L} + \dfrac{\omega L}{|R|}. \end{array}\right. \tag{6.12}$$

The tuning characteristics brought forth are of both cognitive and practical significance. For example, when a negative conductance is connected up to an oscillating circuit for the purpose of increasing its quality factor, it is expedient to perform fine tuning with the help of variable resistance.

It is worth noting that a considerable number of tuning characteristics, concerning the different physical quantities that determine the processes in an oscillating circuit, can also be obtained. These include capacitance current and voltage, conductance current, resistance voltage, magnetic flux of inductance, inductance current and voltage, as well as the numerous combinations of these quantities.

Some tuning characteristics differ from each other only by a constant multiplier. Such is, for example, the capacitance tuning characteristic related to capacitor charge and current, since the charge amplitude differs from the current amplitude by multiplier ω.

At the same time, there is an essential distinction between the capacitance tuning characteristics related to capacitor charge and voltage. For instance, if $x = C$, $y = U_m$ we obtain $x_y = C_r = \dfrac{L}{R^2 + \omega^2 L^2}$ and

$$y_x = U_{mr} = \frac{U}{\dfrac{R}{\sqrt{R^2+\omega^2L^2}} + G\sqrt{R^2+\omega^2L^2}},$$ which differ from the above quoted parameters $x_y = C_r$ and $y_x = q_{mr}$, when $x = C$ and $y = q_m$.

When real experimental investigations are carried out, one should bear in mind the peculiarities of the measuring system. For example, when identifying the capacitance or inductance tuning characteristics, it is necessary to take into account the changes in the value of G and R respectively.

The methods mentioned above with respect to the generalized series oscillating circuit can be used for establishing in an analogous way the parameters of the tuning characteristics of the generalized parallel oscillating circuit.

To conclude this section, we should point out that as a rule capacitors and inductance coils in real systems have distributed volume losses. This fact can be accounted for in first approximation through a respective complication of the equivalent scheme of the generalized oscillating circuit, as shown in Fig.6.3.

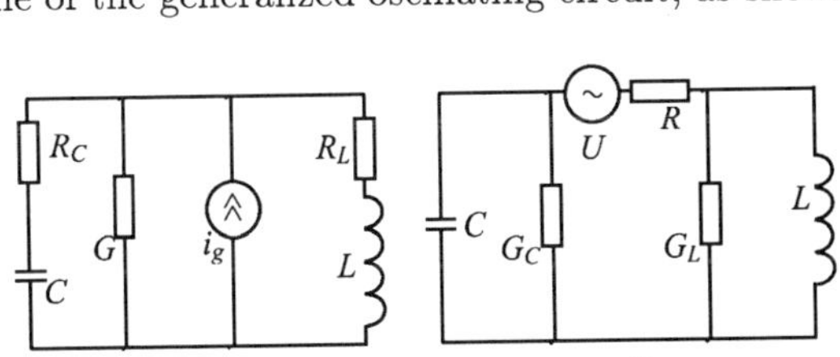

Fig. 6.3. Equivalent circuits of a generalized oscillating circuit accounting for the additional losses of capacitance (a) and inductance (b)

The differential equations and the free oscillation frequencies in these two cases are written down as follows:

$$\frac{d^2q}{dt^2}+\left[\frac{R_L}{L}+\frac{1}{1+GR_C}\left(\frac{G}{C}+\frac{R_C}{L}\right)\right]\frac{dq}{dt}+\frac{1}{LC}\frac{1+GR_L}{1+GR_C}q=\frac{1}{1+GR_C}\left(\frac{di_g}{dt}-\frac{R_L}{L}i_g\right),$$

$$\omega^2_{fa)}=\left(\frac{\omega_0}{1+GR_C}\right)^2-\frac{1}{4}\left[\frac{1}{1+GR_C}\left(\frac{R_C}{L}-\frac{G}{C}\right)+\frac{R_L}{L}\right]^2,$$

$$\frac{d^2q}{dt^2}+\left[\frac{G_C}{C}+\frac{1}{1+RG_L}\left(\frac{G_L}{C}+\frac{R}{L}\right)\right]\frac{dq}{dt}+\frac{1}{LC}\frac{1+RG_C}{1+RG_L}q=\frac{1}{1+RG_L}\left(\frac{U_g}{L}-G_L\frac{dU_g}{dt}\right),$$

$$\omega^2_{fb)}=\frac{\omega_0^2}{1+RG_L}\left[1-\frac{RG_L}{4}\frac{1+\rho^2\frac{G_L}{R}}{1+RG_L}-\frac{1}{4}\left(\rho G_C\sqrt{1+RG_L}-\frac{R}{\rho\sqrt{1+RG_L}}\right)^2\right].$$

When $R_C=0$ and $G_L=0$, the resultant equations and expressions are similar to those derived above for the generalized oscillating circuit in Fig.6.1.

6.2. Energy balance in a generalized oscillating circuit

The generalized oscillating circuit in Fig.6.1. is considered as one-port an in the case of $i_g\equiv 0$ and $U_g=U_m\cos\omega t$ [251].

The complex input resistance of the oscillating circuit is of the following type:

$$\dot Z=R+\frac{G}{G^2+\omega^2C^2}+j\omega\left(L-\frac{C}{G^2+\omega^2C^2}\right).$$

There are two possible varieties of the resonance in the one-port oscillating circuit: a) phase resonance, when the phases of the input current and voltage coincide; b) amplitude resonance, when the amplitude of the input current is maximal. The respective resonance frequencies are equal:

$$\omega_F=\sqrt{\frac{1}{LC}-\frac{G^2}{C^2}}\quad\text{and}\quad\omega_A=\sqrt{\frac{\sqrt{1+2G(R+\rho^2G)}}{LC}-\frac{G^2}{C^2}},\quad\rho=\sqrt{\frac{L}{C}}.$$

We shall limit our consideration to the phase resonance, and for that purpose we shall write the complex current and voltage amplitudes for all circuit elements:

$$\dot I_R=I_R=I_L=I=\frac{U}{R+\rho^2G},\ \dot U_R=U_R=\frac{RU}{R+\rho^2G},\ \dot U_C=\dot U_G=\frac{\rho U}{R+\rho^2G}e^{-j\varphi_C},$$

$$\dot U_L=j\omega_rL\dot I=\frac{\rho\sqrt{1-\rho^2G}}{R+\rho^2G}Ue^{j\frac{\pi}{2}},\quad \mathrm{tg}\,\varphi_C=\sqrt{\frac{1}{\rho^2G^2}-1},$$

$$\dot{I}_C = j\frac{\sqrt{1-\rho^2G^2}}{R+\rho^2G}Ue^{-j\varphi_C}, \quad \dot{I}_G = \frac{\rho G U}{R+\rho^2 G}e^{-j\varphi_C}.$$

With the help of these formulae, we identify expressions for the instantaneous power consumed by the circuit elements:

$$\left| \begin{aligned} P_G &= \frac{1}{2}\frac{\rho^2 G}{(R+\rho^2G)^2}U^2[1+\cos 2(\omega_F t-\varphi_C)], \\ P_R &= \frac{R}{2(R+\rho^2G)^2}U^2(1+\cos 2\omega_F t), \\ P_C &= -\frac{1}{2}\frac{\rho\sqrt{1-\rho^2G^2}}{(R+\rho^2G)^2}U^2\sin 2(\omega_F t-\varphi_C), \\ P_L &= -\frac{1}{2}\frac{\rho\sqrt{1-\rho^2G^2}}{(R+\rho^2G)^2}U^2\sin 2\omega_F t, \\ P_{CL} &= P_C+P_L = \frac{\rho\sqrt{1-\rho^2G^2}}{(R+\rho^2G)^2}U^2\cos\varphi_C\cos(2\omega_F t+\frac{\pi}{2}-\varphi_C). \end{aligned} \right. \tag{6.13}$$

The power in the active elements consists of two components: a constant component (average power) and a harmonic component with frequency $2\omega_F$ (oscillating power), so that the average power is equal to the amplitude of the oscillating power. The active elements consume power in a kick-like way at double frequency. The power in the reactive elements is oscillating, i.e. reactances both receive and transmit electric energy. It is worth noting that for the ordinary series oscillating circuit $P_{CL}=0$, i.e. in a resonance situation, reactances have peculiar autonomy and exchange energy without „noticing" either any losses or the excitation source. Both excitation source and losses are involved in the generalized series oscillating circuit and the bigger the losses, the more considerable the involvement (in a loss-free circuit $\varphi_C = \frac{\pi}{2}$).

The instantaneous energies accumulated in the reactances are as follows:

$$\left| \begin{aligned} W_C &= \frac{L}{4}\frac{U^2}{(R+\rho^2G)^2}[1+\cos 2(\omega_F t+\varphi_C)], \\ W_L &= \frac{L}{4}\frac{U^2}{(R+\rho^2G)^2}(1+\cos 2\varphi_F t), \\ W_{CL} &= W_C+W_L = \frac{L}{2}\frac{U^2}{(R+\rho^2G)^2}[1-\cos\varphi_C\cos(2\varphi_F t+\varphi_C)]. \end{aligned} \right. \tag{6.14}$$

The instantaneous energies in the reactances are related to the instantaneous consumed power through the following dependencies:

$$P_C = \frac{dW_C}{dt}, \quad P_L = \frac{dW_L}{dt}, \quad P_{CL} = \frac{dW_{CL}}{dt},$$

i.e. the power received by the reactances is expended for increasing their energy reserve.

The major conclusion from the analysis is that the instantaneous power generated by the source is not equal to the instantaneous power absorbed by the active elements of the generalized oscillating circuit. Such equality is observed only with respect to average power: $P_a = (P_{GR} + P_{CL})_a = P_{GR_a}$. Indeed, using (6.13) we obtain:

$$P = P_{GR} + P_{CL} = \frac{U^2}{2(R+\rho^2 G)}(1 + \cos 2\omega_F t),$$

$$P_{GR}=P_G+P_R=\frac{U^2}{2(R+\rho^2 G)}\left[1+\sqrt{1-4\rho^2 RG(R+\rho^2 G)^{-2}\sin^2\varphi_C}\cdot\cos(2\omega_F t-2\alpha)\right],$$

$$\operatorname{tg} 2\alpha = -\frac{\rho^2 G \sin 2\varphi_C}{R+\rho^2 G\cos 2\varphi_C}.$$

In conclusion, we shall mention yet another interesting particular case, when $G = \dfrac{1}{\rho}$. Then the following conditions are met: $\omega_F = 0$, $\varphi_C = 0$, $\alpha = 0$. In this case the resonance tuning can be conducted by changing the conductance. Moreover, there is a possibility for „resonance" tuning even with respect to an external generator of constant voltage.

CHAPTER 7.

GENERAL ANALYSIS OF THE PARAMETRIC PHENOMENA IN LINEAR OSCILLATING SYSTEMS WITH PARAMETERS CHANGING IN TIME

The analysis of linear oscillating systems with variable parameters, and particularly of such with parameters that remain periodical in time, is of fundamental importance for the investigation of oscillating systems in a general form. The parameters of nonlinear oscillating systems depend on the voltage applied and on the currents that flow through them, and these, in their turn, are functions of time. Thus, in the long run, nonlinear systems are also systems with variable parameters. In this connection, a principle of linear linkage is formulated in mathematics [112]. It is related to the idea that the phenomena and properties of nonlinear systems can be realized (in the sense of simulated) for each specific (particular) case in the respective linear systems with variable parameters.

This chapter reveals a general method for analyzing linear systems with periodic and almost periodic parameters [78, 252–258].

Qualitative analysis assumes considerable importance in the investigation of complex oscillating systems, since it allows of identifying the most general features of system behavior. The most essential qualitative analysis task is to establish the degree of stability of the system. Here we formulate theorems, as well as differential equations specifying the structure of the system, on the basis of a sufficient number of conditions ensuring stability and asymptotic steadiness of a generalized oscillating circuit with periodic parameters.

7.1. Qualitative analysis of the free processes in a generalized linear oscillating circuit with periodic parameters

The generalized linear oscillating circuit (of the type of the one shown in Fig.6.1) with periodic parameters is called to play an important part in the Theory of nonlinear oscillations and in Radiophysics. Its direct significance is that it is used in Radiophysics as a single oscillating circuit parametric amplifier or generator, in the process of oscillation modulation in generator systems, etc. Its indirect significance is conditioned by the fact that it is a heuristic model of nonlinear autonomous and non-autonomous second-order systems employed in the analysis of process stability in such systems.

This section provides an analysis of the free processes in an oscillating circuit from the point of view of their bounded or unbounded nature, or, in other words, it studies issues of the stability and instability of the oscillating circuit. The difficulty of the task is predetermined by its most general set-up, which requires the application of complex mathematical techniques. Yet such a general perspective of the task makes it interesting from a practical point of view, since the parametric oscillating circuit is quite rich in particular cases, but in principle no specific particular case can reveal the overall diversity of possibilities for the oscillating circuit.

Further down, we have quoted a basic system of two linear differential equations of the generalized parametric oscillating circuit, as well as some particular cases derived from the basic system through variable substitution. The attention is mostly focused on the canonical second-order system, to which the basic system of circuit equations is reduced. Mathieux's equation is equivalent to a rather particular case of a canonical system with periodic coefficients. It is well-known that Mathieux's various equations can be classified within a definite set of classes by zones of stability and instability. These zones can be presented in a two-dimensional plane as areas with sufficiently complex form which intertwine and overlap in a complex way. It turns out that the canonical systems of a general type have analogous properties but the respective stability and instability areas are fixed in cylindrical space obtained by rotating the plane round an axis lying in this plane. The respective results are obtained by employing incomparably more complex methods than in the case of Mathieux's equation and a broader system of mathematical concepts.

7.1.1. Structure of the differential equations describing linear oscillating systems with positive parameters

Linear oscillating circuits can be described by applying a first-order vector linear differential equation

$$\frac{d}{dt}\mathbf{Z} = \mathbf{A}(t)\mathbf{Z} + \mathbf{f}(t), \tag{7.1}$$

where $\mathbf{Z}$ is a n-dimensional vector, whose elements can represent capacitor charges, magnetic flux running through inductances, etc., $\mathbf{A}(t)$ is an $n \times n$-dimensional matrix with elements that can be expressed by circuit parameters (inductances, capacitances, resistances), $\mathbf{f}(t)$ is a free n-dimensional vector, whose components are determined by the electromotive forces connected to the circuit and by the parameters of the system.

In order to identify the structure of equation (7.1), we shall initially analyze the following equation of the free processes with „frozen" (time-independent) parameters:

$$\frac{d}{dt}\mathbf{x} = \mathbf{A}\mathbf{x}, \tag{7.2}$$

where $\mathbf{A} = \text{const}$. If all the parameters of the system are positive, the solution will satisfy the condition: $\lim_{t\to\infty} \mathbf{x}(t) = 0$.

It is obvious that the latter equation meets the condition that $Sp\mathbf{A} < 0$ ($Sp\mathbf{A}$ is the sum total of the main diagonal terms of matrix $\mathbf{A}$).

Lemma 7.1. Any radiophysical system with constant positive parameters, containing active resistances with currents flowing through them, is described by a system of differential equations with constant coefficients, whose matrix includes non-positive main diagonal elements, at least one of which is negative.

Lemma 7.2. The main diagonal in the matrix of the first-order vector differential equation of a radio circuit with constant positive parameters, containing only real reactances (with losses), consists only of negative elements.

The condition that $Sp\mathbf{A} = 0$ and the following Lemma are valid for circuits made up of ideal reactances with constant positive parameters.

Lemma 7.3. The matrix of the vector differential equation of a radio circuit containing only ideal reactances with constant positive parameters has a zero main diagonal.

The Lemmas formulated above are also valid for a certain subclass of linear radio circuits (we shall term it a structurally stationary one) with variable positive parameters. They are characterized by the fact that matrix $\mathbf{A}(t)$ in (7.2) contains no derivatives of the circuit parameters with respect to the time. The following theorem can be formulated in this connection.

Theorem 7.1. Any radio circuits with variable parameters containing no capacitive loops or inductive nodes are structurally stationary.

The proof of Theorem 7.1 is based on Kirchhoff's laws. It demonstrates that the derivatives of the circuit parameters with respect to time emerge only after excluding one of the charges in a capacitive loop, or one of the magnetic fluxes in an inductive node.

With a view to extending the scope of action of the Lemmas on structurally stationary circuits considered above, the latter can be reformulated as separate theorems.

Two coupled oscillating circuits with intrinsic capacitive coupling can serve as an example of a structurally stationary circuit. Two connected oscillating circuits with external capacitive coupling cannot be regarded as a structurally stationary circuit, since in this case the three capacitances form a capacitive loop.

7.1.2. Vector differential equation describing a linear oscillating circuit with time-dependent parameters

The free process in a linear generalized oscillating circuit with changing parameters (of the type shown in Fig.6.1) is described by a linear vector differential equation

$$\frac{d}{dt}\mathbf{x} = \mathbf{A}(t).\mathbf{x}, \tag{7.3}$$

$\mathbf{x} = \text{colon}(x_1, x_2)$, $x_1 = \dfrac{q}{q_{00}}$ – normalized charge of the capacitor,

$x_2 = \dfrac{\phi}{\phi_{00}}$ – normalized magnetic flux of the inductance,

$$\mathbf{A}(t) = \{a_{ij}(t)\},\ ij = 1,2,\ a_{11}(t) = -t_{00}\frac{G(t)}{C(t)} < 0,\ a_{12}(t) = -\frac{t_{00}\phi_{00}}{q_{00}L(t)} < 0,$$

$$a_{21}(t) = \frac{t_{00}q_{00}}{\phi_{00}C(t)} > 0,\ a_{22}(t) = -\frac{t_{00}R(t)}{L(t)} < 0.$$

If x_2 is excluded from system (7.3), the result will be a scalar differential equation with regard to x_1. A scalar equation with respect to x_2 can be obtained in an analogous way. These are equations of the type

$$\frac{d^2x}{dt^2} + a_1(t)\frac{dx}{dt} + a_2(t)x = 0, \tag{7.4}$$

where:

— in the case of $x = x_1$, $a_1(t) = \dfrac{d}{dt}\ln L(t) + t_{00}\left[\dfrac{G(t)}{C(t)} + \dfrac{R(t)}{L(t)}\right]$,

$$a_2(t) = \frac{d}{dt}\left[\frac{t_{00}G(t)}{C(t)}\right] + \frac{t_{00}G(t)}{C(t)}\frac{d}{dt}\ln L(t) + \frac{t_{00}^2}{L(t)C(t)}[1 + R(t)G(t)].$$

— in the case of $x = x_2$, respectively $a_1(t) = \dfrac{d}{dt}\ln C(t) + t_{00}\left[\dfrac{R(t)}{L(t)} + \dfrac{G(t)}{C(t)}\right]$,

$$a_2(t) = \frac{d}{dt}\left[\frac{t_{00}R(t)}{L(t)}\right] + \frac{t_{00}R(t)}{L(t)}\frac{d}{dt}\ln C(t) + \frac{t_{00}^2}{L(t)C(t)}[1 + G(t)R(t)].$$

A substitution of variable (7.4) $x = \exp\left[-\frac{1}{2}\int_0^t a_1(S)dS\right] y$ yields

$$\frac{d^2y}{dt^2} + P(t)y = 0,\quad P(t) = a_2(t) - \frac{1}{4}a_1^2(t) - \frac{1}{2}\frac{da_1(t)}{dt}. \tag{7.5}$$

Given the assumption that in (7.4) $x = x_1$, then

$$y = y_1 = x_1\sqrt{\frac{L(t)}{L(0)}}\exp\left\{\frac{t_{00}}{2}\int_0^t\left[\frac{G(S)}{C(S)} + \frac{R(S)}{L(S)}\right]dS\right\},$$

$$P(t) = \frac{t_{00}^2}{LC}(1 + RG) + \frac{1}{2}\frac{d}{dt}\left(\frac{t_{00}G}{C} - \frac{t_{00}R}{L}\right)^2 + \frac{1}{2}\left(\frac{t_{00}G}{C}\right.$$
$$\left. - \frac{t_{00}R}{L}\right)\frac{d}{dt}\ln L + \frac{1}{4}\left(\frac{d}{dt}\ln L\right)^2 - \frac{1}{4}\left(\frac{t_{00}G}{C} + \frac{t_{00}R}{L}\right)^2 - \frac{1}{2L}\frac{d^2L}{dt^2}.$$

In the case of $x = x_2$ in (7.4), it follows that

$$y = y_2 = x_2\sqrt{\frac{C(t)}{C(0)}}\exp\left\{\frac{t_{00}}{2}\int_0^t\left[\frac{R(S)}{L(S)}+\frac{G(S)}{C(S)}\right]dt\right\},$$

$$P(t) = \frac{t_{00}^2}{CL}(1+GR) + \frac{1}{2}\frac{d}{dt}\left(\frac{t_{00}R}{L} - \frac{t_{00}G}{C}\right)^2 + \frac{1}{2}\left(\frac{t_{00}R}{L} - \frac{t_{00}G}{C}\right)\frac{d}{dt}\ln C + \frac{1}{4}\left(\frac{d}{dt}\ln C\right)^2 - \frac{1}{4}\left(\frac{t_{00}R}{L}+\frac{t_{00}G}{C}\right)^2 - \frac{1}{2C}\frac{d^2C}{dt^2}.$$

Equation (7.3) can be reduced to a vector equation of a canonical type analogous to (7.5) by carrying out the following substitution:

$$\mathbf{X} = \mathbf{Z}\exp\left\{\frac{1}{2}\int_0^t[a_{11}(t)+a_{22}(t)]dt\right\}.$$

We obtain

$$\frac{d}{dt}\mathbf{Z} = \mathbf{B}(t)\mathbf{Z}, \tag{7.6}$$

$\mathbf{Z} = \mathrm{colon}(Z_1, Z_2)$, $\mathbf{B}(t) = \{b_{ij}(t)\}$, $i,j = 1,2$

$b_{11} = -b_{22} = \frac{1}{2}(a_{11} - a_{22})$, $b_{12} = a_{12}$, $b_{21} = a_{21}$.

The condition that $Sp\mathbf{B}(t) = 0$ is indicative of the canonical character of equation (7.6). In the case under consideration, it is in a rather simple form.

If we introduce a Hamiltonian function, i.e. square form of the system

$$H(t, Z_1, Z_2) = \frac{1}{2}b_{21}(t)Z_1^2 - b_{11}(t)Z_1Z_2 - \frac{1}{2}b_{12}(t)Z_2^2. \tag{7.7}$$

Equation (7.6) can be written in the form of the following system

$$\frac{dZ_1}{dt} = -\frac{\partial H}{\partial Z_2}, \quad \frac{dZ_2}{dt} = \frac{\partial H}{\partial Z_1}. \tag{7.8}$$

Let us compare the canonical system (7.6) with the equation of a general type (7.3). It is evident that the elements of the second diagonal of the matrixes are identical, while the elements of the main diagonals differ. They are presented in an extended form:

$$b_{11} = -t_{00}\frac{G_{eqv}}{C}, \quad G_{eqv} = \frac{1}{2}\left(G - \frac{R}{\rho^2}\right),$$

$$b_{22} = -t_{00}\frac{R_{eqv}}{L}, \quad R_{eqv} = -\rho^2 G_{eqv}, \quad \rho = \sqrt{\frac{L(t)}{C(t)}}.$$

Hence the conclusion that the necessary and sufficient condition for the oscillating circuit to be described by a canonical equation is

$$R_{eqv} = -\rho^2 G_{eqv}.$$

It is obvious that any conservative oscillating circuit ($R = 0, G = 0$) is described by a canonical vector differential equation.

It is not difficult to verify that if a Hamiltonian oscillating circuit is described by Eq.(7.8), the coefficient of the first derivative has an average value equal to zero, since $a_1 = \frac{d}{dt}\ln L$ or $a_1 = \frac{d}{dt}\ln C$.

Let us express the Hamiltonian function of system (7.8) by using the circuit parameters

$$H(t, Z_1, Z_2) = \frac{t_{00}}{2}\left(\frac{q_{00}}{\phi_{00}C}Z_1^2 + \frac{\rho G - \dfrac{R}{\rho}}{\sqrt{LC}}Z_1 Z_2 + \frac{\phi_{00}}{q_{00}L}Z_2^2\right). \tag{7.9}$$

When analyzing the canonical system, it is important to identify the non-negative condition, $H(t, Z_1, Z_2) \geq 0$, for all values of the arguments, for which the solutions of equations Z_1 and Z_2 are unknown. The bracketed middle term in (7.9) is a serious impediment to such an estimate. That is why we use the inequality $-\frac{Z_1^2 + Z_2^2}{2} \leq Z_1 Z_2 \leq \frac{Z_1^2 + Z_2^2}{2}$ to obtain the following bilateral estimate

$$\begin{aligned} &\frac{t_{00}}{2}\left(1 - \frac{1}{2}\left|\rho G - \frac{R}{\rho}\right|\right)\left(\frac{q_{00}}{\phi_{00}C}Z_1^2 + \frac{\phi_{00}}{q_{00}L}Z_2^2\right) \leq H(t, Z_1, Z_2) \leq \\ &\leq \frac{t_{00}}{2}\left(1 + \frac{1}{2}\left|\rho G - \frac{R}{\rho}\right|\right)\left(\frac{q_{00}}{\phi_{00}C}Z_1^2 + \frac{\phi_{00}}{q_{00}L}Z_2^2\right). \end{aligned} \tag{7.10}$$

As a result of these inequalities, in the absence of any dissipative losses ($G \equiv 0, R \equiv 0$), the following equation is obtained

$$H(t, Z_1, Z_2) = \frac{t_{00}}{2}\left(\frac{q_{00}}{\phi_{00}C}Z_1^2 + \frac{\phi_{00}}{q_{00}L}Z_2^2\right).$$

This is the normalized instantaneous energy accumulated in the reactances of the oscillating circuit.

Canonical systems and their equivalent equations can be either stable or unstable, but they cannot be asymptotically stable.

Let us consider some other forms of the canonical vector system (7.4), for example the following one:

$$\frac{d}{dt}\mathbf{Z} = \mathbf{JHZ}, \tag{7.11}$$

$$\mathbf{J}=\begin{pmatrix} 0 & 1 \\ -1 & 0 \end{pmatrix}, \ \mathbf{H}=\begin{pmatrix} b_{21} & -b_{11} \\ -b_{11} & b_{12} \end{pmatrix}=\begin{pmatrix} \dfrac{t_{00}q_{00}}{\phi_{00}C(t)} & \dfrac{t_{00}}{2}\left[\dfrac{G(t)}{C(t)}-\dfrac{R(t)}{L(t)}\right] \\ \dfrac{t_{00}}{2}\left[\dfrac{G(t)}{C(t)}-\dfrac{R(t)}{L(t)}\right] & \dfrac{t_{00}\phi_{00}}{q_{00}L(t)} \end{pmatrix},$$

where $\mathbf{H}$ is a symmetrical matrix of the Hamiltonian function $H(t, Z_1, Z_2)$.

Since $\mathbf{J}^{-1} = -\mathbf{J}$, Eq.(7.11) can also be written in the form:

$$\mathbf{J}\frac{d}{dt}\mathbf{Z} = \mathbf{HZ}. \tag{7.12}$$

Sometimes it is convenient to present the initial system (7.3) in the form of an integral system of equations. If, for instance, $\mathbf{X}(t) = \{X_{ij}(t)\}$, $i, j = 1, 2$ is a fundamental system of solutions to (7.3), it will satisfy the integral equation

$$\mathbf{X}(t) = \mathbf{X}(t_0) + \int_{t_0}^{t} \mathbf{A}(S)\mathbf{X}(S)dS,$$

where $\mathbf{X}(t_0)$ is the matrix of the initial conditions.

This form of the oscillating circuit equation is convenient when using the successive approximations method, in the case of seeking a solution in the form of $\mathbf{X}(t) = \mathbf{X}(t_0) + \sum_{k=1}^{\infty} \mathbf{X}_k(t)$, where

$$\mathbf{X}_k(t) = \int_{t_0}^{t} \mathbf{A}(S)\mathbf{X}_{k-1}(S)dS, \quad k = 1, 2, \ldots, \quad \mathbf{X}_0 = \mathbf{X}(t_0).$$

This recurrent formula allows of consistently identifying all series terms in the solution. It would not be difficult to show that this series is absolutely and evenly convergent over an arbitrary finite interval.

Further on we shall demonstrate a multistage transformation approach. If the variables in (7.3) are substituted in the following way

$$\mathbf{x} = \mathbf{B}(t)\mathbf{y}, \quad \mathbf{B}(t) = \begin{pmatrix} \dfrac{1}{a_{21}(t)} & 0 \\ 0 & 1 \end{pmatrix}, \tag{7.13}$$

the equation will assume the form

$$\frac{d}{dt}\mathbf{y} = \mathbf{C}(t)\mathbf{y}, \tag{7.14}$$

$$\mathbf{C}(t) = \begin{pmatrix} c_{11}(t) & a_{12}(t)a_{21}(t) \\ 1 & a_{22}(t) \end{pmatrix}, \quad c_{11} = a_{11} + \left(\frac{d}{dt}a_{21}\right)\frac{1}{a_{21}}.$$

A comparison between (7.14) and (7.3) shows that Eq.(7.14) describes an oscillating circuit with constant capacitance or constant inductance, while all other respective parameters vary in time.

Let us continue the transformation by performing the following substitution in (7.14)

$$\mathbf{y} = \mathbf{Z} \exp\left\{\frac{1}{2}\int_{t_0}^{t} [C_{11}(S) + a_{22}(S)]dS\right\}.$$

The result is

$$\frac{d}{dt}\mathbf{Z} = \mathbf{D}(t)\mathbf{Z}, \tag{7.15}$$

$$\mathbf{D}(t) = \begin{pmatrix} d_{11}(t) & a_{12}(t)a_{21}(t) \\ 1 & -d_{11}(t) \end{pmatrix}, d_{11}(t) = \frac{1}{2}\left(a_{11} - a_{22} + \frac{da_{21}}{dt}\frac{1}{a_{21}}\right).$$

Obviously (7.15) is a simplified canonical system, since one of the matrix elements is constant.

Finally, if we carry out the following substitution in Eq.(7.15),

$$\mathbf{Z}(t) = \mathbf{R}(t)\mathbf{U}(t), \quad \mathbf{R}(t) = \begin{pmatrix} d_{11}(t) & 1 \\ 1 & 0 \end{pmatrix},$$

we shall obtain

$$\frac{d}{dt}\mathbf{U} = \mathbf{M}(t)\mathbf{U}, \quad \mathbf{M}(t) = \begin{pmatrix} 0 & 1 \\ r(t) & 0 \end{pmatrix}, \tag{7.16}$$

$$r(t) = a_{12}a_{21} + d_{11}^2 - \frac{d}{dt}d_{11}.$$

System (7.16) describes an oscillating circuit without losses, which contains one time-dependent reactance.

System (7.16) is much simpler than the initial system (7.3), and the solutions of both equations are related in the following way:

$$\mathbf{x}(t) = k(t, t_0)\mathbf{N}(t)\mathbf{U}, \tag{7.17}$$

where

$$k(t, t_0) = \exp\left\{\frac{1}{2}\int_{t_0}^{t}\left[a_{11}(\tau) + a_{22}(\tau) + \frac{da_{21}(\tau)}{d\tau}\frac{1}{a_{21}(\tau)}\right]d\tau\right\},$$

$$\mathbf{N}(t) = \begin{pmatrix} \frac{1}{2a_{21}}\left(a_{11} - a_{22} + \frac{1}{a_{21}}\frac{d}{dt}a_{21}\right) & \frac{1}{a_{21}} \\ 1 & 0 \end{pmatrix}.$$

The investigation of the free processes in linear oscillating circuits with time dependent parameters of a general type is connected with enormous mathematical difficulties conditioned by the general character of the problem. Indeed, it is necessary to examine a huge number of equations (7.3) characterized by four functions – the elements of the matrix $a_{i,j}$, $i, j = 1, 2$.

The canonical systems lend themselves to a sufficiently accurate and clear classification by the boundedness or boundlessness of their solutions. The possible canonical systems make up a set that can be visualized as a set of points in a three-dimensional cylindric space. The cylindric space breaks down into a set of alternating areas of stability and instability. Certain investigations have been carried out and they have yielded a sufficiently comprehensive picture of the general properties of the free processes in a linear oscillating circuit with periodic parameters.

7.1.3. Classification of the free processes in Hamiltonian oscillating circuits

Again, the canonical systems lend themselves to a sufficiently distinct and clear classification by the boundedness or boundlessness of their solutions. It is possible to create a comprehensive picture of the general regularities governing the free processes in a linear oscillating circuit with periodic parameters. For the sake of visualization, we shall use a three-dimensional cylindric space which is broken down into a set of alternating areas of stability and instability.

Consider a canonical system with periodic coefficients in the following vector form

$$\frac{d}{dt}\mathbf{Z} = \mathbf{JHZ}. \tag{7.18}$$

The fundamental matrix of the solutions of the vector equation (7.18), given the initial conditions $\mathbf{Z}(0) = \mathbf{I}$, where $\mathbf{I}$ is a unit matrix, is presented as

$$\mathbf{Z}(t) = \mathbf{P}(t)e^{\mathbf{K}t}. \tag{7.19}$$

The properties of the matrixes in (7.19) are as follows:

a) $\mathbf{P}(t)$ is a real matrix-function, which is either periodic, i.e. $\mathbf{P}(t+T) = \mathbf{P}(t)$ or anti-periodic $\mathbf{P}(t+T) = -\mathbf{P}(t)$, T – period, $\mathbf{P}(0) = \mathbf{I}$, $\frac{d}{dt}\mathbf{P}$ is piece-wise continuous, i.e. it exists almost everywhere and is summable;

b) $\mathrm{Det}\mathbf{P}(t) \equiv 1$;

c) $\mathbf{K}$ is a real square matrix with constant elements, $Sp\mathbf{K} = 0$.

The investigation of the properties of matrix $\mathbf{K}$ is of crucial importance in the stability analysis. If the matrix function of the vector equation (7.18) is known, matrix $\mathbf{K}$ will be determined in the following way:

$$\mathbf{K} = \frac{1}{T}\ln[\pm\mathbf{Z}(T)]. \tag{7.20}$$

The alternative sign in (7.20) is chosen in such a way as to secure a real matrix $\mathbf{K}$.

Let us dwell more elaborately on matrix sets $\{\mathbf{K}\}$ and $\{\mathbf{P}(t)\}$. For the sake of visualization we shall use geometrical terms.

1) *Structure of the set* $\{\mathbf{K}\}$.

We shall present the set of various constant second-order matrixes $\mathbf{K}$ with zero spur in the following form

$$\mathbf{K} = \begin{pmatrix} -x & y-z \\ y+z & x \end{pmatrix},$$

where x, y, z are various values along the axes of the Cartesian coordinate system $Oxyz$. This presentation yields an ordinary three-dimensional (Euclidean) space. A completely defined matrix $\mathbf{K}$ corresponds to each point in this space.

The characteristic equation of this matrix,

$$\mathrm{Det}(\mathbf{K} - \lambda \mathbf{I}) = 0, \tag{7.21}$$

has the following form: $\lambda^2 = x^2 + y^2 - z^2$.

Given a real λ, this characteristic equation represents a single-band hyperboloid, whose canonical equation is in the following form: $\frac{x^2}{\lambda^2} + \frac{y^2}{\lambda^2} - \frac{z^2}{\lambda^2} = 1$.

The graphic representation of the single-band hyperboloid is shown in Fig.7.1.a, where $a = \lambda$ and $b = \lambda$ are the real semi-axes and $c = j\lambda$ is the imaginary semi-axis.

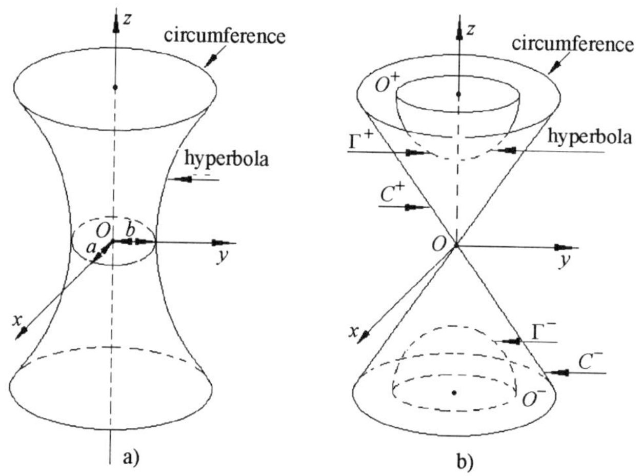

Fig. 7.1. One-belt hyperboloid illustrating the characteristic equation of matrix $\mathbf{K}$ in the case of a real λ (a); Two-belt hyperboloid illustrating the same in the case of an imaginary λ (b)

If λ is imaginary, (7.21) determines a double-band hyperboloid: $\frac{x^2}{\lambda^2} + \frac{y^2}{\lambda^2} - \frac{z^2}{\lambda^2} = -1$. The graphic representation of the double-band hyperboloid is given in Fig.7.1.b, where $a = j\lambda$ and $b = j\lambda$ are imaginary semi-axes, and $c = \lambda$ is a real semi-axis.

When $\lambda = 0$, the result is a borderline case of a cone: $x^2 + y^2 = z^2$, separating the two families of the above-mentioned hyperboloids. Conventionally, we shall ascribe a plus sign to the cone and hyperboloid tops, and a minus sign to their bottoms. The set of the points on the cone surface will be denoted as C^+ and C^- respectively. The set of the points in the interior of the cone ($x^2 + y^2 < z^2$) will be denoted as O (O^+ and O^-), while that of the exterior points ($x^2 + y^2 > z^2$) will be designated as H.

It turns out that systems (7.18) with periodic or anti-periodic solutions correspond to the set C, systems with bounded solutions correspond to the set O, while systems with unbounded solutions are in congruence with the set H. We shall denote the surface points of the double-band hyperboloid family as Γ (Γ^+ and Γ^-). The three-dimensional space constructed in this way will be designated as R^3.

The elucidation of the structure of the matrixes $\mathbf{K}$ and $e^{\mathbf{K}t}$ is of particular significance for the stability investigation of the canonical systems.

The matrix $\mathbf{K}$ can be presented in one of the following forms:

a) $\mathbf{K} = \mathbf{S}\begin{pmatrix} \lambda & 0 \\ 0 & -\lambda \end{pmatrix}\mathbf{S}^{-1}$ (set H),

b) $\mathbf{K} = \mathbf{S}\begin{pmatrix} 0 & \varepsilon \\ 0 & 0 \end{pmatrix}\mathbf{S}^{-1}$ (set C),

c) $\mathbf{K} = \mathbf{S}\begin{pmatrix} 0 & -\beta \\ \beta & 0 \end{pmatrix}\mathbf{S}^{-1} = \beta\mathbf{S}\mathbf{J}\mathbf{S}^{-1}$ (set O),

where $\mathbf{S}$ is a real matrix, λ, ε and β are real numbers.

It follows from equation $\mathbf{S}^{-1}e^{\mathbf{K}t}\mathbf{S} = e^{\mathbf{S}^{-1}\mathbf{K}\mathbf{S}}$ that:

when $\mathbf{K} \in \mathbf{H}$ $\quad e^{\mathbf{K}t} = \mathbf{S}\begin{pmatrix} e^{\lambda t} & 0 \\ 0 & e^{-\lambda t} \end{pmatrix}\mathbf{S}^{-1}$,

when $\mathbf{K} \in \mathbf{C}$ $\quad e^{\mathbf{K}t} = \mathbf{S}\begin{pmatrix} 1 & \varepsilon t \\ 0 & 1 \end{pmatrix}\mathbf{S}^{-1}$,

when $\mathbf{K} \in O$ $\quad e^{\mathbf{K}t} = \mathbf{S}e^{\mathbf{J}\beta t}\mathbf{S}^{-1} = \mathbf{S}(\mathbf{I}\cos\beta t + \mathbf{J}\sin\beta t)\mathbf{S}^{-1}$.

Hence, in case a) both solutions of the fundamental system (7.18) are unbounded, if the whole time axis is taken into account $t \in [-\infty, \infty]$. If the object of consideration is the semi-axis $t \in [0, \infty]$, one of the solutions of the fundamental system should be regarded as unbounded (exponentially increasing), and the other one as bounded (exponentially decreasing).

In case b) one solution under initial conditions $\mathbf{Z}_1(0) = \mathbf{S}\begin{pmatrix} 0 \\ 1 \end{pmatrix}$ is periodic or anti-periodic, while the other one – under initial conditions $\mathbf{Z}_2(0) = \mathbf{S}\begin{pmatrix} 1 \\ 0 \end{pmatrix}$ – is linearly increasing. In case c) all solutions are bounded.

Let us once again refer to formula (7.20) and consider the issue of the single-valuedness of the matrix $\mathbf{K}$.

For the sake of simplifying the presentation, we shall introduce the following denotations: $\mathbf{K}T = \mathbf{Y}$, $\mathbf{Z}(T) = \mathbf{B}$. Then equation (7.20) takes the following form:

$$e^{\mathbf{Y}} = \pm\mathbf{B}. \tag{7.22}$$

As a result of a relevant transformation, the matrix $\pm\mathbf{B}$ can be reduced to one of the following forms:

(a') $\begin{pmatrix} \mu & 0 \\ 0 & \dfrac{1}{\mu} \end{pmatrix}$, $\mu > 0$, $\mu \neq 1$;

(b') $\mathbf{I}+\begin{pmatrix} 0 & 1 \\ 0 & 0 \end{pmatrix}$;

(c') $\begin{pmatrix} \alpha & -\beta \\ \beta & \alpha \end{pmatrix}$, $\alpha^2+\beta^2=1$

If the matrix $\pm\mathbf{B}$ can be reduced to form (a') or (b'), there is a solution only in the case when the characteristic numbers of matrix $\pm\mathbf{B}$ are positive. The sign in (7.22) is chosen on the basis of this condition as well. This is the only solution and it is obtained by using the following formulae:

in case (a') – $\mathbf{Y}=\ln(\pm\mathbf{B})=\dfrac{2\mu\ln\mu}{\mu^2-1}(\pm\mathbf{B})-\dfrac{\mu^2+1}{\mu^2-1}\ln\mu\mathbf{I},$ (7.23)

in case (b') – $\ln(\pm\mathbf{B})=\pm\mathbf{B}-\mathbf{I}.$ (7.24)

If $\mathbf{B}=\pm\mathbf{I}$, we arrive at the solution by applying the formula:

$$\ln(\pm\mathbf{I})=n\pi\mathbf{S}^{-1}\mathbf{J}\mathbf{S}\quad(n=0,\pm1,\pm2,\ldots), \tag{7.25}$$

where n is an even number when a plus sign is ascribed and a negative number when a minus sign is attributed.

When $\pm\mathbf{B}$ can be reduced to form (c') and $\mathbf{B}\neq\pm\mathbf{I}$, all values $\ln(\pm\mathbf{B})$ are yielded by the formula

$$\ln(\pm\mathbf{B})=\frac{\varphi+n\pi}{\sin\varphi}\mathbf{B}-(\varphi+n\pi)\operatorname{ctg}\varphi-\mathbf{I},\quad(n=0,\pm1,\pm2,\ldots), \tag{7.26}$$

where $e^{\pm j\varphi}$ are the characteristic numbers of the matrix $\mathbf{B}$ $(0<\varphi<\pi)$; n is an even number when a plus sign is ascribed and a negative number when the sign is minus.

It follows from (7.22) that $\mathrm{Det}e^{\mathbf{Y}}=e^{Sp\mathbf{Y}}=1$, $Sp\mathbf{Y}=0$, i.e. all solutions $\mathbf{Y}$ of Eq.(7.22) fall in space R^3.

For matrixes $\mathbf{Y}$ and $\pm\mathbf{B}$ it is possible to have simultaneous occurrence of either cases (a) and (a'), or cases (b) and (b'), or cases (c) and (c'). So, we can conclude that in the case of unbounded increase in the solution, i.e. (a') and (b'), matrix $\mathbf{K}=\dfrac{1}{T}[\pm\mathbf{Z}(T)]$ has a single value and $\mathbf{K}\in H\cup C^+\cup C^-$.

In the case of a bounded solution (c') the matrix $\mathbf{K}$ is not a single-valued one. There are two options here: a non-trivial one, when $\mathbf{K}\in O$, and a trivial one, for $\mathbf{K}=0$ (in the second option the matrix degrades to a zero one).

Let us use Γ_0^+ and Γ_0^- to denote the top and bottom of the hyperboloid: $x^2+y^2-z^2=-\dfrac{\pi^2}{4T^2}$; O_0^+ and O_0^- to designate respectively the upper and lower area between the cone: $x^2+y^2-z^2=0$ and this hyperboloid; O_1^+ and O_1^- to mark respectively the areas between the top and the bottom of the cone: $x^2+y^2-z^2=0$ and the hyperboloid: $x^2+y^2-z^2=-\dfrac{\pi^2}{T^2}$.

If we take into account the fact that the matrix $\mathbf{P}(t)$ has a single value determined by the matrix $\mathbf{K}$ in (7.19), it is obvious that in order to ensure a one-value fundamental matrix of solutions (7.19) it would be necessary and sufficient to select the matrix $\mathbf{K}$ falling in the area

$$R_0^3 = H \cup C \cup O_0^+ \cup O_0^- \cup \Gamma_0^+.$$

2) *Structure of set* $\{\mathbf{P}(t)\}$.

We shall use Θ_2 to denote the set of second-order matrixes with constant real elements and with a determinant equal to one. Let us assume that $\mathbf{X}$ is a matrix belonging to this set. It can always be presented in the form

$$\mathbf{X} = \mathbf{a}|\mathbf{b}, \tag{7.27}$$

where $\mathbf{a}$ is a vector corresponding to the first column of the matrix, and $\mathbf{b}$ is a vector corresponding to the second column of the matrix. The elements of matrix $\mathbf{X}$ are expressed through the projections of vectors $\mathbf{a}$ and $\mathbf{b}$: $\mathbf{X} = \begin{pmatrix} a_x & b_x \\ a_y & b_y \end{pmatrix}$. The angle between the vectors is established according to the formula:

$$\alpha = \arccos \frac{(\mathbf{a}, \mathbf{b})}{|\mathbf{a}||\mathbf{b}|} = \arccos \frac{a_x b_x + a_y b_y}{\sqrt{(a_x^2 + a_y^2)(b_x^2 + b_y^2)}},$$

$0 < \alpha < \pi$.

It turns out that the determinant of the matrix $\mathbf{X}$ can be presented in the form

$$\mathrm{Det}\mathbf{X} = |\mathbf{a}||\mathbf{b}| \sin\alpha = 1. \tag{7.28}$$

Let us examine solution (7.19) of the canonical system of two equations. Given a fixed t, the matrix $\mathbf{Z}(t)$ is an element of the set Θ_2; in the case of changing time, this matrix is continuously converted from one element (matrix) of the set Θ_2, into another element of the same set, i.e. a trajectory is described in Θ_2, which can be visualized as a curve situated within a torus (fig. 7.2). We are interested in the normalized matrix of the solutions when the initial point of the trajectory is a unit matrix, and the final one is the matrix $\mathbf{Z}(t)$.

Fig.7.2 can be treated in another way as well. Let us consider a section of the torus with plane $Z = 0$. The result is a circuit in the plane Oxy. Each point on this circuit corresponds to a constant vector $\mathbf{a} = \begin{pmatrix} x \\ y \end{pmatrix}$. Hence it is obvious that Fig.7.2 shows one of the solutions $\mathbf{Z}(t)$, corresponding to the initial condition $\mathbf{e}_1 = \begin{pmatrix} 1 \\ 0 \end{pmatrix}$. φ is the angle of rotation of the vector $\mathbf{X}(t)$ for the time interval $[0, t]$. The angle

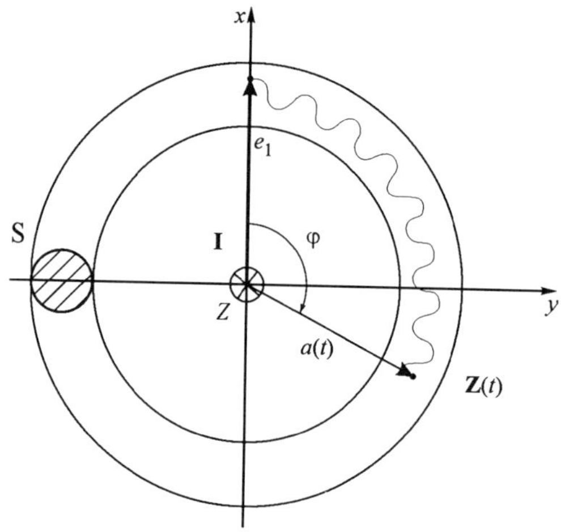

Fig. 7.2. Map of the solution $\mathbf{Z}$ (t) as a trajectory within the torus

of rotation over the time span of one period T, i.e. for the interval $[t, t+T]$, is of particular interest.

It follows from (7.19) that $\mathbf{Z}(T) = \mathbf{P}(T)\mathbf{a}$, $\mathbf{a}$ is a constant vector.

As a result of the condition that $\mathbf{P}(t+T) = \pm\mathbf{P}(t)$, over a time interval of $\Delta t = T$ the vector $\mathbf{Z}(t)$ turns at an angle divisible by π, i.e.

$$\varphi_T = n\pi (n = 0, \pm 1, \pm 2, \ldots), \tag{7.29}$$

where n is an even number for a periodic matrix $\mathbf{P}(t)$ and an odd number for an anti-periodic matrix. We have to point out that n does not depend on the initial conditions, i.e. two arbitrary vector solutions belonging to the relevant canonical system rotate in synchronism so that the angle between them remains unchanged. Hence, n in (7.29) is a system characteristic. Moreover, it proves to be quite an important characteristic from the viewpoint of the stability theory.

Formula (7.29) allows of classifying canonical systems according to the values of n. Indeed, each particular canonical system is characterized by its own matrix $\mathbf{Z}(t)$, hence its own matrix $\mathbf{P}(t)$ as well, determining the number of semi-revolutions n_p over an interval $\Delta t = T$ for an arbitrary vector-solution of the canonical system under consideration.

Let us carry out a qualitative examination of the behaviour of the solutions of canonical systems. Let us assume that there is a canonical system

$$\frac{d}{dt}\mathbf{Z} = \mathbf{A}(t)\mathbf{Z}, \tag{7.30}$$

$Sp\mathbf{A}(t) = 0.$

We shall consider the set L^3, which is made up of triads of periodic piece-wise continuous functions (elements of matrix $\mathbf{A}(t)$) $a_{11}(t), a_{12}(t), a_{21}(t)$. We introduce the following norm in L^3:

$$\int_0^T [|a_{11}(t)| + |a_{12}(t)| + |a_{21}(t)|]dt.$$

Then L^3 is transformed into a complete linear normalized space (Banach space).

Each matrix $\mathbf{A}(t)$ has its corresponding matrix function $\mathbf{Z}(t)$, which, in its turn, is juxtaposed to a couple of matrixes $(\mathbf{P}(t), \mathbf{K})$, where $\mathbf{P}(t) \in \Omega$, Ω is metric space, $\mathbf{K} \in H \cup C \cup O_1^+$. The correspondence indicated above is written in the following way: $L^3 = \Omega \times (H \cup C \cup O_1^+)$.

We use o to denote the beginning of the coordinate system in R^3. Then $C = C^+ \cup o \cup C^-$. We break down space Ω into a countable set of subspaces, $\Omega = \sum_{n=-\infty}^{\infty} \Omega_n$, and introduce the following designations:

$$H_n = \Omega_n \times H, \quad C_n^{*+} = \Omega \times C^+,$$
$$O_n = \Omega_n \times O_1^+, \quad C_n^{*-} = \Omega_n \times C^-,$$
$$C_n = \Omega_n \times C, \quad C_n^{*0} = \Omega_n \times O.$$

The product of the sets should be seen in the following way: $\mathbf{A}(t) \in H_n$, if $\mathbf{P}(t) \in \Omega_n$, $\mathbf{K} \in H$.

The following denotations are used: $H = U_{n=-\infty}^{\infty} H_n$, $O = U_{n=-\infty}^{\infty} O_n$, $C = U_{n=-\infty}^{\infty} C_n$, $C^{*\pm} = U_{n=-\infty}^{\infty} C_n^{\pm}$, $C^{*o} = U_{n=-\infty}^{\infty} C_n^{o}$, $C^* = C^+ \cup C^-$.

The fundamental matrices $\mathbf{Z}(t)$ related to the canonical systems belonging to different areas of stability have different representations:

$$\mathbf{A}(t) \in H \to \mathbf{Z}(t) = \mathbf{P}(t)\mathbf{S}\begin{pmatrix} e^{\lambda t} & 0 \\ 0 & e^{-\lambda t} \end{pmatrix}\mathbf{S}^{-1}, \ (\lambda > 0),$$

$$\mathbf{A}(t) \in O \to \mathbf{Z}(t) = \mathbf{P}(t)\mathbf{S}\begin{pmatrix} \cos\beta t & -\sin\beta t \\ \sin\beta t & \cos\beta t \end{pmatrix}\mathbf{S}^{-1}, \left(0 < \beta < \frac{\pi}{T}\right),$$

$$\mathbf{A}(t) \in C^{*\pm} \to \mathbf{Z}(t) = \mathbf{P}(t)\mathbf{S}\begin{pmatrix} 1 & \pm t \\ 0 & 1 \end{pmatrix}\mathbf{S}^{-1},$$

$$\mathbf{A}(t) \in C^{*o} \to \mathbf{Z}(t) = \mathbf{P}(t).$$

The open connected sets O_n are referred to as stability (boundedness) areas, the open linked sets H_n are respectively termed areas of instability, C_n are the boundaries between the areas of stability and instability (the boundaries themselves are referred to the unstable areas, in accordance with another law – that of the growing unbounded solutions).

Let us consider the issue of the composition of space L^3 which is important for applied problems. The matrix function of the fundamental system (7.19) can be viewed as a trajectory in torus Θ_2. Each canonical system has its corresponding matrix function. Therefore one intuitively arrives at the assertion that the set of Hamiltonian systems (7.30), or the set of matrixes $\{\mathbf{A}(t)\}$, which is the same, has a corresponding set of matrix functions $\{\mathbf{Z}(t)\}$. For the purpose of proving this correspondence, set $\{\mathbf{Z}(t)\}$ should be arranged in such a way as to form a space analogous to space L^3 for matrix $\{\mathbf{A}(t)\}$. It should also be shown that the correspondence $\mathbf{A}(t)\gtrless\mathbf{Z}(t)$ is reciprocally continuous. Seeking to arrange set $\{\mathbf{Z}(t)\}$, we introduce the following intervals: $\rho(\mathbf{Z}_1,\mathbf{Z}_2) = \sup\limits_{0\leq t\leq T} ||\mathbf{Z}_1(t) - \mathbf{Z}_2(t)|| + \int_0^T ||\frac{d}{dt}[\mathbf{Z}_1(t)-\mathbf{Z}_2(t)]||dt \ (0 \leq t \leq T)$. Then set $\{\mathbf{Z}(t)\}$ (like L^3 for $\{\mathbf{A}(t)\}$) becomes a complete linear normalized space. These two spaces can be treated as identical and denoted in the same way – as L^3. The dependence of the matrix function $\mathbf{Z}(t)$ on the matrix $\mathbf{A}(t)$ is continuous. Moreover, there is observance of this continuity both by norm and by interval.

As shown above, the points on torus Θ_2 are unimodular matrices (matrices with a determinant equal to one). The points in the space R^3 for matrices $\mathbf{K}$ represent matrices with a trace equal to zero. For the purpose of achieving a single value, it would be sufficient to select such points for matrixes $\mathbf{K}$ that lie between the upper and lower bottom of the two-band hyperboloid: $z^2 - x^2 - y^2 = \frac{\pi^2}{4T^2}$, and to include one of the two boundaries in this part of space R^3. The bottoms of the two-band cone: $x^2 + y^2 = z^2$, are situated between the bottoms of this hyperboloid. The space between the cone bottoms corresponds to the unstable area. The space between the respective bottoms of the cone and the hyperboloid corresponds to the stable area. The surface of the cone is the boundary between the stable area and the unstable one, but it is treated as belonging to the unstable area. Any point belonging to torus Θ_2, i.e. any unimodular matrix with constant parameters $\mathbf{Z}$ can always be presented in the form: $\mathbf{Z} = \pm e^{\mathbf{K}t}$. In this way the stable and unstable areas, which are entirely determined by matrix $\mathbf{K}$, can be transferred to torus Θ_2. It has already been shown that the matrix function Θ_2 of system (7.30) „winds" round the torus Θ_2, and the number of windings for a time interval of $\Delta t = T$ is $n/2$ (when the canonical system is in the n-th stable or unstable area). In the case of stability, the end of the winding at the close of interval $[t, t+T]$ proves to be at the same distance from the torus center as in the beginning of the interval. In the case of instability, the end of the winding proves to be further away from the torus center than its beginning. Let us replace the torus containing $n/2$ windings with $n/2$ identical tori, each containing one winding of the solution. Let us cut each torus at one and the same place along its section and then shift the ends of the section in different directions along axis z (the result is a spiral winding). Then we stick together the section ends of the different tori so as to link up the solution windings and obtain the complete solution. Thus, instead of $n/2$ tori, the result is a spiral area with $n/2$ windings. By deforming the spiral area directly one can obtain a cylindrical space whose section is shown in Fig.7.3. The space itself can be obtained by rotating the figure round the hatch-dotted line.

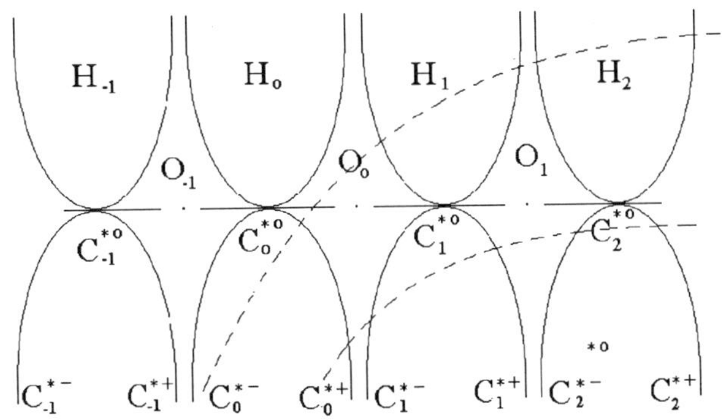

Fig. 7.3. Universal three-dimensional cylindric space whose points represent elements of the set of canonical systems of second order

The cylindric space R^3 encompasses various canonical systems. A particular case of such systems is, for instance, Mathieux's equation (the dotted line in Fig.7.3 indicates the area corresponding to Mathieux's equation).

7.1.4. Phase plane of a linear oscillating circuit with periodic parameters

Let us consider the canonical system [255]

$$\frac{d}{dt}\mathbf{x} = \mathbf{A}(t)\mathbf{x}, \tag{7.31}$$

$$\mathbf{A}(t) = \begin{pmatrix} -\beta(t) & -\gamma(t) \\ \alpha(t) & \beta(t) \end{pmatrix}, \quad \mathbf{A}(t+T) = \mathbf{A}(t).$$

The solution of the canonical system (7.31) can be presented in the form

$$\mathbf{x}(t) = \mathbf{P}(t)e^{\mathbf{K}t}\mathbf{x}(0), \tag{7.32}$$

where $\mathbf{P}(t)$ is a periodic matrix ($\mathbf{P}(t+T) = \mathbf{P}(t)$) or an anti-periodic matrix ($\mathbf{P}(t+T) = -\mathbf{P}(t)$), $\mathbf{K}$ is a constant matrix with $Sp\mathbf{K} = 0$.

Matrix $\mathbf{K}$ contains information on the stability of system (7.31), while matrix $\mathbf{P}(t)$ provides information on precisely what area of stability or instability system (7.31) refers to.

Consider the case when system (7.31) refers to the n-th area of instability, i.e. $\mathbf{A}(t) \in H_n$. Then $\mathbf{P}(t) \in \Omega_n$, $\mathbf{K} \in H \in R^3$ (matrix $\mathbf{K}$ corresponds to some point situated between the bottoms of the cone in Fig.7.1). Let $\mathbf{a}_+$ and a_- be eigenvectors to matrix $\mathbf{K}$, i.e. $\mathbf{K}\mathbf{a}_+ = \lambda\mathbf{a}_+$, $\mathbf{K}\mathbf{a}_- = \lambda\mathbf{a}_-$, λ and $-\lambda$ are numbers proper to matrix $\mathbf{K}$. Then exponent properties prompt that

$$e^{\mathbf{K}t}\mathbf{a}_+ = e^{\lambda t}\mathbf{a}_+, \quad e^{\mathbf{K}t}\mathbf{a}_- = e^{\lambda t}\mathbf{a}_-.$$

An arbitrary initial vector can be presented in the form:

$$\mathbf{x}(0) = \alpha_1\mathbf{a}_+ + \alpha_2\mathbf{a}_-.$$

Let us draw up the following vector function

$$\mathbf{y}(t) = e^{\mathbf{K}t}\mathbf{x}(0) = \alpha_1 e^{\lambda t}\mathbf{a}_+ + \alpha_2 e^{-\lambda t}\mathbf{a}_-. \tag{7.33}$$

This vector is a solution to the system of type (7.31) with constant coefficients in the case when this system is unstable. As time goes by, the module of vector $\mathbf{y}(t)$ increases, while its direction draws closer to that of vector $\mathbf{a}_+$. It is well-known that the phase plane of such a system contains a peculiar point of the „saddle" type.

Solution (7.32) can be presented in the form

$$\mathbf{x}(t) = \mathbf{P}(t)\mathbf{y}(t). \tag{7.34}$$

In this case matrix $\mathbf{P}(t)$ plays the role of an operator transforming vector $\mathbf{y}$ into vector $\mathbf{x}$. It is a unimodular operator. This implies an equal degree of „increase" or „decrease" of vectors $\mathbf{x}(t)$ and $\mathbf{y}(t)$. For example, if vectors $\mathbf{y}_1$ and $\mathbf{y}_2$ form a parallelogram, vectors $\mathbf{x}_1$ and $\mathbf{x}_2$ form a parallelogram with the same area. In addition, the modules of vectors $\mathbf{x}_{1,2}$ and the angle between them may vary from those of vectors $\mathbf{y}_{1,2}$. If the phase plane for $\mathbf{y}(t)$ is known, operator $\mathbf{P}(t)$ rotates this phase plane unevenly, so that it performs $n/2$ semi-revolutions over period T, and the rotation is accompanied by periodic unimodular deformation of the phase plane.

The solution to system (7.31) situated in the n-th area of instability H_n can be presented in the following way. Let us take a horizontal opaque white plane (screen) and, connected to it, a transparent plane, which we fix to an axis in point O above the opaque plane. The whole system is illuminated from above by a flat parallel light beam perpendicular to the planes. The transparent plane rotates periodically round the axis, performing n semi-rotations over period T. A transforming black point moves on the transparent plane (solution $\mathbf{y}(t)$) along trajectories forming a „saddle" (Fig.7.4). During its rotation the transparent plane is deformed unimodularly and periodically with period T. Then the shade of the representing point on the lower plane represents the representing point of solution $\mathbf{x}(t)$. As regards solutions $\mathbf{x}_1(t) = e^{\lambda t}\mathbf{P}(t)\mathbf{a}_+$ and $\mathbf{x}_2(t) = e^{-\lambda t}\mathbf{P}(t)\mathbf{a}_-$, the rotation angles for a time period

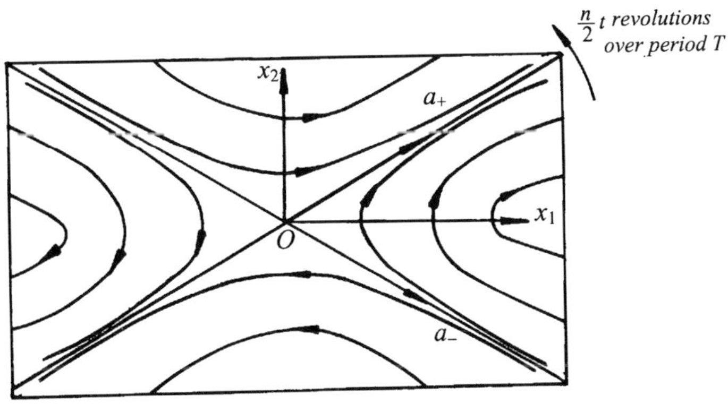

Fig. 7.4. Trajectories of the set of vector functions, when a linear oscillating circuit with periodic parameters is situated in an instability area

T are $\varphi_{x_1} = \varphi_{x_2} = n\pi$. For any other solution, the rotation angle φ_x falls within the following limits: $n\pi < \varphi_x < (n+1)\pi$ or $(n-1)\pi < \varphi_x < n\pi$.

If the canonical system (7.31) belongs to the nth area of stability, Q_n, the difference between the visual model and the previous one will be that point o will be a peculiar point of a „center" type. Then the phase trajectories of the transparent plane will be in the form of concentric ellipses (Fig.7.5). In this case, for a period T, each solution rotates at an angle φ_x within the following range: $n\pi < \varphi < (n+1)\pi$.

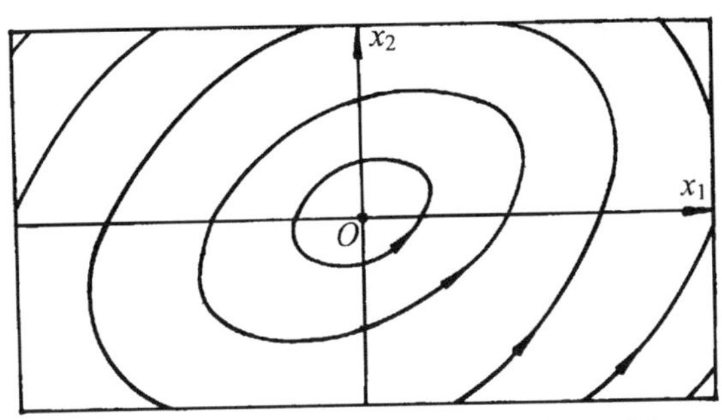

Fig. 7.5. Trajectories of motion in a canonical system situated in an stability area

Let us consider the case when the canonical system (7.31) is on the boundary between the stable and unstable areas. There are two typical boundaries: C_n^{*-} – the boundary between O_{n-1} and H_n, and C_n^{*+} – the boundary between H_n and O_n. These two boundaries intersect and the result is boundary C_n^{*0} running between boundaries C_n^{*-} and C_n^{*+}. Let us initially consider boundary C_n^{*-}. We introduce matrix $\mathbf{J}_1 = \begin{pmatrix} 0 & 1 \\ 0 & 0 \end{pmatrix}$. The following belonging exists in space of R^3 for $\mathbf{K}$: $\mathbf{J}_1 \in C^-$. With respect to the boundary under consideration, $\mathbf{K}$ can be presented in the following way:

$$\mathbf{K} = \mathbf{S}\mathbf{J}_1\mathbf{S}^{-1}, \ \mathrm{Det}\mathbf{S} > 0.$$

In this case there are two linearly independent vectors $\mathbf{a}_1$ and $\mathbf{a}_2$, so that $\mathbf{K}\mathbf{a}_1 = 0$, $\mathbf{K}\mathbf{a}_2 = a_1$. Then

$$e^{\mathbf{K}t}\mathbf{a}_1 = \left(\mathbf{I} + \mathbf{K}t + \frac{1}{2!}\mathbf{K}^2t^2 + \ldots\right)\mathbf{a}_1 = \mathbf{a}_1,$$

$$e^{\mathbf{K}t}\mathbf{a}_2 = \mathbf{a}_2 + t\mathbf{a}_1, \quad \text{since } \mathbf{K}^2 = \mathbf{K}^3 = \ldots = 0.$$

If $\mathbf{d} = \alpha_1\mathbf{a}_1 + \alpha_2\mathbf{a}_2$ is an arbitrary vector, then $e^{\mathbf{K}t}\mathbf{d} = \mathbf{d} + \alpha_2 t\mathbf{a}_1$. It is obvious that the phase trajectories on the transparent plane are straight lines parallel to vector $\mathbf{a}_1$ (Fig.7.6). The rotation angle of solution $x(t)$ satisfies the inequalities: $(n-1)\pi < \varphi_x < n\pi$.

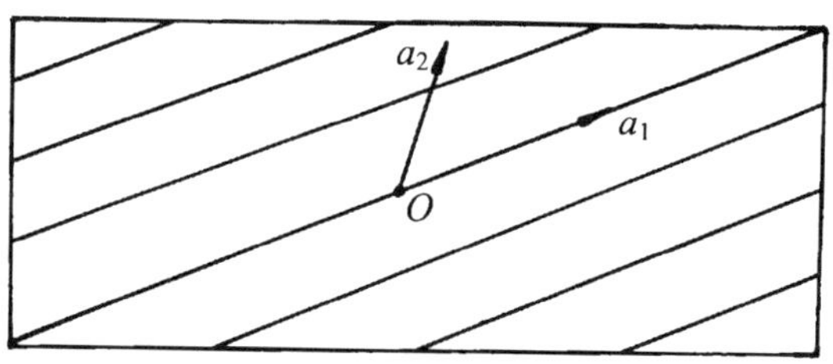

Fig. 7.6. Trajectories of the solution of a canonical system at the boundary of the area of stability

If the canonical system is situated on boundary C_n^{*+}, the picture will differ from the previous one only in that the direction of the trajectories will change to its opposite. The rotation angle of the solution satisfies the following inequalities: $n\pi \leq \varphi_x \leq (n+1)\pi$

Finally, if the canonical system (7.31) is located on boundary C_n^{*0}, the representing point on the transparent plane remains motionless.

7.1.5. *Stability of the canonical systems*

Let us assume that we have the following canonical system

$$\frac{d}{dt}\mathbf{x} = \mathbf{J}\mathbf{H}(t)\mathbf{x}, \tag{7.35}$$

where $\mathbf{H}(t) = \begin{pmatrix} \alpha & \beta \\ \beta & \gamma \end{pmatrix}$.

An important role in the stability investigation is played by the rotation angle of vector $\mathbf{x}(t) = \mathbf{X}(t)\mathbf{c}$, $\mathbf{c}$ is a constant vector.

We shall denote the rotation angle of a given vector $\mathbf{Z}(t) = \begin{pmatrix} p \\ q \end{pmatrix}$ as $\varphi_Z(t)$. Obviously,

$$\frac{d\varphi_Z(t)}{dt} = \frac{d}{dt}\operatorname{arctg}\frac{q}{p} = \frac{\dot{q}p - \dot{p}q}{p^2+q^2} = \frac{1}{p^2+q^2}\operatorname{Det}\begin{pmatrix} p & \dot{p} \\ q & \dot{q} \end{pmatrix},$$

$$\text{i.e. } \varphi_Z(t) - \varphi_Z(0) = \int_0^t \frac{\operatorname{Det}\mathbf{Z}|\dot{\mathbf{Z}}}{(\mathbf{Z},\mathbf{Z})}. \tag{7.36}$$

By using this formula, as well as the relations $\operatorname{Det}\mathbf{a}|\mathbf{J}\mathbf{b} = (\mathbf{a},\mathbf{b})$, we obtain the following equation for the rotation angle of the canonical system (7.35)

$$\varphi_x = \int_0^T \frac{(\mathbf{H}\mathbf{x},\mathbf{x})}{(\mathbf{x},\mathbf{x})}dt$$

We designate the characteristic numbers of matrix $\mathbf{H}(t)$ as follows: $h_{\max}(t)$ and $h_{\min}(t)$:

$$h_{\substack{\max\\ \min}}(t) = \frac{\alpha+\gamma}{2} \pm \sqrt{\frac{(\alpha-\gamma)^2}{4} + \beta^2}.$$

Since

$$h_{\min}(t) \le \frac{(\mathbf{H}\mathbf{x},\mathbf{x})}{(\mathbf{x},\mathbf{x})} \le h_{\max}(t),$$

it follows that

$$\int_0^T h_{\min}(t)dt \le \varphi_x \le \int_0^T h_{\max}(t)dt.$$

Let us assume that $\mathbf{H} = \begin{pmatrix} \alpha_0 & \beta_0 \\ \beta_0 & \gamma_0 \end{pmatrix}$, α_0, β_0, γ_0 are arbitrary numbers, satisfying the conditions that $\alpha_0 > 0$, $\gamma_0 > 0$, $\alpha_0\gamma_0 - \beta_0^2 = 1$, while $h_{\min}$, $h_{\max}$ are roots of the equation:

$$\operatorname{Det}[\mathbf{H}(t) - h\mathbf{H}_0] = 0.$$

If inequalities

$$n\pi < \int_0^T h_{\min}(t)dt \le \int_0^T h_{\max}(t)dt < (n+1)\pi \tag{7.37}$$

are satisfied for a definite $n = 0, \pm 1, \pm 2, ...$, Eq.(7.35) will be stable and $\mathbf{H}(t) \in O_n$.

As an illustration of this stability criterion, we shall consider equation

$\frac{d^2y}{dt^2} + p(t)y = 0$, which is a particular case of the canonical system (7.35). Let us set $C > 0$ as a constant, for which: $\frac{n\pi}{T} < C < \frac{(n+1)\pi}{T}$, n is an integer.

Provided that inequalities

$$n\pi < \frac{1}{C}\int_0^T p_C^-(t)dt \leq \frac{1}{C}\int_0^T p_C^+(t)dt < (n+1)\pi, \tag{7.38}$$

where $p_C^+(t) = \begin{cases} p(t), \text{at } p(t) > C^2 \\ C^2, \quad \text{at } p(t) \leq C^2 \end{cases}$,

$$p_C^-(t) = \begin{cases} p(t), \text{at } p(t) \leq C^2 \\ C^2, \quad \text{at } p(t) > C^2 \end{cases},$$

are satisfied, the equation under consideration is stable and $p(t) \in O_n$.

Let us formulate a second stability criterion. Let inequality $[\mathbf{H}(t)\mathbf{C}, \mathbf{C}] \geq 0$ be valid for any t and $\mathbf{C}$. If inequalities $k\pi < m_- \leq M_+ < (k+1)\pi$ are satisfied, Eq.(7.35) belongs to the k-th stability area.

And if inequalities $M_- > k\pi$ and $m_+ < k\pi$ are satisfied, Eq.(7.35) belongs to the k-th instability area.

If inequality $[\mathbf{H}(t)\mathbf{C}, \mathbf{C}] \leq 0$ is valid for any t and $\mathbf{C}$, the following substitutions should be carried out in the previous inequalities: $M_\pm$ should be replaced by $M_\mp$, and $m_\pm$ by $m_\mp$.

Here:

$$M_\pm = \frac{A_\pm + C_\pm}{2} + \sqrt{\frac{(A_\pm - C_\pm)^2}{4} + B_\pm^2},$$

$$m_\pm = \frac{A_\pm + C_\pm}{2} - \sqrt{\frac{(A_\pm - C_\pm)^2}{4} + B_\pm^2},$$

$$A_\pm = \int_0^T \exp\left(\pm\int_0^t g dt\right)\alpha dt, \; B_\pm = \frac{1}{2}\int_0^T \exp\left(\pm\int_0^t g dt\right)\beta dt,$$

$$C_\pm = \int_0^T \exp\left(\pm\int_0^t g dt\right)\gamma dt, \; g(t) = 2\sqrt{\frac{(\alpha-\gamma)^2}{4} + \beta^2}.$$

Sometimes, owing to various reasons, Eq.(7.35) proves to be inconvenient for analyzing its stability. In such cases it is desirable to transform (7.35) into another equation located in the same area of stability or instability. Let us illustrate this option.

The following denotations are introduced as a supplement to (7.35):

$$\beta_0 = \frac{1}{T}\int_0^T \beta(t)dt, \quad \delta(t) = \int_0^t [\beta(t) - \beta_0]dt. \tag{7.39}$$

The variables are substituted as follows:

$$x_1 = e^{-\delta(t)} z_1, \quad x_2 = e^{\delta(t)} z_2.$$

The result is a system of equations:

$$\left| \begin{array}{l} \dfrac{dz_1}{dt} = -\beta_0 z_1 - \gamma_1(t) z_2 \\ \dfrac{dz_2}{dt} = \alpha_1(t) z_1 + \beta_0 z_2. \end{array} \right. \tag{7.40}$$

This is a system of a canonical type with a matrix $\mathbf{H}_1(t) = \begin{pmatrix} \alpha_1(t) & \beta_0 \\ \beta_0 & \gamma_1(t) \end{pmatrix}$, where $\alpha_1(t) = \alpha(t)e^{-2\delta(t)}$ and $\gamma_1(t) = \gamma(t)e^{2\delta(t)}$.

Systems (7.35) and (7.40) are situated in the same areas of stability or instability, for it can be shown that the rotation angles of arbitrary vector solutions $\mathbf{x}(t)$ and $\mathbf{Z}(t)$ for these two systems of equations are identical. System (7.40) however is frequently more convenient for investigation, since two elements in matrix $\mathbf{H}_1(t)$ have proved to be equal and unchanging (constant) in time.

It is always possible to identify two constant matrixes $\mathbf{C}^{\pm}$, so as to satisfy the inequalities: $\mathbf{C}^{-} \leq \mathbf{H}_1(t) \leq \mathbf{C}^{+}$. These two matrixes are determined in a sufficiently simple way as

$$\mathbf{C}^{\pm} = \begin{pmatrix} \alpha_0^{\pm} & \beta_0 \\ \beta_0 & \gamma_0^{\pm} \end{pmatrix}, \tag{7.41}$$

$$\alpha_0^{+} = \max_t \alpha(t), \quad \gamma_0^{+} = \max_t \gamma_1(t),$$

$$\alpha_0^{-} = \min_t \alpha(t), \quad \gamma_0^{-} = \min_t \gamma_1(t).$$

The constant matrix $\mathbf{C}^{\pm}$ is referred to the n-th stability zone, i.e. $\mathbf{C}^{\pm} \in 0_n$, if $\dfrac{n^2\pi^2}{T^2} < \mathrm{Det}\mathbf{C}^{\pm} < \dfrac{(n+1)^2\pi^2}{T^2}$.

On this basis, the stability criterion can be formulated in the following manner. If the inequalities

$$\frac{n^2\pi^2}{T^2} \leq \alpha_0^{\pm}\gamma_0^{\pm} - \beta_0^2 \leq \frac{(n+1)^2\pi^2}{T^2} \tag{7.42}$$

are satisfied and $\alpha_0^{+} + \gamma_0^{+} > 0$, $\alpha_0^{-} + \gamma_0^{-} > 0$ Eq.(7.35) will be stable and $\mathbf{H}(t) \in O_n$ $(n > 0)$.

Provided that inequalities (7.42) are satisfied, but $\alpha_0^{+} + \gamma_0^{+} < 0$ and $\alpha_0^{-} + \gamma_0^{-} < 0$, Eq.(7.35) will also be stable and $\mathbf{H}(t) \in O_{-n-1}$ $(n > 0)$.

In the case of $\alpha_0^{+}\gamma_0^{+} - \beta_0^2 < 0$ and $\alpha_0^{-}\gamma_0^{-} - \beta_0^2 < 0$, $\mathbf{H}(t)$ will belong to the zero area of instability.

Finally we shall dwell on the following practical issue. Let us assume that there is a particular canonical system of the type of (7.35). How should we answer the question which area of stability or instability it belongs to? In order to come up with an answer, we should approximate the matrix of system $\mathbf{H}(t)$ with two piece-wise constant matrixes $\mathbf{H}^-(t)$ and $\mathbf{H}^+(t)$, so that $\mathbf{H}^-(t) \leq \mathbf{H}(t) \leq \mathbf{H}^+(t)$. As the approximation accuracy increases, one of the following two conclusions is ultimately arrived at: (a) Matrices $\mathbf{H}^-(t)$ and $\mathbf{H}^+(t)$ fall in one area of stability or instability, which will also encompass matrix $\mathbf{H}(t)$; (b) These matrices never fall in the same area: one of them is situated in the stable area, while the other one is located in the adjacent unstable zone. Then obviously matrix $\mathbf{H}(t)$ lies on the boundary between these two areas.

The outlined methods require an ability to determine which area of stability or instability the piece-wise constant matrixes belong to. We divide up the interval $[O, T]$ into smaller sub-intervals like this: $0 = t_0 < t_1 < t_2 < \ldots < t_n = T$, $t_i - t_{i-1} = \tau_i$. We set $\mathbf{A}(t) = \mathbf{K}_i$ for $t_{i-1} < t < t_i$. Then $\mathbf{X}(t_i) = e^{\mathbf{K}_i\tau_i} \bullet \ldots \bullet e^{\mathbf{K}_2\tau_2} e^{\mathbf{K}_1\tau_1}$ and in particular

$$\mathbf{X}(T) = e^{\mathbf{K}_n\tau_n} \bullet \ldots \bullet e^{\mathbf{K}_2\tau_2} e^{\mathbf{K}_1\tau_1}. \tag{7.43}$$

It is convenient to calculate the matrix exponent $e^{\mathbf{K}}$ according to the formula:

$$e^{\mathbf{K}} = \operatorname{ch}\mu\mathbf{I} + \frac{1}{\mu}\operatorname{sh}\mu\mathbf{K},$$

$\pm\mu$ are characteristic figures of $\mathbf{K}$. In the particular case of $\mu = 0$, one can determine $e^{\mathbf{K}} = \mathbf{I} + \mathbf{K}$.

The characteristic equation regarding $\mathbf{X}(T)$ is of the form

$$\lambda^2 - 2a\lambda + 1 = 0,$$

where $2a = Sp\mathbf{X}(T)$. Provided that $|a| < 1$, it follows that $\mathbf{A}(t) \in O$ (all solutions are bounded), and when $|a| > 1$ $\mathbf{A}(t) \in H$ is valid (there are unbounded solutions as well). With a view to identifying the number of the stable or unstable range, the rotation angle of the solution should be computed.

Let us take $\mathbf{x}_0 = \begin{pmatrix} 1 \\ 0 \end{pmatrix}$ as an initial vector. Then for $t = t_1$ the solution will take the form $\mathbf{x}_1 = e^{\mathbf{K}_1\tau_1}\begin{pmatrix} 1 \\ 0 \end{pmatrix}$. If $\mathbf{K}_1$ is unstable (i.e. $\mathrm{Det}\mathbf{K}_1 \leq 0$), the rotation angle for a period of time τ_1 will fall within the range $-\pi \leq \tau_1 < \pi$. If $\mathbf{K}_1$ is stable (i.e. $\mathrm{Det}\mathbf{K}_1 > 0$), the matrix column can be used to determine the rotation angle only with an accuracy up to the term $m\pi$ (m is an integer). For the purpose of establishing the rotation angle accurately, it is necessary to set temporarily $T = \tau_1$ and to determine the stable range to which $\mathbf{K}_1$ belongs. Let us assume that m is the number of this stable range, while φ_1 is the rotation angle over a period of time τ. Then $m\pi < \varphi_1 < (m+1)\pi$ and the angle can be determined accurately.

For $t = t_2$ the solution will be $\mathbf{x}_2 = e^{\mathbf{K}_2\tau_2}e^{\mathbf{K}_1\tau_1}\mathbf{x}_0$; once again the rotation angle φ_2 is determined and the total resultant rotation angle is summed up: $\varphi = \varphi_1 + \varphi_2$. The complete rotation angle is: $\varphi = \varphi_1 + \varphi_2 + \ldots + \varphi_n$. If $|a| < 1$ and $m\pi < \varphi < (m+1)\pi$, $\mathbf{A}(t) \in O_m$. And if $|a| > 1$, the solution will be unstable: $\mathbf{A}(t) \in H_m$. We should determine m. As it was shown above, there is only one matrix $\mathbf{K}$, satisfying the equation $e^{\mathbf{K}T} = \pm\mathbf{B}$, $\mathbf{B} = e^{\mathbf{K}_n\tau_n}e^{\mathbf{K}_{n-1}\tau_{n-1}} \bullet \ldots \bullet e^{\mathbf{K}_2\tau_2}e^{\mathbf{K}_1\tau_1}$ (out of the two possible signs we select the one for which $\mathbf{K}$ is real). Let us assume that a_+ and a_- are the natural vectors of matrix $\mathbf{K} : \mathbf{K}\mathbf{a}_\pm = \pm\mu\mathbf{a}_\pm$ $(\mu > 0)$. We have already shown that all vectors lying in quadrants I and III (this refers to the natural vectors) turn at an angle φ over time T: $m\pi < \varphi < (m+1)\pi$, and the vectors situated in quadrants II and IV turn at an angle φ: $(m-1)\pi < \varphi < m\pi$. The problem of determining the rotation angle boils down to plotting vectors $\mathbf{a}_+$ and $\mathbf{a}_-$, clarifying which quadrant vector $\mathbf{X}_0 = \begin{pmatrix} 1 \\ 0 \end{pmatrix}$ lies in, and determining the angle by using one of the inequalities given above.

Thus, by employing simple algebraic operations one can always determine the stable or unstable area to which the interval-constant matrixes $\mathbf{H}^-(t)$ and $\mathbf{H}^+(t)$ belong.

7.1.6. Stability criteria of a generalized linear resonance circuit

We shall call up some major points of Lyapounov's second method [259] which we shall use as a mathematical basis for investigating the stability of a generalized resonance circuit.

The real scalar function $V(\mathbf{x}, t)$, defined and continuous in the area $S_t = \{\mathbf{x}, t\}$, $||\mathbf{x}|| \le S^0$, $t \ge 0$, t – time, and meeting the condition

$$V(0, t) = 0, \quad t \ge 0$$

is termed Lyapounov's function.

The derivative of the scalar function $V(\mathbf{x}, t)$ multiplied by the vector argument $\mathbf{x}$ is called a n-dimensional vector-function,

$$\frac{\partial V(\mathbf{x}, t)}{\partial t} = \text{colon}\left(\frac{\partial V}{\partial x_1}, \frac{\partial V}{\partial x_2}, \ldots, \frac{\partial V}{\partial x_n}\right).$$

Lyapounov's function will be qualified as sign-constant in the range S_t, if $V(\mathbf{x}, t) \ge 0$ or $V(\mathbf{x}, t) \le 0$, $\mathbf{x}, t \in S_t$. In the first case, Lyapounov's function is a sign-constant positive or non-negative one, and in the second case it is a sign-constant negative or non-positive function.

Lyapounov's function $V(\mathbf{x})$, which is explicitly independent of time, is termed a function of a fixed sign in the range $S = \{\mathbf{x}\}$, $||\mathbf{x}|| \leq S$ if for all $\mathbf{x} \in S$ are valid $V(\mathbf{x}) > 0$ or $V(\mathbf{x}) < 0$ except for $||\mathbf{x}|| \leq 0$.

In the first case the function $V(\mathbf{x})$ is called definitely positive, and in the second case definitely negative.

Function $V(\mathbf{x}, t)$, which is explicitly dependent on time, is termed definitely positive in the range S_t, if there is a function S, which is definitely positive in $W(x)$ and explicitly independent of time, so that $V(\mathbf{x}, t) \geq W(\mathbf{x})$, $\mathbf{x}, t \in S_t$.

If the inequality $V(\mathbf{x}, t) \leq -W(\mathbf{x})$ is satisfied, the function $V(\mathbf{x}, t)$ is called definitely negative.

The function $V(\mathbf{x}, t)$ is qualified as allowing an infinitely small upper boundary in the range S_t, if for an arbitrary number $C > 0$ one can find such a number $\delta > 0$, that $|V(\mathbf{x}, t)| < C$ when $||\mathbf{x}|| < \delta$, $t \geq 0$.

The objective of Lyapounov's second method is to provide a qualitative analysis of systems of n differential equations of first order, usually presented in a vector form,

$$\frac{d}{dt}\mathbf{x} = \mathbf{f}(\mathbf{x}, t), \tag{7.44}$$

where $\mathbf{f}(\mathbf{x}, t)$ is a continuous vector-function defined in the range $S_t^* = \{\mathbf{x}, t\}$, $||\mathbf{x}|| < S^*$, $t \geq 0$, $S^* = \text{const} > 0$.

The motion $\mathbf{x} = 0$ for $t \geq 0$ is termed unperturbed. If, at a certain initial moment $t_0 \geq 0$, the equality $\mathbf{x}_0 = \mathbf{x}(t_0) \neq 0$ is satisfied, the motion $\mathbf{x}(\mathbf{x}_0, t)$, caused by the initial perturbation $\mathbf{x}_0$, is called perturbed motion (the term „free process" is more commonly used in Radiophysics).

Lyapounov's first theorem. The unperturbed motion is stable with respect to variables $x_1, x_2, \ldots, x_n$, if there is a function $V(x, t)$, definitely positive in S_t, whose derivative, $\dot{V}(\mathbf{x}, t) = \left(\frac{\partial V}{\partial \mathbf{x}}, \mathbf{f}(\mathbf{x}, t)\right) + \frac{\partial V}{\partial t}$, counted in accordance with the perturbed Eq.(7.44) is a sign-constant negative function in this area.

Definition. The function $V(\mathbf{x}, t)$ has an infinitely small upper boundary at $\mathbf{x} \to 0$, if, for a definite $t_0 > a$, $|V(\mathbf{x}, t)| \to 0$ is satisfied when $\mathbf{x} \to 0$ for t falling in the interval $[t_0, \infty)$.

Lyapounov's second theorem. An unperturbed motion is asymptotically stable if there is a definitely positive function $V(\mathbf{x}, t)$, allowing an infinitely small upper boundary in S_t, whose derivative, counted by virtue of the unperturbed motion Eq.(7.44), proves to be a function definitely negative in S_t.

Lyapounov's third theorem. Let there be a function $V(\mathbf{x}, t)$ for system (7.44), continuously differentiable with respect to $\mathbf{x}$ and t, allowing an infinitely small upper boundary for $\mathbf{x} \to 0$ and possessing a derivative of a fixed sign $\dot{V}(\mathbf{x}, t)$ by virtue of system (7.44). If at some $t_0 > 0$ a point $(\mathbf{x}_0, t_0)$ is found in an arbitrary area $||\mathbf{x}|| < \Delta$, for which the sign of the function $V(\mathbf{x}, t)$ is identical with the sign of its derivative, the trivial solution of the system $(x = 0)$ is unstable at $t \to \infty$.

We shall consider a generalized linear oscillating circuit of the type shown in Fig.6.1, assuming that for $t \geq 0$ its parameters change in accordance with an arbitrary continuous law, yet they remain positive:

$$C(t), G(t), L(t), R(t) > 0$$

The free process is described by the following system of differential equations concerning the charge q of the capacitor and the magnetic flux ϕ of the inductance:

$$\left| \begin{aligned} \frac{dx_1}{dt} &= -\frac{G}{C}x_1 - \frac{r}{L}x_2 \\ \frac{dx_2}{dt} &= \frac{1}{rC}x_1 - \frac{R}{L}x_2 \end{aligned} \right. , \tag{7.45}$$

where $x_1 = \dfrac{q}{q_{00}}$, $x_2 = \dfrac{\phi}{\phi_{00}}$, $r = \dfrac{\phi_{00}}{q_{00}}$; q_{00}, ϕ_{00}, r – constants.

The linear system (7.45) is a particular case of the system (7.44), where $\mathbf{x} = \mathrm{colon}(x_1, x_2)$,

$$\mathbf{f}(\mathbf{x}, t) = \mathbf{A}(t)\mathbf{x}, \quad \mathbf{A}(t) = \begin{pmatrix} -\dfrac{G(t)}{C(t)} & -\dfrac{r}{L(t)} \\ \dfrac{1}{rC(t)} & -\dfrac{R(t)}{L(t)} \end{pmatrix}.$$

Two criteria (sufficient conditions) for oscillating circuit stability will be retained.

Criterion 7.1. The oscillating circuit will be stable according to Lyapounov's definition, if a positive constant r can be selected, for which, given $t \geq 0$, the following inequalities will be satisfied:

$$\left| \begin{aligned} \frac{G}{C} \pm \frac{1}{2}\left(\frac{r}{L} - \frac{1}{rC}\right) \geq 0, \\ \frac{R}{L} \pm \frac{1}{2}\left(\frac{r}{L} - \frac{1}{rC}\right) \geq 0. \end{aligned} \right. \tag{7.46}$$

In order to prove this we select a definitely positive Lyapounov function [93, 260–264] in the form:

$$V(\mathbf{x}) = \frac{1}{2}(\mathbf{x}, \mathbf{x}) = \frac{1}{2}(x_1^2 + x_2^2). \tag{7.47}$$

Its derivative in compliance with system (7.46) is

$$\dot{V}(\mathbf{x}) = x_1\dot{x}_1 + x_2\dot{x}_2 = -\frac{G}{C}x_1^2 - \frac{R}{L}x_2^2 - \left(\frac{r}{L} - \frac{1}{rC}\right)x_1x_2.$$

The last term can be estimated by employing the inequality:

$$-\frac{x_1^2+x_2^2}{2} \le x_1x_2 \le \frac{x_1^2+x_2^2}{2}.$$

Therefore, $\dot{V}(\mathbf{x})$ is situated between two functions conditioned by the different signs of the expression

$$-\left[\frac{G}{C} \pm \frac{1}{2}\left(\frac{r}{L}-\frac{1}{rC}\right)\right]x_1^2 - \left[\frac{R}{L} \pm \left(\frac{r}{L}-\frac{1}{rC}\right)\right]x_2^2.$$

We arrive at the conclusion that if inequalities (7.46) are satisfied, the derivative of Lyapounov's function is non-positive, which, in this case, is the condition of oscillating circuit stability according to Lyapounov's definition.

Given a satisfaction of the strict inequalities (7.46), i.e.

$$\left|\begin{array}{l} \dfrac{G}{C} \pm \dfrac{1}{2}\left(\dfrac{r}{L}-\dfrac{1}{rC}\right) > 0 \\ \dfrac{R}{L} \pm \dfrac{1}{2}\left(\dfrac{r}{L}-\dfrac{1}{rC}\right) > 0, \end{array}\right. \tag{7.48}$$

Lyapounov's conditions for asymptotic stability are met.

Consequence: The oscillating circuit with positive parameters, where $G(t)$ and $R(t)$ are arbitrary time functions, while C and L are constant, is asymptotically stable according to Lyapounov's definition. Indeed, in this case parameter r can be selected so that $r = \sqrt{\frac{L}{C}}$. Then the bracketed expressions in (7.48) will be nullified. The same consequence can be arrived at by using the energy conservation law as a starting point. Then we have to take into account that in the case of constant reactances there is no energy input in the circuit. The energy is continuously dissipated in the active elements of the circuit at varying speed.

Criterion 7.2. The oscillating circuit will be stable, if, for an arbitrary t, the following system of inequalities is satisfied:

$$\left|\begin{array}{l} R - \dfrac{k-1}{2}\dfrac{dL}{dt} - \dfrac{l}{2}\dfrac{L}{C}\dfrac{dC}{dt} \ge 0 \\ G - \dfrac{l-1}{2}\dfrac{dC}{dt} - \dfrac{k}{2}\dfrac{C}{L}\dfrac{dL}{dt} \ge 0 \ , \end{array}\right. \tag{7.49}$$

where k and l are an arbitrary couple of integers taken from the set: $0, \pm 1, \pm 2, ..., \pm\infty$.

In order to prove this, we set $r = 1$ ($q_{00} = 1$ and $\phi_{00} = 1$) in (7.45) and choose a definitely positive Lyapounov function in the form

$$V = L^{k-1}C^l\phi^2 + L^kC^{l-1}q^2. \tag{7.50}$$

Then, provided that inequality (7.49) is satisfied, its total derivative

$$\frac{dV}{dt} = -2L^{k-2}C^{l}\left[R - \frac{k-1}{2}\frac{dL}{dt} - \frac{l}{2}\frac{L}{C}\frac{dC}{dt}\right]\phi^2$$
$$- 2L^{k}C^{l-2}\left[G - \frac{l-1}{2}\frac{dC}{dt} - \frac{k}{2}\frac{C}{L}\frac{dL}{dt}\right]q^2,$$

will be non-positive, i.e. once again Lyapounov's condition concerning the stability of the specific oscillating circuit under consideration is met. Analogously, provided that the strict inequalities (7.49) are satisfied, Lyapounov's criteria of asymptotic stability of the oscillating circuit are met.

Consequence: The oscillating circuit will be stable, if, for $t \geq 0$, the following system of inequalities is satisfied:

$$R + \frac{1}{2}\frac{dL}{dt} \geq 0, \quad G + \frac{1}{2}\frac{dC}{dt} \geq 0.$$

These inequalities are yielded by (7.49) at $k = l = 0$. In this case Lyapounov's function (7.50) acquires a clear physical meaning, since it represents the instantaneous energy stored in the circuit reactances.

Let us now consider the case when $R(t)$, $G(t) > 0$ for $t \geq 0$, and let us assume that the continuously changing reactances of the generalized oscillating circuit (Fig.7.1) can take both positive and negative values. In reality, a similar situation occurs in the case of Josephson superconducting junctions, whose equivalent inductance takes negative values during a part of the changing period. In a more general treatment, this is a system where the effect of a single-frequency non-degenerate parametric regeneration is manifested [189, 171]. This effect has been considered from the viewpoint of the modulation-parametric interactions in section 2.1.

The free process in the generalized oscillating circuit (Fig.6.1), excluding sources i_g and U_g, can be described by the following system of differential equations concerning the voltage U at the capacitor and the current i flowing through the inductance:

$$\left|\begin{aligned} \frac{dx_1}{dt} &= -\frac{1}{C}\left(G + \frac{dC}{dt}\right)x_1 - \frac{1}{rC}x_2 \\ \frac{dx_2}{dt} &= \frac{r}{L}x_1 - \frac{1}{L}\left(R + \frac{dL}{dt}\right)x_2 \quad , \end{aligned}\right. \tag{7.51}$$

where $x_1 = \dfrac{U}{U_{00}}$, $x_2 = \dfrac{i}{i_{00}}$, $r = \dfrac{U_{00}}{i_{00}}$; U_{00}, i_{00}, r are constants.

Lyapounov's function is set in the form

$$V = \operatorname{sign} C\frac{rC}{2}x_1^2 + \operatorname{sign} L\frac{L}{2r}x_2^2, \tag{7.52}$$

where, for example, $\operatorname{sign} C = \left| \begin{array}{l} 1, \text{for } C > 0 \\ 0, \text{for } C = 0\,. \\ 1, \text{for } C < 0 \end{array} \right.$

Hence function (7.52) is positively defined. Its derivative by virtue of (7.51) is

$$\frac{dV}{dt} = -\operatorname{sign} Cr\left(G + \frac{1}{2}\frac{dC}{dt}\right)x_1^2 - \operatorname{sign} L\frac{1}{r}\left(R + \frac{1}{2}\frac{dL}{dt}\right)x_2^2$$
$$- (\operatorname{sign} C - \operatorname{sign} L)x_1 x_2.$$

The bracketed expression in the last term can take the following values: -2, -1, 0, 1, 2. If the extreme values are considered, it becomes obvious that $\frac{dV}{dt}$ is always located between the following two functions:

$$-\operatorname{sign} Cr\left(G + \frac{1}{2}\frac{dC}{dt}\right)x_1^2 - \operatorname{sign} L\frac{1}{r}\left(R + \frac{1}{2}\frac{dL}{dt}\right) \pm 2x_1 x_2. \tag{7.53}$$

Since evidently $-(x_1^2 + x_2^2) \leq \pm 2x_1x_2 \leq (x_1^2 + x_2^2)$, it can be seen that the two functions (7.30) are located between the functions

$$-\left[\operatorname{sign} Cr\left(G + \frac{1}{2}\frac{dC}{dt}\right) \pm 1\right]x_1^2 - \left[\operatorname{sign} L\frac{1}{r}\left(R + \frac{1}{2}\frac{dL}{dt}\right) \pm 1\right]x_2^2. \tag{7.54}$$

Lyapounov's theorems (from Lyapounov's second method) and Exp.(7.54) allow obtaining criteria of the stability or instability of the oscillating circuit in this case.

Stability criterion. The oscillating circuit will be stable, if the following system of non-strict inequalities is satisfied in the interval $[t_0, \infty)$ and given a positive r:

$$\left| \begin{array}{l} \operatorname{sign} Cr\left(G + \frac{1}{2}\frac{dC}{dt}\right) - 1 \geq 0 \\ \operatorname{sign} L\frac{1}{r}\left(R + \frac{1}{2}\frac{dL}{dt}\right) - 1 \geq 0. \end{array} \right. \tag{7.55}$$

The oscillating circuit will be asymptotically stable, if the system of strict inequalities (7.55) is satisfied under the same conditions.

Instability criterion. The oscillating circuit will be unstable, if the following system of strict inequalities is satisfied at $t \to \infty$ and a positive r:

$$\left| \begin{array}{l} \operatorname{sign} Cr\left(G + \frac{1}{2}\frac{dC}{dt}\right) + 1 < 0 \\ \operatorname{sign} L\frac{1}{r}\left(R + \frac{1}{2}\frac{dL}{dt}\right) + 1 < 0. \end{array} \right.$$

It follows from system (7.28) that if the signs of all oscillating circuit parameters are changed to their opposites, the result will be a system of the same type. This implies that if all the parameters of the oscillating circuit change in time according to arbitrary laws, yet in such a way that their signs change to their opposites simultaneously, from the viewpoint of stability the oscillating circuit will be equivalent to another oscillating circuit, whose parameters change in time in compliance with laws equal to the modules of the respective laws governing the changes of the initial oscillating circuit.

7.1.7. Stability of an oscillating circuit with a piece-wise linear volt-coulomb characteristic

In linear approach, the concept of stability coincides with the concept boundedness of all solutions of the respective differential equation with a zero right-hand part and bounded initial conditions. Having lost its stability, the parametric amplifier is converted into a parametric oscillator.

The equation of a series oscillating circuit containing capacitance $C(t_r)$, periodically changing in time, acquires the following form with respect to the charge q

$$\frac{d^2q}{dt_r^2} + \frac{R}{L}\frac{dq}{dt_r} + \frac{1}{LC(t_r)}q = 0, \tag{7.56}$$

where R and L are the constant resistance and the inductance of the oscillating circuit.

We use the following substitution $q_1 = q\exp\left(-\frac{1}{2}\frac{R}{L}t_r\right)$ to reduce Eq.(7.56) to the form:

$$\frac{d^2q_1}{dt_r^2} + \left[\frac{1}{LC(t_r)} - \frac{1}{4}\frac{R^2}{L^2}\right]q_1 = 0, \tag{7.57}$$

which contains no explicit dissipative term.

The capacitance $C(t_r)$ is regarded as a piece-wise linear time function,

$C = C_1$ for $t_r \in (0, t_{r1})$, and $C = C_2$ for $t_r \in (t_{r1}, T_r)$,

where T_r is the capacitance change period. Thus the capacitance changes twice in one period: when $t_r = t_{r1}$, it leaps from C_1 to C_2, and when $t_r = T_r$, from C_2 to C_1.

Equation (7.57) can be written in the following dimensionless form:

$$\frac{d^2y}{dt^2} + a(t)y = 0, \tag{7.58}$$

where $t = \frac{t_r}{t_{00}}$, $t_{00} = \sqrt{L}\sqrt[4]{C_1C_2}$, $a(t) = \alpha^2 = \sqrt{\frac{C_2}{C_1}}$ at $t \in (0, t_1)$, and $a(t) = \frac{1}{\alpha^2} = \sqrt{\frac{C_1}{C_2}}$ at $t \in (t_1, T)$, $t_1 = \frac{t_{r1}}{t_{00}}$, $T = \frac{T_r}{t_{00}}$.

Equation (7.58) has constant coefficients in the interval $(0, t_1)$, and its solution is a cosine curve. The same can be established for the interval (t_1, T). At moment t_1 the amplitude, the frequency and the initial phase of the cosinusoid change in a leap-like way, yet so that the function turns out to be continuous and sufficiently smooth (with continuous first-order derivative). Equation (7.58) is a particular case of Hill's equation. It follows from the latter's theory that it cannot be asymptotically stable, i.e. if $y(t)$ is a solution of the equation, $\lim_{t\to\infty} y(t) \neq 0$.

Therefore, the stability problem boils down to specifying one of the two cases: a) the case of stability, when the above-mentioned bound is a finite number, and b) the case of instability, when the module of the bound is equal to infinity.

The solution of Eq.(7.58) is determined in the following way:

$$\left| \begin{array}{l} y(t) = y_{01} \cos(\alpha t + \varphi_1) \text{ for } t \in (0, t_1) \\ y(t) = y_{02} \cos\left(\dfrac{1}{\alpha} t + \varphi_2\right) \text{ for } t \in (t_1, T) \end{array} \right. , \tag{7.59}$$

y_{01} and φ_1 are determined by the initial conditions, y_{02} and φ_2 are specified by the condition of continuousness of the function $y(t)$ and its derivative at the moment t_1, when the capacitance changes in a leap-like manner,

$$\varphi_2 = \operatorname{arctg}\left[\alpha^2 \operatorname{tg}(\alpha t_1 + \varphi_1) - \frac{1}{\alpha} t_1\right], \quad y_{02} = \alpha^2 y_{01} \frac{\sin(\alpha t_1 + \varphi_1)}{\sin\left(\dfrac{1}{\alpha} t_1 + \varphi_2\right)}.$$

Equation (7.58) can be presented in a matrix form:

$$\frac{d}{dt}\begin{bmatrix} y_1 \\ y_2 \end{bmatrix} = \begin{bmatrix} 0 & 1 \\ -a(t) & 0 \end{bmatrix} \begin{bmatrix} y_1 \\ y_2 \end{bmatrix}, \quad y_1 = y, \quad y_2 = \frac{dy}{dt},$$

or in short

$$\frac{d}{dt}\mathbf{Y} = \mathbf{A}(t)\mathbf{Y}. \tag{7.60}$$

According to Flocker's theory, the fundamental matrix of the solutions satisfies the condition

$$\mathbf{Y}(t+T) = \mathbf{Y}(t)\mathbf{Y}(T), \tag{7.61}$$

where $\mathbf{Y}(T)$ is a constant matrix, called monodromy matrix. This matrix allows of obtaining Lyapounov's constant $a = Sp\mathbf{Y}(T)$, where Sp is the sum of the elements of the principal diagonal of the matrix.

In accordance with Lyapounov's first method, the following three cases are distinguished for Eq.(7.58): 1) $|a| > 2$, 2) $|a| < 2$ and 3) $|a| = 2$. In the first case the equation is unstable, in the second one it is stable, and in the third case it is determined by the bound between the stable and unstable ranges.

Using solution (7.59) it is not difficult to obtain the fundamental solution (7.60), which can serve for constructing the monodromy matrix, whose elements are obtained in the form

$$\begin{aligned} y_{11} &= \cos\alpha t_1 \cos\frac{1}{\alpha}(T-t_1) - \alpha^2 \sin\alpha t_1 \sin\frac{1}{\alpha}(T-t_1), \\ y_{12} &= \frac{1}{\alpha}\sin\alpha t_1 \cos\frac{1}{\alpha}(T-t_1) + \alpha\cos\alpha t_1 \sin\frac{1}{\alpha}(T-t_1), \\ y_{21} &= -\alpha\cos\alpha t_1 \sin\frac{1}{\alpha}(T-t_1) - \frac{1}{\alpha}\sin\alpha t_1 \cos\frac{1}{\alpha}(T-t_1), \\ y_{22} &= -\frac{1}{\alpha^2}\sin\alpha t_1 \sin\frac{1}{\alpha}(T-t_1) + \cos\alpha t_1 \cos\frac{1}{\alpha}(T-t_1). \end{aligned}$$

This helps derive an expression for Lyapounov's constant:

$$\begin{aligned} a = SpY(T) = y_{11} + y_{22} = 2\cos\alpha t_1 \cos\frac{1}{\alpha}(T-t_1) \\ - \left(\alpha^2 + \frac{1}{\alpha^2}\right)\sin\alpha t_1 \sin\frac{1}{\alpha}(T-t_1). \end{aligned} \tag{7.62}$$

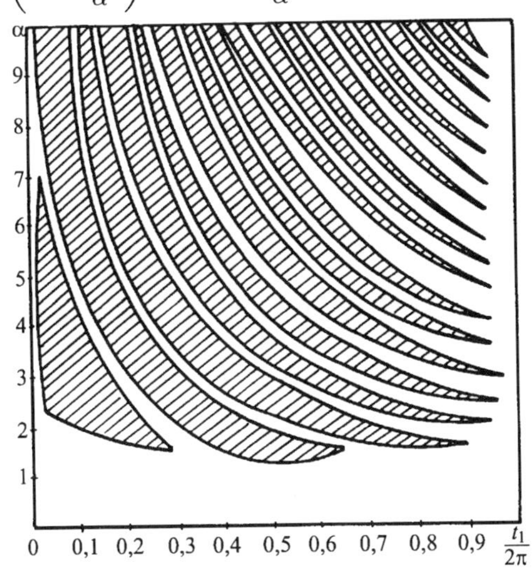

Fig. 7.7. Areas of stability and instability (the latter ones being hatched) of a parametric oscillating circuit with a partially constant 2π-periodic capacitance

This expression allows of identifying which of the above-listed three cases the system refers to in each specific occurrence and in what way the stability problem is to be solved. The areas of stability or instability are determined by three parameters – α, t_1 and T, i.e. in the general case they can be plotted in a three-dimensional space.

Fig.7.7 shows the areas of stability and instability (hatched) on the plane $\left(\alpha, \frac{t_1}{2\pi}\right)$ for the particular case of $T = 2\pi$. Parameter α changes within the range $1 \le \alpha \le 10$. The case for $\alpha < 1$ can be obtained from the case for $\alpha > 1$, given above, by carrying out the following substitution of the parameters: $\beta = \frac{1}{\alpha}$, $\tau_1 = T - t_1$.

7.1.8. Analysis of the free processes in a linear quasi-harmonic oscillating circuit

Object of the analysis in this section is a linear quasi-harmonic oscillating circuit, whose free processes are described by a quasi-harmonic time function, i.e. by a sine function with an amplitude slowly changing in time, and first derivative of the phase by time. The major application area of such oscillating circuits is that of harmonic signal modulators.

Let us consider the general case of a series resonance circuit, where the inductance L, capacitance C and resistance R are time-dependent. The free process in such a circuit is described by the equation

$$\frac{d^2x}{dt^2} + 2\alpha(t)\frac{dx}{dt} + \omega_0^2(t)x = 0, \tag{7.63}$$

where $x = \frac{q}{q_{00}}$, $t = \frac{t_r}{t_{00}}$ – normalized capacitor charge and normalized time respectively,

$$\alpha = \frac{\frac{dL}{dt} + t_{00}R}{2L}, \quad \omega_0(t) = \frac{t_{00}}{\sqrt{LC}}.$$

The general solution of Eq.(7.63) can be presented in the form

$$x = A(t,\tau)\sin\varphi(t,\tau), \tag{7.64}$$

where

$$\left|\begin{aligned} A(t,\tau) &= \frac{1}{\sqrt{\omega_0(t)}}\frac{1}{\sqrt{\omega_0(\tau)}}e^{-\beta(t,\tau)} \\ \beta(t,\tau) &= \int_\tau^t [\alpha + \left(\alpha + \frac{1}{2}\frac{d}{dt}\ln\omega_0\right)\cos 2\varphi(t,\tau)]dt \\ \varphi(t,\tau) &= \int_\tau^t [\omega_0 + \left(\alpha + \frac{1}{2}\frac{d}{dt}\ln\omega_0\right)\sin 2\varphi(t,\tau)]dt \end{aligned}\right. \tag{7.65}$$

Here the constant τ determines an arbitrary initial point of time.

Equation (7.65) given above yields an indicator of the integrability of Eq.(7.63) in quadratures:

$$\alpha(t) + \frac{d}{dt}\ln\omega_0(t) = 0.$$

The first and second equations of (7.65) can produce an indicator of the asymptotic stability of the oscillating circuit –

$$\frac{d\omega_0(t)}{dt} < 0 \text{ for } t > t', \quad t' - \text{ an arbitrary moment.}$$

So far we did not impose any limitations on the laws of time-dependent oscillating circuit parameters.

Let us assume that the oscillating circuit satisfies the quasi-harmonic condition, i.e. we assume that the first co-multiplier in the right-hand part of (7.64) is a slowly changing time-dependent function, while the second one is a quickly oscillating time-dependent function. Under these conditions the approximate extremums of the solution are $\varphi(t_k, \tau) = (2k+1)\frac{\pi}{2}$, $k = 0, \pm 1, \pm 2, \dots$, respectively the zeroes of the solution are $\varphi(t_k^0, \tau) = k\pi$.

The logarithmic damping decrement is determined by using two adjacent extremums of the solution

$$\nu(t_k) = \ln \frac{A(t_{k-1}, \tau)}{A(t_k, \tau)} = \int_{t_{k-1}}^{t_k} (\alpha + \frac{1}{2}\frac{d}{dt}\ln\omega_0)\cos^2\varphi dt$$

The non-negativity of the logarithmic damping decrement for any t_k is an evidence of stability, and its positiveness is an evidence of asymptotic stability of the quasi-harmonic oscillating circuit.

The last expression can yield an analytical indicator of stability of the quasi-harmonic oscillating circuit:

$$\alpha = \frac{1}{2}\frac{d}{dt}\ln\omega_0 \geq 0 \tag{7.66}$$

If we introduce the parameter of instantaneous characteristic resistance of the oscillating circuit $\rho = \sqrt{\frac{L}{C}}$, (7.46) can be used to obtain a simple criterion of asymptotic stability – $\rho = \text{const}$.

We introduce the concept instantaneous quality factor of the oscillating circuit $Q = \frac{\rho}{R}$, and as a result the formula of the logarithmic damping decrement acquires the form

$$\nu(t_k) = \int_{k\pi-\frac{\pi}{2}}^{k\pi+\frac{\pi}{2}} \frac{\cos^2\varphi}{Q + 0.5\sin 2\varphi}(d\varphi + Q d\ln\rho).$$

Let $Q = \text{const}$. Then

$$\nu(t_k) = \int_{k\pi-\frac{\pi}{2}}^{k\pi+\frac{\pi}{2}} \frac{\cos^2\varphi d\varphi}{Q + 0.5\sin 2\varphi} + \int_{\ln\rho(t_{k-1})}^{\ln\rho(t_k)} \frac{Q\cos^2\varphi}{Q + 0.5\sin 2\varphi} d(\ln\rho).$$

The first integral is equal to $\pi(4Q^2-1)^{-\frac{1}{2}}$. Hence, the stability criterion can be written as follows:

$$\int_{\ln\rho(t_{k-1})}^{\ln\rho(t_k)} \frac{Q\cos^2\varphi}{Q+0.5\sin 2\varphi} d(\ln\rho) < \frac{\pi}{\sqrt{4Q^2-1}}.$$

Since $\max\limits_{\varphi} \dfrac{Q\cos^2\varphi}{2Q+\sin 2\varphi} = \dfrac{2Q^2}{4Q^2-1}$, substituting the subintegral expression for its maximum value, we obtain:

$$\frac{4Q^2}{4Q^2-1}\ln\frac{\rho(t_k)}{\rho(t_{k-1})} < \frac{\pi}{\sqrt{4Q^2-1}}.$$

This inequality allows of obtaining the following simple stability expression at $Q = \text{const}$:

$$\ln\frac{\rho_{\max}}{\rho_{\min}} < \frac{\pi}{2Q}.$$

For a more detailed investigation of the free processes in an oscillating circuit with slowly changing parameters, it is desirable to obtain an approximate solution of the last integral equation in (7.65). It is presented in the following way:

$$\varphi = \gamma+\varepsilon,\ \gamma = \int_\tau^t \omega_0 dt,\ \varepsilon = \int_\tau^t \left(\alpha + \frac{1}{2}\frac{d}{dt}\ln\omega_0\right)\sin 2\varphi dt.$$

The inequality $\gamma \gg \varepsilon$ is valid for many cases of practical importance, since the subintegral function of the first integral has a constant sign, while that of the second one is quickly oscillating. After differentiating the last equation in (7.65), we obtain:

$$\frac{d\varepsilon}{dt} = \left(\alpha + \frac{1}{2}\frac{d}{dt}\ln\omega_0\right)(\sin 2\gamma\cos 2\varepsilon\alpha\cos 2\gamma\sin 2\varepsilon).$$

In linear approximation ($\cos 2\varepsilon \approx 1$, $\sin 2\varepsilon \approx 2\varepsilon$), this equation is solved in quadratures:

$$\varepsilon = [2\int_\tau^t \mu\cos 2\gamma dt]\int_\tau^t \mu\sin 2\gamma\exp[-2\int_\tau^t \mu\cos 2\gamma]dt,$$

$\mu = \alpha + \dfrac{1}{2}\ln\omega_0$.

After obtaining ε, Eq.(7.65) can be computed in quadratures and in this way the problem is completely solved.

7.2. Stationary regime in linear radiophysical systems with periodic and almost periodic parameters

It is assumed here that the investigated system is asymptotically stable, hence the stationary regime coincides with the forced oscillations in the system [256].

Let us consider an oscillating circuit with parallel connection of capacitance C, conductance G, inductance L and an ideal source of harmonic current $i = I\cos(\omega t + \varphi)$. We assume that the circuit parameters change in time according to the law

$$P = P_0[1 + m_p\cos(\Omega t + \varphi_p)], \quad P_0 > 0, \tag{7.67}$$

where P is a generalized designation standing for C, G or $\Gamma = \dfrac{1}{L}$.

The circuit equation is of the following form:

$$\frac{d}{dt}\{C_0[1 + m_C\cos(\Omega t + \varphi_C)]U\} + G_0[1 + m_G\cos(\Omega t + \varphi_C)]U$$

$$+\Gamma_0[1 + m_\Gamma\cos(\Omega t + \varphi_\Gamma)]\int_{-\infty}^{t} U\,dt = I\cos(\omega t + \varphi). \tag{7.68}$$

Applying the generalized complex amplitudes method, we write down the determining equation of the system and then proceed with the following infinite system of linear algebraic equations:

$$\begin{array}{|c|c|c|c|c|}
\hline
\dot y_{-2,0} & \dot y_{-2,1} & & & \\
\hline
\dot y_{-1,-1} & \dot y_{-1,0} & \dot y_{-1,1} & & \\
\hline
 & \dot y_{0,-1} & \dot y_{0,0} & \dot y_{0,1} & \\
\hline
 & & \dot y_{1,-1} & \dot y_{1,0} & \dot y_{1,1} \\
\hline
 & & & \dot y_{2,-1} & \dot y_{2,0} \\
\hline
\end{array}
\cdot
\begin{array}{|c|}
\hline
\dot u_{-2} \\ \hline \dot u_{-1} \\ \hline \dot u_0 \\ \hline \dot u_1 \\ \hline \dot u_2 \\ \hline
\end{array}
=
\begin{array}{|c|}
\hline
 \\ \hline \\ \hline \dot I \\ \hline \\ \hline \\ \hline
\end{array}, \tag{7.69}$$

$$\dot y_{k,0} = G + j\left[(\omega + k\Omega)C_0 - \frac{\Gamma_0}{\omega + k\Omega}\right],$$

$$\dot y_{k,-1} = \frac{\overset{*}{m}_G}{2}G_0 + j\left[\frac{\overset{*}{m}_C}{2}(\omega + k\Omega)C_0 - \frac{\overset{*}{m}_\Gamma}{2}\frac{\Gamma_0}{\omega + (k+1)\Omega}\right],$$

$$\dot y_{k,1} = \frac{\dot m_G}{2}G_0 + j\left[\frac{\dot m_C}{2}(\omega + k\Omega)C_0 - \frac{\dot m_\Gamma}{2}\frac{\Gamma_0}{\omega + (k-1)\Omega}\right],$$

$\dot{m}_p = m_p e^{j\varphi_p}$, $\overset{*}{m}_p = m_p e^{-j\varphi_p}$, $\dot{I} = Ie^{j\varphi}$

The matrix equation (7.69) is essentially a generalized version of Ohm's law,

$$\dot{\mathbf{Y}}\dot{\mathbf{U}} = \dot{\mathbf{I}}. \tag{7.70}$$

This matrix equation can be presented in an interesting expanded form:

$$[\dot{\mathbf{G}} + j(\omega\dot{\mathbf{C}} - \dot{\mathbf{\Gamma}}\omega^{-1})]\dot{\mathbf{U}} = \dot{\mathbf{I}},$$

where $\omega = \mathrm{diag}(\ldots, \omega - k\Omega, \ldots, \omega - \Omega, \omega, \omega + \Omega, \ldots, \omega + k\Omega, \ldots)$; $\dot{\mathbf{G}}$, $\dot{\mathbf{C}}$ and $\dot{\mathbf{\Gamma}}$ are infinite three-diagonal matrices of the type of (5.8).

A consistent solution of system (7.70) shows that the vector we are looking for can be expressed by continued fractions in the following way:

$$\dot{U}_0 = \frac{\dot{I}}{\dot{Y}_{0,0}^{(\infty)}}, \quad \dot{U}_1 = -\dot{U}_0 \frac{\dot{Y}_{1,-1}}{\dot{Y}_{1,0}^{(\infty)}}, \quad \dot{U}_{-1} = -\dot{U}_0 \frac{\dot{Y}_{-1,1}}{\dot{Y}_{-1,0}^{(\infty)}},$$

$$\dot{U}_2 = -\dot{U}_1 \frac{\dot{Y}_{2,-1}}{\dot{Y}_{2,0}^{(\infty)}}, \quad \dot{U}_{-2} = -\dot{U}_{-1} \frac{\dot{Y}_{-2,1}}{\dot{Y}_{-2,0}^{(\infty)}},$$

$$\ldots\ldots\ldots\ldots\ldots\ldots\ldots\ldots\ldots\ldots,$$

$$\dot{U}_k = -\dot{U}_{k,-1} \frac{\dot{Y}_{k,-1}}{\dot{Y}_{k,0}^{(\infty)}}, \quad \dot{U}_{-k} = -\dot{U}_{-k+1} \frac{\dot{Y}_{-k,1}}{\dot{Y}_{-k,0}^{(\infty)}},$$

$$\ldots\ldots\ldots\ldots\ldots\ldots\ldots\ldots\ldots\ldots,$$

$$k = 0, 1, 2, \ldots, \infty,$$

$$\dot{Y}_{0,0}^{(\infty)} = \dot{Y}_{0,0}(1 - \dot{K}_1 - \dot{K}_{-1}), \quad \dot{Y}_{1,0}^{(\infty)} = \dot{Y}_{1,0}(1 - \dot{K}_2), \quad \dot{Y}_{-1,0}^{(\infty)} = \dot{Y}_{-1,0}(1 - \dot{K}_{-2}),$$

$$\dot{Y}_{2,0}^{(\infty)} = \dot{Y}_{2,0}(1 - \dot{K}_3), \quad \dot{Y}_{-2,0}^{(\infty)} = \dot{Y}_{-2,0}(1 - \dot{K}_{-3}), \ldots\ldots\ldots,$$

$$\dot{Y}_{k,0}^{(\infty)} = \dot{Y}_{k,0}(1 - \dot{K}_{k+1}), \quad \dot{Y}_{-k,0}^{(\infty)} = \dot{Y}_{-k,0}(1 - \dot{K}_{-k-1}), \ldots\ldots\ldots,$$

$$\dot{K}_1 = \cfrac{\dot{C}_1}{1 - \cfrac{\dot{C}_2}{1 - \ldots}}, \quad \dot{K}_{-1} = \cfrac{\dot{C}_{-1}}{1 - \cfrac{\dot{C}_2}{1 - \ldots}}, \quad \dot{K}_2 = \cfrac{\dot{C}_2}{1 - \cfrac{\dot{C}_3}{1 - \ldots}},$$

$$\dot{K}_{-2} = \cfrac{\dot{C}_{-2}}{1 - \cfrac{\dot{C}_{-3}}{1 - \ldots}}, \ldots\ldots\ldots\ldots, \dot{K}_k = \cfrac{\dot{C}_k}{1 - \cfrac{\dot{C}_{k+1}}{1 - \ldots}},$$

$$\dot{K}_{-k} = \cfrac{\dot{C}_{-k}}{1 - \cfrac{\dot{C}_{-k-1}}{1 - \ldots}}, \ldots\ldots\ldots\ldots\ldots\ldots,$$

$$\dot{C}_k = \frac{\dot{Y}_{k,-1}}{\dot{Y}_{k,0}} \frac{\dot{Y}_{k-1,1}}{\dot{Y}_{k-1,0}}, \quad \dot{C}_{-k} = \frac{\dot{Y}_{-k,1}}{\dot{Y}_{-k,0}} \frac{\dot{Y}_{-k+1,-1}}{\dot{Y}_{-k+1,0}}.$$

The difficult part is to calculate the continued fractions with complex elements. The analysis shows that the following regularities are valid:

$$\dot{K}_{\pm 1} = \frac{\dot{C}_{\pm 1}}{1 - \dot{K}_{\pm 2}} = \frac{\dot{C}_{\pm 1}}{1 - \dfrac{\dot{C}_{\pm 2}}{1 - \dot{K}_{\pm 3}}} = \frac{\dot{C}_{\pm 1}}{1 - \dfrac{\dot{C}_{\pm 2}}{1 - \dfrac{\dot{C}_{\pm 3}}{1 - \dot{K}_{\pm 4}}}} = \ldots \tag{7.71}$$

These continued fractions are of a limiting periodic nature, since they satisfy the bound equations $\lim_{k \to \infty} \dot{C}_k = \lim_{k \to \infty} \dot{C}_{-k} = \dfrac{m_C^2}{4}$.

The continued $\dot{K}_k$ and $\dot{K}_{-k}$ can be computed with an arbitrarily high accuracy for large values of k, likewise

$$\dot{K}_k \approx \frac{\dfrac{m_C^2}{4}}{1 - \dot{K}_k} \approx \dot{K}_{-k}, \tag{7.72}$$

or

$$\dot{K}_k^2 - \dot{K}_k + \frac{m_C^2}{4} \approx 0, \quad \dot{K}_{\pm k} \approx \frac{1}{2}\left(1 \pm \sqrt{1 - m_C^2}\right).$$

Since for $m_G = m_C = m_r = 0$ all $\dot{K}_k$ and $\dot{K}_{-k}$ should be equal to zero, from a physical point of view it would be meaningful to have a minus sign in front of the square root in the previous equation. For large quantities of k, the following approximate equations will be valid

$$\dot{K}_{\pm k} \approx \dot{K}_{\pm(k+1)} \approx \dot{K}_{\pm(k+2)} \approx \ldots \approx \frac{1}{2}\left(1 - \sqrt{1 - m_C^2}\right) \tag{7.73}$$

In this way a forced solution of Eq.(7.69) is obtained. The speed of series convergence is high, because at $m_p < 1$ the coefficients are ideally smooth. Hence, when particular practical problems are to be solved, it is not necessary to calculate a large number of harmonic components.

Let us consider another method of solving the infinite system (7.69). Equation (7.68) is written in a dimensionless form:

$$\frac{d}{dt}\{C_0[1 + m_C \cos(\tau + \varphi_C)]\varkappa\} + g_0[1 + m_g \cos(\tau + \varphi_g)]\varkappa$$
$$+ \gamma_0[1 + m_\Gamma \cos(\tau + \varphi_\Gamma)] \int_{-\infty}^{\tau} \varkappa d\tau = \chi \cos(\omega' \tau + \varphi), \tag{7.74}$$

where $\tau = \dfrac{t}{t_{00}}$, $\chi = \dfrac{\Gamma}{i_{00}}$, $\varkappa = \dfrac{U}{U_{00}}$, $C_0 = C\dfrac{U_{00}}{t_{00} i_{00}}$, $g_0 = G_0 \dfrac{U_{00}}{i_{00}}$, $\gamma_0 = \Gamma_0 \dfrac{t_{00} U_{00}}{i_{00}}$,

$\omega' = \omega t_{00} = \dfrac{\omega}{\Omega}$.

Substituting in (7.74) a solution of the equation in the form

$$\hat{\varkappa} = \sum_{k=-\infty}^{\infty} \dot{K}_k e^{j(\omega'+k)\tau},$$

we obtain an infinite system of algebraic equations:

$$\begin{pmatrix} 1 & \theta_{-2,-1} & & & \\ \theta_{-1,-2} & 1 & \theta_{-1,0} & & \\ & \theta_{0,-1} & \boxed{1} & \theta_{0,1} & \\ & & \theta_{1,0} & 1 & \theta_{1,2} \\ & & & \theta_{2,1} & 1 \end{pmatrix} \begin{pmatrix} \vdots \\ \vdots \\ \boxed{\dot{K}_0} \\ \vdots \\ \vdots \end{pmatrix} = \begin{pmatrix} \vdots \\ \vdots \\ \boxed{\dot{\eta}} \\ \vdots \\ \vdots \end{pmatrix}, \qquad (7.75)$$

or $\mathbf{\Theta K} = \mathbf{H}$,

$$\Theta_{k,k-1} = \frac{1}{2}\frac{\dot{m}_g g_0 + j[(\omega'+k)\dot{m}_C C_0 - \dot{m}_\Gamma \gamma_0(\omega'+k-1)^{-1}]}{g_0 + j[(\omega'+k)C_0 - \gamma(\omega'+k)^{-1}]},$$

$$\Theta_{k,k+1} = \frac{1}{2}\frac{\overset{*}{m}_g g_0 + j[(\omega'+k)\overset{*}{m}_C C_0 - \overset{*}{m}_\Gamma \gamma_0(\omega'+k+1)^{-1}]}{g_0 + j[(\omega'+k)C_0 - \gamma(\omega'+k)^{-1}]},$$

$$k = -\infty, \ldots, -1, 0, 1, \ldots, \infty, \qquad \dot{\eta} = \frac{\chi e^{j\varphi}}{g_0 + j\left(\omega' C_0 + \dfrac{\gamma_C}{\omega'}\right)}.$$

The infinite system has one main diagonal. The aim is to reduce it to a normal form, i.e. to a semi-infinite system with a single principal diagonal. To this end we transform the denotations $\dot{K}_0 = x_1$, $\dot{K}_{-k} = x_{2k}$, $\dot{K} = x_{2k+1}$ and change the order of the equations respectively. The result is a semi-infinite system $\mathbf{Ax} = \mathbf{b}$,

$$\begin{pmatrix} 1 & a_{12} & a_{13} & & & & \\ a_{21} & 1 & & a_{24} & & & \\ a_{31} & & 1 & & a_{35} & & \\ & a_{42} & & 1 & & a_{46} & \\ & & a_{53} & & 1 & & a_{57} \\ & & & a_{64} & & 1 & \\ & & & & a_{75} & & 1 \end{pmatrix} \begin{pmatrix} x_1 \\ x_2 \\ x_3 \\ x_4 \\ x_5 \\ x_6 \\ x_7 \end{pmatrix} = \begin{pmatrix} b_1 \\ \\ \\ \\ \\ \\ \\ \end{pmatrix} \qquad (7.76)$$

The elements of the matrix (7.76) are expressed by using the elements of the matrix (7.75) and substituting the indexes:

$$2l \to -l, \quad 2l-1 \to l-1, \quad l = 1,2,3,\ldots$$

For example: $a_{3,6} = \Theta_{1,-3}$, $x_{1,3} = \dot{K}_6$, $b_1 = \dot{\eta}$.

System (7.76) can be solved approximately by applying the following reduction method. The system is limited to a definite number of n equations (n is an odd number). The following solution is obtained: $\mathbf{x}_n = \mathbf{A}_n^{-1}\mathbf{b}_n$. If the approximation is insufficient, the number of the equations is increased to $n+m$ (m is an even number). Obviously

$$\mathbf{A}_{n+m} = \begin{bmatrix} \mathbf{A}_n & \mathbf{U} \\ \mathbf{V} & \mathbf{F}_m \end{bmatrix},$$

where $\mathbf{U}$, $\mathbf{V}$, $\mathbf{F}_m$ are determined by the structure of the matrix in (7.76). Then the following is obtained:

$$\mathbf{A}_{n+m}^{-1} = \begin{bmatrix} \mathbf{A}_n^{-1} - \mathbf{A}_n^{-1}\mathbf{U}(\mathbf{V}\mathbf{A}_n^{-1}\mathbf{U} - \mathbf{F}_m)^{-1}\mathbf{A}_n & (\mathbf{U}\mathbf{F}_m\mathbf{V} - \mathbf{A}_n)^{-1}\mathbf{U}\mathbf{F}_m^{-1} \\ (\mathbf{V}\mathbf{A}_n^{-1}\mathbf{U} - \mathbf{F}_m)^{-1}\mathbf{V}^{-1} & \mathbf{F}_m^{-1} - \mathbf{F}_m^{-1}\mathbf{V}(\mathbf{U}\mathbf{F}_m^{-1}\mathbf{V} - \mathbf{A}_n)^{-1}\mathbf{U}\mathbf{F}_m^{-1} \end{bmatrix}.$$

Let us consider a series oscillating RLC-circuit driven by a harmonic e.m.f. $\varepsilon = \mathcal{E}\cos(\omega t + \gamma)$ in the situation when its parameters change according to an almost periodic law

$$P = P_0[1 + m_{P1}\cos(\Omega_1 t + \varphi_{P1}) + m_{P2}\cos(\Omega_2 t + \varphi_{P2})]. \tag{7.77}$$

The generalized designation P should be substituted for the respective denotations of the circuit elements R, L, $S = \dfrac{1}{C}$.

The equation of the series circuit is

$$L\frac{di}{dt} + \left(R + \frac{dL}{dt}\right)i + S\int i dt = \mathcal{E}\cos(\omega t + \gamma), \tag{7.78}$$

and the current in a stationary regime is of the form:

$$\hat{i} = \sum_{k=-\infty}^{\infty}\sum_{l=-\infty}^{\infty} \dot{I}_{k,l}e^{j(\omega + k\Omega_1 + l\Omega_2)t}. \tag{7.79}$$

By transforming Eq.(7.78) according to the generalized complex amplitudes method (Section 5.1) and substituting (7.79) in it, the following is obtained

$$\sum_{k=-\infty}^{\infty}\sum_{l=-\infty}^{\infty}\Bigg\{R_0\left(\dot{I}_{k,l}+\frac{\dot{m}_1\dot{I}_{k-1,l}+\overset{*}{m}_1\dot{I}_{k+1,l}}{2}+\frac{\dot{m}_2\dot{I}_{k,l-1}+\overset{*}{m}_2\dot{I}_{k,l+1}}{2}\right)$$
$$+j(\omega+k\Omega_1+l\Omega_2)L_0\Bigg[\dot{I}_{k,l}+\frac{\dot{m}_{L_1}}{2}\dot{I}_{k-1,l}\left(1-\frac{\Omega_1}{\omega+k\Omega_1+l\Omega_2}\right)$$
$$+\frac{\overset{*}{m}_{L_1}}{2}\dot{I}_{k+1,l}\left(1+\frac{\Omega_1}{\omega+k\Omega_1+l\Omega_2}\right)+\frac{\overset{*}{m}_{L_2}}{2}I_{k,l-1}\left(1-\frac{\Omega_2}{\omega+k\Omega_1+l\Omega_2}\right)$$
$$+\frac{\overset{*}{m}_{L_2}}{2}\dot{I}_{k,l+1}\left(1+\frac{\Omega_2}{\omega+k\Omega_1+l\Omega_2}\right)-\frac{jS_0}{\omega+k\Omega_1+l\Omega_2}\Bigg[\dot{I}_{k,l}$$
$$+\frac{\dot{m}_{S_1}}{2}\dot{I}_{k-1,l}\left(1+\frac{\Omega_1}{\omega+(k-1)\Omega_1+l\Omega_2}\right)+\frac{\overset{*}{m}_{S_1}}{2}\dot{I}_{k+1,l}\left(1-\frac{\Omega_1}{\omega+(k+1)\Omega_1+l\Omega_2}\right)$$
$$+\frac{\dot{m}_{S_2}}{2}\dot{I}_{k,l-1}\left(1+\frac{\Omega_2}{\omega+k\Omega_1+(l-1)\Omega_2}\right)+\frac{\overset{*}{m}_{S_2}}{2}\dot{I}_{k,l+1}\left(1-\frac{\Omega_2}{\omega+k\Omega_1+(l+1)\Omega_2}\right)\Bigg]\Bigg]\Bigg\}$$
$$\times e^{j(\omega+k\Omega_1+l\Omega_2)t}=\dot{\mathcal{E}}e^{j\omega t}, \tag{7.80}$$

where

$$\dot{m}_1=m_1e^{j\varphi_1},\quad \dot{m}_2=m_2e^{j\varphi_2},$$

$$m_1=\sqrt{m_{R_1}^2+m_{L_1}^2\left(\frac{\Omega_1L_0}{R_0}\right)^2+2m_{R_1}m_{L_1}\frac{\Omega_1L_0}{R_0}\sin(\varphi_{R_1}-\varphi_{L_1})},$$

$$m_2=\sqrt{m_{R_2}^2+m_{L_2}^2\left(\frac{\Omega_2L_0}{R_0}\right)^2+2m_{R_2}m_{L_2}\frac{\Omega_2L_0}{R_0}\sin(\varphi_{R_2}-\varphi_{L_2})},$$

$$\varphi_1=\operatorname{arctg}\frac{m_{R_1}\sin\varphi_{R_1}+m_{L_1}\dfrac{\Omega_1L_0}{R_0}\cos\varphi_{L_1}}{m_{R_1}\cos\varphi_{R_1}-m_{L_1}\dfrac{\Omega_1L_0}{R_0}\sin\varphi_{L_1}},$$

$$\varphi_2=\operatorname{arctg}\frac{m_{R_2}\sin\varphi_{R_2}+m_{L_2}\dfrac{\Omega_2L_0}{R_0}\cos\varphi_{L_2}}{m_{R_2}\cos\varphi_{R_2}-m_{L_2}\dfrac{\Omega_2L_0}{R_0}\sin\varphi_{L_2}},$$

$$\hat{R}+\frac{d\hat{L}}{dt}=R_0(1+\dot{m}_1e^{j\Omega_1t}+\dot{m}_2e^{j\Omega_2t}).$$

Terms with equal frequencies are selected and an infinite system of equations is obtained which can be presented in a matrix form with two-index multiplication:

$\dot{\mathbf{Z}} \bullet \bullet \dot{\mathbf{I}} = \dot{\mathcal{E}}$, $\dot{\mathbf{Z}} = \{\dot{Z}_{k,l,\varepsilon,\mu}\}$ – a four-dimensional infinite matrix, $\dot{\mathbf{I}} = \{\dot{I}_{\varepsilon,\mu}\}$, $\dot{\mathcal{E}} = \{\dot{\mathcal{E}}_{k,l}\}$ – two-dimensional infinite matrices,

$$\dot{Z}_{k,l,k,l} = R_0 + j\left[(\omega + k\Omega_1 + l\Omega_2)L_0 - \frac{S_0}{\omega + k\Omega_1 + l\Omega_2}\right],$$

$$\dot{Z}_{k,l,k-1,l} = \frac{1}{2}\left\{R_0\dot{m}_1 + j[\omega + (k-1)\Omega_1 + l\Omega_2]L_0\dot{m}_1 - \frac{jS_0\dot{m}_{S_1}}{\omega + (k-1)\Omega_1 + l\Omega_2}\right\},$$

$$\dot{Z}_{k,l,k+1,l} = \frac{1}{2}\left\{R_0\overset{*}{m}_1 + j[\omega + (k+1)\Omega_1 + l\Omega_2]L_0\overset{*}{m}_{L_1} - \frac{jS_0\overset{*}{m}_{S_1}}{\omega + (k+1)\Omega_1 + l\Omega_2}\right\},$$

$$\dot{Z}_{k,l,k,l-1} = \frac{1}{2}\left\{R_0\dot{m}_2 + j[\omega + k\Omega_1 + (l-1)\Omega_2]L_0\dot{m}_{L_2} - \frac{jS_0\dot{m}_{S_2}}{\omega + k\Omega_1 + (l-1)\Omega_2}\right\},$$

$$\dot{Z}_{k,l,k,l+1} = \frac{1}{2}\left\{R_0\overset{*}{m}_2 + j[\omega + k\Omega_1 + (l+1)\Omega_2]L_0\overset{*}{m}_{L_2} - \frac{jS_0\overset{*}{m}_{S_2}}{\omega + k\Omega_1 + (l+1)\Omega_2}\right\}.$$

The outlined method of analyzing the stationary process when the oscillating circuit is asymptotically stable can also be used for analyzing the free process in an oscillating circuit, when the latter occurs either in a stable range or in one of the unstable ones. In both cases, the free process can be respectively presented as a sum of „exponentially damping sinusoids", or as „a sum of exponentially increasing sinusoids", where the exponential terms contain complex frequencies.

Consider the free oscillations in a circuit described by the following equation concerning the capacitor charge

$$L\frac{d^2q}{dt^2} + \left(R + \frac{dL}{dt}\right)\frac{dq}{dt} + Dq = 0, \tag{7.81}$$

where $D = \dfrac{1}{C}$, L, R change according to the law (7.67).

We are seeking a solution of Eq.(7.81) of the form

$$\hat{q} = \sum_{k=-\infty}^{\infty} \dot{q}_k e^{j(\dot{\omega}_C + k\Omega)t} = e^{j\dot{\omega}_C t}\sum_{k=-\infty}^{\infty} \dot{q}_k e^{jk\Omega t}, \tag{7.82}$$

where $\dot{\omega}_C = \omega_C + j\alpha$ is a complex frequency of the free oscillations.

It is noteworthy that when the complex amplitudes method is used for analyzing the free processes, its meaning changes slightly. While in the analysis of the forced oscillations the differential equation is substituted for a non-equivalent one by adding an imaginary component, constructed according to definite rules, to the right-hand part, in the analysis of the free processes this is attained by adding an imaginary component to the initial conditions and by superposing the initial conditions.

Equation (7.81) is transferred to the complex plane and solution (7.82) is substituted in it to obtain the following determining equation

$$-L_0\left(1+\frac{\dot m_L}{2}e^{j\Omega t}+\frac{\overset{*}{m}_L}{2}e^{-j\Omega t}\right)\sum_{k=-\infty}^{\infty}(\dot\omega_C+k\Omega)^2\dot q_k e^{j(\dot\omega_C+k\Omega)t}$$

$$+R_0\left(1+\frac{\dot m_\Gamma}{2}e^{j\Omega t}+\frac{\overset{*}{m}_\Gamma}{2}e^{-j\Omega t}\right)\sum_{k=-\infty}^{\infty}j(\dot\omega_C+k\Omega)\dot q_k e^{j(\dot\omega_C+k\Omega)t}$$

$$+D_0\left(1+\frac{\dot m_D}{2}e^{j\Omega t}+\frac{\overset{*}{m}_D}{2}e^{-j\Omega t}\right)\sum_{k=-\infty}^{\infty}\dot q_k e^{j(\dot\omega_C+k\Omega)t}=0, \tag{7.83}$$

where $\dot m_\Gamma=\dot m_R+\frac{j\Omega L_0}{R_0}\dot m_L$, $\overset{*}{m}_\Gamma=\overset{*}{m}_R-\frac{j\Omega L_0}{R_0}\overset{*}{m}_L$.

Equation (7.83) is used for deriving an infinite system of algebraic equations

$$\begin{array}{|c|c|c|c|c|c|c|}\hline \dot a_{-2,-3} & \dot a_{-2,-2} & \dot a_{-2,-1} & & & & \\ \hline & \dot a_{-1,-2} & \dot a_{-1,-1} & \dot a_{-1,0} & & & \\ \hline & & \dot a_{0,-1} & \boxed{\dot a_{0,0}} & \dot a_{0,1} & & \\ \hline & & & \dot a_{1,0} & \dot a_{1,1} & \dot a_{1,2} & \\ \hline & & & & \dot a_{2,1} & \dot a_{2,2} & \dot a_{2,3} \\ \hline \end{array}\cdot\begin{array}{|c|}\hline \dot q_{-3}\\ \hline \dot q_{-2}\\ \hline \dot q_{-1}\\ \hline \boxed{\dot q_0}\\ \hline \dot q_1\\ \hline \dot q_2\\ \hline \dot q_3\\ \hline \end{array}=0, \tag{7.84}$$

where

$$\dot a_{k,k}=-(\dot\omega_C+k\Omega)^2L_0+j(\dot\omega_C+k\Omega)R_0+D_0,$$

$$\dot a_{k,k+1}=-[\dot\omega_C+(k+1)\Omega]^2\frac{\overset{*}{m}_L}{2}L_0+j[\dot\omega_C+(k+1)\Omega]\frac{\overset{*}{m}_\Gamma}{2}R_0+\frac{\overset{*}{m}_D}{2}D_0,$$

$$\dot a_{k,k-1}=-[\dot\omega_C+(k-1)\Omega]^2\frac{\dot m_L}{2}L_0+j[\dot\omega_C+(k-1)\Omega]\frac{\dot m_\Gamma}{2}R_0+\frac{\dot m_D}{2}D_0,$$

$$k=-\infty,\dots,-1,0,1,\dots,\infty.$$

System (7.84) has a non-trivial solution when its determinant Δ is zero,

$$\Delta=0. \tag{7.85}$$

Equation (7.85) allows determining the complex frequencies $\dot\omega_C+k\Omega$, which yield the imaginary part α. If $\alpha>0$, the circuit is asymptotically stable, and vice-versa: if $\alpha<0$, the circuit is unstable.

If the method of Gauss is used for computing determinants through a reduction to a triangular form, Eq.(7.85) can be written in the form:

$$\dot{a}_{00}(1 - \dot{K}_{+0} - \dot{K}_{-0}) \prod_{\substack{n=-\infty \\ n\neq 0}}^{\infty} (1 - \dot{K}_n)\dot{a}_{nn} = 0, \tag{7.86}$$

where $\dot{K}_{+0}$, $\dot{K}_{-0}$, $\dot{K}_n$ are continued fractions:

$$\dot{K}_k = \frac{\dot{C}_k}{1 - \frac{\dot{C}_{k+1}}{1- .}}, \qquad \dot{K}_{-k} = \frac{\dot{C}_{-k}}{1 - \frac{\dot{C}_{-k-1}}{1-...}},$$

$$\dot{C}_k = \frac{\dot{a}_{k,k-1}}{\dot{a}_{k,k}} \frac{\dot{a}_{k+1,k}}{\dot{a}_{k+1,k+1}}, \qquad \dot{C}_{-k} = \frac{\dot{a}_{-k,-k+1}}{\dot{a}_{-k,-k}} \frac{\dot{a}_{-k-1,-k}}{\dot{a}}.$$

Given a sufficiently large k, one can obtain $\dot{K}_k = \frac{1}{2}(1 - \sqrt{1 - m_L^2})$ in a way analogous to that in (7.71) - (7.73).

It follows from (7.86) that the determinant Δ is equal to zero, when one of the following equations is satisfied:

$$1 - \dot{K}_{+0} - \dot{K}_{-0} = 0 \quad , \quad 1 - \dot{K}_n = 0. \tag{7.87}$$

The infinite set of Eqs.(7.87) contains information on the infinite set of frequencies of the free oscillations $\dot{\omega} + k\Omega$, whose imaginary part can be used for judging the stability of the circuit.

The approach presented above is applicable to analyses of differential equations with changing coefficients of the n-th order. In reality one usually writes n equations of the first order, which are sometimes difficult to transform into one equation of the n-th order.

We shall show that it is possible to avoid the transformation of the vector differential equation of the first order with an $n \times n$- dimensional matrix into a differential equation of the n-th order.

Let the oscillating circuit with periodic parameters in the complex plane be described by the following differential equations:

$$\frac{d}{dt}\hat{\mathbf{x}} = \hat{\mathbf{A}}(t) \circ \hat{\mathbf{x}} + \hat{\mathbf{f}}(t), \tag{7.88}$$

where $\hat{\mathbf{f}}(t) = \dot{\mathbf{F}}e^{j\omega t}$, $\dot{\mathbf{F}} = \text{colon}(\dot{F}_1, \dot{F}_2)$.

If the solution is presented in the form

$$\hat{\mathbf{x}} = \sum_{k=-\infty}^{\infty} \dot{\mathbf{x}}_k e^{j(\omega + k\Omega)t}$$

and substituted in (7.88), the following determining equation is obtained:

$$\sum_{k=-\infty}^{\infty}\left[j(\omega+k\Omega)\dot{\mathbf{x}}_k-\sum_{n=-\infty}^{\infty}\dot{\mathbf{A}}_n\dot{\mathbf{x}}_{k-n}\right]e^{jk\Omega}=\dot{\mathbf{F}}. \tag{7.89}$$

Equation (7.89) determines an infinite system of linear algebraic equations

$$\begin{bmatrix} j(\omega-\Omega)\mathbf{I}-\dot{\mathbf{A}}_0 & -\dot{\mathbf{A}}_{-1} & -\dot{\mathbf{A}}_{-2} \\ -\dot{\mathbf{A}}_1 & j\omega\mathbf{I}-\dot{\mathbf{A}}_0 & -\dot{\mathbf{A}}_{-1} \\ -\dot{\mathbf{A}}_2 & \dot{\mathbf{A}}_1 & j(\omega+\Omega)\mathbf{I}-\dot{\mathbf{A}}_0 \end{bmatrix} \begin{bmatrix} \dot{\mathbf{X}}_{-1} \\ \dot{\mathbf{X}}_0 \\ \dot{\mathbf{X}}_1 \end{bmatrix} = \begin{bmatrix} \\ \dot{\mathbf{F}} \\ \\ \end{bmatrix}, \tag{7.90}$$

the principal diagonal of whose matrix is of the form:

$$\begin{aligned} &\operatorname{diag}[\ldots,j(\omega-k\Omega)\mathbf{I}-\dot{\mathbf{A}}_0,\ldots,j(\omega-\Omega)\mathbf{I}-\dot{\mathbf{A}}_0,j\omega\mathbf{I}-\dot{\mathbf{A}}_0,\\ &\quad,j(\omega+\Omega)\mathbf{I}-\dot{\mathbf{A}}_0,\ldots,j(\omega+k\Omega)\mathbf{I}-\dot{\mathbf{A}}_0,\ldots]. \end{aligned}$$

The analysis conducted above reveals the following general regularities.

When the complex amplitudes method is used for investigating a scalar differential equation of an arbitrary order with periodic coefficients, the result is an infinite system of algebraic equations with a matrix made up of complex elements. If the same method is applied to a vector differential equation of the second order (2×2), the result is an infinite system of algebraic equations with a cellular matrix, each element of which is a matrix of the second order (2×2) with complex figures. The unknown column vector is also cellular. It is obvious that if the initial vector differential equation has a matrix of the n-th order, each cell in the matrix of the infinite system of algebraic equations will also be a matrix of the n-th order.

When examining a linear vector differential system with an almost periodic coefficient, whose matrix is of the n-th order, the complex amplitudes method leads to an infinite system of algebraic equations with a spatial cellular matrix, the number N of whose dimensions depends on the number of the incommensurable frequencies within the spectrum of the almost periodic coefficients in the initial differential system. Each cell in the spatial matrix is an N-dimensional matrix of the n-th order.

The form of the infinite linear system of algebraic equations yields a peculiar classification providing a complete qualitative picture of the entire set of linear oscillating systems with periodic and almost periodic parameters.

Let us consider the more general case of a linear oscillating circuit with an arbitrary ideally smooth law of parameter change in time. In this case, the coefficients and the solution of the respective differential system can be presented in the form of a Taylor series expansion by time. By equating the coefficients at the equal powers of the argument (time) we obtain a linear semi-infinite system of algebraic equations with real coefficients.

Let the differential equation of the oscillating circuit be of the form

$$\frac{d}{dt}\mathbf{x} = \mathbf{A}(t)\mathbf{x} + \mathbf{f}(t), \tag{7.91}$$

$\mathbf{A}(t)$ is a matrix of the n-th order, whose elements are ideally smooth time functions, $\mathbf{f}(t)$ is a column vector with analogous elements.

We are seeking a solution of Eq.(7.91) in the form of an infinite power series

$$\mathbf{x}(t) = \sum_{k=0}^{\infty} \mathbf{x}_k t^k.$$

By expanding the matrix function and the vector function in a Taylor series, we present Eq.(7.91) in the following form

$$\frac{d}{dt}\sum_{k=0}^{\infty} \mathbf{x}_k t^k = \left(\sum_{k=0}^{\infty} \mathbf{A}_k t^k\right)\left(\sum_{k=0}^{\infty} \mathbf{x} t^k\right) + \sum_{k=0}^{\infty} \mathbf{f}_k t^k,$$

$$\mathbf{A}_k = \frac{1}{k!}\frac{d^k}{dt^k}\mathbf{A}(t)|_{t=0}, \quad \mathbf{f}_k = \frac{1}{k!}\frac{d^k}{dt^k}\mathbf{f}(t)|_{t=0}.$$

After differentiation and multiplication of the respective series, we obtain:

$$\sum_{k=0}^{\infty}(k+1)\mathbf{x}_{k+1} = \sum_{k=0}^{\infty} t^k \sum_{l=1}^{k} \mathbf{A}_l \mathbf{x}_{k-l} + \sum_{k=0}^{\infty} \mathbf{f}_k t^k. \tag{7.92}$$

By selecting the terms with an equal multiplier t^k, a semi-infinite system of linear algebraic equations is derived from (7.92)

$$(k+1)\mathbf{x}_{k+1} = \sum_{l=0}^{k} \mathbf{A}_l \mathbf{x}_{k-l} + \mathbf{f}_k,$$

or, presented in an expanded form

$$\begin{aligned}
\mathbf{x}_1 &= \mathbf{A}_0\mathbf{x}_0 + \mathbf{f}_0, \\
2\mathbf{x}_2 &= \mathbf{A}_1\mathbf{x}_0 + \mathbf{A}_0\mathbf{x}_1 + \mathbf{f}_1, \\
&\cdots\cdots\cdots\cdots\cdots\cdots \\
n\mathbf{x}_n &= \mathbf{A}_{n-1}\mathbf{x}_0 + \mathbf{A}_{n-2}\mathbf{x}_1 + \ldots + \mathbf{A}_0\mathbf{x}_{n-1} + \mathbf{f}_{n-1}.
\end{aligned}$$

This is a semi-infinite recurrent system of linear algebraic equations. The initial condition $\mathbf{x}_0$ serves for determining $\mathbf{x}_1$. Once $\mathbf{x}_0$ and $\mathbf{x}_1$ are known, it is not difficult to obtain $\mathbf{x}_2$, etc. As the necessary number of $\mathbf{x}_n$ figures is determined, the approximate solution is written in the form:

$$\mathbf{x}(t) \approx \sum_{k=0}^{n} \mathbf{x}_k t^k.$$

This approach can also be applied in the case when the coefficients of the vector differential equations are periodic or semi-periodic time function. From an algorithmic point of view, the problem proves to be simpler. This is convenient when certain quantitative computations are required. But when it is necessary to conduct an analysis of the internal structure of the signals and to provide general characteristics of the solution properties, the harmonic approach outlined above is more productive.

7.3. Parametric resonance in a linear oscillating circuit with periodic parameters

So far the phenomenon resonance has been investigated comparatively fully in the oscillating systems with constant parameters, less so in the linear systems with parameters that remain periodic in time. Fundamental papers related to the second issue first came out over 65 years ago [86]. Currently the possibilities for the realization of linear non-stationary systems are by far greater and this is the reason why certain fundamental works from the 1930s, in particular those written by G.S. Gorelik [86], are more topical nowadays than they were at the time of their appearance.

In this section we shall delve into the theory and the physical interpretation of the resonance in a linear circuit with periodic parameters [265–268] by resorting to geometrical categories and to an n-dimensional ($n \leq 4$) Euclidean space.

It is known that an adequate class of describing functions for the resonance systems with constant parameters are the harmonic (sine) time functions. When the dependence of the oscillating process on time is periodic but not harmonic, it is most often presented as a sum of sine-like terms (if the process is periodic, these are Fourier series). This privileged role of the sine-like functions is not determined by the fact that they are „the simplest" oscillations, but rather by the fact that the harmonic resonators with constant parameters used in physics and technology have the property of separating precisely such sine-like components from the composition of an arbitrary external force. From the viewpoint of the class of oscillating systems with periodic time-dependent parameters, harmonic resonators are a rather particular case. The adequate class of describing functions for this more general class of systems is much wider – these are the almost periodic time functions.

The phenomenon resonance in a linear periodic oscillating circuit can be manifested in various ways, therefore a conventional classification is necessary. The analysis of the written sources [86] and the investigations conducted so far [265–267] show that there are three major cases of resonance manifestation in a linear periodic oscillating circuit, which we shall conventionally term resonance 1, resonance 2 and resonance 3. If we take the equation of Hill concerning the natural oscillations of the circuit as a classification basis, we can establish the following correspondence:

Resonance 1 – Hill's equation is in the stability zone;

Resonance 2 – the equation concerning the natural oscillations belongs to the boundary between the ranges of stability and instability;

Resonance 3 – Hill's equation is in the instability zone.

It should be pointed out that since Hill's equation is unstable on the boundary, it is only natural that the boundary should be termed boundary of instability. If a dissipative term (even an infinitesimally small one) is added to the equation of Hill, the equation becomes stable on the bound, and it would be appropriate to term the latter „bound of stability".

The properties of resonances 1, 2, and 3 are widely at variance. Resonance 2 cannot be realized in a pure form. But it is of considerable significance because of two reasons: first, as a linkage between Resonance 1 and Resonance 3, which are of greater practical importance, and second, Resonance 2 can be approximately realized not precisely on the instability boundary but in a narrow band including this boundary.

7.3.1. Resonance 1. Geometric meaning of the resonance in a linear oscillating circuit with periodic parameters

Let us consider a series oscillating circuit with periodically changing parameters R, C and L, whose equation concerning the capacitor charge q in a dimensionless form is [265]

$$\frac{d^2x}{d\tau^2} + 2\alpha(\tau)\frac{dx}{d\tau} + \omega_0^2(\tau)x = f(\tau), \tag{7.93}$$

where $x = \dfrac{q}{q_{00}}$, $\tau = \dfrac{t}{t_{00}}$, $\alpha = \dfrac{1}{2}\left(t_{00}\dfrac{R}{L} + \dfrac{d}{d\tau}\ln L\right)$, $\omega_0(\tau) = \dfrac{t_{00}^2}{\sqrt{LC}}$, $f(\tau) = \varepsilon(\tau)\dfrac{t_{00}^2}{q_{00}L}$, $\varepsilon(\tau)$ is a driving electromotive force.

If we represent $\alpha(\tau) = \alpha_0 + \alpha_1(\tau)$ and introduce a new variable $x = ye^{-\int \alpha_1 d\tau}$, Eq.(7.93) acquires the form:

$$\frac{d^2y}{d\tau^2} + 2\alpha_0\frac{dy}{d\tau} + \left(\omega_0^2 - 2\alpha_0\alpha_1 - \alpha_1^2 - \frac{d\alpha_1}{d\tau}\right)y = f(\tau)e^{\int \alpha_1(\tau)d\tau}. \tag{7.94}$$

The assumption is that $f(\tau)$ is an almost periodic function. Obviously Eq.(7.94) is of the same type as the initial Eq.(7.93), with the only difference that Eq.(7.94) has a constant damping factor. Hence, without any loss of generality in our investigation, we can further analyze the following equation

$$\frac{d^2x}{d\tau^2} + 2\alpha\frac{dx}{d\tau} + \omega_0^2(\tau)x = f(\tau), \tag{7.95}$$

where α is a constant damping factor.

We assume that the free oscillations equation

$$\frac{d^2x}{d\tau^2} + 2\alpha\frac{dx}{d\tau} + \omega_0^2(\tau)x = 0 \tag{7.96}$$

is stable according to Lyapounov's criteria, while the natural oscillations equation

$$\frac{d^2x}{d\tau^2} + \omega_0^2(\tau)x = 0 \tag{7.97}$$

may be situated in the stability range, on the instability bound and in the range of instability.

In accordance with the adopted conventional classification, we shall consider here Resonance 1, corresponding to the first case – Eq.(7.97) is in the area of instability.

The fundamental system of solutions of Eq.(7.97) consists of two linearly independent quasi-periodic functions U and V called Hill's functions. These functions can be presented in Fourier series:

$$U = \sum_{k=1}^{\infty} a_k \cos[(\nu + k\Omega)\tau + \varphi_k] \quad \text{and} \quad V = \sum_{k=1}^{\infty} b_k \cos[(\nu + k\Omega)\tau + \psi_k];$$

$\Omega = \dfrac{2\pi}{T}$, $\omega_0(\tau + T) = \omega_0(\tau)$, T is the change period of the coefficient $\omega_0^2(\tau)$ in Eq.(7.97).

When analyzing an arbitrary particular Eq.(7.97), we obtain a set of quasi-periodic functions, representing its solutions. Each solution can be presented in the form

$$x(\tau) = aU + bV, \tag{7.98}$$

where a and b are constants completely determined for a given solution. As we substitute various real numbers for a and b, we obtain the whole set of solutions of Eq.(7.97). Hence the set of solutions of Eq.(7.97) is an Euclidean plane, all points of which are identified with completely determined solutions of this equation. The analogy between the solutions plane and the ordinary plane is completely transparent. Each Eq.(7.97) or, more precisely, each function $\omega_0^2(\tau)$ has its corresponding Euclidean plane of solutions (7.98). It is natural to choose the trivial solution $x(\tau) = 0$ as an origin of the coordinates. It is expedient to have functions

U and V represented by an orthonormalized fundamental system of solutions, for which

$$(U,U)=\lim_{T\to\infty}\frac{1}{T}\int_0^T U^2(t)dt=1,\quad (V,V)=\lim_{T\to\infty}\frac{1}{T}\int_0^T V^2(t)dt-1$$

and

$$(U,V)=\lim_{T\to\infty}\frac{1}{T}\int_0^T U(t)V(t)dt=0.$$

The natural question of a choice of the initial conditions with a view to obtaining orthogonal solutions is posed. Let us first consider the particular case when $\omega_0^2(\tau)$ = const. Then the initial conditions of the orthogonal system of solutions can be set in the following way

$$U(0)=A,\quad \dot U(0)=0,\quad V(0)=0,\quad \dot V(0)=B. \tag{7.99}$$

Indeed, in this case the solutions are: $U(\tau)=A\cos\omega_0\tau$ and $V(\tau)=\dfrac{B}{\omega_0}\sin\omega_0\tau$. The orthogonality is obvious. In order to ensure orthonormalization, one should postulate $A=\sqrt{2}$ and $B=\sqrt{2}\omega_0$.

It can be shown that in the general case, too, when $\omega_0^2(\tau)=\omega_0^2(\tau+T)$, two orthogonal solutions can be identified by setting initial conditions (7.99). Later, multiplied by constants, these solutions can be normalized and an orthonormalized system can be obtained. In this case, the initial conditions (7.99) set the direction of the unit vectors U, V.

Let us assume that the unit vectors U, V are chosen ones. It is known from the theory of differential equations that the Wroskian of Eq.(7.99) is constant,

$$W(U,V)=\det\begin{vmatrix}U(\tau) & V(\tau)\\ \dot U(\tau) & \dot V(\tau)\end{vmatrix}=U(\tau)\dot V(\tau)-\dot U(\tau)V(\tau)=\text{const}=2w. \tag{7.100}$$

w is used to denote a constant completely defined for the given solutions U and V. By averaging the foregoing equation, we obtain:

$$(U,\dot V)-(\dot U,V)=2w. \tag{7.101}$$

Besides, by averaging the identity $U\dot V+\dot U V=\dfrac{d}{d\tau}(UV)$ and accounting for the fact that there is an almost periodic function to the right of the bracket, we obtain:

$$(U,\dot V)+(\dot U,V)=0. \tag{7.102}$$

It follows from (7.101) and (7.102) that $(U,\dot V)=w$, $(\dot U,V)=-w$.
In this way we obtain

$$U\perp V,\quad U\perp\dot U,\quad V\perp\dot V, \tag{7.103}$$

U is not perpendicular to $\dot{V}$, V is not perpendicular to $\dot{U}$.

The employment of geometric categories in the analysis of Eq.(7.97) allows of attaining certain flexibility, since it is possible to compare different solutions by geometric properties. For example, let us choose a particular solution of Eq.(7.97): $x(\tau) = aU(\tau) + bV(\tau)$. Here a and b are completely defined figures. This specific solution is represented on the Euclidean plane of the solutions by a radius-vector with module $\mathrm{mod}x(\tau) = \rho_x = \sqrt{(x,x)} = \sqrt{a^2+b^2}$ and argument $\arg x(\tau) = \alpha_x = \mathrm{arctg}\,\frac{b}{a}$, which is read from the unit-vector U in a counter-clockwise direction. The argument can also be determined through a scalar product – $\alpha_x = \arccos \dfrac{(x,U)}{\sqrt{(x,x)}}$ $= \arcsin \sqrt{1 - \frac{(x,U)^2}{(x,x)}}$.

If two solutions $x_1(\tau)$ and $x_2(\tau)$ are given, it is possible to determine the extent to which they are away from each other by introducing the concept of interval between the solutions $d(x_1, x_2)$. This is a positive figure equal to the square root of the scalar square of the difference between the solutions, $d = \sqrt{(x_1 - x_2, x_1 - x_2)}$.

If the operator $L_0 = \dfrac{d^2}{d\tau^2} + \omega_0^2(\tau)$ in Eq.(7.97) is an even operator (to this end it is necessary and sufficient that the function should be even), it can be shown that U is an even function, while V is an odd one, i.e. $U = \sum\limits_{k=1}^{\infty} \alpha_k \cos(\nu + k\Omega)\tau$, $V = \sum\limits_{k=1}^{\infty} \alpha_k \sin(\nu + k\Omega)\tau$.

Here α_k are such coefficients for which the normalization of the functions $(U,U) = 1$ and $(V,V) = 1$ is fulfilled.

When investigating phenomena in an oscillating circuit described by Eq.(7.93), it is not sufficient to analyze only the Euclidean plane of the solutions of Eq.(7.97). For this purpose it is necessary to consider in addition yet another Euclidean space, which we shall term expanded hyperplane of the solutions of Eq.(7.97). The basis of this hyperplane is provided by the vectors U, V, $\dot{U}$, $\dot{V}$ – it is a Euclidean space determined by these four vectors. Let us show that this is a four-dimensional space. To this end, we shall consider two planes: UV and $\dot{U}\dot{V}$. We shall analyze the common points of these two planes. The first plane contains all the solutions of Eq.(7.97). In order to arrive at the equation, whose solutions belong to the second plane $\dot{U}\dot{V}$, we differentiate Eq.(7.97) by τ to obtain

$$\frac{d^2\dot{x}}{d\tau^2} + \omega_0^2(\tau)\dot{x} + 2\omega_0(\tau)\frac{d\omega_0(\tau)}{d\tau}\int \dot{x}\,d\tau = 0. \tag{7.104}$$

It is obvious that Eq.(7.104) concerning $\dot{x}$ differs from Eq.(7.97) due to the existence of the last term. This means that out of all solutions of Eqs.(7.97) and (7.104) only the trivial ones can coincide. The plane of the solutions UV and the plane of the derivatives of these solutions $\dot{U}\dot{V}$ have just one point in common –

the origin of the coordinates. It is well-known that two planes situated in a three-dimensional space cannot intersect in just one point. This comes to prove that the expanded hyperplane is a four-dimensional Euclidean space. An interesting point to make is that since the parameter $\omega_0(\tau) = \text{const}$ for the oscillating circuit with constant parameters in Eq.(7.97), the last term in (7.104) is equal to zero and Eqs. (7.97) and (7.104) coincide. This implies that the circuit with constant parameters is a particular case of the more general system with periodic parameters, for which the solutions plane and the expanded hyperplane of the solutions of Eq.(7.97) coincide and represent a Euclidean plane.

Let us visualize the expanded hyperplane. A Cartesian system of coordinates with unit vectors i, j, k is taken. The unit vector k is turned round the unit vector i at a definite angle. The result is $i \perp j$, $i \perp k$, j is not perpendicular to k. This is an approximate model of the expanded hyperplane (the model is three-dimensional, while the expanded hyperplane is four-dimensional; this model will help us in our further geometric interpretations).

Let us construct a more precise, though less pictorial model. We assume that we have a four-dimensional Cartesian system of coordinates with unit vectors i, j, k, l. As an example of such a system, one can refer to the space-time system of coordinates which is widely used in the relativity theory. Unit vector k is turned round unit vector j at a definite angle in one direction (for example counter-clockwise, if viewed from the end of unit vector i), while unit vector l is rotated round unit vector j at the same angle in the opposite direction. Unit vectors i, j are substituted for unit vectors U, V, and unit vectors k, l are substituted for non-unit vectors $\dot{U}, \dot{V}$. The result is the set of properties (7.103) and, besides, the property that $\dot{V}$ is not perpendicular to $\dot{U}$.

Having considered the pattern of the solutions of Eq.(7.97), we shall now proceed with the investigation of the resonance phenomena in an oscillating circuit described by Eq.(7.95), given the assumption that Eq.(7.97) is stable. First and foremost we have to clarify what the resonance phenomenon consists of. If we substitute various almost periodic functions with equal intensity $I_f = (f, f)$ in the right-hand part of Eq.(7.95), the intensity of the quasi-periodic oscillations in the circuit $I_x = (x, x)$ will also vary for each almost periodic function $f(\tau)$ in the stationary mode. At that, it will be possible to find such an almost periodic function $f(\tau) = f_r(\tau)$, for which the intensity of the oscillations $(x_r, x_r) = I_{x_{\max}}$ will have a maximum value. If this happens, the circuit will be set in resonance. A question crops up here: to which function f_r does the circuit respond with the specific phenomenon resonance? The theory of resonance in an oscillating circuit with constant parameters does not allow to identify the exact function to which the circuit responds with resonance in this case. Since in the normal case free oscillations are harmonic (sine) time functions, it is indeed impossible to establish whether the circuit responds to its natural oscillations, or to the n-th derivative by the time of its natural oscillations, or to an integral (which can be multifold) by the time of its natural oscillations, for sine functions have the property of multiple reproduction in situations of differentiation and integration. The question posed above is answered in a general form in [50, 86], and for the case under consideration the following can be written:

$$f_r(\tau) = P\dot{U}(\tau) + Q\dot{V}(\tau), \quad P, Q = \text{const.} \tag{7.105}$$

Furthermore, the solution of Eq.(7.95) regarding the stationary mode is in the following form:

$$x(\tau) = \frac{1}{2}[PU(\tau) + QV(\tau)]. \tag{7.106}$$

In this way the oscillations in a situation of resonance are directly related to the solutions $U(\tau)$, $V(\tau)$ of Eq.(7.97). This result can be understood if $f(\tau)$ is replaced with $\alpha f(\tau)$ in Eq.(7.95) and one moves on to equation

$$\frac{d^2x}{d\tau^2} + \omega_0^2(\tau)x = \alpha\left(f - 2\frac{dx}{d\tau}\right). \tag{7.107}$$

The solution of Eq.(7.107) is denoted as $x(\tau, \alpha)$, and that of equation (7.97) as $x(\tau)$. It follows from the principle of superposition that

$$x(\tau) = \frac{x(\tau, \alpha)}{\alpha}. \tag{7.108}$$

After a substitution of (7.105) and (7.106) in (7.107), the resultant equation is (7.97). The forced oscillations of the circuit in a state of resonance coincide with the free oscillations of the same circuit without resistance, or, to put it differently, with the natural oscillations of this circuit, if $f(\tau) = 2\dfrac{dx(\tau, \alpha)}{d\tau}$. The oscillating circuit responds with resonance, when the external acting force contains a derivative by the time of the natural oscillations of the circuit. It follows from (7.108) that when $\alpha \to 0$, the oscillations in a state of resonance increase infinitely, which is also observed with the common harmonic resonators. The provided interpretation of resonance [265] can serve as a basis for determining whether a certain almost periodic function will lead to resonance in (7.95) or not in the following way. For this purpose, the function is presented in the form

$$f(\tau) = P\dot{U} + Q\dot{V} + g(\tau), \quad g(\tau) \perp U(\tau), \quad g(\tau) \perp V(\tau), \tag{7.109}$$

$$P = -\frac{(V, f)}{w}, \quad Q = \frac{(U, f)}{w}.$$

If $P = Q = 0$, there is no resonance. Otherwise the resonance is there. Expansion (7.109) is unusual in form. $P\dot{U} + Q\dot{V}$ is separated from the composition of the exciting force, moreover the remainder $g(\tau)$ is not orthogonal with respect to $\dot{U}$ and $\dot{V}$, as it is usually believed, but it is orthogonal with respect to U and V. The unusual expansion may lead to difficulties in a number of particular cases.

For example, let us set $f(\tau)$ in the form $f = P\dot{U} = RU$, $P, R = \text{const}$. In this case it is not possible to believe that the first term of $f(\tau)$ causes resonance, since the second term is not orthogonal with respect to $U(\tau)$. It is even more astonishing that $\dot{U} \perp U$. Many such examples can be quoted.

We shall try to explain this case by using the geometric basis presented above. It follows from (7.103) that U and V can be presented as follows:

$$U = g\dot{V} + g_U(\tau), \quad V = P\dot{U} + g_V(\tau), \tag{7.110}$$

and in the general case $g(\tau) = g_U(\tau) + g_V(\tau)$.

Fig.7.8 shows the positions of the terms U, V, belonging to the composition of the external influencing force. We can imagine that vectors U, V are projected on vectors $\dot{U}, \dot{V}$ respectively, but this is an unusual projection. In the situation of a normal projection $g_U \perp \dot{V}$ and $g_V \perp \dot{U}$, while in this case $g_U \perp U$ and $g_V \perp V$. The normal projection is shown in Fig.7.8 by a dotted line. In the case under consideration, the result is oxygonal projection. In the example given above the external influencing force should be presented as follows: $f(\tau) = P\dot{U} + RU = P\dot{U} + Rq\dot{V} + g_U(\tau)$. The resonance component of the influencing force is $f_r(\tau) = P\dot{U} + Rq\dot{V}$.

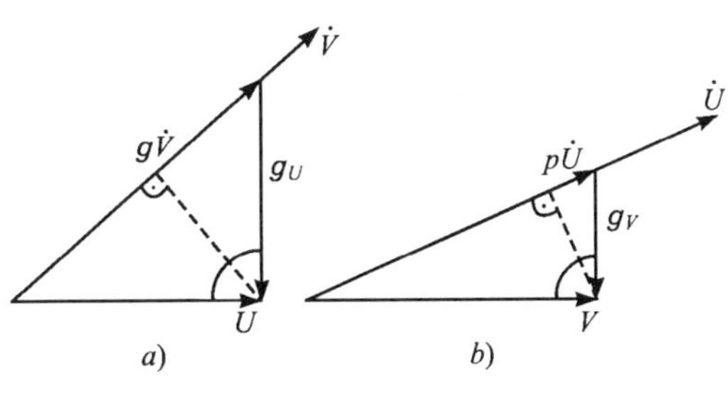

Fig. 7.8. Resolution of the external acting force $f(\tau)$ by a linear periodic oscillating circuit at Resonance 1: a) $f(\tau) = U$, b) $f(\tau) = V$

The analysis of the resonance phenomena in a linear periodic oscillating circuit, when the equation of its natural oscillations is in the stability range, poses a number of mathematical problems. While the normal harmonic resonator identifies the normal sine time functions as the most simple ones and responds to them, the linear periodic resonator identifies Hill's functions – quasi-periodic functions with a frequency spectrum $(\nu \pm k\Omega)$, $k = 0, 1, 2, 3, \ldots$, as belonging to the simplest type. It is known that an arbitrary periodic and an almost periodic function can be presented in the form of a sum of harmonic functions, and the mathematical technique for such an expansion – Fourier series, is highly developed. In exactly the same way, an arbitrary periodic and almost periodic function (including a sine function) can be expanded by using Hill's finite or infinite functions but the respective mathematical technique has not been sufficiently developed so far. Hence the phenomenon resonance in periodic resonators cannot be comprehensively described by employing an adequate mathematical technique that would be the simplest and most appropriate for the considered group of phenomena. We are compelled to make partial use of the language of sine time functions, which is unnatural from the viewpoint of periodic resonators, and this is the reason why some properties of periodic resonators seem strange and amazing at first glance.

7.3.2. Resonance 2. First and second power resonance of a linear oscillating circuit with periodic parameters

In accordance with the classification adopted above, we shall consider here Resonance 2 corresponding to the case when the equation of the natural oscillations (7.97) is on the boundary of the unstable range [266].

The linearly independent solutions of equation (7.97) are in the following form:

$$x_1 = U, \quad x_2 = \gamma U\tau + V, \quad \gamma = \text{const}, \tag{7.111}$$

U and V are periodic functions with period π or 2π, the constant γ is the resonance determining parameter. The initial conditions can be selected in such a way that functions U and V will be ortho-normalized, i.e. $(U, V) = 0$, $(U, U) = 1$, $(V, V) = 1$. Here, once again, $(\cdot, \cdot)$ stands for a scalar product of functions, which can be defined as a period average of a product of periodic functions, for example: $(U,V) = \frac{1}{2\pi}\int_{\tau}^{\tau+2\pi} UV\,d\tau$, $(U,U) = \frac{1}{2\pi}\int_{\tau}^{\tau+2\pi} U^2\,d\tau$. The following equations are valid for the periodic functions: $(U, \dot{U}) = 0$, $(V, \dot{V}) = 0$. The Wroskian (7.100), as we know, is constant, so it follows that $(U, \dot{V}) = -(\dot{U}, V) = W = \text{const}$.

The periodic functions U and V are the basis of the resonance theory in the system (7.97). Let us assume that in (7.95) $f(\tau) = \dot{U}$. Then a direct substitution can yield the forced oscillations

$$x = \frac{U}{2\alpha}. \tag{7.112}$$

This is resonance, since for $\alpha \to 0$, $x \to \infty$.

In the case of $f(\tau) = \dot{V}$, we obtain

$$x = \frac{\gamma U}{2\alpha^2} + \frac{V}{2\alpha}. \tag{7.113}$$

This is also resonance according to the same property. These two types of resonance differ qualitatively by their selectivity. In the first case, α in the denominator of (7.112) is to the first power. This is a first power resonance. In the second case the denominator of one of the terms (7.113) contains α square. This is a second power resonance.

As elucidated above, in the general case of an arbitrary almost periodic function in (7.65), the circuit expands this function in the following way:

$$f(\tau) = P\dot{U} + Q\dot{V} + g, \quad P, Q = \text{const}, \quad (\mathrm{U}, \mathrm{g}) = 0, \quad (\mathrm{V}, \mathrm{g}) = 0. \tag{7.114}$$

This is an extraordinary expansion. The terms proportional to $\dot{U}$ and $\dot{V}$ are separated and the remainder is orthogonal not with respect to $\dot{U}$ and $\dot{V}$, but with respect to U and V. By conducting scalar multiplication of (7.104) by U and V we can convince ourselves that the expansion is one-valued, at that $P = -\dfrac{(V, f)}{W}$, $Q = \dfrac{(U, f)}{W}$. The exciting force $f(\tau)$ causes forced oscillations in the circuit,

$$x = \frac{\gamma Q}{2\alpha^2} + \frac{PU + QV}{2\alpha} + Z. \tag{7.115}$$

The first term in the x expression in (7.105) yields a second power resonance, and the second one yields a first power resonance. Z is a non-resonance component; it can be shown that it remains finite at $\alpha \to 0$.

Let us consider the geometric meaning of Resonance 2. In many aspects it coincides with the case of Resonance 1, but it also has its peculiarities. It is obvious that each solution of (7.97) is expressed by the fundamental system (7.111)

$$x = Ax_1 + Bx_2, \quad A, B = \text{const}. \tag{7.116}$$

To each solution of (7.116) we juxtapose a periodic function

$$y = AU + BV. \tag{7.117}$$

By setting all sorts of initial conditions and thereby changing the constants A and B, we obtain a set of periodic functions $\{y\}$, typical of the given oscillating circuit. We provide unit vectors corresponding to the functions and retain the same denotations for them – U and V. Then the Euclidean plane determined by these unit vectors will correspond to the set $\{y\}$ in the sense that each point of this plane corresponds to definite solution x (7.106) of Eq.(7.97). Following the same principle, we introduce a four-dimensional Euclidean space determined by the four vectors U, V, $\dot{U}$, $\dot{V}$. These vectors form an oxygonal coordinate system, since $U \perp V$, $U \perp \dot{U}$, $V \perp \dot{V}$, but in the remaining three pairs $U\dot{V}$, $\dot{U}V$, $\dot{U}\dot{V}$ the vectors are not perpendicular.

We shall illustrate expansion (7.114) in a four-dimensional space with an oxygonal coordinate system $UV\dot{U}\dot{V}$. The expansion can be presented in three stages: 1. We separate the components proportional to $\dot{U}, \dot{V}$, in the ordinary way (the remainder is orthogonal with respect to $\dot{U}, \dot{V}$): $f = A\dot{U} + B\dot{V} + g_{\dot{U}\dot{V}}$, $g_{\dot{U}\dot{V}} \perp \dot{U}$, $g_{\dot{U}\dot{V}} \perp \dot{V}$, $A, B = \text{const}$. 2. By applying the ordinary procedure, we separate from the remainder a term, which is proportional to U, then we expand this term in a non-standard way as shown in Fig.7.9: $g_{\dot{U}\dot{V}} = aU + g_1 = b\dot{V} + g_U + g_1$, $g_1 \perp U$, $g_U \perp U$. 3. Proceeding in the normal way, we separate from its remainder a term, which is proportional to V, then we expand this term in a non-standard way as shown in Fig.7.9: $g_U + g_1 = CV + g_2 = d\dot{U} + g_{UV} + g_2$, $g_{UV} \perp U$, $g_{UV} \perp V$, $g_2 \perp U$, $g_2 \perp V$. Uniting all terms proportional to $\dot{U}$ and $\dot{V}$ and comparing the result to (7.114) we obtain:

$$P = A + a, \quad Q = B + b, \quad g = g_{UV} + g_2.$$

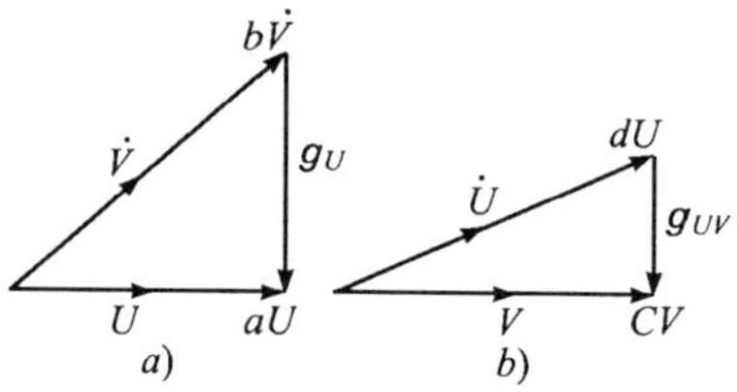

Fig. 7.9. Illustration of the three-stage expansion of the external acting force in four-dimensional space with an oxygonal coordinate system at Resonance 2

Let us consider the following case of an oscillating circuit described by the equation: $\ddot{y} + 2\delta\dot{y} + (a + m\cos 2t)y = \cos(t - \varphi)$.

The equation concerning the natural oscillations of this circuit is

$$\ddot{y} + (a + m\cos 2t)y = 0, \tag{7.118}$$

and it is a Mathieu equation. Using the small perturbations method, we compute the ortho-normalized solutions U, V, and the parameter γ. A necessary and sufficient condition for Eq.(7.118) to belong to the left-hand or right-hand stability boundary is the equation

$$a = 1 - \frac{m}{2} - \frac{m^2}{32} + \frac{m^3}{512} + \dots$$

The following expressions are yielded for the left-hand boundary:

$$U = \sqrt{2}\left(1 - \frac{m^2}{512} + \dots\right)\left(\cos t + \frac{m}{16}\cos 3t + \dots\right),$$

$$V = \sqrt{2}\left(1 - \frac{7m^2}{512} + \dots\right)\left(\sin t + \frac{m}{16}\sin 3t + \dots\right),$$

$$\gamma = -\frac{m}{2}\left(1 - \frac{3m^2}{256} + \dots\right), \quad W = 1 + \frac{3}{128}m^2 + \dots$$

Analogously, the following are related to the right-hand boundary:

$$U = \sqrt{2}\left(1 - \frac{m^2}{512} + \ldots\right)\left(\sin t + \frac{m}{16}\sin 3t + \ldots\right),$$

$$V = \sqrt{2}\left(1 - \frac{7m^2}{512} + \ldots\right)\left(\cos t + \frac{m}{16}\cos 3t + \ldots\right),$$

$$\gamma = -\frac{m}{2}\left(1 - \frac{3m^2}{256} + \ldots\right), \quad W = 1 - \frac{3}{128}m^2 + \ldots$$

Given a sufficiently small m, the following approximations can be regarded as valid:

— for the left-hand boundary

$$U = \sqrt{2}\cos t, \quad V = \sqrt{2}\sin t, \quad \gamma = -\frac{m}{2}, \quad W = 1;$$

— for the right-hand boundary

$$U = \sqrt{2}\sin t, \quad V = \sqrt{2}\cos t, \quad \gamma = -\frac{m}{2}, \quad W = 1.$$

It is expedient to present the external influencing force in the following form:
— for the left-hand boundary

$$\cos(t - \varphi) = -\frac{\sin\varphi}{\sqrt{2}}(-\sqrt{2}\sin t) + \frac{\cos\varphi}{\sqrt{2}}(\sqrt{2}\cos t) = P\dot{U} + Q\dot{V},$$

$$P = -\frac{\sin\varphi}{\sqrt{2}}, \quad Q = \frac{\cos\varphi}{\sqrt{2}};$$

— for the right-hand boundary

$$\cos(t - \varphi) = \frac{\cos\varphi}{\sqrt{2}}(\sqrt{2}\cos t) - \frac{\cos\varphi}{\sqrt{2}}(\sqrt{2}\sin t) = P\dot{U} + Q\dot{V},$$

$$P = \frac{\cos\varphi}{\sqrt{2}}, \quad Q = -\frac{\sin\varphi}{\sqrt{2}}.$$

If, for example, $\varphi = 0$, a second power resonance is realized on the left-hand boundary, and a first power resonance is realized on the right-hand one. If $\varphi = \frac{\pi}{2}$, the opposite is true: a first power resonance is realized on the left-hand boundary, and a second power resonance is realized on the right-hand one. If we change the initial phase φ gradually, the first and second power resonance will alternate on the two boundaries, moreover the relative weight of the former with respect to the latter will also change gradually.

7.3.3. Resonance 3. Equation of the natural oscillations in the instability range

Let us consider the case of resonance in a linear periodic oscillating circuit, when the equation of the natural oscillations is in the unstable range (Resonance 3, according to the adopted classification (See 7.3 and [267]). This type of resonance is of practical significance, since, like Resonance 1, it can be practically realized in a relatively simple way. On the other hand, while Resonance 1 can be viewed as a direct generalization of the „ordinary" resonance, Resonance 3, is characterized by sufficiently extraordinary properties and it is quite far from the ordinary resonance in its tuning and resonance characteristics. In the case of Resonance 3 the oscillating circuit may have quite high selectivity, given a sufficiently high damping factor – a property that is not observed in an oscillating circuit with positive constant parameters.

The equations of the oscillating circuit in a state of Resonance 3 do not differ in type from the analogous equations for the other types of resonance (7.95) – (7.97). In the case of Resonance 3, Eq.(7.97) should be located in the instability range, while the free oscillations Eq.(7.96) should be in the stability range. Thus the damping factor α „shifts" the equation from the instability range to the stability one. The parameter α can be presented as

$$\alpha = h + \delta, \tag{7.119}$$

where h reflects such losses, for which, given $\alpha = h$, Eq.(7.95) turns out to be on the boundary between the stability and instability ranges (at non-zero h this boundary belongs to the stability range, so it is expedient to refer to it as a stability boundary in this case); δ stands for any additional losses, and further on we shall take it that $\delta \ll h$. When considering Resonance 3, parallel with Eq.(7.95) the following equation is also of great importance:

$$\ddot{y} + 2h\dot{y} + \omega_0^2(\tau)y = 0. \tag{7.120}$$

The solution of Eq.(7.97) can be presented in the form

$$y = AU + Be^{-2ht}V, \tag{7.121}$$

A and B are constants selected in such a way that functions U and V should be normalized: $(U,U) = 1$, $(V,V) = 1$. It turns out, however, that the orthogonality condition $(U,V) = 0$ cannot always be satisfied; this condition is met only when Eq.(7.120) concerns the lowest point of the stability boundary.

By way of illustration Fig.7.10 shows the lowest point A of the stability boundary of the equation

$$\ddot{y} + 2h\dot{y} + (a + m\cos 2t)y = 0, \text{ where } a \approx 1 - \frac{m^2}{32},$$

$$h \approx \frac{m}{4} - \frac{3}{1024}m^3,\ U \approx \left(1 - \frac{m^2}{512}\right)\left[\cos t - \sin t + \frac{m}{16}(\cos 3t - \sin 3t)\right],$$

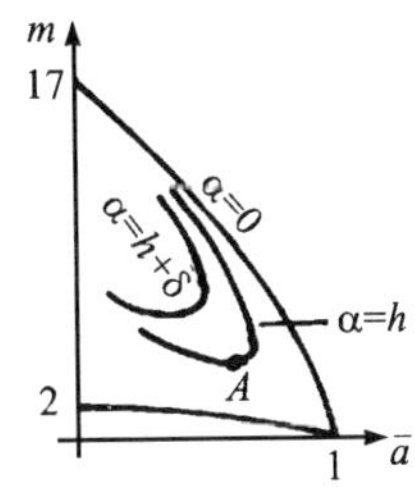

Fig. 7.10. Illustration of the areas of stability and instability of Equation $\ddot{y} + 2h\dot{y} + (a + m\cos 2t)y = 0$

$V \approx \left(1 - \frac{m^2}{512}\right)\left[\cos t + \sin t + \frac{m}{16}(\cos 3t + \sin 3t)\right]$. The case when Eq.(7.120) corresponds to the lowest point of the stability boundary is convenient since it provides a possibility to compare Resonance 3 with Resonance 1 and 2. That is why we shall limit ourselves to this case, i.e. we shall consider that the condition $(U, V)=0$ is satisfied, and therefore the periodic functions U, V are ortho-normalized.

In compliance with the approach presented above, the resonance term of the external influencing force $f(\tau)$ in (7.95) is proportional to $\dot{U}$.

Indeed, assuming $f = \dot{U}$, we obtain the solution

$$x = \frac{U}{2\delta}. \tag{7.122}$$

This is first-power resonance, yet it has a peculiar characteristic. While in the case of first power Resonance 1 and 2 the denominator of the right-hand part of (7.122) contains the damping factor α, in the case of Resonance 3 this is the additional damping factor δ, which can be much smaller than the damping factor α.

In the general case the force $f(\tau)$ in (7.95) can be expanded:

$$f = P\dot{U} + g; \quad (g, V) = 0, \quad P = \text{const}. \tag{7.123}$$

At $\delta \to 0$ the first term tends to infinity and the second one remains finite. This expansion is the only one and $P = \frac{(f, V)}{(\dot{U}, V)}$. Resonance emerges under the condition $P \neq 0$

If $f = \dot{V}$,

$$x = \frac{V}{4h + 2\delta} \approx \frac{V}{4\alpha}. \tag{7.124}$$

This is the ordinary first power resonance that can be referred to as quasi-resonance because it is much less pronounced than the peculiar first-power resonance (7.123). The function g in (7.124) can be expanded as $g = Q\dot{V} + g_1$, $Q = \text{const}$, $(g_1, U) = 0$, $(g_1, V) = 0$. Then $P\dot{U}$ yields resonance, $Q\dot{V}$ yields quasi-resonance, and g_1 is a non-resonance term. It can be demonstrated that the resonance and quasi-resonance terms change their roles (in the same way as the functions U and V do) for the equation conjugate with respect to (7.120). This equation is of the following form: $\ddot{y} - 2h\dot{y} + \rho(t)y = 0$.

Let us provide a visualized geometric interpretation of Resonance 3. A periodic function $z = AU + BV$ is juxtaposed to solution (7.121). A family of such functions at various constant values for A, B and given a norm $||z|| = \sqrt{(z, z)}$ is an Euclidean plane, to which an ordinary plane of coplanar vectors with common origin can be juxtaposed. Then a definite vector of that family corresponds to each solution (7.121). We shall designate the solution and the vector corresponding to it by one and the same letter. Then a Cartesian coordinate system with unit vectors U and V will automatically be set on our plane. Besides the above-mentioned Euclidean plane, we shall also consider a four-dimensional hyperplane, determined by the vectors U, V, $\dot{U}$, $\dot{V}$, which form an oxygonal coordinate system, since out of the six possible pairs of vectors only three – UV, $\dot{U}U$, $\dot{V}V$, are orthogonal. Two planes are set aside - $\dot{U}V$ and $U\dot{V}$. They are cross-sections of the hyperplane. The first plane can be termed a resonance one, and the second plane can be qualified as quasi-resonance.

The external influencing force $f(\tau)$ refers to a sufficiently broad class of almost periodic functions. This class „cannot be accommodated" in the four-dimensional hyperplane, hence function $f(\tau)$ can be expanded to a principal part, located on the hyperplane, and a multidimensional remainder. The projection of the principal part on the plane $\dot{U}V$ contains a resonance member which can be separated in three stages, as shown in Fig.7.11, and which is equal to $(A + C)\dot{U}$. A quasi-resonance term can be separated in an analogous way. For this purpose, the projection of the principal part of the function $f(\tau)$ on the plane $U\dot{V}$ should be considered.

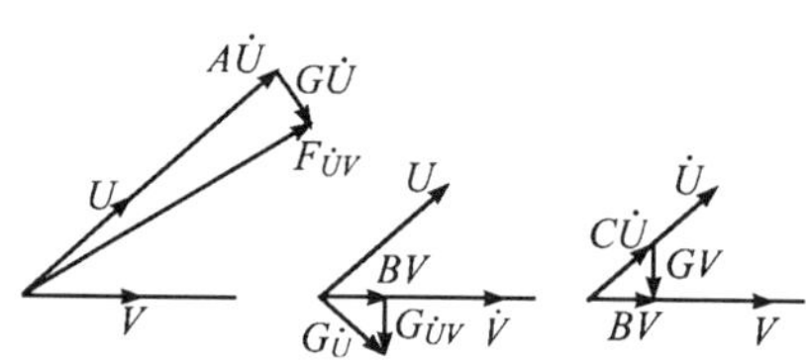

Fig. 7.11. Illustration of the staged separation of the main part of the external acting force in the hyperplane U, V, $\dot{U}$, $\dot{V}$

So in the case of Resonance 3, the linear periodic oscillating circuit responds to the resonance term with a very high degree of sensitivity, to the quasi-resonance term with a high degree of sensitivity, and to the non-resonance term with a low degree of sensitivity. In the case of Resonance 2 we also identified three degrees – second and first power resonance and non-resonance, even though here the difference between a second power resonance and a first power one is by far smaller than the distinction between resonance and quasi-resonance in the situation of Resonance 3. In the case of Resonance 1 there is U-resonance and V-resonance which are at the same level. In the Resonance 3 group there is U-resonance and V-quasi-resonance.

According to this property, Resonance 3 is close to Resonance 1, while in the case of Resonance 2 the first and second power resonances are U-resonances.

Let us set the equation of the circuit where Resonance 3 is excited to be in the form

$$\ddot{y} + 2(h+\delta)\dot{y} + [\rho_0 + m\rho(\tau)]y = f(\tau), \tag{7.125}$$

where ρ_0 = const; $\rho(\tau)$ is a periodic function with period π, zero average value and a module of its maximum value equal to one; m is a modulation index, $f(\tau)$ – an almost periodic function. In the absence of modulation, $m \equiv 0$, the force $f_1 = a\cos\sqrt{\rho_0}t + b\sin\sqrt{\rho_0}t$ yields resonance, while the force $f_2 = a\cos t + b\sin t$ fails to yield resonance. In the presence of modulation, i.e. when $m > 0$, these two forces change roles – force f_2 becomes resonance creating, while force f_1 is non-resonance creating. The modulation of a power-accumulating (reactive) parameter in the circuit changes the period of the external forces, to which the circuit responds with the emergence of resonance.

For the purpose of conducting quantitative comparison, we introduce the concept of sensitivity of the circuit to the external acting forces. It is determined by the following number:

$$\sigma = \frac{(y,y)}{(f,f)}.$$

At $(f,f) = 1$, $\sigma = (y,y)$.

For the harmonic resonator ($m = 0$), the maximum sensitivity in the case of resonance is

$$\sigma_0 = \frac{1}{4\rho_0(h+\delta)^2} = \frac{1}{4\omega_0^2(h+\delta)^2},$$

where $\omega_0 = \sqrt{\rho_0}$ is the own frequency of the resonator.

In the case of modulation of power-accumulating (reactive) elements in the circuit ($m \neq 0$), the maximum sensitivity in the presence of resonance is

$$\sigma_1 = \frac{1}{4\delta^2(\dot{U},V)^2}.$$

In the case of quasi-resonance the sensitivity is approximately

$$\sigma_2 = \frac{1}{16h^2(\dot{U},V)^2}.$$

It turns out that in the presence of resonance in a linear periodic resonator, the sensitivity of the latter can exceed by far the sensitivity of the harmonic resonator,

$$\frac{\sigma_1}{\sigma_0} = \frac{\omega_0^2}{(\dot{U},V)^2}\frac{h^2}{\delta^2}, \quad \delta \ll h.$$

In the situation of quasi-resonance, the sensitivity is of the same order as that of the harmonic resonator: $\dfrac{\sigma_2}{\sigma_0} = \dfrac{1}{4}\dfrac{\omega_0^2}{(\dot{U},V)^2}.$

Let us now consider the impact of detuning on the processes in the oscillating circuit. In the case of Resonance 3, detuning means a degree of remoteness from the stability boundary. Fig.7.12 provides an illustration of the detuning concept. The oscillating circuit situated on the stability boundary, for example in points M, N, P, is regarded as tuned. Depending on the detuning, these points can be shifted to the left or to the right (points M_1, N_1, P_1 and M_2, N_2, P_2).

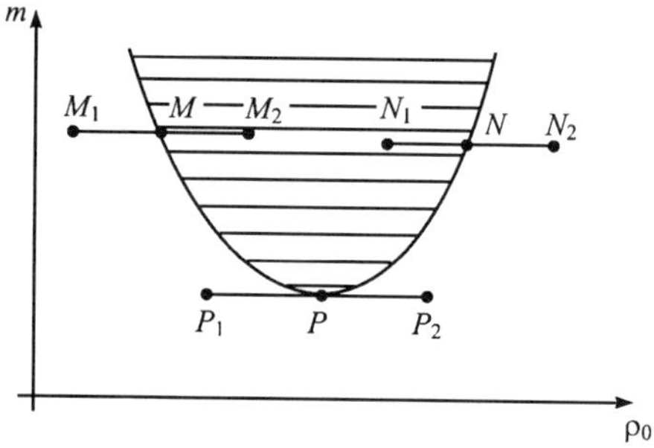

Fig. 7.12. Illustration of one of the areas of instability (hatched) in a periodic oscillating circuit. M, N, P – points of tuning. M_i, N_i, P_i – points corresponding to positive or negative detuning, $(i = 1, 2)$

The detuning ξ = const is supposed to be additive, and the detuned oscillating circuit is described by the following equation concerning the free oscillations:

$$S[y] + \xi y = 0, \tag{7.126}$$

where the operator $S[\cdot] = \dfrac{d}{d\tau^2} + 2h\dfrac{d}{d\tau} + \rho(\tau)$.

Two linearlly independent solutions of Eq.(7.126) can be presented as:

$$y_1 = e^{\alpha\xi t}(U + \xi\varphi_1), \quad y_2 = e^{-(2h+\alpha\xi)t}(V + \xi\Psi_1), \tag{7.127}$$

where $\alpha = -\dfrac{(U, V)}{2[h(U, V) - (\dot{U}, V)]} = -\dfrac{(U, V)}{2[h(U, V) + (U, \dot{V})]}$, φ_1 and Ψ_1 are periodic functions with the same period as U and V. They are determined by the small perturbations method. If the tuned circuit is located on the left-hand boundary, for example at point M, at $\xi < 0$ this point shifts to the stable range – point M_1, at $\xi > 0$ it moves respectively to the unstable range at point M_2. This behaviour occurs when $\alpha > 0$. It can be shown that analogous processes take place on the right-hand stability boundary (for example point N), in the case of $\alpha < 0$, and at the lowest point P at $\alpha = 0$.

Let us have a closer look at this meaning of detuning as a degree of drifting away from resonance. Here we are mostly interested in the case when the straight line P_1P_2 in Fig.7.12 lies entirely in the stability range. This case is described by the following differential equation:

$$S[y] + 2\delta\dot{y} + \xi y = f, \tag{7.128}$$

where $\delta > 0$ is the additional damping caused by the fact that the straight line P_1P_2 is situated below the stability boundary.

Let us assume that resonance occurs when $\xi = 0$. Obviously it should be expected that the free oscillations equation

$$S[y] + 2\delta\dot{y} + \xi y = 0$$

has lineally independent solutions close to those for $\delta = 0$. By using these known solutions and applying the small perturbations method, we identify the following two lineally independent solutions:

$$y_1 = e^{(-\delta - \beta\xi^2)t}(U + \delta\varphi_1),\ y_2 = e^{-(\delta - \beta\xi^2 + 2h)t}(V + \xi\Psi_1), \tag{7.129}$$

where $\beta = -\dfrac{(U, \Psi_1)}{2w} = -\dfrac{(V, \varphi_1)}{2w} > 0$.

Taking into account (7.129) we find a solution of Eq.(7.128) at the following initial conditions $t = 0, y = 0, \dot{y} = 0$:

$$y = \frac{1}{2w}\left\{-e^{-(\delta + \beta\xi^2)t}(U + \xi\varphi_1)\int_0^t e^{(\delta + \beta\xi^2)t} f(V + \xi\Psi_1)dt\right.$$

$$\left. + e^{-(\delta - \beta\xi^2 + 2h)t}(V + \xi\Psi_1)\int_0^t e^{(2h + \delta - \beta\xi^2)t} f(U + \xi\varphi_1)dt\right\}. \tag{7.130}$$

Let us consider two cases: $(f, V) \neq 0$ and $(f, V) = 0$.

Let us posit $(f, V) \neq 0$. Then the major term in (7.130) is in the following form:

$$Y = -\frac{1 - e^{-(\delta + \beta\xi^2)t}}{2w(\delta + \beta\xi^2)}(f, V)U.$$

In the absence of detuning ($\xi = 0$) we have a typical case of first-power resonance. The resonance characteristic is bell-shaped. The so-called „strong" resonance occurs.

Let $(f, V) = 0$. The major term in (7.130) is

$$Y = \frac{\xi(f, \Psi_1)}{2w}\frac{1 - e^{-(\delta + \beta\xi^2)t}}{\delta + \beta\xi^2}U.$$

In the particular case when $f = U$, $(U, \Psi_1) = -2w\beta$, this expression takes the form:

$$Y = \frac{\beta\xi}{\delta + \beta\xi^2}\left[1 - e^{-(\delta + \beta\xi^2)t}\right]U.$$

The following expression is valid for the standard:

$$Y_\infty = \frac{\xi}{\delta' + \xi^2} U, \tag{7.131}$$

where $\delta' = \dfrac{\delta}{\beta}$ is the reduced damping.

It is obvious that (7.131) has an extremum with regard to the detuning ξ, which is determined by the following optimal values:

$$\xi_{opt} = \pm\sqrt{\delta'}. \tag{7.132}$$

By substituting (7.132) in (7.131) we obtain a formula for the oscillations with maximum amplitude:

$$Y_{\max} = \pm \frac{\sqrt{\beta}}{\sqrt{\delta}} U.$$

This corresponds to a two-humped tuning curve: at $\xi = 0$ there are no oscillations (actually they are not entirely absent, since they are not determined only by the major term of the solution which is analyzed here); when $\xi = \pm\sqrt{\delta'}$, the amplitude of the oscillations reaches its maximum value; a further increase in the module of detuning leads to monotonous decline in the intensity of the oscillations. This is the so-called „weak" resonance.

The „weak" resonance in a linear periodic resonator is a peculiar phenomenon that has no analogue in the ordinary harmonic resonators.

Let us consider the more general case, when the external force can be presented in the form $f = f_1 + f_2$, $(f_1, V) \neq 0$, $(f_2, V) = 0$, $(f_2, \Psi_1) \neq 0$.

It is obvious that the force f_1 yields „strong" resonance, while the force f_2 causes „weak" resonance.

In the typical particular case, when $f_1 = AV$, $f_2 = BU$, the major term in a stationary mode is of the form:

$$Y_\infty = \left(\frac{A}{2} + B\beta\xi\right) \frac{U}{\delta + \beta\xi^2}.$$

The extreme points with respect to the detuning ξ are identified as: $\xi_1 = \dfrac{B}{A}\delta$, $\xi_2 = -\dfrac{A}{\beta B} - \dfrac{B}{A}\delta$.

When $\xi = \xi_1$, a sharp and highly pronounced maximum is obtained:

$$Y_{\max} = \frac{A}{2\delta}\left(1 + \frac{B^2}{A^2}\beta\delta\right) U.$$

In the case of $\xi = \xi_2$, depending on the correlation between A and B, the result is either a less pronounced maximum or a less pronounced minimum.

Let us consider a relatively simple particular case:

$$\ddot{y} + 2h\dot{y} + (1 + m\cos 2t)y = 2\cos(1+\varepsilon)t, \quad \varepsilon \ll 1. \tag{7.133}$$

At $m \ll 1$, the natural oscillations of this equation are determined according to the formulae: $U = \cos t - \sin t$, $V = \cos t + \sin t$.

In compliance with the general theory of Resonance 3, the external force should be expanded in the following way: $f = P\dot{U} + g$, $(g, V) = 0$; g should be orthogonal with respect to V, and in our particular case this is equivalent to orthogonality of g with respect to $\dot{U}$.

The right-hand part of (7.133) is presented in the form

$$2\cos(1+\varepsilon)t = (\cos\varepsilon t - \sin\varepsilon t)(\cos t + \sin t) + (\cos\varepsilon t + \sin\varepsilon t)(\cos t - \sin t). \tag{7.134}$$

Since the first multipliers of both terms change slowly in time ($\varepsilon \ll 1$), they can be conventionally regarded as constant. Also it is not difficult to see that the functions $\cos t + \sin t$ and $\cos t - \sin t$ are orthogonal. This provides grounds for representing the expansion (7.134) as $P\dot{U} + Q\dot{V}$, where $P = \cos\varepsilon t - \sin\varepsilon t$, $Q = \cos\varepsilon t + \sin\varepsilon t$. The denotation $g = Q\dot{v}$ can be introduced, since $\dot{V}$ is always orthogonal with respect to V.

Thus the resonance term has been separated, it yields the following oscillation:

$$y_r = \frac{P}{2\delta}U = \frac{\cos\varepsilon t - \sin\varepsilon t}{2\delta}(\cos t - \sin t) = \frac{\sqrt{2}}{2\delta}\sin\left(\varepsilon t - \frac{\pi}{2}\right)(\cos t - \sin t)$$

As can be seen, at one point of time the resonance appears at maximum power with an effective oscillations value $\overline{y}_r = \sqrt{(y_r, y_r)} = \dfrac{\sqrt{2}}{2\delta}$, while at another point of time it fully disappears as $\overline{y}_r = 0$ in accordance with a periodic law with a period of $\dfrac{2\pi}{\varepsilon}$. The second term of the external force $Q\dot{V}$ is a quasi-resonance one. It causes oscillations:

$$y_{qr} = \frac{QV}{4h} = \frac{\cos\varepsilon t + \sin\varepsilon t}{4h}(\cos t + \sin t) = \frac{\sqrt{2}\cos\left(\varepsilon t - \frac{\pi}{4}\right)}{4h}(\cos t + \sin t).$$

Similarly, at one point of time the quasi-resonance appears at maximum power with an effective oscillations value $\overline{y}_{qr} = \dfrac{\sqrt{2}}{4h}$, while at another point of time it fully disappears in accordance with a periodic law with the same period. Moreover, when the resonance disappears, the quasi-resonance appears at maximum power and vice versa: when the quasi-resonance disappears, the resonance appears at maximum power. Since $\delta \ll h$, the resonance oscillations are by far more intensive than the quasi-resonance ones.

CHAPTER 8.

NONLINEAR OSCILLATING SYSTEMS WITH PARAMETERS CHANGING IN TIME

8.1. The principle of linear connection in the analysis of forced oscillations in a nonlinear oscillating circuit

The principle of linear connection postulates that each solution of a nonlinear differential equation can be matched with a linear equation of such form that its solution, given initial conditions coinciding with the initial conditions of the original equation, will be identical to the solution of the given nonlinear equation [112]. Unfortunately, formulated in this way, the principle only informs about such a possibility without providing recommendations for its practical implementation. A similar method is used below for analyzing a nonlinear oscillating circuit by using the method of complex amplitudes.

Let us consider the nonlinear differential equation with polynomial coefficients (See 4.3) [187, 188, 269]:

$$\frac{d^2x}{dt^2}+\left(a_0+\sum_{i=1}^{m}a_ix^i\right)\frac{dx}{dt}+\left(b_0+\sum_{i=1}^{n}b_ix^i\right)x=A\cos(\Omega t+\alpha)+a\cos(\omega t+\beta),\ a\ll A. \tag{8.1}$$

This equation describes the processes in a generator with synchronizing action $A\cos(\Omega t+\alpha)$, or in a nonlinear oscillating circuit in a forced regime to which a signal with small intensity is sent: $a\cos(\omega t+\beta)$. The first term in the right-hand part of the equation plays a double role: a) it changes the equation coefficients in time, and b) performs the functions of a free term (or external excitation). The second term should be regarded only as a free term by virtue of its small intensity with respect to the first term. Hence the response evoked by the second term will be analogous to the one corresponding to a linear equation with variable coefficients, which are changed in time by the first term in the right-hand part. Therefore Eq.(8.1) is solved in two stages. During the first stage the equation is solved without taking into account the second term on the right-hand side of the equation:

$$\frac{d^2x_0}{dt^2}+\left(a_0+\sum_{i=1}^{m}a_ix_0^i\right)\frac{dx_0}{dt}+\left(b_0+\sum_{i=1}^{n}b_ix_0^i\right)x_0=A\cos(\Omega t+\alpha). \tag{8.2}$$

Then the solution of the initial equation (8.1) is presented in the form:

$$x(t)=x_0(t)+\delta(t), \tag{8.3}$$

where the condition $(\delta,\delta) \ll (x_0, x_0)$ is satisfied.

Substituting (8.3) in (8.1) we obtain

$$\frac{d^2\delta}{dt^2} + \left(a_0 + \sum_{i=1}^{m} a_i x_0^i\right)\frac{d\delta}{dt} + \left(b_0 + 2\sum_{i=1}^{n} b_i x_0^i \right.$$

$$\left. + \sum_{i=1}^{m} a_0 x^{i-1}\frac{dx_0}{dt}\right)\delta = a\cos(\omega t + \beta). \tag{8.4}$$

This is a linear differential equation whose coefficients are periodic time functions with angular frequency Ω.

In order to solve Eq.(8.2) we use the generalized method of complex amplitudes (See 5.1). We write

$$\frac{d^2\hat{x}_0}{dt^2} + \left[a_0 + \sum_{i=1}^{m} a_i (Re\hat{x}_0)^i\right]\frac{d\hat{x}_0}{dt} + \left[b_0 + \sum_{i=1}^{n} b_i (Re\hat{x}_0)^i\right]\hat{x}_0 = \dot{A}e^{j\Omega t}. \tag{8.5}$$

Equation (8.5) is solved in accordance with the method of successive approximations. The following equation is zero-order approximation of Eq.(8.5)

$$\frac{d^2\hat{x}^{(0)}}{dt^2} + a_0\frac{d\hat{x}_0^{(0)}}{dt} + b_0\hat{x}_0^{(0)} = \dot{A}e^{j\Omega t}. \tag{8.6}$$

Its solution is

$$\hat{x}_0^{(0)} = \dot{X}_0^{(0)}e^{j\Omega t}, \quad \dot{X}_0^{(0)} = \frac{\dot{A}}{b_0 - \Omega^2 + ja_0\Omega}.$$

Now that we have this solution, we draw up the first approximation of Eq.(8.5):

$$\frac{d^2\hat{x}_0^{(1)}}{dt^2} + \left[a_0 + \sum_{i=1}^{m} a_i (Re\hat{x}_0^{(0)})^i\right]\frac{d\hat{x}_0^{(1)}}{dt} + \left[b_0 + \sum_{i=1}^{n} b_i (Re\hat{x}_0^{(0)})^i\right]\hat{x}_0^{(1)} = \dot{A}e^{j\Omega t}. \tag{8.7}$$

The function $\hat{x}_0^{(0)}$ is known: $Re\hat{x}_0^{(0)} = X_0^{(0)}\cos(\Omega t + \varphi^{(0)})$, $\operatorname{tg}\varphi^{(0)} = \dfrac{-a_0\Omega}{b_0 - \Omega^2}$. The Eq.(8.7) is a linear differential equation with periodic coefficients with a period $T = \dfrac{2\pi}{\Omega}$. Its solution is of the form

$$\hat{x}_0^{(1)} = \sum_{k=-\infty}^{\infty} \dot{X}_k^{(1)}e^{jk\Omega t}, \quad \dot{X}_k^{(1)} = X_k^{(1)}e^{j\varphi_k^{(1)}}.$$

The values $X_k^{(1)}$ and $\varphi_k^{(1)}$ (or their equivalent values $\dot{X}_k^{(1)}$), $k = -\infty, \ldots, -1, 0, 1, \ldots, \infty$ should be determined by the solution of Eq.(8.7). The finite set of these values can be approximately determined in the way described below. Going on with the solving process, we draw up the following approximation for equation (8.5). Thus we arrive at the l-th approximation

$$\frac{d^2\hat{x}_0^{(l)}}{dt^2} + \left[a_0 + \sum_{i=1}^{m} a_i (Re\hat{x}_0^{(l-1)})^i\right] \frac{d\hat{x}_0^{(l)}}{dt} + \left[b_0 + \sum_{i=1}^{n} b_i (Re\hat{x}_0^{(l-1)})^i\right] \hat{x}_0^{(l)} = \dot{A}e^{j\Omega t}. \quad (8.8)$$

It is noteworthy that the methods of solving all approximations of Eq.(8.5) are of the same type, except for the zero-order approximation (8.6). That is why we shall elaborate on the l-th approximation on the condition that the solution of approximation $(l-1)$ is known:

$$\left| \begin{array}{l} \hat{x}_0^{(l-1)} = \sum\limits_{k=-\infty}^{\infty} X_k^{(l-1)} e^{jk\Omega t}; \quad \dot{X}_k^{(l-1)} = X_k^{(l-1)} e^{j\varphi_k^{(l-1)}} \\ x_0^{(l-1)} = Re\hat{x}_0^{(l-1)} = \sum\limits_{k=-\infty}^{\infty} X_k^{(l-1)} \cos(k\Omega t + \varphi_k^{(l-1)}). \end{array} \right. \quad (8.9)$$

On this condition, Eq.(8.9) is a linear differential equation with periodic coefficients. Its solution is of the form:

$$\hat{x}_0^{(l)} = \sum_{k=-\infty}^{\infty} X_k^{(l)} e^{jk\Omega t}; \quad \dot{X}_k^{(l)} = X_k^{(l)} e^{j\varphi_k^{(l)}}. \quad (8.10)$$

So the task consists in finding, strictly speaking, an infinite set of complex numbers $\dot{X}_k^{(l)}$, $k = -\infty, \ldots, -1, 0, 1, \ldots, \infty$. Only in this case the series (8.10) will be strictly and accurately determined. Naturally, in reality it is never required that a periodic function should be computed precisely; it is sufficient to identify a certain number of complex amplitudes. The expansion (8.10) contains one constant component $\dot{x}_0^{(0)}$, two first harmonics with complex amplitudes $\dot{x}_1^{(l)}$, $\dot{x}_{-1}^{(l)}$ and frequencies Ω and $-\Omega$, two second harmonics with frequencies 2Ω, -2Ω, etc., two q-th harmonics with frequencies $q\Omega$, $-q\Omega$, where q is a positive integer. So it is necessary to determine $\dot{x}_k^{(1)}$, $k = -q, \ldots, -1, 0, 1, \ldots, q$.

Bearing in mind that the solution for approximation $(l-1)$ is known, we shall present equation (8.8) in the following way:

$$\frac{d^2\hat{x}_0^{(l)}}{dt^2} + \left(\sum_{p=-\infty}^{\infty} \dot{A}_p e^{jp\Omega t}\right) \frac{d\hat{x}_0^{(l)}}{dt} + \left(\sum_{p=-\infty}^{\infty} \dot{B}_p e^{jp\Omega t}\right) \hat{x}_0^{(l)} = \dot{A}e^{j\Omega t}, \quad (8.11)$$

$$\dot{A}_{-p} = \overset{*}{A}_p, \quad B_{-p} = \overset{*}{B}_p.$$

The coefficients in (8.11) are presented as Fourier series.

When (8.10) is substituted in (8.11) and the respective sums are multiplied, the following is obtained:

$$\sum_{k=-\infty}^{\infty}\left\{-k^2\Omega^2\dot{x}_k^{(l)} + \sum_{p=-\infty}^{\infty}[j(k-p)\Omega\dot{A}_p + \dot{B}_p]\dot{x}_{k-p}^{(l)}\right\} e^{jk\Omega t} = \dot{A}e^{j\Omega t}. \tag{8.12}$$

As terms with identical imaginary indices are selected and the common exponential multiplier is cancelled, a linear algebraic summation of an infinite number of equations is obtained

$$\mathbf{A}.\mathbf{x} = \mathbf{b}, \tag{8.13}$$

$\mathbf{A} = \{a_{ij}\}$, $\mathbf{x} = \{X_i^{(l)}\}$, $\mathbf{b} = \{b_i\}$, $i, j = -\infty, \ldots, 1$.

$b_1 = \dot{A}$, $b_i = 0$ at $i \neq 1$, $\mathbf{x}$ and $\mathbf{b}$ are vector-columns.

All elements of matrix $\mathbf{A}$ can be expressed by using two formulae:

$a_{k,k} = -k^2\Omega^2 + jk\Omega\dot{A}_0 + \dot{B}_0$, $\quad a_{k,l} = jl\Omega\dot{A}_{k-l} + \dot{B}_{k-l}$, $k \neq l$,

$k, l = -\infty, \ldots, -1, 0, 1, \ldots, \infty$.

In an extended form the infinite system (8.13) can be written in the following way:

$$\begin{bmatrix} a_{-2,-2} & a_{-2,-1} & a_{-2,0} & a_{-2,1} & a_{-2,2} \\ a_{-1,-2} & a_{-1,-1} & a_{-1,0} & a_{-1,1} & a_{-1,2} \\ a_{0,-2} & a_{0,-1} & a_{0,0} & a_{0,1} & a_{0,2} \\ a_{1,-2} & a_{1,-1} & a_{1,0} & a_{1,1} & a_{1,2} \\ a_{2,-2} & a_{2,-1} & a_{2,0} & a_{2,1} & a_{2,2} \end{bmatrix} \begin{bmatrix} \dot{x}_{-2}^{(l)} \\ \dot{x}_{-1}^{(l)} \\ \dot{x}_{0}^{(l)} \\ \dot{x}_{1}^{(l)} \\ \dot{x}_{2}^{(l)} \end{bmatrix} = \begin{bmatrix} \\ \\ \\ \dot{A} \\ \\ \end{bmatrix}.$$

This system can be solved by using the reduction method. In the first approximation we limit ourselves to a system of 3 equations, whose matrix contains 9 elements of the matrix of system (8.13) with a central element a_{00}, while the vector columns contain 3 elements each. In the second approximation, 5 such equations are taken and in the q-th approximation, the number of the equations is $(2q+1)$. As we solve the system in the q-th approximation, we obtain:

$$\hat{x}_0^{(l)} = \sum_{k=-q}^{q} \dot{X}_k^{(l)} e^{jk\Omega t}; \quad \dot{X}_k^{(l)} = X_k^{(l)} e^{j\varphi_k^{(l)}}.$$

This is used to find the solution of the initial equation (8.2):

$$x^{(l)} = Re\hat{x}_0^{(l)} = \sum_{k=-q}^{q} X_k^{(l)} \cos(k\Omega t + \varphi_k^{(l)}). \tag{8.14}$$

The issue of the evaluation of the obtained l-th approximation crops up. Since the difference $x_0^{(l)} - x_0^{(l-1)}$ is a periodic function, it is desirable that the evaluation should be carried out on the basis of a definite positive number. For that purpose the concept of a scalar square of this difference is once again used and in this case it is written as follows:

$$\begin{aligned}\Delta_l^2 &= (x_0^{(l)} - x_0^{(l-1)}, x_0^{(l)} - x_0^{(l-1)}) \\ &= \frac{1}{2}\sum_{k=-q}^{q} [X_k^{(l)^2} + X_k^{(l-1)^2} - 2X_k^{(l)}X_k^{(l-1)}\cos(\varphi_k^{(l)} - \varphi_k^{(l-1)})].\end{aligned}$$

The positive number Δ_l should be possibly smaller, for example it should be one order smaller than the admissible error.

The method described above allows obtaining an approximate solution (8.14) of Eq.(8.2), while the error is due to: a) the rejection of the higher harmonics with numbers $-\infty, \ldots, -q-2, -q-1, q+1, q+2, \ldots, \infty$; b) the inaccurate computation of the harmonics with numbers $-q, \ldots, -1, 0, 1, \ldots, q$ included in the solution.

Having obtained the solution (8.14) of Eq.(8.2), we proceed to solve equation (8.4) in the l-th approximation. If $x_0(t)$ is known, the Eq.(8.4) in the l-th approximation is a linear differential equation with periodic coefficients with period T. Therefore it can be solved in an analogous way, but some peculiarities should be taken into account. While in (8.2) the periods of the coefficients and the right-hand part of the equation coincide, in (8.4) the periods of the coefficients $T = \dfrac{2\pi}{\Omega}$ and those of the right-hand part $T_1 = \dfrac{2\pi}{\omega}$ differ. Therefore the solution of Eq.(8.2) is a periodic function, while the solution of Eq.(8.4) is a quasi-periodic function

$$\delta(t) = Re\hat{\delta}(t) = Re \sum_{k=-\infty}^{\infty} \dot{D}_k e^{j(\omega+k\Omega)t}, \quad \dot{D}_k = D_k e^{j\Psi_k}.$$

The problem, as before, leads to an infinite system of equations, analogous to (8.13). Solving it approximately, we obtain

$$\hat{\delta}(t) \approx \sum_{k=-q}^{q} \dot{D}_k e^{j(\omega+k\Omega)t} \tag{8.15}$$

In an analogous way, in order to evaluate the obtained p-th approximation, we construct the scalar square of the difference $\delta^{(p)} - \delta^{(p-1)}$:

$$\Delta_p^2 = (\delta^{(p)} - \delta^{(p-1)}, \delta^{(p)} - \delta^{(p-1)})$$
$$= \frac{1}{2} \sum_{k=-q}^{q} [D_k^{(p)^2} + D_k^{(p-1)^2} - 2D_k^{(p)} D_k^{(p-1)} \cos(\Psi_k^{(p)} - \Psi_k^{(p-1)})].$$

The obtained expressions (8.14) and (8.15) are used for setting up the approximate solution of Eq.(8.1):

$$x(t) = Re\dot{x}(t) = Re \sum_{k=-q}^{q} (\dot{x}_k^{(l)} + \dot{D}_k e^{j\omega t}) e^{jk\Omega t}.$$

The proposed method of solving the nonlinear equation (8.1) does not go beyond the limits of the linear theory. The convergence of the iteration process of the solution, as well as the convergence of the infinite system (8.13) are entirely determined by the properties of the coefficients of Eq.(8.1). If these coefficients represent sufficiently smooth time functions and their sign remains unchanged, the solution is convergent. Usually in reality the coefficients are ideally smooth (they have all derivatives). Hence the doubtful cases should be treated as exceptions, furthermore it would be expedient to quote convergence evidence only when such peculiar specific problems are tackled.

The described analytical approach allows substituting the complex task of analyzing nonlinear systems with a number of simpler tasks of analyzing linear systems with time dependent parameters. The well-known mathematical principle postulating this possibility has not yet become popular in Radiophysics and Mechanics. The method of complex amplitudes allows us to vest this principle in a form suitable for research. Moreover, the use of matrix algebra provides a possibility for compact written presentation and orderly transformations.

We shall present here yet another possible version of the realization of the linear connection principle.

In order to be more specific, we shall consider a system of two linked oscillating circuits with internal capacitance connection (Fig.8.1). We choose as major variables the voltages of the capacitances U_C, U_{C_1}, U_{C_2} and the currents of the inductances i_{L_1} and i_{L_2}. For the sake of attaining more general results, all elements of the system presented in Fig.8.1. will be considered as nonlinear. Further on, the index „*s*" will be used to denote the static parameters of the elements, while index „*d*" will be employed to designate their dynamic parameters. The system is described by the following five first-order differential equations.

$$\left|\begin{array}{l} C_{d_1}(U_{C_1})\dfrac{dU_{C_1}}{dt} + G_{s_1}(U_{C_1})U_{C_1} - i_{L_1} = 0 \\ L_{d_1}(i_{L_1})\dfrac{di_{L_1}}{dt} + U_{C_1} + R_{s_1}i_{L_1} + U_C = U_g \\ C_d(U_C)\dfrac{dU_C}{dt} - i_{L_1} + G_s(U_C)U_C + i_{L_2} = 0 \\ C_{d_2}(U_{C_2})\dfrac{dU_{C_2}}{dt} + G_{s_2}(U_{C_2})U_{C_2} - i_{L_2} = 0 \\ L_{d_2}(i_{L_2})\dfrac{di_{L_2}}{dt} - U_C + U_{C_2} + R_{s_2}(i_{L_2})i_{L_2} = 0, \end{array}\right. \tag{8.16}$$

where U_g is the exciting voltage.

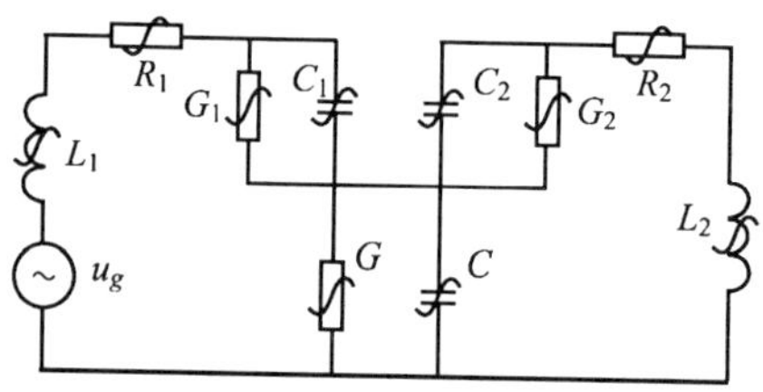

Fig. 8.1. A system of two coupled oscillating circuits with internal capacitance connection

We proceed with normalized variables: $\tau = \dfrac{t}{t_{00}}$, $x_1 = \dfrac{U_{C_1}}{U_{00}}$, $x_2 = \dfrac{i_{L_1}}{i_{00}}$, $x_3 = \dfrac{U_C}{U_{00}}$, $x_4 = \dfrac{U_{C_2}}{U_{00}}$, $x_5 = \dfrac{i_{L_2}}{i_{00}}$. A scale resistance is also introduced: $r_{00} = \dfrac{U_{00}}{i_{00}}$. The parameters of the elements can be presented in the following normalized form:

$$C_1(x_1) = C_{d_1}(x_1)\frac{r_{00}}{t_{00}}, \qquad g_1(x_1) = G_{s_1}(x_1)r_{00},$$

$$l_1(x_2) = L_{d_1}(x_2)\frac{1}{t_{00}r_{00}}, \quad r_1(x_2) = \frac{R_{s_1}(x_2)}{r_{00}},$$

$$C(x_3) = C_d(x_3)\frac{r_{00}}{t_{00}}, \qquad g(x_3) = r_{00}G_s(x_3),$$

$$C_2(x_4) = C_{d_2}(x_4)\frac{r_{00}}{t_{00}}, \qquad g_2(x_4) = G_{s_2}(x_4)r_{00},$$

$$l_2(x_5) = L_{d_2}(x_5)\frac{1}{t_{00}r_{00}}, \quad r_2(x_5) = \frac{R_{s_2}(x_5)}{r_{00}},$$

$$f_2(\tau) = \frac{U_g(\tau)}{U_{00}}.$$

As these designations are taken into account, the system (8.16) can be written in a vector form:

$$
\begin{array}{|c|c|c|c|c|}
\hline
C_1(x_1) & & & & \\ \hline
 & l_1(x_2) & & & \\ \hline
 & & C_1(x_1) & & \\ \hline
 & & & C_2(x_4) & \\ \hline
 & & & & l_2(x_5) \\ \hline
\end{array}
\;\frac{d}{d\tau}\;
\begin{array}{|c|}
\hline x_1 \\ \hline x_2 \\ \hline x_3 \\ \hline x_4 \\ \hline x_5 \\ \hline
\end{array}
$$

$$
+\begin{array}{|c|c|c|c|c|}
\hline
g_1(x_1) & -1 & & & \\ \hline
1 & r_1(x_2) & & & \\ \hline
 & -1 & g(x_3) & & 1 \\ \hline
 & & & g_2(x_4) & -1 \\ \hline
 & & -1 & 1 & r_2(x_5) \\ \hline
\end{array}
\;
\begin{array}{|c|}
\hline x_1 \\ \hline x_2 \\ \hline x_3 \\ \hline x_4 \\ \hline x_5 \\ \hline
\end{array}
=
\begin{array}{|c|}
\hline \\ \hline f_2(\tau) \\ \hline \\ \hline \\ \hline \\ \hline
\end{array}
\tag{8.17}
$$

Parallel with the vector-column $\mathbf{X} = \mathrm{colon}(x_1, x_2, \ldots, x_5)$, we also introduce a diagonal matrix containing the same elements: $\mathbf{x} = \mathrm{diag}(x_1, x_2, \ldots, x_5)$. Now we can present system (8.17) in a concise form:

$$
\mathbf{A}(\mathbf{X})\frac{d}{d\tau}\mathbf{x} + \mathbf{B}(\mathbf{X})\mathbf{x} = \mathbf{f}(\tau). \tag{8.18}
$$

The matrix elements depending on the elements of the unknown vector-column are expanded in a Taylor series and the matrixes are transformed in the following way:

$$
\mathbf{B}(\mathbf{X})=
\begin{array}{|c|c|c|c|c|}
\hline
g_1^{(0)}+g_1^{(1)}x_1+g_1^{(2)}x_1^2+\cdots & -1 & & & \\ \hline
1 & r_1^{(0)}+r_1^{(1)}x_2+r_1^{(2)}x_2^2+\cdots & 1 & & \\ \hline
 & -1 & g^{(0)}+g^{(1)}x_3+g^{(2)}x_3^2+\cdots & & 1 \\ \hline
 & & & g_2^{(0)}+g_2^{(1)}x_4+g_2^{(2)}x_4^2+\cdots & \\ \hline
 & & -1 & 1 & g_2^{(0)}+r_2^{(1)}x_5+r_2^{(2)}x_5^2+\cdots \\ \hline
\end{array}
=
\begin{array}{|c|c|c|c|c|}
\hline
g_1^{(0)} & -1 & & & \\ \hline
1 & r_1^{(0)} & 1 & & \\ \hline
 & -1 & g^{(0)} & & 1 \\ \hline
 & & & g_2^{(0)} & -1 \\ \hline
 & & -1 & 1 & r_2^{(0)} \\ \hline
\end{array}
$$

$$+\begin{array}{|c|c|c|c|c|}\hline g_1^{(1)}x_1 & & & & \\ \hline & r_1^{(1)}x_2 & & & \\ \hline & & g^{(1)}x_3 & & \\ \hline & & & g_2^{(1)}x_4 & \\ \hline & & & & r_2^{(1)}x_5 \\ \hline \end{array} + \begin{array}{|c|c|c|c|c|}\hline g_1^{(2)}x_1^2 & & & & \\ \hline & r_1^{(2)}x_2^2 & & & \\ \hline & & g^{(2)}x_3^2 & & \\ \hline & & & g_2^{(2)}x_4^2 & \\ \hline & & & & r_2^{(2)}x_5^2 \\ \hline \end{array} + \cdots$$

$$= \mathbf{B_0} + \mathbf{B_1 X} + \mathbf{B_2 X^2} + \cdots$$

The result is an expansion in a matrix Taylor series – the matrix $\mathbf{B}$ by the powers of matrix $\mathbf{X}$. This expansion can be used for presenting the vector equation (8.18) in the form

$$(\mathbf{A}_0 + \mathbf{A}_1\mathbf{X} + \mathbf{A}_2\mathbf{X}^2 + \ldots)\frac{d}{d\tau}\mathbf{x} + (\mathbf{B}_0 + \mathbf{B}_1\mathbf{X} + \mathbf{B}_2\mathbf{X}^2 + \ldots)\mathbf{x} = \mathbf{f}(\tau). \qquad (8.19)$$

This type of vector differential equation is convenient for the iteration procedures. In radiophysical problems, it is particularly convenient to construct iterations on the basis of the method of complex amplitudes. Usually in these problems the elements of the vector columns $\mathbf{f}$ and $\mathbf{x}$ can be presented in the form of summations of harmonic components, i.e. these are almost periodic functions or their particular cases (quasi-periodic, periodic functions). Then it is convenient to use the extended generalized method of complex amplitudes (Chapter 5, It. 5.1).

The transform of (8.19) is of the form:

$$\left[\mathbf{A}_0 + \mathbf{A}_1\frac{\hat{\mathbf{X}} + \check{\mathbf{X}}}{2} + \mathbf{A}_2\left(\frac{\hat{\mathbf{X}} + \check{\mathbf{X}}}{2}\right)^2 + \ldots\right]\frac{d}{d\tau}\hat{\mathbf{x}}$$

$$+\left[\mathbf{B}_0 + \mathbf{B}_1\frac{\hat{\mathbf{X}} + \check{\mathbf{X}}}{2} + \mathbf{B}_2\left(\frac{\hat{\mathbf{X}} + \check{\mathbf{X}}}{2}\right)^2 + \ldots\right]\hat{\mathbf{x}} = \hat{\mathbf{f}}(\tau). \qquad (8.20)$$

The iteration method should be regarded as a natural general method for solving such equations:

0) Zero approximation of Eq.(8.20) is the equation obtained in the case when $\mathbf{A}_k = 0$, $\mathbf{B}_k = 0$, $k = 1, 2, \ldots$, i.e.

$$\mathbf{A}_0 \frac{d}{d\tau} \hat{\mathbf{x}}^{(0)} + \mathbf{B}_0 \hat{\mathbf{x}}^{(0)} = \hat{\mathbf{f}}(\tau).$$

This is a vector differential equation with constant coefficients. As we solve it, we determine $\hat{\mathbf{x}}^{(0)}$ and $\hat{\mathbf{X}}^{(0)}$, $\check{\mathbf{X}}^{(0)}$. By substituting these matrices in (8.20) we obtain

1) First approximation, i.e.

$$\left(\mathbf{A}_0 + \mathbf{A}_1 \frac{\hat{\mathbf{X}}^{(0)} + \check{\mathbf{X}}^{(0)}}{2} + \ldots\right) \frac{d}{d\tau} \hat{\mathbf{x}}^{(1)} + \left(\mathbf{B}_0 + \mathbf{B}_1 \frac{\hat{\mathbf{X}}^{(0)} + \check{\mathbf{X}}^{(0)}}{2} + \ldots\right) \hat{\mathbf{x}}^{(1)} = \mathbf{f}(\tau). \tag{8.21}$$

Here the coefficients represent known functions with argument τ, i.e. this is a linear vector equation with variable time-dependent coefficients. As we solve it, we determine $\hat{\mathbf{x}}^{(1)}$ and $\hat{\mathbf{X}}^{(1)}$, $\check{\mathbf{X}}^{(1)}$. By substituting the latter matrices in (8.20), we obtain a second approximation, etc. In the k-th approximation, the coefficients of an equation of the type of (8.21) contain matrices $\hat{\mathbf{X}}^{(k-1)}$, $\check{\mathbf{X}}^{(k-1)}$ and a vector-column $\hat{\mathbf{x}}^{(k)}$. A vector equation with variable coefficients is solved at each stage, and this is simpler than the solution of the original nonlinear equation (8.18). In the k-th approximation, we determine the vector function $\mathbf{x}^{(k)}(\tau) = Re\hat{\mathbf{x}}^{(k)}(\tau)$, which should be close to the vector function $\mathbf{x}(\tau)$ that we are trying to identify. In the $(k-1)$-th approximation, the vector function $\mathbf{x}^{(k-1)}(\tau)$ is obtained with effective meaning $x_{eff}^{(k-1)} = \sqrt{(x^{(k-1)}, x^{(k-1)})}$; in the k-th approximation, respectively $\mathbf{x}^{(k)}(\tau)$ and $x_{eff}^{(k)}$. The effective difference is a positive number $\Delta_k = \left| x_{eff}^{(k-1)} - x_{eff}^{(k)} \right|$.

Normally the iteration procedure is terminated when the effective difference is one order smaller than the error given in advance.

8.2. „Strong" and „weak" resonance in a nonlinear system with explicitly time-dependent parameters

The regenerated oscillating systems (underexcited generators) containing active devices with the respective feedbacks are of particular interest for the functional radiophysical systems [270]. Most often the regenerated oscillating systems belong to the class of systems with linear periodic resistance and nonlinear reactive element described by the following differential equation:

$$\frac{d^2U}{dt^2} + (k - p\cos 2\omega t)\frac{dU}{dt} + \left[\omega_0^2 + \frac{q}{2}\frac{d(U^2)}{dt}\right] U = \omega_0^2 \varepsilon \cos(\omega t + \gamma), \tag{8.22}$$

where U is the voltage on the capacitor of the oscillating circuit, $\varepsilon\cos(\omega t+\gamma)$ is the external acting signal, k, p, q and ω_0 are parameters. If, for example, the voltage U changes in time with angular frequency ω, the reactive element containing a term proportional to $\dfrac{dU^2}{dt}$ will change with angular frequency 2ω. In this case either first stability zone or first instability zone will be realized [89].

Driven by general physical considerations, we write the solution of Eq.(8.22) in the form:

$$U=\sum_{k=0}^{\infty}U_{2k+1}\cos[(2k+1)\omega t+\varphi_{2k+1}]\leftarrow\hat{U}=\sum_{k=-\infty}^{\infty}\dot{U}_{2k+1}e^{j(2k+1)\omega t}.$$

Usually in practice the parameters p, q, ε are sufficiently small, so we can confine ourselves to the first harmonic:

$$\hat{U}=\dot{U}_1e^{j\omega t}+\dot{U}_{-1}e^{-j\omega t}.$$

Transforming (8.22) by applying the method of complex amplitudes, we obtain the determining equation, which can be broken down into the following two nonlinear algebraic equations with respect to the terms with positive and negative frequencies:

$$(\omega_0^2-\omega^2+j\omega k)\dot{U}_1+\frac{j\omega p}{2}\dot{U}_{-1}-\frac{j\omega q}{2}(U_1^2+\overset{*}{U}_{-1}\overset{*}{U}_1+\frac{1}{2}\dot{U}_1U_{-1})\dot{U}_1=\omega_0^2\dot{\varepsilon},\quad(8.23)$$

$$(\omega_0^2-\omega^2-j\omega k)\dot{U}_{-1}-\frac{j\omega p}{2}\dot{U}_1+\frac{j\omega q}{4}U_1^2\overset{*}{U}_1=0.\qquad(8.24)$$

After determining $\dot{U}_{-1}$ from (8.24) and substituting the expression in (8.23), we obtain:

$$\begin{aligned}&(\omega_0^2-\omega^2)\dot{U}_1+j\omega\left(k+\frac{q}{2}U_1^2\right)\dot{U}_1-\frac{\omega^2p^2}{4}\frac{\omega_0^2-\omega^2+j\omega k}{(\omega_0^2-\omega^2)^2+\omega^2k^2}\dot{U}_1\\&-\frac{\omega_0^2-\omega^2+j\omega k}{(\omega_0^2-\omega^2)^2+\omega^2k^2}\frac{\omega^2}{8}pq\dot{U}_1^3+\frac{3(\omega_0^2-\omega^2)-j\omega k}{(\omega_0^2-\omega^2)^2+\omega^2k^2}\frac{\omega^2pq}{8}U_1^2\overset{*}{U}_1\\&-\frac{\omega^2q^2}{16}\frac{\omega_0^2-\omega^2+j\omega k}{(\omega_0^2-\omega^2)^2+\omega^2k^2}=\omega_0^2\dot{\varepsilon}.\end{aligned}\qquad(8.25)$$

If the real and imaginary parts are equated separately and the terms $\sim q^2$, $\sim p^2$, $\sim qp$ are ignored, the following system of two algebraic equations will be obtained:

$$\left|\begin{aligned}&(\omega_0^2-\omega^2)U_1\cos\varphi_1-\frac{\omega p}{2}U_1\sin\varphi_1-\omega U_1\left(k+\frac{q}{4}U_1^2\right)\sin\varphi_1=\omega_0^2\varepsilon\cos\gamma\\&(\omega_0^2-\omega^2)U_1\sin\varphi_1-\frac{\omega p}{2}U_1\cos\varphi_1+\omega U_1\left(k+\frac{q}{4}U_1^2\right)\cos\varphi_1=\omega_0^2\varepsilon\sin\gamma.\end{aligned}\right.\qquad(8.26)$$

When the trigonometric functions $\cos\varphi_1$ and $\sin\varphi_1$ are excluded from system (8.26), the following is obtained:

$$U_1^2 = \varepsilon^2 \frac{\xi^2 + \dfrac{\overline{\omega}^2\overline{p}^2}{4} + \overline{\omega}^2(\overline{k} + \dfrac{1}{4}\overline{q}U_1^2)^2 + \overline{\omega}\overline{p}\xi\sin 2\gamma - \overline{\omega}^2\overline{p}(\overline{k} + \dfrac{1}{4}\overline{q}U_1^2)\cos 2\gamma}{[\xi^2 + \overline{\omega}^2(\overline{k} + \dfrac{1}{4}\overline{q}U_1^2)^2 - \dfrac{1}{4}\overline{\omega}^2\overline{p}^2]^2}, \quad (8.27)$$

where $\xi = \dfrac{\omega_0^2 - \omega^2}{\omega^2}$, $\overline{\omega} = \dfrac{\omega}{\omega_0}$, $\overline{k} = \dfrac{k}{\omega_0}$, $\overline{p} = \dfrac{p}{\omega_0}$, $\overline{q} = \dfrac{q}{\omega_0}$.

The expression (8.27) is a polynomial of the fifth power with respect to U_1^2. By introducing the denotation $\nu = k + \dfrac{1}{4}qU_1^2$ this polynomial can be presented in a nominal form:

$$\nu^5 - \overline{k}\nu^4 + 2\left(\frac{\xi^2}{\overline{\omega}^2} - \frac{1}{4}\overline{p}^2\right)\nu^3 - \left[2\overline{k}\left(\frac{\xi^2}{\overline{\omega}^2} - \frac{1}{4}\overline{p}^2\right) - \frac{\overline{q}}{4\overline{\omega}^2}\varepsilon^2\right]\nu^2$$
$$+ \left[\left(\frac{\xi^2}{\overline{\omega}^2} - \frac{1}{4}\overline{p}^2\right)^2 + \frac{\varepsilon^2}{4}\frac{\overline{pq}}{\overline{\omega}^3}\cos 2\gamma\right]\nu - \left[\overline{k}\left(\frac{\xi^2}{\overline{\omega}^2} - \frac{1}{4}\overline{p}^2\right)^2\right.$$
$$\left. + \frac{\varepsilon^2}{4\overline{\omega}^2}\overline{q}\left(\frac{\xi^2}{\overline{\omega}^2} + \frac{\overline{p}\xi}{\omega}\sin 2\gamma + \overline{p}^2\right)\right] = 0.$$

Let us consider the case of $\varepsilon = 0$ in the absence of external excitation. It follows from (8.27) that this is a case when a trivial process $U_1^2 = 0$ exists, but it is not the only one, since the sufficient condition for its realization is a zero denominator in the right-hand part. If this condition is not satisfied, other processes may also occur in the system. As the denominator is equated to zero, the following is obtained:

$$U_1^2 = \frac{4}{q}\left(\sqrt{\frac{\overline{p}^2}{4} - \frac{\xi^2}{\overline{\omega}^2}} - \overline{k}\right). \quad (8.28)$$

Equation (8.26) is used for determining the initial phase of this process

$$\operatorname{tg}\varphi_1 = \sqrt{\frac{\overline{\omega}\overline{p} - \sqrt{\overline{\omega}^2\overline{p}^2 - 4\xi^2}}{\overline{\omega}\overline{p} + \sqrt{\overline{\omega}^2\overline{p}^2 - 4\xi^2}}} = \frac{2\xi}{\overline{\omega}\overline{p} + \sqrt{\overline{\omega}^2\overline{p}^2 - 4\xi^2}}.$$

It is typical that the more pronounced the nonlinearity, the smaller the susceptibility of the oscillating circuit to driving (U_1^2 is $\sim \dfrac{1}{q}$).

Let us consider the case when $\varepsilon \neq 0$, $\overline{q} = 0$ (linear case). Equation (8.27) is used to obtain

$$U_1^2 = \varepsilon^2 \frac{\xi^2 + \dfrac{\overline{\omega}^2\overline{p}^2}{4} + \overline{\omega}^2\overline{k}^{2tet53d+}\,\overline{\omega}\overline{p}\xi\sin 2\gamma - \overline{\omega}^2\overline{p}\overline{k}\cos 2\gamma}{\left(\xi^2 + \omega^2\overline{k}^2 - \dfrac{\overline{\omega}^2\overline{p}^2}{4}\right)^2}. \quad (8.29)$$

The resonance is manifested at $\omega = \omega_0$, i.e. $\xi = 0$, $\overline{\omega} = 1$. For this frequency formula (8.29) assumes the form

$$U_1^2 = \varepsilon^2 \frac{\dfrac{\overline{p}^2}{4} + \overline{k}^2 - \overline{p}\overline{k}\cos 2\gamma}{\left(\overline{k}^2 - \dfrac{\overline{p}^2}{4}\right)^2}. \tag{8.30}$$

Depending on the initial phase γ, two types of resonance are possible at this frequency – „strong" resonance, when $\cos 2\gamma = -1$, and „weak" resonance in the case of $\cos 2\gamma = 1$. Further on we shall use the index „s" for the „strong" resonance and the index „w" for the „weak" one.

In the case of the „strong" resonance

$$\overline{U_{1s}} = \frac{U_{1\varepsilon}}{\varepsilon} = \frac{1}{\left|\overline{k} - \dfrac{\overline{p}}{2}\right|}, \tag{8.31}$$

and in the case of the „weak" resonance $\overline{U}_{1w} = \dfrac{U_{1w}}{\varepsilon} = \dfrac{1}{\left|\overline{k} + \dfrac{\overline{p}}{2}\right|}$. Obviously $\dfrac{1}{\overline{U_{1s}}} + \dfrac{1}{\overline{U_{1w}}} = \dfrac{2}{\overline{U_1}^{(0)}}$, where $\overline{U_1}^{(0)}$ is the amplitude for $\overline{p} = 0$ (ordinary resonance with constant parameters).

In the range of the „strong" resonance, Eq.(8.29) can be presented in this way

$$\overline{U_1}^2(\xi) = (\xi + 1)\frac{\xi^2(\xi + 1) + \left(\overline{k} + \dfrac{\overline{p}}{2}\right)^2}{\left[\xi^2(\xi + 1) + \left(\overline{k}^2 - \dfrac{\overline{p}^2}{4}\right)\right]^2} \approx \frac{\xi^2 + \left(\overline{k} + \dfrac{\overline{p}}{2}\right)^2}{\left(\xi^2 + \overline{k}^2 - \dfrac{\overline{p}^2}{4}\right)^2}. \tag{8.32}$$

In an analogous way, the following is obtained for the „weak" resonance

$$\overline{U_1}^2(\xi) = (\xi + 1)\frac{\xi^2(\xi + 1) + \left(\overline{k} - \dfrac{\overline{p}}{2}\right)^2}{\left[\xi^2(\xi + 1) + \left(\overline{k}^2 - \dfrac{\overline{p}^2}{4}\right)\right]^2} \approx \frac{\xi^2 + \left(\overline{k} - \dfrac{\overline{p}}{2}\right)^2}{\left[\xi^2 + \left(\overline{k}^2 - \dfrac{\overline{p}^2}{4}\right)\right]^2}. \tag{8.33}$$

The form of the resonance curve will be determined by investigating the extremum functions (8.32) and (8.33).

The condition $\dfrac{dU_1^2(\xi)}{d\xi}=0$ leads to the following polynomial equation

$$\xi^6-\xi^5-4\xi^4+(2\overline{k}^2+kp-2)\xi^3-\left(\overline{k}^2+\frac{9}{4}\overline{p}^2+5\overline{k}\overline{p}\right)\xi^2$$
$$-2\left(\overline{k}^2+\frac{3}{4}\overline{p}^2+2\overline{k}\overline{p}\right)\xi+\left(\overline{k}+\frac{\overline{p}}{2}\right)^3\left(\overline{k}-\frac{\overline{p}}{2}\right)=0. \tag{8.34}$$

Depending on the coefficients of this equation, it can have from zero to six real solutions, determining the extreme points which account for the form of the resonance curve.

In the general case, when the resonance is „strong", the resonance curve may contain up to three peaks, and the central peak coincides, as in the case with the ordinary resonance, with the zero detuning point.

In the case of „weak" resonance, the condition $\dfrac{d\overline{U}_1^2(\xi)}{d\xi}=0$ also leads to a polynomial equation analogous to (8.34)

$$\xi^6-\xi^5-4\xi^4-(2+\overline{p}\overline{k}-2\overline{k}^2)\xi^3+\left(5\overline{k}\overline{p}-\overline{k}^2-\frac{9}{4}p^2\right)\xi^2$$
$$-\left(2\overline{k}^2-4\overline{k}\overline{p}+\frac{3}{2}\overline{p}^2\right)\xi+\left(\overline{k}-\frac{\overline{p}}{2}\right)^3\left(\overline{k}+\frac{\overline{p}}{2}\right)=0. \tag{8.35}$$

Analogously to (8.34), Eq.(8.35) has six real roots. Therefore, in the case of „weak" resonance, the resonance curve is also of a multiextreme character. Depending on the values of the coefficients, the number of the troughs may be from zero to three.

Fig.8.2 shows the resonance curves for „strong" and „weak" resonance in the case of a maximum number of extremums. The resonance characteristic of a „weak" resonance has two maximums, symmetrically positioned with respect to $\xi=0$ at detunings $\xi_W^m=\pm\sqrt{\dfrac{\overline{p}^2}{2}-(\overline{k}-\overline{p})^2}$. The meanings of the arguments in the maximum points are the same and equal to $\overline{U}_{1W\,\max}^2(\xi_W^m)=\dfrac{1}{\overline{p}}\dfrac{\overline{k}-\dfrac{\overline{p}}{4}}{\left(2\overline{k}-\dfrac{3}{4}\overline{p}^2\right)}$.

For the sake of completeness, we shall point out that the extremums are not realized exactly when $\xi=0$. In the case of „strong" resonance the maximum is shifted slightly to the right of $\xi=0$, while in the case of „weak" resonance it is the other way round – the maximum is shifted slightly to the left of the point for $\xi=0$.

Winding up the analysis of the linear case, we should point out that in compliance with the expression (8.31) the „strong" resonance excitation threshold is reached in the case when the equation $\overline{k}=\dfrac{\overline{p}}{2}$ is satisfied.

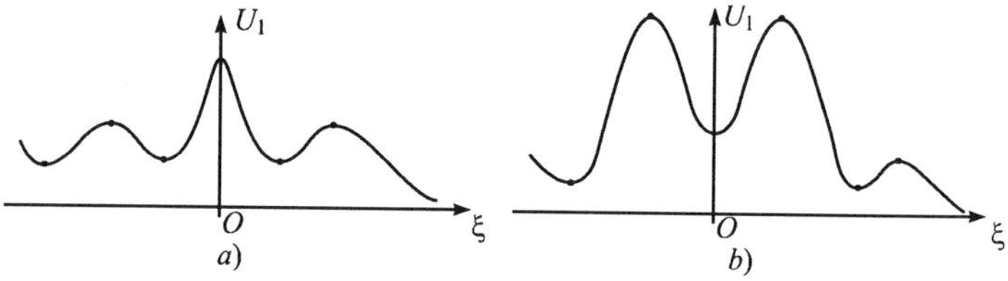

Fig. 8.2. Resonance characteristics for „strong" a) and „weak" b) resonance in the case of a maximum number of extremums

Let us now consider the „strong" and „weak" resonance by taking into account the nonlinear nature of Eq.(8.22). Under the „strong" resonance condition $\cos 2\gamma = -1$, the expression for the resonance characteristic (8.27) takes the form

$$U_1^2(\xi) = \varepsilon^2(\xi+1)\frac{\dfrac{\xi^2(\xi+1)}{\left(\overline{k}+\dfrac{1}{4}\overline{q}\overline{U}_1^2+\dfrac{\overline{p}}{2}\right)^2}+1}{\left[\dfrac{\xi^2(\xi+1)}{\left(\overline{k}+\dfrac{1}{4}\overline{q}\overline{U}_1^2\right)^2-\dfrac{\overline{p}^2}{4}}+1\right]\left(\overline{k}+\dfrac{1}{4}\overline{q}\overline{U}_1^2-\dfrac{\overline{p}}{2}\right)^2}. \tag{8.36}$$

In the case of resonance, $\xi = 0$ and Eq.(8.36) is presented as

$$U_{1s}^2 = \varepsilon^2\frac{1}{\left(\overline{k}+\dfrac{1}{4}\overline{q}\overline{U}_{1s}^2-\dfrac{\overline{p}}{2}\right)^2}. \tag{8.37}$$

Equation (8.37) is of the third power with respect to U_{1s}^2. Its real solution is

$$U_{1s}^2 = \frac{8}{3\overline{q}}\left(\overline{k}-\frac{\overline{p}}{2}\right)\left(\sqrt{3}\,\mathrm{ch}\left\{\frac{1}{3}\mathrm{arch}\left[\frac{1}{3\sqrt{3}}+\frac{3\sqrt{3}}{8}\frac{q\varepsilon^2}{\left(\overline{k}-\dfrac{\overline{p}}{2}\right)^3}\right]\right\}-1\right) \tag{8.38}$$

for

$$D = \frac{64}{\overline{q}^6}\left\{\left[\frac{8}{27}\left(\overline{k}-\frac{\overline{p}}{3}\right)^3+q\varepsilon^2\right]^2-\frac{64}{27}\left(\overline{k}-\frac{\overline{p}}{2}\right)^6\right\}>0$$

and

$$U_{1s}^2 = \frac{8}{3\overline{q}}\left(\overline{k}-\frac{\overline{p}}{2}\right)\left(\sqrt{3}\cos\left\{\frac{1}{3}\arccos\left[\frac{1}{3\sqrt{3}}+\frac{3\sqrt{3}}{8}\frac{q\varepsilon^2}{\left(\overline{k}-\dfrac{\overline{p}}{2}\right)^3}\right]\right\}-1\right),$$

when $D < 0$.

An interesting point worth noting is that at $\overline{k} = \dfrac{\overline{p}}{2}$

$$U_{1s} = \sqrt[3]{\frac{4\varepsilon}{q}}. \tag{8.39}$$

It follows from (8.39) that the nonlinearity whose power here is determined by q, deteriorates the power properties of the circuit. Indeed, while in the linear case ($q = 0$) the condition $\overline{k} = \dfrac{\overline{p}}{2}$ determined an excitation mode, $R_{1s} \to \infty$, in the nonlinear case this is not a threshold regime and the amplitude U_{1s} is so much smaller as the parameter q is larger.

In the case of the „weak" resonance characterized by the condition $\cos 2\gamma = 1$, the resonance characteristic is of the form

$$U_1^2(\xi) = \varepsilon^2(\xi+1)\frac{\dfrac{\xi^2(\xi+1)}{\left(\overline{k}+\dfrac{1}{4}\overline{q}\overline{U}_1^2-\dfrac{\overline{p}}{2}\right)^2}+1}{\left\{\dfrac{\xi^2(\xi+1)}{\left[\left(\overline{k}+\dfrac{1}{2}\overline{q}\overline{U}_1^2\right)^2-\dfrac{1}{4}\overline{p}^2\right]^2}+1\right\}^2\left(\overline{k}+\dfrac{1}{2}\overline{q}\overline{U}_1^2+\dfrac{\overline{p}}{2}\right)^2}.$$

In a resonance mode, $\xi = 0$, the expression for the amplitude in the case of „weak" resonance takes the form

$$U_{1W}^2 = \frac{\varepsilon^2}{\left(\overline{k}+\dfrac{1}{4}qU_{1W}^2+\dfrac{\overline{p}}{2}\right)^2}.$$

This is a cubic equation with respect to U_{1W}^2, which has a single real solution

$$U_{1W}^2 = \frac{8}{3q}\left(\overline{k}+\frac{\overline{p}}{2}\right)\left(\operatorname{ch}\left\{\frac{1}{3}\operatorname{arch}\left[1+\frac{27}{8}\frac{\overline{q}\varepsilon^2}{\left(\overline{k}+\dfrac{\overline{p}}{2}\right)^3}\right]\right\}-1\right).$$

Under the condition $\dfrac{\overline{p}}{2} = \overline{k}$ the following is obtained:

$$U_{1W}^2 = \frac{16}{3}\frac{\overline{k}}{\overline{q}}\left\{\operatorname{ch}\left[\frac{1}{3}\operatorname{arch}\left(1+\frac{27}{64}\frac{\overline{q}\varepsilon^2}{\overline{k}^3}\right)\right]-1\right\}.$$

The analysis indicates that the phenomena of „strong" and „weak" resonance are retained in the considered non-linear system, yet they are not so distinctly manifested as in the linear case considered before.

8.3. Quasi-periodic oscillations in an auto-generator with an oscillating circuit containing nonlinear reactance

Currently, transistors are increasingly used at higher frequencies, when the basic frequency in the spectra of voltages and currents is higher, and even by far higher, than the typical boundary frequency of the transistor [271]. It turns out that the transistor „works" under such conditions, too, but a number of regime peculiarities are observed. In practice such problems arise in the microwave range or in lower ranges close to it.

On this issue, theory lags behind practice and, as a rule, goes no further than explaining mostly experimental results. We shall hereby provide a theoretical analysis of the peculiarities in the operation of transistor generators at high frequencies, when the interelectrode capacitances of the transistors are comparable to the capacitances of the oscillating circuits in the system.

Let us consider an ordinary transistor generator (Fig.8.3a) with a transformer feed-back. We shall work on the assumption that the interelectrode capacitances of the transistor play an essential part. If the transistor is replaced with a Giakoletto equivalent circuit, the result is the equivalent replacement circuit of the generator, presented in Fig.8.3b. The generator contains five nonlinear parameters: $C = C(q)$, $G = G(q)$, $C_{bc} = C_{bc}(q_{bc})$, $G_{bc} = G_{bc}(q_{bc})$, $S = S(q)$ is the slope of the transistor characteristic. The generator presented in this way is described by a system of 4 differential equations of the first order with respect to the charges of the capacitances and the magnetic fluxes of the inductances. This system can be transformed into the following system of two differential equations of the second order

$$\frac{d^2\mathbf{x}}{dt^2} + \mathbf{A}(\mathbf{x})\frac{d\mathbf{x}}{dt} + \mathbf{B}(\mathbf{x})\mathbf{x} = 0, \tag{8.40}$$

where $\mathbf{A} = \{a_{i,j}\}$, $\mathbf{B} = \{b_{i,j}\}$, $i, j = 1, 2$, $\mathbf{x} = \text{colon}(q, q_{bc})$,

$$a_{11} = \frac{1 - k\sqrt{\dfrac{L}{L_{fb}}}}{1 - k^2}\frac{R}{L} + q\frac{d}{dq}\left(\frac{G - S}{C}\right) + \frac{G - S}{C},$$

$$a_{12} = \frac{1}{1 - k^2}\left[-\frac{R}{L} + \frac{R_{fb}}{L_{fb}} + \frac{k}{\sqrt{LL_{fb}}}(R - R_{fb})\right],$$

$$a_{21} = -q\frac{dS}{dq}\frac{1}{C} + \frac{S}{C}, \quad a_{22} = \frac{1}{1 - k^2}\frac{R_{fb}}{L_{fb}}$$

$$+\frac{k}{1-k^2}\sqrt{\frac{L}{L_{fb}}}\frac{R}{L}-q_{bc}\frac{dG_{bc}}{dq_{bc}}\frac{1}{C_{bc}}-\frac{G_{bc}}{C_{bc}},$$

$$b_{11}=\frac{1}{1-k^2}\left(\frac{1+RG}{L}+\frac{1+SR_{fb}}{L_{fb}}-k\frac{2+GR+SR_{fb}}{\sqrt{LL_{fb}}}\right)\frac{1}{C},$$

$$b_{12}=\frac{1}{1-k^2}\left[\frac{1}{L_{fb}}+\left(\frac{R}{L}-\frac{R_{fb}}{L_{fb}}\right)G_{bc}-k\frac{1+(R-R_{fb})G_{bc}}{\sqrt{LL_{fb}}}\right]\frac{1}{C_{bc}},$$

$$b_{21}=\frac{1}{1-k^2}\left(\frac{1+SR_{fb}}{L_{fb}}-k\frac{1+RG}{\sqrt{LL_{fb}}}\right)\frac{1}{C},$$

$$b_{22}=\frac{1}{1-k^2}\left(\frac{1+G_{bc}R_{fb}}{L_{fb}}-k\frac{RG_{bc}}{\sqrt{LL_{fb}}}\right)\frac{1}{C_{bc}},\quad k=\frac{M}{\sqrt{LL_{fb}}}\,.$$

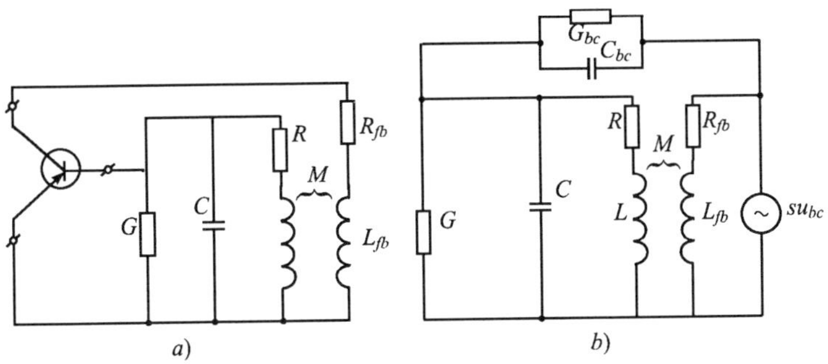

Fig. 8.3. Transistor generator with transformer feedback a) and its equivalent circuit after Giakoletto b)

The nature of the dependence of matrices **A**, **B** upon the vector **x** is the following: the first columns of the matrices depend only on the first line q of the unknown vector, while the second ones depend on the second line q_{bc} of the same vector. This structure of the matrixes allows introducing a diagonal matrix $\mathbf{X}=\mathrm{diag}(q,q_{bc})$ and expanding the matrices into a matrix Taylor series introduced in a previous section, i.e.

$$\frac{d^2\mathbf{x}}{dt^2}+(\mathbf{A}_0+\mathbf{A}_1\mathbf{X}+\dots)\frac{d}{dt}\mathbf{x}+(\mathbf{B}_0+\mathbf{B}_1\mathbf{X}+\dots)\mathbf{x}=0. \tag{8.41}$$

If the nonlinearities are ignored, the result would be an equation of a double-frequency system analogous to the system of two coupled circuits with transformer connection.

The method of successive approximations developed in 8.1 can be used to indicate that in the stationary mode the solution of Eq.(8.41) is of the form

$$\mathbf{x}(t) = \sum_{k,l=0}^{\infty} \text{colon}\left\{x_{kl}^{(1)} \text{colon}[(k\omega_1 \pm l\omega_2)t + \varphi_{kl}^{(1)}],\right.$$

$$\left., x_{kl}^{(2)} \text{colon}[(k\omega_1 \pm l\omega_2) + \varphi_{kl}^{(2)}]\right\}. \qquad (8.42)$$

This expression makes it clear that the spectral composition of the oscillations is sufficiently complex, moreover the result is a Fourier series, which is typical of the almost periodic functions. It turns out that at sufficiently high frequencies the ordinary transistor generator of sine oscillations is converted into a generator of almost periodic oscillations. This phenomenon is caused by the circumstance that the nonlinear interelectrode capacitances and conductances at sufficiently high frequencies are included in the oscillating system of the generator. The resonance circuit of the generator becomes nonlinear and its selective properties are widely at variance with those of the ordinary circuit.

The considered phenomenon of transforming transistor generators of sine oscillations into generators with almost periodic oscillations at increased generated frequency is essentially of a general nature and does not depend on the type of the feed-back in the generator. Generators with almost periodic oscillations can be realized in a microwave range, as well as in a low-frequency range. The requirement is that the oscillating system should contain nonlinear capacitances and (or) inductances, and the generator should have at least two degrees of freedom.

By employing the mathematical induction method, the result obtained (8.42) can be generalized for equations of the type of (8.41) with arbitrary order matrices. For example, if the emitter-base circuit in the generator presented in Fig.8.3 includes a more complex oscillating system, for instance a system of two coupled oscillating circuits, the resultant equation will be of the type of (8.41) but its matrices will be of the sixth or seventh order (depending on the type of the connection between the circuits).

The number of the base frequencies n is determined by the number of the freedom degrees of the generator as a half of the differential system order (8.41). Two options are possible: the number of the degrees of freedom ν is either an integer or a half-integer. In the first case $n = \nu$, and in the second $n = \nu - \frac{1}{2}$. In the general case the determining vector $\mathbf{x}$ of the generator is presented as time-dependent function in the following form:

$$\mathbf{x}(t) = \sum_{k_1,k_2,\ldots,k_n=0}^{\infty} \text{colon}\left[x_{k_1,k_2,\ldots,k_n}^{(1)} \cos\left(\sum_{i=1}^{n} \pm k_i\omega_i t + \varphi_{k_1,k_2,\ldots,k_n}^{(1)}\right)\right.$$

$$\left., \ldots, x_{k_1,k_2,\ldots,k_n}^{(n)} \cos\left(\sum_{i=1}^{n} \pm k_i\omega_i t + \varphi_{k_1,k_2,\ldots,k_n}^{(n)}\right)\right].$$

This formula shows that as the number of the freedom degrees of the generator grows, the spectrum of its oscillations becomes richer.

8.4. Thermoparametric oscillations

This section will be a brief presentation of the principle of operation of additional non-traditional energy sources, i.e. thermoparametric generator systems for converting thermal energy from the heat supply mains located in plants and enterprises into electric power.

Fig.8.4 shows a model system consisting of mass m, suspended in the middle of an imponderable metallic string, while the latter is connected to a circuit of alternating current with frequency ω [139]. Under certain conditions the string can vibrate both in the plane of the drawing (Fig.8.4) and in that perpendicular to it (around the $0_1 0_2$ axis). Let us first of all consider the particular case when the oscillations are in the plane of the drawing. The equation describing the oscillations of the mass in this case is of the form:

$$m\ddot{x} = mg - 2F\sin\beta - h\dot{x}, \tag{8.43}$$

where F is the strain force of the string, $\sin\beta = \dfrac{x}{l} = \dfrac{x}{\sqrt{x^2 + \tilde{l}_0^2}}$, $2\tilde{l}_0$ is the distance between the supports $0_1 0_2$, h is the friction coefficient, g is the acceleration of gravity.

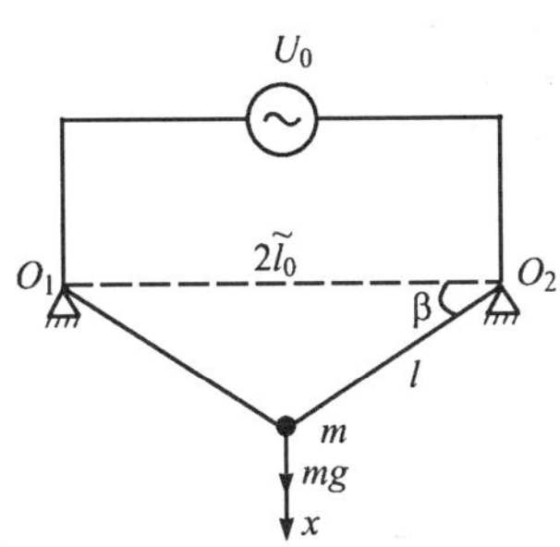

Fig. 8.4. A model system for investigating thermoparametric oscillations

The strain force of the string depends on the degree to which it is heated:

$$F = ES\frac{l - l_0(1 + \alpha T)}{l_0}, \tag{8.44}$$

where E is Young's modulus, S is the cross section of the string, T is the difference between the temperature of the string and that of the surrounding environment, α

is the linear expansion coefficient, $l_0 = \sqrt{x_0^2 + \tilde{l}_0^2}$ is half the length of the string in an unstrained condition, given $T = 0$ and $m = 0$, x_0 is the deflection distance of the middle point of the string under the same conditions.

We shall assume that $l - l_0 \ll l_0$. Then, as we take into account the relation between l_0 and $\tilde{l}_0$, we obtain $l = l_0 + \dfrac{(x^2 - x_0^2)}{2l_0}$ and

$$2F \sin\beta = ES \left[\frac{x^2 - x_0^2}{l_0^2} - 2\alpha T \right] \frac{x}{l_0}. \tag{8.45}$$

The equation concerning the temperature T of the string will be written under the assumption that the heat conductivity of the string is sufficiently high for all string points to reach the same temperature over one oscillation period and the cooling of the string is governed by Newton's law, the convective heat transfer coefficient being q and depending on the speed module:

$$q = q_0 + q_1 \sqrt{|\dot{x}|}, \tag{8.46}$$

where $q_0 = 1,3.10^{-3} S_s \dfrac{W}{(^\circ)}$, S_s is the lateral surface of the string. Then the equation concerning T assumes the form

$$mc\dot{T} = -q_0 T - q_1 \sqrt{|\dot{x}|} T + \frac{U_0^2}{2R}(1 + \cos 2\omega t), \tag{8.47}$$

where c is the heat accumulation capacity of the material of the ball with mass m, R is the resistance of the string. As the string is heated and elongated, the resistance R increases as a result of the tensor-resistive effect, so that

$$R = R_0 \left(1 + \beta_1 T + \beta_2 \frac{l - l_0}{l_0} \right), \tag{8.48}$$

where $\beta_1 \simeq 6.10^{-3} \text{grad}^{-1}$, $\beta_2 \simeq 0,2$ (the quantity β_2 is determined by Poisson's coefficient).

We shall assume it that over the oscillation period of the current in the string $\dfrac{2\pi}{\omega}$ the temperature T is not changed substantially, while the frequency ν lies beyond the areas of parametric excitation of the string. Then the term $\dfrac{U_0^2}{2R}\cos 2\omega t$ in Eq.(8.47) can be ignored.

The stationary solution x_{st} of Eqs. (8.43) and (8.47), obtained by taking into account (8.45) and (8.48), is determined from the equation

$$mg = 2ES \frac{x_{st}}{l_0} \left[\frac{x_{st}^2 - x_0^2}{l_0^2} - 2\alpha T_{st} \right], \tag{8.49}$$

where

$$T_{st} = \frac{1}{2\beta_1}\left(\sqrt{1+\frac{2U_0^2\beta_1}{R_0 q_0}}-1\right) \tag{8.50}$$

(in this case the ignored term is $\beta_2 \dfrac{l_{st}-l_0}{l_0}$ as compared to unity).

We introduce deviations from the stationary solution $\xi = \dfrac{x - x_{st}}{l_0}$, $v = \dfrac{T - T_{st}}{T_{st}}$ and write Eqs.(8.43) and (8.47) concerning these deviations in the following form

$$\left|\begin{array}{l} \ddot{\xi} + \nu^2\xi = \varepsilon\left[-2\delta\dot{\xi} - 3\dfrac{ESx_{st}}{ml_0^2}\xi^2 - \dfrac{ES}{ml_0}\xi^2 + K\left(1+\dfrac{l_0}{x_{st}}\xi\right)\nu\right] \\ \dot{v} + \gamma v + a\xi\left(1+\dfrac{l_0}{2x_{st}}\xi\right) + \gamma_1\sqrt{|\dot{\xi}|} = -\varepsilon\gamma_1\nu\sqrt{|\dot{\xi}|}, \end{array}\right. \tag{8.51}$$

where $\nu^2 = \dfrac{g}{2x_{st}} + 2\dfrac{ES}{ml_0^3}x_{st}^2$, $2\delta = \dfrac{h}{m}$, $K = 2ES\dfrac{\alpha T_{st}x_{st}}{ml_0^2}$, $r = \dfrac{q_0}{mc}\left(1+\dfrac{\beta_1 T_{st}}{1+\beta_1 T_{st}}\right)$,
$a = \dfrac{q_0 x_{st}\beta_2}{mcl_0(1+\beta_1 T_{st})}$, $\gamma_1 = \dfrac{q_1 T_{st}\sqrt{l_0}}{mc}$, ε is a small parameter, which should be equated to unity in the final expressions.

System (8.51) refers to the class of oscillating systems with an inertial excitation. The condition for self-oscillation in such systems is of the form

$$Ka \geq 2\delta(\nu^2 + 2\delta\gamma + \gamma^2). \tag{8.52}$$

Therefore, reasons for possible self-oscillation in the case under consideration can be the dependence of the resistance of the string on its deformation ($a \neq 0$) and the sufficient inertia in the temperature changes in the string (it follows from (8.52) that the inertia parameter γ^{-1} should be bigger than
$\gamma_c^{-1} = \left(\sqrt{\delta^2 + K\dfrac{a}{2}\delta - \nu^2} - \delta\right)^{-1}$).

In order to check whether the excitation of the oscillations is of a soft or hard nature, we shall work out an equation concerning the oscillation amplitude of the variable ξ. In addition, we shall provide an approximate solution to (8.51) by using the asymptotic method of Krilov-Bogolyubov-Mitropolski [97] and limiting ourselves to a cubic term of the amplitude under consideration. For that purpose we set

$$\xi = A\cos(\nu t + \varphi_1) + \varepsilon U_1 + \ldots,$$

$$v = \frac{a}{\sqrt{\nu^2+\gamma^2}} A\cos(\nu t + \varphi_1 - \varphi_2) - \frac{al_0}{4x_{st}\gamma} A^2$$
$$- \frac{4\gamma_1\sqrt{2\pi\nu}}{\gamma\Gamma^2(\frac{1}{4})}\sqrt{A} - \frac{al_0}{4x_{st}\sqrt{4\nu^2+\gamma^2}} A^2\cos(2\nu t + 2\varphi_1 - \varphi_3)$$
$$+ \frac{8\gamma_1\sqrt{2\pi\nu}}{\sqrt{4\nu^2+\gamma^2}\Gamma^2(\frac{1}{4})}\sqrt{A}\cos(2\nu t + 2\varphi_1 - \varphi_3) + \ldots + \varepsilon\nu_1^2 + \ldots,$$

where $\operatorname{tg}\varphi_2 = \frac{\nu}{\gamma}$, $\operatorname{tg}\varphi_3 = \frac{2\nu}{\gamma}$, $\Gamma\left(\frac{1}{4}\right)$ being a gamma function.

The type of the solution concerning the function v in zero approximation by ε is obtained from the second equation of (8.50) through a harmonic expansion of $\sqrt{|\dot{\xi}|} = \sqrt{A\nu|\sin(\nu t + \varphi)|}$. Moreover, only the terms containing zero and second harmonic have been ejected from the series.

In the first approximation by ε, we obtain the following equation about the amplitude of oscillations A:

$$\dot{A} = (\mu_1 + \mu_2\sqrt{A} - \mu_3 A^2)A, \tag{8.53}$$

where $\mu_1 = \frac{Ka}{2(\nu^2+\gamma^2)} - \delta$, $\mu_2 = \frac{4\gamma_1 K l_0\sqrt{2\pi\nu}}{(4\nu^2+\gamma^2)x_{st}\Gamma^2\left(\frac{1}{4}\right)}$, $\mu_3 = \frac{Kal_0^2}{8x_{st}^2(4\nu^2+\gamma^2)}$.

Since $\mu_2 > 0$, it follows that the excitation of the oscillations is hard. Hence, a stationary oscillation mode is possible even for a limited range of negative values for μ_1.

It is worth noting that the self-oscillation condition $\mu_1 > 0$, following from Eq.(8.53), coincides with (8.52), if $\delta\gamma \ll \nu^2 + \gamma^2$. This inequality is also one of the conditions for applicability of the asymptotic method employed.

Now, let us consider the more general case, when the string can not only shift vertically, but also oscillate around the $0_1 0_2$ axis. We introduce a rotation angle φ and write equations concerning the motion of the mass m and the temperature T

$$\left|\begin{aligned} & m\ddot{x} = mg\cos\varphi - 2F\sin\beta - h\dot{x} + mx\dot{\varphi}^2 \\ & m\frac{d^2(x^2\dot{\varphi})}{dt} + Hx\dot{\varphi} + mgx\sin\varphi = 0 \\ & mc\dot{T} = -q_0 T - q_1(\dot{x}^2 + x^2\dot{\varphi}^2)^{\frac{1}{4}} + \frac{U_0^2}{2R}, \end{aligned}\right. \tag{8.54}$$

where F is determined by formula (8.44) and R – by formula (8.42).

In the equilibrium state $\varphi = 0$ and, both x_{st} and T_{st} are the same as those in the particular case considered above. In a linear approximation, the equations concerning the deviation $\xi = x - x_{st}$ and $v = T - T_{st}$ do not depend on φ. Therefore, everything mentioned above applies fairly for them as well. If the oscillation frequency of the variable x lies in one of the areas of parametric resonance of the variable φ, the excitation of vertical oscillations of the mass necessarily leads to oscillation of the rotation angle φ as well. If this condition is not met, the excitation of oscillations of the variable φ is possible only in the hard way at the expense of the fact that the oscillations of the temperature T will contain a second harmonic in the oscillations φ, which will cause modulation of the variable x with the same frequency. The latter, in its turn, will entail parametric energy input in the oscillations of the variable φ. Equations (8.54) make it possible to calculate the stationary amplitude of the oscillations in such a process.

In the model system under consideration – Fig.8.4, the role of the high frequency voltage source is reduced mostly to heating the moving element of the system. Obviously, the heating source can also be of a different origin – for example the main heat lines of plants and factories or other thermal sources. In addition, it is possible to envisage a mechanism for inducing an electromotive force so that the flowing current increases the temperature further on and enhances the motion of the moving element of the system. In this way, after an initial mechanical drive, the feedback in the system closes and it is converted into an autoparametric generating system, whose power supply source is a thermal one. This can be used as a basis to design additional non-traditional electric power sources.

CHAPTER 9.

GROUPING OF COUPLED OSCILLATING SYSTEMS IN STABLE ELECTROMECHANICAL FORMATIONS

9.1. Introduction

In recent years the opinion has prevailed that the entropy of systems is not as such universal a value as traditionally presented [5, 17, 19, 22, 27, 32–35, 38]. Thermodynamics permits the existence, in closed systems, of spontaneous processes of entropy reduction and, respectively, a local order promotion in the form of fluctuations, whose value and frequency abide by the probability laws of classical statistics.

Currently, the research on local organization in open systems develops extensively regardless of the „entropy" notions [32]. These are studies on the processes of self-organization or, in a certain specific aspect, of synergetics (synergism is a Greek word denoting a „joint action"). The task of synergetics as a branch of science is to identify the general laws governing the processes of self-organisation that bring about the formation of systems with more orderly space and temporal structures. In general, the research under way has two aspects: 1/ a search for more general and universal criteria and qualitative parameters of order, as compared to the classical concept of entropy, suitable for characterizing open systems, and 2/ identification of the mechanisms of self-organization, the issue of major interest (from a methodological point of view) being the macroscopic processes of self-organization, whose mechanisms are well manifested.

As a model example of a self-organizing system, we could mention the phenomenon of metal dust arrangement in bizarre forms under the action of an electromagnetic SHF wave. This is also a method of visualizing and photographing the SHF field, when the metal dust is applied in a thin film directly on photographic paper. The totality of particles can be regarded as a large number of linear or quasi-linear resonators with varying detuning, both in the ordinary sense of the word used in Radiophysics and in the geometric sense. Empiric data show that a portion of the particles, grouped in couples or larger sets, prove to be tuned and capable of absorbing the energy of the SHF field, emitting sparks as it does so (a process of sparking). As one shakes slightly the entire volume of metal dust, the number of the tuned resonator systems (hence, the number of the sparks, as well), given a finite time interval, can be excessive. The exposure to sufficiently powerful radiation causes incessant glow in the entire diffusion area of the SHF waves. It is possible to identify glowing resonators in a square sized 5 x 5 mm within a 3 cm microwave range. The ejection of some of the particles in such a tuned complex causes the extinction of the spark. In contrast, a slight shaking of the dust-like mass causes more intensive glowing since it helps surmount the friction at rest and promotes the ponderomotive forces by increasing the mobility of the particles and their efficient grouping and tuning. Enhanced grouping performance is also observed when using highly modulated SHF radiation, as the broad spectrum conditions an effective absorption of the SHF energy in a situation of considerable deviations from the resonance conditions. There is „competitive struggle" between the „relevant" (resonance) complexes sets and the „irrelevant" ones (those wide apart from the resonance conditions). Moreover, the picture is not statistically frozen and invariable: it is dynamic. The dynamism and „competitive struggle" are boosted by impact action and vibrations of varying nature.

To wind up, we can formulate briefly some of the conditions for a manifestation of a self-organisation process:

(i) Existence of a disorderly open system with subsystems of certain assimilating properties and a medium that feeds it to a limited extent but with sufficient variety (for instance with a disorder typical of „noise").

(ii) Availability of „a range of possibilities" as a system of potentially possible orderly conditions in a certain medium and nonlinear oscillation in the system.

(iii) Consistence between the „assortment" of the feeding and the assimilation capacities of the subsystems. This condition can be modified as follows: a capacity for selective assimilation of the separate components of the disorderly feeding („noise").

(iv) A low level of dissipative couplings imposed on the subsystems.

(v) Competition among the subsystems in their struggle to assimilate the available feeding.

In his fundamental works on the foundations of synergetics, H. Haken introduces, as a parameter of the „degree of order" of the self-organizing system, the existence of a set of modes of the oscillation amplitude in the system and a possibility for competition among them [32].

It is worth noting that the tendency of self-organization of the macroscopic systems is usually suppressed by the highly pronounced dissipative couplings. This fact seems to explain the lack of evolution in a vertical direction and a transition to a qualitatively higher level. The major reason in this case is manifested in the damage under the action of the destructive dissipative factors. From a methodological perspective, the investigation of the self-organizing systems in the inanimate nature acquires particular importance. What matters most in this case is to surmount the metaphysical view of matter as an inert, passive, fossilized and frozen mass, whose natural state is absolute rest.

This Chapter 9, as well as the next Chapter 10 analyze some synergetic aspects from the viewpoint of oscillating processes in radiophysical and electromechanical oscillating systems with a limited number of degrees of freedom that come together in stable stationary formations. Issues of ponderomotive (mechanical) equilibrium in grouped dynamic systems are examined. A model of two interacting resonators with a mechanical degree of freedom, i.e. with a possibility to move one of them in relation to the other, has been used as a basis to demonstrate that such a system can be grouped in definite stable formations.

The mathematical description of self-organization is reduced to the establishment of a link between the effect of a certain action and its cause depending on time. If the equation of self-organization is to be obtained, the forces that are external to the system have to be ejected from the written equation. From a mathematical perspective, it is simpler to expand the system in such a way that the external forces included in the self-organization equation should become internal. In our particular case – the system of two coupled resonators – we shall present the external exciting oscillator as an emerging „internal" non-equilibrium ponderomotive force acting as a cause and „indicator" of the process of grouping in a stable dynamic structure.

9.2. Generalized conditions for grouping in stable electromechanical formations

This section investigates the ponderomotive interactions of coupled mobile resonators under the action of electromagnetic waves (or equivalent e.m.f.) in the process of their grouping in stable formations [272, 273].

In the general case, it is assumed that a change in the distance between the resonators alters the inductive, capacitive and dissipative coupling, which entails a change in the ponderomotive forces between them. Each of the resonators is formed by linear reactive and active elements.

In the situation of a slow alteration of the parameters of the mutual couplings, the processes in the depicted system are described by the following system of differential equations:

$$\left| \begin{array}{l} \beta_{11}\ddot{x}_1 + \varepsilon_{11}\dot{x}_1 + \alpha_{11}x + \beta_{12}\ddot{x}_2 + \varepsilon_{12}\dot{x}_2 + \alpha_{12}x_2 = \Phi_1(x_1, \dot{x}_1, x_2, \dot{x}_2, t) \\ \beta_{22}\ddot{x}_2 + \varepsilon_{22}\dot{x}_2 + \alpha_{22}x_2 + \beta_{21}\ddot{x}_1 + \varepsilon_{21}\dot{x}_1 + \alpha_{21}x_1 = \Phi_2(x_1, \dot{x}_1, x_2, \dot{x}_2, t) \\ \beta_{33}\ddot{x}_3 + \varepsilon_{33}\dot{x}_3 + \dfrac{\partial \beta_{12}}{\partial x_3}(\dot{x}_1\dot{x}_2) + \dfrac{\partial \alpha_{12}}{\partial x_3}(x_1x_2) = 0, \end{array} \right. \tag{9.1}$$

where x_1 and x_2 are the charges of the first and second resonator, x_3 is the distance between the resonators; α_{ik}, β_{ik}, ε_{ik} $(i, k = 1, 2)$ are the partial and mutual coefficients of capacitance, inductance and resistance respectively, β_{33} and ε_{33} are the mass and the frictional coefficient of the mobile resonator, $\dfrac{\partial \beta_{12}}{\partial x_3}(\dot{x}_1\dot{x}_2)$ and $\dfrac{\partial \alpha_{12}}{\partial x_3}(x_1x_2)$ are the ponderomotive forces; Φ_1 and Φ_2 reflect the action of the electromagnetic waves.

The aggregate action of the electromagnetic waves on the resonators standing at a certain distance can be replaced with a concentrated source of e.m.f. which has an identical effect. Hence, further on we shall consider that there is a source of e.m.f. E with frequency Ω acting in just one of the resonators. We also assume that the resonators are identical, i.e. $\alpha_{11} = \alpha_{22}$, $\beta_{11} = \beta_{22}$, $\varepsilon_{11} = \varepsilon_{22}$, $\alpha_{12} = \alpha_{21}$, $\beta_{12} = \beta_{21}$, $\varepsilon_{12} = \varepsilon_{21}$, $n^2 = n_1^2 = n_2^2 - \dfrac{\alpha_{11}}{\beta_{11}} - \dfrac{\alpha_{22}}{\beta_{22}}$. Under these conditions the solution of the system of Eqs. (9.1) can be presented in the form

$$x_1 = Ae^{j\omega t}, \quad x_2 = Be^{j\omega t}. \tag{9.2}$$

As the assumptions are taken into account, Eq.(9.1) is transformed as follows:

$$\left| \begin{aligned} &[(n^2 - \omega^2) + 2j\delta\omega]x_1 + [(\gamma_2 n^2 - \gamma_1\omega^2) + 2j\delta\omega\varkappa]x_2 = Ee^{j\omega} \\ &[(\gamma_2 n^2 - \gamma_1\omega^2) + 2j\delta\omega\varkappa]x_1 + [(n^2 - \omega^2) + 2j\delta\omega]x_2 = 0 \\ &\beta_{33}\ddot{x} + \varepsilon_{33}\dot{x}_3 + \beta_{11}\frac{\partial\gamma_1}{\partial x_3}(\dot{x}_1\dot{x}_2) + \alpha_{11}\frac{\partial\gamma_2}{\partial x_3}(x_1x_2) = 0, \end{aligned} \right. \tag{9.3}$$

where $\gamma_1 = \dfrac{\beta_{12}}{\beta_{11}}$, $\gamma_2 = \dfrac{\alpha_{12}}{\alpha_{11}}$, $\varkappa = \dfrac{\varepsilon_{12}}{\varepsilon_{11}}$ and $\delta = \dfrac{\varepsilon_{11}}{2\beta_{11}}$ are reduced coefficients of mutual couplings, which change in the ranges given below, while the coordinate x_3 alters:

$$-1 \le \gamma_1 \le 1, \quad 0 \le \varkappa \le 1, \quad 0 \le \gamma_2 \le 1.$$

The amplitudes of the coordinates x_1 and x_2 are determined from the first two equations of the system

$$\left| \begin{aligned} &A = N[(n^2 - \omega^2)\mathbf{i} + 2\delta\omega\mathbf{k}] \\ &B = -N[(n^2\gamma_2 - \gamma_1\omega^2)\mathbf{i} + 2\delta\omega\varkappa\mathbf{k}], \end{aligned} \right.$$

where $N = \dfrac{E^2}{[(n^2 - \omega^2) + j2\delta\omega]^2 - [(n^2\gamma_2 - \omega^2\gamma_1) + j2\delta\varkappa\omega]^2}$ is a coefficient depending on the parameters of the resonators as well as on the amplitude and frequency of the force exerting the action, **i**, **k** – orthogonal singleton basis.

Given $\dfrac{\partial\gamma_2}{\partial x_3} = 0$ (a constant ohmic coupling), the equilibrium position x_{30} of the resonators in relation to each other is determined by the equilibrium condition: zero ponderomotive force, i.e.

$\dfrac{\partial\gamma_1}{\partial x_3}(\dot{x}_1\dot{x}_2) = 0$. Since $\dfrac{\partial\gamma_1}{\partial x_3} \ne 0$, $(\dot{x}_1, \dot{x}_2) = 0$ or $(x_1x_2) = 0$. This means that the system of two coupled resonators will be in equilibrium, when the phase difference between $\dot{x}_1$ and $\dot{x}_2$ corresponds to a quarter of the period of the external force.

The period average ponderomotive force of interaction between the resonators is determined by the scalar product of the current amplitudes, where

$$\cos(\dot{x}_1, \dot{x}_2) = \frac{(n^2 - \omega^2)(\gamma_1\omega^2 - n^2\gamma_2) - 4\delta^2\omega^2\varkappa}{\sqrt{(n^2 - \omega^2)^2 + 4\delta^2\omega^2}\sqrt{(n^2\gamma_2 - \gamma_1\omega^2)^2 + 4\delta^2\omega^2\varkappa^2}}. \tag{9.4}$$

The condition for resonator equilibrium representing a condition for orthogonality of the coordinates x_1 and x_2 is written in the form

$$(n^2 - \omega^2)(\gamma_1\omega^2 - n^2\gamma_2) = 4\delta^2\omega^2\varkappa. \tag{9.5}$$

For the purpose of investigating stability, we shall formulate a linearized equation regarding small deviations ξ from the state of equilibrium. A change in the state of the resonators causes an alteration in the coefficients of inductive, capacitive and dissipative couplings. This leads to a change in the currents and the outcome is the occurrence of ponderomotive forces between the resonators. As the change in state is expressed through a change in the currents and charges, the following is obtained:

$$\beta_{33}\ddot{\xi} + \varepsilon_{33}\dot{\xi} + K\xi = 0, \tag{9.6}$$

where K is a coefficient depending on the parameters of the resonators, the amplitudes, the frequency of the external electromagnetic action, as well as on the laws governing the alteration of the inductive, capacitive and dissipative couplings along the coordinate x_3 (if x_{30} is set in advance).

It follows from Eq.(9.6) that the equilibrium state is always stable, when $K \geq 0$. Given values for $K \geq \dfrac{\varepsilon_{33}^2}{4\beta_{33}}$ or $\dfrac{\varepsilon_{33}^2}{4\beta_{33}} > K > 0$ the equilibrium position is described on the phase plane by a peculiar point of the type of a stable focus or stable node. A condition for the existence of this state is the presence of a dissipative coupling ($\varkappa \neq 0$) and damping ($\varepsilon_{11} \neq 0$) in the resonators. The stable equilibrium position is determined by the frequency Ω of the external e.m.f. or the respective length of the electromagnetic wave λ and it is not dependent on their amplitude. If the equilibrium position is stable, the resonator system represents a common stable formation.

As the length of the electromagnetic wave or the frequency of e.m.f. source, whose action is identical, change, the coupling parameters in the resonator system alters. This causes the occurrence of ponderomotive forces. Under their action, the resonators assume a new stable equilibrium position consistent with the new wavelength.

There are definite values of the coefficients of capacitive, inductive and dissipative coupling that correspond to each frequency value of the external action. When the frequency of the external force alters, the resonators assume a new equilibrium position. Hence, this is a way for definite congruence to be maintained between the changes in the system itself, the resonance properties and the frequency of the external action.

9.3. Peculiarities of the processes of interaction between coupled oscillating systems

Seeking to determine the resonance frequencies ω_1 and ω_2 of the electromechanical system of resonators, we shall consider the free oscillations in the coupled system under the assumption that the resonators are positioned at a definite distance from each other, which is conditioned by the wavelength or by the respective frequency ω_0 of the equivalent acting e.m.f. source. Under condition (9.5) there are certain values of the coupling coefficients γ_{10}, γ_{20}, $\varkappa_0$ that correspond to this frequency of the external acting force. Given the assumption that the resonators have a high quality factor, the resonance frequencies of the system are determined by the following expressions:

$$\omega_1 = n\sqrt{\frac{1-\gamma_{20}}{1-\gamma_{10}}}, \qquad \omega_2 = n\sqrt{\frac{1+\gamma_{20}}{1+\gamma_{10}}}.$$

Condition (9.5) will be presented in a different form, where η will stand for the ratio between the frequency of the partial system n and the frequency ω of the external force: $\eta = \dfrac{n}{\omega}$. Then

$$Q_0^2(\gamma_{10} - \gamma_{20}\eta^2)(\eta^2 - 1) = \eta^2\varkappa_0.$$

It is not hard to show that for $\eta = 1$ $\omega_2^2 = \omega_0^2\dfrac{\eta^2 + \gamma_{20}}{1+\gamma_{10}}$. This means that $\omega_0 \leq \omega_2 \leq n \leq \omega_1$. In this case, when $0 \leq \eta \leq 1$, the following inequality holds: $\omega_0 \geq \omega_2 \geq n \geq \omega_1$.

The ratios that are obtained express a law of correspondence between one of the frequencies of the electromechanical resonator system and the frequency of the external e.m.f. or the respective length of the acting electromagnetic wave.

Fig.9.1 displays the dependence of the natural resonance frequencies of the electromechanical system of resonators upon its mechanical coordinate x_3. As the distance between the resonators x_3 increases, the first natural resonance frequency decreases non-monotonically. This is of considerable importance, when the resonators are grouped in a single mechanically stable formation, since this effect occurs under the condition $\omega < n$. Knowing the alteration of the natural frequency of the system, one can predict the resonance phenomena for all values of the frequency of external action.

In the general case, the distance between the resonators, the coupling coefficients and the quality factor (the non-resonance frequencies ω_1 and ω_2) are continuous functions of the length of the electromagnetic wave or the frequency of the external force. A leap of the first order occurs at some peculiar points of these dependencies. One of these particular points is determined by the value of the partial frequency of the resonators.

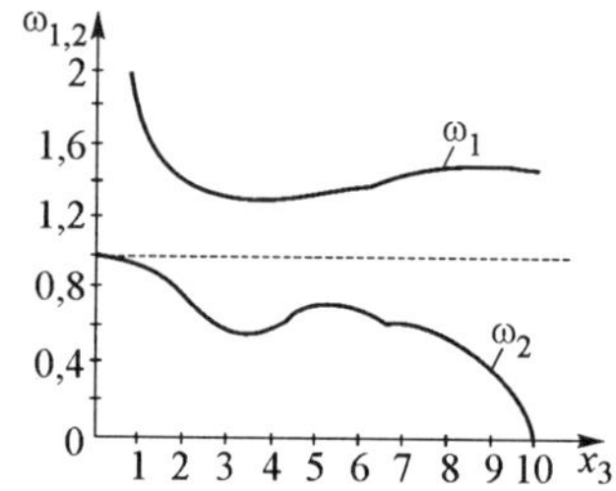

Fig. 9.1. Dependence of the natural resonance frequencies of an electromechanical system of two coupled resonators on the distance between them

The energy of the ponderomotive interaction of the resonators can be determined in the following way:

$$U = -\int F(x_3)dx_3 = N^2\left[\frac{n^2-\omega^2}{2}\left(\frac{k_1^2\beta_{11}\omega^4}{x_3^2} + k_2^2\alpha_{11}n^2x_3^2\right)\right.$$
$$\left. - 4\delta^2 \varkappa\omega^2\left(\frac{k_1\beta_{11}\omega^2}{x_3} + k_2\alpha_{11}x_3\right)\right], \tag{9.7}$$

where F is the ponderomotive force of interaction of the resonators.

In the course of the calculations, we set $\gamma_1 = \dfrac{k_1}{x_3}$, $\gamma_2 = -k_2x_3$, $\varkappa = \text{const}$, where k_1 and k_2 depend on the geometry of the resonators and the properties of the environment where the interactions take place.

Fig.9.2 shows the dependence of the potential energy of the ponderomotive interaction between coupled resonators on the mechanical coordinate. The dependence is characterized by several minimum values, which correspond to one and the same frequency of external action. This means that it is possible to achieve stable mechanical equilibrium of the electromechanical system for one and the same exciting frequency at several values of the energy of the coupled resonators.

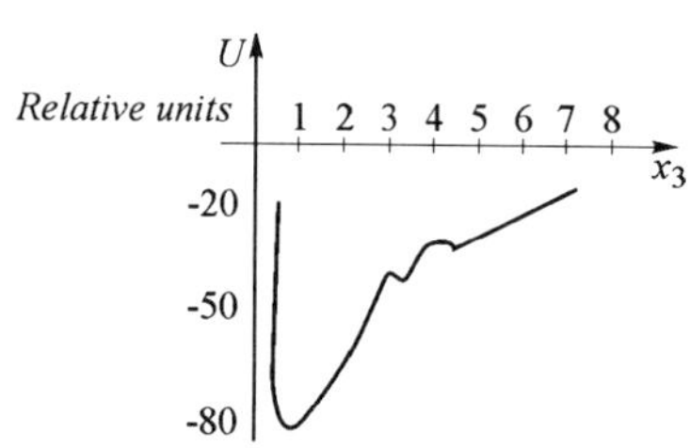

Fig. 9.2. Dependence of the potential energy of the interaction of two coupled resonators on the distance between them

It is worth noting that the stable equilibrium position of the resonators is characterized by a non-zero coefficient of the dissipative coupling. If the coefficient of the dissipative coupling is equal to zero, stable equilibrium is never attained, since for an arbitrary distance between the resonators the average ponderomotive force for the period does not zero. Nevertheless, due to their inertia and depending on the frequency of the external force, the resonators can, in this case, maintain certain geometry with respect to each other that we shall call „quasi-equilibrium".

The expression (9.7) yields the following equation for the equilibrium distances:

$$(n^2-\omega^2)k_2^2\alpha_{11}n^2x_{30}^4 - 4\delta^2\varkappa\omega^2 k_2\alpha_{11}x_{30}^3 + 4\delta^2\varkappa\omega^4\beta_{11}k_1x_{30} - (n^2-\omega^2)k_1^2\beta_{11}\omega^4 = 0,$$

whose solution is

$$x_{30}^{(1)} = \frac{\omega}{n}\sqrt{\frac{k_1}{k_2}},$$
$$x_{30}^{(2)} = \frac{R_c - \sqrt{R_c^2 - 4R_cR_L}}{2},$$
$$x_{30}^{(3)} = \frac{R_c + \sqrt{R_c^2 - 4R_cR_L}}{2},$$

R_L and R_c are the equilibrium distances for resonators inductively coupled with ohmic and capacitive-ohmic bonds respectively,

$$R_L = \frac{k_1(n^2-\omega^2)}{4\delta^2\varkappa}, \qquad R_c = \frac{4\delta^2\varkappa\Omega^2}{k_2 n^2(n^2-\omega^2)}.$$

The equilibrium position $x_{30}^{(1)}$ is the basic one because it does not depend on any additional conditions. The principal minimum of the potential energy of the coupling between the resonators (See Fig.9.2) corresponds to this equilibrium state of the system. The former is determined according to the following formula:

$$U_{\min}^{(1)} = N^2\alpha_{11}\omega^2(k_1k_2)^{\frac{1}{2}}[(n^2-\omega^2)\sqrt{k_1k_2} - 8\delta\varkappa\frac{\omega}{n}]. \tag{9.8}$$

The condition for the existence of two other equilibrium states that are realized only in the event of definite ratios between the coupling coefficients and the parameters of the system, is expressed as follows:

$$R_c > 4R_L. \tag{9.9}$$

For some quantities of the coupling coefficients, the depth of the local minimums on the curve of the potential energy can be comparable to the depth of the main minimum. In this case the electromagnetic system may have three stable equilibrium states.

In the absence of fluctuations the system, being in one of its stationary states, cannot transcend to another one without the application of external action. Any small random actions may cause inconsiderable fluctuation oscillations of the system close to one of its stationary states and it can pass from one equilibrium state to another.

Having obtained the analytical expression for the potential energy and regarding the electromechanical system of oscillators as similar to the harmonic oscillator in the region of the equilibrium point, we can determine the frequency of the mechanic oscillations in the first approximation:

$$\Omega_0 = 2Nk_2n^2\sqrt{\frac{\beta_{11}}{2\beta_{33}}}\sqrt{(n^2-\omega^2)-(k_1k_2)^{-\frac{1}{2}}2\delta^2\varkappa\frac{\omega}{n}}. \tag{9.10}$$

It should be pointed out that the maximum absorption of electromagnetic energy by a stable mechanical system of resonators corresponds to frequencies that are close to the natural resonance frequencies of the system ω_1 and ω_2. This fact is important, for example, for elucidating the mechanism of acoustic wave generation in a solid body under the action of electromagnetic radiation.

9.4. Electromagnetic tracking system

It is worth noting that there is a difference of principle between the „frequency impacts" in the system of interacting resonators described above and the „frequency impacts" in the case of the highly popular phenomena of entrainment and synchronization of the frequencies [273]. In the latter case, there is a definite phase relation between the external force and the generalized coordinates of the system. Moreover, the parameters of the system remain unchanged. In the case discussed above the „frequency actions" are manifested in a change in the system parameters under the action of external forces. Furthermore, the frequency (spectrum) of absorption corresponds to one of the natural frequencies of the system of coupled resonators, though they are not exactly equal. Thus it is evident that the tendency of the resonators of varying nature to group together in stable formations is manifested in a shift of these resonators under the action of mutual ponderomotive forces towards an equilibrium state. At that these forces are nullified and their action disappears.

The two cases described below are particularly interesting from a pragmatic point of view:

(i) A system of resonators with a changing inductive and constant ohmic coupling (there is no capacitive coupling);

(ii) A system of resonators with a changing capacitive and constant ohmic coupling (there is no inductive coupling)

Let us consider the first case, i.e. when $\beta_{12} = \text{const}$, $\alpha_{12} = 0$, in brief. The reduced coupling coefficient depends on the distance between the resonators

$$\gamma_1 = \frac{\beta_{12}}{\beta_{11}} = \frac{k_1}{x_3}, \quad \text{where} \quad k_1 = \text{const}. \tag{9.11}$$

The solution of system (9.1) is presented in a way analogous to that of (9.2)

$$x_1 = A_1 e^{j\omega t}, \qquad x_2 = B_1 e^{j\omega t}. \tag{9.12}$$

The amplitudes A_1 and B_1 in (9.12) are determined from (9.1) as follows:

$$\left| \begin{aligned} A_1 &= N[(n^2 - \omega^2)\mathbf{k} + 2\delta\omega\mathbf{i}] \\ B_1 &= -N[-\gamma_1\omega^2\mathbf{k} + 2\varkappa\delta\omega\mathbf{i}]. \end{aligned} \right. \tag{9.13}$$

The average ponderomotive force of interaction between the resonators is proportional to the scalar product of the amplitudes and it is of the following form:

$$F_L = \frac{\partial\beta_{12}}{\partial x_3}(-\omega^2)(A_1 B_1) = \frac{\partial\beta_{12}}{\partial x_3}\omega^2 N^2[-(n^2 - \omega^2)\gamma_1\omega^2 + 4\varkappa\delta^2\omega^2]. \tag{9.14}$$

The potential energy of the ponderomotive interaction can be determined, after taking into account (9.11), by integrating (9.14):

$$U_L = k_1\beta_{11}N^2\omega^4\left[\frac{k_1(n^2 - \omega^2)}{2x_3^2} - \frac{4\delta^2\varkappa}{x_3}\right]. \tag{9.15}$$

As (9.15) is analyzed, given $\omega < n$, for the equilibrium states of the system of resonators, the dependence of the coordinate x_3 on the frequency of the external harmonic action can be determined:

$$x_{3L}^{(\omega)} = \frac{k_1(n^2 - \omega^2)}{4\delta^2\varkappa}. \tag{9.16}$$

It follows from (9.16) that when ω is increased to n, the respective equilibrium distance between the resonators decreases. The minimum value of the potential energy (at absolute quantity) is determined from the relation

$$|U_{\min}| = \frac{8\beta_{11}\delta^4\varkappa^2 N^2\omega^4}{(n^2 - \omega^2)} \tag{9.17}$$

and it corresponds to the stable equilibrium position of the system.

The analysis shows that when only one coupling changes (in this case the inductive one) in the presence of a constant ohmic coupling, it is possible to group the system of related resonators in stable formations. There is a definite dependence of the potential energy on the mechanical coordinate between the resonators that corresponds to each value of the acting frequency, so that the increase in the frequency of the external action correlates with a rise in the energy of the coupling between the resonators.

The following sections of Chapter 9: 9.5 and 9.6 deal with the second case – a changing capacitive and constant ohmic coupling.

9.5. A tracing system with a self-tuning capacitance

9.5.1. Conditions for grouping two oscillating systems with changing capacitive and constant ohmic coupling in stable formations

Object of consideration is the interaction of two linear oscillating systems with a capacitive and constant (invariable) dissipative coupling – Fig.9.3 [274–279]. The circuit of one of the systems includes a generator of electric oscillations. The capacitance of the conventionally separated coupling capacitor C_t can change as a result of a mechanical shift of one of the plates. The essence of the phenomenon under review consists in that the mobile plate of the coupling capacitor C_t can assume a strictly defined position conditioned by the frequency value of the forcing generator, i.e. the capacitor traces the frequency of the exciting generator by changing the distance between the plates.

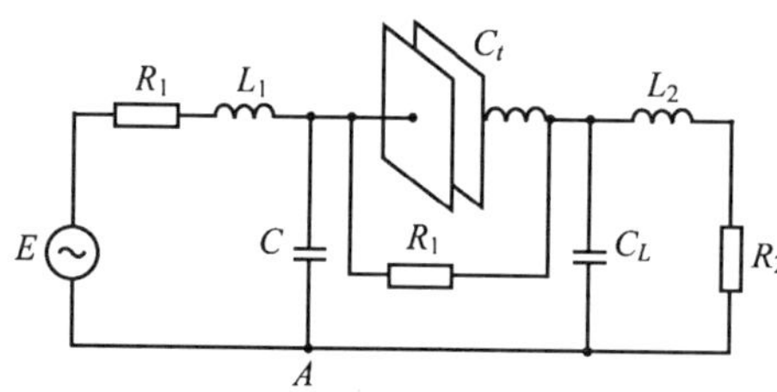

Fig. 9.3. A system of two linear radiophysical oscillating circuits with capacitive and dissipative coupling and an exciting generator

The phenomenon of adaptive self-tuning by altering the capacitance in line with the frequency of the generator is both heuristically significant in the investigation of various synergetic phenomena and effectively applicable in the practice of instrument engineering, automation and radio engineering.

The differential equations describing the electric processes in a coupled system of two identical electric oscillating circuits and the mechanical process of shifting the mobile capacitor plate are of the following form:

$$\left| \begin{aligned} &\ddot{x}_1 + 2\delta\dot{x}_1 + n^2 x_1 + 2\delta\varkappa\dot{x}_2 + \gamma n^2 x_2 = \frac{1}{\beta_{11}} E\cos\omega t \\ &2\delta\varkappa\dot{x}_1 + \gamma n^2 x_1 + \ddot{x}_2 + 2\delta\dot{x}_2 + n^2 x_2 = 0 \\ &M\ddot{r} + \beta_{11}\frac{\partial\gamma_1}{\partial r} n^2 \overline{x_1 x_2} = 0, \end{aligned} \right. \tag{9.18}$$

where x_1 and x_2 are the charges of the first and second oscillating circuits of the system; $\delta = \frac{\varepsilon_{11}}{2\beta_{11}}$, $n^2 = \frac{a_{11}}{\beta_{11}}$; β_{11}, ε_{11}, a_{11} are the coefficients of the inductance, capacitance and damping in the oscillating circuits, $\gamma = \frac{a_{12}}{a_{11}}$, $\varkappa = \frac{\varepsilon_{12}}{\varepsilon_{11}}$, a_{12} and ε_{12} are the coefficients of capacitive and dissipative coupling; M is the mass of the mobile plate of the capacitor, r is the coordinate of the mechanical shifting of the mobile plate, $\frac{\partial a_{12}}{\partial r}\overline{x_1 x_2}$ is the average ponderomotive force of the interaction between the capacitor plates, $E\cos\omega t$ is a generator of electric oscillations, $0 < \varkappa < 1$, $-1 < \gamma < 1$.

The solution of the first two equations in the system (9.18) is presented in the form:

$$x_1 = A_2 e^{j\omega t}, \qquad x_2 = B_2 e^{j\omega t}, \tag{9.19}$$

where A and B are complex values of the amplitudes.

The amplitudes A_2 and B_2 of the charges x_1 and x_2 are expressed from (9.18) after taking into account (9.19):

$$A_2 = N[(n^2 - \omega^2)\mathbf{i} + 2\delta\omega\mathbf{k}], \quad B_2 = -N[n^2\gamma\mathbf{i} + 2\delta\varkappa\omega\mathbf{k}], \tag{9.20}$$

where N is a coefficient depending on the parameters of the system, on the amplitude and frequency of the acting force. After a process of averaging for the period $T = \frac{2\pi}{\omega}$, the ponderomotive force of interaction between the capacitor plates is determined from the scalar product of the charge amplitudes

$$\cos(\widehat{x_1 x_2}) = \frac{(\omega^2 - n^2)n^2\gamma - 4\delta^2\omega^2\varkappa}{\sqrt{(n^2-\omega^2)^2 + 4\delta^2\omega^2}\sqrt{n^4\gamma^2 + 4\delta^2\omega^2\varkappa^2}}. \tag{9.21}$$

In the case of a ponderomotive force equal to zero, the capacitor plates will be in equilibrium. The plate equilibrium condition is

$$(\omega^2 - n^2)n^2\gamma - 4\delta^2\omega^2\varkappa = 0. \tag{9.22}$$

If the equilibrium of the capacitor plates is stable, then, in line with Eq.(9.22), the reduced coefficient of the capacitive coupling ω assumes a strictly defined value for the given frequency of the external force γ. Given a sufficiently broad frequency range and a slow and monotonous change in the frequency of the external force, there is a monotonous and continuous alteration of the parameter of the capacitive coupling and, respectively, of the distance between the capacitor plates.

9.5.2. Ponderomotive forces and interaction energy in connected resonance systems with self-tuning capacitance

Let us assume that in the event of a changing distance between the mobile and immobile capacitor plate the coefficient of capacitive coupling changes in accordance with the law

$$a_{12} = \frac{h}{r}, \tag{9.23}$$

where h is a constant depending on the area of the capacitor plates.

As (9.23) is taken into account, the following is obtained for the average value of the ponderomotive force per period:

$$\begin{aligned} F &= \frac{\partial a_{12}}{\partial r}\overline{x_1 x_2} = D\frac{1}{2}\frac{\partial a_{12}}{\partial r}[(\omega^2 - n^2)Q_0^2\gamma - \omega^2 \varkappa] \\ &= D\left[\frac{hQ_0^2\left(\frac{n^2}{\omega^2} - 1\right)}{r^7} - \frac{\varkappa}{r^4}\right] = F_1 - F_2, \end{aligned} \tag{9.24}$$

where $Q_0 = \frac{n}{2\delta}$ is the quality factor of the oscillating system, D is a coefficient depending on the amplitude and frequency of the external force.

Fig.9.4 presents the plot of the terms F_1 and F_2 in accordance with equation (9.24) and as depending on the distance between the capacitor plates. The middle curve represents the dependence of the aggregate average ponderomotive force F. At a definite value of the distance between the capacitor plates, the average ponderomotive force is equal to zero. If there is a minimum of the energy of interaction of the plates at this point, the equilibrium of the plates will be stable.

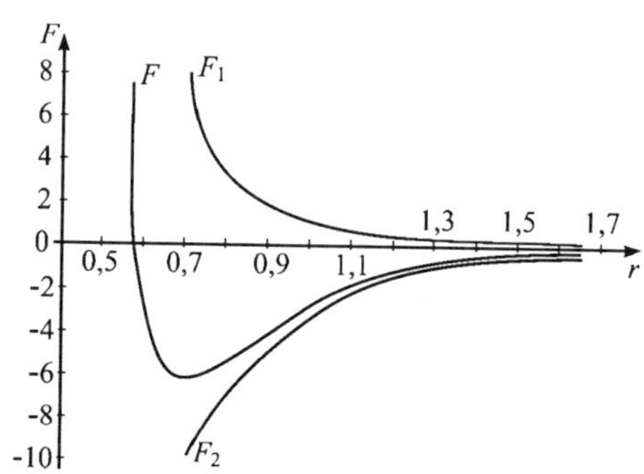

Fig. 9.4. Theoretical dependence of the ponderomotive force (F) and its components (F_1, F_2) on the distance between the capacitor plates

The analytical expression for the energy of the ponderomotive interaction between the capacitor plates is of the following form:

$$W = -\int F dr = D\left[\frac{hQ_0^2\left(\frac{n^2}{\omega^2}-1\right)}{6r^6} - \frac{\varkappa}{3r^3}\right] + C. \qquad (9.25)$$

Fig.9.5 shows the dependence of the interaction energy of plates W on the distance r between them. The interaction energy has its minimum at the equilibrium point, hence the ponderomotive equilibrium of the plates is stable.

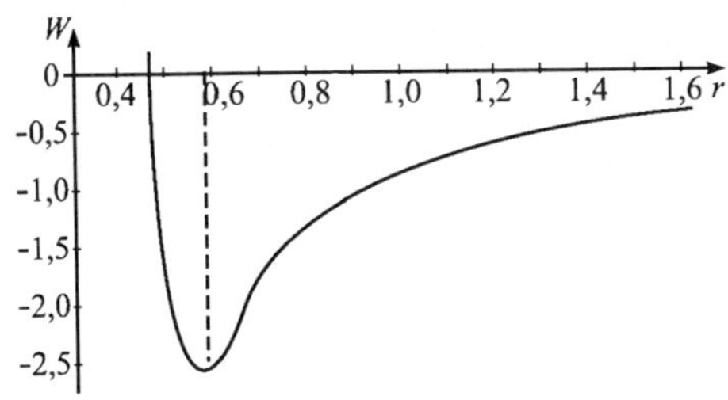

Fig. 9.5. Theoretical dependence of the energy of the interaction between the capacitor plates on the distance between them

The coefficient and frequency values used in plotting the graphs (Fig.9.4 and Fig.9.5) are as follows: $DhQ_0^2 = 0,6$; $\frac{n^2}{\omega^2} = 2$; $C = 0$; $D\varkappa = 3$.

The stable equilibrium position of the capacitor plates is characterized by a zero ponderomotive force. It is determined by the frequency of the external force. As the frequency of the exciting generator changes, the parameters of the couplings also alter and this results in the occurrence of the respective ponderomotive forces. The latter change the distance between the capacitor plates and thereby bring them to a new stable equilibrium position.

The phenomenon described in Sections 9.5 and 9.6 and the analytical results shown above are important for the solution of certain problems in instrument engineering, automation and radio engineering. They make it possible to formulate some new principles and methods of implementing levitation, electrodynamic supports, systems for automatic frequency regulation, separation of the real component from the imaginary one, when measuring complex dielectric permittivity, etc. These principles can be employed to create simple and reliable sensors for technical vision and sensing in robotics, precise manipulators, including such for aggressive environments, etc.

9.6. Grouping of coupled dipole resonators under the action of an external electromagnetic wave

Let us consider the interaction between an electromagnetic wave and a system of weakly dissipative identical linear dipole resonators coupled with a quasielastic (coulomb) force [277].

Let us assume that the distance between the electric dipoles is r and that they lie on axis x. The mass of each of the charges is m, while the distance between the opposite charges $\pm e$ of the dipoles is respectively x_1, x_2. Let the electromagnetic waves propagate at the angle of φ in the direction towards axis x. We also take it that the size of the dipoles is respectively $L_1 = L_2 = L$ and that they are by far smaller than the length of the electromagnetic wave λ: $L \ll \lambda$. Hence the influence of the diffraction, reflection and scattering of the waves on the parameters of the external field is neglected.

The differential equations describing the processes of interaction of a coupled system of identical dipole resonators with an electromagnetic wave can be presented in the following form:

$$\left| \begin{aligned} &\beta\ddot{p}_1 + \varepsilon\dot{p}_1 + \alpha p_1 + \alpha_{12}p_2 = \Phi_0 \cos\omega t \\ &\beta\ddot{p}_2 + \varepsilon\dot{p}_2 + \alpha p_2 + \alpha_{12}p_1 = \Phi_0 \cos(\omega t + \psi), \end{aligned} \right. \tag{9.26}$$

where $p_1 = ex_1$, $p_2 = ex_2$ are the dipole moments of the resonators; β, ε, α are respectively the own inertia, dissipation and quasi-elasticity coefficients of the resonators; α_{12} is the coefficient of quasi-elastic coupling; Φ_0 is the amplitude of external action; ψ is the electric angle of the phase shift of the external wave action. The system of Eqs. (9.26) can be written in the form:

$$\left| \begin{aligned} &\ddot{p}_1 + 2\delta\dot{p}_1 + \omega_0^2 p_1 + \gamma p_2 = f_0 e^{j\omega t} \\ &\ddot{p}_2 + 2\delta\dot{p}_2 + \omega_0^2 p_2 + \gamma p_1 = f_0 e^{j(\omega t+\psi)}, \end{aligned} \right. \tag{9.27}$$

where $2\delta = \dfrac{\varepsilon}{2\beta} = \dfrac{2e^2\omega_0^2}{3c^3 m 4\pi\varepsilon_0}$; $\omega_0^2 = \dfrac{\alpha}{\beta}$; $\gamma = \dfrac{\alpha_{12}}{\beta} = \dfrac{2e^2}{4\pi\varepsilon_0 r^3 m}$; $f_0 = \dfrac{\Phi_0}{\beta}$.

The solution of the system of equations (9.27) is once again sought in the form

$$P_1 = Ae^{j\omega t}, \qquad P_2 = Be^{j\omega t}, \tag{9.28}$$

where A and B are the complex amplitudes of the dipole moments.

After taking into account (9.27) and (9.28), the system of equations (9.26) is represented as

$$\left| \begin{aligned} &[(\omega_0^2 - \omega^2) + j2\delta\omega]A + \gamma B = f_0 \\ &\gamma A + [(\omega_0^2 - \omega^2) + j2\delta\omega]B = f_0 e^{j\psi}. \end{aligned} \right. \tag{9.29}$$

The following notation is introduced: $a = \omega_0^2 - \omega^2$, $b = 2\delta\omega$, $c = \gamma$. Expressions for the amplitudes A and B are obtained from (9.29):

$$\left| \begin{aligned} A &= f_0 \frac{(a - c\cos\psi) + j(b - c\sin\psi)}{a^2 - b^2 - c^2 + j2ab} \\ B &= f_0 \frac{(a\cos\psi - b\sin\psi - c) + j(a\sin\psi + b\cos\psi)}{a^2 - b^2 - c^2 + j2ab}. \end{aligned} \right. \tag{9.30}$$

The potential of the coulomb attraction of two dipole resonators characterized by moment shifts x_1 and x_2 is determined from the expression:

$$V = \left[-\frac{e^2}{r - x_1} - \frac{e^2}{r + x_2} + \frac{e^2}{r + x_2 - x_1} + \frac{e^2}{r} \right] \frac{1}{4\pi\varepsilon_0} \simeq -\frac{2e^2 x_1 x_2}{r^3} \frac{1}{4\pi\varepsilon_0} = -\frac{2P_1P_2}{r^3} \frac{1}{4\pi\varepsilon_0} \tag{9.31}$$

given $|x_1|$, $|x_2| \ll r$.

When (9.31) is taken into account, the ponderomotive force of interaction acquires the form:

$$\mathcal{F} = -\frac{\partial V}{\partial r} = -\frac{3P_1P_2}{2\pi\varepsilon_0 r^4}.$$

The period average value of the ponderomotive force with a view to (9.28) is determined as

$$\overline{\mathcal{F}} = -\frac{3(A, B)}{2\pi\varepsilon_0 r^4}. \tag{9.32}$$

The scalar product (A, B) using (9.30) is written as follows:

$$(A, B) = N[\cos\psi(a^2 + b^2 + c^2) - 2ac], \tag{9.33}$$

where $N = \dfrac{f_0^2}{[(a^2 - b^2 - c^2)^2 + 4a^2b^2]^2}$.

Accounting for (9.33), the expression for the ponderomotive force (9.32) assumes the form

$$\overline{\mathcal{F}} = -\frac{M}{r^{10}} + \frac{G}{r^7} - \frac{K}{r^4}, \tag{9.34}$$

where $M = N\dfrac{3}{2\pi\varepsilon_0}\cos\psi c^2 r^6$, $G = N\dfrac{3\varkappa}{2\pi\varepsilon_0}$, $\varkappa = \dfrac{ae^2}{\pi\varepsilon_0 m}$, $K = N\dfrac{3}{2\pi\varepsilon_0}(a^2 + b^2)\cos\psi$.

Let us identify the conditions under which the functions $N(r, \omega)$ and $\cos\psi(r, \omega)$ are actually independent of r. For the function $N(r, \omega)$ this condition is derived from the following inequalities:

$$a^2 - b^2 - c^2 > 4a^2b^2, \qquad a^2 - b^2 > c^2 \tag{9.35}$$

and it is of the form

$$r > \sqrt{\frac{2e^2}{4\pi\varepsilon_0 m\sqrt{a^2 - b^2 - 4a^2b^2}}}.$$

As dissipation is disregarded, the following is obtained:

$$r > \sqrt{\frac{e^2}{2\pi\varepsilon_0 m(\omega_0^2 - \omega^2)}}. \tag{9.36}$$

The quantity $\cos\psi$, included in expression (9.34) is determined in the following way:

$$\cos\psi = \cos\left[\frac{2\pi r}{\lambda}\cos\varphi\right] = \cos\left[\frac{r\omega}{c}\cos\varphi\right]. \tag{9.37}$$

In the case when the wavelength of the external emission λ is by far bigger than the relative distance between the resonators r, i.e.

$$r \ll \lambda, \tag{9.38}$$

the value ψ tends to zero, while the function $\cos\psi$ tends to its maximum. In the area of the maximum, the function $\cos\psi$ can be regarded as virtually constant.

As (9.34), (9.36) and (9.38) are taken into account, the energy of the ponderomotive interaction is presented in the following way:

$$\overline{V} = -\frac{H}{r^9} + \frac{L}{r^6} - \frac{P}{r^3}, \tag{9.39}$$

where $H = \dfrac{M}{9}$, $L = \dfrac{G}{6}$, $P = \dfrac{K}{3}$.

Fig.9.6 and Fig.9.7 present the theoretical dependencies of the force and energy of the interaction of dipole resonators on the distance between them for the area determined by (9.36). The following are taken as numeric values of the coefficients: $G = 77.6210^{-58}$, $K = 6.5610^{-28}$, $H = 23.0510^{-88}$, $L = 12.9410^{-58}$, $M = 207.4510^{-88}$, $P = 2.1810^{-28}$.

In a state of stable equilibrium the ponderomotive force is equal to zero, while the energy of interaction has a minimum.

Expression (9.34) and inequality (9.35) yield a condition for ponderomotive equilibrium:

$$\cos\psi(a^2 + b^2)r^6 - \varkappa r^3 + \left(\frac{\varkappa}{2a}\right)^2 \cos\psi = 0. \tag{9.40}$$

The dependence of the distance between the resonators, in accordance with (9.40), is determined as

$$r_{1,2} = \sqrt{\frac{\varkappa \pm \sqrt{\varkappa^2 - 4(a^2 + b^2)\left(\dfrac{\varkappa}{2a}\right)^2 \cos^2\psi}}{2(a^2 + b^2)\cos\psi}}. \tag{9.41}$$

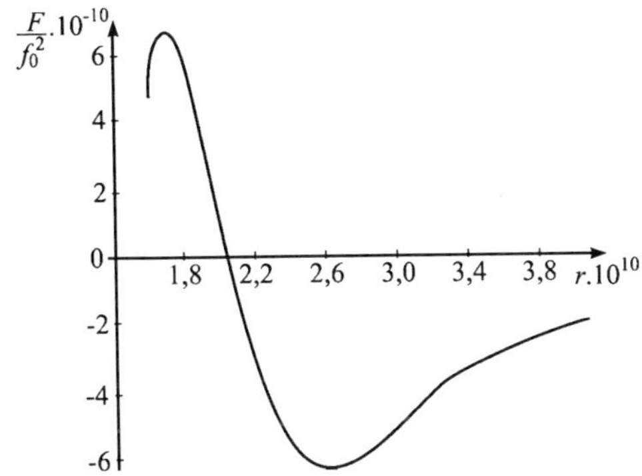

Fig. 9.6. Theoretical dependence of the ponderomotive force of interaction between dipole resonators on the distance between them

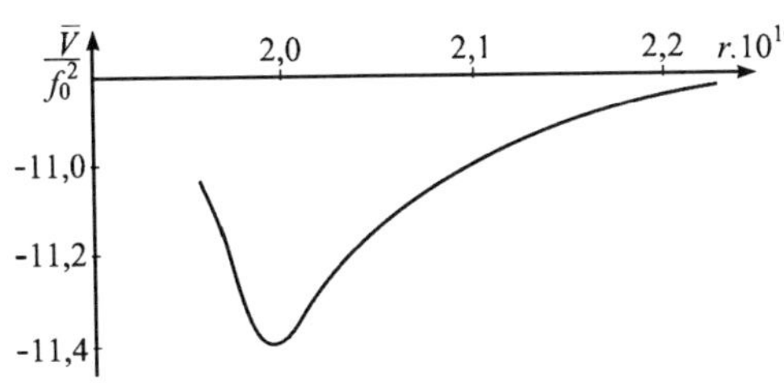

Fig. 9.7. Theoretical dependence of the energy of the interaction between dipole resonators on the distance between them

As the own dissipation of the resonators is ignored, the following is obtained from (9.41):

$$r_1 = \frac{1}{a}\sqrt{\frac{\varkappa}{2}\left[\frac{1+\sin\psi}{\cos\psi}\right]}, \tag{9.42}$$

$$r_2 = \frac{1}{a}\sqrt{\frac{\varkappa}{2}\left[\frac{1-\sin\psi}{\cos\psi}\right]}. \tag{9.43}$$

In order to determine the stable equilibrium state, we investigate the sign of the second derivative of the energy of interaction of the resonators. Taking into account (9.38), we obtain:

$$\overline{V''(r)} = -\overline{\mathcal{F}'(r)} = -N\frac{3}{2\pi\varepsilon_0}\frac{1}{r^{11}}[4r^6a^2\cos\psi - 7\varkappa r^3 + 10\left(\frac{\varkappa}{2a}\right)^2\cos\psi].$$

It is not difficult to notice that $\overline{V''}_{1,2}(r_1) > 0$ and $\overline{V''}_{1,2}(r_2) < 0$.

Given the conditions considered above and the set value range of the frequency ω, the dependence (9.42) corresponds to the stable equilibrium state, while the dependence (9.43) corresponds to the unstable one.

As Expr.(9.39) is integrated, after taking into consideration (9.37), by an arbitrary closed area encompassing the connected system of resonators, the following is obtained

$$\int_s \cos\psi ds = \frac{\varkappa r^3}{(a^2+b^2)r^6 + \left(\frac{\varkappa}{2a}\right)^2} \int_s ds, \tag{9.44}$$

where s is the area of integration.

As the function $\cos\psi$ is expanded in powers of a small parameter and the expression (9.44) is transformed, the following is obtained:

$$\left(r^8 \frac{\omega^2}{4c^2} - r^6\right)(a^2+b^2) + r^3\varkappa + r^2\left(\frac{\omega\varkappa}{4ac}\right)^2 - \left(\frac{\varkappa}{2a}\right)^2 = 0. \tag{9.45}$$

Equation (9.45) represents an implicit dependence of the relative distance r on the frequency ω. Its solution is associated with certain difficulties and an approximation of it can be obtained by using numeric methods.

The analysis indicates that the tendency of two coupled resonators to be grouped into stable formations is manifested by the changing of their relative distance in accordance with the frequency and phase shift of the external action on the system.

The paper [279] describes the established phenomenon of frequency attraction. This is a phenomenon that differs in quality and principle from the widely popular phenomenon of entrainment and synchronization of frequencies. During the interaction of two linear oscillating circuits with a capacitive and resistive coupling, the capacitive coupling alters in accordance with the change in frequency of the external exciting generator. The alteration in the capacitive coupling causes a change in the natural resonance frequencies of the system. There are two natural resonance frequencies that correspond to each frequency value of the exciting generator. At that, one of them – the lower one – follows the frequency of the exciting generator and its value is only slightly different from that frequency, i.e. one of the natural resonance frequencies „catches up" with the frequency of the exciting generator by incessantly self-tuning its value to that frequency.

CHAPTER 10.

A PHENOMENON OF EXCITATION OF CONTINUOUS OSCILLATIONS WITH A DISCRETE SET OF STABLE AMPLITUDES ("QUANTIZED" OSCILLATION EXCITATION)*

10.1. Introduction (Major model notions)

Synergetics, being one of the new concepts laying out the road to the development of a single theory of system self-organization, shares many traits with the systems approach, which is well-known one in cybernetics. As far back as the 1940s, the eminent US researcher L. von Bertalanfey wrote his fundamental work: „General Systems Theory: A New Approach to Uniting Sciences", where he expressed the idea that what mattered was not so much superficial and vague analogies between phenomena, but isomorphism in the mathematical sense of the word, i.e. the strict consistence of all elements of the compared systems. Today this thesis underlies the synergetic approach to system analysis based on generalized models as well. It is worth noting that most frequently the general mathematical models are equations describing oscillating phenomena in a broad class of systems. A central issue raised in relation to the majority of synergetic models is the investigation of what are known as „resonance excitations". As already emphasized in Chapter 9, the existence of nonlinear oscillating motion in open systems is an essential condition for the manifestation of synergetic phenomena of self-organization.

This chapter presents the phenomenon of excitation of continuous oscillations with a possible discrete set of stable amplitudes under the action of an external periodic force that is nonlinear along the coordinate. This is a phenomenon of a highly general nature manifested in various dynamic systems. What is meant here is the display of peculiar „quantization" by the parameter of intensity of the excited oscillations, i.e. given unchanging conditions, it is possible to excite oscillations with a strictly defined discrete set of amplitudes; the rest of the amplitudes are „forbidden". The realization of oscillations with certain amplitude from the „permitted" discrete set of amplitudes is determined by the initial conditions. The occurrence of this unusual property is determined by the new initial conditions, i.e. nonlinear action of the external exciting force with respect to the coordinate of the system subjected to excitation.

It is well-known fact that the Theory of Nonlinear Oscillations considers mostly the action of external periodic forces on oscillating systems. These forces are either independent of the coordinate of the system, or linear with respect to

* An investigation supported by the Bulgarian National Council „Scientific Research" under Contract No. H3-1106/01

that coordinate (the latter are in essence the classical parametric systems – See Chapters 5 and 7). Recent years also saw investigations of nonlinear parametric phenomena occurring in dynamic systems described by equations with polynomial nonlinearity [79].

The phenomenon under consideration is characterized by other initial conditions, i.e. nonlinearity of the external acting force with respect to the coordinate of the system that is being excited. The result is the occurrence of qualitatively new properties typical of both a linear and a nonlinear oscillating system under the action of a nonlinear force. Naturally, we need to point out that in this case the „linear" system is actually pseudo-linear, since the oscillator can indeed be linear, but the external force depends in a nonlinear way upon the coordinate of its motion.

Let us give the reader an initial idea of the class of systems that will be considered in this chapter, and then we shall go back and review the historical prerequisites for this analysis and some previous investigations of kindred phenomena.

The motion occurring in oscillating systems of varying physical nature under the action of an external periodic force, which is nonlinear along the coordinate of the system, can be described, in the general case, by the following equation:

$$\ddot{x} + 2\delta_0\dot{x} + \omega_0^2 x + f(x) = F(x, t_r), \tag{10.1}$$

where x is a generalized coordinate of the system, a dot signifies differentiation by time, i.e. $\dot{x} = \dfrac{dx}{dt_r}$, t_r is real time, δ_0 is a coefficient reflecting the dissipative properties of the system, ω_0 is the natural resonance frequency given an infinitesimal amplitude of the oscillations, $f(x)$ is a function characterizing the nonlinearity of the system, $F(x, t_r)$ is the external periodic force, nonlinear along the coordinate of the system.

This chapter will be mostly devoted to the consideration of the periodic solutions of the equations describing the motion of a pendulum under the action of an external periodic force that is nonlinear along the coordinate:

$$\ddot{x} + 2\delta_d\dot{x} + \sin x = F_\nu(x, t_r), \tag{10.2}$$

where x is the angle of displacement of the pendulum from the lower equilibrium point, $\dot{x} = \dfrac{dx}{dt}$, t is dimensionless time, whose scale is the reciprocal resonance frequency $\dfrac{1}{\omega_0}$, $2\delta_d = \dfrac{2\delta_0}{\omega_0}$ is the damping relation, $F_\nu(x,t) = \dfrac{F(x, t_r)}{\omega_0^2}$.

The acting force – nonlinear along x and (almost) periodic along t – $F_\nu(x,t)$ can be presented, for example, in the following form:

$$F_\nu(x,t) = \varepsilon(x) F_0 \sin \nu t, \tag{10.3}$$

where F_0 and ν are constants representing the amplitude and frequency of the external harmonic force, the function $\varepsilon(x)$ yields the dependence on the coordinate x, which can be analytically expressed in various ways, such as:

$$\varepsilon(x) = \begin{cases} 1, & \text{when } |x| \leq d' \\ 0, & \text{when } |x| > d' \end{cases} \tag{10.4}$$

or $\varepsilon(x) = e^{-\frac{x^2}{2d'^2}}$, $\varepsilon(x) = \begin{cases} \frac{1}{2}[1+\cos(\frac{\pi}{d'}x)], & \text{at } |x| \leq d' \\ 0 & \text{at } |x| > d' \end{cases}$ and so on, $d' \ll 1$ is a parameter limiting the external action on a small part of the trajectory of motion in the system.

Equations (10.2), (10.3) and (10.4) reveal the obvious characteristics of the initial conditions set in the class of systems under consideration: (i) the impact of the external force is short-lived in comparison with the period of the oscillations that are excited, and (ii) the dependence on the coordinate presupposes the existence of a peculiar internal self-adaptive feedback. These conditions lead to the occurrence of the phenomenon of "quantization" of the dynamic states and strong adaptive stability of the stationary modes. As we shall show further on, these regularities can also be observed in the case of the action of an incident plane wave upon an oscillator, i.e. in the case of absence of any artificial organization of external nonlinear action.

There are oscillating processes and phenomena based on non-homogeneous interactions, on the utilization of the inertia-driven properties of charged particles and the like [78, 125] that are tackled by SHF Electronics and Physical Electronics, by charged particles accelerating technology, by Optics, Mechanics, Acoustics, Chemical Kinetics, Biology and other areas of science and technology. The mechanism of interaction has different manifestations in each specific case and mode (as phase selection, as particle grouping, as self-modulation, etc. [42, 78, 121, 122, 125, 132, 137, 160, 280]). A single principle can be traced as underlying all these mechanisms. According to it, a high frequency external force has nonlinear (non-homogeneous) action with respect to the coordinate of the system motion. The term „high-frequency force" refers, both here and further on, to a force, whose frequency is by far higher than the frequency of the oscillations that are excited. For example, all cyclic accelerators of charged particles abide by the relation $\nu = N\omega$, where ν is the frequency of the accelerating high-frequency electromagnetic field, ω is the frequency of the cycle of charged particles, N is an integer (called acceleration ratio), which reaches values in the order of scores and hundreds. The acceleration process is accompanied by an increase in ν (phasotron), in N (microtron), in ω (synchrotron) or in ν and ω (synchrophasotron). The excitation of stationary modes is based on processes of self-modulation, grouping and phase selection. The stationary oscillations (or rotatory mode) of the pendulum under the action of a high-frequency force are obviously analogous to the motion of a „resonance" particle in a cyclic accelerator.

The basic model considered in this Chapter shares many common features with the well-known problem investigated by Fermi [281]. As an explanation of the origin of cosmic rays, Fermi proposed a mechanism for accelerating charged particles by means of collisions with moving magnetic field structures. A great number of works deal with the simplest model case – what is known as the model of Fermi-Ulam [23]. In the setup of the Fermi-Ulam scattering problem a ball is made to fly and impact dissipatively on a single sinusoidally vibrating surface under the influence of gravitational acceleration, which reverses the flight. The amplitude of the surface vibration of the cosine type and the coefficient of restitution between the ball and the surface control the ball dynamics. In recent years, there have even been discussions in principle on the possibility of using the mechanism of Fermi to accelerate space rockets in the gravitational fields of planets and stars and to create a model of what is known as „gravitational engine“. The part of a vibrating surface can be played, for example, by the field of a rotating binary stars.

In summary, a class of phenomena and systems with specific excitation can be formed. It can be most generally termed a class of kick-excited self-adaptive dynamical systems. Kick-excitation is represented by a short (compared with the basic period of the oscillations) impact of an external periodic force. The self-adaptivity consists in the self-tuning of the system to the external kick excitation, which conditions the super stability of the oscillations. The Fermi problem as well as the model setup of Fermi-Ulam can be identified as one of the basic phenomena covered by the constructed class of kick-excitable systems and phenomena. Analogous phenomena occur in other systems, too, and they can also be referred to this class, such as those in periodically kicked hard oscillators, in kicked rotators, in driven impact oscillators, in ice-structure interaction models [78, 282–284], etc. This class also includes the above-mentioned cyclotron accelerator of charged particles, self-modulation phenomena, particle grouping and phase selection phenomena.

As early as 1919 the works of H. Barkhausen and K. Kurtz outlined methods for generating SHF oscillations based on entirely new principles [285]. In their first announcement concerning their generator, N. Barkhausen and K. Kurtz pointed out the conditions for generation maintenance: „Certain order should be introduced in the motion of the charges so that the majority of the electrons will have the same phase of motion. In addition, the order should be maintained by the very motion of the electrons“. The order referred to by N. Barkhausen and K. Kurtz implies nothing other than a grouping of the electrons, while the mechanism for maintaining this order consists in a relevant change in the speed (time) of motion of the grouped electrons during their interaction with the external field. Virtually all SHF electronic devices that came about later use the mechanism of the grouping of electron flows. The energy of the phase selection leads to an increase in the energy of the SHF field in the case when „irregular“ phase electrons are quickly removed and the „regular phase“ electrons prevail. From a common single perspective, it can be concluded that the interaction mechanism in SHF electronic devices is characterized by the nonlinear nature of the applied external SHF force with respect to the coordinate of motion of the charged particles.

A striking example of the so formed class of kick-excited self-adaptive phenomena and systems is the model of a pendulum influenced by quasi-periodic short-term actions, as considered in this Chapter. Instead of the flow of charged particles with modulated initial speed, typical of the cyclic accelerators of charged particles, there is a macro-physical oscillating object, such as a pendulum. A high-frequency periodic force acts on a small part of the motion trajectory of the pendulum. The latter is localized in a certain area, which we shall further on term area (or zone) of interaction. The existence of stable periodic motions with a discrete set of possible stationary amplitudes in such macro-physical oscillating systems has been established and proven [78, 132, 138, 139, 155–158, 280, 286–316].

The pendulum is a well-known phenomenon intensively studied for over 300 years. Currently, the pendulum is rightly considered as one of the most general models of nonlinear dynamics [317–321]. We can express the biased opinion that if a certain phenomenon is observed in the case of the pendulum, it has a considerable degree of generality. Phenomena such as „resonance", „enhancement, synchronization and stabilization of the frequency", etc. have been discovered in the „pendulum" type of systems. In the beginning of the 1950s N.N. Bogolyubov and P.L.Kapitza discovered the possibility to stabilize the top equilibrium point of the pendulum by using light high-frequency modulation applied to the point of suspension. This is a phenomenon that was later used in the experiments to establish control on the thermonuclear reaction for the purpose of stabilizing heated plasma [317, 321–323]. Nowadays the pendulum is also widely used as a basic paradigm to analyze, both theoretically and experimentally, the phenomena of excitation of complex irregular and chaotic oscillations [322, 323]. Extensive numerical and analog simulations have shown that this simple, low-dimensional system exhibits complex behavior, including frequency and phase locking, intermittency and fractal basin boundaries.

The inexhaustibility of the pendulum as a general model is once again supported by the phenomenon presented in this chapter, i.e. the excitation of continuous oscillations with an amplitude belonging to a discrete set of possible stable amplitudes. Fig.10.1 offers a schematic presentation of a diagram of the pendulum system under consideration. The deviation of the pendulum from the lower equilibrium position is denoted by x. The external high-frequency force $F = F_0 \sin \nu t$, where $F_0 = \text{const}$ acts in a limited area $[-d', d']$ of the motion trajectory of the pendulum situated symmetrically around the lower equilibrium point. This is the meaning of the concept „inhomogeneous action" with respect to the motion trajectory of the pendulum. The same can be expressed by means of the concept „nonlinear harmonic force" which implies nonlinear dependence of its amplitude on the motion coordinate of the system subject to excitation, i.e. the pendulum. The direction of the external force action is parallel (periodical along or opposite) to the direction of the pendulum motion. Initially, when the

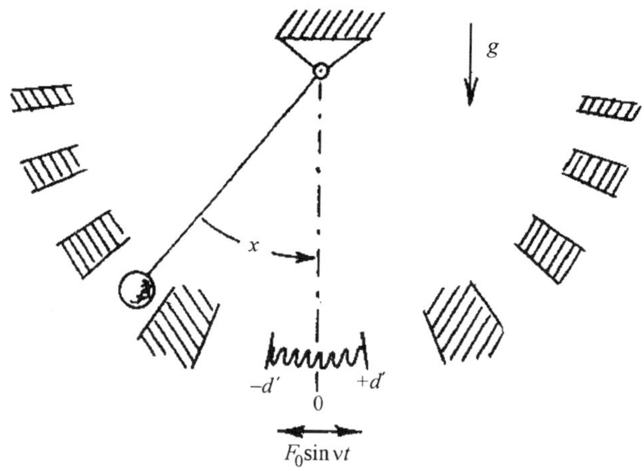

Fig. 10.1. Illustration of a kick-excited pendulum under the action of an external periodic force, located in a narrow area around the lower equilibrium point. The major attraction areas to the possible discrete series of stable amplitudes are shaded

pendulum is diverted from the equilibrium position beyond the $[-d', d']$ area and allowed to oscillate, it passes periodically through the $[-d', d']$ area and gets exposed to the action of the external force $F = F_0 \sin \nu t$. Under these conditions it is possible to establish a stationary oscillation mode of the pendulum with quasi-constant amplitude located within one of the shaded areas of attraction in Fig.10.1. The specific stationary amplitude of the pendulum motion is determined by the initial coordinate deviation and the initial speed (i.e. by the initial conditions). There can be different modes of pendulum motion depending on the initial conditions: it either adheres to one of the possible stationary stable orbits, or its motion damps quickly. This constitutes the heuristic value of the phenomenon – the existence of a possible discrete set of stationary amplitudes, i.e. of peculiar „quantization“ of the pendulum motions by intensity parameter. Parallel with that, there are „forbidden“ areas of initial conditions with respect to which the motion is only damping. Obviously, there is a phenomenon of excitation of „quantized“ oscillations, of „quantization“ of the dynamic states of the macro system. The excitation of one amplitude or another depends on the initial conditions, given unchanged other parameters and conditions. In this case we regard the pendulum as a self-oscillating system with a high-frequency source of energy (unlike the generally accepted concept that self-oscillating systems should have a constant source of energy [48–52]). Quantization (the idea of quanta, photons, phonons, gravitons) is postulated in Quantum Mechanics, while Relativity theory does not derive quantization from geometric considerations. In the case of the established phenomenon, the „quantized nature“ of portioned energy transfer stems directly from the mechanisms of the process and has a precise mathematical description.

The quasi-harmonic oscillator obeys the classical laws to a greater extent than any other system. A number of problems, related to quasi-harmonic oscillators, have the same solution in classical and quantum mechanics.

The fundamental problem, the analytical study of which goes back to Huygens [44, 53], is isomorphic with respect to a variety of physical phenomena, particularly such as radio-frequency driven Josephson junctions and charge-density wave transport [140]. This correspondence, recognized over a quarter of a century ago, has led to various studies of phenomena related to the Josephson effect by means of mechanical analogs.

A well-known phenomenon is that of J. Bethenod [44, 45], which is essentially as follows: If one provides a physical pendulum with a piece of soft iron and places the pendulum above a coil, the pendulum starts oscillating and reaches a stationary state with a constant amplitude. Several theories of this phenomenon were discarded before Y. Rocard formulated the differential equation, describing a kind of parametric excitation [44]. Presumably the parametric excitation of oscillations is a consequence of the change in the inductance of the solenoid as the pendulum passes over it. We show below that even in these conditions the physical nature of the pendulum excitation may be quite different, and the closing section clearly outlines the distinctions between the parametric mechanism of excitation and that presented here. The action on the pendulum is not associated with the fact that it changes the equivalent inductance of the solenoid. Rather it has to do with the fact that it simply passes periodically through its alternating magnetic field, thus acquiring the respective impetus. The same will occur if the interaction between the pendulum and the external force was purely mechanical – for example a feather may oscillate at high frequency at the lower equilibrium point and as the pendulum passes through the oscillation zone it is pushed „back and forth", and under certain conditions there is a summation resultant kick forward. 3.1 shows (See Fig.3.2) that another non-parametric mechanism of oscillation excitation, associated with a respective change in the inductance value of the solenoid, is also possible. In this case, the solenoid is complemented by a capacitor to form an oscillating circuit tuned to the frequency of the power supplying current. The pendulum motion is maintained by the occurrence of a Lorenz force of attraction at the moments when it gets close to the solenoid. The change in the inductance of the solenoid plays the part of a self-adapting controlling relay element ensuring that the process of pendulum acceleration prevails over the process of its retardation by the Lorenz force, i.e. guaranteeing an effective input of energy into the oscillating process.

Our main purpose here is to consider a case when an external periodic source acts upon a part of the trajectory of a moving pendulum in the absence of parametric influence manifested in the changing of a reactive parameter. A general picture of the pendulum motion under inhomogeneous excitation under different conditions is presented. The major dynamic regularities in the motion of the pendulum under consideration are revealed by using numerical simulations technique. An analytical proof of the phenomenon under consideration is also provided.

10.2. Numerical experiment of excitation of „quantized" pendulum oscillations

The inhomogeneously AC driven, damped pendulum system is described by the following set of three first order autonomous differential equations

$$\left|\begin{array}{l} \dot{x} = y \\ \dot{y} = -2\delta_d\dot{x} - \sin x + \varepsilon(x)F_0 \sin z \ , \\ \dot{z} = \nu \end{array}\right. \tag{10.5}$$

where x is the angle of elevation of the pendulum, $y = \dfrac{dx}{dt}$ is its angular velocity; the driven torque is a sinusoidal torque with amplitude F_0, frequency ν, and phase $z = \nu t + \varphi$, φ is the initial phase; δ_d is the damping; a point denotes the operation of differantiation by non-dimensional time $t = \omega_0 t_r$, where ω_0 is the natural resonance frequency of the pendulum for oscillations with an infinitesimal amplitude, the frequency of the external periodic source is expressed in ω_0 units; the case under consideration is $\nu \gg 1$.

We accept that the function $\varepsilon(x)$ determining the nonlinearity of the external periodic force in relation to the coordinate of the system which is excited should be expressed as

$$\varepsilon(x) = \begin{cases} 1, & \text{for } |x| \le d' \\ 0, & \text{for } |x| > d', \end{cases} \tag{10.6}$$

where the parameter d' thereby determines a symmetric zone of action in the area of the lower equilibrium position, $d' \ll 1$. The function $\varepsilon(x)$ is different from zero within a narrow range on the x coordinate, by far narrower than the typical oscillation amplitude of the pendulum. The function $\varepsilon(x)$ can be conventionally termed a feeding function, while the range within which it is different from zero can be called an active zone of the system. At this point we shall emphasize once again the link with the name of the already formed class of kick-excited self-adapting phenomena and systems: the external force is linked, through the function $\varepsilon(x)$, with the motion coordinate in an adaptive mode, and at the same time it exerts action in the form of short impulses, much shorter than the oscillation period of the system. There is free motion in the system during most of the oscillation period.

Equations (10.5) and (10.6) imply that an almost symmetric solution is an almost periodic solution with a period T, which is an odd-integer multiple of the driven period $\dfrac{2\pi}{\nu}$: $T = (2n+1)\dfrac{2\pi}{\nu}$, $n = 1, 2, 3, \ldots$.

A fourth-order Runge-Kutta routine can be employed to compute a numerical solution of Eqs.10.5 and 10.6. Generally it is good to choose the integration time step as 0.001 of the natural period. For every round of calculations we set, for example, that the discarding points should be 500 thousand and the calculating points should be 250 thousand. A comparison between the analytic and computed solutions of the linearized form of Eqs.(10.5) indicates that this technique yields numerical precision of seven decimal digits over one natural cycle. A substantial issue from the viewpoint of the precision of the numeric integration is the fact that the right-hand part of the second equation in (10.5) is subject to a leap at $|x| = d'$, and this leap influences the precision of the calculations. A technique that can help eliminate the influence of this leap is to use the Runge-Kutta method with a variable step. The approach is as follows: let us have a certain moment t on the trajectory where the step is dt. The solution for the moment $t + dt$ is obtained in two independent ways: by means of one integration with step dt and by means of two consecutive integrations with step $\frac{dt}{2}$. These two solutions at moment $t + dt$ may differ by some small magnitude. But if the absolute value of the difference between the two solutions is greater than some preset threshold value (for instance 10^{-8}), the integrating procedure assumes that the precision has not been sufficient, it reduces the step by a half and resumes the integration at moment t with the new step. This splitting of the step may continue until the set threshold value of precision is reached. The opposite is also possible: if the difference between the two solutions is by far smaller than the accuracy threshold (for example 5 to 10 times), the step may automatically double. This computing technique makes it possible for the magnitude of the step to vary depending on the nature of the function $\varepsilon(x)$ on the right-hand side of the second differential equation in (10.5). In the particular case, when the solution reaches the leap point $|x| = d'$, the step will split in such a way that the trajectory around the leap will be calculated precisely; then it will automatically increase again, so that no operating uptime should be lost for calculations with an unnecessarily small step in the regular section of the trajectory. During the next step there will be fragmentation again, etc. This is exactly the method used to make the numerical calculations presented in this Chapter.

Equation (10.5) constitutes a flow in a three-dimensional phase space with dynamic variables x, $y = \dot{x}$ and a drive phase z. The control parameters F_0 and δ_d and the initial conditions x_0 and $y_0 = \frac{dx}{dt}$ determine the pendulum motion. When the point of departure is the physical excitation mechanism, which will be explained at length further and which is related to a frequency lock-on and phase synchronization, the frequency ν of the external acting force during the numerical experiment should be constant. As it shall become clear further on, the initial phase φ, determining the condition of the external acting force at the time of pendulum entry into the action zone, plays an essential part in the adapting maintenance of the pendulum oscillations. At the same time, it is established that at the time of the initial release of the pendulum the phase φ has a value of equal probability within the range $(0 - 2\pi)$, which implies that the pendulum makes its entry into the action zone of the external force with an equally probable value of the initial phase φ. The phase φ is a quantity varying from one period to the other, mainly during the transition process, and it plays a crucial part in the adaptive self-sustained pendulum oscillations. In all calculations the initial value of the initial phase φ is selected to be zero: $\varphi = \varphi_0 = 0$. It is established that after the end of the transition period a mode of automatic adaptive self-tuning of the initial phase around some value $\varphi = \varphi_{st}$ sets in. It is typical of each mode and the respective totality of parameters.

Computer solutions of Eqs.10.5 are obtained and the resulting data are analyzed by using different diagnostic tools: time series of the angular x and the angular velocity $\dot{x}$, phase-plane plots ($\dot{x}$ vs. x), bifurcation characteristics (e.g. the oscillation amplitude vs. the controlled parameters) etc. We obtain numerical solutions for F_0 values in the range of $0.1 \leq F_0 \leq 50.$, for δ_d values of the damping in the range of $0.0001 \leq \delta_d \leq 0.5$, for d' values in the range of $0.001 \leq d \leq 0.05$, for fixed values of the driving frequency $\nu = 51.; 97.; ...$ and starting from the initial conditions $y_0 = \dfrac{dx_0}{dt} = 0$, $x_0 =$ vary. In all cases, in order to eliminate transients, the solutions are run through at least 1000 periods of the driving force before any actual data-taking begins.

Further on, some results of the numerical experiment are presented in a time continuous form for the following values of the parameters: $\beta = 0.01$; $\nu = 51.$; $F_0 = 2.$; $d' = 0.025$, $y_0 = 0.$, $x_0 =$ vary.

Fig.10.2 and Fig.10.3 show, respectively, the time series and the joint phase portraits of stationary steady motion of the pendulum at four different initial conditions: $x_0 = 0.25$, 1.05, 1.35; 1.5. Fig.10.2a presents the time series of the coordinate x, and Fig.10.2b – the time series of the angular speed $\dot{x}$. Both in Fig.10.2b, and in Fig.10.3 one can easily discern the drastic changes in the speed of the pendulum motion in the narrow zone of action of the external force. Fig.10.2 and Fig.10.3 have been used in an endeavour to illustrate different possible characteristic modes and cases. Given an initial condition $x_0 = 1.5$, periodic oscillations are excited that are very close to the harmonic ones with a stationary amplitude ~ 1.45. The initial condition $x_0 = 1.05$ determines the stationary amplitude ~ 1.1. In both cases the value of the initial condition has been selected in such a way as to be very close to the possible („permitted“) amplitude values. The area of attraction in relation to the initial conditions of each of the possible stationary amplitudes (See the shaded areas in Fig.10.1) varies, for instance between 15% and 7%, with the increase in the absolute value of x_0 respectively from 0.25 to 1.5. When the initial condition is set between these areas of attraction, different modes are possible: either the oscillations are rapidly damping, or the motion is „caught“ and stabilized at one of the possible, for example the lower, („permitted“) orbits. The latter possibility is illustrated in Fig.10.4 and Fig.10.5, which present the transition processes of establishing stationary motion for the same four initial conditions as those specified with respect to Fig.10.2 and Fig.10.3. Given the initial condition $x_0 = 1.35$ (clearly visible in Fig.10.5) – a value falling in the range between the „permitted“ values of the stationary amplitudes ~ 1.1 and ~ 1.45, the pendulum „becomes heavier“ and passes in transit through the possible stationary orbit with amplitude ~ 1.1, and it is „caught“ in an orbit with an amplitude ~ 0.75. Another typical peculiarity of the presented data on the pendulum behavior is that while the

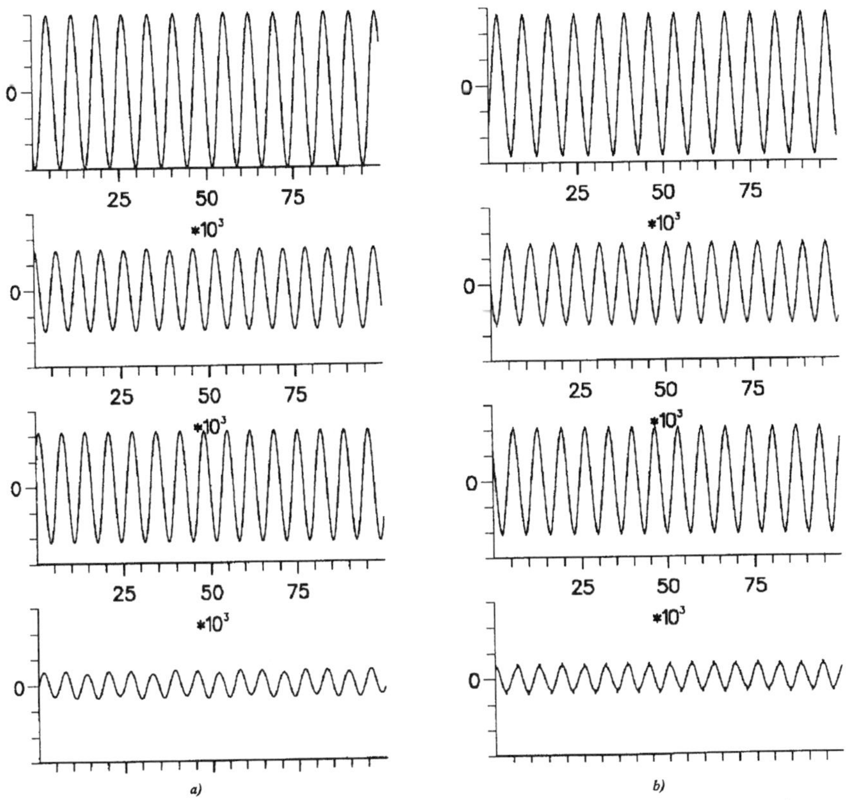

Fig. 10.2. Time series of the coordinate (Fig.2a) and the velocity (Fig.2b) in steady-state stationary regime of the pendulum's motion with the following initial conditions: $x_0 = 1.5$; 1.35; 1.05 and 0.25. The values provided the rest unchanging parameters are: $y_0 = 0$, $F_0 = 2$, $v = 51$, $\beta = 0.01$ and $d' = 0.025$

nature of its motion along one of the orbits, in this case with an amplitude ~ 0.25, is sufficiently close to the harmonic law of motion, in the event of motion with other amplitudes (in this case of lower values), amplitude-two and amplitude-three modulated motion can be observed. This is particularly typical of the case, when the orbit has an amplitude ~ 0.25 and represents amplitude-three motion.

On the whole, Figs.10.2, 10.3, 10.4 and 10.5 illustrate the most important general characteristic of the system under review: discrete nature („quantization") of the possible steady motions by the intensity parameter, where the specific oscillation amplitude is determined by the initial conditions. The occurrence of a definite sequence of possible stable and steady stationary amplitudes has to do with the condition $\nu \gg 1$. It is also regulated by the conditions of catching phase φ and attaining phase synchronization between the pendulum motion and

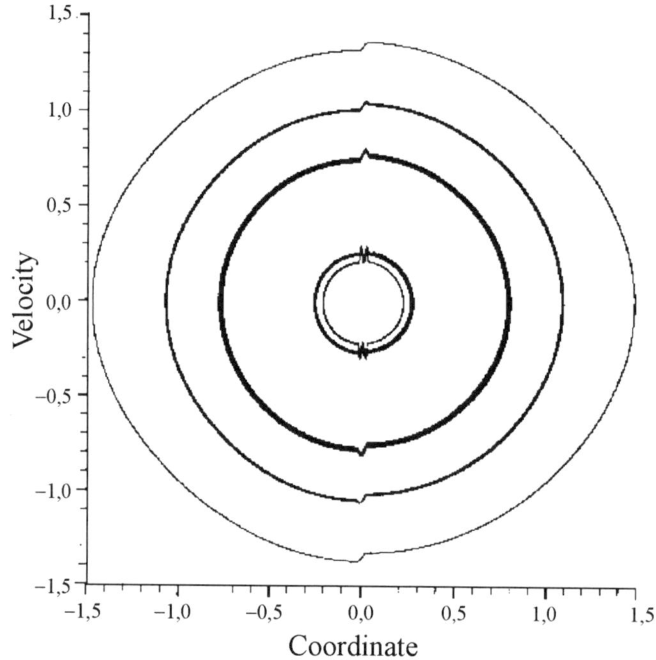

Fig. 10.3. Joint phase portraits of stationary steady motions of the pendulum (coordinate vs. velocity) for the same initial conditions and with the same values of the remaining parameters as in Fig.10.2

the external periodic force. The physical mechanism of phase adaptivity and its role in sustaining unchanging pendulum oscillations, while a number of parameters and conditions alter substantially, will be elucidated below and in the Sections to come.

For every stationary oscillation amplitude of the pendulum there exists a threshold value $F_0 = F_{th}$ for the amplitude of the force of external action. In this way, for example, for an amplitude $\sim$ 1.1 the threshold value is: $F_{th} = 1.0$. The pendulum oscillations stay close to harmonic with increasing the value so that: $F_0 > F_{th}$. At a value of the excitation amplitude of $F_0 \sim 2.8$, a bifurcation of tripling the period occurs (see Fig.10.6). While for $F_0 < 2.8$ the pendulum excitations have a single-valued unit amplitude (Fig.10.6a). At a minor increase of the value of F_0 amplitude-3 oscillations are established as a result of a bifurcation (see Fig.10.6b). These oscillations undergo some increase, without changing their nature, up to a value of $F = 3.2595$ (see Fig.10.6c). With a further minor increase of the value of the excitation amplitude F_0, a new bifurcation occurs and the oscillations in the system become strongly irregular (see Fig.10.6d).

It is well-known that Poincaré's cross section method is a major tool for

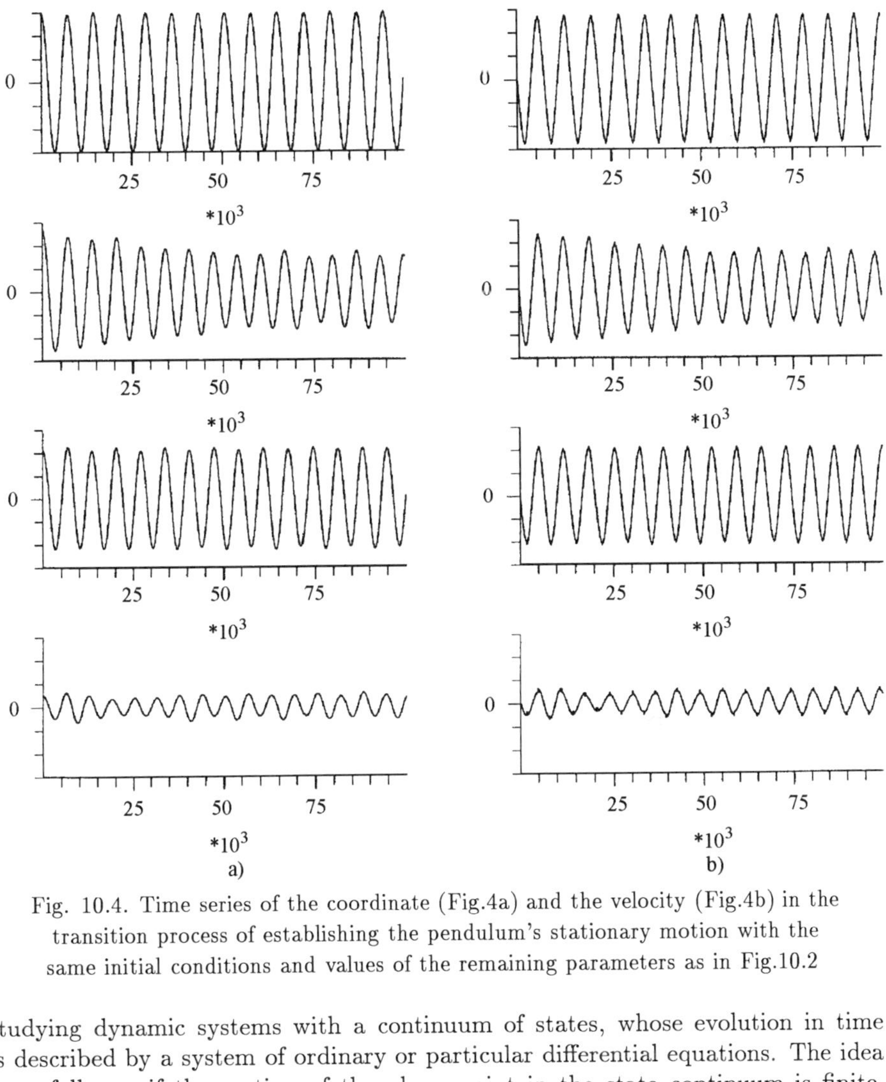

Fig. 10.4. Time series of the coordinate (Fig.4a) and the velocity (Fig.4b) in the transition process of establishing the pendulum's stationary motion with the same initial conditions and values of the remaining parameters as in Fig.10.2

studying dynamic systems with a continuum of states, whose evolution in time is described by a system of ordinary or particular differential equations. The idea is as follows: if the motion of the phase point in the state continuum is finite, then some (hyper) surface can be selected in the phase space, such as its dimension be smaller than the dimension of the space itself by unity. In addition, it should be selected in such a way, that the phase trajectory should cross it during the motion. The consecutive intersections of the trajectory with this plane, called Poincaré's cross section, in a certain direction (for example from left to right), form a sequence of points on it $(x_1, x_2, x_3, ..., x_n)$. In this way, an actual transition from a continuous to a discrete problem takes place. Indeed, the evolution of the system on

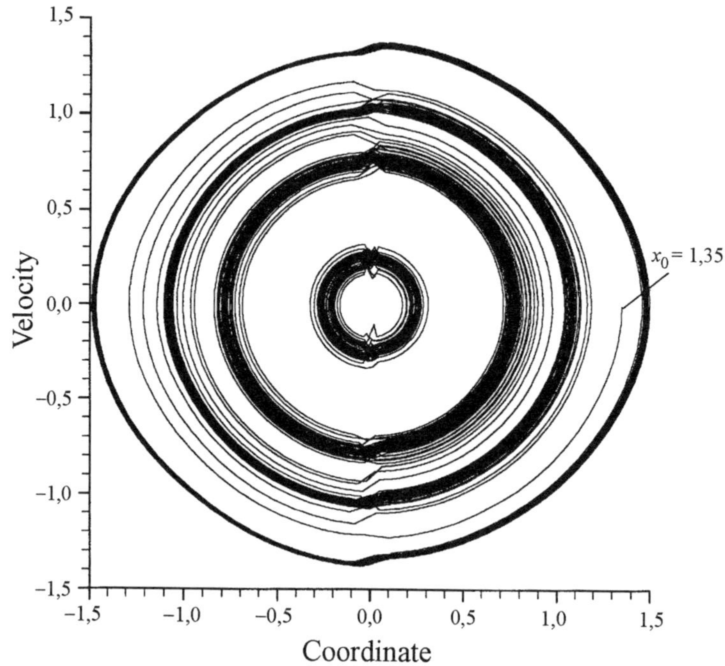

Fig. 10.5. Phase portraits (coordinate vs. velocity) of the pendulum's motion in the transition process of establishing stationary dynamical regime with the same values of the parameters as in Fig.10.2

the Poincaré cross section can be presented by several (initially unknown) discrete maps P: $x_{n+1} = p(x_n)$, where the components of the vector x are fewer than the phase space dimension by one.

It is interesting that the nature of the sequence $\{x_n\}$ can serve as a basis for drawing some important qualitative conclusions on the dynamic mode, in which the system under review operates. For example, let us assume that the phase space is three-dimensional (such is the case with the above-considered kick-excited pendulum). Therefore the Poincaré cross section is a two-dimensional plane. Several different motion modes can exist in a three-dimensional space: the trajectory may gravitate towards some particular point (focus or node), towards a limit cycle, it may be coiled on a two-dimensional torus, or it may tend towards some non-trivial attracting set – a strange attractor. The case of a steady singular point corresponds to damping oscillations. The issue that evokes greatest interest is limit cycle, which intersects with the Poincaré plane only at a certain number of points depending on the number of its periods – if it is a simple cycle, the point of intersection is just one, if it is a period – 2 cycle, they are two, if it is a period

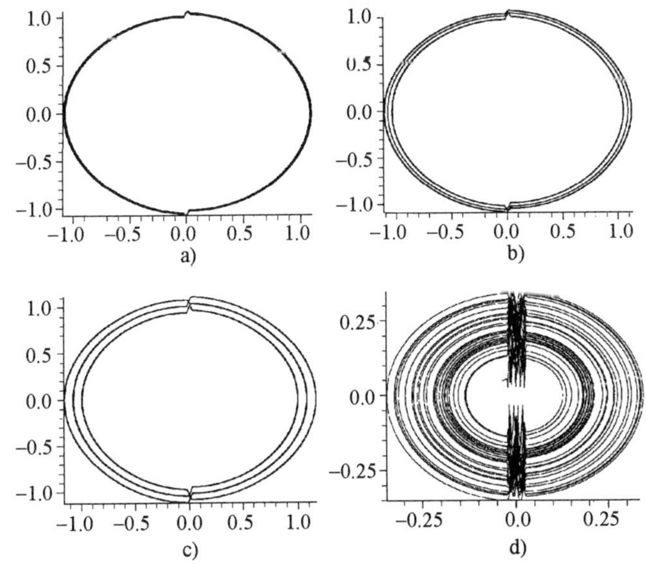

Fig. 10.6. Phase portraits of the motion in the pendulum system at $x_0 = 0.5$ and $F_0 = 2.775$ (Fig.10.6a), $F_0 = 2.8$ (Fig.10.6b), $F_0 = 3.2595$ (Fig.10.6c), $F_0 = 3.26$ (Fig.10.6d) and the same values for the remaining parameters as in Fig.10.2

– 3 one, they are three, and so on. The limit torus is a more complicated case: a set of points appear in the Poincaré cross section to form a closed curve with the topology of a circle, or several such closed curves, if the axis of the torus intersects with the Poincaré surface at several periodic points. If the motion takes place on a strange attractor, the set of intersection points will be of a fractal (Cantor) nature of the type, for example, of the famous attractor in the Hénon map, or they will simply be stochastically dispersed in some finite area of the cross section surface – such is the case with the non-integrable conservative, or very close to conservative, systems. Hence, a mere glance at the set of intersection points will be sufficient for the experimenter to identify qualitatively what type of a motion mode they are dealing with: whether it is a limit cycle and what its period is, whether it is a limit torus, whether it is a strange attractor of a Cantor type or if there is stochastic wandering of the trajectory in a finite area in the phase space.

Let us now proceed with the choice of a Poincaré cross section for the kick-excitable pendulum under examination here. In this case the extended phase space is $(x,\, \dot{x},\, t)$; the left-hand side of (10.2) however is periodically dependent on time t and in order to limit the motion of the phase point, we can consider motion in cylindrical space $(x, \dot{x}, \varphi = \nu t \pmod{2\pi})$. It is immediately obvious that the very geometry of the problem we are trying to solve, as described by the specified equation, offers a relevant way of choosing a Poincaré cross section. The existence of an active zone limited by the two planes $x = \pm d'$ indicates that the natural location of the Poincaré cross section is on one of these planes: either at the point of entry or at the point of exit of the trajectory from the active zone. It is possible, for example, to choose the first option: to accept that the Poincaré cross section is the plane through which the phase point enters the active zone in the course of its motion (See Fig.10.7):

$$x = \begin{cases} -d', & \dot{x} > 0 \\ +d', & \dot{x} < 0 \end{cases}. \tag{10.7}$$

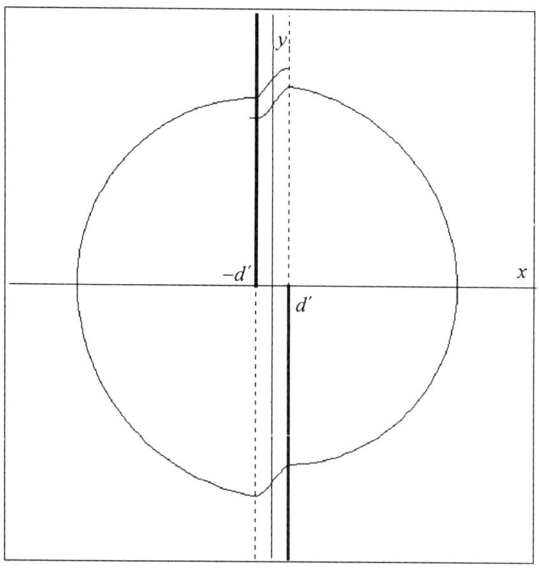

Fig. 10.7. Motion trajectory of the kick pendulum on the phase plane $(x, \dot{x})$. The dotted line indicates the boundaries of the active zone, and the bold line – the selected Poincaré section

The two variables of the Poincaré map are the velocity $v = \dot{x}$ and the phase of the external force $\varphi = \nu t \pmod{2\pi}$ at the moment of the n-th intersection. It can be seen that the first variable is linked with the total energy of the system, as

$$E_0 = T + U = \frac{V^2}{2} + (1 - \cos d') \simeq \frac{V^2}{2} + \frac{d'^2}{2}, \tag{10.8}$$

where E_0 is the total energy of the system, T and U are respectively the kinetic and potential energy of the system. In this sense the Poincaré map $(V, \varphi_{n+1}) = P(V, \varphi_n)$ proves to be chosen in „energy-angle“ variable.

Fig.10.7 shows that if one accepts as map iterations only those points where the trajectory crosses the plane of the cross section in a definite direction (for example from left to right), as this should be done in the general case, the consecutive iterations will be separated in time by an oscillation period, and during that time the phase point crosses the active zone twice: once in the upper half-plane and once in the lower one ($V \gtrless 0$). On the other hand, Eq.(10.2) indicates that the system under investigation can have certain symmetry. Its left-hand part is an odd function of x, respectively of $\dot{x}$; with regard to these variables, the right-hand side of the equation is an even function, but it depends on the phase variable φ as $\sin\varphi$. It follows from this that the equation of the kick excitable pendulum is invariant with respect to the change of the variables $(x, \dot{x}, \varphi) \to (-x, -\dot{x}, \varphi + \pi)$. Due to this symmetry of the system it is possible to use the Poincaré map T, where the sequential iterations are divided by a half-period, rather than by a whole period; this new map will include the points of intersection with the plane of the cross section in the cases of both positive and negative velocities. In this case the Poincaré map will be presented as double map of one period: $P = T^2$. For the purpose of obtaining the map T, the phase variable should be shifted by π and a change should be introduced in the sign of the velocity at the intersection points corresponding to a negative velocity, i.e. the latter should be taken by absolute value. So, the representation with reduced symmetry acquires the following form:

$$
\left|\begin{array}{l}
(V,\varphi)_{n+1} = T(V,\varphi)_n \\
V = |\dot{x}|; \quad \varphi = \begin{cases} \nu t \ (\mathrm{mod}\, 2\pi), & \text{at } \dot{x} > 0 \\ \nu t + \pi \ (\mathrm{mod}\, 2\pi), & \text{at } \dot{x} < 0 \end{cases}
\end{array}\right. \qquad (10.9)
$$

Further on the results of the numerical integration of the system (10.5) are presented in the form of a Poincaré cross section in coordinates defined by (10.9). In these coordinates, the stationary attractors are represented by fixed points. Thus, for example, in the case of $F_0 = 1.5$, the total number of the fixed points in the Poincaré map is five and they have the following coordinates:

Table 10.1

N	V^0	φ^0
1	0.2260	1.131
2	0.7506	1.077
3	1.0135	1.263
4	1.1924	1.187
5	1.3256	0.939

The next step is to determine the basins of these stationary attractors (points), or, to put it differently, the set of initial conditions that evolve toward each of the listed fixed points. It is appropriate that the initial conditions should also be set on a sufficiently fine rectangular grid in the Poincaré cross section in the variables (V, φ), at that the results should be presented in a color image, where each pixel corresponds to an initial condition at a grid point and its color differs depending on which of the fixed points the consecutive iterations from the Poincaré representation are drawing closer to in the course of time. Such a map for $F_0 = 1.5$ is shown in Fig.10.8. Its resolution power is 640×480 points, as velocity V, in the range of 0 to 2, is plotted along the abscissa (according to the relations (10.8) this means that the motion of the pendulum is finite, i.e. no full turns round the point of suspension are considered), and the phase variable φ in the range of $-\pi$ to π is plotted along the ordinate. One can clearly see the five fixed (stationary) points marked by crosses and their basins, each in a different color. The typical structure of the basins definitely stands out: they have a broad oval part surrounding the fixed point, with a long and narrow „tail“ jutting out. Away from the fixed points, the adjacent pixels have complexly interwoven colors, i.e. quite close initial conditions can be attracted by quite different attractors. This complex, intertwining pattern of the basins shows that their boundaries are fractal in nature. Moreover, the considerable fragmentation of the basins in areas situated away from the fixed points leads to extremely strong dependence of the ultimate state of the system on the initial conditions. Additionally, even the slightest changes in the initial conditions may result in an ultimate state that switches from one attractor to another.

Besides the basins of the five fixed points, there is yet another distributed area, marked in dark blue, that can be seen in Fig.10.8. The initial conditions of this area substantiate attraction of the motion towards a different stationary mode, which, in the phase space situation, is a stable limit cycle situated entirely between the two planes of the Poincaré cross section $x = \pm d'$. Thus the trajectories attracted to this limit cycle cannot be observed on the discrete map chosen by us, as they do not cross the plane of the Poincaré cross section at all. Under this mode, the equation of the kick-excited pendulum (10.2) simply degenerates to the following equation: $\ddot{x} + 2\delta_d \dot{x} + \sin x = F_0 \sin \nu_t$. This is the case of a pendulum experiencing the action of a high frequency external sine force. It is noteworthy that this stationary solution can be observed in the case of a different Poincaré cross section, for example, when a plane $x = 0$ cuts across the ordinate in Fig.10.7. Under this mode the oscillations are rapid, with small amplitude and frequency equal to the frequency of the external force. The important fact in the case under consideration is the existence of such a solution, which attracts a substantial part of the initial conditions to form the basin marked in dark blue.

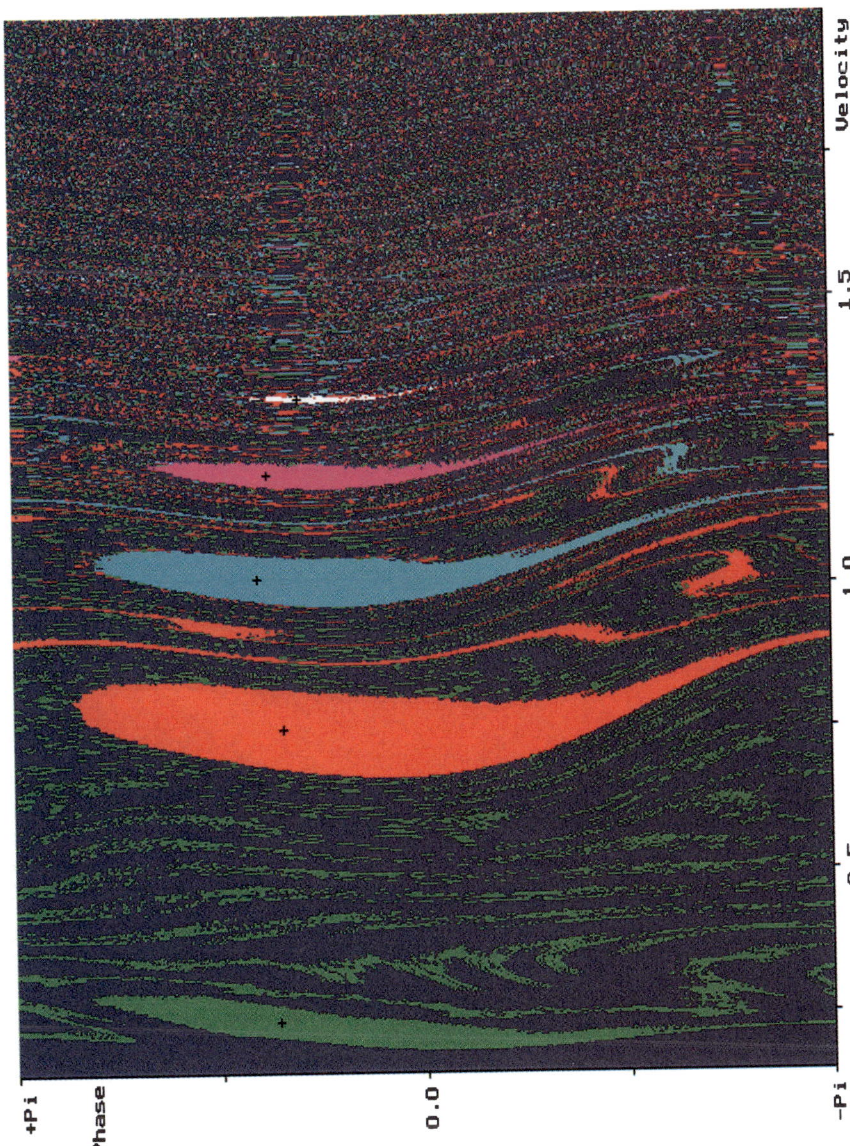

Fig.10.8. Basins of the initial conditions in the Poincare section corresponding to a discrete set of stationary periodic solutions of the system of equations (10.5) at $F_0 = 1.5$. The five stationary (fixed) points (cf. Table 10.1) are marked by crosses

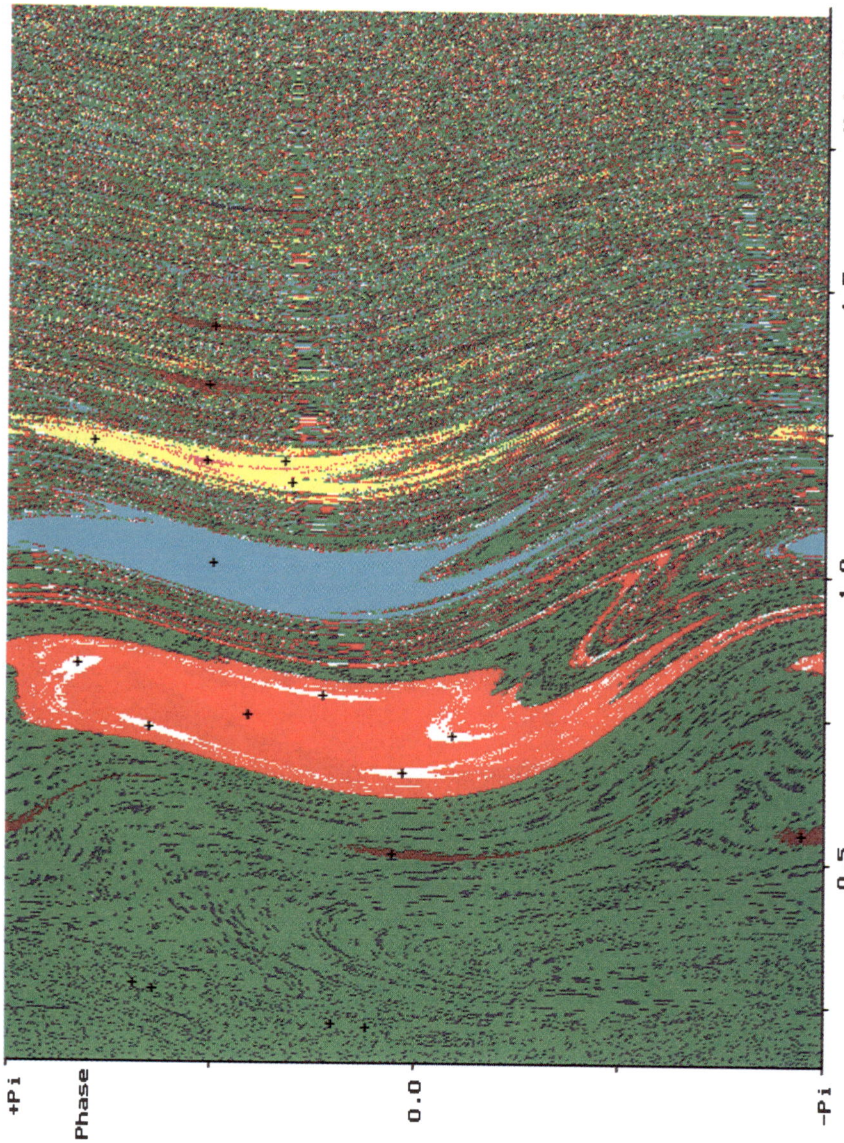

Fig. 10.9. Evolution of the basins of initial conditions from Fig.10.8 with increase of the amplitude of the external driving force to $F_0 = 2.5$

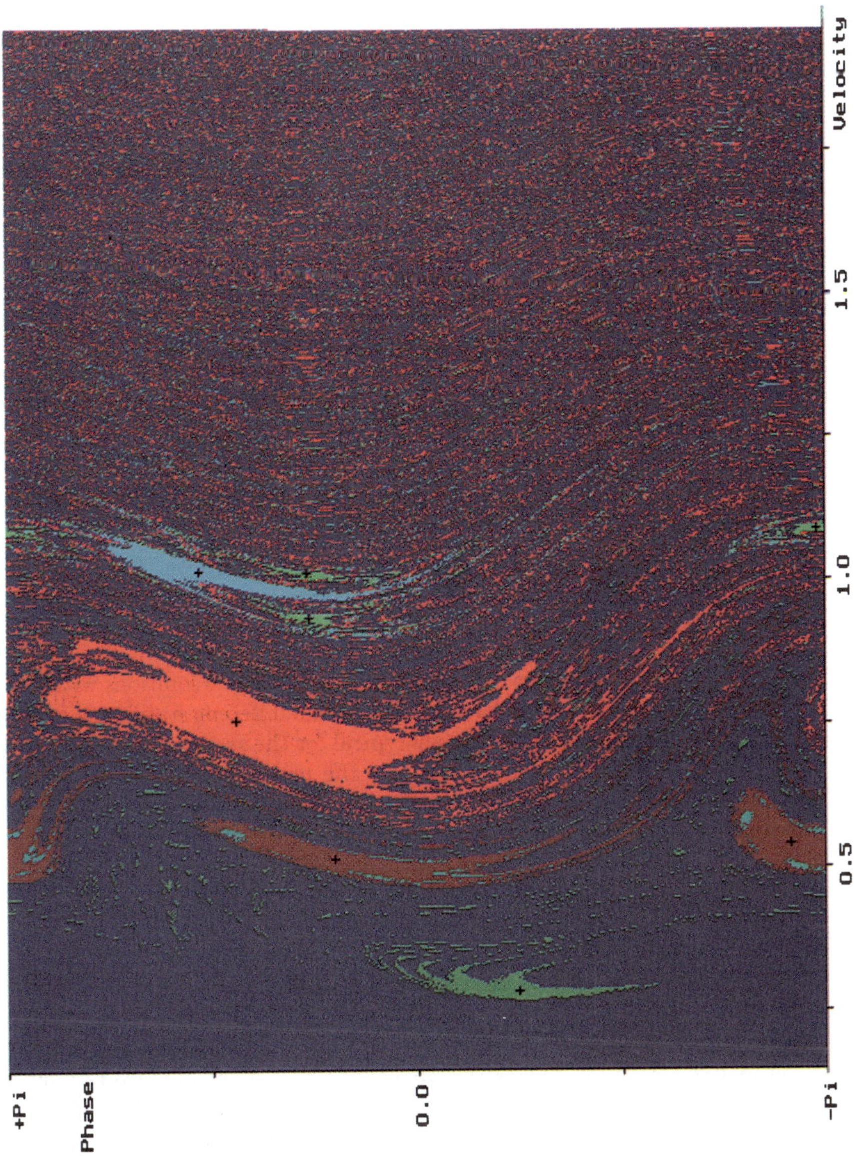

Fig. 10.10. Evolution of the basins of initial conditions from Fig.10.8 with increase of the amplitude of the external driving force to $F_0 = 3.4$

Our next task is to study the behavior of the stationary mode, while changing the parameter F_0 – the amplitude of the external acting force. Figs.10.9 and 10.10 show color images of the basins of the different attractors in the case of $F_0 = 2.5$ and $F_0 = 3.4$ respectively. Firstly, Fig.10.9 displays the basins of the first stationary point and of the orbit, all of which lies in the active area, as completely interwoven. The first fixed point has undergone two successive bifurcations of doubling, and it is actually a periodic point with period 4. Further on, a solution with period 5 has emerged around the second fixed point, and a similar one, this time with period 3, has appeared with respect to the fourth one. The latter solution has almost fully „appropriated“ the entire original basin of the basic fixed point, leaving only a small area in the center and three narrow strips coming out of it for that point. Thirdly, new fixed points have emerged after the fifth one (also marked by crosses), which was last in the case of $F_0 = 1.5$. Their basins however are relatively small in size. And finally, yet another stationary solution with period 2 has appeared between the first and the second fixed points. Fig.10.10 shows a following stage in the evolution of the system, given a change in the control parameter F_0. In the case of $F_0 = 3.4$, the stationary solutions for energies of greater magnitude have become unstable and have disappeared together with the basins at this orbit. Around the third fixed point there is a solution with period 3. The basins by this orbit have become by far narrower in comparison with the previous ones with lower values for F_0. And a new stable orbit with period 2×4 has emerged around the independent solution with period 2.

It can be seen from all three color maps that the periodic solutions with periods $M = 3, 4, 5, ...$, appearing in the surroundings of the different stationary points for certain quantities of the force F_0, are very typical for the system. Periodic solutions are also observed in their vicinities, but now the period is $M \times N$, where N can also assume values $N = 3, 4, 5, ...$ etc. For example, one can observe solutions with periods 4×3, 8×3 and others with relatively narrow basins. The reason for the existence of such periodic points around the stationary orbits has to do mostly with the small damping δ_d. As the shrinkage of the phase volume per unit of time $\Lambda = \frac{\dot{V}}{V} = -\delta_d = \text{const}$ is small, one can expect that the Jacobian of the Poincaré representation will be close to unity ($J = \exp(-\delta_d T) = 1-)(\beta)$, where T is the semi-period of oscillation close to π). This indicates that the Poincaré representation can be viewed as approximately conservative, but perturbed with small dissipative correction. The structure of the phase plane in maps retaining the phase area is well known (See for example [23]). It contains areas of regular motion occupied by invariant KAM orbits as well as regular islands around the elliptical fixed points of the map. The regions of chaotic motion occupy the areas around the hyperbolic fixed points. The elliptic points, in their turn, are surrounded by Birkhoff M-chains made up of alternating elliptic and hyperbolic points with period M; the M-periodic elliptic points, in their turn, are surrounded by even smaller Birkhoff $M \times N$-chains, etc. This complex structure of the phase space however is not stable with respect to arbitrarily small dissipative perturbations. Even in the cases of only slightly dissipative maps, the areas of chaotic motion around the hyperbolic points are destroyed while the elliptic points become stable focuses. Therefore the Birkhoff M-chains are transformed into M-periodic points surrounding the basic fixed point. The periodic solution in the kick-pendulum system corresponds to that very type of Birkhoff chains. For the purpose of illustrating the relation between the conservative and slightly dissipative dynamics, Figs 10.11 and 10.12 present the structure of the motions in the Poincaré plane for the kick-pendulum system under consideration in the conservative case of $\delta_d = 0$ and values of the external force $F_0 = 1.5$ and $F_0 = 2.5$ respectively. A comparison with Fig.10.8 and Fig.10.9 promptly reveals the direct relation between the different periodic modes and the Birkhoff islands.

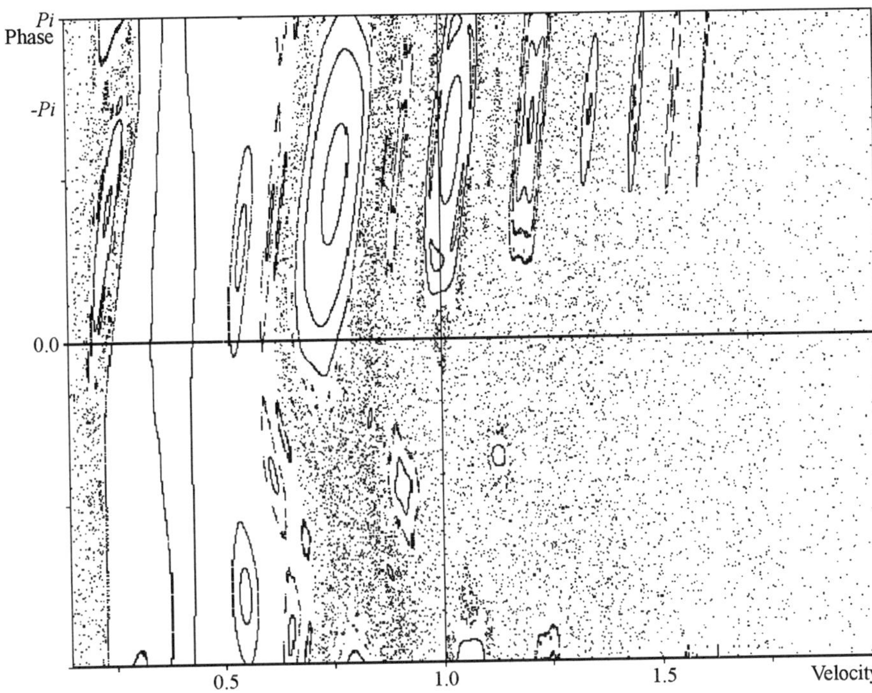

Fig.10.11. Structure of the Poincaré map of the motion of conservative kicked pendulum ($\delta_d = 0$) at $F = 1.5$

The next step in the numerical analysis is to build a bifurcation diagram of the system, i.e. to trace the evolution of the different attractors of the system, for example, when the external force F_0 varies gradually within a broad range. Such a diagram, built for the energy variable V (the velocity of the motion) is

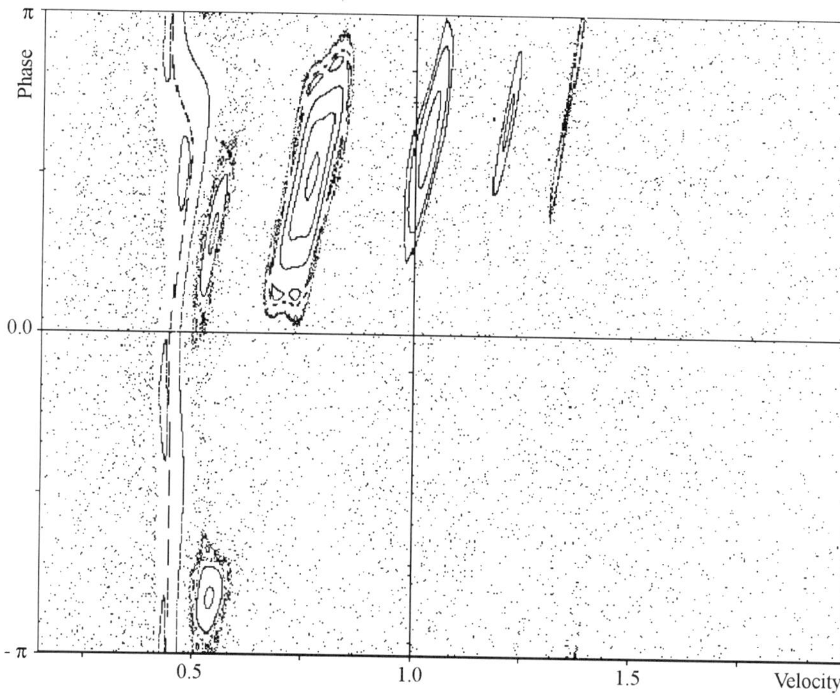

Fig.10.12. Structure of the Poincaré map of the motion of conservative kicked pendulum ($\delta_d = 0$) at $F = 2.5$

shown in Fig.10.13. It reveals a whole family of fixed points (they are at ten). The light gray lines indicate the periodic solutions with period 3 around the first several stationary points as well as the solution with period 2 between the first and the second orbit. Fig.10.14 provides a better view of the typical branches of the bifurcation diagram. The fixed and periodic points are brought about by tangential or saddle-focal bifurcation, and as parameter F_0 goes up they all, regardless of their period, undergo cascades of bifurcations of doubling of the period. We can see that the solutions with period 3 are engendered in close proximity to the basic orbit around which they are situated. Yet, it should be pointed out that the basic orbit remains stable; the process of three-furcation, during which the periodic orbit is born, does not affect its stability in any way: the predominant part of the basin seems to be simply „ceded“ to the stable solution that have come about in its surroundings (See Figs.10.8, 10.9, and 10.10). In actual fact, as F_0 increases, the periodic points quickly move away from the basic orbit, their basins become narrower (See Fig.10.10), and they get into a cascade of doubling bifurcations of the same universal type as those that affect the basic orbits.

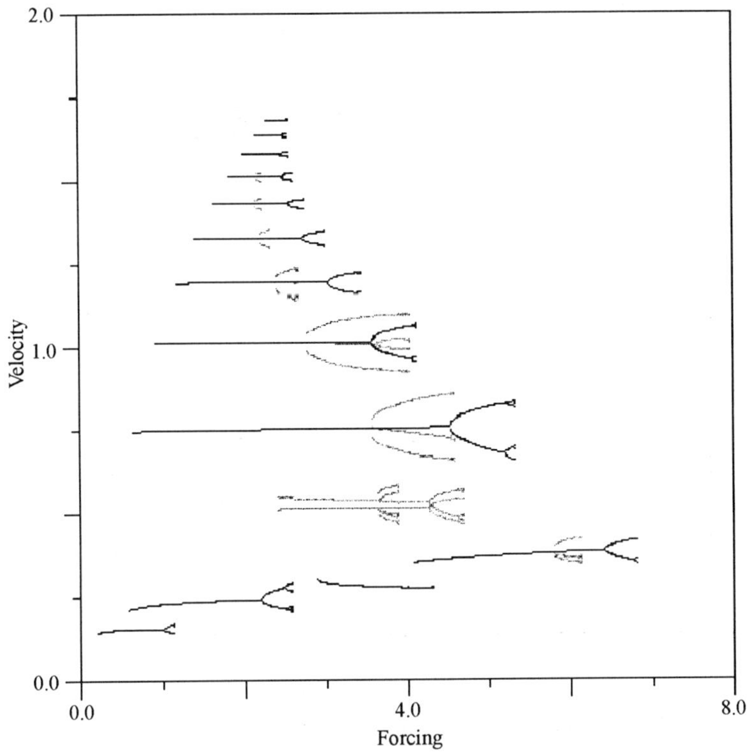

Fig. 10.13. Bifurcation diagram for the whole family of fixed points (black) and period-3 points surrounding each of them (grey)

When considering the consecutive doublings, one should remember that the object of examination is the Poincaré map for a half-period and not for a whole period. In this sense the first doubling of the map corresponds to bifurcation of a loss of symmetry of the orbit in the three-dimensional phase space, and only the second doubling of the map becomes first doubling of the orbit. To put it differently: the initial three-dimensional system (10.2) has additional symmetry and before reaching the stage of a cascade of doubling bifurcations and orbits with even periods, a bifurcation of loss of symmetry has to take place. This additional symmetry implies that the map for one period P can be presented as a second iteration of some other map: $P = T^2$. And if at the time of the first doubling of T the value of the critical multiplier is -1, then this multiplier will be equal to +1 for the doubling map P, and this value corresponds to bifurcation of loss of symmetry rather than to a period doubling. Or, to phrase it in yet another way: when the Poincaré map is designed for a half-period instead of a period, the

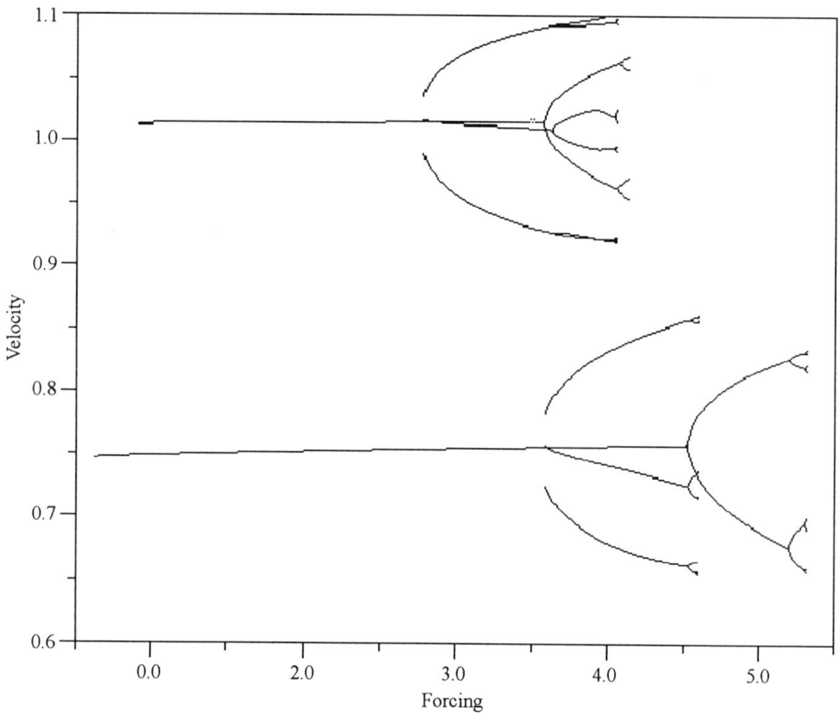

Fig. 10.14. Bifurcation diagram for the second and third fixed point and the corresponding period-3 solution

symmetry of the system is removed, for the map for a half-period T is no more equipped with the additional symmetry of the initial system. Fig.10.13 clearly reveals not only the above-mentioned characteristic peculiarity of the process of rendering discrete the possible stationary amplitudes (a peculiar „quantization" by intensity of motion in the system), but the second very important regularity of the system under consideration as well, i.e. the independence of the established stationary amplitude of pendulum oscillation from the change in the amplitude of the external high-frequency acting force in a broad range: the percentage amounting to scores and even hundreds. It can be seen that the energy of the oscillations remains almost constant throughout the interval, as regards parameter F_0, between the moment of birth and that of doubling of the period of the respective fixed point with changes smaller than 0.1%. The system exhibits an unusually pronounced self-adaptivity and extreme stability of the stationary oscillations excited once in it; this is valid to a larger or lesser degree for each of the possible stationary amplitudes.

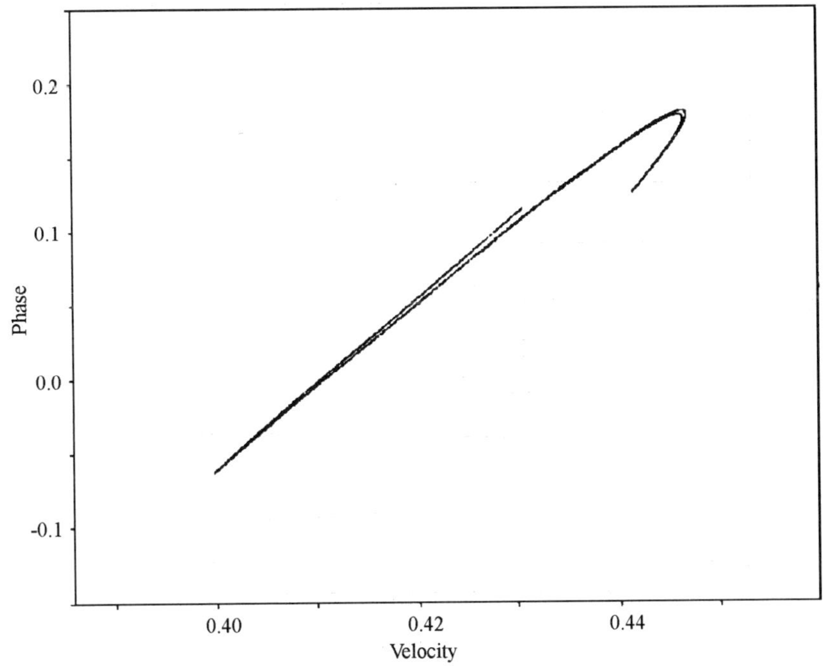

Fig.10.15. A strange attractor of the kicked pendulum for parameters $\delta_d = 0.2$ and $F_0 = 11.9$

Fig.10.13 strikes one with the fact that the cascade of bifurcations of period doubling does not end with a noticeable area of chaotic motion, i.e. it does not end with a strange attractor as is the case with two-dimensional maps and with most of the dissipative systems with three-dimensional phase space. Instead, the attractor seems to suddenly disappear somewhere around the parameter $F_0 = F_\infty$, corresponding to the point of compression of the sequence of bifurcations. As the quantity of the controlling parameter F_0 exceeds this value, the solutions corresponding to the attractor that has disappeared are attracted by some of the other attractors in the system (which, as we have seen, constitute a set) and the stationary mode of the system changes in a leap-like manner. Since numerical simulations are inevitably accompanied by calculation „noise", which is intrinsic to each computation procedure of the type of the Runge Kutta one, the numerical analysis of the bifurcational characteristic close to the transition to chaos proves to be possible only with limited accuracy. As a consequence of this noise, the successive iterations of the Poincaré map do not gravitate towards a particular geometric point but fill in a minute area around the stationary point. The numerical experiment shows that in most cases this area is in the form of an ellipsoid strongly drawn in the direction of the own vector of Lyapunov's greater indicator. The situation

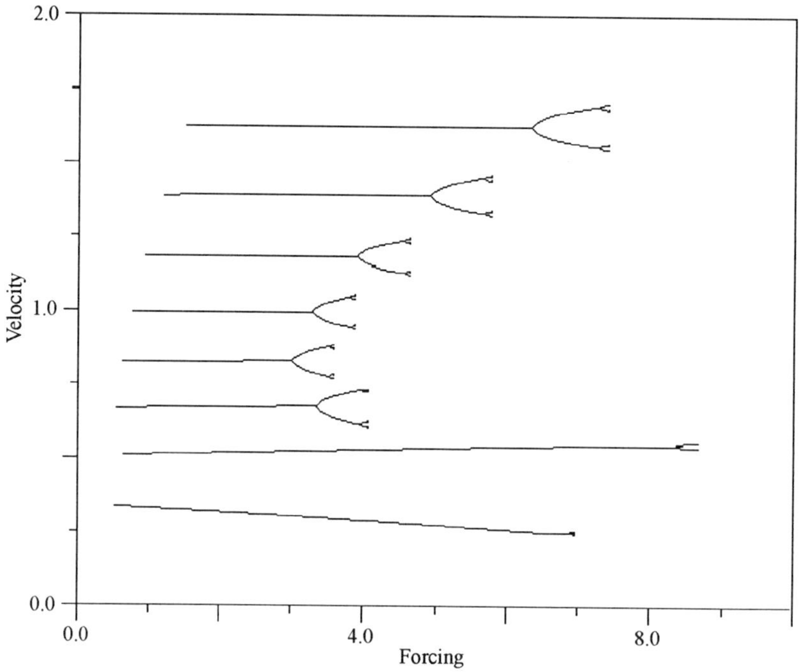

Fig.10.16. Bifurcations diagram of a kick-excited Duffing oscillator. The attenuation, frequency and the active zone width are taken to be as in the kick-pendulum numeric experiment: $\delta_d = 0.01$, $v = 51$ and $d' = 0.025$

is further aggravated by the fact that for parameter values approaching F_∞ from below, the higher Lyapunov's indicator, though remaining negative, does not stay very close to zero. This means that a random fluctuation of the trajectory, which is a consequence of the computation noise, would damp very slowly and meanwhile there will be superposition of new noise fluctuations to it. Practically only the initial three or four doublings can be detected - then the trajectory becomes too „blurred“ as a result of the noise and the higher Lyapunov's indicator, which is close to zero. The following doublings that occur at an ever smaller scale cannot be distinguished on the Poincaré map. It can only be seen that at one moment, removed by a parameter of about 10^{-4}, for example after the fourth bifurcation of doubling, the attractor suddenly ceases to exist. A possible explanation of this effect is that as the transition to chaos comes closer, the basin of the doubling attractor quickly shrinks and reduces its size; at a certain point of time it can become smaller than the area where the trajectory wanders as a consequence of the noise, and the trajectory may abandon it and reach some of the other attractors.

It is worth noting that if a relatively large damping is selected in the system under consideration (10.5), typical strange attractors will also be observed. They will be obtained after the cascade of reverse doubling bifurcations; a single segment one is shown in Fig.10.15.

As our aim is to reiterate the commonness of the condition that the external exciting force should be nonlinear – a condition pooling a broad circle of systems into a single class and conditioning the existence of the two most typical regularities: discreteness of the oscillations by amplitude and pronounced adaptive stability, let us consider a dynamic system with another type of a potential well. To this end Fig.10.16 presents the bifurcational diagram of the Duffing oscillator with a single potential well: $u(x) = \frac{x^2}{2} + \frac{x^4}{4}$. In this case, unlike the pendulum one, the oscillation period T drops as the energy E goes up. Besides, the dependence $T(E)$ has no peculiarities of the type of shooting of the period toward infinity at some extreme quantity of the energy due to proximity to the separatrix: merely, in this case the potential has but one extreme at $x = 0$, which is a minimum, and this precludes the existence of hyperbolic peculiar points. Despite all these particular differences between the two systems, Fig.10.16 shows clearly that the kick-excitable Duffing oscillator also indicates distinctly the characteristic features of the established class of systems: „quantization" of the stationary energies and independence of the stationary amplitudes from the change in the amplitude of the external exciting force within a very wide range.

10.3. Analytical proof of the existence of „quantized" kick-pendulum oscillations

10.3.1. An approach used in the case of small amplitudes of pendulum oscillation

Let us write down the equation of pendulum motion in the following form [156]:

$$\ddot{x} + \omega_0^2 x = f(x, \dot{x}) + \varepsilon(x) F_0 \sin \nu t, \tag{10.10}$$

where x is the angle of pendulum deviation from the vertical, ω_0 is the frequency of the small free oscillations of the pendulum, $f(x, \dot{x}) = \omega_0^2(x - \sin x) - \delta_d \dot{x}$ is a small function accounting for the non-linearity of the free oscillations and the friction, F_0 and ν are respectively the amplitude and the frequency of the external force ($\nu \gg \omega_0$), the function $\varepsilon(x)$ is hereafter assumed to be an even function represented by expression (10.4).

The solution of the nonlinear equation (10.10) in an approximation to the small slowly changing amplitudes is presented in the form:

$$x = a \sin \theta, \quad \theta = \omega t + \alpha,$$
$$a = a(t), \quad \alpha = \alpha(t)$$

as we set

$$\dot{x} = \omega a \cos\theta \text{ and } \dot{a}\sin\theta + a\dot{\alpha}\cos\theta = 0.$$

As $\ddot{x}$ is formed and a substitution is effected in (10.10), the following system of equations is obtained:

$$\left|\begin{array}{l} \dot{a} = -\dfrac{\omega_0^2 - \omega^2}{\omega} a \sin\theta\cos\theta + \dfrac{1}{\omega} f(a\sin\theta, \omega a\cos\theta)\cos\theta + \dfrac{F_0}{\omega}\varepsilon(a\sin\theta)\cos\theta\sin\nu t \\ \dot{\alpha} = \dfrac{\omega_0^2 - \omega^2}{\omega}\sin^2\theta - \dfrac{1}{\omega a} f(a\sin\theta, \omega a\cos\theta)\sin\theta - \dfrac{F_0}{\omega a}\varepsilon(a\sin\theta)\sin\theta\sin\nu t. \end{array}\right. \tag{10.11}$$

Taking into account that the function $\varepsilon(a\sin\theta)$ is periodic along θ with a period π, we can expand it in the following way:

$$\varepsilon(a\sin\theta) = A_0(a) + 2\sum_{m=1}^{\infty} A_{2m}(a)\cos 2m\theta,$$

$$A_{2m}(a) = \frac{2}{\pi}\int_0^{\pi/2} \varepsilon(a\cos\theta)\cos 2m\theta s\theta. \tag{10.12}$$

As indicated by the product

$$\varepsilon(a\sin\theta)\sin\nu t = A_0\sin\nu t + \sum_{m=1}^{\infty} A_{2m}[\sin(\nu t - 2m\theta) + \sin(\nu t + 2m\theta)]$$

given that

$$\nu = n\omega, \quad n = 2m+1, \quad \text{i.e. } |\nu - n\omega| \ll \omega, \tag{10.13}$$

there is a force in the right-hand part of the motion equation (10.10), which is synchronous (at $\omega \approx \omega_0$) to the pendulum's own oscillations.

As the averaging method [97] is applied to the system of equations (10.11), the following shortened equations are obtained:

$$\left|\begin{array}{l} \dot{a} = -\dfrac{\delta_d}{2} a + \dfrac{F_0}{2\omega}(A_{n-1} + A_{n+1})\sin\varphi \\ \dot{\varphi} = \nu - n\omega - n\left[\dfrac{\omega_0^2 - \omega_a^2}{2\omega} - \dfrac{F_0}{2\omega a}(A_{n-1} + A_{n+1})\cos\varphi\right] \end{array}\right., \tag{10.14}$$

where

$$\left|\begin{array}{l} \varphi = \nu t - n\theta = (\nu - n\omega)t - n\alpha \\ \dot{\varphi} = \nu - n(\omega + \dot{\alpha}) \end{array}\right., \tag{10.15}$$

$$\omega_a^2 = \frac{2J_1(a)}{a}\omega_0^2 = \left(1 - \frac{a^2}{8} + \ldots\right)\omega_0^2, \tag{10.16}$$

at that ω_a is the frequency of the free pendulum oscillations with a finite amplitude a [97].

The system of equations (10.14) is suitable for analyzing both stationary oscillations and the process of establishing a stationary mode.

In a stationary oscillation mode the conditions $\dot{a} = 0$ and $\dot{\varphi} = 0$ should be fulfilled, and the frequency of oscillation

$$\omega + \dot{\alpha} = \frac{\nu}{n}$$

should be an odd subharmonic of the frequency of the external exciting force. Under these conditions, as well as under the obvious condition $\dot{\alpha} = 0$, the system of equations (10.14) yields the following expressions:

$$\left| \begin{array}{ll} F_0 \sin\varphi = \omega_0^2 a \dfrac{\xi_0}{A_{n-1} + A_{n+1}}, & \xi = \dfrac{\delta_d \omega}{\omega_0^2} \\ F_0 \cos\varphi = \omega_0^2 a \dfrac{\xi}{A_{n-1} - A_{n+1}}, & \xi = \dfrac{\omega^2 - \omega_a^2}{\omega_0^2} \end{array} \right. , \tag{10.17}$$

determining the amplitude a and the phase $\varphi = -n\alpha$ of the stationary oscillations.

In the case of $n \gg 1$, the small changes in a lead to large changes in the magnitude of φ determining the phase of the high-frequency force at the moment, when the pendulum passes through the equilibrium position ($\varphi = \nu t$ at $\theta = 0$). In accordance with the first expression in (10.17), $\varphi \to 0$ at $\delta_d \to 0$ or $F_0 \to \infty$.

(10.16) yields the formula of the relation between the amplitude of the exciting force F_0 and the amplitude of the oscillations a:

$$F_0 = \omega_0^2 a \sqrt{\left(\frac{\xi_0}{A_{n-1} + A_{n+1}}\right)^2 + \left(\frac{\xi_0}{A_{n-1} - A_{n+1}}\right)^2}. \tag{10.18}$$

The function (10.4) is substituted in the integral (10.12) to determine:

$$A_n = \frac{2}{\pi}\frac{\sin n\theta}{n}, \qquad \theta = \arcsin\frac{d'}{a} \simeq \frac{d'}{a} \text{ at } \frac{d'}{a} \ll 1.$$

The analysis shows that, first, for $\theta_0 \ll 1$ and $n\theta \geq 1$, the coefficients A_{n-1} and A_{n+1} differ slightly and that is why in expressions (10.17) and (10.18) their difference is substantially smaller than their sum, and second, the value ξ depends on the amplitude a, and in accordance with formula (10.15) we have

$$\xi \approx \frac{a^2 - a_0^2}{8} \simeq \frac{a_0}{4}(a - a_0), \quad a_0 \approx \sqrt{8\left(1 - \frac{\omega^2}{\omega_0^2}\right)^2} \simeq 4\sqrt{1 - \frac{\omega}{\omega_0}}.$$

That is why for high values of F_0 the dependence of F_0 on a is almost linear:

$$F_0 \approx \omega_0^2 a_0 \frac{a_0(a-a_a)}{4|A_{n-1}(a_0)-A_{n+1}(a_0)|}. \tag{10.19}$$

Actually, in practice, we are interested in the dependence of the amplitude a of the excited pendulum oscillations on the value of the amplitude F_0 of the external exciting force. This dependence is quite small: it would not be hard to calculate from either (10.18) or (10.19) that if the magnitude of F_0 is increased, say, fifteen times, the amplitude a would go up by less than 0.3%. Thus the theoretical consideration in an approximation to small amplitudes also proves the existence of high adaptive stability of the pendulum system under consideration.

For the purpose of stability analysis of the stationary oscillations defined by relations (10.17) and (10.18), the system of equations (10.14) is written in the form: $\dot{a} = f(a,\varphi)$, $\dot{\varphi} = g(a,\varphi)$. f_a, f_φ, g_a and g_φ are used to denote the derivatives $\frac{df}{da}$, $\frac{df}{d\varphi}$, $\frac{dg}{da}$ and $\frac{dg}{d\varphi}$ respectively, taken at constant values for a and φ, corresponding to the stationary solution. The stability condition can be written in the form:

$$\mathrm{Re} P_{1,2} < 0,$$

where

$$P_{1,2} = \frac{f_a + g_\varphi}{2} \pm \sqrt{\left(\frac{f_a - g_\varphi}{2}\right)^2 + f_\varphi g_a}. \tag{10.20}$$

(10.20) takes into account that the small deviations of a and φ from the stationary values can be expressed by the formulae $\delta a + A_1 e^{p_1 t} + A_2 e^{p_2 t}$, $\delta\varphi = B_1 e^{p_1 t} + B_2 e^{p_2 t}$, at that A_1, A_2, B_1 and B_2 are constant.

In practice the system works at high values for F_0 and low ones for φ. Under these conditions one can set

$$f_a = -\frac{\delta_d}{2}, \quad f_\varphi = \frac{F_0}{2\omega}(A_{n-1} + A_{n+1}), \quad g_a = -\frac{n\omega a}{8}, \quad g_\varphi = 0,$$

and the condition for stability of the dynamic pendulum system is

$$f_\varphi > 0, \quad A_{n-1} + A_{n+1} > 0. \tag{10.21}$$

Condition (10.21) reveals the physical essence of the mechanism of the system adaptive stability. In the case of $A_{n-1} + A_{n+1} > 0$ and $\varphi > 0$, the pendulum reduces its amplitude and in accordance with formula (10.16) its oscillation period increases. That is why the pendulum returns to the active zone (the action zone) $|x| \leq d'$ with a delay, i.e. with a smaller phase φ. At $A_{n-1} + A_{n+1} < 0$ and $\varphi > 0$, the pendulum increases its amplitude and enters the zone $|x| \leq d'$ with a growing phase φ. Having used an analytical approach, at this point we arrive at the conclusions made in section 10.2 during the numerical experiment, i.e. that the high self-adaptive stability of the system is associated with the self-tuning of the phase of the external exciting force at the moment, when the pendulum crosses the boundary of the active action zone.

Let us consider the dependence of the value of the stationary amplitudes on the relative frequency detuning in the system. Given small deviations from the equilibrium position, the pendulum equation is presented in the form [138]:

$$\ddot{x} + 2\delta_d \dot{x} + \omega_0^2(1 - \gamma x^2)x = \varepsilon(x) F_0 \cos \nu t,$$

where the function $\varepsilon(x)$ is again adopted in the form (10.4), γ is a constant.

As the solution of the above equation is again adopted in the form

$$x = a(t)\cos[\omega t + \varphi(t)]$$

and its right-hand side is presented as [139]

$$\varepsilon(x) F \cos \nu t = v\left(t - t_n + \frac{d'}{a\omega}\right) v\left(t_n + \frac{d'}{a\omega} - t\right) F_0 \cos \nu t,$$

where $\nu(.)$ is a Heavyside function, and t_n designates the moments of pendulum passage through the equilibrium position, in the first approximation, while applying the averaging method [97], the following system of equations is obtained with regard to a and φ:

$$\dot{a} = -\delta_d a - \frac{2F_0}{\pi\nu}(-1)^n \sin\frac{\nu\pi}{2\omega} \sin\frac{\nu d'}{\omega a} \sin\frac{\nu[(n+1)\pi - \varphi]}{\omega},$$

$$\dot{\varphi} = -\Delta(a) - \frac{F_0 d'}{\pi\nu a^2}(-1)^n \sin\frac{\nu\pi}{2\omega}\left(\frac{\omega a}{\nu d'} \sin\frac{\nu d'}{\omega a} - \cos\frac{\nu d'}{\omega a}\right) \cos\frac{\nu[(n+1)\pi - \varphi]}{\omega}.$$

Here

$$\Delta(a) = \omega - \omega_0(a) \ll \omega, \quad \omega_0(a) \approx \omega_0\left[1 - \frac{3\gamma a^2}{8}\right].$$

The condition for existence of a stationary solution is independence of the „slow“ time t_n, i.e. of the number n. To this end it is sufficient to set $\omega = \dfrac{\nu}{2m+1}$, where m is an integer.

The dependence of the stable (continuous lines) and unstable (dotted lines) values of the stationary amplitudes on the relative detuning $\dfrac{\nu - M\omega_0}{\delta_d}$, where $M = 2m_0 + 1$ is an odd number, is shown in Fig.10.17. The dependencies have been calculated for the following values:

$$B_m = \frac{2F_0}{\pi d' \delta_d \omega_0 M^2} = 1, \quad \frac{3\omega_0 d'^2 M^3}{8\delta_d}\gamma = 20, \quad \frac{\omega_0}{\delta_d} = 10.$$

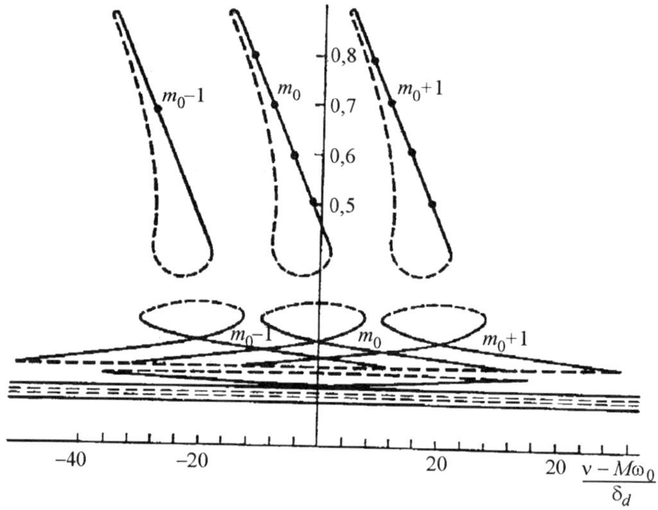

Fig. 10.17. Dependence of the stable (continuous lines) and unstable (dotted lines) values of the stationary amplitudes from the relative frequency detuning

The presented dependencies reveal the almost periodic nature of a number of leaves corresponding to different values of the possible oscillation amplitudes.

The „period" is $\dfrac{2\omega_0}{\delta_d}$ but the sequence is not strictly periodic due to the change in the number m. For each set of values for the parameters of the external force F_0 and ν, there is a definite series of stable oscillation amplitudes in the system.

The initial conditions determine whether the oscillating system will fall into one regime or another. The magnitudes of the possible oscillation amplitudes, starting with a definite number s, are determined approximately as [139]

$$a_{m,s} \simeq \frac{2d'(2m+1)}{\pi(2s+1)},$$

where s is an integer.

The time of interaction of the pendulum with the external force in the event of stationary oscillations is determined as:

$$\tau_s = \frac{2d'}{a_{m,s}\omega} = \frac{2d'(2m+1)}{a_{m,s}\nu} = \frac{d'(2m+1)}{\pi a_{m,s}} T_\nu \simeq (s+\frac{1}{2})T_\nu,$$

where $T_\nu = \dfrac{2\pi}{\nu}$ is the period of external action.

We can add the following touch to the physical explanation of the described effect. The average value of the external exciting force for the interaction time is different from zero and depends on the phase φ, which, in its turn, is determined by the value B_m. The system selects the value of the phase adaptively in such a way that the external force should have an accelerating impact on the pendulum. When the pendulum moves in the opposite direction and once again enters the interaction zone, the direction of the average force is reversed as a result of the fact that for one period of pendulum oscillation an odd number of external force oscillations take place. That is why during each act of interaction, a definite portion of energy is communicated to the pendulum, which compensates for the damping losses.

In the case of gradual change in the frequency of the acting force ν, the amplitude and the frequency of the pendulum oscillation may leap. At that, it is also possible for „hysteresis" to occur, i.e., the values of the amplitude and frequency of the pendulum oscillation when increasing ν may not coincide with the same values when decreasing ν.

We shall note here, that even though the frequency of the pendulum oscillations should be an odd number of times smaller than the frequency of the action (given evenness of the external acting force), its value always remains close to the own resonance frequency. The latter is achieved by means of a self-adaptive selection of the number m on the part of the oscillating system. That is why the dependence of the oscillation frequency ω on the frequency of the external exciting action ν is actually very weak, and therefore an important conclusion can be drawn: that the sign required for referring the system under consideration to the self-oscillating systems is in place.

10.3.2. Spectrum of the possible oscillation amplitudes of a pendulum under the action of an external nonhomogeneous force

Once again we use equations (10.2) while taking into account (10.3) and (10.4).

We consider the case of pendulum oscillations without turning, i.e., when the coordinate falls within the range $[-\pi, \pi]$.

Conventionally, we number the time moments, determined by the action zone, as shown in Fig.10.18.

The pendulum motion in the intervals $[4n, 4n+1]$, $[4n+2, 4n+3]$, ... (out of the action zone) can be described by an unperturbed equation

$$\ddot{x} + \sin x = 0. \tag{10.22}$$

Multiplying Eq.(10.22) by $\dot{x}$ and integrating, we find

$$\frac{1}{2}\dot{x}^2 - \cos x = W - 1, \tag{10.23}$$

where W is an integration constant corresponding to the full system energy.

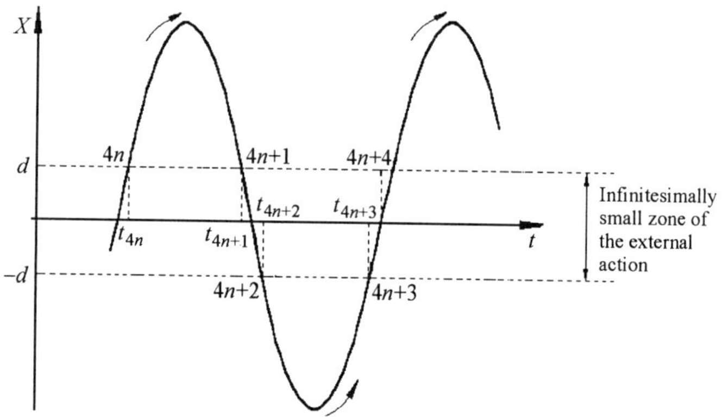

Fig. 10.18. Illustration in relation to the analysis of the pendulum's interaction with an external periodic force in the active zone

From Eq.(10.23) we obtain

$$\dot{x} = \pm\sqrt{2W - 4\sin^2\frac{x}{2}}. \tag{10.24}$$

Introducing the designation $u = \dfrac{x}{2}$ and $\sin u = z$ and considering Eq.(10.24), we can write

$$t - \alpha = \int \frac{dz}{\pm\sqrt{(1-z^2)\left(\dfrac{W}{2} - z^2\right)}},$$

where α is a constant.

Further on, we use the incomplete normal elliptic integral of first kind $F(.,.)$, so

$$t - \alpha = \pm a \int_0^z \frac{dz}{\sqrt{(a^2 - z^2)(z^2 - b^2)}} = F(\varphi, k), \tag{10.25}$$

where the amplitude $\varphi = am(t - \alpha, k)$, $m = k^2$, k is the modulus of the elliptic function, m is the parameter of the elliptic function.

In the case under consideration $a^2 = 1$, $b^2 = \dfrac{W}{2} < 1$ (in correspondence with the condition $-\pi < x < \pi$),

$$k = \sqrt{\frac{W}{2}}, \quad m = \frac{W}{2}, \quad \sin\varphi = \frac{z}{\sqrt{\dfrac{W}{2}}} = \frac{\sin\dfrac{x}{2}}{k}. \tag{10.26}$$

The solution of the equation (10.22) can be presented in the following form

$$x = 2\arcsin[k\,\mathrm{sn}(t-\alpha)], \tag{10.27}$$

where sn(.) is the sine of the amplitude (Jacobi's elliptic function).

Taking into account the dissipation, Eq.(10.26) becomes

$$\ddot{x} + \sin x = -2\delta_d \dot{x}. \tag{10.28}$$

Multiplying Eq.(10.28) by $\dot{x}$ and integrating, we find

$$\frac{d}{dt}\left[\frac{1}{2}\dot{x}^2 - \cos x\right] = -2\delta_d \frac{d}{dt}\int \dot{x}^2\,dt,$$

or

$$\frac{dW}{dt} = \frac{d}{dt}[-2\delta_d \int \dot{x}^2\,dt]. \tag{10.29}$$

For a half of the period, we obtain the following from (10.26) and Eq.(10.29)

$$2\Delta m = \Delta W = -2\delta_d \int \dot{x}^2\,dt.$$

Using (10.27), we can write

$$\dot{x} = 2k\,\mathrm{cn}(t-\alpha) \tag{10.30}$$

and

$$\int \dot{x}^2\,dt = 4k^2 \int \mathrm{cn}^2(t-\alpha)dt,$$

where cn(.) is the cosine of the amplitude (Jacobi's elliptic function).

Noting that $\int \mathrm{cn}^2(t-\alpha)dt = \frac{1}{k^2}[E(am(t-\alpha),k) - (1-k^2)(t-\alpha)]$ and $am[t-\alpha+2K(k),k] = am(t-\alpha,k)+\pi$, $E(\varphi+\pi,k) = E(\varphi,k)+2E(k)$, where $E(.,.)$ is incomplete elliptic integral of the second kind, hence

$$\int_0^{t+2K(k)} \mathrm{cn}^2(t-\alpha)dt = \frac{1}{k^2}[2E(k) - (1-k^2)2K(k)].$$

For half a period we have

$$2\Delta m = \Delta W = -16\delta_d[E(k) - (1-k^2)K(k)], \tag{10.31}$$

$$\Delta k = -\frac{4\delta_d}{k}[E(k) - (1-k^2)K(k)]. \tag{10.32}$$

In the case of small k, $0 < k \ll 1$, we can find

$$\int_t^{t+2K(k)} \mathrm{cn}^2(t-\alpha)dt \simeq \int_0^\pi \cos^2(t-\alpha)dt = \frac{\pi}{2}. \tag{10.33}$$

Combining Eqs.(10.30), (10.31) and (10.33), we obtain for half a period

$$\Delta m \simeq -2\pi\delta_d m, \quad \Delta k \simeq -\pi\delta_d k. \tag{10.34}$$

Let us introduce the following designations:

$$\Delta t_{4n} = t_{4n+1} - t_{4n}, \quad \Delta t_{4n+2} = t_{4n+3} - t_{4n+2}.$$

The border points are $x = \pm d'$ and the semi-periods are symmetrical with respect of the time points $t_{4n,\max}$ and $t_{4n+2,\max}$ (see Fig.10.17).

For $\left\{ \begin{array}{c} t = t_{4n,\max} \\ t = t_{4n+2,\max} \end{array} \right\}$ we have $\varphi = \left\{ \begin{array}{c} + \\ - \end{array} \right\} \dfrac{\pi}{2}$.

Using (10.26) we can determine

$$\varphi = \arcsin\left(\frac{\sin\frac{x}{2}}{k}\right). \tag{10.35}$$

Combining Eqs.(10.25) and (10.35) we find

$$\begin{aligned} \Delta t_{4n} &= 2\left[F\left(\frac{\pi}{2}, k\right) - F\left(\arcsin\frac{\sin\frac{d'}{2}}{k}, k\right)\right] \simeq \left[K(k) - F\left(\frac{d'}{2k}, k\right)\right] \\ &\simeq 2\left[K(k) - \frac{d'}{2k}\right], \end{aligned} \tag{10.36}$$

$$\Delta t_{4n+2} = 2\left[F\left(\frac{\pi}{2}, k\right) - F\left(\arcsin\frac{\sin\frac{d'}{2}}{k}, k\right)\right] = \Delta t_{4n} \simeq \left[K(k) - \frac{d'}{2k}\right]. \tag{10.37}$$

The expressions (10.36) and (10.37) are valid when $k > \sin\dfrac{d'}{2}$.

Further on, we use the approach developed in [305] on the basis of joining the solutions.

In the region $|x| < d'$, noting that $d' \ll 1$, we can use the linear approximation of the Eq.(10.22), i.e. $\ddot{x} + 2\delta_d\dot{x} + x \simeq \dfrac{F_0}{2d'}\sin\nu t$ and its solution in the form

$$x = \mathrm{R}e^{-\delta_d t}\sin[\omega(t-\gamma)] + \frac{\frac{F_0}{2d'}}{\sqrt{(\nu^2-1)^2 + (2\nu\delta_d)^2}}\sin(\nu t + \varphi_\nu),$$

where $\omega = \sqrt{1-\delta_d^2}$.

Let us assume that $\nu > 1$, then $\varphi_\nu = \operatorname{arctg} \dfrac{2\nu\delta_d}{\nu^2 - 1} + \pi$.

When $0 < \delta_d \ll 1$ and $\nu \gg 1$, the frequency $\omega \simeq 1$ and

$$x \simeq R\sin(t-\gamma) + \frac{\dfrac{F_0}{d'}}{1-\nu^2}\sin\nu t, \quad \dot{x} \simeq R\cos(t-\gamma) + \nu\frac{\dfrac{F_0}{2d'}}{1-\nu^2}\cos\nu t.$$

Now, let us consider the region out of the action zone $[-d', d']$, but close to that zone, i.e. $|x| > d'$, $|x| \simeq d'$.

Under these conditions we can write:

$$x = 2\arcsin[k\,\mathrm{sn}(t-\alpha)] = 2\arcsin\{k\,\mathrm{sn}[2K(k)-(t-\alpha)]\} \simeq 2k[2K(k)-(t-\alpha)].$$

It follows that the moment t_{4n+1} can be found from the equation

$$2k[2K(k)-(t_{4n+1}-\alpha)] \simeq d' \tag{10.38}$$

when $\dot{x} \simeq -2k$.

From the condition of lacking x and $\dot{x}$ interruption in the point $t = t_{4n+1}$, it follows that

$$R\sin(t_{4n+1} - \gamma + \frac{\dfrac{F_0}{2d'}}{1-\nu^2}\sin\nu t_{4n+1} \simeq d', \tag{10.39a}$$

$$R\cos(t_{4n+1}-\gamma) + \nu\frac{\dfrac{F_0}{2d'}}{1-\nu^2}\cos\nu t_{4n+1} \simeq -2k_{4n+1}. \tag{10.39b}$$

Solving the system (10.39) we can obtain formulae for $R = R_{4n+1}$ and $\gamma = \gamma_{4n+1}$. Analogously, when going out of the action zone, i.e. for the point $t = t_{4n+2} = t_{4n+1} + \Delta t_{4n+1}$, where

$$\Delta t_{4n+1} = t_{4n+2} - t_{4n+1}, \tag{10.40}$$

we can write

$$R\sin(t_{4n+1}+\Delta t_{4n+1}-\gamma) + \frac{\dfrac{F_0}{2d'}}{1-\nu^2}\sin[\nu(t_{4n+1}+\Delta t_{4n+1})] \simeq -d', \tag{10.41a}$$

$$R\cos(t_{4n+1}+\Delta t_{4n+1}-\gamma) + \gamma\frac{\dfrac{F_0}{2d'}}{1-\nu^2}\cos[\nu(t_{4n+1}+\Delta t_{4n+1})] \simeq -2k_{4n+2}. \tag{10.41b}$$

If

$$\nu\Delta t_{4n+1} \ll 1, \tag{10.42}$$

the Eqs.(10.39) give $\Delta t_{4n+1} \simeq \dfrac{2d'}{-R\cos(t_{4n+1}-\gamma)+\nu\dfrac{\frac{F_0}{2d'}}{1-\nu^2}\cos\nu t_{4n+1}}$.

Taking into account Eq.(10.39b), Expr. (10.40) becomes

$$\Delta t_{4n+1} \simeq \frac{d'}{k_{4n+1}}. \tag{10.43}$$

Let us introduce the designation

$$\Delta k_{4n} \simeq k_{4n+1} - k_{4n}. \tag{10.44}$$

A comparison of (10.44) and (10.34) reveals $\Delta k_{4n} \simeq -\pi\delta_d k_{4n}$. Considering (10.36), we can write

$$\Delta t_{4n} \simeq 2K(k_{4n}) - \frac{d'}{k_{4n}}. \tag{10.45}$$

Using Eq.(10.41b), under the condition (10.42), we find

$$k_{4n+2} \simeq \frac{1}{2}\left\{-R\cos(t_{4n+1}-\gamma) + R\Delta t_{4n+1}\sin(t_{4n+1}-\gamma) + \nu\frac{\frac{F_0}{2d'}}{\nu^2-1}\cos\nu t_{4n+1}\right.$$
$$\left. -\nu^2\frac{\frac{F_0}{2d'}}{\nu^2-1}\Delta t_{4n+1}\sin\nu t_{4n+1}\right\}. \tag{10.46}$$

Substituting Eqs.(10.39) and (10.43) in Eq.(10.46) we obtain

$$k_{4n+2} \simeq k_{4n+1} - \frac{F_0}{4k_{4n+1}}\sin\nu t_{4n+1}. \tag{10.47}$$

Analogously we can write the following equations

$$\Delta t_{4n+2} \simeq 2K(k_{4n+2}) - \frac{d'}{k_{4n+2}}, \tag{10.48}$$

$$\Delta k_{4n+2} \simeq -\pi\delta_d k_{4n+2}, \tag{10.49}$$

$$k_{4n+3} = k_{4n+2} + \Delta k_{4n+2}, \tag{10.50}$$

$$t_{4n+3} = t_{4n+2} + \Delta t_{4n+2}.$$

For the region $4n+3 \to 4n+4$ (see Fig.10.18) we have ($R = R_{4n+3}, \gamma = \gamma_{4n+3}$)

$$R\sin(t_{4n+3} - \gamma) - \frac{\frac{F_0}{2d'}}{\nu^2 - 1} \sin \nu t_{4n+3} \simeq -d', \tag{10.51a}$$

$$R\cos(t_{4n+3} - \gamma) - \nu \frac{\frac{F_0}{2d'}}{\nu^2 - 1} \sin \nu t_{4n+3} \simeq 2k_{4n+3}, \tag{10.51b}$$

$$t_{4n+4} = t_{4n+3} + \Delta t_{4n+3},$$

$$R\sin(t_{4n+3} + \Delta t_{4n+3} - \gamma) - \frac{\frac{F_0}{2d'}}{\nu^2 - 1} \sin \nu(t_{4n+3} + \Delta t_{4n+3}) \simeq d', \tag{10.52a}$$

$$R\cos(t_{4n+3} + \Delta t_{4n+3} - \gamma) - \nu \frac{\frac{F_0}{2d'}}{\nu^2 - 1} \cos \nu(t_{4n+3} + \Delta t_{4n+3}) \simeq 2k_{4n+4}. \tag{10.52b}$$

Assuming, that

$$\nu \Delta t_{4n+3} \ll 1 \tag{10.53}$$

and combining Eqs.(10.51) and (10.52a) we find

$$\Delta t_{4n+3} \simeq \frac{2d'}{R\cos(t_{4n+3} - \gamma) - \nu \frac{\frac{F_0}{2d'}}{\nu^2 - 1} \cos \nu t_{4n+3}}.$$

From (10.51b) it follows

$$\Delta t_{4n+3} \simeq \frac{d'}{k_{4n+3}}, \quad t_{4n+4} = t_{4n+3} + \Delta t_{4n+3}. \tag{10.54}$$

Taking into account Eqs.(10.49) and (10.47), we can write

$$k_{4n+2} \simeq k_{4k} - \pi \delta_d k_{4n} - \frac{F_0}{4k_{4n}} \sin \nu t_{4n+2}.$$

Considering the condition (10.53), Eq.(10.52b) can be rewritten

$$k_{4n+4} \simeq \frac{1}{2} \left\{ R\cos(t_{4n+3} - \gamma_{4n+3}) - \nu \frac{\frac{F_0}{2d'}}{\nu^2 - 1} \cos \nu t_{4n+3} \right.$$
$$\left. - R\Delta t_{4n+3} \sin(t_{4n+3} - \gamma) + \nu^2 \frac{\frac{F_0}{2d'}}{\nu^2 - 1} \Delta t_{4n+3} \sin \nu t_{4n+3} \right\}. \tag{10.55}$$

Using Eqs.(10.51) and (10.55) we can write

$$k_{4n+4} \simeq k_{4n+3} + \frac{F_0}{4k_{4n+3}} \sin \nu t_{4n+3}. \tag{10.56}$$

A comparison between (10.56), on the one hand, and (10.49) and (10.50), on the other hand, reveals

$$\Delta k_{4n+4} \simeq k_{4n+2} - \pi\delta_d k_{4n+2} + \frac{F_0}{4k_{4n+2}} \sin \nu t_{4n+4}.$$

Combining Eqs.(10.43) and (10.45) we find

$$t_{4n+2} \simeq \left[2K(k_{4n+2}) - \frac{d'}{k_{4n}}\right] + \frac{d'}{k_{4n+1}} + t_{4n+1} \simeq 2K(k_{4n}) + t_{4n}.$$

Drawing an analogy with Eqs.(10.48) and (10.54) we can obtain

$$t_{4n+4} \simeq \left[2K(k_{4n+2}) - \frac{d'}{k_{4n+2}}\right] + \frac{d'}{k_{4n+3}} + t_{4n+2} \simeq 2K(k_{4n+2}) + t_{4n+2}.$$

In the long run we have obtained the following system of equations:

$$t_{4n+2} \simeq t_{4n} + 2K(k_{4n}) \tag{10.57a}$$

$$k_{4n+2} \simeq k_{4n} - \pi\delta_d k_{4n} - \frac{F_0}{4k_{4n}} \sin \nu t_{4n+2} \tag{10.57b}$$

$$t_{4n+4} \simeq t_{4n+2} + 2K(k_{4n+2}) \tag{10.57c}$$

$$\Delta k_{4n+4} \simeq k_{4n+2} - \pi\delta_d k_{4n+2} + \frac{F_0}{4k_{4n+2}} \sin \nu t_{4n+4}. \tag{10.57d}$$

The spectrum of possible stationary amplitudes of continuous oscillations is determined by the expression:

$$\nu(t_{4n+2} - t_{4n}) = 2\pi N, \tag{10.58}$$

where $N = 1, 2, 3, ...$ is the ratio of frequency division.

The Eq.(10.58) can also be written in the form:

$$\nu[K(k_{2n}) + K(k_{2n+2})] = \pi N.$$

Below, we show that N has to be an odd number.

Designating five successive time points as t_0, t_1, t_2, t_3 and t_4 and corresponding values of k as k_0, k_1, k_2, k_3 and k_4 (by analogy with the steps in Fig.10.18), we can write the following conditions for the stationary mode:

$$k_4 = k_0, \tag{10.59a}$$
$$\nu(t_4 - t_0) = 2\pi N. \tag{10.59b}$$

The Eq.(10.59b) follows from the condition of oscillation synchronization with the external excitement.

The Eqs.(10.57) can be rewritten

$$k_2 = k_0 - \pi\delta_d k_0 - \frac{F_0}{4k_0}\sin\nu t_2, \qquad k_4 = k_2 - \pi\delta_d k_2 + \frac{F_0}{4k_0}\sin\nu t_0,$$
$$t_2 = t_0 + 2K(k_0), \qquad t_4 = t_2 + 2K(k_2).$$

If we consider the condition of symmetry between the upper $\{4n \to 4n+2\}$ and lower $\{4n+2 \to 4n+4\}$ periods, we can find that it is possible to have the symmetry only if N is an odd number, i.e. $N = 2l+1$, $l = 0, 1, 2, 3, ...$ and if the following equation is fulfilled: $\sin\nu[t + 2K(k)] = -\sin\nu t$..

From Eqs.(10.38) and (10.39) it follows $2\nu K(k_0) = 2\pi(l + \frac{1}{2})$ and

$$\sin\nu t_0 = -\sin\nu t_2, \tag{10.60a}$$

$$\cos\nu t_0 = -\cos\nu t_2. \tag{10.60b}$$

Combining Eqs.(10.32), (10.59a) and (10.60) we can write

$$\frac{16\delta_d}{F_0}[E(k_0) - (1 - k_0^2)K(k_0)] - \sin\nu t_0 = 0, \tag{10.61}$$

which, in the case of $k \to 0$, becomes

$$\frac{4\pi\delta_d k_0^2}{F_0} - \sin\nu t_0 = 0. \tag{10.62}$$

Solving Eq.(10.62) we can determine the initial phase t_0.

Apparently there exists an excitation threshold,

$$|F_0| > 4\pi\delta_d k^2. \tag{10.63}$$

For the F_0 values above the value $F_0 = 4\pi\delta_d k^2$, a discretization of the possible stationary oscillating amplitudes appears.

As we have assumed the solution symmetry, for the sake of system stability examination, it is enough to consider only a half of the period.

From Eq.(10.57b) we determine the variation

$$\delta k_{4n+2} = \left(1 - \pi\delta_d + \frac{F_0}{4k_{4n}^2}\sin\nu t_{4n+2}\right)\delta k_{4n} - \nu\frac{F_0}{4k_{4n}}(\cos\nu t_{4n+2})\delta t_{4n+2}.$$

Taking into account the relation [324] $\dfrac{dK(k)}{dk} = \dfrac{1}{k}\left[\dfrac{E(k)}{1-k^2} - K(k)\right]$ as well as the approximate formulae $E(k)=\frac{\pi}{2}\left(1-\frac{k^2}{4}\right)+O(k^4)$, $K(k)=\frac{\pi}{2}\left(1+\frac{k^2}{4}\right)+O(k^4)$ we can write

$$\frac{dK(k)}{dk} = \frac{\pi}{4}k + O(k^3). \tag{10.64}$$

Considering (10.64), we can find the following variation from (10.57c):

$$\delta t_{4n+4} = \delta t_{4n+2} + 2\frac{dK(k)}{dk}\delta k_{4n+2} = \frac{\pi}{2}k_{4n+2}\left(1 - \pi\delta_d + \frac{F_0}{4k_{4n}^2}\sin\nu t_{4n+2}\right)\delta k_{4n}$$
$$+ \left(1 - \frac{\pi}{8}\nu F_0\frac{k_{4n+2}}{k_{4n}}\cos\nu t_{4n+2}\right)\delta t_{4n+2}.$$

Taking into account that $k_{4n+2} = k_{4n} = k_0$, $\sin\nu t_{4n+2} = -\sin\nu t_0$, $\cos\nu t_{4n+2} = -\cos\nu t_0$, it follows from Eq.(10.61) that $\dfrac{F_0\sin\nu t_{4n+2}}{4k_{4n}^2} = -\dfrac{F_0\sin\nu t_0}{4k_0^2}$ $= -\pi\delta_d$, so, we can write

$$\left|\begin{array}{l} \delta k_{4n+2} = (1-2\pi\delta_d)\delta k_{4n} + \nu\dfrac{F_0}{4k_0}(\cos\nu t_0)\delta t_{4n+2} \\[2ex] \delta t_{4n+4} = \dfrac{\pi}{2}k_0(1-2\pi\delta_d)\delta k_{4n} + \left(1 + \dfrac{\pi}{8}\nu F_0\cos\nu t_0\right)\delta t_{4n+2} \end{array}\right. .$$

Let us assume that $\delta k_{4n+2} = \lambda\delta k_{4n}$ and $\delta t_{4n+4} = \lambda\delta t_{4n+2}$.
Hence we can write

$$\left|\begin{array}{l} (1-2\pi\delta_d-\lambda)\delta k_{4n} + \nu\dfrac{F_0}{4k_0}(\cos\nu t_0)\delta t_{4n+2} = 0 \\[2ex] \dfrac{\pi}{2}k_0(1-2\pi\delta_d)\delta k_{4n} + \left(1 + \dfrac{\pi}{8}\nu F_0\cos\nu t_0 - \lambda\right)\delta t_{4n+2} = 0 \end{array}\right. .$$

The characteristic equation is in the form

$$\lambda^2 - \lambda(2 - 2\pi\delta_d + \frac{\pi}{8}\nu F_0\cos\nu t_0) + (1-2\pi\delta_d) = 0$$

and its solution is

$$\lambda_{1,2} = 1 - \pi\delta_d + \frac{\pi}{16}\nu F_0\cos\nu t_0 \pm \sqrt{\left(1 - \pi\delta_d + \frac{\pi}{16}\nu F_0\cos\nu t_0\right)^2 - 1 + 2\pi\delta_d}.$$

The stability condition is: $|\lambda_{1,2}| < 1$.

Apparently, the solution is stable when the following condition is satisfied: $F_0 \nu \cos \nu t_0 < 0$.

Generally, we have proved that oscillations with an amplitude taken from a possible set of stable amplitudes can be excited in the system under consideration.

The spectrum of the symmetrical solution amplitudes can be expressed as $2\nu K(k_0) = 2\pi \left(l + \frac{1}{2}\right)$, $l = 0, 1, 2, 3...$, which yields the spectrum of amplitudes k_0, $K(k_0) = \left(l + \frac{1}{2}\right)\frac{\pi}{\nu}$, $l = 0, 1, 2, 3...$ and an odd ratio of frequency division $N = 2l + 1$, $l = 0, 1, 2, 3...$

10.3.3. Rotator under non-homogeneous action

A phenomenon of rotation excitation with a strictly determined discrete set of possible rates is presented [313]. The model considered is that of a pendulum rotating in the field of an external periodic and non-homogeneous influence, which corresponds to the condition $\frac{W}{2} > 1$ (See (10.23)).

Once again we shall consider first the non-perturbed equation (10.22). The constants of the integral (10.25) may be determined here as

$$a^2 = \frac{W}{2}, \quad b^2 = 1, \quad m = \frac{b^2}{a^2} = \frac{2}{W}, \quad k = \sqrt{m} = \sqrt{\frac{2}{W}}, \quad \frac{W}{2} = \frac{1}{m} = \frac{1}{k^2}, \quad a = \frac{1}{k}$$

(See [324]) and, respectivly

$$a \int_0^z \frac{dz}{\sqrt{\left(\frac{W}{2} - z^2\right)(1 - z^2)}} = F(\varphi, k),$$

$$z = \sin\varphi,$$

$$t - \alpha = \int_0^z \frac{dz}{\pm\sqrt{\left(\frac{W}{2} - z^2\right)(1 - z^2)}},$$

$$t - \alpha = kF(\varphi, k).$$

Since $\sin u = z = \sin\varphi$, it follows that

$$u = \varphi,$$

$$t - \alpha = kF(u, k),$$

$$u = am\left(\frac{t-\alpha}{k}, k\right),$$

$$x = 2u = 2am\left(\frac{t-\alpha}{k}, k\right).$$

Considering Eq.(10.28), which accounts for the dissipation, we can write, besides Expr.(10.29), that

$$\Delta W = -2\delta_d \int \dot{x}^2 dt$$

$$\dot{x}^2 = \frac{4}{k^2} du^2 \left(\frac{t-\alpha}{k}\right). \tag{10.65}$$

Since $\int dn^2 u du = E[am(u,k),k]$, then $\int \dot{x}^2 dt = \frac{4}{k^2}\int dn^2\left(\frac{t-\alpha}{k}\right) dt$ and finally

$$\int \dot{x}^2 dt = \frac{4}{k} E\left[am\left(\frac{t-\alpha}{k}\right), k\right].$$

The quantity $\dot{x}$ is periodic, with a period: $\Delta t = 2kK(k)$. Taking into account this fact and having in mind that $am\left[\frac{t+2kK(k)-\alpha}{k}\right] - am\left(\frac{t-\alpha}{k}\right) = \pi$, $E(\varphi+\pi,k) - E(\varphi,k) = 2E(k)$ and $\int_t^{t+2kK(k)} \dot{x}^2 dt = \frac{8}{k}E(k)$ we find

$$\Delta W = -\frac{8\delta_d}{k} E\left[am\left(\frac{t-\alpha}{k}\right), k\right]_t^{t+2kK(k)} = -\frac{16\delta_d}{k} E(k).$$

Considering that $W = \frac{2}{k^2}$, then: $\Delta W = -\frac{4}{k^3}\Delta k$. Hence, for a single period we obtain

$$\Delta k = 4\delta_d k^2 E(k). \tag{10.66}$$

For $k \to 0$, considering that $E(k) \to \frac{\pi}{2}$, we can find the following approximate form: $\Delta k \simeq 2\pi\delta_d k^2$. For $x = 2\pi n$, $n = 0, \pm1, \pm2, ...$ we have $dn\left(\frac{t-\alpha}{k}\right) \simeq 1$, hence, regarding Expr. (10.65) we may write

$$\dot{x} \approx \frac{2}{k}. \tag{10.67}$$

Suppose that the derivative $\dot{x}$ varies slightly during the transition $\{2n+1 \to 2n+2\}$ (See Fig.10.17), and k is invariable. For the purpose of distinguishing the rotator mode the typical points in the action zone in Fig.(10.17) are denoted as

$2n+1,\ 2n+2, \ldots, t_{2n+1}, t_{2n+2}, \ldots$ instead of $4n+1,\ 4n+2, \ldots, t_{4n+1}, t_{4n+2}, \ldots$ Then we can write the following for the difference $\Delta \dot{x}$ of the values of $\dot{x}$ before the transition and after the transition

$$\Delta \dot{x} = \int_{-\Delta x}^{\Delta x} \delta(x) F_0 \sin vt \frac{dt}{dx} dx = \frac{k}{2} F_0 \sin vt. \tag{10.68}$$

Since the external exciting force acts only once in a period of rotation and considering Exprs.(10.67) and (10.68)

$$\dot{x}_{2n+1} - \dot{x}_{2n} \approx \frac{k_{2n}}{2} F_0 \sin vt, \quad \frac{2}{k_{2n+1}} - \frac{2}{k_{2n}} \approx \frac{k_{2n}}{2} F_0 \sin vt$$

or

$$k_{2n+1} - k_{2n} \approx -\frac{k_{2n}^3}{4} F_0 \sin vt_{2n}. \tag{10.69}$$

Using Rel.(10.66), we obtain $k_{2n+2} = k_{2n+1} + 4\delta_d k_{2n+1}^2 E(k)$, or, on the basis of Rel.(10.69):

$$k_{2n+2} = k_{2n} + 4\delta_d k_{2n}^2 E(k) - \frac{k_{2n}^3}{4} F_0 \sin vt_{2n}. \tag{10.70}$$

For $k \to 0$ we have approximately

$$k_{2n+2} = k_{2n} + 2\pi \delta_d k_{2n}^2 - \frac{k_{2n}^3}{4} F_0 \sin vt_{2n}. \tag{10.71}$$

The rotating period is determined by the expression

$$t_{2n+2} = t_{2n} + 2K_{2n} K(k_{2n}). \tag{10.72}$$

The spectrum of the rates of rotation can be derived from the equation

$$v(t_{2n+2} - t_{2n}) = 2\pi N, \tag{10.73}$$

where $N = 1, 2, 3, \ldots$ is a coefficient of division of the external action frequency.

Combining (10.72) and (10.73) we can write an equation for the periodic solution, $2vkK(k) = 2\pi N$, thus obtaining a formula for the discrete k values

$$kK(k) = \frac{\pi}{v} N, \quad N = 1, 2, 3, \ldots$$

Since $k_{2n+2} = k_{2n}$ for the periodic solution, we obtain the following from Eq.(10.70)

$$\frac{16\delta_d}{F_0 k_{2n}} E(k_{2n}) - \sin vt_{2n} = 0.$$

For $k_{2n} \sim k_0 \to 0$ we get from Eq.(10.71) an equation for the initial phase $t_0 \sim t_{2n}$, i.e.

$$\frac{8\pi\delta_d}{k_0 F_0} = \sin vt_0. \tag{10.74}$$

Obviously, the threshold value of the amplitude of the external impact can be determined as $|F_0| \geq \dfrac{8\pi\delta_d}{k_0}$.

However, surpassingly, this limit leads to rotation of the pendulum and to a discretization of its stationary dynamical states.

Let us analyze the stability of the solution obtained.

Using the relation and approximations that have led to the Eq.(10.64) and considering Eqs. (10.71) and (10.72) we can produce the following variations:

$$\left| \begin{array}{l} \delta k_{2n+2} = \left(1 + 4\pi\delta_d k_{2n} - \dfrac{3}{4}k_{2n}^2 F_0 \sin vt_{2n}\right)\delta k_{2n} - \dfrac{k_{2n}^3}{4} vF_0 \cos vt_{2n} \delta t_{2n} \\ \delta t_{2n+2} = \left[2K(k) + 2k\dfrac{\pi}{4}\right]\delta k_{2n} + \delta t_{2n} \end{array} \right. . \tag{10.75}$$

When we use the approximation $K(k) \simeq \dfrac{\pi}{2}\left(1 + \dfrac{k^2}{4}\right)$, Eqs. (10.75) assume the form

$$\left| \begin{array}{l} \delta k_{2n+2} = \left(1 + 4\pi\delta_d k_0 - \dfrac{3}{4}k_0^2 F_0 \sin vt_0\right)\delta k_{2n} - \dfrac{k_0^3}{4} vF_0 \cos vt_0 \delta t_{2n} \\ \delta t_{2n+2} = \pi\left(1 + \dfrac{3}{4}k_0^2\right)\delta k_{2n} + \delta t_{2n} \end{array} \right. .$$

Let $\delta k_{2n+2} = \lambda \delta k_{2n}$ and $\delta t_{2n+2} = \lambda \delta t_{2n}$.
Then

$$\left| \begin{array}{l} \left(1 + 4\pi\delta_d k_0 - \dfrac{3}{4}k_0^2 F_0 \sin vt_0 - \lambda\right)\delta k_{2n} - \dfrac{k_0^3}{4} vF_0 \cos vt_0 \delta t_{2n} = 0 \\ \pi\left(1 + \dfrac{3}{4}k_0^2\right)\delta k_{2n} + (1-\lambda)\delta t_{2n} = 0 \end{array} \right. .$$

The nonzero solution is determined by the characteristic equation:

$$\begin{aligned} \lambda^2 - \left(2 + 4\pi\delta_d k_0 - \frac{3}{4}k_0^2 F_0 \sin vt_0\right)\lambda + 1 + 4\pi\delta_d k_0 \\ - \frac{3}{4}k_0^2 F_0 \sin vt_0 + \frac{\pi}{4}k_0^3\left(1 + \frac{3}{4}k_0^2\right) vF_0 \cos vt_0 = 0. \end{aligned} \tag{10.76}$$

Considering Rel.(10.74), Eq.(10.76) can be written in the form:

$$\lambda^2 - 2\lambda(1 - \pi\delta_d k_0) + 1 - 2\pi\delta_d k_0 + \frac{\pi}{4}k_0^3\left(1 + \frac{3}{4}k_0^2\right)vF_0 \cos vt_0 = 0.$$

Its solutions are:

$$\lambda_{1,2} = 1 - \pi\delta_d k_0 \pm \sqrt{(1 - \pi\delta_d k_0)^2 - 1 + 2\pi\delta_d k_0 - \frac{\pi}{4}k_0^3\left(1 + \frac{3}{4}k_0^2\right)vF_0 \cos vt_0}.$$

The stability condition given by the inequality $|\lambda_{1,2}| < 1$ is fulfilled under the following general condition:

$$vF_0 \cos vt_0 > 0.$$

10.3.4. General conditions for pendulum oscillation excitation under the action of an external nonlinear force

The following equation is subject to analysis:

$$\frac{d^2x}{dt^2} + 2\delta_d\frac{dx}{dt} + f(x) = \varepsilon(x)F_0 \sin vt, \qquad (10.77)$$

where the nonlinearity of the oscillation system is expressed in a general form by the function $f(x) : f(0) = 0,\ f(-x) = -f(x)$. The nonlinearity of the external exciting harmonic force, as regards the coordinate x of the system under consideration, is accounted for by the function

$$\varepsilon(x) = \begin{cases} 1, & |x| \le d' \\ 0, & |x| > d' \end{cases}, \quad d' \ll 1 \text{ is a constant.}$$

It is clear that the external action is assumed to be pulsating in nature and appearing around the point $x = 0$. At the same time the function $\varepsilon(x)$ presupposes the existence of self-adaptivity of the excited system with respect to the external exciting force.

The solution is sought in the interval of x values for which $xf(x) \ge 0$. The nonlinearity of Eq.(10.77) is not limited, i.e. the function $f(x)$ can be essentially a nonlinear one, but the analysis is carried out for the condition of small perturbations. In this case Eq.(10.77) can be represented in the form

$$\frac{d^2x}{dt^2} + f(x) = \mu h_1 + \mu h_2, \qquad (10.78)$$

where μ is a small parameter, $0 \le \mu \ll 1$, $\mu h_1 = -2\delta_d\frac{dx}{dt}$, $\mu h_2 = \varepsilon(x)F_0 \sin vt$.

Drawing on the basic ideas of the method of nonlinear transformation of variables (see Section 2.2.1 and [185]), a new coordinate y is set according to the expression

$$y = \sqrt{2V(x)}, \quad \text{where } V(x) = \int_0^x f(x')dx',$$

and a new nonlinear time is introduced by the relation: $\dfrac{dt}{d\tau} = G(y) = \dfrac{dx}{dy} = \dfrac{y}{f(x)}$.

Eq.(10.78) becomes

$$\frac{d^2y}{d\tau^2} + \xi^2 y = \mu\Phi + (\xi^2 - 1)y, \tag{10.79}$$

where ξ is the possible frequency detuning, $\mu\Phi = (\mu h_1 + \mu h_2)G(y)$.

The solution of Eq.(10.79) can be sought in the form: $y = R\cos\Psi$, $\Psi = \xi\tau + \gamma$, where R and γ are the amplitude and the initial phase respectively.

The average (shortened) perturbation equation related to R and γ are:

$$\left| \begin{array}{l} \left\langle \dfrac{dR}{d\tau} \right\rangle = -\left[\left\langle \dfrac{\mu\Phi}{\xi} \sin\Psi \right\rangle \right] \\ \left\langle \dfrac{d\gamma}{d\tau} \right\rangle = -\left\langle \dfrac{\mu\Phi}{\xi} \dfrac{\cos\Psi}{R} \right\rangle - \left\langle \dfrac{\xi^2 - 1}{2\xi} \right\rangle \end{array} \right. , \tag{10.80}$$

where the sign $<>$ denotes the averaging procedure by Ψ.

We consider the influence of the two perturbation actions μh_1 and μh_2 separately.

Assuming that $\mu\Phi = \mu h_1 G(y)$, the functional expression in Eq.(10.80) can be successively clarified:

$$\left\langle \frac{\mu\Phi}{\xi} R\sin\Psi \right\rangle = \frac{2\beta}{\xi^2}\frac{\xi}{2\pi} \int_0^{2\pi/\xi} \left(\frac{dx}{dt}\right)^2 G(y)d\tau = \frac{2\delta_d}{2\pi}\frac{T}{2\pi}\left[\frac{1}{T}\int_0^T \left(\frac{dx}{dt}\right)^2 dt\right],$$

where $T = \displaystyle\int_0^{2\pi/\xi} G(y)d\tau$ is the period of oscillation.

The energy in the system, determined as $W = \displaystyle\int_0^{x_{\max}} f(x')dx'$, can be derived from the expression

$$\left\langle \frac{dW}{dt} \right\rangle = \frac{\left\langle \dfrac{dW}{d\tau} \right\rangle}{\left\langle \dfrac{dt}{d\tau} \right\rangle} = \frac{-\left\langle \dfrac{\mu\Phi}{\xi} R\sin\Psi \right\rangle}{\langle G(y) \rangle} = -\frac{2\delta_d}{\xi^2}\left[\frac{1}{T}\int_0^T \left(\frac{dx}{dt}\right)^2 dt\right].$$

In the case of small amplitudes, $\left\langle \frac{dW}{dt} \right\rangle = -2\delta_d W$.

Further on, taking into account that $G(y)$ is an even function,

$$\left\langle \frac{\mu\Phi}{\xi}\frac{\cos\Psi}{R} \right\rangle = -\frac{2\delta_d}{\xi R}\langle(-\xi R\sin\Psi)G(R\cos\Psi)\cos\Psi\rangle = 0.$$

Hence it follows that $\frac{d\gamma}{d\tau} = -\left(\frac{\xi^2-1}{2\xi}\right)$.

Now, let us analyze the influence of the external pulse action on the system manifested at $|x| \le d'$, $d' \ll 1$. First let us consider the change of the energy in the system. The dissipative term $2\delta_d \frac{dx}{dt}$ in Eq.(10.77) can be neglected for the short time of pulse action. Integrating Eq.(10.77) we can write

$$\frac{d}{dt}\left[\frac{1}{2}\left(\frac{dx}{dt}\right)^2 + V(x)\right] = \varepsilon(x)\frac{dx}{dt}F_0\sin vt, \tag{10.81}$$

where t_0 corresponds to the moment when $x = 0$.

Evaluating Eq.(10.81) in the interval $[t_0 - \Delta t_0, t_0 + \Delta t_0]$ corresponding to the pulse action one can write $\int_{t_0-\Delta t_0}^{t_0+\Delta t_0} dW = \int_{t_0-\Delta t_0}^{t_0+\Delta t_0} \varepsilon(x)\frac{dx}{dt}F_0\sin vt_0\,dt$.

Hence $\Delta W = W(t_0 + \Delta t_0) - W(t_0 - \Delta t_0) = 2d'F_0\sin vt_0$.

Then the pulse influence on the phase γ can be revealed as

$$\begin{aligned}\left\langle \frac{d\gamma}{d\tau} \right\rangle &= \left\langle \frac{\mu\Phi}{\xi}\frac{\cos\Psi}{R} \right\rangle = -\frac{1}{2\pi}\int_0^{2\pi}\frac{1}{\xi}G(y)\varepsilon(x)F_0\sin vt\frac{\cos\Psi}{R}d\Psi \\ &\simeq F_0\sin vt_0\left(-\frac{1}{2\pi\xi}\int_{-\pi}^{\pi}G(y)\frac{\cos\Psi}{R}\varepsilon(x)d\Psi\right).\end{aligned} \tag{10.82}$$

Taking into account that $G(y)$ is an even function, it can be shown that the integral in (10.82) is equal to zero. Hence, the pulse action does not influence the initial phase.

Now let us consider the Poincaré section at the points $x = \pm d'$ – see Fig.10.19, where a motion portrait at the phase plane $(x, dx/dt)$ is presented.

In the moment t_n: $y = R\cos\Psi_n = 0$, i.e. $\cos(\xi\tau_n + \gamma_n) = 0$ and

$$\xi\tau_n + \gamma_n = -\frac{\pi}{2} + l\pi, \quad l = 0, \pm, \pm 2, \pm 3, \ldots$$

In the moment t_{n+1}: $y = R\cos\Psi_{n+1} = 0$, i.e. $\cos(\xi\tau_{n+1} + \gamma_{n+1}) = 0$ and $\xi\tau_{n+1} + \gamma_{n+1} = \frac{\pi}{2} + l\pi$. Hence: $\tau_{n+1} - \tau \simeq \pi + (\xi - 1)\pi$.

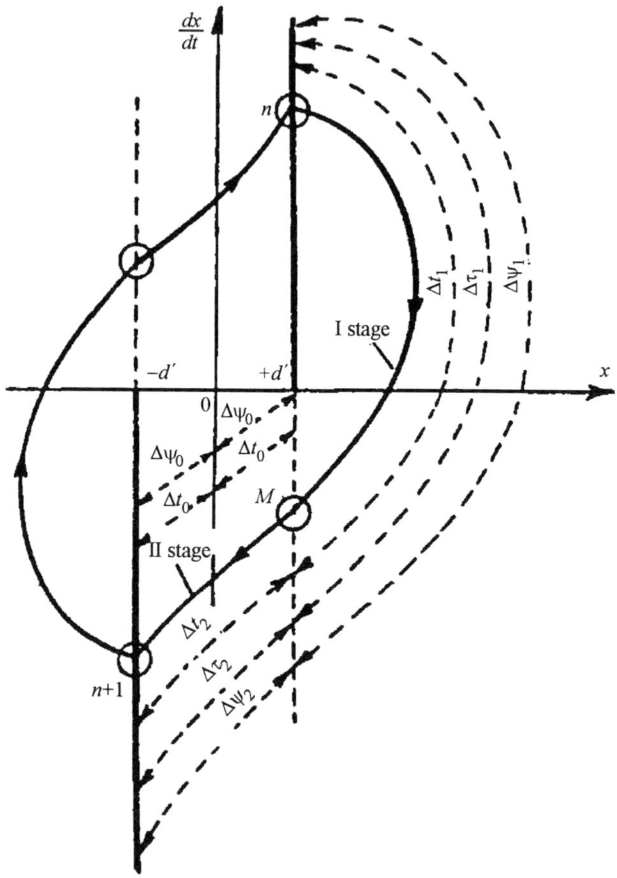

Fig. 10.19. Illustration in relation to the analysis of the energy change during the semi-cycles of pendulum motion

Then, $t_{n+1}-t_n = \dfrac{T}{2}+G(0)(\tau_{n+1}-\tau_n-\dfrac{\pi}{\xi}) \simeq \dfrac{T}{2}+\dfrac{(\xi-1)\pi}{\omega_0}$, where the following is taken into account: $G(0) = \dfrac{y}{f(0)} = \dfrac{1}{\omega_0}$, ω_0 is the natural resonance frequency of the system at infinitesimal amplitudes.

The change of the energy W during the semicycles is considered stage by stage as it is illustrated in Fig.10.19.

Stage 1. $\Delta W_1 \approx \left\langle \frac{dW}{dt} \right\rangle \frac{T}{2} = -\frac{\delta_d T}{\xi^2}\left[\frac{1}{T}\int_0^T \left(\frac{dx}{dt}\right)^2 dt\right]$ and for small amplitudes $\Delta W_1 \approx -\delta_d WT$.

Stage 2. $\Delta W_2 = 2d' F_0 \sin vt_{n+1}$.

As a whole, for the semicycle (the transition) $n \to n+1$ we have received:

$t_{n+1} = t_n + \frac{T}{2} + \frac{(\xi-1)\pi}{\omega_0}$, $\xi = \frac{\Pi(W_n)}{T}$, where $\Pi(W) = \int_0^{2\pi} G(y)d\Psi$ is the period of the autonomous oscillation at time t,

$W_{n+1} = W_n - \frac{\delta_d T}{\xi^2}\left[\frac{1}{T}\int_0^T \left(\frac{dx}{dt}\right)^2 dt\right] + 2d' F \sin vt_{n+1}$, or in the case of small amplitudes $W_{n+1} = W_n - \delta_d W_n T + 2d' F_0 \sin vt_{n+1}$.

When considering symmetric stationary solutions the following conditions are to be satisfied: $t_{n+1} = t_n + \frac{T}{2}$, $W_{n+1} = W_n = W$, $\xi = 1$ and $\Pi(W) = T$,

$$2d' F_0 \sin v_0 t_{n+1} - \delta_d WT = 0 \tag{10.83}$$

(this is in the case of small amplitudes).

It follows From Eq.(10.83) that there is a limitation „from below" caused by the value of the amplitude F_0 necessary for the excitation of stationary oscillations, i.e. there exists a threshold amplitude value $|F_0| \geq \frac{\delta_d WT}{2d'}$.

The oscillation amplitudes $R = \sqrt{2W}$ do not depend on the value of F_0. That creates conditions for system self-adaptivity within a wide range of values of the amplitude of the external exciting force.

For the purpose of solution stability examination, we proceed to the variational system of equations as follows:

$$\delta t_{n+1} = \delta t_n + \frac{1}{2}\frac{d\Pi}{dW}\delta W_n, \quad \delta W_{n+1} = H\delta t_{n+1} + (1-\delta_d T)\delta W_n,$$

where $H = 2d' vF_0 \cos vt$.

Let $\delta t_{n+1} = \lambda \delta t_n$ and $\delta W_{n+1} = \lambda \delta W_n$. The stability condition $|\lambda| < 1$ can be verified by using the characteristic equation

$$\lambda^2 - (2 - \delta_d T + \frac{1}{2}H\frac{d\Pi}{dW})\lambda + 1 - \delta_d T = 0,$$

whose solution is

$$\lambda_{1,2} = 1 - \frac{\delta_d T}{2} + B \pm \sqrt{\left(1 - \frac{\delta_d T}{2} + B\right)^2 - 1 + \delta_d T}, \tag{10.84}$$

where $B = \frac{1}{2} d' v \frac{d\Pi}{dW} F_0 \cos vt_{n+1}$.

We consider two cases.

Case 1. The quantity under the square root mark in (10.84) has a value greater than or equal to zero, i.e. $\lambda_{1,2}$ are real quantities. Taking into account that $|\delta_d T| \ll 1$, we can formulate two sub-cases.

Sub-case 1.1. At $B \geq 0$, $\lambda_{1,2} \simeq 1 - \frac{\delta_d T}{2} + B \pm \sqrt{2|B|}$. The inequality $|B| > \frac{\delta_d T}{2}$ is valid and the stability condition $\lambda_{1,2} < 1$ is not satisfied.

Sub-case 1.2. For the quantity under the square root mark in (10.84) to be greater than zero, the following condition should be satisfied: $B < \delta_d T - 2 < 0$. Then $\lambda_{1,2} \simeq 1 - \frac{\delta_d T}{2} + B \pm \sqrt{|B|(|B| - 2)}$, where $B < -2$. However under the condition $B < -2$ it follows that one of the roots e.g. $\lambda_1 < -1$, so the solution does not satisfy the stability conditions.

Case 2. Let the quantity under the square root mark in (10.84) be less than zero, i.e. $B(2 - \delta_d T + B) < 0$. Since $|\delta_d T| \ll 1$ we can regard that $-2 < B < 0$. Under these conditions the two roots of the characteristic equation are complex conjugated ones, $\lambda_{1,2} = \rho e^{\pm i\alpha}$, and $\rho^2 = \lambda_1 \lambda_2 = 1 - \delta_d T$, i.e. $\rho < 1$ and $|\lambda_{1,2}| < 1$. It is seen that the stability condition is satisfied.

Taking into account that $|\delta_d T| \ll 1$, it follows from the condition (10.83) that at greater values of F_0, $B \simeq \frac{1}{2} d' v |F_0| \frac{d\Pi}{dW}$, i.e. there is a limitation „from above" also with respect to the value of the amplitude F_0 of the external exciting force.

Finally, the general conditions for stability of systems excited oscillations can be written down as follows:

$$-2 < B < 0 \quad \text{and} \quad \frac{8}{v\left|\frac{d\Pi}{dW}\right|} \geq 2d'|F_0| \geq \delta_d WT.$$

A distinctive result is that there exist limitations „from below" as well as „from above" with respect to the value of the amplitude F_0 of the external exciting harmonic force.

10.3.5. A proof of the existence of a modulation – parametric channel for energy input in the oscillation process

The analysis below has been made in support of the thesis formulated under Chapter 1 of this book regarding the existence of a parametric channel for energy transformation and input as a consequence of the reversibility of the modulation-parametric interactions whenever signals are manipulated in resonance systems.

Once again we write down the equation of the nonlinear oscillator under the non-homogeneous action of an external periodic force in the following form:

$$\ddot{x} + 2\delta_d \dot{x} + \omega_0^2 \sin x = \varepsilon(x) F_0 \sin \nu t \tag{10.85}$$

(the denotations are the same as introduced in the previous section).

In accordance with the analytical technique developed in Section 1.4, the solution of (10.85) is presented in the form:

$$x = X_0 + \sum_n X_n, \tag{10.86}$$

where X_0 corresponds to the basic oscillations in the system, and $\sum_n X_n$ is the sum of the combined components generated under the non-homogeneous action of the external periodic force.

As (10.86) is substituted in (10.85) and the considerably lower intensity of the combined components $\sum_n x_n$ in comparison with the basic oscillations X_0 is taken into account, the following are written down:

an equation regarding the basic oscillations:

$$\begin{aligned} &\ddot{X}_0 + 2\delta_d \dot{X}_0 + \omega_0^2 \sin X_0 + \left[\omega_0^2 \cos X_0 \sum_n X_n\right]_{X_0} \\ &= [\varepsilon(X_0) F_0 \sin \nu t]_{X_0} + \left[\frac{d\varepsilon(X_0)}{dx} \sum_n X_n F_0 \sin \nu t\right]_{X_0} ; \end{aligned} \tag{10.87}$$

an equation regarding the l-th combined component:

$$\ddot{X}_l + 2\delta_d \dot{X}_l + \left[\omega_0^2 \cos X_0 \sum_n X_n\right]_l = [\varepsilon(X_0) F_0 \sin \nu t]_l + \left[\frac{d\varepsilon(X_0)}{dx} \sum_n X_n F_0 \sin \nu t\right]_l . \tag{10.88}$$

The indices X_0 and l in (10.87) and (10.88) indicate that only the addents respectively to the frequency of the basic oscillations X_0 or the relevant combined frequency are selected from the respective terms of the equations.

We set

$$X_0 = a\cos(\omega t + \varphi) = a\cos\theta, \tag{10.89}$$

$$\sum_n X_n = \sum_{n=-\infty}^{\infty} A_n \cos[(v + n\omega)t + \varphi_n], \tag{10.90}$$

where a, ω and φ are the amplitude, frequency and phase of the basic oscillations; A_n, $v + n\omega$ and φ_n are the amplitude, frequency and phase of the n-th combined component. The frequency ω is close to the resonance frequency ω_0.

Given the condition of (10.89), we use the following harmonic expansions of the functions:

$$\varepsilon(X_0) = \varepsilon(a\cos\theta) = a_0 + 2\sum_{n=1}^{\infty} a_{2n}\cos 2n\theta, \tag{10.91}$$

$$\frac{d\varepsilon(X_0)}{dx} = \frac{d\varepsilon(a\cos\theta)}{dx} = 2\sum_{m=1}^{\infty} b_{2m}\sin 2m\theta. \tag{10.92}$$

The following formulae for the Fourier coefficients are obtained for the function $\varepsilon(x)$ of the type $\varepsilon(x) = \begin{cases} 1, & |x| \le d' \\ 0, & |x| > d' \end{cases}$ in (10.91) and (10.92):

$$a_0 = \frac{3\theta_0}{\pi}, \quad a_{2n} = \frac{1}{\pi n}\sin 2n\theta_0, \quad b_{2m} = \frac{2}{\pi a}\frac{\sin 2m\theta_0}{\sin\theta_0}, \quad \theta_0 = \arccos\frac{d'}{a}. \tag{10.93}$$

As (10.89), (10.90), (10.91) and (10.92) are substituted in (10.87), shortened equations are obtained for establishing the amplitude a and the phase φ of the basic oscillations in the system while accounting for an infinite spectrum of combined frequencies:

$$\dot{a} = -\delta_d a + \frac{G_1}{2\omega} - \frac{F_0}{2\omega}(a_{N-1} - a_{N+1})\cos N\varphi, \tag{10.94}$$

$$\dot{\varphi} = -\frac{\omega^2 - \tilde{\omega}_0^2}{2\omega} + \frac{G_2}{2\omega a} + \frac{F_0}{2\omega a}(a_{N-1} + a_{N+1})\sin N\varphi, \tag{10.95}$$

where

$$\begin{Bmatrix} G_1 \\ G_2 \end{Bmatrix} = \omega_0^2\frac{J_0(a)}{a}\left[A_{1-N}\begin{Bmatrix}\sin\\ \cos\end{Bmatrix}(\varphi - \varphi_{1-N}) + A_{-1-N}\begin{Bmatrix}\sin\\ \cos\end{Bmatrix}(\varphi - \varphi_{-1-N})\right]$$
$$+\,\omega_0^2\sum_{k=1}^{\infty}(-1)^k J_{2k}(a)\left[\sum_{n=\pm1-2k-N}^{*} A_n\begin{Bmatrix}\sin\\ \cos\end{Bmatrix}(\varphi \mp 2k\varphi \mp \varphi_n)\right.$$
$$\left. +\sum_{n=\mp1+2k-N}^{*} A_n\begin{Bmatrix}\sin\\ \cos\end{Bmatrix}(\varphi \mp 2k\varphi \pm \varphi_n)\right]$$

$$
-F_0 \sum_{m=1}^{\infty} b_{2m} \left[- \sum_{n=\mp 1+2m}^{*} A_n \left\{ \begin{matrix} \sin \\ \cos \end{matrix} (\varphi \mp 2m\varphi \pm \varphi_n) \right\} \right.
$$

$$
+ \sum_{n=\pm 1-2m}^{*} A_n \left\{ \begin{matrix} \sin \\ \cos \end{matrix} (\varphi \mp 2m\varphi \mp \varphi_n) \right\} + \sum_{n=\pm 1+2m-2N}^{*} A_n \left\{ \begin{matrix} \sin \\ \cos \end{matrix} (\varphi \pm 2mk\varphi \mp \varphi_n) \right\}
$$

$$
\left. - \sum_{n=\pm 1-2m-2N}^{*} A_n \left\{ \begin{matrix} \sin \\ \cos \end{matrix} (\varphi \mp 2m\varphi \mp \varphi_n) \right\} \right], \tag{10.96}
$$

J is a Bessel function of the first order, $\tilde{\omega}_0^2 = \omega_0^2 \dfrac{2J_1(a)}{a}$ is the change in the resonance frequency of the non-isochronous oscillations with the change in their amplitude, $N = \dfrac{v}{\omega}$ is the ratio of the frequency division in the system, $\sum^*$ means that every term in the sum consists of two addents, respectively for the upper and lower sign of the index n.

In the regime of stationary oscillations $\dot{a} = 0$ and $\dot{\varphi} = 0$. After denoting $\delta_0 = 2\delta_d \dfrac{\omega}{\omega_0}$, $\delta = \dfrac{G_1}{\omega_0^2 a}$, $\xi = \dfrac{\omega^2 - \tilde{\omega}_0^2}{\omega_0^2}$, $\eta = \dfrac{G_2}{\omega_0^2 a}$, the following expressions are obtained for the amplitude and the phase of the stationary oscillations in the system from (10.94) and (10.95):

$$
a = \frac{F_0}{\omega_0} \sqrt{\left(\frac{\delta_0 - \delta}{a_{N-1} - a_{N+1}} \right)^2 + \left(\frac{\xi - \eta}{a_{N-1} + a_{N+1}} \right)^2}, \tag{10.97}
$$

$$
\varphi = \frac{1}{N} \operatorname{arctg} \left(\frac{\xi - \eta}{\delta_0 - \delta} \frac{a_{N-1} - a_{N+1}}{a_{N-1} + a_{N+1}} \right). \tag{10.98}
$$

(10.97) and (10.98) reveal the role of the combined components (expressed through $\dfrac{G_1}{2\omega}$ and $\dfrac{G_2}{2\omega a}$) in (10.94) and (10.95)) in the formation of the respective stationary amplitude and phase.

A complex frequency spectrum conditioned by the highly non-homogeneous action of the external force is generated in the system. As a result of the reversibility of modulation-parametric interactions, the combined components convey energy both to the basic oscillations and to the oscillations in each of the other combined components of the spectrum. The maintenance of stationary oscillations in the system is an integral effect of the summary action of an infinite spectrum of combined components, given an indirect involvement of such a spectrum. Since the excitation of oscillations is of a synchronous nature, as a result of the action of an external high-frequency source, the system under consideration stands out with its strong phase selectivity. Like in the case of cyclic accelerators, where only „balance" particles are phase selected and accelerated, of significant importance here are those combined components X_n, whose phase φ_n

has the most relevant value from the perspective of the optimum energy input in the area of action of the external high-frequency source. This is the essence of the adaptivity of the system and of the self-tuning to a stable stationary mode. To put it differently, the summary action of the combined frequencies contributing in the case of quasi-stationary frequency of the oscillations in the system, phase-tunes these oscillations to the most favourable phase (from an energy point of view) for establishing a stationary mode. This transition process is analyzed by writing down and solving shortened equations analogous to (10.94) and (10.95) for a certain volume of combined components (10.88). The establishment of the stationary amplitude a and the phase φ of the oscillations in the system is of an oscillating, quickly damping nature. One can also observe processes, where the amplitude and the phase, already in a stationary mode, continue „to wander" in a certain range of values around the respective resonance values a and φ. These are quite specific modes with underdeveloped chaotic oscillations. There is a clash of two powerful processes: the process of stochastization of the oscillations in a highly nonlinear system under an external non-homogeneous action and the process of strong adaptive stabilization and attraction to the stable stationary resonance modes. As a whole this is an intricate synergetic formative process of a complex dynamic mode.

It is interesting to point out that there exist l-th combined frequencies satisfying the condition:

$$v + l\omega = \pm\omega, \tag{10.99}$$

which fall in the frequency band of transmission of the oscillating unit in the system. They act „directly" upon the basic oscillations in the system, this without precluding the parametric action caused by the reversibility of the modulation-parametric interactions. Indeed, if out of the entire spectrum of combined frequencies one takes into account only those l-th components, which satisfy condition (10.99), i.e. $l = \pm 1 - N$, the infinite sums in expression (10.96) disappear and it assumes the following form:

$$\begin{aligned}
\begin{Bmatrix} G_1 \\ G_2 \end{Bmatrix} &= \omega_0^2 \left\{ [J_0(a) \pm J_2(a)] A_{1-N} \begin{Bmatrix} \sin \\ \cos \end{Bmatrix} (\varphi - \varphi_{1-N}) \right. \\
&\quad \left. + [J_0(a) \pm J_2(a)] A_{-1-N} \begin{Bmatrix} \sin \\ \cos \end{Bmatrix} (\varphi + \varphi_{-1-N}) \right\} \\
&\quad - \frac{F_0}{2} \left\{ (b_N - b_{N+2}) \left[A_{1-N} \begin{Bmatrix} \sin \\ \cos \end{Bmatrix} [(N+1)\varphi - \varphi_{1-N}] \right. \right. \\
&\quad \left. + A_{-1-N} \begin{Bmatrix} \sin \\ \cos \end{Bmatrix} [(N+1)\varphi + \varphi_{-1-N}] \right] \\
&\quad + (b_{N-2} - b_N) \left[A_{1-N} \begin{Bmatrix} \sin \\ \cos \end{Bmatrix} [(N-1)\varphi + \varphi_{1-N}] \right. \\
&\quad \left. \left. + A_{-1-N} \begin{Bmatrix} \sin \\ \cos \end{Bmatrix} [(N-1)\varphi - \varphi_{-1-N}] \right] \right\}.
\end{aligned} \tag{10.100}$$

In order to determine the stationary amplitudes and phases of the two reported combined components $X_{1-N} = A_{1-N}\cos(\omega t+\varphi_{1-N})$ and $X_{-1-N} = A_{-1-N}\cos(-\omega t+\varphi_{-1-N})$, we write down Eq.(10.88) in a complex matrix form by using the analytical approach developed under section 1.4:

$$\begin{vmatrix} -\omega^2 & 0 \\ 0 & -\omega^2 \end{vmatrix} \begin{vmatrix} X^C_{1-N} \\ X^C_{-1-N} \end{vmatrix} + \begin{vmatrix} j\delta_d & 0 \\ 0 & -j\delta_d\omega \end{vmatrix} \begin{vmatrix} X^C_{1-N} \\ X^C_{-1-N} \end{vmatrix}$$

$$+ \begin{vmatrix} \omega_0^2 J_0(a) & \omega_0^2 J_2(a)e^{-j2\Psi} \\ -\omega_0^2 J_2(a)e^{j2\Psi} & \omega_0^2 J_0(a) \end{vmatrix} \begin{vmatrix} X^C_{1-N} \\ X^C_{-1-N} \end{vmatrix}$$

$$-\frac{F_0}{2} \begin{vmatrix} b_N(e^{jN\varphi}+e^{-jN\varphi}) & b_{N-2}e^{j(N-2)\varphi}+b_{N+2}e^{-j(N+2)\varphi} \\ b_{N+2}e^{j(N+2)\varphi}+b_{N-2}e^{-j(N-2)\varphi} & b_N(e^{jN\varphi}+e^{-jN\varphi}) \end{vmatrix} \begin{vmatrix} X^C_{1-N} \\ X^C_{-1-N} \end{vmatrix}$$

$$= \begin{vmatrix} \omega_0^2 J_1(a)e^{j\varphi} + j\dfrac{F_0}{2}[a_{N+1}e^{j(N+1)\varphi} - a_{N-1}e^{-j(N-1)\varphi}] \\ \omega_0^2 J_1(a)e^{-j\varphi} + j\dfrac{F_0}{2}[a_{N-1}e^{j(N-1)\varphi} - a_{N+1}e^{-j(N+1)\varphi}] \end{vmatrix}, \tag{10.101}$$

where X^C_{1-N} and X^C_{-1-N} are complex quantities.

As parameters $A_{\pm 1-N}$ and $\varphi_{\pm 1-N}$ derived from (10.101) are substituted in (10.100), (10.97) and (10.98) can be used to obtain the possible discrete spectrum of stationary amplitudes and phases of the oscillations in the system.

The analysis shows that besides the basic modulation channel of the „argument" action (See (10.17) and (10.18)), there is also a modulation-parametric channel for energy input in the system. That is why we believe that this type of systems, phenomena, processes and mechanisms make up a separate subclass of dynamic systems: modulation nonlinear-parametric systems with adaptive phase self-tuning. The quantitative analysis conducted on the basis of Eqs.(10.93) – (10.98) indicates the following: The existence of a modulation-parametric channel for energy input makes the mechanism of phase self-tuning underlying the energy exchange and the maintenance of continuous oscillations in the system more flexible and effective. Moreover, on the one hand, there is substantial specification of the discrete order of possible stable amplitudes, and, on the other hand, the oscillations in the system exhibit greater independence from the random changes in the external factors, in the qualitative factor of the oscillating unit and in the amplitude of the acting force in a wide range.

10.3.6. Excitation of continuous oscillations with a discrete set of stable amplitudes in a pseudo-linear oscillating system

The investigation is carried out by using an equation, describing a linear oscillating system under inhomogeneous action, as [325]

$$\ddot{X} + 2\delta_d \dot{X} + \omega_0^2 X = \omega_0^2 F_0 f(\alpha X) \cos vt, \tag{10.102}$$

where X is a generalized dimensionless coordinate; F_0 and v are the external force amplitude and frequency respectively; ω_0 is a system natural frequency; t is real time and $f(\alpha X)$ is a nonlinear function (to be concretized henceforth), α – constant.

We substitute the independent variable $\tau = vt$ and denote $\tilde{\beta} = \dfrac{\delta_d}{v}$, $\tilde{\omega}_0 = \dfrac{\omega_0}{v}$, $\tilde{F} = F_0 \tilde{\omega}_0$. Also, if $\dot{X}_\tau = dX/d\tau$ and $\ddot{X}_\tau = d^2X/d\tau^2$, the following is obtained from (10.102):

$$\ddot{X}_\tau + 2\tilde{\beta}\dot{X}_\tau + \tilde{\omega}_0^2 X = \tilde{\omega}_0^2 \tilde{F} f(\alpha X) \cos \tau. \tag{10.103}$$

The variables in (10.103) are substituted, as follows:

$$X = A\cos(\tilde{\omega}_0 \tau + \Psi), \quad \dot{X}_\tau = -A\tilde{\omega}_0 \sin(\tilde{\omega}_0 \tau + \Psi). \tag{10.104}$$

Denoting $\varphi = \tilde{\omega}_0 \tau + \psi$ and $\tilde{A} = A/\tilde{F}$, we obtain from (10.103), with consideration of (10.104), the following:

$$\left| \begin{array}{l} \dot{\tilde{A}}_\tau \cos\varphi - \tilde{A}\dot{\psi}_\tau \sin\varphi = 0 \\ \tilde{\omega}_0(\dot{\tilde{A}}_\tau \sin\varphi + \tilde{A}\dot{\psi}_\tau \cos\varphi) = -2\tilde{\beta}\tilde{\omega}_0 \tilde{A} \sin\varphi - \tilde{\omega}_0 f(\alpha X) \cos\tau \end{array} \right. . \tag{10.105}$$

Using (10.105) we determine:

$$\left| \begin{array}{l} \dot{\tilde{A}}_\tau = -\tilde{\beta}\tilde{A} - \tilde{\beta}\tilde{A}\cos 2\varphi - f(\alpha X)\cos\tau \sin\varphi \\ \tilde{A}\dot{\psi}_\tau = -\tilde{\beta}\tilde{A}\sin 2\varphi - f(\alpha X)\cos\tau\cos\varphi \end{array} \right. . \tag{10.106}$$

Let us assume that the nonlinear function $f(\alpha X) = f(\alpha\tilde{A}\cos\varphi)$ is an even function: i.e. it can be factorized to Fourier expansion as:

$$f(\alpha X) = C_0(\varkappa) + \sum_{k=1}^{\infty} C_k(\varkappa)[e^{2jk\varphi} + e^{-2jk\varphi}],$$

where $\varkappa = \alpha\tilde{A}$, $\alpha \leq 1$.

The following expansions will be useful for the solution of system (10.106):

$$f(\alpha X)e^{j\varphi} = \sum_{k=0}^{\infty} [C_k(\varkappa)e^{j(2k+1)\varphi} + C_{k+1}(\varkappa)e^{-j(2k+1)\varphi}], \tag{10.107}$$

$$I = f(\alpha X)e^{j\varphi}\cos\tau = \frac{1}{2}\left\{\sum_{k=0}^{\infty} C_k(\varkappa)e^{j[\tau+(2k+1)\varphi]} + C_{k+1}(\varkappa)e^{j[\tau-(2k+1)\varphi]}\right.$$

$$\left. + \sum_{k=0}^{\infty} C_k(\varkappa)e^{-j[\tau-(2k+1)\varphi]} + C_{k+1}(\varkappa)e^{-j[\tau+(2k+1)\varphi]}\right\}. \tag{10.108}$$

Note that $\tau - (2k+1)\varphi = \tau[1-(2k+1)\tilde{\omega}_0] - (2k+1)\psi = \tau\overline{\delta}_k - (2k+1)\psi$, where $\overline{\delta}_k = 1-(2k+1)\tilde{\omega}_0$. Let's determine the value of k_0, at which $\overline{\delta}_k$ reaches its minimum. If $\tilde{\omega}_0 = 1/(2k_0+\Delta)$, where $1 \geq \Delta \geq 0$, then:

$$\overline{\delta}_k = \frac{2(k_0-k)+(\Delta-1)}{2k_0+\Delta}. \tag{10.109}$$

It can be seen from (10.109) that at $\Delta > 0$ the function $\overline{\delta}_k$ is minimized for $k = k_0$, whereas at $\Delta = 0$ $\overline{\delta}_k$ reaches its minimum for two values of k: $k = k_0$ and $k = k_0 + 1$.

Let's discuss the first case, when $\Delta > 0$. We designate $\phi = (2k+1)\psi - \tau\overline{\delta}_{k_0}$. After separating only the „slow" terms from (10.108), we obtain:

$$I_1 = \frac{1}{2}[C_k(\varkappa)e^{j\phi} + C_{k_0+1}(\varkappa)e^{-j\phi}]. \tag{10.110}$$

The following reduced (shortened) differential equations describing system (10.102) processes are obtained from (10.106):

$$\left|\begin{aligned} &\dot{\tilde{A}}_\tau = -\tilde{\beta}\tilde{A} - \mathrm{Im}[\overline{f(\alpha X)\cos\tau e^{j\phi}}] \\ &\tilde{A}\left(\frac{\dot{\phi}_\tau + \overline{\delta}_{k0}}{2k_0+1}\right) = -\mathrm{Re}[\overline{f(\alpha X)\cos\tau e^{j\varphi}}] \end{aligned}\right. . \tag{10.111}$$

where the line above denotes average.

Considering (10.110), system (10.111) is presented as:

$$\left|\begin{aligned} &\dot{\tilde{A}}_\tau = -\tilde{\beta}\tilde{A} - \frac{1}{2}[C_{k_0}(\varkappa) - C_{k_0+1}(\varkappa)]\sin\phi \\ &\tilde{A}\dot{\phi}_\tau = -\overline{\delta}_{k0}\tilde{A} - \frac{1}{2}[C_{k_0}(\varkappa) - C_{k_0+1}(\varkappa)]\cos\phi \end{aligned}\right. . \tag{10.112}$$

If we set $P_{k_0}(\varkappa) = \frac{1}{2}[C_{k_0}(\varkappa) - C_{k_0+1}(\varkappa)]$ and $R_{k_0}(\varkappa) = \frac{1}{2}[C_{k_0}(\varkappa) + C_{k_0+1}(\varkappa)]$, (10.112) is formally presented as:

$$\dot{\tilde{A}}_\tau = -\tilde{\beta}\tilde{A} - P_{k_0}(\alpha)\sin\varphi, \tag{10.113a}$$

$$\tilde{A}\dot{\phi}_\tau = -\overline{\delta}_{k_0}\tilde{A} - R_{k0}(\alpha)\cos\phi. \tag{10.113b}$$

Let us consider in details the most interesting case, when $\overline{\delta}_{k_0} = 0$: i.e., $\Delta = 1$. In a physical context it means that the impact force frequency is exactly equal to the odd harmonic of the resonance frequency of the excited oscillating system, i.e., the transformation products of the impact frequency fall exactly on a central tuning frequency of the oscillating system. If the above conditions are complied with, system (10.113) becomes

$$\dot{\tilde{A}}_\tau = -\tilde{\beta}\tilde{A} - P_{k_0}(\varkappa)\sin\varphi, \tag{10.114a}$$

$$\tilde{A}\dot{\phi}_\tau = -R_{k_0}(\varkappa)\cos\phi. \tag{10.114b}$$

The condition for the established (stationary) mode is $\tilde{A} = \text{const}$, $\varkappa = \alpha\tilde{A} = \text{const}$, and $\dot{\phi}_\tau = 0$, i.e.

$$R_{k_0}(\alpha)\cos\phi = 0. \tag{10.115}$$

It follows from (10.114b) and (10.115) that the stable equilibrium state of variable $\phi(\tau)$ at $\tilde{A} = \text{const}$ is determined as

1. $\phi(\tau) = \dfrac{\pi}{2}$, if $R_{k_0}(\varkappa) < 0$,
2. $\phi(\tau) = -\dfrac{\pi}{2}$, if $R_{k_0}(\varkappa) > 0$.

If we consider condition

$$\sin\phi = -\operatorname{sign} R_{k_0}(\varkappa), \tag{10.116}$$

Eq.(10.114a) assumes the form

$$\dot{\tilde{A}}_\tau = -\tilde{\beta}\tilde{A} + P_{k_0}(\varkappa)\operatorname{sign} R_{k_0}(\varkappa) = \Phi(\tilde{A}). \tag{10.117}$$

Equation (10.117) allows us: (i) to examine the transition processes of establishing the oscillations amplitude in the discussed system; (ii) to determine the stable and unstable meanings of $\tilde{A}$ in the system.

In the case when $\overline{\delta}_k \neq 0$, the following can be obtained from (10.113) by analogy with (10.117):

$$\cos\phi = -\frac{\overline{\delta}_{k_0}\tilde{A}}{R_{k_0}(\varkappa)}, \tag{10.118a}$$

$$\sin\phi = -[\operatorname{sign} R_{k_0}(\varkappa)]\sqrt{1 - \left(\frac{\overline{\delta}_{k_0}\tilde{A}}{R_{k_0}(\varkappa)}\right)^2}. \tag{10.118b}$$

Equations (10.118) are exact for $\tilde{A}$ values, complying with condition $|\bar{\delta}_{k_0}\tilde{A}| < |R_{k_0}(\varkappa)|$. Hence, the maximum possible meaning of $\tilde{A}$ can be obtained.

From (10.118) and (10.113a) it follows that

$$\dot{\tilde{A}}_\tau = -\tilde{\beta}\tilde{A} + P_{k0}(\varkappa)]\sqrt{1 - \left(\frac{\bar{\delta}_{k0}\tilde{A}}{R_{k0}(\varkappa)}\right)^2}\,\operatorname{sign} R_{k0}(\varkappa). \tag{10.119}$$

Case $\Delta = 0$ is not particularly interesting and is omitted here.

Considering three forms of function $f(\alpha X)$, let us find analytical expressions for functions $R_{k_0}(\varkappa)$ and $P_{k_0}(\varkappa)$.

$$1. f(\alpha X) = e^{-\alpha X^2} = e^{-\varkappa^2/2 - \varkappa^2/2}\cos 2\varphi = e^{-\varkappa^2/2}\left[I_0\left(\frac{\varkappa^2}{2}\right) + 2\sum_{k=1}^{\infty}\left(\frac{\varkappa^2}{2}\right)\cos 2k\varphi\right],$$

where $I_k(.)$ is a modified Bessel function of the k-th order.

Apparently, in that case

$$C_0(\varkappa) = e^{-\varkappa^2/2} I_0\left(\frac{\varkappa^2}{2}\right),$$

and

$$C_k(\varkappa) = e^{-\varkappa^2/2} I_k\left(\frac{\varkappa^2}{2}\right),$$

and

$$P_{k_0}(\varkappa) = \frac{1}{2}e^{-\varkappa^2/2}\left[I_{k_0}\left(\frac{\varkappa^2}{2}\right) - I_{k_0+1}\left(\frac{\varkappa^2}{2}\right)\right] > 0,$$

$$R_{k_0}(\varkappa) = \frac{1}{2}e^{-\varkappa^2/2}\left[I_{k_0}\left(\frac{\varkappa^2}{2}\right) - I_{k_0+1}\left(\frac{\varkappa^2}{2}\right)\right] > 0.$$

Besides, in the $\tilde{A} \leq \tilde{A}_{\max}$ values area, where $\bar{\delta}_{k_0}\tilde{A} < R_{k_0}(\varkappa)$, it is possible to establish the oscillations with definite amplitude. The transitional process of amplitude establishing is described by Eq.(10.119), in our case in the form:

$$\dot{\tilde{A}}_\tau = -\tilde{\beta}\tilde{A} + P_{k_0}(\varkappa)\sqrt{1 - \left[\frac{\bar{\delta}_{k_0}\tilde{A}}{R_{k_0}(\varkappa)}\right]^{-2}} = \Phi(\tilde{A}). \tag{10.120}$$

If $\bar{\delta}_{k_0} = 0$ at $k_0 \gg 1$ the system phase portrait, described by Eq.(10.120), is in the form demonstrated in Fig.10.20. It can be seen that three equilibrium states are presented in the system, where $\tilde{A}_1$ is unstable, whereas $\tilde{A}_0$ and $\tilde{A}_2$ are stable. However, $\tilde{A}_1 < \tilde{A}_2$; i.e., the discussed system is characterized by stiff excitation.

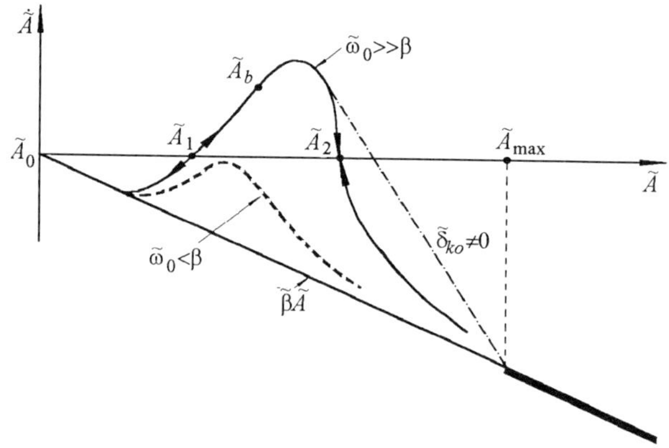

Fig. 10.20. Phase portrait of a resonance system under the action of an external non-linear force (Eq.(10.120))

It is necessary to bring the system out of equilibrium state, when the given initial amplitude $\tilde{A}_b > \tilde{A}_1$ is reached. Further, the energy insertion mechanism from an external source is activated and stationary oscillations are set.

The transition period duration is determined by using the formula

$$T = \int_{\tilde{A}_b}^{\gamma \tilde{A}^2} \frac{dA}{\Phi(\tilde{A})},$$

where $\gamma = 1 - \varepsilon$, $\varepsilon \ll 1$ is a selected number, determining the degree of proximity of $\tilde{A}$ to the stationary value $\tilde{A}_2$.

$$2. f(\alpha X) = \cos \alpha X = \cos(\varkappa \cos \varphi) = I_0(\varkappa) + 2 \sum_{k=1}^{\infty} (-1)^k I_{2k}(\varkappa) \cos 2k\varphi.$$

In the case $C_k(\varkappa) = (-1)^k I_{2k}(\varkappa)$ and

$$P_{k_0}(\varkappa) = \frac{(-1)^{k_0}}{2} [I_{2k_0}(\varkappa) - I_{2k_0+2}(\varkappa)] = (-1)^{k_0} I'_{2k_0+1}(\varkappa),$$

$$R_{k_0}(\varkappa) = \frac{(-1)^{k_0}}{2} [I_{2k_0}(\varkappa) + I_{2k_0+1}(\varkappa)] = (-1)^{k_0} (2k_0 + 1) \frac{I_{2k_0+1}(\varkappa)}{\alpha}.$$

Here $I_k(.)$ are Bessel functions of the first kind and k-th order, $I'_k(.)$ is the Bessel function derivative.

At $\overline{\delta}_{k_0} = 0$, from (10.119) it follows that

$$\dot{\tilde{A}}_\tau = -\tilde{\beta}\tilde{A} + (-1)^{k_0} I'_{2k_0+1}(\varkappa) \operatorname{sign}\{(-1)^{k_0} I'_{2k_0+1}(\varkappa)\}. \tag{10.121}$$

The phase portrait character, corresponding to Eq.(10.121), is shown on Fig.10.21. The equilibrium amplitudes $\tilde{A}_1$, $\tilde{A}_3$, $\tilde{A}_5$, ... are unstable, whereas $\tilde{A}_2$, $\tilde{A}_4$, $\tilde{A}_6$, ... are stable. For small $\tilde{A}$ values, the equilibrium state is determined by equation

$$I'_{2k_0+1}(\varkappa) = (-1)^{k_0}\tilde{\beta}\tilde{A} = 0,$$

i.e. $A_k = \varkappa_k/\alpha$, where $\varkappa_k$ are equation $I'_{2k_0+1}(\varkappa) = 0$ roots, independent of $\tilde{F}$.

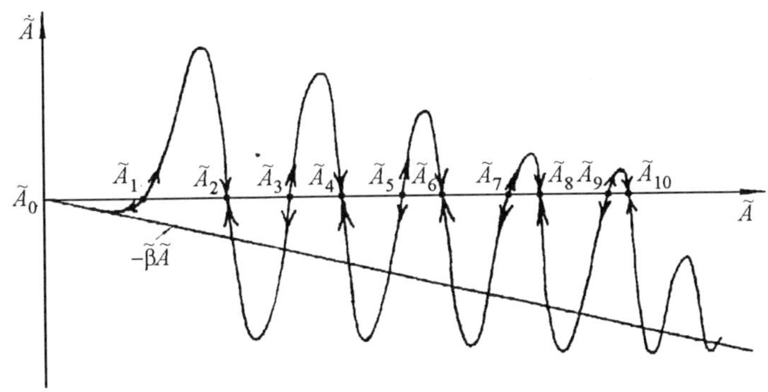

Fig.10.21. Phase portrait of a resonance system under the action of an external non-linear force with periodical non-linearity (Eq.(10.121))

3. $f(\alpha X) = \cos 2k\varphi$.

On condition that $\Delta = 1$, the following are obtained from (10.117): $P_{k_0}(\varkappa) = \frac{1}{2}$, $R_{k_0}(\varkappa) = \frac{1}{2}$ and

$$\dot{\tilde{A}}_\tau = -\tilde{\beta}\tilde{A} - \frac{1}{2}\sin\phi,$$

$$\tilde{A}\dot{\phi}_\tau = -\frac{1}{2}\cos\phi.$$

It follows that

$$\dot{\tilde{A}}_\tau = -\tilde{\beta}\tilde{A} + \frac{1}{2} = \Phi(\tilde{A}). \tag{10.122}$$

The phase portrait, corresponding to Eq.(10.122) is shown in Fig.10.22. The stationary amplitude is $\tilde{A}_{st} = 1/2\tilde{\beta}$, and the transition period duration is determined as

$$T = \int_0^{\gamma\tilde{A}_{st}} \frac{d\tilde{A}}{\Phi(A)} = \frac{1}{\tilde{\beta}}\ln\left(\frac{1}{1-\gamma}\right).$$

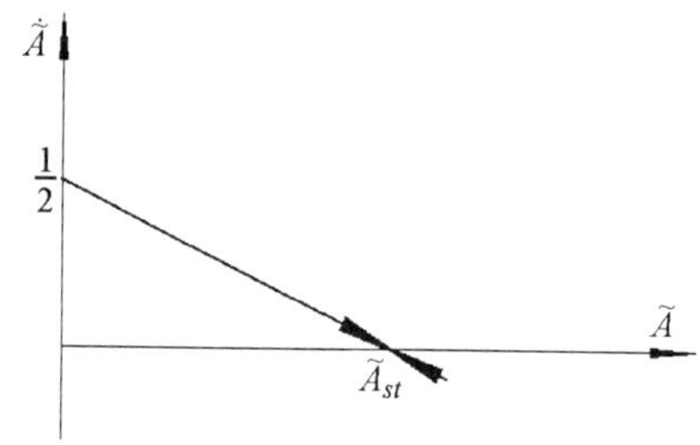

Fig.10.22. Phase portrait of a resonance system, in which a regenerative frequency division of external influence is accomplished (Eq.(10.122))

The above analysis has conclusively demonstrated the efficiency of the invented mechanism for oscillations excitation in resonance systems under the influence of external high-frequency periodic forces that are nonlinear *vis-à-vis* the coordinates.

It should be noted that the relation $v = n\omega_0$ is complied with in all cyclic accelerators; there v is the accelerating high-frequency field frequency, ω_0 is the charged particles rotating frequency, and n is acceleration multiplicity, reaching tens and hundreds. That is why the stationary oscillations discussed above are analogous to the movement of an „equilibrium" particle in a cyclic accelerator. Particles, similar to the equilibrium particle, perform slow phase oscillations in cyclic accelerators. Their analogue in our system is the fluctuating approximation to the stationary values $\tilde{A}_{st}$ and ϕ_{st}. In our problem, the phase oscillations damping is determined by the friction coefficient β, while in charged particles accelerators, damping is the result of radio-emission. Here v, n and ω_0 are constants, however in the cyclic accelerators the process of charged particles acceleration is accompanied by an increase in v (phasotron), n (microtron), ω_0 (synchrotron) or v and ω_0 (synchrophasotron). Injection in acceleration mode (for accelerators) and in stationary oscillations mode (our system) represents a separate problem.

The presented mechanism of continuous oscillations excitation makes it possible to examine, from this perspective, the processes of plasma particles interaction with electromagnetic waves. For example, an equation of the (10.103) form is obtained with a right-hand side $\phi(x,t) = E\cos kx \sin vt$, where $\tilde{\beta} = \omega_i$ is the collision frequency of ion-electron-neutral atoms and $\omega_0 = eB/MC$ is the ion cyclotron frequency, e is electron charge, M is ion mass, C is light velocity, in the case of electromagnetic wave interaction with a particle in cylindric waveguide with longitudinal magnetic field B and E type wave. If, for example, $v = n\omega_0$ UHF oscillation is transformed into low-frequency oscillation ω_0, then the corresponding correlation between E and E_0 (longitudinal electric field) is:

$$E = \frac{e}{M}E_0\frac{\Omega_p}{v} = \frac{e}{M}E_0\frac{C}{V_A},$$

where

$$\Omega_p = 4\pi C^2 \frac{N}{M}$$

is Langmuir plasma frequency,

$$V_a = \frac{B}{\sqrt{4\pi NM}}$$

is Alfven velocity, N is plasma density. The condition for plasma heating is defined as

$$E_0 > \frac{M}{l} \frac{\omega_i \omega_0^2 R_0 n}{\omega_p},$$

where R_0 and l are the waveguide radius and length,

$$n = \frac{v}{\omega_0}, \qquad \omega_p = 4\pi l^2 \frac{N}{M}.$$

The examination of efficiency of energy transformation from the centimetre, IR and optical wavebands in the low-frequency oscillations demonstrates the potential for generation of powerful low-frequency waves in the Solar system near-planet space.

Let us present yet another approach to the analysis of the discrete dynamics of a pseudo-linear oscillating system with a short time interval of interaction with an external energy source.

Let us once again write the equation of a „linear" pendulum in the following form [139]:

$$\ddot{x} + 2\delta_d \dot{x} + \omega_0^2 x = f(x,t) = \nu(d' + x)\nu(d' - x) F_0 \cos vt, \tag{10.123}$$

where $\nu(.)$ is the Heavyside function.

We shall consider such modes of oscillation in the oscillator, where the amplitude a is by far larger than the semi-interval of interaction d'. Given this condition for the interaction interval, the motion in the oscillator can be regarded as uniform motion at the speed of $\pm a\omega$, where ω is the frequency of the excited oscillations. Therefore the force $f(x,t)$ can be regarded as dependent only on the time and amplitude of motion in the oscillator, and it can be set at

$$f(x,t) = F_v(a,t) = \nu(t - t_n + \frac{d'}{a\omega})\nu(t_n + \frac{d'}{a\omega} - t) F_0 \cos vt, \tag{10.124}$$

where t_n are the moments of passage through the position $x = 0$.

The pseudo-linearity of the system is determined by the non-linear dependence $F_v(a,t)$.

The solution of Eq.(10.123), while taking into consideration (10.124), is sought in the following form:

$$x = a(t)\cos[\omega t + \varphi(t)], \tag{10.125}$$

where $a(t)$ and $\varphi(t)$ are the slowly changing amplitude and phase of the oscillations. The role of „slow" time on which a and φ are dependent, can be played by the discrete time

$$t_n = \frac{n\pi - \varphi}{v} + \frac{\pi}{2v}. \tag{10.126}$$

In the case of the first approximation of the averaging method [97] the equations for a and φ have the following form:

$$\left|\begin{array}{l} \dot{a} = -\delta_d a - \dfrac{1}{\omega}\overline{[F_v(a,t)\sin(\omega t + \varphi)]_n} \\ \dot{\varphi} = \Delta - \dfrac{1}{\omega a}\overline{[F_v(a,t)\cos(\omega t + \varphi)]_n} \end{array}\right., \tag{10.127}$$

where $\Delta = \omega_0 - \omega$ is the detuning between the natural frequency of the oscillator and the frequency of its oscillations, the bar above means time averaging, while the index n implies that the averaging is done for the n-th „period" of the oscillations.

After accounting for (10.126), averaging, and taking into consideration the condition $v \gg \omega$, the following is obtained:

$$\left|\begin{array}{l} \dot{a} = -\delta_d a - \dfrac{2F_0}{\pi v}(-1)^n \sin\dfrac{v\pi}{2\omega}\sin\dfrac{vd'}{\omega a}\sin\dfrac{v[(n+1)\pi - \varphi]}{\omega} \\ \dot{\varphi} = \Delta - \dfrac{2F_0 d'}{\pi v a^2}(-1)^n \sin\dfrac{v\pi}{2\omega}\left(\dfrac{\omega a}{vd'}\sin\dfrac{vd'}{\omega a} - \cos\dfrac{vd'}{\omega a}\right) \end{array}\right.. \tag{10.128}$$

For Eqs.(10.128) to have a stationary solution, their right-hand parts should not be dependent on the slow time t_n, i.e. on the number n. This condition corresponds to the ratio $\omega = \omega_m = \dfrac{v}{2m+1}$, where m is an integer corresponding to the minimum value of the detuning $|\Delta|$. Under this condition

$$(-1)^n \sin\frac{v(n\pi - \varphi)}{\omega} = -\sin\frac{v\varphi}{\omega_m}, \quad (-1)^n\cos\frac{v(n\pi - \varphi)}{\omega} = \cos\frac{v\varphi}{\omega_m}.$$

It is convenient to rewrite Eqs.(10.128) in the form:

$$\left|\begin{array}{l} \dot{y}_m = -\delta_d y_m\left(1 + \dfrac{\omega_0}{\omega_m}Q_m \sin\Phi_m \dfrac{\sin y_m^{-1}}{y_m}\right) \\ \dot{\Phi}_m = -\tilde{\Delta} - \dfrac{\omega_0}{\omega_m}\dfrac{\delta Q_m}{y_m^2}\cos\Phi_m(y_m \sin y_m^{-1} - \cos y_m^{-1}) \end{array}\right., \tag{10.129}$$

where $y_m = \frac{\omega_m a}{v d'} = \frac{a}{(2m+1)d'}$, $\Phi_m = \frac{v\varphi}{\omega_m} = (2m+1)\varphi$,

$$Q_m = (-1)^m \frac{2F_0\omega_m^2}{\pi d'\delta_d\omega_0 v^2} = (-1)^m \frac{2F_0}{\pi d'\delta_d\omega_0(2m+1)^2}, \quad \tilde{\Delta} = v - (2m+1)\omega_0.$$

The stationary solution of Eqs.(10.129) is determined from the following system of transcendent equations:

$$\left| \begin{aligned} &\left(\frac{\omega_0^2}{\omega_m^2} Q_m^2 \frac{\sin^2 y_m^{-1}}{y_m^2} - 1\right)\left(\sin y_m^{-1} - \frac{\cos y_m^{-1}}{y_m}\right)^2 = \frac{\tilde{\Delta}^2}{\delta_d^2}\sin^2 y_m^{-1} \\ &\sin\Phi_m = -\frac{\omega_m y_m}{\omega_0 Q_m \sin y_m^{-1}} \\ &\cos\Phi_m = -\tilde{\Delta}\left[\frac{\omega_0}{\omega_m}\frac{\delta_d Q_m}{y_m^2}(y_m \sin y_m^{-1} - \cos y_m^{-1})\right]^{-1} \end{aligned} \right. . \tag{10.130}$$

The stability of the stationary solutions $y_m^{(st)}$ and $\Phi_m^{(st)}$ obtained from (10.130) is investigated by means of linearization of Eqs.(10.129) with respect to small deviations $\xi = y_m - y_m^{(st)}$; $\eta = \Phi_m - \Phi_m^{(st)}$. The linearized equations have the following form:

$$\left| \begin{aligned} &\dot{\xi} = -\delta_d[(1 + \frac{\operatorname{ctg} y_m^{-1}}{y_m})\xi - y_m \operatorname{ctg}\Phi_m \eta] \\ &\dot{\eta} = \frac{\delta_d}{y_m}(1 + \frac{1}{y_m^2} - \frac{\operatorname{ctg} y_m^{-1}}{y_m})\operatorname{ctg}\Phi_m \xi - \delta_d(1 - \frac{\operatorname{ctg} y_m^{-1}}{y_m})\eta \end{aligned} \right. . \tag{10.131}$$

Upon writing the characteristic equation for the system (10.131) the following condition required for the stability of the stationary values of the amplitudes is obtained:

$$1 - \left(\frac{\operatorname{ctg} y_m^{-1}}{y_m}\right)^2 + \left(1 + \frac{1}{y_m^2}\frac{\operatorname{ctg} y_m^{-1}}{y_m}\right)\frac{\tilde{\Delta}^2}{\delta_d^2} y_m^2 \sin^2 y_m^{-1}(y_m \sin y_m^{-1} - \cos y_m^{-1})^{-2} \geq 0. \tag{10.132}$$

The dependence of the stable (continuous lines) and of the unstable (dotted lines) values of the stationary amplitudes on the relative detuning $\frac{\tilde{\Delta}}{\delta_d}$ in the case of $Q_m = 1$ and $\frac{\omega_0}{\delta_d} = 10$ is shown in Fig.10.23.

These dependencies take the form of several leaves. For different values of Q_m these leaves turn out to be embedded in each other so that leaves with a smaller area corresponds to smaller values of Q_m. As the value of Q_m goes down, the upper leaves gradually disappear. The top leaf vanishes when $|Q_m| = \frac{2}{\pi}$.

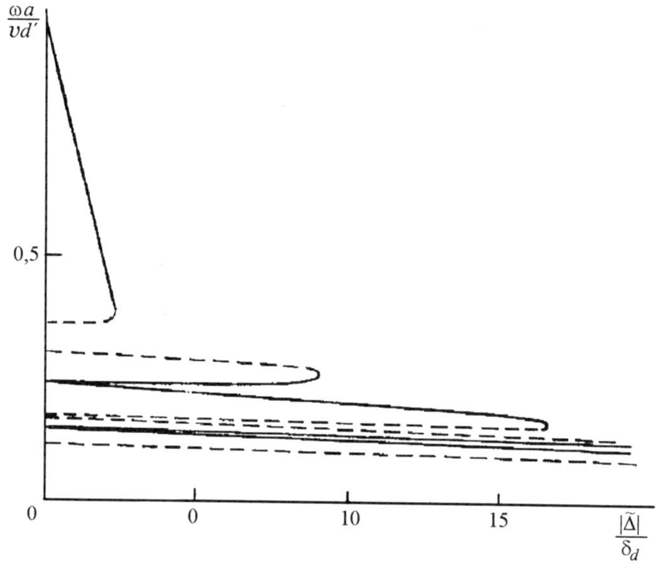

Fig.10.23. Dependence of the stationary amplitudes on the relative asynchronization – stable (continuous lines) and unstable (dotted lines)

Thus it has been proven that in the case of pseudo-linear systems once again there is a discrete series of stable oscillation amplitudes in the oscillator for every set of values of the parameters of the external force F_0 and v. The emergence of one oscillation mode rather than another is determined by the initial conditions. The values of the possible amplitudes a_s, starting from a definite number s (in reality from $s = 1$) are approximately equal to

$$a_s = \frac{2d'(2m+1)}{\pi(2s+1)}. \tag{10.133}$$

It follows from (10.133) and (10.124) that the time of action of the external force on the oscillator in stationary mode is

$$\tau_{ms} = \frac{2d'}{v_m a_s} = \frac{d'(2m+1)T_v}{\pi a_s} = T_v(s + \frac{1}{2}), \tag{10.134}$$

where $T_v = \dfrac{2\pi}{v}$ is the period of external action. Obviously the adaptive mechanism for maintenance of continuous oscillations operates at times of interaction in the order of half a wave of the external action or an odd number of half-waves. In this case the average value of the energy contributed to the oscillation process is positive and it is effectively governed by the phase Φ_m, determined by the parameter Q_m.

In this situation the condition of two energy inputs per oscillation period is also met.

In the case of odd nonlinearity of the external acting force, the function (10.124) can be written in the following form:

$$f(x,t) = \nu(d'^2 - x^2)\Phi(x)F_0 \cos vt, \tag{10.135}$$

where in the simplest case $\Phi(x) = \operatorname{sign} x$, i.e. Eq.(10.135) assumes the form

$$f(x,t) = \nu(t - t_n + \frac{d'}{a\omega})\nu(t_n + \frac{d'}{a\omega} - t)(-1)^n \operatorname{sign}(t - t_n)F_0 \cos vt. \tag{10.136}$$

When (10.136) and (10.128) are taken into account, the following equations are obtained for the amplitude a and the phase φ:

$$\left| \begin{aligned} \dot{a} &= -\delta_d a + \frac{4F_0}{\pi v} \cos\frac{v\pi}{2\omega} \sin\frac{2vd'}{2\omega a} \sin\frac{v(n\pi - \varphi)}{\omega} \\ \dot{\varphi} &= -\Delta - \frac{4F_0 d'}{\pi v a^2} \cos\frac{v\pi}{2\omega} \sin\frac{vd'}{2a\omega}\left(\frac{\omega B}{vd'}\sin\frac{vd'}{2\omega a} - \cos\frac{vd'}{2\omega a}\right)\cos\frac{v(n\pi - \varphi)}{\omega}. \end{aligned} \right. \tag{10.137}$$

Equations (10.137) have a non-trivial stationary solution at $\omega = \omega_m = \dfrac{v}{2m}$, where m is an arbitrary integer. Thus, unlike the situation in the previous case, the frequency of the stationary oscillations should be an even number smaller than the frequency of the acting force. This conclusion is consistent with the results presented in the following paragraph [56]. It is obvious that in the general case, when the function $\Phi(x)$ in (10.134) is asymmetric, both even and odd subharmonics of the external action frequency can be excited, moreover their amplitudes will depend on the odd or even component of the function $\Phi(x)$ respectively.

10.3.7. Pendulum oscillations in case of oddness of the external exciting force

In all previous sections 10.3.1 – 10.3.6 the case under consideration was that of an even external exciting force, where, it was established, the ratio between the external force frequencies and the exciting oscillations should be an odd integer. In the foregoing section it was already noted that in the opposite case – that of an odd exciting force, the ratio between the frequencies should be an even number.

Let us elaborate on the case of an odd external acting nonlinear force [286], where $F_n(x,t)$ in (10.2) is represented by the following nonlinear function:

$$F_v(x,t) = 2F_0\varepsilon(d'^2 - x^2)\sin\frac{x}{2}\sin(vt + \varphi_v), \tag{10.138}$$

where $\varepsilon(z) = \begin{cases} 1, & \text{at } z \geq 0 \\ 0, & \text{at } z < 0 \end{cases}$.

The choice of the type of nonlinearity $2F_0 \sin\frac{x}{2}$ is conditioned only by a consideration of simplicity of the presentation, which can actually be transferred unchanged to another, arbitrarily selected, odd smooth function.

So, the case of oddness of the function (10.138), reflecting the action of an external nonlinear force on the pendulum, leads to the following equation:

$$\ddot{x} + 2\delta_d\dot{x} + \sin x = 2F_0\varepsilon(d'^2 - x^2)\sin\frac{x}{2}\sin(vt + \varphi_v). \tag{10.139}$$

Further on, an expression of the stationary amplitudes of the oscillator described by Eq.(10.139) [286] will be derived, and so will be the condition for the stability of these amplitudes. Once again, it is assumed that the friction coefficient $2\delta_d$ and the parameter d' are small quantities, while the frequency of the external action is $v \gg 1$ (by far greater than the own resonance frequency of the pendulum).

Once again, like in 10.3.2 above, the solution regarding an unperturbed conservative pendulum (10.22) can be taken as a zero approximation when solving equation (10.139).

The solution of equation (10.22) contains the energy of the oscillations W and the related amplitudes k in the capacity of a parameter:

$$\sin\frac{x}{2} = k\,\mathrm{sn}(t,k), \quad \dot{x} = 2k\mathrm{cn}(t,k), \quad W = 2k^2, \quad 1 \geq k \geq 0.$$

The quantity d' is equivalent to two other quantities: S_0 and t_0, which are convenient for the following substitution:

$$\sin\frac{d'}{2} = S_0 = k\sin(t_0, k).$$

An equation determining the change in the pendulum energy $W(t)$ with time will be written. To this end, Eq. (10.139) is multiplied by the derivative by time of the generalized coordinate x. The first and third terms on the left-hand side of the equation will correspond, as a sum total, to the derivative of the energy by time, and the other two terms will determine the dynamics of the pendulum and the change in its energy on a large temporal scale as compared to the oscillation period of an ideal autonomous pendulum, i.e.

$$\dot{W}(t) = -2\delta_d\dot{x}^2 + 2F_0\varepsilon(d'^2 - x^2)\dot{x}\sin(vt + \varphi_v). \tag{10.140}$$

An assumption of small contribution of the terms with $2\delta_d$ and F_0, i.e. of small change in the pendulum energy in the course of one period is made. The solution of Eq.(10.22) in the capacity of first approximation is substituted in (10.140). We obtain

$$\dot{W}(t) = -8\delta_d k^2[\mathrm{cn}(t,k)]^2 + 4F_0k^2\sin(vt + \varphi_v)\mathrm{sn}(t,k)\mathrm{cn}(t,k)\varepsilon\{S_0^2 - k^2[\mathrm{sn}(t,k)]^2\}.$$

As a transfer is made from energy to amplitude the following is obtained:

$$\frac{d\ln k}{dt} = -2\delta_d[\mathrm{cn}(t,k)]^2 + F_0\,\mathrm{sn}(t,k)\,\mathrm{cn}(t,k)\sin(vt+\varphi_v)\varepsilon\{S_0^2 - k^2[\mathrm{sn}(t,k)]^2\}. \tag{10.141}$$

For the stationary amplitudes

$$\left\langle \frac{d\ln k}{dt} \right\rangle_{t\to\infty} = 0.$$

This condition allows a calculation of the possible stationary amplitudes while averaging (10.141) for a large number of periods T_0 of pendulum oscillations, where

$$T_0 = 4\int_0^{\pi/2} [1-k^2(\sin\alpha)^2]^{-1/2}d\alpha = 4K(k).$$

After averaging (10.141) for time nT_0 in the case of $n\to\infty$ the following is obtained:

$$\begin{aligned} 0 = &-2\delta_d[K(k)^{-1}k^{-2}[E(k)-(1-k^2)K(k)] \\ &+ F_0[4nK(k)]^{-1}\sum_{i=0}^{2n-1}\int_{-t_0}^{t_0} \mathrm{sn}(t,k)\,\mathrm{cn}(t,k)\sin\{\omega[t+2iK(k)]+\varphi_v\}dt. \end{aligned}$$

The difference between the two terms on the right-hand part can be nullified when $n\to\infty$, if the sum in the second term is proportional to n, and this requirement is met provided that $2vK(k_m) = 2\pi m$, where m is a natural number. This condition is determined by the amplitude of the stationary oscillations k_m and, therefore, the ratio between the period of the pendulum oscillations T_0 and the period of the external acting force T_v should be, respectively,

$$\frac{T_0}{T_v} = T_0\frac{v}{2\pi} = 2m.$$

Taking into account that $S_0 \ll k$, it is not difficult to derive an equation regarding the phase φ_v of the external acting force at the moments when $x=0$, which is required for stationary oscillations to be established:

$$\cos\varphi_v^{(m)} = \frac{\delta_d v^2[E(k_m)-(1-k_m^2)K(k_m)]}{F_0k_m^2\left(\sin\frac{vS_0}{k_m} - \frac{vS_0}{k_m}\cos\frac{vS_0}{k_m}\right)}.$$

The stability test indicates that the pendulum oscillations will be stable regarding both amplitude and phase under the following condition:

$$\frac{vS_0}{k_m}\,\mathrm{ctg}\,\frac{vS_0}{k_m} < 1.$$

The analysis presented above shows that in the case of odd nonlinearity of the external exciting force, the realization of a discrete set of stationary amplitudes of pendulum oscillations determined by the initial conditions is also possible.

10.3.8. Approach in case of large amplitudes of pendulum oscillation

The equation that describes pendulum swinging caused by the action of a force, nonlinear to the coordinate, can be written in the general form:

$$\ddot{x} + 2\delta_0 \dot{x} + \omega_0^2 x + f(x) = F(x,t), \tag{10.142}$$

where x is the angular distance to equilibrium, δ_0 is a coefficient describing the system's dissipative properties, ω_0 is the resonance frequency of the small oscillations, $t = \omega_0 t_r$, t_r is real time.

In order to integrate the nonlinear Eq.(10.142) using the methods of the Theory of nonlinear oscillations, we introduce a new variable y and nonlinear time τ. So, the strongly nonlinear reactive term $\sin x$ in Eq.(10.142) may be „excluded". The transformation of variables is performed in accordance with the scheme proposed by K.A. Samoylo (see [185] and Section 2.2.1), which, for the case under consideration, yields:

$$y = \operatorname{sign} x \sqrt{2\int_0^x \sin x' dx'} = 2\sin\frac{x}{2}, \tag{10.143}$$

$$\frac{dt}{d\tau} = \frac{dx}{dy} = \frac{y}{\sin[x(y)]} = G(y). \tag{10.144}$$

Functions $x(y)$ and $G(y)$ in Eq.(10.144) are easily expressed, taking into account Expr.(10.143):

$$x(y) = 2\arcsin\left(\frac{y}{2}\right), \quad G(y) = \frac{1}{\sqrt{1-\dfrac{y^2}{4}}}. \tag{10.145}$$

As (10.143) and (10.144) are replaced in (10.142), the following is obtained:

$$\frac{d^2y}{d\tau^2} + y = \left[-2\delta_d \frac{dy}{d\tau} + F(x,\tau)\right] G(y). \tag{10.146}$$

The transmission to the variables makes the system quite similar to a linear conservative system, whose state is represented by a point in phase space moving in a circle with constant angular velocity. The well-known methods of the Theory of nonlinear oscillations may be applied to such a system. It should be mentioned that in terms of the new variables, all the features of the initial system are kept. The transformations (10.144) and (10.145) are right, if the conditions $G(0) = 1$ and

$G(y) > 0$ for all the values of y are fulfilled. The first condition is obviously fulfilled (see Exp.(10.145)). The second one is fulfilled for $-\pi < x < \pi$ or $-2 < y < 1$. Further consideration will be performed for this interval of y values. In physical terms, it means that initial conditions and external action provide a pendulum swing with an angle amplitude smaller than $\pm\pi$.

It is assumed that the solution of Eq.(10.146) is quasiharmonic with nominal frequency $\omega_n = \dfrac{v}{N}$, where $N \gg 1$ is an odd integer, and that $\omega_n \sim \omega_0$. Hence, Eq.(10.146) is rewritten in a different form:

$$\frac{d^2y}{d\tau^2} + \beta^2 y = -2\delta_d \frac{dy}{d\tau} + F(x,\tau)G(y) + (\beta^2 - 1)y, \qquad (10.147)$$

where $2\delta_d = \dfrac{2\delta_0}{\omega_0}$ and $F(x,\tau) = \dfrac{F_0(x,t)}{\omega_0^2}$, $(\beta^2 - 1)$ correspond to the frequency detuning, $\beta \sim 1$.

We assume that the solution of Eq.(10.146) is:

$$y = R\cos\Psi = R\cos(\beta\tau - \varphi_v), \qquad (10.148)$$

where R and φ_v are the oscillation amplitude and phase.

The dependence of the normalized time t on the angle Ψ can be expressed in agreement with (10.144), (10.145) and (10.148) as

$$t = \frac{1}{\beta}\int_0^{\Psi} \frac{d\psi}{\sqrt{1 - \dfrac{R^2}{4}\cos^2\Psi}}. \qquad (10.149)$$

Considering Exp.(10.149), the normalized oscillation period is:

$$T_0 = \frac{1}{\beta}\int_0^{2\pi} \frac{d\psi}{\sqrt{1 - \dfrac{R^2}{4}\cos^2\Psi}} = \frac{4}{R}K\left(\frac{R}{2}\right), \qquad (10.150)$$

where $K\left(\dfrac{R}{2}\right)$ is the complete elliptic integral of first kind.

The coefficient β is derived from (10.150):

$$\beta = \frac{2vK\left(\dfrac{R}{2}\right)}{\pi\omega_0 N}. \qquad (10.151)$$

We proceed to solve Eq.(10.147).

The shortened (averaged) differential equations for amplitude R and phase φ_v may be written as:

$$\left|\begin{array}{l}\left\langle\dfrac{dR}{d\tau}\right\rangle = -\dfrac{1}{2\pi\beta}\displaystyle\int_0^{2\pi} L\sin\Psi d\Psi \qquad (10.152a)\\ \left\langle\dfrac{d\varphi_v}{d\tau}\right\rangle = -\dfrac{1}{2\pi\beta R}\displaystyle\int_0^{2\pi} L\cos\Psi d\Psi, \qquad (10.152b)\end{array}\right.$$

where $L = 2\delta_d\beta R\sin\Psi + F(x,\tau)G(y) + (\beta^2-1)R\cos\Psi$. The sign $<>$ denotes the procedure of averaging by time τ.

Taking into account that

$$\int_0^{2\pi}\sin^2\Psi G(y)d\Psi = \int_0^{2\pi}\frac{\sin^2\Psi}{\sqrt{1-\dfrac{R^2}{4}\cos^2\Psi}}d\Psi = 4K\left(\frac{R}{2}\right) + \frac{16}{R^2}\left[E\left(\frac{R}{2}\right) - K\left(\frac{R}{2}\right)\right],$$

where $K(.)$ and $E(.)$ are the complete elliptic integrals of the first and the second kind, the shortened equations (10.152) take the form:

$$\left|\begin{array}{l}\left\langle\dfrac{dR}{d\tau}\right\rangle = -\dfrac{1}{\pi}\beta R\left\{4K\left(\dfrac{R}{2}\right) + \dfrac{16}{R^2}\left[E\left(\dfrac{R}{2}\right) - K\left(\dfrac{R}{2}\right)\right]\right\}\\ \qquad -\dfrac{1}{2\pi\beta}\displaystyle\int_0^{2\pi} F(x,\tau)G(y)\sin\Psi d\Psi \qquad (10.153a)\\ \left\langle\dfrac{d\varphi_v}{d\tau}\right\rangle = -\dfrac{1}{2\pi\beta R}\displaystyle\int_0^{2\pi} F(x,\tau)G(y)\cos\Psi d\Psi - \dfrac{\beta^2-1}{2\beta}. \qquad (10.153b)\end{array}\right.$$

Let us first consider the external action zone as a limited but finite part of the motion trajectory of the pendulum defined by the function:

$$\varepsilon(x) = \begin{cases} 1, & \text{at } |x| \le d' \\ 0, & \text{at } |x| > d' \end{cases}, \qquad (10.154)$$

d' sets the limits of the external action zone, $d' \ll 1$.

In this case the external force in (10.142) is presented as

$$F(x,t) = \varepsilon[x(y)]F_0\sin\left[\frac{v}{\omega_0}t(\tau) - \varphi_v\right]. \qquad (10.155)$$

Taking into account (10.148), the function $\varepsilon[x(R\cos\Psi)]$ can be written as

$$\varepsilon[x(R\cos\Psi)] = \begin{cases} 1, & \text{at } |\Psi| \le \theta_0 \\ 0, & \text{at } |\Psi| > \theta_0 \end{cases}, \qquad (10.156)$$

where $\theta_0 = \arccos\left(\frac{2}{R}\sin\frac{d'}{2}\right)$.

Besides, the function $L(.)$ in the system of equations (10.152) assumes the form:

$$L[R\cos\Psi, -\beta R\sin\Psi, e(\Psi-\varphi_v)] = 2\delta_d\beta R\sin\Psi$$

$$+ c[x(R\cos\Psi)]\frac{F_0}{\sqrt{1-\frac{R^2}{4}\cos^2\Psi}}\sin\left[\frac{\pi N}{2K\left(\frac{R}{2}\right)}\int_0^{\Psi}\frac{d\Psi}{\sqrt{1-\frac{R^2}{4}\cos^2\Psi}} - \varphi\right]$$

$$+ (\beta^2-1)R\cos\Psi.$$

The following denotations are introduced:

$$\left\{\begin{matrix} I_1 \\ I_2 \end{matrix}\right\} = \int_{z_0}^{K(R/2)} \sin\left\{r\left[K\left(\frac{R}{2}\right)-Z\right]\right\}\left\{\begin{matrix} \operatorname{sn} Z \\ \operatorname{cn} Z \end{matrix}\right\} dZ, \tag{10.157}$$

$$\left\{\begin{matrix} I_3 \\ I_4 \end{matrix}\right\} = \int_{z_0}^{K(R/2)} \cos\left\{r\left[K\left(\frac{R}{2}\right)-Z\right]\right\}\left\{\begin{matrix} \operatorname{sn} Z \\ \operatorname{cn} Z \end{matrix}\right\} dZ, \tag{10.158}$$

where $Z = F(\alpha, \frac{R}{2})$ and $Z_0 = F(\alpha_0, \frac{R}{2})$ are incomplete elliptic integrals of the first kind, $\alpha = \frac{\pi}{2} - \Psi$, $\alpha_0 = \frac{\pi}{2} - \theta_0$, $\operatorname{sn} Z$ and $\operatorname{cn} Z$ are the sine and cosine of the amplitude (Jacobi elliptic functions), $r = \dfrac{\pi N}{2K\left(\frac{R}{2}\right)}$.

As (10.157) and (10.158) are taken into account, the shortened equations regarding the amplitude R and the phase φ_v become

$$\left|\begin{aligned} \frac{dR}{d\tau} &= -\delta_d R - \frac{2}{\pi\beta}F_0(I_1\cos\varphi_v - I_3\sin\varphi_v) && (10.159a)\\ \frac{d\varphi_v}{d\tau} &= -\frac{2}{\pi\beta R}F_0(I_2\cos\varphi_v - I_4\sin\varphi_v) - (\beta-1). && (10.159b) \end{aligned}\right.$$

For the stationary mode $\left(\frac{dR}{d\tau} = 0,\ \frac{d\Psi}{d\tau} = 0\right)$ the following expressions are obtained for the established amplitude R and phase φ_v from the system of equations (10.159):

$$R = \frac{2F_0(I_1I_4 - I_2I_3)}{\pi\beta\delta_d\sqrt{(\sigma I_1 - I_2)^2 + (\sigma I_3 - I_4)^2}}, \tag{10.160}$$

$$\varphi_v = \operatorname{arctg}\frac{\sigma I_1 - I_2}{\sigma I_3 - I_4}, \tag{10.161}$$

where $\sigma = \dfrac{\beta - 1}{\delta_d}$.

The stability of the obtained solutions (10.160) and (10.161) can be analyzed on the basis of the approach developed in 10.3.1.

Fig.10.24 presents the theoretical dependence of the stable roots of Eq.(10.160) on the amplitude of the external acting force F_0 in the case of $\delta_d = 0.01$, $\omega_0 = 3.14$, $N = 111$ and $d' = 0.02$. The regularities characteristic of the system can be clearly seen: in the case of unchanging external action, stable stationary oscillations with a possible discrete set of stable amplitudes R_n can exist; virtually none of the amplitudes of the possible discrete set of stationary amplitudes R_n is dependent on the changes in the amplitude F_0 of the external acting force within a broad range above certain threshold values. Thus a peculiar effect of „quantization" of the possible stationary amplitudes is manifested in a dynamic macrosystem. The realization of oscillations of either one or another amplitude of the possible discrete set is determined by the initial conditions. Moreover, the energy from the outside is provided in such a way that the stationary amplitude R_n $(n = 1, 2, 3, ...)$ of the oscillations and the energy portion absorbed periodically by the system should actually be independent of the changes in the intensity of the external action.

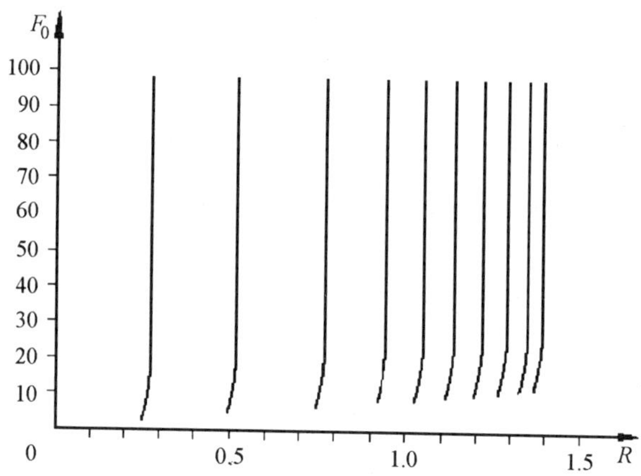

Fig.10.24. Theoretical dependence of possible stable amplitudes R_n on the intensity of the external action force F_0 (Eq.10.160)

The mechanism of establishment and maintenance of the stationary oscillations has to do with a capture and adaptive automatic self-tuning of the phase φ_v, determining the interaction and the input of energy from an external high-frequency source into the excited oscillations. This mechanism is illustrated in Fig.10.25 and Fig.10.26. Fig.10.25 shows the theoretical dependence of the discrete values of the phase φ_v corresponding to the stationary amplitudes R_n of the oscillations in the system. Each stationary amplitude determined by the initial conditions has a corresponding strictly determined energy portion, periodically introduced from the outside, which is set by the respective phase φ_v. A strictly fixed discrete series of stationary phases R_1, R_2, $R_3, \ldots$ can be juxtaposed to the discrete series of stationary amplitudes $\varphi_1^{(1)}$, $\varphi_2^{(2)}$, $\varphi_3^{(3)}, \ldots$ for each value of the amplitude of the acting force F_0. When the amplitude F_0 changes, the oscillations in the system remain unchanged, which is also due to the adaptive self-tuning of the phase φ_v. This case is illustrated by Fig.10.23, which shows the changes in the phase with the change in the amplitude of the acting force F_0, which ensure an invariable and stable amplitude of the stationary oscillations $R = \text{const}$.

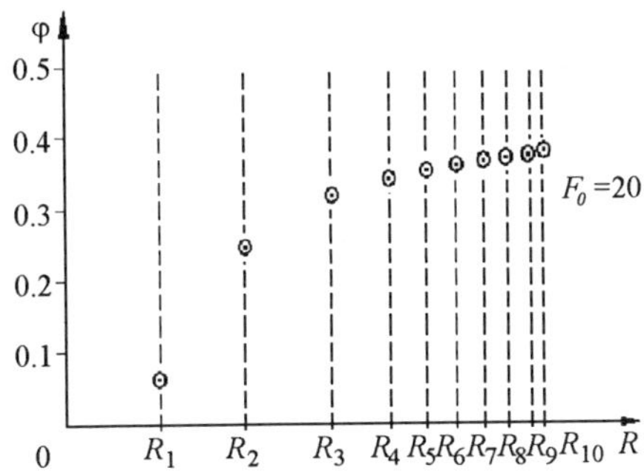

Fig.10.25. Correspondence between stable oscillations amplitudes R_n and stationary initial phases (φ_ν^n) under unchanged intensity of the external acting force (Eq.10.161)

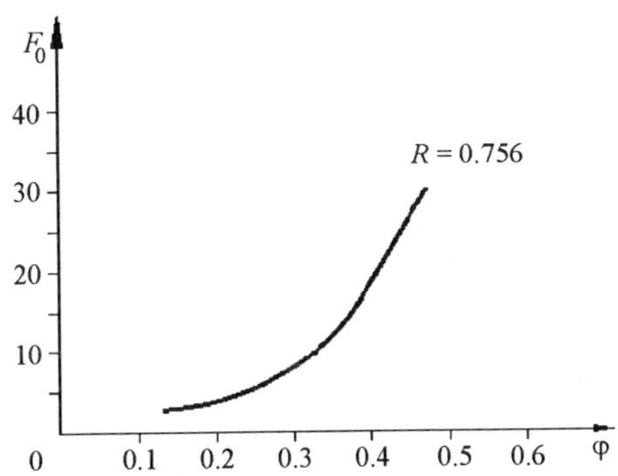

Fig.10.26. Illustration of the theoretical dependence of stationary initial phases on the acting force's intensity alternation keeping the oscillation's amplitudes unchangeable and stable

Now let us render the function $F(x,t)$ concrete as follows:

$$F(x,t) = \delta(x) F_0 \sin vt, \tag{10.162}$$

where $\delta(x)$ is Dirak's δ-function, F_0 and v are the external harmonic force amplitude and frequency respectively. We assume that $v = N\omega_0$, where $N = 1, 2, 3, ...$ Taking into account the solution form (10.148), $\delta(x)$ can be presented in the form

$$\delta(x) = \sum_{t} \left| \frac{d\Psi}{dx} \right| \delta(\Psi - \Psi_{0,i}), \tag{10.163}$$

where the values $\Psi_{0,i}$ are determined from the equation

$$x(\Psi_{0,i}) = 0. \tag{10.164}$$

Considering Eqs.(10.163) and (10.164), the Eqs.(10.153) become

$$\left| \begin{aligned} &\left\langle \frac{dR}{d\tau} \right\rangle = -\frac{1}{\pi}\beta R \left\{ 4K\left(\frac{R}{2}\right) + \frac{16}{R^2}\left[E\left(\frac{R}{2}\right) - K\left(\frac{R}{2}\right) \right] \right\} \\ &\quad -\frac{1}{2\pi\beta}\left[GF_0 \sin vt\left(\frac{\pi}{2}\right) \left| \frac{d\Psi}{dx} \right|_{\frac{\pi}{2}} - GF_0 \sin vt\left(\frac{3\pi}{2}\right) \left| \frac{d\Psi}{dx} \right|_{\frac{3\pi}{2}} \right] && (10.165a) \\ &\left\langle \frac{d\varphi_v}{d\tau} \right\rangle = -\frac{\beta^2 - 1}{2\beta}. && (10.165b) \end{aligned} \right.$$

Noting that $\dfrac{d\Psi}{dx} = \dfrac{d\Psi}{d\tau}\dfrac{d\tau}{dt}\dfrac{dt}{dx} = \dfrac{d\Psi}{d\tau}\dfrac{d\tau}{dt}\dfrac{d\tau}{dy} = -\dfrac{1}{G(y)R\sin\Psi}$ and that $G\left(\dfrac{\pi}{2}\right) = G\left(\dfrac{3\pi}{2}\right) = 1$, Eq.(10.165a) can be rewritten as

$$\left\langle \frac{dR}{d\tau} \right\rangle = -\frac{1}{\pi}\beta R \left\{ 4K\left(\frac{R}{2}\right) + \frac{16}{R^2}\left[E\left(\frac{R}{2}\right) - K\left(\frac{R}{2}\right) \right] \right\}$$
$$-\frac{F_0}{2\pi\beta R}\left[\sin vt\left(\frac{\pi}{2}\right) - \sin vt\left(\frac{3\pi}{2}\right) \right]. \tag{10.166}$$

Introducing the designation $t\left(\dfrac{\pi}{2}\right) = t_1$ and taking into account Exprs.(10.150) and (10.151), we can write

$$t\left(\frac{3\pi}{2}\right) = t_1 + \frac{2}{\beta}K\left(\frac{R}{2}\right), \quad \sin vt\left(\frac{3\pi}{2}\right) = (-1)^N \sin vt_1.$$

Let us now consider two cases:
a) Case of even $N (N = 2l,\ l = 0, 1, 2, 3, ...)$.
In this case $\sin vt\left(\dfrac{\pi}{2}\right) - \sin vt\left(\dfrac{3\pi}{2}\right) = 0$ and there are no stationary solutions (the oscillations are damped);

b) Case of odd $N (N = 2l + 1,\ l = 1, 2, 3, \ldots)$.

In this case $\sin vt\left(\frac{\pi}{2}\right) - \sin vt\left(\frac{3\pi}{2}\right) = 2\sin vt_1$ and

$$\left|\begin{aligned}\left\langle\frac{dR}{d\tau}\right\rangle &= e(R, \varphi_v),\\ \left\langle\frac{d\varphi_v}{d\tau}\right\rangle &= g(R, \varphi_v),\end{aligned}\right.$$

where

$$e(R, \varphi_v) = -\frac{1}{\pi}\beta R\left\{4K\left(\frac{R}{2}\right) + \frac{16}{R^2}\left[E\left(\frac{R}{2}\right) - K\left(\frac{R}{2}\right)\right]\right\} - \frac{F_0}{\pi\beta R}\sin vt_1, \quad (10.167a)$$

$$g(R, \varphi_v) = -\frac{\beta^2 - 1}{2\beta}. \quad (10.167b)$$

For the stationary mode ($e(R, \varphi_v) = 0$ and $g(R, \varphi_v) = 0$) from Eq.(10.167b) we can find the condition $\beta = 1$, which can be rewritten while taking into account Expr.(10.151) as $K\left(\frac{R}{2}\right) = \frac{\pi}{v}\left(l + \frac{1}{2}\right)$, $l = 0, 1, 2, 3, \ldots$

Denoting $k = \frac{R}{2}$ (the module of the elliptic function), from Eq.(10.167a) we can find the second condition of the stationary mode in the form

$$\frac{16\delta_d}{F_0}[E(k) - (1 - k^2)K(k)] - \sin vt_1 = 0. \quad (10.168)$$

When $k \to 0$, Eq.(10.168) is simplified,

$$\frac{4\pi\delta_d k^2}{F_0} + \sin vt_1 = 0, \quad (10.169)$$

and corresponds to the condition

$$|F_0| > 4\pi\delta_d k^2. \quad (10.170)$$

For the sake of stability estimation we can rewrite Eqs.(10.167) under the condition $k \to 0$ as

$$\left|\begin{aligned} e(R, \varphi_v) &= -2\delta_d k - \frac{F_0}{2\pi\beta k}\sin vt_1, & (10.171)\\ g(R, \varphi_v) &= -\frac{\beta^2 - 1}{2\beta}, & (10.172)\end{aligned}\right.$$

where $\beta = 1$.

The characteristic equation can be written as

$$\lambda^2 - \lambda(e_R + g_\varphi) + e_R g_\varphi - e_R g_R = 0,$$

taking the final form

$$\lambda(\lambda - e_R) = 0, \tag{10.173}$$

where e_R, g_φ, g_R are the corresponding partial derivatives.

From Eq.(10.173) we find $\lambda_1 = 0$, $\lambda_2 = e_R$. The stability condition is: $\lambda_2 = e_R < 0$ i.e. $e_k < 0$.

Using Eq.(10.171) we obtain $e_k = -2\delta_d + \dfrac{F_0}{2\pi k^2} \sin vt_1$.

Comparison with (10.169) reveals the stability condition in the form

$$e_k = -4\delta_k < 0. \tag{10.174}$$

As the value $\delta_d > 0$ *apriori,* the inequality (10.174) is fulfilled and the solution for odd N describes a discrete set of stable stationary oscillations.

10.4. „Quantization" of the oscillations of an oscillator under the action of an incoming (falling) wave

The foregoing sections under 10.3 presented the class of the kick-excited self-adaptive dynamic systems. They provided numerical and analytical evidence of the behavior and major properties of the dynamic systems to which an external nonlinear exciting force had been applied. The discussion was based on the general model of a pendulum and the nonlinearity of the external action we introduced by using specially selected functional dependencies.

In this section we shall consider an „oscillator-wave" general model and we shall show that the non-homogeneous external action is realized in a natural way without an artificial creation of special conditions for that purpose. For the sake of consistency of the presentation we shall consider the excitation of oscillations in a nonlinear oscillator of the „pendulum" type under the action of an incident wave. We shall show that under certain conditions continuous oscillations are excited in the system, whose frequency is close to the system's own one and whose amplitude belongs to a definite discrete spectrum of possible amplitudes. The second important feature is also manifested, i.e. self-adaptive stability of the excited oscillations with a determined amplitude given a broad range of alteration of the wave intensity.

10.4.1. A model of the interaction of an oscillator with an electromagnetic wave: an approach in the case of small amplitudes of the oscillations in the system

Let us consider the interaction between an electromagnetic wave and a slightly dissipative nonlinear oscillator. Let us take an electric charge q with mass

m oscillating around a definite static point in the direction of the x-axis under the action of a nonlinear reverse force. The electromagnetic wave also propagates in the direction of the x-axis and it has a longitudinal component in the electric field E. The motion equation of the charge interacting with the wave can be presented in the form:

$$m(\ddot{x} + 2\delta_d \dot{x} + \omega_0^2 \sin x) = Eq \sin(vt - kx), \tag{10.175}$$

where $2\delta_d$ is a damping coefficient, ω_0 is the own frequency of small oscillations of the charge, v is the frequency of the wave and k is the wave number. The case under consideration is that of $v \gg \omega_0$.

It is assumed that when oscillations are excited in the charge under the wave action, the symmetry of its motion in relation to the equilibrium position is not affected significantly and the charge coordinate changes according to the law:

$$x = a \sin\theta, \quad \theta = \omega t + \alpha, \quad a = a(t), \quad \alpha = \alpha(t). \tag{10.176}$$

As solution (10.176) is substituted in the right-hand part of Eq.(10.175), the following is obtained:

$$Eq \sin(vt - ka \sin\theta) = Eq \sum_{n=-\infty}^{\infty} J_n(ka) \sin(vt - n\theta).$$

Setting $\dot{x} = a\omega \cos\theta$, $\dot{a} \sin\theta + \dot{\alpha} a \cos\theta = 0$ and using the method of Krilov-Bogolyubov-Mitropolsky [97], we write the following to a first approximation:

$$\left| \begin{aligned} \frac{da}{dt} &= -2\delta_d a + \frac{F_0}{\omega} \sum_{n=-\infty}^{\infty} J_n(ka) \sin(vt - n\theta) \cos\theta, && (10.177a) \\ \frac{d\alpha}{dt} &= \frac{\Omega^2 - \omega^2}{2\omega} - \frac{F_0}{a\omega} \sum_{n=-\infty}^{\infty} J_n(ka) \sin(vt - n\theta) \sin\theta, && (10.177b) \end{aligned} \right.$$

where Ω is the frequency of the free oscillations of the charge with amplitude a,

$$\Omega^2 = \frac{2J_1(a)}{a}\omega_0^2 \simeq \omega_0^2 \left[1 - \frac{a^2}{8} + O(a^4)\right], \tag{10.178}$$

$J_n(.)$ are Bessel functions of the first kind, $F_0 = \dfrac{Eq}{m}$.

The excitation of undamping oscillations in the oscillator with frequency $\omega = \omega_s$, close to the oscillator's own, is possible under the condition $\dfrac{2\pi}{\lambda} a > 1$, where λ is the length of the acting wave. As a result of the interaction of the oscillator and the wave, frequency components appear in the spectrum of the force, that are sufficiently close to the frequency of its own oscillations. Then the action of these spectral components prevails overwhelmingly over the others. In this case the right-hand part of Eq.(10.175) will be in the form:

$$\frac{1}{m}Eq\sum_{n=-\infty}^{\infty} J_n(ka)\sin(vt-n\theta)$$
$$=\left\{J_{\frac{v}{\omega}-1}(ka)\sin\left[\omega t-\left(\frac{v}{\omega}-1\right)\alpha\right]-J_{\frac{v}{\omega}+1}(ka)\sin\left[\omega t+\left(\frac{v}{\omega}+1\right)\alpha\right]\right\}. \quad (10.179)$$

Provided that $v>\omega$, several spectral components of the exciting wave can fall in the resonance area of the nonlinear oscillator and each of them can excite stationary oscillations of the oscillator with a corresponding amplitude of the possible discrete series of stationary amplitudes. In the case of fixed parameters of the oscillator and the wave, the excitation of oscillations with certain amplitude from the possible series of amplitudes is determined by the initial conditions. In line with relation (10.178), the values of the discrete series of possible stationary amplitudes can be calculated according to the formula:

$$a_{s0}=\sqrt{8(1-\frac{v^2}{s^2\omega_0^2})}\simeq 4\sqrt{1-\frac{v}{s\omega_0}}, s=1,2,3,... \quad (10.180)$$

After averaging the right-hand parts of Eqs.(10.177) and (10.179) the following are obtained:

$$\frac{da_s}{dt}=-\delta_d a+\frac{F_0}{2\omega_s}[J_{s-1}(ka_s)+J_{s+1}(ka_s)]\sin(pt-\gamma_s), \quad (10.181a)$$
$$\frac{d\alpha_s}{dt}=\frac{\Omega^2-\omega_s^2}{2\omega_s}-\frac{F_0}{2\omega_s a_s}[J_{s-1}(ka_s)-J_{s+1}(ka_s)]\cos(pt-\gamma_s), \quad (10.181b)$$

where $p=v-\frac{v}{\omega_s}\ll\omega_s$, $\gamma_s=\frac{v}{\omega_s}\alpha_s$, $\omega_s=\frac{v}{s}$.

In line with certain recurrent ratios related to the Bessel functions, Eqs.(10.181) can be presented in the form:

$$\frac{da_s}{dt}=-\delta_d a_s-\frac{vF_0}{\omega_s^2 ka_s}J_s(ka_s)\sin(pt-\gamma_s), \quad (10.182a)$$
$$\frac{d\alpha_s}{dt}=\frac{\Omega_s^2-\omega_s^2}{2\omega_s}-\frac{F_0}{\omega_s a_s}J_s'(ka_s)\cos(pt-\gamma_s). \quad (10.182b)$$

In the case of a stationary oscillation mode of the oscillator ($\frac{da_s}{dt}=0$ and $\frac{d\alpha_s}{dt}=0$), the following is obtained from Eqs.(10.182):

$$\operatorname{tg}\gamma_s=\frac{2\delta_d a_s\omega_s^2 k}{(\Omega_s^2-\omega_s^2)v}\frac{J_s'(ka_s)}{J_s(ka_s)}.$$

The relation between the intensity of the longitudinal component of the wave and the amplitude of the oscillations in the oscillator are of the form:

$$F_0^2 = \left[\frac{a_{s0}^2 \omega_s^2 \delta_d k}{v J_s(k a_{s0})}\right]^2 + \left[\frac{a_{s0}(\Omega_s^2 - \omega_s^2)}{2 J_s'(k a_{s0})}\right]^2 . \qquad (10.183)$$

For higher intensity of the wave, Eq.(10.183) can be presented as:

$$F_0 \simeq \frac{a_{s0}^2 \omega_s^2 (a_s - a_{s0})}{8 J_s'(k a_s)} .$$

The first term in formula (10.183) is the minimum threshold value F_0 of the wave intensity. If the intensity of the wave is lower than this threshold value, it is possible to excite only forced oscillations in the oscillator with a frequency coinciding with the wave frequency. When the intensity values of the wave excede the threshold value, the motion of the oscillator takes place with one of the amplitudes in the discrete series (10.180). In the case of $v > \omega_0$ depending on the initial conditions, each amplitude is realized at oscillation frequency Ω_s, close to the own frequency of the oscillator. Using the approach recounted in 10.3.1, it is not difficult to demonstrate that in a situation of fixed values of the frequency v and the amplitude F_0 of the external force, the motion in the oscillator with the possible discrete series of amplitudes (10.180) is stable.

The analysis that has been conducted indicates that the continuous wave with a frequency by far higher than the frequency of an oscillator, can excite in it oscillations with a frequency close to oscillator's own, and with an amplitude belonging to a discrete series of possible stable amplitudes. The establishment of either one or another amplitude depends on the initial conditions. In a stationary mode, the magnitude of the amplitude is virtually independent of the wave intensity, when the latter changes in a wide range over a definite threshold value. This is reminiscent of Einstein's explanation of the photoelectric effect when using Planck's quantization hypothesis. In this case the absorption is also independent of the intensity of the incoming wave. Besides, the absorbed frequencies can be expressed as integer multiples of a certain basic frequency reminding one of resonance phenomena.

10.4.2. „Quantized" cyclotron motion

As the history of research development shows, the revelation of generation mechanisms in planetary magnetosphere radio sources is connected with the level of understanding of physical phenomena and the development of concept in modern radiophysics. The most vivid example is the investigation of cyclotron maser processes and the subsequent discovery of similar processes in the nature of all magnetized planets [326].

When an electromagnetic wave interacts with resonators, the effect of „quantization“ of possible stationary stable oscillating amplitudes occurs without satisfying any specially organized conditions (like the inhomogeneous action of an external harmonic force).

An electric charge, moving on a circular orbit in a homogeneous permanent magnetic field is considered. When the charge is irradiated by a flat electromagnetic wave whose length is commensurable with the orbit radius, an effect of discretization („quantization“) of the possible stable orbit radii (or motion velocities) is observed.

A recurrent expression for the possible stable radius values (correspondingly, for the possible rotation speed values) is derived. It is shown, that a radius threshold value exists and that for the values above it, a discretization of the possible stable radius values arises. A general stability investigation is carried out.

Let us consider an electric charge q in a magnetic field $\mathbf{B}$ and an electric field $\mathbf{E}$. The equation of motion in three-dimensional Euclidian space is

$$m\frac{d^2r}{dt^2} = \mathbf{F} = q(\mathbf{E} + \mathbf{V} \times \mathbf{B}) - 2m\beta\mathbf{V}, \qquad (10.184)$$

where m is the mass of the moving charge q, $\mathbf{V} = \frac{dr}{dt}$ is the velocity, β is the coefficient of dissipation.

Considering Eq.(10.184) and assuming that the motion is in the plane $z = 0$, we can write:

$$\left| \begin{array}{l} m\dfrac{dV_x}{dt} = q(E_x + V_y B_0) - 2m\beta V_x \\ m\dfrac{dV_y}{dt} = q(E_y - V_x B_0) - 2m\beta V_y \end{array} \right. . \qquad (10.185)$$

For a constant magnetic field $\mathbf{B} = \mathbf{e}_z B_0 = \text{const}$ the cyclotron frequency

$$\omega_0 = -\frac{qB_0}{m}. \qquad (10.186)$$

Taking into account Eq.(10.186), Eqs.(10.185) take the form

$$\left| \begin{array}{l} \dfrac{dV_x}{dt} = -\dfrac{\omega_0}{B_0}E_x - \omega_0 V_y - 2\beta V_x \\ \dfrac{dV_y}{dt} = -\dfrac{\omega_0}{B_0}E_y + \omega_0 V_x - 2\beta V_y \end{array} \right. . \qquad (10.187)$$

A solution, corresponding to the rotation plus drift, is sought in the form

$$x = R\cos\Psi + at, \quad y = R\sin\Psi + bt, \quad \Psi = \omega t + \varphi, \qquad (10.188)$$

where R, φ, a, b are constants in the stationary regime, $\omega = \text{const}$.

The values a and b can be obtained from the following equations:

$$\left| \begin{array}{l} -\dfrac{\omega_0}{B_0}\langle E_x\rangle - \omega_0 b - 2\beta a = 0, \\ -\dfrac{\omega_0}{B_0}\langle E_y\rangle + \omega_0 a - 2\beta b = 0, \end{array} \right. \tag{10.189}$$

where the sign $\langle\ \rangle$ denotes the averaging by time t.

Integrating (10.187), we can write

$$\left| \begin{array}{l} V_x = -\dfrac{\omega_0}{B_0}\displaystyle\int E_x dt - \omega_0 y - 2\beta x + \mathrm{const}_1, \\ V_y = -\dfrac{\omega_0}{B_0}\displaystyle\int E_y dt + \omega_0 x - 2\beta y + \mathrm{const}_2. \end{array} \right. \tag{10.190}$$

Neglecting $\dfrac{da}{dt}$ and $\dfrac{db}{dt}$ from (10.188), we obtain

$$\left| \begin{array}{l} V_x = \dfrac{dx}{dt} = -\omega R\sin\Psi + \dfrac{dR}{dt}\cos\Psi - \dfrac{d\varphi}{dt}R\sin\Psi + a, \\ V_y = \dfrac{dy}{dt} = -\omega R\cos\Psi + \dfrac{dR}{dt}\sin\Psi + \dfrac{d\varphi}{dt}R\cos\Psi + b. \end{array} \right. \tag{10.191}$$

The substitution of (10.191) into (10.190) gives:

$$\left| \begin{array}{l} \dfrac{dR}{dt}\cos\Psi - \dfrac{d\varphi}{dt}R\sin\Psi = -\dfrac{\omega_0}{B_0}\displaystyle\int E_x dt - a - \omega_0 bt \\ \qquad -2\beta at + (\omega-\omega_0)R\sin\Psi - 2\beta R\cos\Psi + \mathrm{const}_1, \\ \dfrac{dR}{dt}\sin\Psi + \dfrac{d\varphi}{dt}R\cos\Psi = -\dfrac{\omega_0}{B_0}\displaystyle\int E_y dt - b + \omega_0 at \\ \qquad -2\beta bt - (\omega-\omega_0)R\cos\Psi - 2\beta R\sin\Psi + \mathrm{const}_2. \end{array} \right. \tag{10.192}$$

Considering (10.192) it gets clear how const_1 and const_2 have to be determined for the constant part of the Eqs.(10.190) to fall out.

Considering also (10.189) we can write:

$$\dfrac{dR}{dt}\cos\Psi - \dfrac{d\varphi}{dt}R\sin\Psi = -\dfrac{\omega_0}{B_0}(\text{periodical part of} \int E_x dt) + (\omega-\omega_0)R\sin\Psi - 2\beta R\cos\Psi, \tag{10.193a}$$

$$\dfrac{dR}{dt}\sin\Psi + \dfrac{d\varphi}{dt}R\cos\Psi = -\dfrac{\omega_0}{B_0}(\text{periodical part of} \int E_y dt) - (\omega-\omega_0)R\cos\Psi - 2\beta R\sin\Psi. \tag{10.193b}$$

From Eqs.(10.193) we have

$$\left|\begin{aligned} \frac{dR}{dt} &= \frac{\omega_0}{B_0}[-\cos\Psi(\text{periodical part of} \int E_x dt) \\ &\qquad - \sin\Psi(\text{periodical part of} \int E_y dt)] - 2\beta R, \qquad (10.194a) \\ \frac{dR}{dt} &= \frac{1}{R}\frac{\omega_0}{B_0}[\sin\Psi(\text{periodical part of} \int E_x dt) \\ &\qquad - \cos\Psi(\text{periodical part of} \int E_y dt)] - (\omega - \omega_0). \qquad (10.194b) \end{aligned}\right.$$

We consider a plane electromagnetic wave (i.e. $\mathbf{E}\mathbf{k} = 0$, where $\mathbf{k}$ is the wave vector, $E \sim \cos(\tilde{v}t - k_x x - k_y y - k_z z + a)$).

Let us assume that $k_x = k_z = 0$ and $k_y = k$. Then $E_y = E_z = 0$ and

$$E_x = -k\tilde{E}_0 \cos[(\tilde{v} - kb)t - kR\sin\Psi + \alpha]. \qquad (10.195)$$

Assuming that $v = \tilde{v} - kb$ and $-k\tilde{E}_0 = E_0$, Eq.(10.195) can be rewritten in the form

$$E_x = E_0 \cos(vt + \alpha - kR\sin\Psi). \qquad (10.196)$$

We assume $v = N\omega$, $N = 1, 2, 3, \ldots$

For the sake of solving Eqs.(10.194) and considering Eq.(10.196), we derive the following expansions:

$$\begin{aligned} &\cos\Psi[\text{periodic part of} \int \cos(vt - kR\sin\Psi + \alpha)dt] \\ &= \text{finite part} \left\{ \frac{J_0(kR)}{2N\omega} \{\sin[(N-1)\omega t + \alpha - \varphi] + \sin[(N+1)\omega t + \alpha + \varphi]\} \right. \\ &+ \sum_{j=1}^{\infty} J_{2j}(kR) \left\{ \frac{\sin[(2j+N-1)\omega t + (2j-1)\varphi + \alpha] + \sin[(2j+N+1)\omega t + (2j+1)\varphi + \alpha]}{2(2j+N)\varphi} \right. \\ &+ \left. \frac{\sin[(2j-N-1)\omega t + (2j-1)\varphi - \alpha] + \sin[(2j-N+1)\omega t + (2j+1)\varphi - \alpha]}{2(2j-N)\omega} \right\} \\ &+ \sum_{j=1}^{\infty} J_{2j-1}(kR) \left\{ \frac{\sin[(2j-2-N)\omega t + (2j-2)\varphi - \alpha] + \sin[(2j-N)\omega t + 2j\varphi - \alpha]}{2(2j-1-N)\omega} \right. \\ &- \left.\left. \frac{\sin[(2j-2+N)\omega t + (2j-2)\varphi + \alpha] + \sin[(2j+N)\omega t + 2j\varphi + \alpha]}{2(2j-1+N)\omega} \right\}\right\}, \end{aligned} \qquad (10.197)$$

$$\sin\Psi[\text{periodic part of} \int \cos(vt - kR\sin\Psi + \alpha)dt]$$

$$- \text{finite part} \left\{ \frac{J_0(kR)}{2N\omega} \{\cos[(N-1)\omega t + \alpha - \varphi] - \cos[(N+1)\omega t + \alpha + \varphi]\} \right.$$
$$+ \sum_{j=1}^{\infty} J_{2j}(kR) \left\{ \frac{\cos[(2j+N-1)\omega t + (2j-1)\varphi + \alpha] - \cos[(2j+N+1)\omega t + (2j+1)\varphi + \alpha]}{2(2j+N)\varphi} \right.$$
$$\left. + \frac{\cos[(2j-N-1)\omega t + (2j-1)\varphi - \alpha] - \cos[(2j-N+1)\omega t + (2j+1)\varphi + \alpha]}{2(2j-N)\omega} \right\}$$
$$+ \sum_{j=1}^{\infty} J_{2j-1}(kR) \left\{ \frac{\cos[(2j-2-N)\omega t + (2j-2)\varphi - \alpha] - \cos[(2j-N)\omega t + 2j\varphi - \alpha]}{2(2j-1-N)\omega} \right.$$
$$\left.\left. - \frac{\cos[(2j-2+N)\omega t + (2j-2)\varphi + \alpha] - \cos[(2j+N)\omega t + 2j\varphi + \alpha]}{2(2j-1+N)\omega} \right\}\right\}, \tag{10.198}$$

where $J(.)$ are Bessel functions of the first kind.

Using (10.194) we can write the shortened (averaged) equations:

$$\left\langle \frac{dR}{dt} \right\rangle = -\frac{\omega_0}{B_0} E \left\langle \cos\Psi \{\text{periodic part of} [\int \cos(vt - kR\sin\Psi + \alpha) dt]\} \right\rangle$$
$$-2\beta R \tag{10.199a}$$
$$\left\langle \frac{d\varphi}{dt} \right\rangle = \frac{1}{R} \frac{\omega_0 E_0}{B_0} \left\langle \sin\Psi \{\text{periodic part of} [\int \cos(vt - kR\sin\Psi + \alpha) dt]\} \right\rangle$$
$$-(\omega - \omega_0) \tag{10.199b}$$

From (10.197), (10.198) and (10.199) we obtain

$$\left\langle \cos\Psi \{\text{periodic part of} [\int \cos(vt - kR\sin\Psi + \alpha) dt]\} \right\rangle$$
$$= -\frac{1}{\omega} J_N'(kR) \sin(N\varphi - \alpha), \tag{10.200a}$$
$$\left\langle \sin\Psi \{\text{periodic part of} [\int \cos(vt - kR\sin\Psi + \alpha) dt]\} \right\rangle$$
$$= -\frac{N}{\omega kR} J_N(kR) \cos(N\varphi - \alpha), \tag{10.200b}$$

where $J_N'(.)$ is the first derivative of the Bessel function of the first kind.

Taking into account (10.200) Eqs.(10.199) can be rewritten as

$$\left\langle \frac{dR}{dt} \right\rangle = f(R, \varphi) \tag{10.201a}$$
$$\left\langle \frac{d\varphi}{dt} \right\rangle = g(R, \varphi) \tag{10.201b}$$

where

$$\left| \begin{array}{l} f(R,\varphi) = \dfrac{\omega_0}{\omega}\dfrac{E_0}{B_0}J_N'(kR)\sin(N\varphi-\alpha) - 2\beta R, \\ g(R,\varphi) = \dfrac{\omega_0}{\omega}\dfrac{E_0}{B_0}\dfrac{N}{kR^2}J_N(kR)\cos(N\varphi-\alpha) - (\omega-\omega_0). \end{array} \right. \quad \begin{array}{r} (10.202a) \\ (10.202b) \end{array}$$

The stationary solution corresponds to the conditions

$$\left\langle \frac{dR}{dt} \right\rangle = 0, \quad \left\langle \frac{d\varphi}{dt} \right\rangle = 0. \tag{10.203}$$

In order to analyze the system stability we produce the following variations:

$$\left| \begin{array}{l} \delta\dfrac{dR}{dt} = \dfrac{\partial f}{\partial R}\delta R + \dfrac{\partial f}{\partial \varphi}\delta\varphi \\ \delta\dfrac{d\varphi}{dt} = \dfrac{\partial g}{\partial R}\delta R + \dfrac{\partial g}{\partial \varphi}\delta\varphi \end{array} \right. . \tag{10.204}$$

Using f_R, f_φ, g_R and g_φ to denote the derivatives $\dfrac{\partial f}{\partial R}$, $\dfrac{\partial f}{\partial \varphi}$, $\dfrac{\partial g}{\partial R}$ and $\dfrac{\partial g}{\partial \varphi}$ in Eqs.(10.204) for constant (stationary) values of R and φ, corresponding to the steady-state oscillations, the stability condition can be written as

$$\mathrm{Re}(\lambda_{1,2}) < 0, \tag{10.205}$$

where

$$\lambda_{1,2} = \frac{f_R + g_\varphi}{2} \pm \sqrt{\left(\frac{f_R - g_\varphi}{2}\right)^2 + f_\varphi g_R} \tag{10.206}$$

since the time dependence of the small variations of R and φ from their steady-state values is governed by the equations $\delta R = A_1 e^{\lambda_1 t} + A_2 e^{\lambda_2 t}$ and $\delta\varphi = B_1 e^{\lambda_1 t} + B_2 e^{\lambda_2 t}$, where A_1, A_2, B_1 and B_2 are constant.

Considering Eq.(10.206) the condition (10.205) can be rewritten as

$$\left| \begin{array}{l} f_R + g_\varphi < 0, \\ f_R g_\varphi - f_\varphi g_R > 0. \end{array} \right. \quad \begin{array}{r} (10.207a) \\ (10.207b) \end{array}$$

The partial derivatives can be expressed as follows:

$$\left| \begin{array}{l} f_R = \dfrac{\omega_0}{\omega}\dfrac{E_0}{B_0}k\left[-\dfrac{1}{kR}J_N'(kR) + \left(\dfrac{N^2}{k^2R^2} - 1\right)J_N(kR)\right], \\ f_\varphi = \dfrac{\omega_0}{\omega}\dfrac{E_0}{B_0}NJ_N'(kR)\cos(N\varphi-\alpha), \\ g_R = \dfrac{\omega_0}{\omega}\dfrac{E_0}{B_0}\dfrac{Nk^2}{k^2R^2}\left[-\dfrac{2}{kR}J_N(kR) + J_N'(kR)\right]\cos(N\varphi-\alpha), \\ g_\varphi = -\dfrac{\omega_0}{\omega}\dfrac{E_0}{B_0}k\dfrac{N^2}{k^2R^2}J_N(kR)\sin(N\varphi-\alpha). \end{array} \right. \tag{10.208}$$

Considering Eqs.(10.203) and (10.202), Eqs.(10.208) become

$$\left|\begin{aligned} f_R &= -4\beta + \frac{\omega_0}{\omega}\frac{E_0}{B_0}k\left(\frac{N^2}{k^2R^2}-1\right)J_N(kR)\sin(N\varphi-\alpha), \\ f_\varphi &= \frac{\omega_0}{\omega}\frac{E_0}{B_0}NJ_N'(kR)\cos(N\varphi-\alpha), \\ g_R &= -\frac{2}{R}(\omega-\omega_0) + \frac{\omega_0}{\omega}\frac{E_0}{B_0}Nk^2\frac{1}{k^2R^2}J_N'(kR)\cos(N\varphi-\alpha), \\ g_\varphi &= -\frac{\omega_0}{\omega}\frac{E_0}{B_0}k\frac{N^2}{k^2R^2}J_N(kR)\sin(N\varphi-\alpha). \end{aligned}\right. \tag{10.209}$$

Combining, from (10.209) we can write

$$f_R + g_\varphi = -4\beta - F_0 J_N(\rho)\sin\gamma, \tag{10.210}$$

$$\begin{aligned} f_R g_\varphi - f_\varphi g_R &= F_0^2\frac{N^2}{\rho^2}\left[\left(1-\frac{N^2}{\rho^2}\right)J_N^2(\rho) - J_N'^2(\rho)\right] \\ &+ F_0\left[4\beta\frac{N^2}{\rho^2}J_N(\rho)\sin\gamma + 2(\omega-\omega_0)\frac{N}{\rho}J_N'(\rho)\cos\gamma\right] \\ &+ (N^2-\rho^2)(\omega-\omega_0)^2 + 4\beta^2N^2, \end{aligned} \tag{10.211}$$

where the following designations are introduced: $\frac{\omega_0}{\omega}\frac{E_0}{B_0}k = F_0$, $kR = \rho$, $N\varphi - \alpha = \gamma$.

First we consider the case of small amplitudes, i.e. $|\rho| \ll 1$. In this case we can use the following asymptotical expressions for the Bessel functions [52]

$$J_N(\rho) = \left(\frac{\rho}{2}\right)^N \sum_{l=0}^{\infty}\frac{(-1)^l}{l!(N+l)!}\left(\frac{\rho}{2}\right)^{2l} \simeq \frac{1}{N!}\left(\frac{\rho}{2}\right)^N + \ldots, \tag{10.212}$$

$$J_N'(\rho) = \frac{1}{2(N-1)!}\left(\frac{\rho}{2}\right)^{N-1} + \ldots\,. \tag{10.213}$$

From (10.201), (10.202) and (10.203) we find

$$F_0 J_N'(\rho)\sin\gamma = 2\beta\rho, \tag{10.214a}$$

$$F_0\frac{N}{\rho^2}J_N(\rho)\cos\gamma = \omega - \omega_0. \tag{10.214b}$$

Substituting (10.212) and (10.213) into (10.214) we determine

$$\left|\begin{aligned} \operatorname{tg}\gamma &\simeq \frac{2\beta}{\omega-\omega_0}, \\ |\rho|^{N-2} &\simeq \frac{2^N(N-1)!}{|F_0|}\left[(2\beta)^2 + (\omega-\omega_0)^2\right]^{1/2}. \end{aligned}\right. \tag{10.215}$$

From (10.215) it is evident that the spectrum of the possible amplitudes is uninterrupted and in this case there are no conditions for the amplitude discretization.

When $|\rho| \ll 1$ and $N \gg 1$, we find from (10.210) and (10.211)

$$f_R + g_\varphi = -4\beta < 0, \tag{10.216}$$

$$f_R g_\varphi - f_\varphi g_R \simeq N^2(\omega - \omega_0)^2 + 4\beta^2 N^2 > 0, \tag{10.217}$$

i.e. the conditions (10.207) are satisfied and in this case the system motion is stable.

Let us now consider the resonance case, which means

$$\omega - \omega_0 \simeq 0, \tag{10.218}$$

or considering Eq.(10.214b) this is equivalent to

$$|F_0|\frac{N}{\rho^2} \gg |\omega - \omega_0|. \tag{10.219}$$

Two possibilities follow from Eqs.(10.214):

i) $J_N(\rho) = 0$ and $\cos\gamma \neq 0$ or ii) $\cos\gamma = 0$.

We show that when the amplitudes are large ($\rho \gg 1$) the motion in case i) is unstable, while in case ii) it is stable.

Case i),

$$J_N(\rho) = 0 \qquad \text{and} \qquad \cos\gamma \neq 0. \tag{10.220}$$

Then $J'_N(\rho) \neq 0$. In Eq.(10.211) we neglect $(\omega - \omega_0)$ and $J_N(\rho)$ in correspondence with (10.218) and (10.220). We find: $f_R g_\varphi - f_\varphi g_R \simeq -\dfrac{F_0^2 N^2}{2} J_N'^2(\rho) + 4\beta^2 N^2$. However from (10.214a) it follows

$$F_0 \frac{J'_N(\rho)}{\rho} = \frac{2\beta}{\sin\gamma}, \quad \text{i.e.} \quad f_R g_\varphi - f_\varphi g_R \simeq 4\beta^2 N^2 \left(1 - \frac{1}{\sin^2\gamma}\right) < 0 \tag{10.221}$$

and apparently in this case the motion is unstable.

Case b),

$$\cos\gamma = 0 \tag{10.222}$$

or

$$\sin\gamma = (-1)^m, \quad m = 0, 1 \tag{10.223}$$

(the two cases are possible, e.g. with adding π to γ).

As here β is not of essential significance, for the sake of simplification we set

$$\beta \to 0. \tag{10.224}$$

From Eq.(10.214a) it follows that:

$$J'_N(\rho) = 0. \tag{10.225}$$

The condition (10.225) determines the possible discrete spectrum of amplitudes ρ. These amplitudes do not depend on the force E_0 (or F_0).

Taking into account Eqs.(10.218), (10.222), (10.224) and (10.225) we find from Eq.(10.211):

$$f_R g_\varphi - f_\varphi g_R \simeq F_0^2 \frac{N^2}{\rho^2}\left(1 - \frac{N^2}{\rho^2}\right) J_N^2(\rho). \tag{10.226}$$

When

$$\rho > N, \tag{10.227}$$

from (10.226), it follows that the stability condition (10.207b) is satisfied: $f_R g_\varphi - f_\varphi g_R > 0$.

From (10.210) and (10.223) we obtain

$$f_R + g_\varphi \simeq -4\beta - (-1)^m F_0 J_N(\rho).$$

Selecting the values for m, it is also possible to satisfy the second stability condition (10.207a): $f_R + g_\varphi < 0$, or

$$F_0 \sin\gamma J_N(\rho) > 0. \tag{10.228}$$

10.4.3. The wave nature and dynamical quantization of the Solar System

A heuristic model of the mean distances between the Solar-System planets, their satellites and primaries is here proposed. The model is based on: (i) the concept of the wave nature of the Solar System structure; (ii) the micro-mega analogy (MM analogy) of the micro- and mega-system structures, and (iii) the oscillator amplitude „quantization" phenomenon, occurring under wave action, discovered on the basis of the classical oscillations theory (see 10.4.2.).

From the equation, describing the charge rotation under the action of an electromagnetic wave, we obtain an expression for the discrete set of probable stationary motion amplitudes. The discrete amplitudes values – the „quantization" phenomenon – are defined by the argument values at the extreme points of the N-order Bessel functions. Using this expression, the mean related distances are computed from the Solar System planets and the Saturnian, Uranian and Jovian satellites to the primaries.

The commensurability and resonance phenomena of the Solar System motion structure – (including planets, asteroids, planet satellites) – have been an object of detailed discussions and experimental examination in the last years [326–330].

Systematic observations and measurements have been carried out by using all available means. A considerable bulk of empiric data on the Solar System has been obtained from the „Voyager 2“ mission. The understanding of the inevitable resonance character of evolving mature oscillation systems leads to series of interesting concepts on the resonance character of the Solar System motion dynamics. Among the most outstanding of them is Molchanov's hypothesis [326] on the complete resonance character of the large planet in-orbit motion. A.M. Molchanov has noticed that the mean motion of the nine large planets $n_1, ..., n_9$ is related approximately to nine linear homogeneous equations

$$k_1^{(j)} n_1 + ... + k_9^{(j)} n_9 = 0, \quad j = 1, ..., 9,$$

with integer coefficients $k_1^{(j)}, ..., k_9^{(j)}$. The mean motions of the Jovian, Saturnian and Uranian satellites are related to similar equations.

If asteroids are considered to be planets and if Pluto is excluded, it is established [330] that the planetary distances obey the following regularity:

$$\frac{a_{k+1}}{a_k} \simeq 1.75 \pm 0.20,$$

i.e. the ratio of the semi-major orbital axes of neighbouring planets is almost constant.

So far, whole series of phenomena in the Solar System structure have been considered anomalous or obscure. For example, the revolution periods of the celestial bodies vary within a comparatively narrow range of values even if their masses differ considerably.

Those and other phenomena do not fit properly in the pattern of classical celestial mechanics and are actually ignored due to the absence of a traditional explanation.

The development of astronautics and modern astrophysics calls for: (i) an understanding of the anomalies' origin; (ii) filling in the gaps in the existing theoretical models; (iii) clarification of certain phenomena, such as the nature of stability of commensurability, periodicity and resonances and their probable preferability, having in mind that the above-mentioned phenomena are observed not only among celestial bodies, but also among artificial satellites, space ferries and orbital stations.

Virtually, all contemporary scientists come to the conclusion that the new picture of the Universe is based on „closing the ring of Nature“ that leads to close interrelations between microcosm physics and astrophysics. The general data on near-the-Earth, deep space and especially Solar System structure cannot be thoroughly interpreted without a general notion of the wave nature of occurring phenomena.

Since 1982, great interest has arisen in the quantization problem – (in the large) and in the Universe mega-system wave structure [331-335]. The concept and the corresponding idea of the mega-quantum wave structure of an astronomic system has been considered.

According to the wave Universe concept, the large astronomic systems are considered to be wave dynamic systems that are, in a certain sense, analogous to the atomic system [336]. From that point of view the Solar System is regarded as a wave dynamical system. Its components – the celestial bodies (the Sun, the planets, the satellites and the small celestial bodies) and the interplanetary continuum (interplanetary plasma, electromagnetic field, etc.) are described within a common dynamical substance – field context.

The phenomenologic and dynamic description of such systems is connected with the mega-quantum wave dynamics. Assuming that the Solar System is a wave dynamic system and hence, the micro-mega-analogy (MM-analogy) is valid, series of deductions have been drawn on the basis of the dynamic isomorphism of the atomic and Solar System structures [337]. Different quantization versions have been used – according to Bohr, De Broglie, Sommerfeld, Schrödinger, etc. The essence of all those fundamental works is the substantiation of the existence of mega-waves, realizing short-range interactions on a scale commensurable with the systems scale in any Universal mega-system and, in particular, in the Solar System, representing an analog to De Broglie's waves for gigantic astronomic systems. However, the common opinion is that the major problem at present is to carry out a more detailed identification, and make a phenomenologic and dynamic description of the Solar System mega-quantum structure and of the analogous systems, consisting of a central body and satellites [330].

The above-mentioned concepts of the bodies' waves interaction unity in the micro- and macro-world as well as of the basic realization and manifestation behavior closing up in Nature is the essence and grounds for the presented work. For the purpose of commensurability, resonance and dynamic quantization studies of the Solar System body motion structure, different authors use versions of the quantization phenomena within the framework of Quantum Mechanics. Herewith, a heuristic model is proposed of the discrete distribution of Solar System planets and satellites mean distances from the primaries. That mega-quantum resonance-wave model is based on the Solar System structure wave nature concept, as well as on the oscillator amplitude „quantization" phenomenon, occurring under wave action, discovered on the basis of the classical oscillations theory (see 10.4.1.). The analysis of the latter explains and gives analytic grounds for: (i) the existence of a discrete set of stable („allowed") orbits and a set of resonant, but unstable („forbidden") orbits of the planets and their satellites; (ii) the phase locking of the in-orbit motion; (iii) the trajectory stability only under exceptional initial conditions and in strictly defined areas, and (iv) the independence of the stable stationary orbits from the planets' and satellites' masses.

As a rule, the harmonic oscillator is taken as the basis of the field theory interpretations [336] from the perspective of the classical theory of oscillations and the quantum-mechanical approach. The classical approach is of greater heuristic value and efficiency when it comes to the action of (in the classical sense) fully determined forces in an oscillator. This fact, noted by other researchers, is confirmed by the analysis, presented below.

Here, we take as a general model a periodic motion that is the rotation of a charge in the course of which an electromagnetic wave falls along the X-axis.

The charge motion equation is

$$\ddot{\mathbf{r}} + 2\beta\dot{\mathbf{r}} + \omega_0^2\mathbf{r} = e\mathbf{E}\sin(vt - \mathbf{kr}), \tag{10.229}$$

where $\mathbf{r}$ is the charge radius-vector; 2β is the dissipation coefficient; e is the charge value; $\mathbf{k}$ is the wave vector; $\mathbf{E}$ is the wave field intensity and v, t are the frequency and time parameters, respectively.

Plotting the equation along the X-axis, we obtain

$$\ddot{x} + 2\beta\dot{x} + \omega_0^2 x = eE_x\sin(vt - kx), \tag{10.230}$$

where E_x is the X-axis electrical field component and k is the wave modulus.

The solution of Eq.(10.230), describing a lincar oscillator under wave action, is written in the form of a quasi-harmonic function

$$x(t) = a\sin(\omega t + \alpha), \tag{10.231}$$

where $a(t)$ and $\alpha(t)$ are the slowly changing amplitude and phase, $\omega = v/N$ is the charge periodic motion frequency, and $N = 1, 2, 3, ...$ is a whole number.

When using the averaging method [97], shortened version, these equations are

$$\left| \begin{aligned} \dot{a} &= -\beta a + \frac{eE_x N}{ka\omega} J_N(ka)\sin N\alpha \\ \dot{\alpha} &= \frac{\omega_0^2 - \omega^2}{2\omega} - \frac{eE_x}{a\omega} J_N'(ka)\cos N\alpha \end{aligned} \right. , \tag{10.232}$$

where J_N is the N-order Bessel function of the first kind.

In a charge stationary periodic motion mode, i.e. when $\dot{a} = 0$ and $\dot{\alpha} = 0$, Eqs.(10.232) can be written as

$$\beta a = \frac{eE_x N}{ka\omega} J_N(ka)\sin N\alpha \tag{10.233}$$

$$\omega_0^2 - \omega^2 = \frac{2eE_x}{a} J_N'(ka)\cos N\alpha. \tag{10.234}$$

According to the amplitude „quantization" phenomenon theory presented above, the role of the phase α is essential when the excited oscillation synchronization and adaptive stability maintenance under different perturbing actions are considered.

In a resonance mode* , condition $\omega = \omega_0$ is accomplished and Eq.(10.234) is valid provided that

$$J'_N(ka) = 0 \quad \text{and} \quad \cos N\alpha = 0.$$

In order to examine the stationary amplitude and the phase stability, the set of Eqs.(10.232) is rewritten, as follows:

$$\dot{a} = f(a, \alpha), \quad \dot{\alpha} = g(a, \alpha). \tag{10.235}$$

The derivatives $\partial f/\partial a$, $\partial f/\partial \alpha$ and $\partial g/\partial a$, $\partial g/\partial \alpha$ obtained for the constant values of a and α and corresponding to the stationary oscillations are designated by f_a, f_α and g_a, g_α. Hence, the stability condition can be rewritten as follows:

$$\mathrm{Re}P_{1,2} < 0, \text{ where } P_{1,2} = \frac{f_a + g_\alpha}{2} \pm \sqrt{\left(\frac{f_a - g_\alpha}{2}\right)^2 + f_\alpha g_a}, \tag{10.236}$$

$$\frac{f_a + g_\alpha}{2} = -\frac{\beta}{2} - \frac{eE_xN}{2k\omega a^2} \sin N\alpha [J_N(ka) - 2kaJ'_N(ka)],$$

$$\frac{f_a - g_\alpha}{2} = -\frac{\beta}{2} - \frac{eE_xN}{2k\omega a^2} \sin N\alpha J_N(ka),$$

$$f_a g_\alpha = \frac{(eE_xN)^2}{k\omega^2 a^3} \cos^3 N\alpha J_N(ka)[J'_N(ka) - kaJ''_N(ka)].$$

The $P_{1,2}$ quantities are derived from the time-dependent small deviations of the stationary values of a and α, expressed as:

$$\Delta a = A_1 e^{P_1 t} + A_2 e^{P_2 t}, \quad \Delta\alpha = B_1 e^{P_1 t} + B_2 e^{P_2 t},$$

where A_1, A_2, B_1 and B_2 are constants.

The examination of the stability solutions (Eqs.(10.232)) shows that the stable oscillations amplitude a satisfies the condition

$$J'_N(ka) = 0. \tag{10.237}$$

Hence, the stationary charge oscillations can be realized for amplitudes a_i, belonging to a strictly defined set of amplitude values. The a_i values are determined by the Bessel function extremes, and by using Eq.(10.237) it may be presented as

$$ka_i = j_{N,i}, \tag{10.238}$$

* The resonance case is in conformity with the idea about evolutionary mature systems [330].

where $j_{N,i}$ is the N-th order Bessel function J_N argument value ka_i at the i-th extremum point.

In conformity with the general premises, we assume that the oscillating charge system under wave action may be used as a general model to describe the Solar System. The large Earth orbit semi-axis a_E complies with Eq.(10.237) at $N = 8$ and $i = 5$. Those conditions are assumed to be basic: $ka_E = J_{8,5}$.

Further, by using Exprs.(10.238) the following semi-major axes relation is obtained:

$$\frac{a_i}{a_E} = \frac{ka_i}{ka_E} = \frac{j_{N,i}}{j_{8,5}} = \frac{j_{8,i}}{j_{8,5}}. \tag{10.239}$$

At the same time it should be noted that the Solar System planets arrangement has a regular character, expressed on the whole by the Titius-Bode law [338], giving the mean planets-to-Sun distances a_k by equation

$$a_k = 0.4 + 0.3\ 2^k, \tag{10.240}$$

where a_k is expressed in astronomical units, and $k = -\infty$ for Mercury, $k = 0$ for Venus, $k = 1$ for the Earth, etc. For more than 200 years this law has not found a convincing theoretical explanation.

A better expression of the mean distances is proposed by several authors [339–342] in the following form:

$$r_n = r_0 d^n, \tag{10.241}$$

where n is the n-th planet distance from the Sun or n-th satellite distance from the primary, r_0 is a simple normalizing distance, different for each system, d is a constant, characterizing the geometric progression valid for all systems. The integer n for Mercury is equal to 1, for Venus it is equal to 2, etc. Some of the values do not correspond to an existing celestial body, but to a „hole". Exprs.(10.241), compared to the earlier Titius-Bode law (10.240) is less artificial, as the discount value $n = -\infty$ for Mercury is avoided.

The Solar System planets mean distances are presented in Table 10.2. For the sake of comparison the direct astronomic measurements data is given parallel to the result, computed according to the classical Titius-Bode law (10.240) and Eq.(10.239) according to the „oscillator-wave" model, described above. All data are expressed in astronomical units (A.U.).

A good correspondence is observed between the computed (Eq.(10.239)) and astronomically measured radii. The correspondence between the computed and measured radii of Neptune and Pluto is particularly significant. The Titius-Bode law determines the mean distances of those two planets with an error of 23% and 49%, respectively.

The computed data of the mean satellite distances from Saturn, Uranus and Jupiter, as well as the mean ring system distances from Saturn, are given in Table 10.3. The calculations are made on the basis of the „oscillator-wave" model, by

Table 10.2

Mean planet distances in the Solar System

Planets in the Solar System	Data from direct astronomical measurements of planet distances from the Sun [343], A.U.	Titius-Bode Law (10.240), [343], A.U. k	a_k	Computed planet distances using Eq. (10.239) of the "Oscillator-wave" model i	$\frac{a_i}{a_E} = \frac{j_{8,j}}{j_{8,5}}$
Mercury	0.39	$-\infty$	0.4	1	0.392
Venus	0.72	0	0.7	3	0.723
Earth	1.00	1	1.0	5	1.000
Mars	1.52	2	1.6	9	1.530
Asteroids	2.78	3	2.8	19	2.824
Jupiter	5.2	4	5.2	37	5.132
Saturn	9.55	5	10.0	71	9.474
Chiron (Collewll's objects)	13.71			104	13.689
Uranus	19.18	6	19.6	147	19.180
Neptune	30.03	7	38.8	232	30.035
Pluto	39.67	8	77.2	307	39.598

using an expression, similar to Eq.(10.239). For the sake of compariso16n, the data obtained from direct astronomical distance measurements [343] are presented as well. Again, a good correspondence is seen between the calculated and measured mean distances.

The measurement and computed data normalizing is realized relatively to an arbitrary chosen representative body. Obviously, there are no restrictions and any other satellite can be chosen as a representative body for the performance of similar normalizing.

It is necessary to note that, according to the „oscillator-wave" model, there is a certain differentiation between the Solar System (Table 10.2.) and planet-satellite (Table 10.3.) structures. In the first case, the wave action frequency v and Solar System planets revolution frequencies ω_i relation remains invariable ($N = \gamma/\omega_i$ = const). The differences between the planet mean distances from the Sun are determined by the values of the Bessel's function arguments $j_{8,i}$ at the extreme i-th points. In the second case, i.e., for planet satellite systems, the correspondence between the calculation and astronomical measurements data

Table 10.3

Mean Saturnian, Uranian, Jovian satellites and Saturn-ring-system distances from the primaries in conformity with the proposed „oscillator-wave" model

Satellites	Data from direct astronomical measurements of satellite mean distance from the primaries [343], 10^{-3} A.U.	Normalized data from direct astronomical measurement	N	Normalized computed mean satellite distaces from the primaries in conformity with the „Oscillator-Wave" model
Saturnian satellites				$j_{N,1}/j_{10,1}$
Janus	1.060	0.301	2	0.308
Mimas	1.241	0.352	3	0.357
Enceladus	1.592	0.452	4	0.452
Tethys	1.970	0.550	5	0.545
Dione	2.523	0.716	7	0.729
Rhea	3.524	1.000	10	1.000
Titan	8.166	2.317	25	2.328
Hyperion	9.911	2.812	31	2.838
Iapetus	23.718	6.736	77	6.737
Phoebe	86.580	24.569	284	24.560
Uranian Satellites				$j_{N,1}/j_{10,1}$
Miranda	0.872	0.488	4	0.452
Ariel	1.282	0.718	7	0.728
Umbriel	1.786	1.000	10	1.000
Titania	2.930	1.641	17	1.624
Oberon	3.919	2.104	23	2.152
Jovian Satellites				$j_{N,1}/j_{17,1}$
Amalthea	1.209	0.096	1	0.096
Io	2.819	0.224	3	0.220
Europa	4.489	0.357	5	0.336
Ganymede	7.155	0.568	9	0.563
Calisto	12.585	1.000	17	1.000
Saturn Ring System				$j_{N,1}/j_{14,1}$
Crape of C ring: faint	0.498	0.827	11	0.803
Gap: dark	0.595	0.988	13	0.934
Main B ring: very bright	0.602	1.000	14	1.000
Cassini division: dark	0.782	1.299	18	1.261
Outer A ring: moderately bright	0.802	1.332	19	1.326

are observed for different frequency multiplicities ($N = v/\omega_i$ =vary) and the corresponding values of the first Bessel extremes $j_{N,i}$.

10.4.4. Approach in the case of large amplitudes of the oscillations in a nonlinear dynamical system existing under wave action

Let the nonlinear oscillator be an electric charge q with mass m and let it be able to oscillate along the X-axis with a small friction force $2\delta_0\dot{X}$. Let an electromagnetic wave propagating along the X-axis acts upon the oscillating charge. Let us assume that the wave has a longitudinal component of the electric field E_x. The equation of the charge motion becomes*

$$\ddot{X} + 2\delta_0\dot{X} + \omega_0^2 \sin X = P_0 \sin(vt_r - kX - \varphi), \tag{10.242}$$

where ω_0 is the resonant frequency of small amplitude oscillations $P_0 = E_x q/m$; v, φ and k are the frequency, the initial phase and the wave number respectively, t_r is the real time. We will assume $v \gg \omega_0$.

Let us introduce the dimensionless time $t = \omega_0 t_r$. In this case, Eq.(10.242) takes the form

$$\ddot{X} + 2\delta_d\dot{X} + \sin X = F_0 \sin\left(\frac{v}{\omega_0}t - kX - \varphi\right), \tag{10.243}$$

where $2\delta_d = 2\delta_0/\omega_0$, $F_0 = P_0/\omega_0^2$.

In order to integrate Eq.(10.243) by using the methods of the Theory of nonlinear oscillations, we apply the approach developed in 10.3.8. We introduce the new variable y and the nonlinear time τ using Exprs.(10.143) and (10.144). So, the nonlinear reactive term $\sin X$ in Eq.(10.243) may be „excluded". The functions $X(y)$ and $G(y)$ in (10.144) are easily expressed by taking into account (10.143) in the form (10.145). Substituting (10.143) and (10.144) in (10.243) we obtain

$$\frac{d^2y}{d\tau^2} + y = \left\{-2\delta_d\frac{dy}{d\tau} + F_0 \sin\left[\frac{v}{\omega_0}t(\tau) - kX(y) - \varphi\right]\right\} G(y). \tag{10.244}$$

The further consideration will be performed for the following interval of y values: $-2 < y < 2$.

Before the integration of Eq.(10.244) we will mention that the solution will be quasi-harmonic with nominal frequency $\omega_n = v/N$, where $N \gg 1$ is a positive odd number, however $\omega_n \sim \omega_0$. That is why we will write Eq.(10.244) in the following form:

$$\frac{d^2y}{d\tau^2} + \beta^2 y = -2\delta_d\frac{dy}{d\tau} + F_0 \sin\left[\frac{v}{\omega_0}t(\tau) - kX(y) - \varphi\right] G(y) + (\beta^2 - 1)y, \tag{10.245}$$

* Similar equations describe the behaviour of cosmic charged particles in certain conditions, the process in radio-frequency driven, quantum-mechanical Josephson's junctions, charge density wave transport and many more.

where $\beta \sim 1$ corresponds to the deviation from the resonant frequency.

We assume that in excitation of the charge oscillations by the wave its motion is symmetric with respect to the equilibrium and the charge coordinate changes in agreement with

$$y = R\cos\beta\tau = R\cos\Psi. \tag{10.246}$$

The dependence of the normalized time t on the angle Ψ can be expressed in accordance with (10.144), (10.145) and (10.246) in the form (10.149). The normalized period of the oscillations is given in accordance with (10.150). By use of (10.150) the coefficient β is expressed in the form (10.151).

Now we can solve Eq.(10.245).

The shortened (reduced) differential equations for the amplitude R and phase φ may be written as:

$$\frac{dR}{d\tau} = -\frac{1}{2\pi\beta}\int_0^{2\pi} L[R\cos\Psi, -\beta R\sin\Psi, e(\Psi-\varphi)]\sin\Psi d\Psi, \tag{10.247}$$

$$\frac{d\varphi}{d\tau} = -\frac{1}{2\pi\beta}\int_0^{2\pi} L[R\cos\Psi, -\beta R\sin\Psi, e(\Psi-\varphi)]\cos\Psi d\Psi, \tag{10.248}$$

where

$$L[R\cos\Psi, -\beta R\sin\Psi, e(\Psi-\varphi)] = 2\delta_d\beta R\sin\Psi + \frac{F_0}{\sqrt{1-R^2/4\cos^2\Psi}}\sin\left[\frac{\pi N}{2K(R/2)}\right].$$

Let us introduce the following designations:

$$\begin{Bmatrix} H_1 \\ H_2 \end{Bmatrix} = \int_0^{K(R/2)} \sin(rZ)\cos[D(Z)]\begin{Bmatrix} \operatorname{sn} Z \\ \operatorname{cn} Z \end{Bmatrix} dZ, \tag{10.249}$$

$$\begin{Bmatrix} H_3 \\ H_4 \end{Bmatrix} = \int_0^{K(R/2)} \cos(rZ)\sin[D(Z)]\begin{Bmatrix} \operatorname{sn} Z \\ \operatorname{cn} Z \end{Bmatrix} dZ, \tag{10.250}$$

$$\begin{Bmatrix} H_5 \\ H_6 \end{Bmatrix} = \int_0^{K(R/2)} \cos(rZ)\cos[D(Z)]\begin{Bmatrix} \operatorname{sn} Z \\ \operatorname{cn} Z \end{Bmatrix} dZ, \tag{10.251}$$

$$\begin{Bmatrix} H_7 \\ H_8 \end{Bmatrix} = \int_0^{K(R/2)} \sin(rZ)\sin[D(Z)]\begin{Bmatrix} \operatorname{sn} Z \\ \operatorname{cn} Z \end{Bmatrix} dZ, \tag{10.252}$$

where $Z = F(\Psi, R/2)$ is an incomplete elliptic integral of the first kind,

$$D(Z) = 2kE\left[\arcsin\left(\frac{R}{2}\operatorname{cn} Z\right), \frac{2}{R}\right],$$

$E[.,.]$ is an incomplete elliptic integral of the second kind, $\operatorname{sn} Z$ and $\operatorname{cn} Z$ are the sine and cosine of the amplitude (the Jacobi elliptic functions),

$$r = \frac{\pi N}{2K(R/2)}.$$

Taking into account Exprs.(10.249)-(10.252), the shortened Eqs.(10.247) and (10.248) for establishing the amplitude R and phase φ take the form

$$\frac{dR}{d\tau} = -\delta_d R - \frac{F_0}{2\pi\beta}[(H_1 - H_3)\cos\varphi - (H_5 + H_7)\sin\varphi],$$

$$\frac{d\varphi}{d\tau} = -\frac{F_0}{2\pi\beta}R[(H_2 - H_4)\cos\varphi - (H_6 + H_8)\sin\varphi] - (\beta - 1).$$

For the stationary mode $\left(\frac{dR}{d\tau} = 0,\ \frac{d\varphi}{d\tau} = 0\right)$ we obtain the following expressions for the established values of the amplitude R and the phase φ:

$$R=\frac{F_0[(H_1 - H_3)(H_6 - H_8) - (H_2 - H_4)(H_5 + H_7)]}{2\pi\beta\delta_d\sqrt{[\sigma(H_1 - H_3) - (H_2 - H_4)]^2 + [\sigma(H_5 + H_7) - (H_6 + H_8)]^2}} \qquad (10.253)$$

$$\varphi = \operatorname{arctg}\frac{\sigma(H_1 - H_3) - (H_2 - H_4)}{\sigma(H_5 + H_7) - (H_6 + H_8)}, \qquad (10.254)$$

where $\sigma = (\beta - 1)/\delta_d$.

The analysis of Exprs.(10.253) and (10.254) reveals their qualitative similarity with Exprs.(10.160) and (10.161). Dependencies qualitatively analogous to those shown in Fig.10.24, Fig.10.25 and Fig.10.26 are also obtained in the case under consideration.

Thus the peculiarities typical of amplitude discretization are also proper to nonlinear dynamic systems existing under the action of a falling wave.

Since the „oscillator-wave" system may vary in its physical nature, the observed features are of a significantly common character. In the case considered in Section 10.3.8., the interaction non-homogeneity was especially arranged – by restricting the external force action on a small part of the trajectory. In this case of wave action on the oscillator, on the contrary, nothing has been done to ensure the non-homogeneity of the interaction.

The general nature of the oscillator-wave model can be proved by using various examples drawn from nature, science and technology, where similar systems are in abundance. Wherever waves act on oscillators, conditions for exciting oscillations with a possible discrete series of stable amplitudes are created. It was shown earlier, under section 10.4.3, that the oscillator-wave model can be used as a heuristic model of the distribution of the distances between the planets and the satellites in the Solar System.

The presented oscillator-wave model can also serve as a basis for creating heuristic models of the interaction of electromagnetic waves with particles in plasma media in the Earth's ionosphere and magnetosphere, for a heuristic model of the generation of powerful low-frequency waves in the Earth surrounding space in the presence of the cosmic electromagnetic background and many more. Using the oscillator-wave model under consideration as a point of departure, it is possible, by way of generalization for a 3D case, to demonstrate analytically, for instance, the possibility for optic pumping of a microwave SHF emitter and maintenance of a stationary level of the electromagnetic field in the resonator, given sufficiently high intensity of the initial emission [307, 357].

10.4.5. General conditions for transition to irregular behavior in an oscillator under wave action

The model system under consideration is presented by the following system of equations

$$\left|\begin{aligned} &\dot{x}_1 = x_2 \\ &\dot{x}_2 = -\sin x_1 + \mu[F_0 \sin(\theta - \rho x_1) - \delta_d x_2]\ , \\ &\dot{\theta} = v \end{aligned}\right. \tag{10.255}$$

where $[x_1, x_2, \theta(t)] \in \mathrm{R}^3$, the dot denotes an operation of differentiation by the time t, $\theta = vt$, F_0, v and ρ are the amplitude, the frequency and wave number parameters of the external acting wave respectively, δ_d reflects the dissipation in the system, $0 < \mu \ll 1$ is a small parameter.

Proceeding from the non-perturbed system ($\mu = 0$),

$$\left|\begin{aligned} &\dot{x}_1 = x_2 \\ &\dot{x}_2 = -\sin x_1 \end{aligned}\right. ,$$

i.e. from the system with Hamiltonian function $H = \dfrac{x_2^2}{2} + 1 - \cos x_1$, the phase space divides into two domains and a separating boundary (separatrix), where qualitatively different phenomena occur: I domain, $0 < H < 2$; II separatrix, $H = 2$ and III domain (a regime of rotation), $H > 2$.

The solution in the so outlined domains can be presented in the following form.

I domain, notating $\varkappa = \dfrac{H}{2}$:

$$\left|\begin{aligned} &x_1 = 2\arcsin[\varkappa\, \mathrm{sn}(t - t_0, \varkappa)] + 2\pi N, \quad N = 0, \pm 1, \pm 2, \pm 3, ..., \\ &x_2 = 2\varkappa\, \mathrm{cn}(t - t_0, k), \qquad t_0 = \mathrm{const}, \end{aligned}\right.$$

also

$$\left|\begin{aligned} &x_1 = -2\arcsin[\varkappa\, \mathrm{sn}(t - t_0, \varkappa)] + 2\pi N, \\ &x_2 = 2\varkappa\, \mathrm{cn}(t - t_0, \varkappa). \end{aligned}\right.$$

II separatrix:

$$\left| \begin{aligned} x_1 &= 2\arcsin[\mathrm{th}(t - t_0)] + 2\pi N, \\ x_2 &= \frac{2}{\mathrm{ch}(t - t_0)}, \end{aligned} \right.$$

also

$$\left| \begin{aligned} x_1 &= -2\arcsin[\mathrm{th}(t - t_0)] + 2\pi N, \\ x_2 &= -\frac{2}{\mathrm{ch}(t - t_0)}. \end{aligned} \right.$$

III domain (a regime of rotation), notating $\varkappa^2 = \frac{2}{H}$:

$$\left| \begin{aligned} x_1 &= 2\arcsin\left[\mathrm{sn}\left(\frac{t - t_0}{\varkappa}, \varkappa\right)\right] + 2\pi N, \\ x_2 &= \frac{2}{\varkappa}\mathrm{dn}\left(\frac{t - t_0}{\varkappa}, \varkappa\right), \end{aligned} \right.$$

also

$$\left| \begin{aligned} x_1 &= -2\arcsin\left[\mathrm{sn}\left(\frac{t - t_0}{\varkappa}, \varkappa\right)\right] + 2\pi N, \\ x_2 &= -\frac{2}{\varkappa}\mathrm{dn}\left(\frac{t - t_0}{\varkappa}, \varkappa\right). \end{aligned} \right.$$

Let us consider the system dynamics in the domain of the separatrix (the homoclinic trajectory). We work with the solution

$$x_0 = \begin{pmatrix} x_{01} \\ x_{02} \end{pmatrix} = \begin{pmatrix} 2\arcsin[th(t - t_0)] \\ \dfrac{2}{ch(t - t_0)} \end{pmatrix}.$$

According to ([76], §1.4) we can write

$$\frac{d}{dt}x \equiv \begin{pmatrix} \dfrac{dx_1}{dt} \\ \dfrac{dx_2}{dt} \end{pmatrix} = f_0(x) + \mu f_1(x, t) \equiv \begin{pmatrix} x_2 \\ -\sin x_1 \end{pmatrix} + \mu \begin{pmatrix} 0 \\ F_0 \sin(\theta - \rho x_1) - \delta_d x_2 \end{pmatrix},$$

where, in the right-hand side of the equation, the initial approximation stands at $x_0 = \begin{pmatrix} x_{01} \\ x_{02} \end{pmatrix}$ for $x = \begin{pmatrix} x_1 \\ x_2 \end{pmatrix}$.

Hence,

$$\begin{aligned} f_0 \times f_1 \equiv f_0 \wedge f_1 &= \begin{pmatrix} x_{02} \\ -\sin x_{01} \end{pmatrix} \wedge \begin{pmatrix} 0 \\ F_0 \sin(\theta - \rho x_{01}) - \delta_d x_{02} \end{pmatrix} \\ &= \frac{2F_0}{ch(t - t_0)} \sin\{\theta - 2\rho \arcsin[th(t - t_0)]\} - \frac{4\delta_d}{ch^2(t - t_0)}. \end{aligned}$$

For the system under consideration the Melnikov distance ([23], see formula (3.7.31) can be expressed as follows:

$$D(t_0,t_0) = -\int_{-\infty}^{\infty} f_0 \wedge f_1 dt$$
$$= \int_{-\infty}^{\infty} \left\{ \frac{-2F_0}{\operatorname{ch}(t-t_0)} \sin[\theta - 2\rho \arcsin(\operatorname{th}(t-t_0))] + \frac{4\delta_d}{\operatorname{ch}^2(t-t_0)} \right\} dt. \quad (10.256)$$

Irregular (chaotic) behavior occurs for the areas where $D(t_0, t_0)$ passes through zero. The case of $\rho = 1$ is the most interesting one for practical purposes.

Substituting θ with $\theta = vt+\theta_0 = v(t-t_0)+(vt_0+\theta_0)$, the Eq.(10.256) becomes

$$D(t_0,t_0) = \int_{-\infty}^{\infty} \left\{ -\frac{2F_0}{\operatorname{ch}(t-t_0)} \sin[v(t-t_0)+(vt_0+\theta_0)][1-2\operatorname{th}^2(t-t_0)] \right.$$
$$\left. + \frac{2F_0}{\operatorname{ch}(t-t_0)} \cos[v(t-t_0)+(vt_0+\theta_0)] \frac{2\operatorname{sh}(t-t_0)}{\operatorname{ch}^2(t-t_0)} + \frac{4\delta_d}{\operatorname{ch}^2(t-t_0)} \right\} dt$$
$$= -2F_0 \sin(vt_0+\theta_0) \int_{-\infty}^{\infty} \cos v\tau \frac{(1-\operatorname{sh}^2\tau)}{\operatorname{ch}^3\tau} d\tau$$
$$- 2F_0 \sin(vt_0+\theta_0)^2 \int_{-\infty}^{\infty} \sin v\tau \frac{\operatorname{sh}\tau}{\operatorname{ch}^3\tau} d\tau + 4\delta_d \int_{-\infty}^{\infty} \frac{d\tau}{\operatorname{ch}^2\tau}. \quad (10.257)$$

The integrals in (10.257) are evaluated as follows:

$$\int_{-\infty}^{\infty} \cos v\tau \frac{(1-\operatorname{sh}^2\tau)}{\operatorname{ch}^3\tau} d\tau = \int_{-\infty}^{\infty} \cos v\tau\, d\left(\frac{\operatorname{sh}\tau}{\operatorname{ch}^2\tau}\right) = \cos v\tau \frac{\operatorname{sh}\tau}{\operatorname{ch}^2\tau}\Bigg|_{-\infty}^{\infty}$$
$$-v \int_{-\infty}^{\infty} \sin v\tau \frac{\operatorname{sh}\tau}{\operatorname{ch}^2} d\tau = v \int_{-\infty}^{\infty} \sin v\tau \frac{d(\operatorname{ch}\tau)}{\operatorname{ch}^2\tau}$$
$$= -v \frac{\sin v\tau}{\operatorname{ch}\tau}\Bigg|_{-\infty}^{\infty} + v \int_{-\infty}^{\infty} \frac{d(\operatorname{sh} v\tau)}{\operatorname{ch}\tau} = \frac{v^2\pi}{\operatorname{ch}\frac{v\pi}{2}}. \quad (10.258)$$

$$\int_{-\infty}^{\infty} \sin v\tau \frac{\operatorname{sh}\tau}{\operatorname{ch}^3\tau} d\tau = -\frac{1}{2} \int_{-\infty}^{\infty} \sin v\tau\, d\left(\frac{1}{\operatorname{ch}^2\tau}\right) = -\frac{1}{2} \frac{\sin v\tau}{\operatorname{ch}^2\tau}\Bigg|_{-\infty}^{\infty}$$
$$+\frac{1}{2} \int_{-\infty}^{\infty} \frac{d(\sin v\pi)}{\operatorname{ch}^2\tau} = \frac{v}{2} \int_{-\infty}^{\infty} \frac{\operatorname{ch} v\tau\, d\tau}{\operatorname{ch}^2\tau} = \frac{v^2\pi}{2\operatorname{sh}\frac{v\pi}{2}}. \quad (10.259)$$

$$\int_{-\infty}^{\infty} \frac{d\tau}{\operatorname{ch}^2\tau} = 2. \quad (10.260)$$

According to Exprs.(10.258), (10.259) and (10.260), Eq.(10.257) is expressed by

$$D(t_0, t_0) = -2\pi v^2 F_0 \left(\frac{1}{\operatorname{ch} \frac{v\pi}{2}} + \frac{1}{\operatorname{sh} \frac{v\pi}{2}} \right) \sin(vt_0 + \theta_0) + 8\delta_d$$

or, finally,

$$D(t_0, t_0) = -\frac{4\pi v^2 F_0 e^{v\frac{\pi}{2}}}{\operatorname{sh} v\pi} \sin(vt_0 + \theta_0) + 8\delta_d.$$

Under the condition $D(t_0, t_0) = 0$ and taking into account that $|\sin(vt_0 + \theta_0)| < 1$ and $\delta_d > 0$, the general condition for transition to irregular (chaotic) behaviour in a nonlinear oscillator under wave action takes the form:

$$2\delta_d \operatorname{sh}(|v|\pi) < \pi v^2 |F_0| e^{v\frac{\pi}{2}}.$$

Obviously, this condition is fulfilled in some domains of the space (δ_d, F_0, v).

In conclusion, it is necessary to note that the solution of Eq.(10.255) depends strongly on the choice of initial conditions. Besides, the homoclinic bifurcation is one of the first bifurcations that occur in the transition from regular to irregular motion in the system under consideration. The latter is related to the condition for overcoming the strong self-adaptive mechanism of the system's internal stability. It is also necessary to emphasize that the homoclinic tangency implies formation of a very complicated fractal boundary for the basins of attraction.

10.5. Twist dissipative maps as a generalized model and immanent analyzing technique for the class of kick-excited self-adaptive dynamical systems

10.5.1. Energy balance of the system

The energy vested in the system for a single pass of the phase trajectory through the active zone $[-d', d']$ is defined as

$$\Delta E_{in} = \int_{-d'}^{d'} F_0 \sin[vt(x)] dx. \tag{10.261}$$

Introducing phase variable $\Psi = vt$ and denoting by V_a the mean velocity in the active area, Expr.(10.261) is developed into:

$$\Delta E_{in} = \int_{\Psi_{in}}^{\Psi_{out}} \frac{F_0}{v} \dot{x} \sin \Psi d\Psi = \frac{F_0 V_a}{v} \int_{\Psi_{in}}^{\Psi_{out}} \sin \Psi d\Psi = 2F_0 d' \frac{\sin \xi}{\xi} \sin \Psi_0, \tag{10.262}$$

where $\Psi_0 = \dfrac{\Psi_{in} + \Psi_{out}}{2}$, $\xi = \dfrac{\Psi_{out} - \Psi_{in}}{2} \simeq \dfrac{vd'}{V_a}$. Here, Ψ_0 is the mean phase, meaning the phase of the external force at the time when the movement trajectory passes approximately through the middle of the active area, and ξ is the phase half-width of the active zone. Approximated for the case of small phase width where $\sin\xi \simeq \xi$, Expr.(10.262) is reduced merely to

$$\Delta E_{in} \simeq 2F_0 d' \sin\Psi_0. \tag{10.263}$$

The energy losses of the pendulum for one half-period with narrow active area, compared to the oscillation period, can be assumed equal to the losses of a free damping pendulum provided by the expression

$$\Delta E_{out} \simeq 16\delta_d[E(m) - (1-m)K(m)]. \tag{10.264}$$

In (10.264), the following symbols are used: $m = \dfrac{E_0}{2}$, $E_0 = \dfrac{\dot{x}^2}{2} + (1 - \cos x)$ is the pendulum's full energy; $K(m)$ and $E(m)$ are full elliptic integrals of first and second kind, respectively.

10.5.2. Construction of a discrete map

Let us relate the discrete map to the energy-phase variables referring to time $(n-1)$ where the movement trajectory leaves the action zone (See Fig.10.6). Let us use as energy variable the quantity $m = \dfrac{E_0}{2}$.

From Fig.10.7 it can be seen plainly that the change of the phase variable θ_n within a half-period can be expressed by the pendulum's half-period in the following form

$$\theta_{n+1} = \theta_n + \frac{vT(E)}{2} + \pi = \theta_n + 2vK(m_{n+1}) + \pi (\operatorname{mod} 2\pi).$$

The change of the energy variable m_n follows immediately from the pendulum's energy balance

$$m_{n+1} = m_n + \Delta m_{in} - \Delta m_{out},$$

where $\Delta m_{in} = \dfrac{\Delta E_{in}}{2}$ is expressed by (10.263), and $\Delta m_{out} = \dfrac{\Delta E_{out}}{2}$ – by (10.264), accordingly. Thus, we obtain the following 2D discrete map

$$\left| \begin{array}{l} m_{n+1} = m_n - 8\delta_d[E(m_n) - (1-m_n)K(m_n)] + F_0 d' \sin\theta_n \\ \theta_{n+1} = \theta_n + 2vK(m_{n+1}) + \pi \quad (\operatorname{mod} 2\pi) \end{array} \right. . \tag{10.265}$$

If appropriate expansions for the full elliptic integrals [324] are used, i.e.

$$K(m) = \frac{\pi}{2}(1 + \frac{m}{4} + \frac{9}{64}m^2 + ...), \quad E(m) = \frac{\pi}{2}(1 - \frac{m}{4} - \frac{3}{64}m^2 + ...), \tag{10.266}$$

then the dissipative term in the energy equation from (10.265) can be represented as

$$\Delta m_{out} = 2\pi\delta_d m(1 + \frac{m}{8} + ...) = 2\pi\delta_d m D(m), \qquad (10.267)$$

where $D(m)$ is a function close to unit when the values of m are not big.

Accounting for (10.267), map (10.265) assumes the form:

$$\left| \begin{array}{l} m_{n+1} = m_n[1 - 2\pi\delta_d D(m_n)] + F_0 d' \sin\theta_n \\ \theta_{n+1} = \theta_n + 2vK(m_{n+1}) + \pi \quad (\mathrm{mod}\, 2\pi) \end{array} \right. . \qquad (10.268)$$

Written in this form, the map of the kick-pendulum is clear enough, providing for its basic characteristics to be identified.

What could we say about it even with no particular study? First, it is clearly to be seen that the Jacobian of (10.268) is close to unit: $J = 1 - 2\pi\delta_d[D(m) + mD(m)]$, and $\delta_d \ll 1$. Second, the half-period phase shift is function of the energy variable at time $n+1$ only. System (10.268) has the same form as a large class of maps, among which is the model of Fermi-Ulam, for instance, describing the acceleration of space particles as a radial twist map with slight dissipative correction:

$$\left| \begin{array}{l} J_{n+1} = (1 - \delta) J_n + \varepsilon f(\theta_n) \\ \theta_{n+1} = \theta_n + 2\pi\alpha(J_{n+1}) \end{array} \right. .$$

Here, J and θ are variables of the action-angle type, $f(\theta)$ is a 2π-periodic function of the angular variable, $\alpha(J)$ expresses the dependence of the oscillation period on the system's energy, and $\delta \ll 1$ reflects the slight damping in the system.

The map thus obtained (10.265), and especially its form (10.268), is a very important result because it makes it possible to relate the class of kick-excited self-adaptive dynamical systems considered here to the rather common and well-studied class of radial twist maps.

And last, if we replace the elliptic integral $K(m)$ from the phase equation by its expansion ending in the linear term (small amplitude approximation), map (10.268) assumes the form:

$$\left| \begin{array}{l} m_{n+1} = m_n(1 - 2\pi\delta_d) + F_0 d' \sin\theta_n \\ \theta_{n+1} = \theta_n + (v+1)\pi + \dfrac{v\pi}{4} m_{n+1} \quad (\mathrm{mod}\, 2\pi) \end{array} \right. . \qquad (10.269)$$

That is exactly the private form of the so-called dissipative standard map (DSM) that has been given intense consideration in the references (cf. [344, 345, 346])

$$\left| \begin{array}{l} r_{n+1} = br_n - \dfrac{k}{2\pi} \sin(2\pi\theta_n) \\ \theta_{n+1} = \theta_n + \omega + r_{n+1} \quad (\mathrm{mod}\, 1) \end{array} \right. . \qquad (10.270)$$

Indeed, it can easily be seen that the first map (10.269) is transformed into the second one (10.270) upon changing the variables $(m, \theta) \to (r = \frac{vm}{8}, \theta = \frac{\theta}{2\pi})$ and introducing new parameters: $b = 1 - 2\pi\delta_d$, $\omega = \frac{v+1}{2}$, $k = -\frac{16\pi F_0 d'}{v}$.

A number of physical problems result in dissipative standard maps, such as, in particular, the problem for a non-linear rotator under the action of an external force having the form of time-periodic δ-pulses. This problem is described by Zaslavski's map relating directly to DSM.

10.5.3. Fixed stationary map points

The fixed stationary points can be determined from equation system (10.65), which is written here as follows:

$$\left| \begin{array}{ll} F_0 d' \sin\theta^0 = 8\delta_d[E(m^0) - (1 - m^0)K(m^0)] & (10.271) \\ 2vK(m^0) = (2l - 1)\pi. & (10.272) \end{array} \right.$$

Equation (10.272) reveals that, with fixed value of the frequency v, the system features a whole family of values for stationary energy with various values of the integer index l; moreover, the condition $K(m) \geq \frac{\pi}{2}$ imposes the restriction $(2l-1) \geq v$. If here, we consider again only the linear term in the decomposition of $K(m)$ according to (10.266), we shall obtain the approximate discrete series for m_l^0 with amplitudes that are not big:

$$m_l^0 = 4\left[\frac{2l-1}{v} - 1\right]. \tag{10.273}$$

A very important consequence from discretization condition (10.272) for the considered kick-system is that the stationary values m_l^0 do not depend on the amplitude of the external force F_0, but only on frequency; from here follows the conclusion about the independence F_0 of the oscillation's amplitude on the change of at large. Thus, the discrete map so constructed reflects clearly the two important behavioral regularities of the kick-pendulum, described in the previous indents: the discreteness (peculiar „quantization“) of the possible stable stationary energies and the super-stability under changing intensity of the external action at large.

The remarkable fact here is that the same discretization condition can be also derived from the system's synchronization condition, i.e. from the requirement for the oscillator's oscillation period to be divisible by the period of the external force. In the case of the kick-pendulum considered here, which is an odd system, the oscillation half period in stationary state must equal an odd number of half-periods of the external force, i.e.

$$\frac{T}{2} = 2K(m_l^0) = (2l - 1)\frac{\pi}{v},$$

this condition being absolutely identical with the condition from (10.272).

This fact could lead us to the conclusion that, in a kick-pendulum, ordinary synchronization of the system with the frequency of the external force takes place. However, this is not quite true for several reasons. First, the oscillation amplitude of the kick-system can assume a whole series of discrete values for a single fixed frequency v of the external force. Second, the amplitudes of this series depend only on frequency, changing in no way (or sooner, changing quite slightly) under changing intensity of the external force F_0; while with common „forced" synchronization, the amplitude changes smoothly under changing external force. This behavioral difference results from the fact that, in these two cases, the energy balance is maintained in an absolutely different way. With normal synchronization, the amplitude increases with increase of the external force, thus maintaining the energy balance, while with kick-excitation self-adaptable systems, the energy balance is maintained by changing the time when the movement trajectory enters the active area and not by changing the oscillation amplitude, i.e. it is the result of changing the phase variable instead of the energy one. Actually, the expression for the obtained energy (10.263) is proportional to the sine of the mean phase $\Psi_0 = \theta$, which means that it is only the change of this variable that governs, moreover effectively enough, the system's energy balance. It should be also mentioned that, with common synchronization, there is no synchronization threshold for the intensity of the external action force, while with the kick-pendulum there is such a threshold.

The stationary value of θ^0 can be determined directly from (10.271):

$$\sin\theta_l^0 = \frac{8\delta_d[E(m_l^0) - (1 - m_l^0)K(m_l^0)]}{F_0 d'}. \tag{10.274}$$

Within the interval $[0, 2\pi]$, the stationary phase θ_l^0 can assume two values, one of which is always unstable; we shall consider this matter a little further.

From the condition $\sin\theta_l^0 \leq 1$ it follows that

$$8\delta_d[E(m_l^0 - 1) - (1 - m_l^0)K(m_l^0)] \leq F_0 d'. \tag{10.275}$$

Inequality (10.275) provides the a.m. threshold for the intensity of the external force F_0, above which synchronization takes place and stationary oscillations are maintained in the system (with fixed half-width of the action area d').

Using expansions (10.266) ending in the linear term, we can obtain simplified expressions for (10.274) and (10.275) accordingly, namely

$$\sin\theta_l^0 = \frac{2\pi\delta_d m_l^0}{F_0 d'}\left(1 + \frac{m_l^0}{8}\right) \simeq \frac{2\pi\delta_d m_l^0}{F_0 d'}, \tag{10.276}$$

$$F_0 d' \geq 2\pi\delta_d m_l^0\left(1 + \frac{m_l^0}{8}\right) \simeq 2\pi\delta_d m_l^0. \tag{10.277}$$

Inequality (10.277) provides the lower limit for the value $F_0 d'$ of the product for a given stationary state.

If we replace the approximate expression m_l^0 (10.273) in (10.277), we shall obtain a condition, limiting from above the index l of the possible fixed point with fixed value of the product $F_0 d'$:

$$l \leq \frac{v}{2}\left(\frac{F_0 d'}{8\pi\delta_d} + 1\right) + \frac{1}{2}.$$

10.5.4. Stability of the stationary points. Conditions for bifurcation doubling of period

Let us linearize equation system (10.265) in the vicinity of equilibrium point (m_l^0, θ_l^0). An appropriate equation system is obtained with respect to the small perturbations $(\delta m, \delta\theta)$ about the stationary point, which can be written in the following matrix form

$$\begin{bmatrix} \delta m_{n+1} \\ \delta\theta_{n+1} \end{bmatrix} = \begin{bmatrix} 1 - W & B \\ (1 - W)C & 1 + BC \end{bmatrix} \begin{bmatrix} \delta m_n \\ \delta\theta_n \end{bmatrix} = \hat{\mathrm{M}} \begin{pmatrix} \delta m_n \\ \delta\theta_n \end{pmatrix}, \tag{10.278}$$

where the following symbols are introduced:

$$\left| \begin{array}{l} W = 8\delta_d \left[\dfrac{dE}{dm} - (1 - m_l^0)\dfrac{dK}{dm} + K(m_l^0)\right] \\ B = F_0 d' \cos\theta_l^0; \qquad C = 2v\dfrac{dK}{dm} \end{array} \right. . \tag{10.279}$$

The stability of the stationary points is determined by the multipliers of system (10.278): if the module of both multipliers is less than one, the stationery point is stable, otherwise, it is unstable. Actually, these multipliers are the eigenvalues of the matrix $\hat{\mathrm{M}}$; they can be determined if the characteristic equation is solved

$$\lambda^2 - \lambda T_r \hat{\mathrm{M}} + \det \hat{\mathrm{M}} = 0. \tag{10.280}$$

From (10.278) it can be seen that $T_r\hat{\mathrm{M}} = 2 + BC - W$ and $\det\hat{\mathrm{M}} = 1 - W$. From the last expression, an important conclusion can be made: if both multipliers are complex, then the immovable point is stable, since $|\lambda_1\lambda_1^*| = |\lambda_2\lambda_2^*| = |\lambda_1\lambda_2| = (1 - W) < 1$ (let us remember that W reflects the dissipation in the system and it can be easily shown that, with $\delta_d > 0$, it is a small positive quantity).

The roots of Eq.(10.280) are

$$\lambda_{1,2} = 1 + \frac{BC - W}{2} \pm \sqrt{BC + \left(\frac{BC - W}{2}\right)^2}.$$

They are complex, provided

$$-(2-W)-2\sqrt{1-W}<BC<-(2-W)+2\sqrt{1-W},$$

or, accounting that $W\sim\delta_d$ – provided the following simplified condition is available

$$-(4-2W)+\frac{W^2}{4}<BC<-\frac{W^2}{4}. \tag{10.281}$$

As already shown, when the multipliers are complex numbers, the fixed stationary point is always stable.

When the multipliers are real numbers, the stability condition $|\lambda_1\lambda_2|<1$ results in $\lambda_1<1$ and $\lambda_2>-1$ and inequality (10.281) assumes the form

$$-(4-2W)<BC<0. \tag{10.282}$$

From the above it becomes clear that the intervals complying with stability conditions (10.282) and for complex eigenvalues (10.281) coincide almost completely, but for two narrow bands of width $\frac{W^2}{4}$; the stationary point in these two bands is stable, but the multipliers are real. Notwithstanding that their width may be deemed a small second-order quantity (let us remember that $W\ll 1$), they play an important role in quantitative stability considerations. They indicate that, when stability is lost, i.e. when a multiplier crosses the unit circumference, its imaginary part is exactly zero. Really, with $BC=0$, $\lambda_1=1$, while with $BC=-(4-2W)$, $\lambda_2=-1$. The first case ($BC=0$) corresponds to a saddle-node bifurcation, where a couple of stationary points is generated – a stable and an unstable one; while, when the parameters are changed in the reverse direction, the stable point merges with the unstable one and disappears. In the second case ($BC=-(4-2W)$), a period-doubling bifurcation takes place. It is namely these bifurcations that have been witnessed in the numeric experiment and described in 10.2.

If we replace coefficients W, B and C from (10.279) in (10.282), we shall be able to express the stability condition as a direct function of the system's physical parameters:

$$-4+16\delta_d\left[\frac{dE}{dm}-(1-m_l^0)\frac{dK}{dm}+K(m_l^0)\right]<2v\frac{dK}{dm}F_0d'\cos\theta_l^0<0. \tag{10.283}$$

Since $\frac{dK}{dm}>0$, the right inequality yields simply $\cos\theta_l^0<0$. Let us remember, however, that condition (10.274) for the stationery value of the phase variable determines only the value of $\sin\theta_l^0$, which provides the variable θ_l^0 with the option to assume two values within the interval $[0,2\pi]$: with one of these values $\cos\theta_l^0>0$, and with the other, $\cos\theta_l^0<0$. Now, stability analysis reveals that the first of these two stationary values always corresponds to an unstable stationary point. The two stationary points corresponding to a given index l, a stable and an unstable one, merge in the saddle-focus bifurcations point when $\cos\theta_l^0 = 0$. The last condition makes it possible to determine from (10.274) the bifurcation point for the external force F_{sn}, if we assume the other parameters to be fixed,

$$F_{sn}d' = 8\delta_d[E(m_l^0) - (1-m_l^0)K(m_l^0)]. \tag{10.284}$$

It can be seen that this is exactly the lower limit of condition (10.275), which meant that the energy received by the system could not compensate for the damping when the values of F_0 are smaller than F_{sn}.

From the left inequality of (10.283) we can determine another bifurcation value of F_0 at which the period is doubled. Using expansion (10.266) and omitting all terms but the linear ones in the expressions for the full elliptic integral's derivatives, and neglecting the small term proportional to δ_d, inequality (10.283) takes the form

$$(\cos\theta_l^0)^2 = 1 - (\sin\theta_l^0)^2 < \left[\frac{16}{\pi v F_0 d'\left(1 + \frac{9}{8}m_l^0\right)}\right]^2.$$

Using Expr.(10.276) for the stationary value of $\sin\theta_l^0$ in a small-amplitude approximation, we ultimately obtain inequality

$$Fd' < \sqrt{\left[2\pi\beta m_l^0\left(1+\frac{m_l^0}{8}\right)\right]^2 + \left[\frac{16}{\pi v\left(1+\frac{9}{8}m_l^0\right)}\right]^2} = F_{pd}d'. \tag{10.285}$$

Thus, we obtained an approximate expression for the bifurcation value of the external force F_{pd}, at which the period is doubled. For comparison, the theoretical value of calculated using expression (10.285) $F_{pd}^{theor} = 2.96$ (at $v = 51$, $\beta = 0.01$, $d' = 0.025$, $l = 29$, $m_l^0 = 0.378$) complies well with the value of obtained $F_{pd}^{theor} = 3.04$ in the numeric experiment (See 10.2). The theoretical framework outlined in this Section 10.5. can be used successfully for evaluation and analysis of kick-pendulum type system's behavioral regularities.

10.5.5. Generation of complex periodic solutions: multiplications in weakly dissipative maps

The numerical analysis of kick-excited self-adaptive system (cf.10.2) revealed that, around nearly any fixed stationary point, at some value of the parameter, stable low-period periodic points appear (with various periods – 3, 4, 5). Quite often, these periodic points are surrounded by other ones, for instance, with periods 2 × 4, 3 × 4 or the like. We already saw that these stationary periodic solutions in (weakly) dissipative maps correspond to the elliptic points in the centres of Birkhoff's islands, surrounding the stationary points in the relevant conservative map ($J = 1$). Pakarinen & Nieminen [347] have given a sound theoretical consideration to the so-called multibifurcations in conservative maps – this is the name they use to call the generation of Birkhoff's islands in the immediate vicinity of some stationary point of the map. Let us consider briefly their approach to the problem, bearing in mind the idea to apply it to the kick-map already constructed (10.265); actually, the last one is dissipative, but its Jacobian is but a bit different from unit (according to stability analysis, it equals $1 - W$, where $W \sim \beta \ll 1$ is given by Expr.(10.279)).

The stability of a p-cycle $x_1, x_2, ..., x_p$ ($p = 2, 3, 4, ...$) is determined by the multipliers (eigenvalues) of the matrix

$$\begin{pmatrix} \delta x \\ \delta y \end{pmatrix} = \hat{\mathrm{M}} \begin{pmatrix} \delta x \\ \delta y \end{pmatrix}; \quad \hat{\mathrm{M}} = \prod_{j=1}^{p} \hat{\mathrm{L}}(x_j),$$

where $\hat{\mathrm{L}}(x_j)$ is linearized matrix of the considered map P around point x_j of the j-th iteration of the p-cycle. In the case of a conservative map, it can be shown that the stability condition of the p-cycle is

$$\left| \mathrm{Tr}\hat{\mathrm{M}} \right| < 2 \tag{10.286}$$

(a proof of this condition can be found in [23, §3.3], for instance).

Let us assume that the studied map, as well as the linearized matrix, depends on some parameter s and that for some particular value of s the map is surrounded by a chain of Birkhoff's islands; then, stability condition (10.286) is satisfied. Let us now slowly vary the value of parameter s in such a way that Birkhoff's islands converge gradually to the central stationary point (the „mother" fixed point). At the boundary moment, when the p-cycle is born in the immediate proximity of the central point, we have $x_1 = x_2 = ... = x_p = x_0$ (here, we have denoted x_0 by the position of the central point); at this moment, the p-cycle becomes boundary-stable, i.e.

$$\left| \mathrm{Tr}\hat{\mathrm{M}} \right| = \left| \mathrm{Tr} \prod_{j=1}^{p} \hat{\mathrm{L}}(x_j) \right| = \left| \mathrm{Tr} \left[\hat{\mathrm{L}}(x_0) \right]^p \right| = 2,$$

where $\hat{\mathrm{L}}(x_0)$ is the matrix of the linearized map at point x_0.

To calculate $\mathrm{Tr}\hat{\mathrm{M}}$, a property of square 2 × 2 matrices is used:

$$\mathrm{Tr}(\hat{\mathrm{L}}_0)^p = \lambda_1^p + \lambda_2^p. \tag{10.287}$$

Here, $\lambda_{1,2}$ are the eigenvalues of matrix $\hat{\mathrm{L}}_0$. If its characteristic equation is written in the form $\lambda^2 - T\lambda + D = 0$ (here, T and D stand for the trace and determinant $\hat{\mathrm{L}}_0$) and if we take into account that the map is conservative, i.e. $D = \pm 1$, then, using property (10.287), the condition for boundary stability of the p-cycle can be reduced to the form

$$\left|\mathrm{Tr}\hat{\mathrm{M}}\right| = \left|\mathrm{Tr}\left(\hat{\mathrm{L}}_0\right)^p\right| = \left|2(D^{1/2})^p T_p\left(\frac{1}{2}TD^{-1/2}\right)\right|^2,$$

where T is the p-th Chebishev's polynomial of first kind. Using the properties of Chebishev's polynomials, the last condition for the case where $D = 1$ can be written as

$$\left|\mathrm{Tr}\hat{\mathrm{M}}\right| = \left|2T_p\left(\frac{1}{2}T\right)\right| = |C_p(T)| = 2.$$

Here, C stands for the p-th order Chebishev's polynomials of second kind. The roots of the last expression, solved with respect to the trace of the linearized matrix T, are written immediately as

$$T = \mathrm{Tr}\hat{\mathrm{L}}(x_0) = 2\cos\left(\frac{2k\pi}{p}\right). \tag{10.288}$$

It can be seen that, according to the last expression, the boundary values of T can be more than one, depending on the divisibility p of the studied Birkhoff's chain.

Now, let us attempt to apply the above analysis, which actually holds for conservative maps only, to the kick-pendulum's map (10.265) already constructed. We assumed that its dissipation is weak, being of the order of the weak damping δ_d. The applicability of the above calculations, however, is much more affected by the fact whether Birkhoff's chain is generated in the immediate proximity of the major fixed stationary point or in a place not so close to it. The observations from the numerical calculations on the kick-pendulum's behavior (cf.10.2) reveal that, at least in the case of 3-periodic orbits, the latter appear very close to the stationary points generating them. The higher-period periodicals do not feature such a close proximity; generally, it can be ascertained to some extent that the lower the period of the relevant Birkhoff's chain, the smaller the distance to the major fixed point where it is generated in the dissipative case. Therefore, we shall apply expression (10.288) providing the boundary value of the linearized matrix T trace for the tri-periodic Birkhoff's islands, observed with the kick-pendulum, i.e. we assume $p = 3$.

The linearized matrix in the vicinity of some particular fixed point was already determined by expression (10.278) when analyzing the stationary points' stability. We would write out its trace in the form $\mathrm{Tr}\hat{\mathrm{M}} = 2 - W + BC$, where the notations for W, B and C are given by (10.279). Since in either case we apply the conservative system analysis to a (weakly) dissipative one, we could neglect the weak dissipation W, and substituting in (10.288) the expressions for B and C, write out

$$T = 2 + 2\pi v F d' \frac{dK}{dm} \cos\theta_l^0 = 2\cos\left(\frac{2k\pi}{3}\right). \tag{10.289}$$

In the case where $k = 0$ Eq.(10.289) yields the precise condition for the saddle-node bifurcation of the major fixed stationary point (10.284). With $k = 1$ and $k = 2$, the cosine value in the right-hand side of (10.289) equals to $-\frac{1}{2}$ and this is exactly the case we are interested in. Later, we shall apply a similar procedure to the one used to determine the bifurcation value F_{pd}: substituting expressions $\frac{dK}{dm}$, accounting for expansion (10.266) and for the expansion of $\sin\theta_l^0$ from (10.276); we solve the equation with respect to F to finally obtain an expression for the *trifurcation* parameter F_{3f}, whose form resembles very much the expression for the period's doubling:

$$F_{3f}d' < \sqrt{\left[2\pi\beta m_l^0\left(1+\frac{m_l^0}{8}\right)\right]^2 + \left[\frac{12}{\pi v\left(1+\frac{9}{8}m_l^0\right)}\right]^2}. \tag{10.290}$$

Here again, we can verify the extent to which the theoretical forecast of formula (10.290) agrees with the numerical experiment. Let us calculate the theoretical value F_{3f}^{theor} for the same parameters as before ($v = 51$, $\beta = 0.01$, $d' = 0.025$) and again for the fourth stationary orbit. We obtain $F_{3f}^{theor} = 2.31$, which, compared to the value from the numerical experiment, $F_{3f}^{exp} = 2.38$, shows again the good agreement of the theory presented here with the numerical simulations.

10.5.6. On the class of radial twist maps

Above, we showed that, with some simplification, the problem for the kick-pendulum's behavior can be solved analytically in Poincareś section, thus being reduced to a 2D discrete system of the radial twist map type. However, the analysis of many other physical problems results in similar maps, the two best-known examples being Fermi-Ulam's map, modelling space rays' acceleration at the expense of their interaction with moving magnetic fields, and Zaslavski's map, describing a rotator's motion under the action of time-periodic-pulses. Here, we shall consider the physical principles underlying these two models for the purpose of comparing them more thoroughly with our model of a kick-excited self-adaptive system. We shall consider in principle the class of so-called dissipative twist maps (cf., for instance, [348]). Except for being a generalizing model, this class is also remarkable for the fact that it reveals clearly the link of our models with the nearly-integratable Hamiltonian systems.

The model of Fermi-Ulam represents an accelerating space particle as a solid elastic ball, moving freely between two parallel walls, one of which is fixed, and the other, slightly oscillating. With sine oscillations, this model's map has the form [23, §3.4]

$$\left|\begin{array}{l} u_{n+1} = |u_n + \sin\varphi_n| \\ \varphi_{n+1} = \varphi_n + \dfrac{M}{u_{n+1}} \quad (\mathrm{mod}\, 2\pi) \end{array}\right. .$$

Here, $u = \dfrac{v}{V}$ is the particle's dimensionless velocity at the time of its impact with the oscillating wall (V is the amplitude of the wall's velocity), φ is the wall's oscillation phase at the time of impact, $M = \dfrac{1}{16a}$ (a is the wall's oscillation amplitude). The absolute value in the first equation accounts for direction change at $u < 1$.

It would not be difficult to find out several meaningful resemblances between Fermi-Ulam's models and the kick-excited self-adapted system we have defined. First, both maps are written out using variables, equivalent to system energy and to external periodic force phases at the time of introducing energy from outside. Second, in both cases, energy introduction is governed by phase parameter and is accomplished within time period much smaller than this system's motion period. Third, with both models, the period is amplitude-dependent, i.e. these are linear oscillator models. If a small linear dissipation is introduced in Fermi-Ulam's model, as is the case, for instance, with [349; 23, §7.3], then, the last qualitative difference with respect to the kick-pendulum will disappear:

$$\left|\begin{array}{l} u_{n+1} = |(1-\delta)u_n + \sin\varphi_n| \\ \varphi_{n+1} = \varphi + \dfrac{2\pi M}{u_{n+1}} \quad (\mathrm{mod}\, 2\pi) \end{array}\right. . \tag{10.291}$$

Let us consider the stationary solutions of (10.291). From the phase equation, it follows that there is a discrete set of stable energy states: $u_m = \dfrac{M}{m}$, $m = 1, 2, \ldots$ Each value for m has two corresponding values for φ_m, determined as $\sin\varphi_m = \dfrac{\delta M}{m}$. The stability study reveals that solution φ_m for which $\cos\varphi_m > 0$ is always unstable, and the other solution for which $\cos\varphi_m < 0$ is stable at $u > \sqrt{\dfrac{\pi M}{2}}$. If we replace u in the last inequality and rewrite it with respect to the parameter M, we obtain that stability is attained at $M > \dfrac{\pi m^2}{2}$; when reducing M below this value, the respective stationary point is subject to doubling bifurcation, since at bifurcation, the critical multiplier equals -1. Moreover, we have the requirement $\sin\varphi_m \leq 1$, which results in equation $M \leq \frac{m}{\delta}$; in the equality point, saddle-focus bifurcation occurs, as a result of which the couple of solutions for both values of is generated φ_m – one stable and one unstable.

It can be seen that, depending on the parameter, the stationary points' bifurcation behavior is qualitatively identical with the behavior of the kick-pendulum – upon saddle-node bifurcation, a couple of solutions is generated, whereas, at some parameter, the stable solution's period is doubled, and a cascade to chaos follows [350, 23]. There is yet another similarity between Fermi-Ulam's model and the kick-pendulum that should be pointed out – the discrete set of fixed stationary points.

The next model system to be considered is the damping rotator subject to the action of external force having the form of time-periodic δ-pulses. The motion of this rotator is described by equation

$$\ddot{\varphi} + \Gamma\dot{\varphi} = F = Kf(\varphi)\sum_{n=0}^{\infty}\delta(t - nT). \tag{10.292}$$

Here, T is the interval between pulses, Γ is damping, and the periodic function $f(\varphi)$ reveals that the amplitude of a given pulse depends on the rotator's momentary position. If we substitute $x = \varphi$ and $y = \dot{\varphi}$, then, by integration, (10.292) can be reduced to a discrete map with respect to (x, y), taken at the time moment preceding the n-th δ-pulse [351]:

$$\left| \begin{aligned} y_{n+1} &= e^{-\Gamma T}[y_n + Kf(x_n)] \\ x_{n+1} &= x_n + \frac{1 - e^{-\Gamma T}}{\Gamma}[y_n + Kf(x_n)] = X_n + \mu y_{n+1} \end{aligned} \right. . \tag{10.293}$$

It should be pointed out that a map identical in form to (10.293) is also obtained when solving the canonical equations for a mechanical oscillator with one-and-a-half degrees of freedom, written in the form [352, 353]:

$$\left| \begin{aligned} \dot{I} &= -\gamma(I - I_0) + \varepsilon q(I_0, \theta)f(t) \\ \dot{\theta} &= \omega(I) = \omega_0 + \frac{\alpha\omega_0}{I_0}(I - I_0) \end{aligned} \right. .$$

Here, I and θ are action-angle variables for the autonomous oscillator, and function $f(t) = \sum_{n=-\infty}^{\infty}\delta(t - nT)$ expresses the form of the external force, which is of the type of periodic infinitely short pulses. Function $q(I_0, \theta)$ is periodic with respect to the angular variable θ, expressing the dependence of pulse intensity on the oscillator's momentary state. If we assume $q(I_0, \theta) = I_0 \cos\theta$, introduce normalized variables $y = \dfrac{I - I_0}{I_0}$, $\theta = 2\pi x$ and denote $\Omega = \omega_0 T$ and $\Gamma = \gamma T$, we can obtain Zaslavski's map [353, 354]:

$$\left| \begin{array}{l} y_{n+1} = e^{-\Gamma}[y_n + \varepsilon\cos(2\pi x_n)] \\ x_{n+1} = x_n + \dfrac{\Omega}{2\pi} + \dfrac{\alpha\Omega}{2\pi}\dfrac{1 - e^{-\Gamma}}{2\pi}[y_n + \varepsilon\cos(2\pi x_n)] \pmod 1 \end{array} \right. .$$

The last system can be written in a yet simpler form, if new parameters, $\omega = \dfrac{\Omega}{2\pi}$, $b = e^{-\Gamma}$ and $K = \dfrac{\alpha\Omega\varepsilon}{2\pi}\dfrac{1 - e^{-\Gamma}}{\Gamma}$, are introduced in it and new variables are substituted, $(y, x) \to \left(r = -\dfrac{Ky}{\varepsilon b}, \theta = x + \dfrac{1}{2}\right)$. Then, Zaslavski's map will take the form of the dissipative standard map (DSM):

$$\left| \begin{array}{l} r_{n+1} = b r_n - \dfrac{K}{2\pi}\sin(2\pi\theta_n)] \\ \theta_{n+1} = \theta_n + \omega + r_{n+1} \pmod 1 \end{array} \right. . \tag{10.294}$$

Comparing (10.269) to (10.294), it is obvious that the kick-pendulum's discrete map, obtained with some simplifications (the case of big amplitudes, ending the extension of the elliptic integral $K(m)$ up to the linear term, weak dependence of damping on the energy variable) is reduced to a DSM. This is a very meaningful fact. It should be also pointed out that, with the rotator, just as with the kick-pendulum (cf. 10.1), under the action of time-periodic δ-pulses, the external force is represented as the product of two terms: one, responsible for time dependence, and the other – for coordinate dependence. With the rotator, the time-dependent term has pulse form, while the position-dependent function, $K(x)$, is of regular and periodic nature; whereas with the kick-pendulum it is just the opposite – the time-dependent term is a sine function, while the coordinate-dependent term has the form of a very narrow (by X) Π-like pulse. Both settings, however, yield discrete maps of one and the same type, namely, standard dissipative maps.

As pointed out in considering Fermi-Ulam's model, a map with one phase variable yields a discrete series of stationary solutions $r_m{=}m$; $\sin(2\pi\theta_m){=}{-}2\pi(1{-}b)\dfrac{m}{K}$ and, again, only one of the solutions for θ_m is stable within some value interval for parameter K – the solution for which $\cos(2\pi\theta_m) > 0$, while the second solution is unstable. The stability area for a given solution m starts with $K_{sn} = 2\pi(1 - b)m$, when the couple of solutions is generated as a result of saddle-node bifurcation, to end up with $K_{pd} = 2\{[\pi m(1{-}b)]^2{+}(1{+}b)^2\}^{1/2}$, where a doubling bifurcation occurs. The interesting fact here is that, further, the doubled solution, prior to becoming subject to the next doubling bifurcation, becomes subject to a loss-of-symmetry bifurcation. This happens because the doubled map features some symmetry, which is first destroyed and only then a cascade of doublings takes place, providing a chaotic attractor.

Analyzing (10.294), we can define the stability area for given values of K and $\delta = 1 - b$ (here, δ stands for damping) whereas

$$\Delta m = m_{sn} - m_{pd} = \begin{cases} \dfrac{K}{2\pi\delta}, & K \leq 2(2-\delta) \\ \dfrac{K}{2\pi\delta} - \dfrac{K}{2\pi\delta}\left[1 - \dfrac{4(2-\delta)^2}{K^2}\right]^{1/2}, & K > 2(2-\delta) \end{cases} .$$

Here, m_{pd} and m_{sn} denote the boundaries, where period-doubling and saddle-node bifurcation occurs. With small values of K, when none of the fixed stationary points has been doubled yet, Δm increases linearly with force increase. After some particular moment, the doubling bifurcations start annihilating the attractors on the side of small m, while the saddle-node bifurcation generates them with big m; at a subsequent time moment, attractors' annihilation outpaces attractors' generation, until, finally, at very big values of K, the interval Δm decreases very fast. We can see, therefore, that the number of stable fixed points has a maximum at some K, and decreases in the boundary cases – becoming equal to unit with quite a small force (the fixed point with $m = 0$ being always stable), and decreasing to zero with a great force.

If we fix some particular point m and vary the force K, we could also define the *life time* of the fixed point as

$$\Delta K = K_{pd} - K_{sn} = 2\pi\delta m \left[1 + \frac{(2-\delta)^2}{(\pi\delta m)^2}\right]^{1/2} - 2\pi\delta m, \tag{10.295}$$

where K_{pd} and K_{sn} are the parameter's bifurcation values, accordingly, with period doubling or saddle-node generation of the fixed point responsible for a given m. From (10.295) it can be seen clearly that, when damping increases life time decreases; by the way, this appears to be also true for arbitrary-period periodic points, and not for fixed points only [355].

10.5.7. Twist Maps and Hamiltonian Dynamics

The couple of 2D discrete systems with quantitavely identical behavior considered so far can be represented in the following general form:

$$\left| \begin{aligned} J_{n+1} &= (1-\delta)J_n + \varepsilon f(\theta_n) \\ \theta_{n+1} &= \theta_n + 2\pi\alpha(J_{n+1}) \quad (\mathrm{mod}\, 2\pi) \end{aligned} \right. . \tag{10.296}$$

This expression reflects completely the physics of the problems. The energy injected by an external periodic force (the period here is assumed to be equal to 2π, which does not impair the generality of consideration) is expressed by a 2π-periodic function of the phase variable $f(\theta)$; ε means the amplitude of the perturbing (exciting) force. Here, the weak damping of the system is assumed to be linear, i.e. $\delta = \text{const}$; this parameter can (slightly) depend on the energy variable $\delta(J)$. The function of the energy variable, $\alpha(J)$, expresses the system period dependence on the amplitude.

If an autonomous Hamiltonian system is integratable, canonical action-angle variables (J_i, θ_i) can be introduced in it, so that the total energy does not depend on the angular variable; the action variables J_i represent the motion integrals. In the case of two degrees of freedom, $H(J_1, J_2) = E = \text{const}$; from here, either of the variables J can be expressed by the other. The motion in the (4D) phase space is then performed over a 2D torus, located on a 3D energetic hyper-plane, and Hamiltonian equations look as follows:

$$\left| \begin{array}{l} \dfrac{dJ_i}{dt} = -\dfrac{\partial H}{\partial \theta_i} = 0 \\[2mm] \dfrac{d\theta_i}{dt} = \dfrac{\partial H}{\partial J_i} = \omega_i(J) \end{array} \right. .$$

If we introduce a Poincaré surface, crossing this torus with some fixed value of the angular variable $\theta_2 = \text{const}$, the map for the other couple of variables will be of the so-called twist map [23] type:

$$\left| \begin{array}{l} J_{n+1} = J_n \\ \theta_{n+1} = \theta_n + 2\pi\alpha(J_{n+1}) \end{array} \right. . \tag{10.297}$$

Here, $\alpha(J) = \dfrac{\omega_1(J)}{\omega_2(J)}$ is the ratio between the angular frequences for both degrees of freedom n which, generally speaking, depends on J. Let us now consider the case of a system, close to an integrable one, i.e. we introduce a small perturbation into the integrable system. In this case, the Hamiltonian can be represented as the sum of the Hamiltonian of the integrable system with the non-integrable perturbation:

$$H(J_1, J_2, \theta_1, \theta_2) = H_0(J_1, J_2) + \varepsilon H_1(J_1, J_2, \theta_1, \theta_2). \tag{10.298}$$

The twist map will assume its perturbed version [23]:

$$\left| \begin{array}{l} J_{n+1} = J_n + \varepsilon f(J_{n+1}, \theta_n) \\ \theta_{n+1} = \theta_n + 2\pi\alpha(J_{n+1}) + \varepsilon g(J_{n+1}, \theta_n) \end{array} \right. , \tag{10.299}$$

where f and g are periodic functions of θ. This system results from the producing function $F_2 = J_{n+1}\theta_n + 2\pi A(J_{n+1}) + \varepsilon F(J_{n+1}, \theta_n)$, moreover

$$\alpha = \frac{dA}{dJ_{n+1}}, \quad f = -\frac{dF}{d\theta_n}, \quad g = \frac{dF}{dJ_{n+1}}. \tag{10.300}$$

10.6. General Characteristics of the Class of Kick-Excited Self-Adaptive Dynamical Systems. Conclusions

The major properties and regularities characterizing the considered class of kick-excited self-adaptive dynamical systems are the following:

1. Excitation of oscillations with quasi-own (quasi-natural) system frequency and a multitude of discrete stationary amplitudes depending only on the initial conditions, i.e. discreteness of the process of absorption by the system of the energy fed by a high-frequency source. A principally new property here is the possibility for exciting oscillations featuring the system's own frequency under the action of an external high-frequency force in non-perturbed linear and conservative linear or non-linear oscillating systems.

2. Adaptive self-regulation of the energy introduced during the oscillation process, which is manifested in stable maintenance of the system's oscillations' amplitude and frequency with significant change of the external action's amplitude, the oscillating unit's (load's) quality (Q-) factor, and other external actions.

3. The possibility for effective frequency division with a high single conversion rate. Available discrete spectrum of the action force's frequencies that can excite oscillations in the system with close amplitudes and frequencies.

The physical essence of the considered oscillations' excitation mechanism lies in the fact that, while the external high-frequency source acts on the oscillating system, the reverse action of the latter generates a frequency- and phase-modulated force. It is well-known that, in literature, frequency-phase modulation is usually analyzed not accounting for its energy interpretation; to generate it in radio-physical systems, complementary energy sources and special devices, called modulators, are used. In the considered case, the modulation principle manifests itself as a mechanism, providing for the oscillations' excitation, triggering the interaction of the oscillating system with the external high-frequency source. Adaptivity and self-regulation result in changing the value of the system's main argument – the initial phase determining the conditions for energy introduction or exhaustion. The change of the initial phase in one direction or other with respect to its stationary value causes an adequate change in the energy introduced by the external source in the oscillating process, whereat the energy portion compensating for dissipative energy losses in the system for each oscillation period remains unchanged on average. If there was no reverse adaptive action of the system on the source, then, for instance, in linear oscillating systems, it would have been possible to excite forced oscillations featuring only the frequency of the external high-frequency source. The considered method of energy introduction by a high-frequency source enables to excite in linear systems continuous oscillations featuring their own (natural) resonance frequency.

The characteristic argument of the system represented by an adaptively adjusting initial phase provides, from the energetic point of view, the most effective interaction between the excited oscillating system and the high-frequency power source. With non-homogenous action of the external high-frequency powering source, a characteristic mixing of its frequency (or frequency spectrum) with the oscillation frequency of the excited system occurs. Putting it in terms of frequencies, the process is expressed as a generation of an infinite spectrum of combination frequencies. Moreover, the necessary condition for such mixing of both oscillation processes, where spectral components appear featuring frequency close to the frequency of the excited system's own resonance frequency, is the adaptive automatic adjustment to the optimal phase. These spectral components maintain the oscillations in the system unchanged, whereas the conditions for excitation of non-damping oscillations are discrete, being determined by the initial kinematic parameters (initial conditions) of the excited system's oscillations.

The available unusual properties and operation regularities of oscillating systems with limited number of degrees of freedom under non-homogenous interaction with external periodic force provide reasons to argue for certain principles, grouping oscillating systems into stable formations. A boundary can be laid across the properties of oscillating systems, beyond which it could be argued that these systems feature certain elementary self-organization principles. Most probably, this boundary represents the state where the system not only receives energy from the source under a forced regime, i.e. under a regime where the external action imposes on the system its own operation conditions, but where the system itself starts to exert its action on the source, modifying and adjusting it to its own operation regime as well. The considered excitation method of non-damping oscillations represents namely such a manner of interaction. With it, the excited oscillating system exerts reverse adaptive action on the periodic source during its interaction with it, as a result of which frequency- or phase- (or, in the general case, argument-) modulated exciting force is formed. This is one of the reasons why the method is called „argument excitation of continuous oscillations".

Considering Eq. (10.1) from a formal mathematical point of view, it could be assumed that word goes merely for a nonlinear – parametric class of oscillating processes and systems. The thorough analysis, however, reveals some substantial differences between the classical parametric phenomena and the „argument" excitation method of continuous oscillations, which can be generalized in the following way:

1. With parametric phenomena, the external force periodically changes an energy-accumulating (reactive) parameter of the oscillating system. For instance, with parametric excitation of a pendulum, the external force acts in a direction perpendicular to the direction of motion, reducing periodically the pendulum's length or changing periodically its effective mass (the electrical analogue being time-changing capacitance and inductance). With the „argument" method, the exciting force acts in the direction of the pendulum's motion, whereas the argument of the coordinate change delays on average, compared to the argument of the system's acceleration.

2. Parametric excitation necessitates pumping frequency $p=\dfrac{2\omega}{n}$, $=1,2,3,4,...$, while with argument excitation, the frequency of the external action $v = n\omega$,

$n = 1, 3, 5, 7, \ldots$, with even equivalent non-linearity, or $n = 2, 4, 6, 8, \ldots$, with odd equivalent nonlinearity; ω being the frequency of excited oscillations in both cases.

3. In the case of parametric excitation, the ratio $2\omega/p$ is not strictly fixed, but there are rather certain frequency excitation areas. The argument method of energy introduction suggests a strictly fixed ratio v/ω.

4. Parametric oscillations are excited under zero initial conditions, i.e. oscillations are generated out of a state of rest. The argument mechanism is accomplished under non-zero initial conditions, since it is based on the interaction between the system's oscillations and the high-frequency energy source.

Oscillation studies in linear and non-linear „argument" systems with zero or negative resistance of the oscillating system reveal that the initial phase assumes a value at which the introduced energy is compensated for by its exhaustion in the interaction area, since the average overall amount of the energy introduced by the external source is zero or negative, accordingly.

Chapters 1 and 2 of the monograph considered the effects of negative impedance generation using modulation-parametric oscillating systems and excitation of oscillations featuring the input signal frequency as a result of the reversability of modulation-parametric interactions. With the „argument" oscillation excitation method, another type of reverse interaction is accomplished, accompanied by frequency-phase modulation of the acting source itself and automatic adjustment of the consumed energy portion. Putting it in terms of impedance, this means that the introduced negative equivalent impedance is regulated by the excited low-frequency system itself, not depending on a large scale on „pumping" parameters (amplitude, frequency, excitation force form).

The phenomenon of continuous oscillation excitation with amplitude from a discrete value set of possible stationary amplitudes has been demonstrated as well on the basis of a common model – oscillator under wave action. It has been shown that phenomenon manifestation conditions are realized in a natural way in an oscillator system interacting with a continuous wave.

A modeling system of oscillating charge under electromagnetic wave action has been considered. It has been shown that the continuous wave with spectral components, considerably higher than the oscillator charge natural frequency, excites charge oscillations with quasi-natural frequency and amplitude belonging to a discrete value set of possible stationary amplitudes, dependent only on the initial conditions. The considered model may be used for phenomenological investigation of plasma particles with electromagnetic waves interactions and waves in the Earth ionosphere and planetary magnetospheres. Hypothesis of adaptive nonlinear parametric wave generation may be suggested for Solar wind control of Jovian heterometric radiations, Saturn modulated radio emissions and Uranian auroral kilometric radiations. The mechanism is connected with natural interaction inhomogeneity and its type can be defined as cyclotron instability in the generation processes.

There is general agreement between researchers that planetary radiation is emitted in extraordinary mode by a maser cyclotron process, and all celestial bodies with magnetic field and energetic electron source are strong radio-emitters due to cyclotron maser instability. The effect, presented here, may throw a new light and enrich the concept of generation mechanisms.

A mechanism of cyclotron processes has been presented that might prove fundamental considering planetary magnetosphere radio-emission. It can be shown that the mechanism may give rise to radio-emission not only in a narrow range of angles almost perpendicular to the magnetic field in the source region, but any time when a wave packet falls upon the charged particle oscillator.

The potential has been shown for excitation of low-frequency continuous oscillations with discrete amplitude set under the influence of wave with incompatibly higher frequency – in that number fall waves from the ultraviolet band, near and far IR range and the radio-band. Possibly, this mechanism is combined with multiple re-emission with downward frequency transformation, and collision mechanisms are accompanied by radio-emission generation mechanisms due to plasma waves transformation into electromagnetic under the „wave-particle“ and „wave-wave“ interactions.

The mechanism may also be combined with maser cyclotron processes, giving initial excitation (initial conditions) in the presence of a magnetic field, whereas later a wave pumping from electromagnetic background is added.

Radio-emission spectrum characteristics might be determined by the properties of the discussed effect – on one hand, a wave with the same (unchangeable) frequency parameter may excite oscillations in a wide frequency band and different amplitudes; on the other hand – waves with different frequency parameter may excite oscillations with the same frequency (e.g., in gyro-resonance frequency area and local plasma frequency, due to the resonance effects).

A heuristic model of the mean distances between the Solar-system planets, their satellites and the primaries has been proposed. The model is based on: (i) the concept of the Solar system structure wave nature; (ii) the micro-mega analogy (MM analogy) of the micro- and mega-system structures, and (iii) the oscillator amplitude „quantization„ phenomenon, occurring under wave action, discovered on the basis of the classical oscillations theory (see 10.4.2).

From the equation, describing the charge rotation under the action of an electromagnetic wave, an expression is obtained for the discrete set of probable stationary motion amplitudes. The discrete amplitude values – the „quantization“ phenomenon – are defined by the argument values at the extreme points of the N-order Bessel functions. Using this expression, the mean related distances are computed from the Solar system planets and the Saturnian, Uranian and Jovian satellites to the primaries.

The harmonic oscillator obeys the classical laws to a greater extent than any other system. A number of problems, related to harmonic oscillators, have the same solution in classical and quantum mechanics.

Regardless of its simplicity, the „oscillator-wave" model obviously reflects a number of processes in the micro- and macro-world. The model is manifested naturally in different material media and harmonizes with the modern ideas about the world into and out of us as a totality of particles and fields. Also, it takes into account the wide extensions of the oscillating processes in Nature. In the presence of particle flows and fields of different nature, the model realizes (materializes) widely in Nature in a very natural way. In one way or another, the model has been considered by a number of authors, but the most essential feature of behavior – the „quantization" phenomenon, has escaped their attention.

The „oscillator-wave" model confirms the bodies' wave interaction mechanism in the micro- and macro-world, as well as the closing down and convergence of the manifestations of basic properties in Nature. The heuristic importance of the model is drawn on the basis of the remarkably fruitful hypothesis about the Solar system wave dynamical structure. Notwithstanding its simplicity, the proposed model gives a phenomenological and dynamical description of the mega-quantum resonance-wave structure of the Solar system and analogous systems, consisting of a central body and satellites. It seems that the model reconciles two opposite approaches – the one of the classical celestial mechanics of N bodies and that of the quantum-mechanical wave system theory. Based on the ideas of the wave nature of the Universe, the model gives a detailed enough identification of the Solar system resonance structure within the frames of the classical theory. The large number of possible resonances are in complete correspondence, e.g., with the fact of the existence of numerous resonances in the Jupiter and Saturn satellites systems. The model gives the orbit quantization, independent on the initial dynamic conditions and the planets' and satellites' masses. At the same time, by analogy with the atomic system theory, the model defines the ranges of the initial dynamic conditions, determining the possibility for planet or satellite capture, forming the Solar system planet and satellite structure. Hence, the model may trace the space substance zones, determining in the course of evolution the present structure of the Solar system's arrangement of planets and satellites. A more detailed study of the „oscillator-wave" model proposed may reveal all possible stable orbits of the Solar system planet and planet-satellite structure, including the fine Saturn ring system structures etc.

Since the system „oscillator-wave" may have a different physical nature, the observed features have significantly common character. In the case which has been considered in Section 10.3.8, the interaction non-homogeneity has been especially arranged – by restriction of the external force action over a small part of the trajectory. In the case of the wave action over the oscillator, on the contrary, nothing has been done to provide the in-homogeneity of the interaction.

The generality of the „oscillator-wave" model can be proven by a variety of examples from nature, science, or technology, in which these systems abound. Anywhere waves are acting on oscillators, conditions are created for excitation of oscillations with a possible discrete set of stable amplitudes.

The „argument" method of introducing energy into oscillating processes and excitation of „quantized" oscillations in dynamical macro-systems is and will be in the future applied in the process of solving important practical problems related to the development of new methods and mechanisms for excitation and maintenance of continuous oscillations and energy transformation. These methods and mechanisms could provisionally be grouped in the following way:

1. High-effective frequency transformation of signals by dividing frequency tens, hundreds, or thousands of times (by a single division), which is displayed by various radiophysical systems featuring a discrete series of stable operation regimes. Examples of these are : transformers of super-high-frequency (SHF) oscillations into lower frequencies; high-efficiency frequency divisors; wide-range-frequency generators; optic-range frequency modulators; high-efficiency transformers of energy from the optic and near-infrared range (where powerful generation sources are available) into energy of the sub-millimeter range (where powerful electromagnetic waves sources are still lacking), used in implementing controllable processes, such as thermo-nuclear reaction.

2. Transformation of energy from one type into another, such as electric into mechanical or vice versa. Examples of this are electric and electromechanical converters, electric signal generators, wave energy converters, non-conventional methods for transformation of thermal energy into electric energy etc.

3. Stabilization of various parameters during their large-scale changing (e.g., 50–100–300%), inclusive of microprocessor systems voltage stabilizers with a wide range of allowed load variation etc.

4. Development of new base elements intended for dedicated computer units featuring a great number of discrete stable dynamical states.

5. Intensification of various processes through special organization of argument interaction for various oscillation or wave processes, e.g. manipulations such as cavity destruction, purification, emulsification of non-mixing liquids or liquid-phase substances, development of various wave technologies.

6. Modeling of micro- and macro-processes by the methods of classical oscillation theory, explanation and model design for processes of electro-magnetic wave interaction in the Earth's ionosphere and magnetosphere, powerful low-frequency wave generation phenomena in near-to-Earth space against the space electromagnetic background, particles' interaction with electromagnetic waves in plasma environments, magnetosphere radio-emissions of the Solar System's outer planets, design and development of a mega-quantum resonance-wave model of the Solar System etc.

Based on the considered „oscillator-wave" model, through generalization for the 3D case, the author's contribution [356, 78] presents analytically the possibility for optic pumping of a microwave (VHF) emitter and maintenance of a stationary level of the resonator's electromagnetic field with great enough power of output emission. A high-efficiency sub-millimeter emitter, designed on this basis [357, 78], would be quite an adequate instrument for radiophysical plasma heating, e.g., in experiments aimed to achieve controllable thermo-nuclear reaction.

The other publications of the author provide some supplementing analysis, a number of other properties and regularities of the dynamic systems pertaining to the class of kick-excited self-adaptive dynamical systems, namely: (i) On the energetics of the process of stationary oscillations' excitation under the action of a coordinate-non-linear periodic force – cf. [358, 78]; (ii) On the excitation of non-damping oscillations in a pseudo-linear dissipative oscillating system with quasi-periodic adaptive switch over of an external high frequency periodic source – cf. [359, 78]; (iii) On the natural and numeric experiment studying the regularities of the considered method for excitation of „quantized" continuous oscillations in linear and non-linear resonance systems – cf. [78]; (iv) On the applications in radiophysical oscillator systems and the development of wave technologies for intensification of various technological processes – cf. [360, 78].

CONCLUSION

This book sets *three major objective points:*

1. To provide knowledge to physicists, radio engineers, specialists in Theory of nonlinear oscillations, Nonlinear Dynamics and other physical and engineering disciplines with a certain set of effects and phenomena occurring in four major classes of radiophysical and physico-technical oscillating systems: linear, parametric, nonlinear and nonlinear-parametric ones;

2. To present a series of analytical methods adequate to the identified circle of issues;

3. To demonstrate the employment of definite research approaches and techniques in the solution of particular research problems.

We present a general picture of modulation-parametric and nonlinear phenomena occurring in resonance amplifying, generating and converting oscillating systems. A generalized interpretation of the behaviour of linear resonance systems with periodic and almost periodic parameters is provided. There are descriptions of phenomena of adaptive (synergetic) grouping of oscillating systems in stable formations: grouping of coupled oscillating systems into stable electro-mechanical structures; a phenomenon of frequency attraction differing in quality and principle from the phenomenon of frequency entrainment and synchronization; a phenomenon of excitation of continuous oscillations with a discrete set of stable amplitudes given the action of an external periodic force that is nonlinear along the coordinate.

The new evidence that the reader can obtain from the book can be summed up in the following way.

The work presents a theory and develops an analytical tool for studying modulation-parametric and nonlinear phenomena in dynamical systems with external pumping and in generator oscillating systems. This has served as a basis for working out general methods for effective control of equivalent impedances in radiophysical systems. A principle of reversibility of modulation-parametric interactions is formulated; it provides a key to the analysis of the regularities in the manipulation of signals in diverse radiophysical and physico-technical systems and offers precise guidelines for research thought in its quest for various effects occurring in specific oscillating systems. It is shown that under certain conditions modulation-parametric interactions lead to an effective change in the value and sign of the reactive parameters and of dissipation, and to the appearance of new conversion properties and regularities. The oscillating systems are classified according to the conditions for the appearance of forces that tend to change the equivalent reactive parameters and dissipation as a result of modulation-parametric interactions. A model of a parametric modulator is created and a generalized conversion matrix is obtained.

The classical Manley-Row energy relations are generalized for an arbitrary number of signal sources in the conditions of linear and periodic in-time reactance, or nonlinear reactance with explicit time dependence, and this allows a most general analysis of energy input, of signal amplification and conversion, and of the stability of modern radiophysical devices. Methods and devices for obtaining broad-band (in the video frequency band) low-noise one-ports with negative capacitance, resistance and inductance have been developed. The major advantages of the offered methods of realization of one-port systems with negative parameters are the following: a. broad range of the possible absolute values of the negative output parameters; b. low own fluctuation noise; c. possibility for broad-range control of the frequency band of negative parameter existence; d. possibility for work in a situation of high current and voltage amplitudes; e. absolute stability of the system beyond the working frequency range. We present modulation-parametric systems as one-ports with negative parameters that are well-grounded and convenient for the engineering practice criteria of usage. They are required for amplification and frequency adjustment purposes for certain classes of sensors and receiver systems. We have developed a method for increasing the sensitivity of radiophysical receiver systems of the capacitive and inductive video sensor type. It is applicable to a multielectrode modulation meter of interplanetary plasma (Solar wind), to a highly sensitive receiver of gravitational waves, to a photosensor with a video amplifier, to photoparametric amplifiers, pyroelectric receivers, piezoelectric transducers, higher-resistant video cameras, etc.

The analytical tool proposed for studying converting, modulation-parametric, autodyne and other processes and modes in autonomous and non-autonomous generator systems combines the advantages of time-dependent analytical methods, based on nonlinear differential equations that provide a most accurate and comprehensive description of the items and phenomena under investigation, and complex spectral methods, securing a most compact written form of the equations and clear physical treatment of the internal structure of the signals and facilitating mathematical analysis. We derive differential equations of the second and third order describing conversion processes in self-exciting oscillator systems in autonomous and non-autonomous mode. We show that the frequency conversion by main channel in a non-autonomous oscillator can be accompanied by signal amplification, while the signals and noises in the image channel can be substantially suppressed. The transmission impedance of conversion is highly dependent on the oscillation amplitude in the non-autonomous generator, on the parameters and the mode. The magnitude and sign of the video frequency impedance of a one-port represented by a synchronized generator can vary within a broad range in the frequency band of synchronization. When a generator is used as a frequency converter, it is possible to achieve frequency selectivity under the respective asynchronous action on the generator. We propose methods for signal amplification, conversion and processing that are elaborated on the basis of the following effects found in generator radiophysical systems with synchronizing or asynchronous external action: a) The effect of frequency conversion by means of a non-autonomous oscillator characterized by amplification of the signal passing trough the main channel and suppression of the signal passing through the image channel (that is a method of increasing the signal/noise ratio by suppressing the noise in the image channel and amplifying the useful signal); b) The effect of broad-range alteration of the magnitude and sign of the video frequency impedance of a one-port represented by a synchronized generator (that is a method of conversion and amplification of frequency modulated signals; also a method for obtaining parametric elements, for modulation and frequency adjustment); c) The effect of frequency selectivity that is characteristic for a generator with asynchronous action (that is a method of spectral selectivity, also a method for analyzing the velocity of moving objects); d)The effect of amplification and equalization of conversion ratios by main and image channels, when using a non-autonomous generator with an output signal containing combined frequencies (that is a method for stabilizing the mode of short-distance self-detecting Doppler radars with ensured equal conversion ratio given a positive or negative Doppler shift of the frequency).

The developed analytical techniques serve as a basis for providing a theoretical description of processes in SHF short-distance self-detecting Doppler radars. We have conducted a theoretical comparative analysis of the conversion properties and oscillation spectrum in autodyne systems described by second and third order differential equations. Two basic types of SHF autodyne systems are considered: one of them registering the autodyne response in the direct current circuit of the active device, and the other one assisted by an external converter. It is shown that the controversial data quoted in the literature regarding the conversion properties of the different short-distance self-detecting Doppler radars are conditioned by the complex dependence of the conversion ratios on the oscillation amplitude, the degree of regeneration in the oscillator and a number of other parameters. General expressions taking into account all influencing parameters and modes are obtained. A separate analysis tackles the oscillation spectrum in the self-detecting Doppler radar. It takes into account the amplitude and the phase modulation as well as their dependence on the factor of reflection of the signal by a moving object, the degree of regeneration and the load of the oscillating circuit of the generator, the movement parameters of the object, the time constant in the direct current circuit of the active element exerting regenerative influence.

Some basic concepts are introduced and the modern ideas of the occurrence of chaotic oscillations in nonlinear radio engineering systems are outlined. A theoretical generalization of the stochastic instability phenomena in an application to the specific class of radiophysical systems with delayed self-action is presented. A theoretical presentation of the possibility to make the oscillations in a short-distance self-detecting Doppler radar stochastic with a total loss of its informative qualities is provided. This is quite important for engineering practice, where, until recently, reliability was associated only with the probability for elements and devices to show defects, for parasitic feedbacks to emerge and the like, and not with the general possibility for chaotic instabilities to develop in the situation of trouble-free operation of the system. The investigation of the possibility for stochastic instabilities to occur is of particular significance for numerous applications, for example, during the operation of autodynes in safety structures, etc.

A theory has been created concerning diffusion impedance and detection with a semiconductor diode in the case of high signal frequencies and amplitudes, including accounting for the accumulation of nonbasic carrier charge in the diode base. It is shown that the inertial properties of the p-n junction in extreme modes do not preclude the possibility for that junction to be used as a complex parametric one-port. We have established effects of „rectification" of the amplitude frequency characteristic of a complex nonlinear and parametric oscillating circuit and demonstrated in principle the possibility of exclusion of the bifurcational instability zone from the resonance characteristics of complex oscillating systems. An analytical method for obtaining equations regarding high level approximations based on the equivalent linearization method is offered. A general modified complex amplitudes method for analyzing the processes in nonlinear oscillating systems is also developed.

A number of problems from the Theory of parametric systems are considered on the basis of the spectral approach. A general interpretation of the parametric resonance in linear and nonlinear oscillating systems with periodically or almost periodically changing parameters is presented. In addition, visual geometric categories have been employed as well. The potential of the complex amplitudes method for analyzing oscillating systems with periodic and almost periodic parameters and nonlinear oscillating systems is revealed and developed. An adequate analytical tool is presented: linear spatial systems of algebraic equations for investigating linear oscillating systems with almost periodic parameters. A classification of the linear and nonlinear oscillating systems is also offered. It is based on the dimension of the respective algebraic systems of equations divided into one-index, two-index, etc. ones, depending on the dimension of the system matrix. The concept of spatial matrix is introduced and methods for multiple-index multiplication of such matrices are provided. An analytical tool for investigating the stationary mode in oscillating systems with periodic parameters based on continued (chain) fractions is presented. The phenomenon resonance is examined for systems with constant, periodic and almost periodic parameters. The involvement of geometric categories while studying this phenomenon makes it possible to visualize the resonance theory and to impart to it relative completeness. The possible linear oscillating systems with periodic and almost periodic parameters are classified by stability and instability zones of the respective differential equations. Three resonance types are formed. The properties of the elements of a generalized oscillating circuit with periodic and almost periodic time-dependent resistance, capacitance and inductance have been considered and their energy characteristics have been studied. Criteria for stability and instability of linear and nonlinear oscillating systems are obtained. A general analytical approach for implementing the principle of linear connection to an application in nonlinear radiophysical systems is offered.

A general theory of adaptive (synergetic) grouping of coupled oscillating systems in stable electromechanical formations is presented. Proof is presented that the orthogonality of the coordinates in coupled oscillating systems is the general condition for adaptive stability of the electromechanical structure. A theory on natural resonance frequencies of an oscillating system with 2.5 degrees of freedom operating in the conditions of adaptive self-tuning is developed. It is shown that the tendency of two coupled oscillating systems to group in stable electromechanical formations is manifested in the alteration of the distance between them in accordance with the frequency of the external action. Theoretical dependencies of the capacitance and the distance between the plates of the connecting self-tuning capacitor on the frequency of the external exciting oscillator are obtained. The resonance properties of a system of two linear oscillating circuits coupled with a variable capacitance and dissipative connection are studied. Two own resonance frequencies correspond to each frequency value of the external exciting oscillator. The lower own resonance frequency is consistently tuned to the frequency of the exciting generator upon its change. Moreover, the former frequency differs only slightly from the latter one. This is the essence of the „frequency attraction" phenomenon, which differs in quality and principle from the well-known phenomena of frequency entrainment and synchronization.

The phenomena established in coupled oscillating systems with a mechanical degree of freedom are of heuristic significance for the present-day modern studies from the perspective of synergetic grouping of systems in stable formations, formulation of fundamental hypotheses on a number of properties of a solid body, of substances in a liquid or gaseous state, and of elecrophysical systems. These phenomena also have an immediate technical application in the creation of tracing devices, displacement transducers, sensors for technical vision, frequency meters and specific frequency sensors, precise manipulators, etc.

A theory of the currently forming subclass of modulation nonlinear parametric phenomena and systems with adaptive phase self-tuning is developed. The theory includes two major elaborations with a considerable degree of generality: general analysis of the dynamic properties and regularities in a linear or nonlinear oscillating system under the action of an external periodic force that is nonlinear along the coordinate, and in a linear or nonlinear oscillating system under the action of an incident (falling) plane wave. The major regularities and properties of the „argument" method of excitation of continuous oscillations are presented: discretization („quantization") of the possible stable oscillation amplitudes, adaptive maintenance of the established oscillations as unchangeable given broad-range variation of the intensity of the exciting periodic force and of the quality factor (the load). The analysis of the mechanism underlying the phenomenon shows that the discreteness of the modes is conditioned by the discreteness of the phase conditions, for stationary oscillation excitation; in addition, the high adaptive stability is also conditioned by the phase relations determining the interaction between the oscillating system and the external periodic source. When the system subject to excitation is nonlinear, the acceleration of the transition process of adaptive attraction to the resonance mode depends on the degree of nonlinearity. The possibility of exciting oscillations in linear oscillating systems by using the action of a high-frequency harmonious force is demonstrated. The discreteness of the amplitudes of the excited oscillations, both in linear and in nonlinear systems, is retained in the cases of zero or negative friction coefficient as well – in such cases the phase adaptivity ensures nullifying or removal of the „excessive" energy. General principles of conversion by frequency and by signal and energy type are formulated. The possibility of highly effective division of the frequency with a high conversion ratio, given a single act of division, is proved. It is stressed that within the „oscillator-wave" system, the non-homogeneous external action is realized in a natural way, without the establishment of any special conditions for that purpose. As the systems of the „oscillator-wave" type can be of varying physical nature, the regularities that have been observed are essentially of a general nature. The „oscillator-wave" model and a generalization for a three-dimensional case have served as a basis for demonstrating the possibility of optic pumping of a microwave emitter in the SHF range, and of maintenance of a stationary level of the electromagnetic field in the resonator, given sufficiently high power of the output emission. The unusual properties and regularities of the phenomenon, such as „quantization" of parameters and modes, as well as its mechanism of manifestation are confirmed and substantiated by a precise experiment conducted both naturally and numerically. The experiment demonstrates that a principle of interaction can be realized in real oscillating systems. This principle however is regarded as acceptable only for the purposes of accelerating (or slowing down) microphysical objects in such systems as SHF electronic devices, cyclotron accelerators of charged particles, etc. The numerical experiment is a visual illustration of the discrete nature of the amplitudes, of the adaptive mechanism of establishing stationary oscillations in the transition process, and their maintenance in the situation of drastic alteration of a number of parameters not only in dissipative oscillating systems but also in linear or nonlinear conservative systems or in systems with negative friction.

A general systematization of the possible applications of the phenomenon in macro-dynamic systems is provided. The mechanism of excitation, interaction and energy conversion discussed here is observed in a number of areas of science and technology: SHF electronics, phenomena occurring in plasma medium, cyclic accelerators of charged particles, and in various processes based on the inertial properties of charged particles. In these different cases, the phenomenon is usually referred to as self-modulation, grouping of particles, phase selection, etc. But all of these mechanisms are united by a more general regularity of the processes in oscillating systems (linear or nonlinear): to absorb portions of energy under non-homogeneous action. It is possible to realize various oscillating systems with numerous stable conditions by using a high-frequency power source. The most general and most natural manifestation of the phenomenon (without any artificially created conditions) is in the case of interaction of waves with oscillators of various nature, when a stable discrete order of intensity levels is realized. A general block diagram of radiophysical „argument" systems is provided. A generalized block diagram is developed and the following specific devices are offered: frequency converters, regenerator of the carrier frequency of a phase manipulated signal, compensating stabilizer of alternating voltage up to 100% alterations, electric micro-engine with a number of discrete speeds, etc. A microwave submillimetre laser-pumped emitter is presented. Josephson detecting and superheterodyne receivers for millimeter waves are developed and investigated. They are used, for example, in radio engineering systems for reconnaissance purposes. Some other applications for special purposes are also referred to: when electromagnetic emission systems are worked out, during the experiments seeking to realize a controlled thermonuclear reaction, when studying the physics of the interaction between electromagnetic waves and a substance, etc. A brief analysis of the applications of devices intended for land, aviation and space systems and developed and presented in the book is provided: low-noise SHF amplifiers for the system of radars, SHF semiconductor generators for various purposes, including such for pumping SHF parametric amplifiers, SHF receivers (including those based on Josephson's quantum effect) for radio engineering reconnaissance; SHF radiometers, which can be used for obtaining SHF radiometric pictures of objects and works, for space meteorological reconnaissance systems; autodyne systems that can be used to avoid clashes, or the other way round - to produce such clashes between objects situated on the land, in the water or in the air; natural imitators of Doppler frequency shift for special training equipment, for radar countermeasures and for one-band radar systems; various tracing devices, sensors for technical vision, precise manipulators; modulation-parametric devices that are to be applied to the systems for optical television and radio engineering reconnaissance, vidicons for security purposes, pyroelectric receivers for special fire alarm systems, in receivers for seismic reconnaissance, special sensor systems, including inductive sensors for analyzing the vibrospectrum of the body and other elements of machines, aircraft, marine vessels, railway and motor transport, electromagnetic and radiation systems for special purposes, in the systems for realizing controlled thermonuclear reaction, etc.

A number of new research problems that can be subject to further scientific and research work are formulated and substantiated: a) effective control of video frequency impedances and conversion properties by using nonlinear and modulation-parametric phenomena and effects in oscillating systems; b) stochastic instability (chaotization) of the oscillations in generator systems with delayed self-action; c) development of general methods for investigating and interpreting the parametric resonance in oscillating systems with periodic and almost periodic parameters; d) adaptive (synergetic) grouping of coupled oscillating systems in stable electro-mechanical formations; e) excitation of continuous oscillations with a discrete set of stable amplitudes in linear or nonlinear radiophysical and physico-technical oscillation systems, under the action of an external periodic force that is nonlinear along the coordinate of the excited system.

REFERENCES

1. *Great Soviet Encyclopedia,* 3rd Edition, v.21 (Soviet Encyclopedia Publishing House, 1975), p.392, in Russian
2. V.V. Potemkin, *Radiophysics* (Moscow State University Publishers, Moscow, 1988), p.339
3. E.N.Lorenz, *The Essence of Chaos* (UCL press, London, 1993), p.227
4. R.Z.Sagdeev, D.A.Usikov and G.M.Zaslavsky, *Nonlinear Physics: From the Pendulum to Turbulence and Chaos* (Harwood Academic Publishers, New York, 1988), p.675
5. H.G.Shuester, *Deterministic Chaos: An Introduction* (VCH, Weinheim, New York, 1995), p.291
6. N.Hall (Ed.), *The New Scientist: Guide to Chaos* (Penguin Books, London, 1991), p.223
7. R.L. Devaney, *A First Course in Chaotic Dynamical Systems: Theory and Experiment* (Addison – Wesley Publishing Co., Inc., New York, 1992), p.302
8. M.Marek and I.Schreiber, *Chaotic Behaviour of Deterministic Dissipative Systems* (Cambridge University Press, Cambridge, New York, 1991), p.319
9. J.M.T. Thompson and H.B. Stewart, *Nonlinear Dynamics and Chaos: Geometrical Methods for Engineers and Scientists* (John Wiley & Sons, Singapore, 1993), p.376
10. J.Awrejcewicz, *Bifurcation and Chaos in Simple Dynamical Systems* (World Scientific, Singapore, 1999), p.126
11. R.N.Madan (Ed.), *Chua's Circuit: A Paradigm for Chaos* (World Scientific, Singapore, 1993), p.1042
12. S.H.Strogatz, *Nonlinear Dynamics and Chaos, With Applications to Physics, Biology, Chemistry and Engineering* (Addison - Wesley Publishing Co., Singapore, 1994), p.498
13. R.C.Hilborn, *Chaos and Nonlinear Dynamics* (Oxford University Press, Oxford, 1994), p.654
14. P.Glendinning, *Stability, Instability and Chaos: An Introduction to the Theory of Nonlinear Differential Equations* (Cambridge University Press, Cambridge, 1994), p.388
15. T.Kapitaniak, *Chaotic Oscillators: Theory and Applications* (World Scientific, Singapore, New Jersey, London, 1992), p.441
16. S.Ferraz-Mello (Ed.), *Chaos, Resonance and Collective Dynamical Phenomena in the Solar System* (Kluwer Academic Publishers, London, 1992), p.416

17. A.S.Dimitriev and V.Y.Kislov, *Stochastic Oscillations in Radiophysics and Electonics* (Nauka Publisher, Moscow, 1990), p.278, in Russian
18. St. Panchev, *Theory of Chaos, With Examples and Applications* (Academic Publishing House, Sofia, 1996), p.382, in Bulgarian
19. Yu.I.Neymark and P.S. Landa, *Stochastic and Chaotic Oscillations* (Nauka Publisher, Moscow, 1987), p.423, in Russian
20. T.Mullin (Ed.), *The Nature of Chaos* (Clarendon Press, Oxford, 1993), p.314
21. G.M.Zaslavsky, R.Z.Sagdeev, D.A.Usikov and A.A.Chernikov, *Weak Chaos and Quasi-Regular Structures* (Nauka Publisher, Moscow, 1991), p.235, in Russian
22. F.C.Moon, *Chaotic Vibrations: An Introduction for Applied Scientists and Engineers* (Cornell University, Ithaca, New York, 1990), p.312
23. A.J.Lichtenberg and M.A.Liberman, *Regular and Chaotic Motion* (Springer-Verlag, Berlin, 1994), p.501
24. G.L.Baker and J.P.Golub, *Chaotic Dynamics* (Cambridge University Press, Cambridge, 1993), p.298
25. V.S.Anishchenko, *Complicated Oscillations in Simple Systems: Origin Mechanisms, Structures and Properties of the Dynamical Chaos in Radiophysical Systems* (Nauka Publisher, Moscow, 1990), p.312
26. V.S.Anishchenko, *Dynamical Chaos in Physical Systems* (Teubnet – Texte, Leipzig, 1989), p.277
27. I.Prigogine, *La Fin Des Certitudes* (Les Editions Odile Jacob S.A., Paris, 1996), p.130
28. G.Haller, *Chaos Near Resonance.* Applied Mathematical Sciences, v.138 (Springer, New York, 1999), p.427
29. P.Cvitanovič, *Universality in Chaos* (Adam Hilger, New York, 1999), p.631
30. V.Damgov and E.Spedicato, *Deterministic Chaos: A New Paradigm in Nature and Social Sciences* (Proc. University of Bergamo, DMSIA, Anno 2002, No.8, Bergamo), p.16
31. V.Damgov, in *Proc. Int. Symposium "50 Years After Velykovsky's "Worlds in Collision: Classical and New Scenarios of the Evolution of the Solar System"*, eds. E.Spedicato and A.Agriesti, (University of Bergamo, Bergamo, 2002)
32. H.Haken, *Advanced Synergetics: Instability Hierarchies of Self-Organizing Systems and Devices* (Springer – Verlag, New York, 1983), p.419
33. Yu.A.Ermolaev and A.L.Sanin, *Electronical Synergetics* (Leningrad University Publishing House, Leningrad. 1989), p.245, in Russian
34. A.Yu.Loskutov and A.S.Michailov, *Introduction to Synergetics* (Nauka Publisher, Moscow, 1990), p.270, in Russian
35. A.Yu.Loskutov and A.S.Michailov, *Synergetics: Chaos and Structures Far From Equilibrium* (Moscow State University Publishing House, Moscow, 1990), p.298, in Russian

36. S.P.Kapitza, S.P.Kurdymov and G.G.Malinezkii, *Synergetics and Future Prediction* (Nauka Publisher, Moscow, 1997), p.285, in Russian
37. I.Prigogine and I.Stengers, *Order Out of Chaos: Man's New Dialogue With Nature* (Heinemann, London, 1984), p.402
38. M.Bushev, *Synergetics: Chaos, Order, Self-Organization* (World Scientific, Singapore, 1994), p.275
39. V.I. Arnold, *Catastrophe Theory* (Nauka Publisher, Moscow, 1990), p.381, in Russian
40. R.Gilmore, *Catastrophe Theory for Scientists and Engineering* (John Wiley & Sons, New York, 1981), v.1, p.350; v.2, p.285
41. J.M.T. Thompson, *Instabilities and Catastrophes in Science and Engineering* (John Wiley & Sons, New York, 1982), p.254
42. I.I.Blechman, *Synchronization in Nature and Technics* (Nauka Publisher, Moscow, 1991), p.292, in Russian
43. M.I. Rabinovich, *XX Century Physics: Development and Perspectives* (Nauka Publisher, Moscow, 1984), p.362, in Russian
44. N.Minorsky, *Nonlinear Oscillations* (D.Van Nastrand Company Inc., Princeton, New York, 1962), p.211
45. N.Minorsky, *Nonlinear Oscillations* (Robert E. Krieger Publishing Company, New York, 1974), p.714
46. L.I.Mandelshtam, N.D.Papaleksi, A.A.Andronov, A.A.Vilt, G.S.Gorelik and S.E.Khaikin, *New Studies on Nonlinear Oscillations* (Fizmatgiz Publisher, Moscow, 1936), p.348, in Russian
47. L.I.Mandelshtam, *Lectures on the Theory of Nonlinear Oscillations* (Nauka Publisher, Moscow, 1972), p.192, in Russian
48. B.Van-der-Pol, *Nonlinear Theory of Electrical Oscillations* (Svyazizdat Publisher, Moscow, 1935), p.188, in Russian
49. A.A.Andronov, A.A.Vitt and S.E.Khaikin, *Theory of Oscillations* (Fizmatgiz Publisher, Moscow, 1974), p.568, in Russian
50. G.S.Gorelik, *Oscillations and Waves* (Fizmatgiz Publisher, Moscow, 1959), p.312, in Russian
51. V.V.Migulin, V.I.Medvedev, E.R.Mustel and V.N. Parigin, *Basic Theory of Oscillations* (Mir Publisher, Moscow, 1988), p.443
52. Ch.Hayashi, *Nonlinear Oscillations in Physical Systems* (Princeton University Press, Princeton, New Jersey, 1985), p.288
53. G.D.Birkhoff, *Dynamical Systems* (American Mathematical Society, Providence, 1966), p.305
54. A.H.Nayfeh and T.M.Dean, *Nonlinear Oscillations* (John Wiley & Sons, New York, 1979), p.704
55. P.Hagedorn, *Nonlinear Oscillations* (Clarendon Press, Oxford, 1988), p.341

56. P.A.Cook, *Nonlinear Dynamical Systems* (Prentice-Hall Int., London, 1986), p.272
57. G.Nicolis, *Introduction to Nonlinear Science* (Cambridge University Press, Cambridge, 1995), p.254
58. R.E.Mickens, *An Introduction to Nonlinear Oscillations* (Cambridge University Press, Cambridge, 1981), p.224
59. E.Atlee Jackson, *Perspectives of Nonlinear Dynamics* (Cambridge University Press, Cambridge, 1991), v.1, p.496; v.2, p.633
60. A.V.Gaponov-Grekhov and M.I.Rabinovich, *Nonlinearities in Action: Oscillations, Chaos, Order, Fractals* (Springer-Verlag, New York, 1992), p.191
61. H.D.I. Abarbanel, M.I.Rabinovich and M.M.Sushchik, *Introduction to Nonlinear Dynamics for Physicists* (World Scientific, Singapore, 1993), p.158
62. L.O.Chua, *Linear and Nonlinear Circuits* (University of California, Berkeley, 1987), p.333
63. M.V. Kapranov, V.N.Kuleshov and G.M.Utkin, *Theory of Oscillations in Radioengineering* (Nauka Publisher, Moscow, 1984), p.320, in Russian
64. J.M.T. Thompson and W.Schiencen, *Nonlinear Dynamics of Engineering Systems* (Cambridge University Press, Cambridge, 1992), p.453
65. A.V.Kulakov and A.A.Rumyanzev, *Introduction to the Physics of Nonlinear Processes* (Nauka Publisher, Moscow, 1988), p.463
66. V.G.Bayrakhtar, V.M.Chernousenko, N.S.Erokhin, A.G.Sitenko and V.E. Zakharov, *Nonlinear World* (World Scientific, Singapore, 1989), v.1, p.272; v.2, p.314
67. M.I.Rabinovich and D.I.Trubezkov, *Introduction to the Theory of Oscillations and Waves* (Nauka Publisher, Moscow, 1984), p.432, in Russian
68. B.S.Andreev, *Theory of Nonlinear Electrical Circuits* (Radio i Sviaz Publisher, Moscow, 1982), p.282, in Russian
69. B.I.Babizkii and V.L.Krupenin, *Oscillations in Strong Nonlinear Systems* (Nauka Publisher, Moscow, 1985), p.283, in Russian
70. R.Bishop, *Oscillations* (Nauka Publisher, Moscow, 1986), p.491, in Russian
71. N.N.Butenin, Yu.I. Neymark and N.A.Fufaev, *Introduction to the Theory of Nonlinear Oscillations* (Nauka Publisher, Moscow, 1987), p.384
72. V.Klimov, *Dynamical Systems: Phase Portraits, Stability* (Zinatne Publisher, Vilnjus, 1991), p.271, in Russian
73. P.S.Landa, *Self-Oscillations in Systems With Finite Number of Degrees of Freedom* (Nauka Publisher, Moscow, 1980), p.248, in Russian
74. F.Verhulst, *Nonlinear Differential Equations and Dynamical Systems* (Springer-Verlag, New York, 1990), p.452
75. H.Kawakami (Ed), *Bifurcation Phenomena in Nonlinear Systems and Theory of Dynamical Systems* (World Scientific, Singapore, 1990), p.344

76. N.B.Tufillaro, T.Abbott and J.Reilly, *An Experimental Approach to Nonlinear Dynamics and Chaos* (Addison-Wesley Publishing Co., New York, 1992), p.340
77. E.Mosekilde, *Topics in Nonlinear Dynamics: Applications to Physics, Biology and Economics* (World Scientific, Singapore, 1996), p.380
78. V.N.Damgov, *Nonlinear and Parametric Phenomena in Radiophysical Systems* ("Prof.Marin Drinov" Publishing House, Sofia, 2000), p.450, in Bulgarian
79. M.Cartmell, *Introduction to Linear, Parametric and Nonlinear Vibrations* (Chapman and Hall, London, 1990), p.242
80. F.Kouril and K,Vrba, *Nonlinear and Parametric Circuits: Principles, Theory and Applications* (John Wiley & Sons, New York, 1988), p.413
81. A.Tondl, T.Ruijgrok, F.Verhulst and R.Nabergoj, *Autoparametric Resonance in Mechanical Systems* (Cambridge University Press, Cambridge, New York, 2000), p.211
82. A.Tondl, V.Kotek and C.Kratochvil, *Vibration Quenching of Pendulum Type Systems by Means of Absorbers* (CERM Akademické Nakladtelstvi s.r.o., Brno, 2001), p.102
83. Sh. Sastry, *Nonlinear Systems: Analysis, Stability, and Control* (Springer, New York, 1999), p.667
84. S.P.Strelkov, *Introduction to the Theory of Oscillations* (Nauka Publisher, Moscow, 1964), p.421, in Russian
85. V.S.Etkin (ed), *Semiconductor Parametric Amplifiers and Convertors* (Radio i Sviaz Publisher, Moscow, 1983), p.343, in Russian
86. G.S.Gorelik, *Journal of Technical Physics* **4** (1934) 1781; **5** (1935) 195, 489
87. V.M.Starjinskii, *Applied Methods of Nonlinear Oscillations* (Nauka Publisher, Moscow, 1977), p.292, in Russian
88. V.A.Taft, *Spectral Methods for Study Nonstationary Circuits and Systems* (Energia Publisher, Moscow, 1978), p.412, in Russian
89. V.A.Yakubovich and V.M.Starjinskii, *Parametric Resonance in Linear Systems* (Nauka Publisher, Moscow, 1987), p.319, in Russian
90. A.Poincaré, *About the Curves Determined by Differential Equations* (Gostekhizdat Publisher, Moscow, 1947), p.142, in Russian
91. A.Poincaré, *Les Méthodes Nouvelles de la Mecanique Celeste* (Gouthier-Villarg, vol.1-1892, vol.2-1893, vol.3-1899)
92. A.Poincaré, *Selected Transactions,* in 3 volumes (Nauka Publisher, Moscow, 1974), p.292, p.212, p.272, in Russian
93. A.M.Lyapunov, *General Problem of Motion Stability* (Fizmatgiz Publisher, Moscow, 1959), p.492, in Russian
94. J.Mawhin, *Applied Mathematics and Mechanics* (ZAMM) **73** (1993) 554
95. V.M.Matrosov and R.N.Kozlov (Eds), *The Methods of Lyapunov's Functions in Dynamics of Nonlinear Systems* (Nauka Publisher, Novosibirsk, 1983), p.413, in Russian

96. V.N.Belikh, *Qualitative Methods of the Theory of Nonlinear Oscillations* (Nijni Novgorod University Publisher, Nijni Novgorod, 1980), p.321, in Russian
97. N.N.Bogoljubov and Yu.A.Mitropolskii, *Asymptotical Methods of the Theory of Nonlinear Oscillations* (Nauka Publisher, Moscow, 1974), p.431, in Russian
98. B.I.Shakhtarin, *Quasi-Harmonic Method and Its Applications to the Analysis of Nonlinear Phase Systems* (Energoatomizdat Publisher, Moscow, 1987), p.296, in Russian
99. K.F. Teodorchik, *Self-Oscillating Systems* (Gostechizdat Publisher, Moscow, 1952), p.214
100. V.F. Jhuravlev and D.M.Klimov, *Applied Methods of the Theory of Nonlinear Oscillations* (Nauka Publisher, Moscow, 1988), p.339, in Russian
101. V.M.Volosov and B.I.Morgunov, *The Averaging Method of the Theory of Nonlinear Oscillating Systems* (Moscow State University Publishing House, Moscow, 1971), p.229, in Russian
102. L.I.Manevich, Yu.V.Mihlin and V.N.Pilipchuk, *Method of Normal Oscillations for Strong Nonlinear Systems* (Nauka Publisher, Moscow, 1989), p.348, in Russian
103. V.P.Maslov, *Complex Method VKB for Nonlinear Equations* (Nauka Publisher, Moscow, 1977), p.296, in Russian
104. R.Clarkson (Ed), *Applications of Analytic and Geometric Methods to Nonlinear Differential Equations* (Proc. NATO ARW, Exter, UK, 1992), p.477
105. J.Dennis and R.Shnabel, *Numerical Methods for Absolute Optimization and Nonlinear Equation Solution* (Mir Publisher, Moscow, 1988), p.412, in Russian
106. A.Van-der-Liele, *Noise and Measurements* (Mir Publisher, Moscow, 1979), p.192, in Russian
107. I.N.Bobrov, *SHF Parametric Amplifiers and Converters* (Technika Publisher, Kiev, 1969), p.329, in Russian
108. J.Marie, *Low-Frequency Parametric Amplifiers With Orthogonal Magnetic Control* (Mir Publisher, Moscow, 1983), p.298, in Russian
109. V.I.Medvedev and A.A.Belov, *Parametric and Self-Oscillating Systems* (Moscow State University Publishing House, Moscow, 1990), p.267, in Russian
110. I.M.Angelov, V.N.Damgov, A.Ya Spasov, L.K.Urshev and A.S.Yanev, *Electropromishlenost i Priborostroene* **17** (1981) 366, in Bulgarian
111. E.Gershenzon, *Autodynes* (Nauka Publisher, Moscow, 1990), p.281, in Russian
112. V.F.Bilov, R.E.Vinograd, D.M.Grobman and V.V.Nemizkii, *Theory of Lyapunov's Indices: Applications to the Problem of Stability* (Nauka Publisher, Moscow, 1966), p.279, in Russian
113. E.A.Grebennikov, *Introduction to the Theory of Resonance Systems* (Moscow State University Publishing House, Moscow, 1987), p.231, in Russian
114. A.Kunizin and L.Tashimov, *On the Stability of Resonance Systems* (Nauka Publisher, Kazahstan, 1990), p.203, in Russian

115. N.D.Papaleksi, *Soviet Physics – Uspehi* **31** (1947) 447
116. B.V.Chirikov, *Interaction of Nonlinear Resonances* (Novosibirsk State University Publishing House, Novosibirsk, 1978), p.193, in Russian
117. A.Shcherbakov, *Matter Self-Organization in the Inanimate Nature* (Moscow State University Publishing House, Moscow, 1990), p.241, in Russian
118. L.I. Mandelshtam, *Complete Works* (Russian Academy of Sciences, Moscow, 1948-1050), v.1, p.406; v.2, p.519; v.3, p.497, in Russian
119. M.Plank, *Complete Works* (Russian Academy of Sciences, Moscow, 1975), p.728, in Russian
120. B.M.Akulin, *Intensive Resonance Interactions in Quantum Electronics* (Nauka Publisher, Moscow, 1987), p.279, in Russian
121. A.A.Alifov and K.B.Frolov, *Interactions Between Nonlinear Oscillating Systems and Energy Sources* (Nauka Publisher, Moscow, 1985), p.319, in Russian
122. M.Ya.Amusia, *Brake Radiation* (Energoatomizdat Publisher, Moscow, 1990), p.264, in Russian
123. I.I.Blechman, *What is the Vibration Ability* (Nauka Publisher, Moscow, 1988), p.333, in Russian
124. M.A.Braun, *Physics at the Edge of New Discoveries* (Sankt Petersburg State University Publishing House, Sankt Petersburg, 1990), p.443, in Russian
125. G.Bruk, *Cyclic Accelerators of Charged Particles* (Atomizdat Publisher, Moscow, 1990), p.292, in Russian
126. V.Burdakov, *Electric Energy from the Space* (Energoatomizdat Publisher, Moscow, 1990), p.221, in Russian
127. A.G.Butkovsky and Yu.A.Samoylenko, *Control of Quantum-Mechanical Processes* (Nauka Publisher, Moscow, 1984), p.414, in Russian
128. V.A.Vanke, *Cosmic Energetical Systems* (Mashinostroenie Publisher, Moscow, 1990), p.192, in Russian
129. V.Volosov and A.Kalinzev, *Nonlinear-Optical Frequency Converters* (Mashinostroenie Publisher, Moscow, 1991), p.261, in Russian
130. Ch.A.Vo, *Theory of the Parametric Influence of High Power Electromagnetic Radiation on Solid States* (Nauka Publisher, Moscow, 1985), p.239, in Russian
131. N.I.Vulfson, *Vibrations of Machines With Cyclic Motion Mechanisms* (Mashinostroenie Publisher, Moscow, 1990), p.276, in Russian
132. R.F.Ganiev and L.E.Ukrainsky, *Dynamics of Particle Subjected to Vibration Actions (Naukova Dumka, Kiev, 1975), p.289, in Russian*
133. I.F.Goncharevich, Vibration: A Non-Standard Way (Nauka Publisher, Moscow, 1986), p.179, in Russian
134. V.I.Gulaev, *Applied Problems of the Theory of Nonlinear Vibrations of Mechanical Systems* (Vishaia Shkola, Moscow, 1989), p.156, in Russian

135. A.Jigliavsky and A.Jilinskas, *Methods for Search of Global Extremum* (Nauka Publisher, Moscow, 1991), p.428, in Russian
136. V.Jilkov, *Measurement Pondermotive SHF Converters: Theory and Applications* (Vishaia Shkola, Kharkov, 1989), p.249, in Russian
137. V.Kanavez, Yu.Mozgovoi and A.Slepkov, *Powerful Electron Flows Emission From the Resonance Delaying Systems* (Moscow State University Publishing House, Moscow, 1991), p.272, in Russian
138. P.S.Landa, *Mashinovedenie* **42** (1988) 90, in Russian
139. P.S.Landa and Ya.B.Douboshinsky, *Soviet Physics – Uspehi* **158** (1989) 729
140. K.K.Likharev, *Introduction to the Dynamics of Josephson Elements* (Nauka Publisher, Moscow, 1985), p.426, in Russian
141. P.Mack-Komik, *Wave Energy Conversion* (Mir Publisher, Moscow, 1988), p.292, in Russian
142. N.N.Prenzlau, V.M.Dimitriev and V.D.Borbishev, *Radiotekhnika i Elektronika* **24** (1979) 1702, in Russian
143. B.N.Rodimov, *Self-Oscillating Quantum Mechanics* (Nauka Publisher, Moscow, 1976), p.371, in Russian
144. E.Sanchez-Palensia, *Non-Homogeneous Media and the Theory of Oscillations* (Mir Publisher, Moscow, 1987), p.199, in Russian
145. Yu.M.Svirejev, *Non-Linear Waves, Dissipative Structures and Catastrophes in Ecology* (Nauka Publisher, Moscow, 1987), p.444, in Russian
146. A.P.Suchorukov, *Nonlinear Wave Interactions in Optics and Radiophysics* (Nauka Publisher, Moscow, 1988), p.331, in Russian
147. Yu.P.Timofeev, S.A.Fridman and M.V.Fok, *Light Conversation* (Nauka Publisher, Moscow, 1985), p.369
148. G.Yu.Treder, *Evolution of the Basic Physical Ideas* (Naukova Dumka, Kiev, 1989), p.416, in Russian
149. F.L.Chernousko, L.D.Akulenko and B.I.Sokolov, *Oscillation Control* (Nauka Publisher, Moscow, 1980), p.289, in Russian
150. V.N.Damgov, *Bulgarian J. Physics* **5** (1978) 519
151. V.N.Damgov, *Elektropromishlenost i Priborostroene* **16** (1980) 100; **16** (1980) 339, in Bulgarian, in *Proceedings 2nd Intern. Conference on Electronic Circuits ICKC* (Chekhoslovakia, Prague, 1979)
152. V.N.Damgov, *Elektropromishlenost i Priborostroene* **17** (1981) 305, in Bulgarian
153. V.N.Damgov, *Ukrainian Physical J.* **42** (1997) 582, in Russian
154. V.B.Braginsky and A.B.Manukin, *Measurement of Weak Forces in Physical Experiments* (Nauka Publisher, Moscow, 1974), p.248, in Russian
155. V.N.Damgov, D.B.Douboshinsky and Y.B.Douboshinsky, *Elektropromishlenost i Priborostroene* **25** (1989) No 4, 2

156. L.A.Vaynshtein and Y.B.Douboshinsky, *J. Technical Physics* **48** (1978) 1321, in Russian
157. V.N.Damgov, *Comptes Rendus de l'Académie Bulgare des Sciences* **43** (1990) No 10, 43
158. M.D.Karasev, *Vestnik of Moscow State University, Ser. Physics & Astronomy* **18** (1977) No 4, 109, in Russian
159. A.Lebedev and A.Shalnov, *Basic Foundations of Accelerators Physics and Techniques* (Energoatomozdat Publisher, Moscow, 1990), p.329, in Russian
160. R.Miroshin and I.Chalidov, *Theory of Local Interactions* (Sankt Petersburg State University Publishing House, Sankt Petersburg, 1991), p.431, in Russian
161. H.Gibbs, *Optical Bistability* (Mir Publisher, Moscow, 1988), p.229, in Russian
162. V.N.Belikh and N.N.Verichev, *Izv.VUZ, Radiophysics* **31** (1988) 921, in Russian
163. G.A.Tyulin, *Dynamics of the Space Vehicles and Space Exploration* (Mashinostroenie Publisher, Moscow, 1986), p.481, in Russian
164. A.Lebedev and O.Nesterenko, *Space Systems of Observation* (Mashinostroenie Publisher, Moscow, 1991), p.521, in Russian
165. G.Petrov, *Airmechanics of High Speeds and Space Exploration* (Radio i Sviaz, Moscow, 1991), p.338, in Russian
166. J.Gonda, *IEEE Trans.* **MTT-25** (1977) 343
167. G.N.Korostelev and L.S.Sotov, *Radiotechnika i Electronika* **34** (1989) 1925
168. K.S.Rzhevkin, *Physical Principles of Semiconductor Devices Operation* (Moscow State University Publishing House, Moscow, 1986), p.311, in Russian
169. A.M.Samson, L.A.Kotomtzeva and N.A.Loyko, *Self-Oscillations in Lasers* (Nauka i Technika Publisher, Minsk, 1990), p.227, in Russian
170. V.N.Damgov, *IEE Proceedings, Pt.G.: Electronic Circuits and Systems* **131** (1984) 9
171. V.N. Damgov, *Comptes Rendus de l'Académie Bulgare des Sciences* **46** (1993) No 2, 36
172. J.M.Manley and H.E.Row, *Proc. IRE* **44** (1956) 904; 47 (1959) 2115
173. J.M.Manley, *IEEE Trans. on Circuit Theory* **CT-7** (1960) 275
174. N.D.Birjuk and V.N.Damgov, *Comptes Rendus de l'Académie Bulgare des Sciences* **38** (1985) 567
175. A.C.Scott, *Proc. IEEE* **61** (1973) 1443
176. N.D.Birjuk and V.N.Damgov, *Izv. VUZ, Radiophysics* **28** (1985) 665, in Russian
177. N.N.Moyseev, *Asymptotical Methods of the Nonlinear Mechanics* (Nauka Publisher, Moscow, 1981), p.403, in Russian
178. G.Murey, *Asymptoticle Analysis* (Nauka Publisher, Moscow, 1999), p.391, in Russian

179. T.G.Strijak, *Asymptotical Method of Normalization* (Vishcha Shkola, Kiev, 1984), p.196, in Russian
180. A.N.Bruevich and S.I.Evtianov, *Approximation of Nonlinear Characteristics and Spectrums Provoked by Harmonic Actions* (Soviet Radio, Moscow, 1965), p.328, in Russian
181. G.Leonov, I.Burkin and A.Shepeliavii, *Frequency Method in the Theory of Oscillations* (Sankt Petersburg University Publishing House, Sankt Petersburg, 1991), p.356, in Russian
182. J.P.Oben, *Applied Nonlinear Analysis* (Mir Publisher, Moscow, 1988), p.406, in Russian
183. E.N.Rozenvasser and S.K.Volovodov, *Operational Methods and Oscillation Processes* (Nauka Publisher, Moscow, 1985), p.299, in Russian
184. R.Edvards, *Fourier Series in Modern Presentation* (Mir Publisher, Moscow, 1985), v.1, p.298; v.2, p.348
185. K.A.Samoylo, *Method for Analysis of Oscillating Systems of Second Order* (Soviet Radio, Moscow, 1976), p.295, in Russian
186. V.N.Damgov, *Bulgarian J. Physics* **4** (1979) 482
187. N.D.Birjuk and V.N.Damgov, *Electrichestvo* **27** (1988) No 8, 46, in Russian
188. N.D.Birjuk and V.N.Damgov, in *MTA Sz TAKI Proceedings* (Computer and Automation Institute, Hungarian Academy of Sciences, 1988)
189. A.N.Vistavkin, V.N.Gubankov, L.S.Kuzmin, K.K.Likharev and V.V.Migulin, *Radiotechnika i Electronika* **26** (1981) 1706, in Russian
190. U.Johnes and V.Tron, *Continuous Fractions: Analytical Theory and Applications* (Mir Publisher, Moscow, 1985), p.347, in Russian
191. V.N.Damgov, P.G.Georgiev and G.A.Russeva, *Radiotekhnika i Electronika* **29** (1984) 465, in Russian; in *IEEE Proceedings of Int. Symposium on Circuits and Systems* (Italy, Rome, 1982)
192. V.N.Damgov and M.N.Zenov, *Bulgarian J. Physics* **11** (1984) 312; in *Proceedings IX Intern. Conference on Nonlinear Oscillations* (Naukova Dumka Publisher, Kiev, 1984)
193. V.N.Damgov, *Radiotechnika i Electronika* **30** (1985) 524, in Russian; in *26 Midwest Symposium on Circuits and Systems* (Mexico, Puebla, 1983)
194. V.N.Damgov, *Comptes Rendus de l'Académie Bulgare des Sciences* **36** (1983) 607
195. V.N.Damgov, *Comptes Rendus de l'Académie Bulgare des Sciences* **36** (1983) 1163
196. V.Bogachev, V.Lisenko and S.Smolsky, *Transistor Generators and Autodynes* (Energoatomizdat Publisher, Moscow, 1990), p.396, in Russian
197. I.M.Kogan, *Short-Range Radiolocation* (Soviet Radio, Moscow, 1983), p.341, in Russian

198. V.Ya.Noskov, *Izv. VUZ, Radioelectronika* **34** (1991), No 10, 44, in Russian
199. Yu.Alekseev, *Izv. VUZ, Radioelectronika* **33** (1990), No 10, 78, in Russian
200. Y.L.Khotunzev and D.Ya.Tamarchak, *Synchronized Generators and Autodynes at Semiconductor Devices* (Radio i Sviaz, Moscow, 1992), p.319, in Russian
201. V.A.Ivanov and S.M.Smolsky, *in MEI Proceedings,* vol.522 (MEI Publisher, Moscow, 1981) 86, in Russian
202. A.F.Tereshchenko, *Radiotechnika* **41** (1986) No 11, 57, in Russian
203. Ya.S.Izkhoki, *Nonlinear Radiotechnics* (Soviet Radio, Moscow, 1955), p.427
204. V.N.Damgov, *Radiotechnika* **42** (1987) No 5, 6, in Russian
205. V.N.Damgov, *Radiotechnika* **45** (1990) No 4, 5, in Russian
206. V.N.Damgov, *Radiotechnika i Electronika* **33** (1988) 1207, in Russian
207. V.N.Damgov and A.P.Popov, *in Proceedings of 3rd National Symposium „Physics and Electronization"* (NTS Publisher, Sofia, 1985) v.3, 669, in Bulgarian
208. P.T.Zubov and Yu.L.Khotunzev, *Radiotechnika i Electronika* **29** (1984) 69
209. J.Engberg and T.Larsen, *Noise Theory of Linear and Nonlinear Circuits* (John Wiley & Sons, Singapore, 1995), p.296
210. V.N.Damgov, *Electrotechnika i Electronika* **28** (1992) No1-2, 31, in Bulgarian
211. V.N.Damgov, *in Proceedings of Xth Intern. Conference on Noise in Physical Systems* (Hungarian Academy of Sciences, Budapest, 1990)
212. V.N.Damgov, *Electropromishlenost i Priborostroene* **18** (1982) 488, in Bulgarian
213. M.E.Gerzenshtein and F.A.Levinzon, *Radiotechnika* **35** (1980) 6, in Russian
214. V.N.Damgov and P.G.Georgiev, *Inductive Sensor - Spectrum Analyser* (Patent of Republic of Bulgaria No 30009), p.27
215. V.Damgov, *Aerospace Research in Bulgaria* **13** (1997) 83
216. V.Damgov and E.Spedicato, *in Proceedings of the University of Bergamo* (DMSIA, Bergamo, 2003)
217. V.N.Damgov, *in Proceedings of IEEE Intern. Symposium on Circuits and Systems* (Rome, Italy, 1982)
218. V.N.Damgov, *Methods and Devices for Implementation Low-Noise Wide-Band Negative Resistances, Negative Capacitances and Negative Inductances* (Patents of Republic of Bulgaria, No 25959, No 25960, No 25961, No 28971, No 29260 and No 30008)
219. M.D.Akseenko, M.V.Baranochnikov and O.V.Smolin, *Microelectronic Photoreceivers* (Energoatomizdat, Moscow, 1984), p.249, in Russian
220. V.N.Damgov, *Method and Device for Measuring Biological Membranes* (Patent of Republic of Bulgaria, No 29993)
221. V.N.Damgov, *Multielectrode Modulation Device for Measuring the Interplanetary Plasma* (Patent of Republic of Bulgaria, No 45821)

222. V.N.Damgov, *Advances of Space Research* **11** (1991) 405; *Aerospace Research in Bulgaria* 9 (1992) 9
223. V.N.Damgov, *Short-Range Autodyne System* (Patent of Republic of Bulgaria, No 45520)
224. V.N.Damgov, *Short-Range Radar* (Patent of Republic of Bulgaria, No 47466)
225. V.N.Damgov and D.V.Stoyanov, *One-Band Modulator* (Patent of Republic of Bulgaria, No 35731)
226. A.A.Arefev, E.N.Baskakov and L.N.Stepanova, *Radioengineering Devices at Transistor Equivalents of the p-n-p-n Structure* (Radio i Sviaz Publisher, Moscow, 1982), p.336, in Russian
227. B.V.Chirikov, *Nonlinear Resonance* (Novosibirsk State University, Novosibirsk, 1987), p.386, in Russian
228. V.N.Damgov, *Bulgarian J. Physics* **4** (1977) 420
229. V.N.Damgov, *Electropromishlenost i Priborostroene* **16** (1980) 452, in Bulgarian
230. V.N.Damgov, *Vestnik of Moscow State University, Series Physics – Astronomy* **18** (1977) No 3, 74, in Russian
231. V.N.Damgov, *Bulgarian J. Physics* **5** (1978) 213
232. V.N.Damgov, Y.B.Douboshinsky and M.I.Kozakov, *Linear Reciprocating Electric Motor* (Patent of Rebublic of Bulgaria, No 44194)
233. V.N.Damgov, *Electropromishelnost i Priborostroene* **20** (1984) No 4, 11; in *IEEE Proceedings of Intern. Symposium on Circuits and Systems* (Italy, Rome, 1982); in *Proceedings IX Intern. Conference on Nonlinear Oscillations* (Naukova Dumka Publisher, Kiev, 1984)
234. V.N.Damgov, *IEE Proceedings, Part G: Electronic Circuits and Systems* **131** (1984) 24
235. Yu.A.Mitropolsky and A.A.Molchanov, *Computer Analysis of Nonlinear Resonance Circuits* (Naukova Dumka Publisher, Kiev, 1981), p.361, in Russian
236. M.D.Karasev, *Soviet Physics – Uspehi* **69** (1959) 217
237. V.N.Damgov and K.S.Rzhevkin, *Bulgarian J. Physics* **5** (1978) 613
238. V.N.Damgov and K.S.Rzhevkin, *Izv. VUZ, Radioelectronika* **26** (1983) No 12, 80, in Russian
239. V.N.Damgov, *Electropromishlenost i Priborostroene* **17** (1981) 350, in Bulgarian; in *Proceedings IX Intern. Conference on Nonlinear Oscillations* (Naukova Dumka Publisher, Kiev, 1984)
240. V.N.Damgov and K.S.Rzhevkin, in *Proceedings of the 4th Int. Symposium on Network Theory* (Ljubliana, Yugoslavia, 1979)
241. V.N.Damgov and K.S.Rzhevkin, *Bulgarian J. Physics* **6** (1979) 87
242. G.A.Russeva and V.I.Medvedev, *Bulgarian J. Physics* **16** (1989) 219

243. D.M.Vavriv and V.B.Ryabov, *Radiotechnika i Electronika* **36** (1991) 2148, in Russian
244. V.N.Damgov, P.S.Landa, S.M.Perminov and G.G.Shatalova, *Radiotechnika i Electronika* **31** (1986) 730, in Russian
245. V.N.Damgov, *Electrotechnika i Electronika* **39** (1991) No 8, 9, in Bulgarian
246. V.N.Damgov and T.S.Christov, in *Proceedings of Xth Intern. Conference on Noise in Physical Systems* (Hungarian Academy of Sciences, Budapest, 1990)
247. V.N.Damgov, *Radiotechnika* **48** (1993) No 2, 3, in Russian
248. V.N.Damgov, in *Proceedings 1989 SBMO Int. Microwave Symposium* (Brasil, São Paulo, 1989)
249. N.D.Birjuk and V.N.Damgov, *Technical Ideas* **39** (2002) No 1-2, 39
250. V.N.Damgov and P.G.Georgiev, *Aerospace Research in Bulgaria* **16** (2001) 124; 17 (2002) 34
251. N.D.Birjuk and V.N.Damgov, *Izv. VUZ, Radioelectronics* **28** (1985) No 5, 82
252. N.D.Birjuk and V.N.Damgov, *Aerospace Research in Bulgaria* **13** (1997) 59
253. N.D.Birjuk and V.N.Damgov, *Aerospace Research in Bulgaria* **14** (1998) 20
254. N.D.Birjuk and V.N.Damgov, *Aerospace Research in Bulgaria* **15** (1999) 55
255. N.D.Birjuk and V.N.Damgov, *Comptes Rendus de l'Académie Bulgare des Sciences* **37** (1984) 1027
256. N.D.Birjuk, V.N.Damgov and Yu.F.Epifanzev, *Electrichestvo* **30** (1991) No 2, 63, in Russian
257. N.D.Birjuk and V.N.Damgov, in *Proceedings of Xth Int. Conference on Nonlinear Oscillations* (Bulgarian Academy of Sciences, Varna, 1984)
258. N.D.Birjuk and V.N.Damgov, in *Proceedings 27th Midwest Symposium on Circuits and Systems* (West Virginia, Morgantown, 1984)
259. V.D.Furasov, *Motion Stability, Evolution and Stabilization* (Nauka Publisher, Moscow, 1977), p.299, in Russian
260. K.G.Valiev and G.S.Finin, *Lyapunov Functions Formation* (Naukova Dumka Publisher, Kiev, 1982), p.227, in Russian
261. V.I.Zubov, *Motion Stability: Lyapunov Methods Application* (Vishaya Shkhola Publisher, Moscow, 1984), p.331, in Russian
262. I.G.Malkin, *Lyapunov and Poincaré Methods in the Theory of Nonlinear Oscillations* (Gostechizdat Publisher, Moscow, 1949), p.312, in Russian
263. V.M.Matrosov and P.I.Kozlov (Eds), *Method of Lyapunov Functions in the Dynamics of Nonlinear Systems* (Nauka Publisher, Novosibirsk, 1983), p.412, in Russian
264. J.M.Skowroushi, *Nonlinear Lyapunov Dynamics* (University of Southern California, 1990), p.347
265. N.D.Birjuk and V.N.Damgov, *Izv. VUZ, Radioelectronika* **28** (1985) No 1, 48, in Russian

266. N.D.Birjuk and V.N.Damgov, *Izv. VUZ, Radioelectronika* **30** (1987) No 1, 95, in Russian

267. N.D.Birjuk and V.N.Damgov, *Izv. VUZ, Radioelectronika* **30** (1985) No 3, 93, in Russian

268. N.D.Birjuk and V.N.Damgov, in *Proceedings 28th Midwest Symposium on Circuits and Systems* (Kentucky, Luisville, 1985)

269. N.D.Birjuk and V.N.Damgov, *Radiotechnika i Electronika* **38** (1993) 481, in Russian

270. N.D.Birjuk and V.N.Damgov, *Technical Ideas* **39** (2002) No1-3, 25; No 3–4, 8

271. N.D.Birjuk and V.N.Damgov, *Radiotechnika i Electronika* **30** (1985) 1025, in Russian

272. V.N.Damgov and Y.B.Douboshinsky, *Comptes Rendus de l'Académie Bulgare des Sciences* **36** (1983) 1295

273. V.N.Damgov, D.B.Douboshinsky and T.D.Tagarev, *Technical Ideas* **38** (2001) No1-2, 28, in Bulgarian

274. V.N.Damgov, Y.B.Douboshinsky and D.M.Chekarkov, *Journal of Technical Physics* **55** (1985) 1048, in Russian

275. V.N.Damgov, D.B.Douboshinsky and Y.B.Douboshinsky, in *Proceedings 27th Midwest Symposium on Circuits and Systems* (West Virginia, Morgantown, 1984)

276. V.N.Damgov, D.B.Douboshinsky and Y.B.Douboshinsky, in *Proceedings X-Ith Int. Conference on Nonlinear Oscillations* (János Bolyai's Mathematical Society, Budapest, 1987)

277. V.N.Damgov and D.B.Douboshinsky, *Comptes Rendus de l'Académie Bulgare des Sciences* **45** (1992), No7, 17

278. V.N.Damgov, D.B.Douboshinsky and Y.B.Douboshinsky, in *Proceedings 5th Intern. Symposium on Network Theory* (Jugoslavia, Saraevo, 1984)

279. V.N.Damgov, D.B.Douboshinsky and Y.B.Douboshinsky, *Comptes Rendus de l'Académie Bulgare des Sciences* **40** (1987) No 5, 49

280. L.A.Vaynshtein and D.E.Vakman, *Frequency Division in the Theory of Oscillations and Waves* (Nauka Publisher, Moscow, 1983), p.293, in Russian

281. F.Fermi, *Phys. Rev.* **75** (1949) 392

282. G.A.Cecchi, D.L.Gonzales, M.O.Magnasco, G.B.Mindlin, O.Piro and A.J. Santillan, *Chaos* **3** (1993) No 1, 51

283. H.M.Isomдki, J.von Boehm and R.Raty, *Phys. Lett* **A107** (8) 343

284. K.Y.Tung, *JSME Int.J., Series III* **35** (3), 378

285. A.W.Troesch, D.G.Karr and K.-P.Beier, *Int. J. Bifurcation and Chaos* **2** (3) 607

286. H.Barkhausen and K.Kurz, *Phys.Zs.* **21** (1920) No 1, 534

287. F.B.Bunkin, *J.Technical Physics* **58** (1988) No 11, 2241, in Russian

288. Y.A.Voshtinsky, Y.A.Kravzov, A.S.Magarshak, A.Y.Sinkevich and D.M. Shteingradt, *Cybernetics and Computations* **91** (1991) 75, in Russian
289. V.Damgov and Pl.Trenchev, *Chaos, Solitons and Fractals* **17** (2003) 11
290. V.Damgov, Pl.Trenchev and K.Sheiretsky, *Chaos, Solitons and Fractals* **17** (2003) 41
291. V.N.Damgov and I.G.Popov, *Discrete Dynamics in Nature and Society* **3** (1999) No 6, 1
292. V.N.Damgov, in *Progress Theor. Phys. Suppl.* **No139** (2000) 344
293. V.N.Damgov, in *Dynamical Systems and Chaos 2,* Eds. J.Aizawa, S.Saito and K.Shiraiwa (World Scientific, London, 1995) 558
294. V.N.Damgov and P.G.Georgiev, in *Dynamical Systems and Chaos 2,* Eds. J.Aizawa, S.Saito and K.Shiraiwa (World Scientific, London, 1995) 304
295. V.N.Damgov, *Ukrainian J.Phys.* **38** (1993) No 3, 470, in Russian
296. V.N.Damgov and I.G.Popov, *Comptes Rendus de l'Academie Bulgare des Sciences* **52** (1999) No 5-6, 31
297. V.N.Damgov and I.G.Popov, *Comptes Rendus de l'Academie Bulgare des Sciences* **52** (1999) No 7-8, 29
298. V.N.Damgov and P.G.Georgiev, *Comptes Rendus de l'Academie Bulgare des Sciences* **52** (1999) No 9-10, 22
299. V.N.Damgov and A.V.Holden, *Aerospace Research in Bulgaria* **12** (1996) 5
300. V.N.Damgov, *Aerospace Research in Bulgaria* **12** (1996) 27
301. V.N.Damgov and P.G.Georgiev, *Comptes Rendus de l'Academie Bulgare des Sciences* **48** (1995) No 5, 19
302. V.N.Damgov, *Comptes Rendus de l'Academie Bulgarie des Sciences* **48** (1995) No 7, 29
303. V.N.Damgov and P.G.Georgiev, *Aerospace Research in Bulgaria* **11** (1994) 3
304. M.A.Bikhovsky and V.N.Damgov, *Comptes Rendus de l'Academie Bulgare des Sciences* **46** (1993) No 2, 33
305. V.N.Damgov, *Comptes Rendus de l'Academie Bulgarie des Sciences* **44** (1991) No 9, 37
306. V.N.Damgov and G.S.Grinberg, *Comptes Rendus de l'Academie Bulgare des Sciences* **44** (1991) No 9, 33, in Russian
307. V.N.Damgov, D.B.Douboshinsky and Y.B.Douboshinsky, *Comptes Rendus de l'Academie Bulgare des Sciences* **44** (1991) No 12, 17, in Russian
308. V.N.Damgov, D.B.Douboshinsky and Y.B.Douboshinsky, *Comptes Rendus de l'Academie Bulgare des Sciences* **44** (1991) No 1, 23
309. V.N.Damgov, D.B.Douboshinsky and Y.B.Douboshinsky, *Comptes Rendus de l'Academie Bulgare des Sciences* **43** (1990) No 10, 43, in Russian
310. V.N.Damgov, D.B.Douboshinsky and Y.B. Douboshinsky, *Comptes Rendus de l'Academie Bulgare des Sciences* **40** (1987) No 4, 57, in Russian

311. V.N.Damgov, D.B.Douboshinsky and Y.B. Douboshinsky, *Comptes Rendus de l'Academie Bulgare des Sciences* **39** (1986) No 9, 47, in Russian
312. V.N.Damgov, D.B.Douboshinsky and Y.B. Douboshinsky, *Comptes Rendus de l'Academie Bulgare des Sciences* **39** (1986) No 10, 63, in Russian
313. V.N.Damgov and I.G.Popov, *J.Theor. & Appl. Mech.* **30** (2000) No 2, 3
314. V.N.Damgov and P.G.Georgiev, *Comptes Rendus de l'Academie Bulgare des Sciences* **49** (1996) No 3, 37
315. V.N.Damgov and D.B.Douboshinsky, *Earth, Moon and Planets* **56** (1992) 233
316. V.N.Damgov, *Earth, Moon and Planets* **61** (1993) 87
317. T.G.Strijak, *Method for Investigation of Dynamical Systems of Pendulum Type* (Nauka Publisher, Alma Ata, 1981), p.224, in Russian
318. A.D.Morozov, in *Research Reports in Physics: Nonlinear Waves,* eds. A.V. Gaponov-Grekhov et al. (Springer-Verlag, Berlin, 1990) 276
319. S.J.Liao, *J.Applied Mechanics, Transactions of the ASME* **59** (1992) 970
320. J.Mawhin, *Zeitschrift für Angewantte Mathematik und Mechanik (ZAMM)* **73** (4-5) T54
321. H.Heng, R.Doerner, R.Hüblinger and W.Martienssen, *Int. J. Bifurcation and Chaos* **4** (1994) 751
322. J.A. Blackburn, H.J.T. Smith and N. Groubech-Jensen, *Am. J. Physics* **60** (1992) No 10, 422
323. S.C.Yip and F.Dimaggio, *Computers and Structures* **46** (1993) No 4, 725
324. S.P. Hastings and J.B. McLeod, *American Math. Monthly* **100** (1993) No 6, 563
325. M.I. Abramowitz and I.J. Stegun (Eds), *Handbook of Mathematical Functions* (Dover Publ., N.Y., 1989), p.592
326. A.M.Molchanov, in *Modern Problems of Celestial Mechanics and Astrodynamics* (Nauka Publisher, Moscow, 1973), p.111, in Russian
327. V.V.Beletsky, *Articles on Space Body Motion* (Nauka i Iskustvo, Sofia, 1977), 214p, in Bulgarian
328. I.I.Blehmann, *Synchronization in Nature and Technique* (Nauka Publ., Moscow, 1981), 331p, in Russian
329. N.N.Gorkavy and A.A. Fridmann, *Soviet Physics - Uspehi* **160** (1990) 169
330. A.M.Tchetchelnitsky, *Extremes, Stability and Resonances in the Astrodynamics and Astronautics* (Mashinostroene, Moscow, 1986), p. 308
331. R. Louise, *The Moon and the Planets* **27** (1982) 59
332. R. Louise, *Astrophys. Space Sci.* **86** (1982) 505
333. R.Wayte, *The Moon and the Planets* **26** (1982) 11
334. D.W.Greenberger, *Foundation of Physics* **13** (1983) 903
335. H.L.F. Houpis and D.A. Mendis, *The Moon and the Planets* **29** (1983) 39

336. A.B.Pippard, *The Physics of Vibration. The Simple Vibration in Quantum Mechanics* (Cambridge Univ. Press, London, 1983), p.392
337. R.L. Oldershaw, *Int. J. General Systems* **8** (1982) 1
338. M.M. Neto, *The Titius-Bode Law of Planetary Distances: Its History and Theory* (Pergamon Press, Oxford, 1972), p.114
339. V.Pletser, *Earth, Moon and Planets* **46** (1989) 285
340. V.Lalaja et al. *Earth, Moon and Planets* **44** (1989) 105
341. R. Neuhanser and J.V. Feitzinger, *Astron. Astrophysics* **170** (1986) 174
342. J.J. Rawal, *Earth, Moon and Planets* **44** (1989) 295
343. C.-W. Allen, *Astrophysical Quantities* (The Athlone Press, University of London, London, 1973), p.429
344. T.Bohr, P.Bak and M.Jensen, *Phys. Rev.* **A30** (1984) 1970
345. G. Schmidt and B.W. Wang, *Phys. Rev.* **A30** (1985) 2994
346. M.Casdagli, *Ergodic Theory & Dynamical Systems* **7** (1987) 165
347. P.Pakarinen and R.Nieminen, *J. Phys.* **16** (1983) 2105
348. M.Casadagli, *Physica* **D 29** (1988) 365
349. M.A. Liberman and K.Y. Tsang, *Phys. Rev. Lett.* **55** (1985) 908
350. A.J. Lichtenberg, M.A.Liberman and R.H.Cohen, *Physica* **D 29** (1980) 291
351. H.G.Schuster, *Deterministic Chaos* (Physik-Verlag, Weinheim, 1984), p. 451
352. G.M.Zaslavskii and B.V.Chirikov, *Soviet Physics – Uspehi* **105** (1972) 3
353. G.M.Zaslavskii and H.I.Rachko, *J. Experimental and Theor. Physics* **76** (1979) 2052
354. G.M.Zaslavskii, H.I.Rachko and Kh.R.Ya, *Soviet Phys. JETP* **49** (1978) 1039
355. U.Feudel, C.Grebogi, B.Hunt and J.Yorke, *Phys. Rev. E* **54** (1996) 71
356. V.N.Damgov, D.B.Douboshinsky and Y.B. Douboshinsky, *Comptes Rendus de l'Academie Bulgare des Sciences* **44** (1991) No1, 21
357. V.N.Damgov, D.B.Douboshinsky and Y.B. Douboshinsky, *Microwave Emitter* (Patent of Republic of Bulgaria, No 44197)
358. V.N.Damgov, D.B.Douboshinsky and Y.B. Douboshinsky, *Comptes Rendus de l'Academie Bulgare des Sciences* **40** (1987) No 4, 57, in Russian
359. V.N.Damgov and G.S.Grinberg, *Comptes Rendus de l'Academie Bulgare des Sciences* **44** (1991) No 9, 33, in Russian
360. V.N.Damgov, *Technical Ideas* **30** (1995) No 2, 11, in Bulgarian

SUBJECT INDEX